Diffeology

Mathematical
Surveys
and
Monographs

Volume 185

Diffeology

Patrick Iglesias-Zemmour

American Mathematical Society
Providence, Rhode Island

2010 *Mathematics Subject Classification.* Primary 53Cxx, 53Dxx, 55Pxx, 55P35, 55Rxx, 55R65, 58A10, 58A40, 58Bxx.

For additional information and updates on this book, visit
www.ams.org/bookpages/surv-185

Library of Congress Cataloging-in-Publication Data

Iglesias-Zemmour, Patrick, 1953–
Diffeology / Patrick Iglesias-Zemmour.
pages cm. — (Mathematical surveys and monographs ; volume 185)
Includes bibliographical references.
ISBN 978-0-8218-9131-5 (alk. paper)
1. Global differential geometry. 2. Symplectic geometry. 3. Algebraic topology. 4. Differentiable manifolds. I. Title.

QA670.I35 2013
516.3′62—dc23

2012032894

∞ The paper used in this book is acid-free and falls within the guidelines
established to ensure permanence and durability.
Visit the AMS home page at http://www.ams.org/

10 9 8 7 6 5 4 3 2 1 18 17 16 15 14 13

לֹא עָלֶיךָ הַמְּלָאכָה לִגְמוֹר וְאֵין אַתָּה בֶן־חוֹרִין לִבָּטֵל מִמֶּנָּה

רבי טרפון, משנה אבות ב טז

Contents

Preface

At the end of the last century, differential geometry was challenged by theoretical physics: new objects were displaced from the periphery of the classical theories to the center of attention of the geometers. These are the irrational tori, quotients of the 2-dimensional torus by irrational lines, with the problem of quasi-periodic potentials, or orbifolds with the problem of singular symplectic reduction, or spaces of connections on principal bundles in Yang-Mills field theory, also groups and subgroups of symplectomorphisms in symplectic geometry and in geometric quantization, or coadjoint orbits of groups of diffeomorphisms, the orbits of the famous Virasoro group for example. All these objects, belonging to the outskirts of the realm of differential geometry, claimed their place inside the theory, as full citizens. Diffeology gives them satisfaction in a unified framework, bringing simple answers to simple problems, by being the right balance between rigor and simplicity, and pushing off the boundary of classical geometry to include seamlessly these objects in the heart of its concerns.

However, diffeology did not spring up on an empty battlefield. Many solutions have been already proposed to these questions, from functional analysis to noncommutative geometry, via smooth structures à la Sikorski or à la Frölicher. For what concerns us, each of these attempts is unsatisfactory: functional analysis is often an overkilling heavy machinery. Physicists run fast; if we want to stay close to them we need to jog lightly. Noncommutative geometry is uncomfortable for the geometer who is not familiar enough with the $\mathbf{C}^*$-algebra world, where he loses intuition and sensibility. Sikorski or Frölicher spaces miss the singular quotients. Perhaps most frustrating, none of these approaches embraces the variety of situations at the same time.

So, what's it all about? Roughly, a diffeology on an arbitrary set X declares, which of the maps from $\mathbf{R}^n$ to X are *smooth*, for all integers n. This idea, refined and structured by three natural axioms, extends the scope of classical differential geometry far beyond its usual targets. The smooth structure on X is then defined by all these *smooth parametrizations*, which are not required to be injective. This is what gives plenty of room for new objects, the quotients of manifolds for example, even when the resulting topology is vague. The examples detailed in the book prove that diffeology captures remarkably well the smooth structure of singular objects. But quotients of manifolds are not the sole target of diffeology, actually they were not even the first target, which was spaces of smooth functions, groups of diffeomorphisms. Indeed, these spaces have a natural *functional diffeology*, which makes the category Cartesian closed. But also, the theory is closed under almost all set-theoretic operations: products, sums, quotients, subsets etc. Thanks to these nice properties, diffeology provides a fair amount of applications and examples and offers finally a renewed perspective on differential geometry.

Also note the existence of a convenient *powerset diffeology*, defined on the set of all the subsets of a diffeological space. Thanks to this original diffeology, we get a clear notion of what is a smooth family of subsets of a diffeological space, without needing any model for the elements of the family. This powerset diffeology «encodes genetically» the smooth structure of many classical constructions without any exterior help. The set of the lines of an affine space, for example, inherits a diffeology from the powerset diffeology of the ambient space, and this diffeology coincides with its ordinary manifold diffeology, which is remarkable.

Moreover, every structural construction (homotopy, Cartan calculus, De Rham cohomology, fiber bundles etc.) renewed for this category, applies to all these derived spaces (smooth functions, differential forms, smooth paths etc.) since they are diffeological spaces too. This unifies the discourse in differential geometry and makes it more consistent, some constructions become more natural and some proofs are shortened. For example, since the space of smooth paths is itself a diffeological space, the Cartan calculus naturally follows and then gives a nice shortcut in the proof of homotopic invariance of the De Rham cohomology.

What about standard manifolds? Fortunately, they become a full subcategory. Then, considering manifolds and traditional differential geometry, diffeology does not subtract anything nor add anything alien in the landscape. About the natural question, "Why is such a generalization of differential geometry necessary, or for what is it useful?" the answer is multiple. First of all, let us note that differential geometry is already a generalization of traditional Greek Euclidean geometry, and the question could also be raised at this level. More seriously, on a purely technical level, considering many of the recent heuristic constructions coming from physics, diffeology provides a light formal rigorous framework, and that is already a good reason. Two examples:

Example 1. For a space equipped with a closed 2-form, diffeology gives a rigorous meaning to the moment maps associated with every smooth group action by automorphisms. It applies to every kind of diffeological space, it can be a manifold, a space of smooth functions, a space of connection forms, an orbifold or even an irrational torus. It works that way because the theory provides a unified coherent notion of differential forms, on all these kinds of spaces, and the tools to deal with them. In particular, such a general diffeological construction clearly reveals that the status of moment maps is high in the hierarchy of differential geometry. It is clearly a categorical construction which exceeds the ordinary framework of the geometry of manifolds: every closed 2-form on a diffeological space gets naturally a *universal moment map* associated with its group of automorphisms.

Example 2. Every closed 2-form on a simply connected diffeological space[1] is the curvature of a connection form on some diffeological principal bundle. The structure group of this bundle is the diffeological *torus of periods* of the 2-form, *i.e.*, the quotient of the real line by the group of periods of the 2-form. This construction is completely universal and applies to every diffeological space and to every closed 2-form, whether the form is integral or not. The only condition is that the group of periods is diffeologically discrete, that is, a strict subgroup of the real numbers. The construction of a prequantization bundle corresponds to the special case when the periods are a subgroup of the group generated by the Planck constant h or,

[1]The general case is a work in progress.

if we prefer, when the group of periods is generated by an integer multiple of the Planck constant.

The crucial point in these two constructions is that the quotient of a diffeological group — the group of momenta of the symmetry group by the holonomy for the first example, and the group of real numbers by the group of periods for the second — is naturally a nontrivial diffeological group whose structure is rich enough to make these generalizations possible. In this regard, the *contravariant approaches* — Sikorski or Frölicher differentiable spaces — are globally helpless because these crucial quotients are trivial, and this is irremedible. By respecting the internal (nontrivial) structure of these quotients, diffeology leads one to a good level of generality for such general constructions and statements. The reason is actually quite simple, the contravariant approaches define smooth structures by declaring which maps from X to **R** are smooth. Doing so, they capture only what looks like **R** — or a power of **R** — in X, killing everything else. The quotient of a manifold may not resemble **R** at all, if we wanted to capture its singularity, we would have to compare it with all kinds of standard quotients. *A contrario*, diffeology as a *covariant approach* assumes nothing about the resemblance of the diffeological space to some Euclidean space. It just declares what are the smooth families of elements of the set, and this is enough to retrieve the local aspect of the singularity, if it is it what we are interested in.

Another strong point is that diffeology treats simply and rigorously infinite-dimensional spaces without involving heavy functional analysis, where obviously it is not needed. Why would we involve deep functional analysis to show, for example, that every symplectic manifold is a coadjoint orbit of its group of automorphisms? It is so clear when we know that it is what happens when a Lie group acts transitively, and the group of symplectomorphisms acts transitively. In this case, and maybe others, diffeology does the job easily, and seems to be, here again, the right balance between rigor and simplicity. Recently A. Weinstein et al. wrote "For our purposes, spaces of functions, vector fields, metrics, and other geometric objects are best treated as diffeological spaces rather than as manifolds modeled on infinite-dimensional topological vector spaces" [**BFW10**].

NOTE A. The axiomatics of *Espaces différentiels*, which became later the diffeological spaces, were introduced by J.-M. Souriau in the beginning of the eighties [**Sou80**]. Diffeology is a variant of the theory of *differentiable spaces*, introduced and developed a few years before by K.T. Chen [**Che77**]. The main difference between these two theories is that Souriau's diffeology is more differential geometry oriented, whereas Chen's theory of differentiable spaces is driven by algebraic geometry considerations.

NOTE B. I began to write this textbook in June 2005. My goal was, first of all, to describe the basics of diffeology, but also to improve the theory by opening new fields inside, and by giving many examples of applications and exercises. If the basics of diffeology and a few developments have been published a long time ago now [**Sou80**] [**Sou84**] [**Don84**] [**Igl85**], many of the constructions appearing in this book are original and have been worked out during its redaction. This is what also explains why it took so long to complete. I chose to introduce the various concepts and constructions involved in diffeology from the simple to the complex, or from the particular to the more general. This is why there are repetitions, and some constructions, or proofs, can be shortened, or simplified. I included sometimes these simplifications as exercises at the end of the sections. In the examples treated, I tried to clearly separate what is the responsibility of the category and what is specific. I hope this will help for a smooth progression in the reading of this text.

NOTE C. By the time I wrote these words, and seven years after I began this project, a few physicists or mathematicians have shown some interest in diffeology, enough to write a few papers [**BaHo09**] [**Sta10**] [**Sch11**]. The point of view adopted in these papers is strongly categorical. Diffeology is a Cartesian closed category, complete and cocomplete. Thus, diffeology is an « interesting beast » from a pure categorical point of view. However, if I understand and appreciate the categorical point of view, it does not correspond to the way I apprehended this theory. I may not have commented clearly enough, or exhaustively, on the categorical aspects of the constructions and objects appearing there because my approach has been guided by my habits in classical differential geometry. I made an effort to introduce a minimum of new vocabulary or notation, to give the feeling that studying the geometry of a torus or of its group of diffeomorphisms, or the geometry of its quotient by an irrational line, is the same exercise, involving the same concepts and ideas, the same tools and intuition. I believe that the role of diffeology is to bring closer the objects involved in differential geometry, to treat them on an equal footing, respecting the ordinary intuition of the geometer. All in all, I no longer see diffeology as a replacement theory, but as the natural field of application of traditional differential geometry. But I judged, at the moment when I began this textbook, that diffeology was far enough from the main road to avoid moving too far away. Maybe it is not true anymore, and it is possible that, in a future revision of this book, I shall insist, or write a special chapter, on the categorical aspects of diffeology.

CONTENTS OF THE BOOK

Throughout its nine chapters, the contents of the book try to cover, from the point of view of diffeology, the main fields of differential geometry used in theoretical physics: differentiability, groups of diffeomorphisms, homotopy, homology and cohomology, Cartan differential calculus, fiber bundles, connections, and eventually some comments and constructions on what wants to be *symplectic diffeology*.

Chapter 1 presents the abstract constructions and definitions related to diffeology: objects are diffeologies, or diffeological spaces, and morphisms are *smooth maps*. This part contains all the categorical constructions: sums, products, subset diffeology, quotient diffeology, functional diffeology.

In Chapter 2 we shall discuss the local properties and related constructions, in particular: D-topology, generating families, local inductions or subductions, dimension map, modeling diffeology, in brief, everything related to local properties and constructions.

In Chapters 3 and 4, we shall see the notion of *diffeological vector spaces*, which leads to the definition of *diffeological manifolds*. Each construction is illustrated with several examples, not all of them coming from traditional differential geometry. In particular the examples of the infinite-dimensional sphere and the infinite-projective space are treated in detail.

Chapter 5 describes the diffeological theory of homotopy. It presents the definitions of connectedness, Poincaré's groupoid and fundamental groups, the definition of higher homotopy groups and relative homotopy. The exact sequence of the relative homotopy of a pair is established. Everything relating to functional diffeology of iterated spaces of paths or loops finds its place in this chapter.

Chapter 6 is about Cartan calculus: exterior differential forms and De Rham constructions, their generalization to the context of diffeology. Differential forms are defined and presented first on open subsets of real vector spaces, where everything is clearly explicit, and then carried over to diffeologies. Then, we shall see exterior derivative, exterior product, generalized Lie derivative, generalized Cartan formula, integration on chain, De Rham cohomology on diffeology, chain homotopy operator and obstructions to exactness of differential forms. We shall also see a very useful formula for the variation of the integral of differential forms on smooth chains. In particular, the generalization of Stokes' theorem; the homotopic invariance of De Rham cohomology, and the generalized Cartan formula are established by application of this formula.

Chapter 7 talks about *diffeological groups* and gives some constructions relative to objects associated with diffeological groups, for instance the space of its momenta, equivalence between right and left momenta, etc. Smooth actions of diffeological groups and natural coadjoint actions of diffeological groups on their spaces of momenta are defined.

Chapter 8 presents the theory of *diffeological fiber bundles*, defined by *local triviality along the plots* of the base space (not to be confused with the local triviality of topological bundles). It is more or less a rewriting of my thesis [**Igl85**]. We shall define principal and associated bundles, and establish the exact homotopy sequence of a diffeological fiber bundle. The construction of the universal covering and the construction of coverings by quotient is also a part of the theory, as well as the generalization of the monodromy theorem in the diffeological context. We shall also see, in this general framework, how we can understand connections, reductions, construction of the holonomy bundle and group. In the same vein, we shall represent any closed 1-form or 2-form on a diffeological space by a special structured fiber bundle, a groupoid.

In Chapter 9 we discuss *symplectic diffeology*. It is an attempt to generalize to diffeological spaces the usual constructions in symplectic geometry. This construction will use an essential tool, the *moment map*, or more precisely its generalization in diffeology. We have to note first that, if diffeology is perfectly adapted to describe covariant geometry, *i.e.*, the geometry of differential forms, pullbacks etc., it needs more work when it comes to dealing with contravariant objects, for example vectors. This is why it is better to introduce directly the space of momenta of a diffeological group, the diffeological equivalent of the dual of the Lie algebra, without referring to some putative Lie algebra. Then, we generalize the moment map relative to the action of a diffeological group on a diffeological space preserving a closed 2-form. This generalization also extends slightly the classical moment map for manifolds. Thanks to these constructions, we get the complete characterization of homogeneous diffeological spaces equipped with a closed 2-form ω. This theorem is an extension of the well-known Kirillov-Kostant-Souriau theorem. It applies to every kind of diffeological spaces, the ones regarded as singular by traditional differential geometry, as well as spaces of infinite dimensions. It applies to the exact/equivariant case as well as the not-exact/not-equivariant case, where exact here means *Hamiltonian.* In fact, the natural framework for these constructions is some equivariant cohomology, generalized to diffeology. This theory locates pretty well all the questions related to exactness versus nonexactness, equivariance versus nonequivariance, as well as the so-called Souriau *symplectic cohomology* [**Sou70**]. Incidentally, this definition of the moment map for diffeology gives a way for defining *symplectic diffeology,* without considering the *kernel* of a 2-form for a diffeological space, what can be problematic because of the contravariant nature of the kernel of a form. They are defined as diffeological spaces X, equipped with a closed 2-form ω which are homogeneous under some subgroup of the whole group of diffeomorphisms preserving ω, and such that the moment map is a covering. This definition can be considered as strong, but it includes a lot of various situations.[2] For example every connected symplectic manifold is symplectic in this meaning. Some refinements are needed to deal with some nonhomogeneous singular spaces like orbifolds for example, but this is still a work in progress. Many questions are still open in this new framework of symplectic diffeology. I discuss some of them when they appear throughut the book.

On the structure of the book

The book is made up of numbered chapters, each chapter is made of unnumbered sections. Each section is made of a series of numbered paragraphs, with a title which summarizes the content. Throughout the book, we refer to the numbered paragraphs as (art. X). Paragraphs may be followed by notes, examples, or a proof if the content needs one. This structure makes the reading of the book easy, one can decide to skip some proofs, and the title of each paragraph gives an idea about what the paragraph is about. Moreover, at the end of most of the sections there are one or more exercises related to their content. These exercises are here to familiarize the reader with the specific techniques and methods introduced by diffeology. We are forced, sometimes, to reconsider the way we think about things and change our methods accordingly. The solutions of the exercises are given at the end of the book in a special chapter. Also, at the end of the book there is a list of the main notations used. There is no index but a table of contents which includes the title of each paragraph, so it is easy to find the subject in which one is interested in, if it exists.

Acknowledgment

Thanks to Yael Karshon and François Ziegler for their encouragements to write this textbook on diffeology. I am especially grateful to Yael Karshon for our many discussions on orbifolds and smooth structures, and particularly for our discussions on « symplectic diffeology » which pushed me to think seriously about this subject.

I am very grateful to The Hebrew University of Jerusalem for its hospitality, specials thanks to Emmanuel Farjoun for his many invitations to work there. I started, and mainly carried out, this project in Jerusalem. Without my various

[2] I have often discussed the question of symplectic diffeological spaces with J.-M. Souriau. The definition given here seems to be a good answer to the question, because this moment map and the related construction of *elementary spaces* include the complete case of symplectic manifolds even if the action of the group of symplectomorphisms is not Hamiltonian, or the orbit is not linear but affine. But the few discussions I have had with Yael Karshon about the status of *symplectic orbifolds* will maybe lead to a refinement of the concept of symplectic diffeological space. It is however too early to conclude.

stays in Israel I would never have finished it. Givat Ram campus is a wonderful place where I enjoyed the studious and friendly atmosphere, and certainly its peaceful and intense scientific environment. Thanks also to David Blanc from the University of Haifa, to Misha Polyak and Yoav Moriah from the Technion, to Misha Katz from Bar-Ilan University, to Leonid Polterovich and Paul Biran, from Tel-Aviv University, for their care and invitations to share my ideas with them. Thanks also to Yonatan Barlev, Erez Nesrim and Daniel Shenfeld, from the Einstein Institute, for their participation in the 2006 seminar on Diffeology. Working together has been very pleasant and their remarks have been helpful.

Thanks to Paul Donato, my old accomplice in diffeology, who checked some of my claims when I needed a second look. Thanks also to the young guard: Jordan Watt, Enxin Wu and Guillaume Tahar, their enthusiasm is refreshing and their contribution somehow significant. Thanks also to the referee of the manuscript for his support and his clever suggestions which made me improve the content.

Thanks, obviously, to the CNRS that pays my salary, and gives me the freedom necessary to achieve this project. Thanks particularly to Jérôme Los for his support when I needed it.

Thanks also to Liliane Périchaud who helped me fixing misprints, to David Trotman for trying to make my English look English. Thanks eventually to the AMS publishing people for their patience and care, in particular to Jennifer Wright-Sharp for the wonderful editing job she did with the manuscript.

Patrick Iglesias-Zemmour
Jerusalem, September 2005 — Aix-en-Provence, August 2012

Laboratoire d'Analyse, Topologie et Probabilités du CNRS
39, rue F. Joliot-Curie
13453 Marseille, cedex 13
France
&
The Hebrew University of Jerusalem
Einstein Institute
Campus Givat Ram
91904 Jerusalem
Israel

http://math.huji.ac.il/~piz/
piz@math.huji.ac.il

CHAPTER 1

Diffeology and Diffeological Spaces

A *diffeology* on a set X declares, for all $n \in \mathbf{N}$ and for all open sets $U \subset \mathbf{R}^n$, which maps, from U to X, are *smooth*. A *diffeological space* is a set equipped with a diffeology. The elements of a diffeology are called the *plots* of the diffeological space. To suit the consensual meaning of the word "smooth", the plots of a diffeological space satisfy three natural axioms which are the basis of the theory. These are the axioms of *covering*, *locality*, and *smooth compatibility* (art. 1.5).

Relationships between diffeological spaces are defined through *smooth maps*. A smooth map from a diffeological space to another is a map exchanging the plots of the spaces (art. 1.14). Diffeological spaces together with smooth maps form the category {Diffeology}, whose isomorphisms are called *diffeomorphisms*. This category generalizes the ordinary category {Manifolds} insofar as {Manifolds} is a full subcategory of {Diffeology}. But this is not its specificity, since there exist other extensions, and this is not its sole interest.

One of the most striking properties of the category {Diffeology} is its stability under almost all natural set-theoretic constructions. Every subset of a diffeological space carries a natural *subset diffeology*, induced by the ambient space (art. 1.33), and defined by the *pullback of diffeologies* (art. 1.26). As well, every quotient of a diffeological space carries a natural *quotient diffeology* (art. 1.50), defined by the *pushforward of diffeologies* (art. 1.46). There is also a natural diffeology on every product of diffeological spaces (art. 1.55), and on every sum (coproduct) of diffeological spaces (art. 1.39).

Another important diffeological construction is the *functional diffeology*, defined on the set of smooth maps between diffeological spaces (art. 1.57). This diffeology is a key for many other diffeological constructions, for example, in the theory of homotopy of diffeological spaces, but also in the definition of differential forms and bundles, diffeological groups etc. A *powerset diffeology*, defined on the powerset $\mathfrak{P}(X)$ of every diffeological space X, is another example of a general and important construction (Exercise 63, p. 61). But it is more an attempt than a final conclusion.

Linguistic Preliminaries

Every theory needs a precise vocabulary, a fixed set of notations and conventions to be transmitted. In this section, we introduce the basic vocabulary and objects used in diffeology: finite dimensional real vector spaces, domains, and parametrizations.

1.1. Real vector spaces and domains. We denote by $\mathbf{R}$ the field of real numbers. We call n-*domain* every subset of $\mathbf{R}^n$, open for the standard topology. Let us recall that a set $U \subset \mathbf{R}^n$ is open, for the standard topology, if there exist a set $\mathcal{J}$ of indices (possibly empty) and a family $\{B(x_i, r_i)\}_{i \in \mathcal{J}}$ of open balls in $\mathbf{R}^n$, such that $U = \bigcup_{i\in\mathcal{J}} B(x_i, r_i)$. We will denote generally by

$$B(x,r) = \{x' \in \mathbf{R}^n \mid \|x' - x\| < r, \text{ with } x \in \mathbf{R}^n \text{ and } r > 0\}$$

the open ball centered at x with radius r. The set of all n-domains will be denoted by Domains($\mathbf{R}^n$). We shall also denote by Domains the sum of all the Domains($\mathbf{R}^n$) when n runs over the integers. An element of Domains will be called a *real domain*, or simply a *domain*, when there is no possible confusion. But note that the real vector space of which the domain is a subset is implicit.

1.2. Sets, subsets, maps etc. Let X be some set. As usual we call *subset* of X any set A whose elements are elements of X, and we denote $A \subset X$. We denote by $\mathfrak{P}(X)$ the powerset of X, that is, the set of all the subsets of X,

$$\mathfrak{P}(X) = \{A \mid A \subset X\}.$$

Now, let A be a subset of X, every subset $B \subset X$ containing A is called a *superset* of A. By extension, a *superset of a point* $x \in X$ is any superset of the singleton $\{x\}$. For any topological space X, an *open superset* is a superset which is open for the topology of X. A *neighborhood* of $x \in X$ is a superset of an open superset of x. An *open superset* of x or an *open neighborhood* of x are the same thing. For a map f, its domain of definition is denoted by def(f), and the set of its values by val(f). Thus, if $f : X \to Y$, def(f) $= X$ and val(f) $= f(X) \subset Y$. The set of all maps from X to Y will be denoted by Maps(X, Y). Also, for every set X the identity map will be denoted by $\mathbf{1}_X$.

1.3. Parametrizations in sets. Let X be a nonempty set, we call *parametrization in* X every map $P : U \to X$, where U is a domain. If U is an n-domain, we also say that P is an n*-parametrization*. If U is a nonempty n-domain, we say that the *dimension* of P is n, and we denote $\dim(P) = n$. The empty parametrization has, by convention, dimension zero. We shall denote by Param(U, X) the set of all the parametrizations in X defined on U, and by Param(X) the set of all parametrizations in X. If necessary, we shall denote with a star, that is, Param$^\star(U, X)$ or Param$^\star(X)$, the subset of nonempty parametrizations. To avoid useless discussions, let us note that the parametrizations in X form a well defined set. Note first that, for every integer n, for every n-domain U, Param(U, X) is a set. Indeed, Param(U, X) is a subset of the powerset $\mathfrak{P}(U \times X)$, since maps are just some special relations. Now, let us denote

$$\mathrm{Param}_n(X) = \bigcup_{U \in \mathrm{Domains}(\mathbf{R}^n)} \mathrm{Param}(U, X).$$

But since the domains of $\mathbf{R}^n$ are a subset of $\mathfrak{P}(\mathbf{R}^n)$, Param$_n(X)$ is a family of sets, indexed by a set. Therefore, Param$_n(X)$ is itself a set. Then,

$$\mathrm{Param}(X) = \bigcup_{n \in \mathbf{N}} \mathrm{Param}_n(X)$$

is a family of sets indexed by the integers and thus a set. Also note that the nonempty 0-parametrizations in X are naturally identified with the elements of X. In other words, Param$_0^\star(X) \simeq X$, where Param$_n^\star(X)$ denotes the set of nonempty n-parametrizations in X. For every element x in X, we denote by the bold letter

$$\mathbf{x} : \{0\} \to X, \quad \text{with} \quad \mathbf{x}(0) = x,$$

the unique 0-parametrization with value x.

NOTE. It happens that we write this kind of sentence: "Let $P : U \to X$ be a parametrization, let V be a superset of $r \in U$, and let $P \restriction V$ be a parametrization satisfying such condition ...". This implies that V is an open neighborhood of r, since a parametrization is always defined on a domain.

1.4. Smooth parametrizations in domains. Let us recall some basic material on which diffeology is founded. Let U be an n-domain, and let V be an m-domain. Let $f : U \to V$ be a map. Let f be continuous, f is said to be *differentiable* if there exists a map $D(f) : U \to L(\mathbf{R}^n, \mathbf{R}^m)$ such that

$$\text{for all } x \in U \text{ and all } u \in \mathbf{R}^n, \quad D(f)(x)(u) = \lim_{\varepsilon \to 0} \frac{f(x + \varepsilon u) - f(x)}{\varepsilon}.$$

The map $D(f)$ is called *the derivative* of f. For all $x \in U$, the map $D(f)(x)$ is called the *tangent linear map* or simply the *tangent map* of f at x. It belongs to the space $L(\mathbf{R}^n, \mathbf{R}^m)$ of linear maps from $\mathbf{R}^n$ to $\mathbf{R}^m$. The tangent map is made of the partial derivatives of f, and is represented by the $n \times m$ matrix

$$D(f)(x) = \begin{pmatrix} \dfrac{\partial y_1}{\partial x_1} & \cdots & \dfrac{\partial y_1}{\partial x_n} \\ \vdots & \ddots & \vdots \\ \dfrac{\partial y_m}{\partial x_1} & \cdots & \dfrac{\partial y_m}{\partial x_n} \end{pmatrix}, \text{ where } f : x = \begin{pmatrix} x_1 \\ \vdots \\ x_n \end{pmatrix} \mapsto y = \begin{pmatrix} y_1 \\ \vdots \\ y_m \end{pmatrix}.$$

The partial derivatives are a notation for

$$\frac{\partial y_i}{\partial x_j} = D(x_j \mapsto f_i(x_1, \ldots, x_j, \ldots, x_n))(x_j) \quad \text{where} \quad y_i = f_i(x_1, \ldots, x_n).$$

The tangent map $D(f)(x)$ is denoted sometimes by $D(f)_x$ or Df_x or even df_x. If we use the variable mapping notation $f : x \mapsto y$, we may also write $D(x \mapsto y)(x)$ for $D(f)(x)$. We may also use the partial derivative shortcut,

$$\frac{\partial y}{\partial x} \quad \text{or} \quad \frac{\partial f(x)}{\partial x} \quad \text{for} \quad D(f)(x).$$

The map f is said to be *of class* $\mathcal{C}^k$ if it satisfies the following conditions:

a) $k = 0$, where f is continuous.
b) $0 < k < \infty$, where f is continuous, differentiable and its derivative

$$D(f) : U \to L(\mathbf{R}^n, \mathbf{R}^m) \simeq \mathbf{R}^{n \times m}$$

is of class $\mathcal{C}^{k-1}$.
c) $k = \infty$, where f is of class $\mathcal{C}^n$ for every $n \in \mathbf{N}$.

The set of $\mathcal{C}^k$ mappings from U to V is denoted by $\mathcal{C}^k(U, V)$. For $k = \infty$, we say that f is *infinitely differentiable* or *smooth*. The map f is smooth if and only if all its partial derivatives are smooth. Let V be a domain, we call *smooth parametrization in* V every smooth (infinitely differentiable) map $f : U \to V$, where U is some domain. We denote by $\mathcal{C}^\infty_n(V)$ the set of all smooth n-parametrizations in V, and sometimes by $\mathcal{C}^\infty_\star(V)$ the set of all smooth parametrizations in V.

Axioms of Diffeology

A diffeology *on an arbitrary set is defined by declaring which parametrizations in the set are smooth. These parametrizations must satisfy three axioms:* covering, locality *and* smooth compatibility. *They are required to ensure the coherence with the usual smooth parametrizations in real domains. A set equipped with a diffeology becomes then a* diffeological space. *After the introduction of the axiomatics, we give a few simple examples to familiarize the reader with the basics of this theory.*

1.5. Diffeologies and diffeological spaces. Let X be a nonempty set, a *diffeology* of X is any set $\mathcal{D}$ of parametrizations of X, such that the three following axioms are satisfied:

D1. *Covering.* The set $\mathcal{D}$ contains the constant parametrizations $\mathbf{x} : r \mapsto x$ defined on $\mathbf{R}^n$, for all $x \in X$ and all $n \in \mathbf{N}$.

D2. *Locality.* Let $P : U \to X$ be a parametrization. If for every point r of U there exists an open neighborhood V of r such that $P \upharpoonright V$ belongs to $\mathcal{D}$, then the parametrization P belongs to $\mathcal{D}$.

D3. *Smooth compatibility.* For every element $P : U \to X$ of $\mathcal{D}$, for every real domain V, for every F in $\mathcal{C}^\infty(V, U)$, $P \circ F$ belongs to $\mathcal{D}$.

A *diffeological space* is a nonempty set equipped with a diffeology. Formally, a diffeological space is a pair $(X, \mathcal{D})$ where X is the *underlying set* and $\mathcal{D}$ its *diffeology*, but we shall denote most often the diffeological space by one letter, X for example. Some spaces have a natural implicit diffeology and are just denoted by the letter denoting the underlying set, for example $\mathbf{R}$, $\mathbf{C}$, $\mathcal{C}^\infty(\mathbf{R})$, etc. It will be specified when it will not be the case.

1.6. Plots of a diffeological space. Let X be a diffeological space. The elements of the diffeology $\mathcal{D}$ of X are called the *plots of the space* X. In other words, to be a plot of a diffeological space means to be an element of its diffeology. We shall say an *n-plot* of X for a plot defined on an n-domain. We shall denote by $\mathcal{D}(U, X)$ or $\mathrm{Plots}(U, X)$ the set of the plots of X defined on the domain U, by $\mathcal{D}_n(X)$ or $\mathrm{Plots}_n(X)$ the set of the n-plots of X, and sometimes by $\mathcal{D}_\star(X)$ or $\mathrm{Plots}_\star(X)$ the set $\mathcal{D}$ of all the plots of X. We call *global plot* of X every plot defined on a whole space $\mathbf{R}^n$. The set of global n-plots of X is just $\mathcal{D}(\mathbf{R}^n, X)$, it will also be denoted by $\mathrm{Paths}_n(X)$ (art. 1.64).

1.7. Diffeology or diffeological space? The distinction between a diffeology, as a structure, and a diffeological space, as a set equipped with a diffeology, is purely formal. Every diffeology $\mathcal{D}$ of a set X contains the underlying set as the set of nonempty 0-plots (art. 1.3). The difference between a diffeological space and its diffeology is psychological. Sometimes it is more convenient to think in terms of diffeology rather than in terms of diffeological space, and vice versa.

1.8. The set of diffeologies of a set. Since a diffeology $\mathcal{D}$ of a set X is a subset of $\mathrm{Param}(X)$ (art. 1.3), the set of the diffeologies of a set X, that is,

$$\mathrm{Diffeologies}(X) = \{\mathcal{D} \subset \mathrm{Param}(X) \mid \mathcal{D} \text{ satisfies D1, D2 and D3}\},$$

is a subset of the powerset of $\mathrm{Param}(X)$,

$$\mathrm{Diffeologies}(X) \subset \mathfrak{P}(\mathrm{Param}(X)).$$

1.9. Real domains as diffeological spaces. The set of all smooth parametrizations $\mathcal{C}^\infty_\star(\mathrm{U})$ (art. 1.4) in a domain U is a diffeology. It will be called the *usual diffeology*, or the *standard diffeology*, or the *smooth diffeology* of the domain U. The three axioms (art. 1.5) are obviously satisfied. These are precisely the three properties of smooth parametrizations of real domains which have been chosen to define, by extension, diffeologies. Real domains, equipped with the standard diffeology, are the first and basic examples of diffeological spaces.

1.10. The "wire diffeology". This is an example of an unexpected diffeology. Let us consider the smooth parametrizations $\mathrm{P} : \mathrm{U} \to \mathbf{R}^n$ which factorize locally through $\mathbf{R}$, that is, for every r in U, there exist an open neighborhood V of r, a smooth parametrization $\mathrm{Q} : \mathrm{V} \to \mathbf{R}$ and a smooth 1-parametrization F in $\mathbf{R}^n$ such that $\mathrm{P} \upharpoonright \mathrm{V} = \mathrm{F} \circ \mathrm{Q}$. We check immediately that these parametrizations form a diffeology. This diffeology, called the *wire diffeology*, is characterized by the 1-plots and does not coincide with the standard diffeology of $\mathbf{R}^n$. Indeed, the identity map of $\mathbf{R}^n$ is a plot for the standard diffeology but not a plot for the wire diffeology.[1] We can imagine, of course, other diffeologies of this type: those defined by plots which factorize through $\mathbf{R}^2$, $\mathbf{R}^3$, etc. These examples are related to the construction of diffeologies by means of generating families, (art. 1.66).

1.11. A diffeology for the circle. Let us consider the circle S^1 as the subset of complex numbers,

$$\mathrm{S}^1 = \{z \in \mathbf{C} \mid \bar{z}z = 1\}.$$

The parametrizations $\mathrm{P} : \mathrm{U} \to \mathrm{S}^1$ satisfying the following condition are a diffeology:

(♠) For all r_0 in U, there exist an open neighborhood V of r_0 and a smooth parametrization $\varphi : \mathrm{V} \to \mathbf{R}$ such that $\mathrm{P} \upharpoonright \mathrm{V} : r \mapsto \exp(2i\pi\varphi(r))$.

This diffeology is the quotient diffeology of $\mathbf{R}$ by the exponential (art. 1.50).

PROOF. Let us check the three axioms of a diffeology (art. 1.5).
Axiom D1: Let z be a point of S^1, and let $\mathbf{z} : r \mapsto z$ be the associated global constant n-parametrization. Since the exponential is surjective from $\mathbf{R}$ onto S^1, there exists a real number θ such that $\exp(2i\pi\theta) = z$. Then, for every r_0 we choose $\mathrm{V} = \mathbf{R}^n$ and $\varphi(r) = \theta$.
Axiom D2: By definition, a parametrization satisfying the condition (♠) is local.
Axiom D3: Let $\mathrm{F} : W \to \mathrm{U}$ be a smooth parametrization. Let s_0 be a point of W, let $r_0 = \mathrm{F}(s_0)$, let V and φ as described in the condition (♠). Since φ is a parametrization, V is a domain. Since F is a smooth parametrization, it is continuous, hence $\mathrm{V}' = \mathrm{F}^{-1}(\mathrm{V})$ is a domain and $\varphi \circ \mathrm{F} : \mathrm{V}' \to \mathbf{R}$ is a smooth parametrization. Now, by construction, V' contains s_0. Thus, for every point s_0 of W there exist an open neighborhood V' of s_0 and a smooth parametrization $\varphi' = \varphi \circ \mathrm{F} : \mathrm{V}' \to \mathbf{R}$ such that $(\mathrm{P} \circ \mathrm{F}) \upharpoonright \mathrm{V}' : s \mapsto \exp(2i\pi\varphi'(s))$. Therefore, the parametrization $\mathrm{P} \circ \mathrm{F}$ satisfies the condition (♠). □

1.12. A diffeology for the square. Let us consider the square $\mathrm{Sq} = [0,1] \times \{0,1\} \cup \{0,1\} \times [0,1] \subset \mathbf{R} \times \mathbf{R}$. The set of the parametrizations of the square which,

---

[1] This diffeology has been introduced, as a significant example, by J.-M. Souriau. He named it the *spaghetti diffeology*. But because we use it afterwards, we prefer a more neutral, less culinary, name.

regarded as parametrizations of $\mathbf{R} \times \mathbf{R}$, are smooth, is a diffeology. This diffeology is the subset diffeology, inherited from $\mathbf{R}^2$ (art. 1.33).

PROOF. Let us check the three axioms of a diffeology (art. 1.5).
Axiom D1: Every constant parametrization, regarded as a parametrization in $\mathbf{R}\times\mathbf{R}$, is smooth.
Axiom D2: A parametrization in the square which, regarded as a parametrization in $\mathbf{R} \times \mathbf{R}$, is locally smooth at each point of its domain, is smooth.
Axiom D3: The composite of a plot of the square with any smooth parametrization in the source of the plot does not change its set of values and, regarded as a parametrization in $\mathbf{R} \times \mathbf{R}$, is smooth. □

1.13. A diffeology for the sets of smooth maps. Let A and B be two, nonempty, real domains. Let us consider the set $\mathcal{C}^\infty(A, B)$ of all the smooth maps from A to B. The set $\mathcal{D}$ of parametrizations $P : U \to \mathcal{C}^\infty(A, B)$ satisfying the following condition is a diffeology:

$$(\diamondsuit)\quad \mathbf{P} : (r, s) \mapsto P(r)(s) \in \mathcal{C}^\infty(U \times A, B).$$

This diffeology is the functional diffeology of $\mathcal{C}^\infty(A, B)$ (art. 1.57).

PROOF. First of all, note that $U \times A$ is a domain. Thus, it makes sense to consider the set of smooth parametrizations $\mathcal{C}^\infty(U \times A, B)$. Let us check now the three axioms of a diffeology (art. 1.5).
Axiom D1: Every constant parametrization $P : r \mapsto f$, with $f \in \mathcal{C}^\infty(A, B)$, satisfies

$$\mathbf{P} = [(r, s) \mapsto P(r)(s) = f(s)] = f \circ \mathrm{pr}_2 \in \mathcal{C}^\infty(U \times A, B),$$

where $\mathrm{pr}_2 : U \times A \to A$ is the projection onto the second factor. Then, $\mathbf{P}$ is the composite of two smooth maps, therefore $\mathbf{P}$ is smooth.
Axiom D2: Let us consider a parametrization $P : U \to \mathcal{C}^\infty(A, B)$ such that for all $r_0 \in U$ there exists an open neighborhood V of r_0 such that $P \restriction V$ is a parametrization in $\mathcal{C}^\infty(A, B)$ satisfying $(\diamondsuit)$. Since $P \restriction V$ is a parametrization, V is a domain and $\mathbf{P} \restriction V \times A$ belongs to $\mathcal{C}^\infty(V \times A, B)$. Hence, for all (r_0, s_0) in $U \times A$ there exists an open neighborhood $W = V \times A$ such that $\mathbf{P} \restriction W$ is a smooth parametrization in B. Therefore $\mathbf{P}$ is a smooth parametrization in B, that is, it satisfies $(\diamondsuit)$.
Axiom D3: Let $P : U \to \mathcal{C}^\infty(A, B)$ be a parametrization satisfying the condition $(\diamondsuit)$, and let $F : V \to U$ be a smooth parametrization. Let $P' = P \circ F$, the parametrization $\mathbf{P}' = [(t, s) \mapsto P(F(t), s)] = [(t, s) \mapsto (r = F(t), s) \mapsto P(r)(s)]$, as the composite of two smooth parametrizations, is a smooth parametrization. Thus, $P \circ F$ satisfies the condition $(\diamondsuit)$. Therefore, $\mathcal{D}$ is a diffeology of $\mathcal{C}^\infty(A, B)$. □

Exercises

✎ EXERCISE 1 (Equivalent axiom of covering). Check that the first axiom of diffeology (art. 1.5) can be replaced by Axiom D1′: The values of the elements of $\mathcal{D}$ cover X, that is, $\bigcup_{P\in\mathcal{D}} \mathrm{val}(P) = X$.

✎ EXERCISE 2 (Equivalent axiom of locality). Let X be a set, and $\mathcal{D} \subset \mathrm{Param}(X)$. Show that the axiom D2 of diffeology can be replaced by Axiom D2′: For all integers n, for all families $\{P_i : U_i \to X\}_{i\in\mathcal{J}}$ of n-parametrizations such that $r \in U_i \cap U_j$ implies $P_i(r) = P_j(r)$, if all the P_i belong to $\mathcal{D}$, then the following parametrization

P belongs to $\mathcal{D}$:

$$P : U = \bigcup_{i \in \mathcal{I}} U_i \to X, \quad \text{with} \quad P(r) = P_i(r) \quad \text{if} \quad r \in U_i.$$

NOTE. The family $\{P_i : U_i \to X\}_{i \in \mathcal{I}}$ is said to be a *compatible family* of parametrizations, and P is called the *supremum* of the family. The axiom D2′ expresses itself this way: *the supremum of a compatible family of elements of* $\mathcal{D}$ *belongs to* $\mathcal{D}$.

✎ EXERCISE 3 (Global plots and diffeology). Let X be a set, and let $\mathcal{D}$ and $\mathcal{D}'$ be two diffeologies of X. Show that if for every integer n, $\mathcal{D}(\mathbf{R}^n, X) = \mathcal{D}'(\mathbf{R}^n, X)$ (art. 1.6), then $\mathcal{D}$ and $\mathcal{D}'$ coincide. In other words, prove that the global plots characterize the diffeology.

Smooth Maps and the Category Diffeology

We consider now the notion of smooth maps *between diffeological spaces. They are maps which transform, by composition, the plots of the source into plots of the target. Diffeological spaces, together with smooth maps, define a category denoted by* {Diffeology}. *The isomorphisms of this category are called* diffeomorphisms.

1.14. Smooth maps. Let X and X′ be two diffeological spaces. A map $f : X \to X'$ is said to be *smooth* if for each plot P of X, $f \circ P$ is a plot of X′ (Figure 1.1). The set of smooth maps from X to X′ is denoted by $\mathcal{D}(X, X')$. In other words, denoting by $\mathcal{D}$ and $\mathcal{D}'$ the diffeologies of X and X′,

$$\mathcal{D}(X, X') = \{f \in \mathrm{Maps}(X, X') \mid f \circ \mathcal{D} \subset \mathcal{D}'\}.$$

1.15. Composition of smooth maps. The composition of two smooth maps is a smooth map. Diffeological spaces together with the smooth maps define a category denoted by {Diffeology}.

PROOF. The fact that the composition of smooth maps is smooth results directly from the associativity of the composition of maps. Let X, X′ and X″ be three diffeological spaces whose diffeologies are denoted by $\mathcal{D}$, $\mathcal{D}'$ and $\mathcal{D}''$. Let $f : X \to X'$ and $g : X' \to X''$ be two smooth maps, that is, $f \circ \mathcal{D} \subset \mathcal{D}'$ and $g \circ \mathcal{D}' \subset \mathcal{D}''$. Then, $(g \circ f) \circ \mathcal{D} = g \circ f \circ \mathcal{D} \subset g \circ \mathcal{D}' \subset \mathcal{D}''$. Therefore, $g \circ f$ is smooth. □

1.16. Plots are smooth. Let X be a diffeological space and U be a nonempty real domain (art. 1.9). The smooth maps from U to X are exactly the plots of X defined

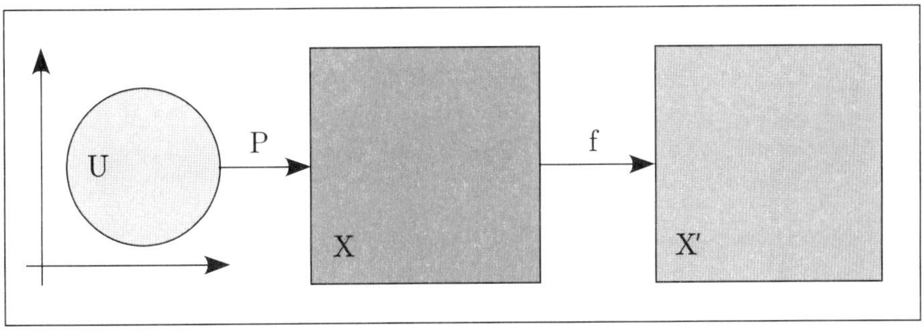

FIGURE 1.1. A smooth map $f \in \mathcal{D}(X, X')$.

on U, and only them. This property, which writes $\mathcal{D}(U, X) = \mathcal{D}(U)$ (art. 1.6), is the very principle of diffeology. The plots of a space X are the parametrizations of X we want to regard as smooth. Now, let $X = V$ be another real domain, then $\mathcal{D}(U, V) = \mathcal{C}^\infty(U, V)$. The smooth maps from U to V, from the diffeological point of view, are exactly the usual smooth maps. This justifies the equivalent notation $\mathcal{C}^\infty(U, X)$ for $\mathcal{D}(U, X)$. By extension, we also denote, for every pair X and X' of diffeological spaces,

$$\mathcal{C}^\infty(X, X') = \mathcal{D}(X, X').$$

PROOF. Let $f : U \to X$ be a smooth parametrization. Thanks to the axiom D3 (art. 1.5), the composite $f \circ \mathbf{1}_U$, where $\mathbf{1}_U$ is the identity of U, is a plot of X. Thus $\mathcal{D}(U, X) \subset \mathcal{D}(U)$. Now, let $P : U \to X$ be a plot. Thanks again to the axiom D3, for every smooth parametrization F in U, the parametrization $P \circ F$ belongs to $\mathcal{D}$. Hence, P is smooth (art. 1.14). Thus, $\mathcal{D}(U) \subset \mathcal{D}(U, X)$. Therefore, $\mathcal{D}(U) = \mathcal{D}(U, X)$. ☐

1.17. Diffeomorphisms. Let X and X' be two diffeological spaces. A map $f : X \to X'$ is called a *diffeomorphism* if f is bijective and if both f and f^{-1} are smooth. Diffeomorphisms are the isomorphisms of the category {Diffeology}. The set of diffeomorphisms from X to X' is denoted by $\mathrm{Diff}(X, X')$. Note that for $X = X'$ we denote by $\mathrm{Diff}(X)$, instead of $\mathrm{Diff}(X, X)$, the diffeomorphisms of X. It is a group for the composition.

PROOF. By definition, the isomorphisms of the category {Diffeology} are the invertible morphisms, that is, invertible smooth maps whose inverses also are smooth. This is what we called diffeomorphisms. ☐

Exercises

✎ EXERCISE 4 (Diffeomorphisms between irrational tori). Let α be some irrational number, $\alpha \in \mathbf{R} - \mathbf{Q}$. Let T_α be the quotient set $\mathbf{R}/(\mathbf{Z} + \alpha\mathbf{Z})$, that is, the quotient of $\mathbf{R}$ by the equivalence relation: $x \sim x'$ if and only if there exist $n, m \in \mathbf{Z}$ such that $x' = x + n + \alpha m$. The set T_α is an Abelian group. We say that T_α is an *irrational torus*. Let $\pi_\alpha : \mathbf{R} \to T_\alpha$ be the canonical projection. Let $\mathcal{D}$ be the set of parametrizations $P : U \to T_\alpha$ such that

(✱) for all $r_0 \in U$ there exist an open neighborhood V of r_0 and a smooth parametrization $Q : V \to \mathbf{R}$ such that $\pi_\alpha \circ Q = P \restriction V$.

1) Check that $\mathcal{D}$ is a diffeology of T_α (art. 1.11).

2) Check that $\mathcal{C}^\infty(T_\alpha, \mathbf{R})$ is reduced to the constants, that is, $\mathcal{C}^\infty(T_\alpha, \mathbf{R}) \simeq \mathbf{R}$.

3) Let α and β be two irrational numbers. Let $f : T_\alpha \to T_\beta$ be a smooth map. Show that there exist an interval $\mathcal{J}$ of $\mathbf{R}$ and some affine map $F : \mathcal{J} \to \mathbf{R}$, such that $\pi_\beta \circ F = f \circ \pi_\alpha \restriction \mathcal{J}$, use the fact that $\mathbf{Z} + \alpha\mathbf{Z}$ is dense in $\mathbf{R}$. Then, show that F can be extended to the whole $\mathbf{R}$ in an affine map. Deduce that $\mathcal{C}^\infty(T_\alpha, T_\beta)$ does not reduce to the constant maps if and only if there exist four integers a, b, c and d such that

$$\alpha = \frac{a + \beta b}{c + \beta d}\cdot$$

4) Show that T_α and T_β are diffeomorphic if and only if α and β are *conjugate* modulo $\mathrm{GL}(2, \mathbf{Z})$, that is, the four integers a, b, c, d of the question 1) satisfy $ad - bc = \pm 1$.

✎ Exercise 5 (Smooth maps on $\mathbf{R}/\mathbf{Q}$). Let $E_{\mathbf{Q}}$ be the quotient of $\mathbf{R}$ by $\mathbf{Q}$. Let $\pi : \mathbf{R} \to E_{\mathbf{Q}}$ be the projection. Since the set $\mathbf{R}$ is a vector space over $\mathbf{Q}$, and $\mathbf{Q}$ is a $\mathbf{Q}$-vector subspace of $\mathbf{R}$, the quotient $E_{\mathbf{Q}}$ inherits the structure of quotient vector space over $\mathbf{Q}$:

- for all $\tau, \tau' \in E_{\mathbf{Q}}$, if $\tau = \pi(x)$ and $\tau' = \pi(x')$, then $\tau + \tau' = \pi(x) + \pi(x') = \pi(x + x')$,
- for all $q \in \mathbf{Q}$ and all $\tau \in E_{\mathbf{Q}}$, if $\tau = \pi(x)$ then $q \cdot \tau = \pi(qx)$.

Let us equip the quotient vector space $E_{\mathbf{Q}}$ with the diffeology defined, *mutatis mutandis*, by the condition (✱) of Exercise 4, p. 8.

1) Show that the only smooth maps from $E_{\mathbf{Q}}$ to $E_{\mathbf{Q}}$ are the $\mathbf{Q}$-affine maps:

$$\mathcal{C}^\infty(E_{\mathbf{Q}}) = \{\tau \mapsto q \cdot \tau + \tau' \mid q \in \mathbf{Q} \text{ and } \tau' \in E_{\mathbf{Q}}\}.$$

2) Show that the only smooth maps from any irrational torus T_α (Exercise 4, p. 8) to $E_{\mathbf{Q}}$ are the constant maps, $\mathcal{C}^\infty(T_\alpha, E_{\mathbf{Q}}) = E_{\mathbf{Q}}$.

3) Also show that $\mathcal{C}^\infty(E_{\mathbf{Q}}, \mathbf{R}) = \mathbf{R}$ and $\mathcal{C}^\infty(E_{\mathbf{Q}}, T_\alpha) = T_\alpha$.

✎ Exercise 6 (Smooth maps on spaces of maps). Let $\mathcal{C}^\infty(\mathbf{R}) = \mathcal{C}^\infty(\mathbf{R}, \mathbf{R})$, equipped with the functional diffeology defined in (art. 1.13).

1) Let f be an element of $\mathcal{C}^\infty(\mathbf{R})$, and let $f^{(k)}$ denote its k-th derivative. Show that, for every integer k, the following map is smooth:

$$\frac{d^k}{dx^k} : \mathcal{C}^\infty(\mathbf{R}) \to \mathcal{C}^\infty(\mathbf{R}) \quad \text{defined by} \quad \frac{d^k}{dx^k}(f) = f^{(k)}.$$

2) Show that, for every real number x, the map $\hat{x} : f \mapsto f(x)$, from $\mathcal{C}^\infty(\mathbf{R})$ to $\mathbf{R}$, is smooth. Deduce that, for every real number x, for every integer k, the following map, called the k-*jet* or the *jet of order* k, is smooth:

$$D^k_x : \mathcal{C}^\infty(\mathbf{R}) \to \mathbf{R}^{k+1} \quad \text{defined by} \quad D^k_x(f) = (f(x), f'(x), \ldots, f^{(k)}(x)).$$

3) Show that, for any pair of real numbers a and b, the following map $I_{a,b}$ is smooth, where the sign $\int$ denotes the Riemann integral:

$$I_{a,b} : \mathcal{C}^\infty(\mathbf{R}) \to \mathbf{R} \quad \text{with} \quad I_{a,b}(f) = \int_a^b f(t)\, dt.$$

4) Let $\mathcal{C}^\infty_0(\mathbf{R})$ be the space of smooth real maps f such that $f(0) = 0$. Check that the parametrizations of $\mathcal{C}^\infty_0(\mathbf{R})$ satisfying the condition (♢) of (art. 1.13) are still a diffeology. Then, using question 3 above, deduce that the derivative $d/dx : f \mapsto f'$, defined from $\mathcal{C}^\infty_0(\mathbf{R})$ to $\mathcal{C}^\infty(\mathbf{R})$, is a diffeomorphism.

Comparing Diffeologies

The inclusion of diffeologies, regarded as subsets of the powerset of all the parametrizations of a set, is a partial ordering. This relation, called fineness, *is a key for the categorical constructions of the theory.*

1.18. Fineness of diffeologies. Let X be a set, a diffeology $\mathcal{D}$ of X is said to be *finer* than another $\mathcal{D}'$ if $\mathcal{D} \subset \mathcal{D}'$. This relation, called *fineness*, is a partial order. We say equivalently that $\mathcal{D}$ is finer than $\mathcal{D}'$ or $\mathcal{D}'$ is *coarser* than $\mathcal{D}$. In other words, $\mathcal{D}$ is finer than $\mathcal{D}'$ if $\mathcal{D}$ has fewer elements than $\mathcal{D}'$. If X' denotes the set X equipped with the diffeology $\mathcal{D}'$, we also denote $X \preceq X'$ to mean that $\mathcal{D}$ is finer than $\mathcal{D}'$.

1.19. Fineness via the identity map. Let us denote by X_1 and X_2 the same set X equipped with two diffeologies $\mathcal{D}_1$ and $\mathcal{D}_2$. The diffeology $\mathcal{D}_1$ is finer than $\mathcal{D}_2$ if and only if the identity map $1_X : X_1 \to X_2$ is smooth.

1.20. Discrete diffeology. Let X be a set. The *locally constant parametrizations* of X are defined as follows:

(♣) A parametrization $P : U \to X$ is said to be *locally constant* if for all $r \in U$ there exists an open neighborhood V of r such that $P \restriction V$ is constant.

The locally constant parametrizations in X form a diffeology called the *discrete diffeology*. The set X, equipped with the discrete diffeology, will be denoted by $X_\circ$, and the discrete diffeology itself by $\mathcal{D}_\circ(X)$. A set equipped with the discrete diffeology will be called a *discrete diffeological space*.

NOTE. For every set X, the discrete diffeology is the finest diffeology of X, that is, finer than any other diffeology of X. Every diffeology contains the discrete diffeology. Moreover, let X and X′ be any two sets equipped with the discrete diffeology. Every map f from X to X′ is smooth, that is, $\mathcal{C}^\infty(X_\circ, X'_\circ) = \text{Maps}(X, X')$. In other words, the correspondence ∘ defined from the category {Set} into the category {Diffeology} by $\circ(X) = X_\circ$, and $\circ(f) = f$, is a full faithful functor called the *discrete functor*.

PROOF. The axioms of diffeology are satisfied by the locally constant parametrizations. For the covering axiom: every constant parametrization is locally constant. For the locality axiom: it is satisfied by the very definition of locally constant parametrizations. For the smooth compatibility: the composite of a locally constant parametrization by a smooth parametrization in its domain is again locally constant. Now, let $\mathcal{D}$ be any other diffeology of X. Let n be any integer, and $P : U \to X$ be a locally constant n-parametrization. Let $r \in U$ and $x = P(r)$. Let $\mathbf{x} : \mathbf{R}^n \to X$ be the constant n-parametrization with value x. By the axiom D1 of diffeology, the parametrization $\mathbf{x}$ is a plot for the diffeology $\mathcal{D}$. Since P is locally constant, there exists an open neighborhood V of r such that $P \restriction V$ is a constant parametrization. Let us denote by $j_V : V \to \mathbf{R}^n$ the natural inclusion, it is a smooth parametrization in $\mathbf{R}^n$. By axiom D2, $\mathbf{x} \circ j_V = P \restriction V$ belongs to $\mathcal{D}$. Hence, P belongs locally to $\mathcal{D}$ everywhere. Thus, by axiom D3, P belongs to $\mathcal{D}$. Therefore, the discrete diffeology $\mathcal{D}_\circ(X)$ is contained in every diffeology $\mathcal{D}$ of X. It is the finest diffeology of X. Next, consider a map $f : X \to X'$ where X and X′ are equipped with the discrete diffeology. Let us recall that f is said to be smooth if the composite of f with any plot of X is a plot of X′ (art. 1.14). The composite of any map with any locally constant parametrization is locally constant. Indeed, let $P : U \to X$ be a locally constant parametrization. Let $r \in U$ and $x = P(r)$. Let V be an open neighborhood of r such that $P \restriction V$ is the constant parametrization defined on V with value x. Hence, $f \circ P \restriction V$ is the constant parametrization in X′, defined on V, with value f(x). Therefore $f \circ P$ is locally constant and $f \in \mathcal{C}^\infty(X_\circ, X'_\circ)$. □

1.21. Coarse diffeology. Let X be a set. The set of all the parametrizations in X is a diffeology. This diffeology, coarser than every other one, is called the *coarse diffeology*. The set X equipped with the coarse diffeology will be denoted by $X_\bullet$, and the coarse diffeology itself by $\mathcal{D}_\bullet(X)$, that is, $\mathcal{D}_\bullet(X) = \text{Param}(X)$. A set equipped with the coarse diffeology will be called a *coarse diffeological space*. Moreover, let X and X′ be two sets equipped with the coarse diffeology. Every map f from X to X′ is smooth, that is, $\mathcal{C}^\infty(X_\bullet, X'_\bullet) = \text{Maps}(X, X')$. In other words, the correspondence

• from the category {Set} to the category {Diffeology} by $\bullet(X) = X_\bullet$ and $\bullet(f) = f$, is a full faithful functor called the *coarse functor*.

NOTE. Combining the existence of the coarse diffeology on every set X with the existence of the discrete diffeology (art. 1.20), we conclude that every diffeology $\mathcal{D}$ of X is somewhere in between,

$$\mathcal{D}_\circ(X) \subset \mathcal{D} \subset \mathcal{D}_\bullet(X).$$

PROOF. For the set Param(X) the two first axioms of diffeology are obviously satisfied. For the third one, just remark that the composite of a parametrization in X with a smooth parametrization in its source is still a parametrization, since to be a parametrization is just to be defined on a real domain. Now, the composite of any map $f : X \to X'$ with any parametrization in X is a parametrization in X', for the same reason as above. Hence, every map $f : X \to X'$ belongs to $\mathcal{C}^\infty(X_\bullet, X'_\bullet)$, therefore $\mathcal{C}^\infty(X_\bullet, X'_\bullet) = \text{Maps}(X, X')$. ☐

1.22. Intersecting diffeologies. Let X be any set. Let $\mathbf{D} = \{\mathcal{D}_i\}_{i\in\mathcal{I}}$ be any family of diffeologies of X, indexed by some set $\mathcal{I}$. The intersection of the elements of the family, denoted by

$$\mathcal{D} = \bigcap_{i\in\mathcal{I}} \mathcal{D}_i,$$

is still a diffeology of X, finer than each element of the family $\mathbf{D}$. Note that we may also denote just by $\cap\mathbf{D}$ the intersection $\bigcap_{i\in\mathcal{I}} \mathcal{D}_i$.

PROOF. Let us first check the three axioms of diffeology (art. 1.5).
Axiom D1: Since every diffeology contains the constant parametrizations, they are contained in each element $\mathcal{D}_i$ of $\mathbf{D}$. Hence, the constant parametrizations also are contained in the intersection $\bigcap_{i\in\mathcal{I}} \mathcal{D}_i$.
Axiom D2: Let $P : U \to X$ be a parametrization in X such that for every point r in U there exists an open neighborhood V of r for which $P \upharpoonright V$ is a parametrization belonging to the intersection $\bigcap_{i\in\mathcal{I}} \mathcal{D}_i$, that is, for each i in $\mathcal{I}$, $P \upharpoonright V \in \mathcal{D}_i$. Thus, by application of the axiom D2 to the diffeology $\mathcal{D}_i$, P is a plot for $\mathcal{D}_i$. Therefore, for all $i \in \mathcal{I}$, $P \in \mathcal{D}_i$, that is, $P \in \bigcap_{i\in I} \mathcal{D}_i$.
Axiom D3: Let $P \in \bigcap_{i\in\mathcal{I}} \mathcal{D}_i$ and let F be a smooth parametrization in the domain of P. For each i in $\mathcal{I}$, $P \in \mathcal{D}_i$. Since each $\mathcal{D}_i$ is a diffeology: $P \circ F$ belongs to $\mathcal{D}_i$, hence $P \circ F$ belongs to the intersection $\bigcap_{i\in\mathcal{I}} \mathcal{D}_i$. Therefore, $\mathcal{D} = \bigcap_{i\in\mathcal{I}} \mathcal{D}_i$ is a diffeology of X. By construction, this diffeology is contained in every $\mathcal{D}_i$, that is, $\mathcal{D}$ is finer than every element of the family $\mathbf{D}$. ☐

1.23. Infimum of a family of diffeologies. Let X be a set and $\mathbf{D}$ be some family of diffeologies of X. The family $\mathbf{D}$ has an infimum for the fineness partial ordering (art. 1.18), this is the intersection of the elements of the family $\mathbf{D}$ (art. 1.22).

$$\inf(\mathbf{D}) = \bigcap_{\mathcal{D}\in\mathbf{D}} \mathcal{D} = \{P \in \text{Param}(X) \mid \text{for all } \mathcal{D} \in \mathbf{D}, P \in \mathcal{D}\}.$$

It is the coarsest diffeology contained in every element of the family $\mathbf{D}$.

PROOF. Let us recall that the infimum of $\mathbf{D}$, if it exists, is the greatest lower bound of $\mathbf{D}$. A diffeology $\mathcal{D}'$ is a lower bound of $\mathbf{D}$ if and only if $\mathcal{D}' \subset \mathcal{D}$ for all $\mathcal{D} \in \mathbf{D}$. The set of lower bounds of $\mathbf{D}$ is not empty, since it contains the discrete diffeology (art. 1.21). Since $\cap\mathbf{D} = \bigcap_{\mathcal{D}\in\mathbf{D}} \mathcal{D}$ is a diffeology (art. 1.22) and since for all $\mathcal{D}' \in \mathbf{D}$,

$\bigcap_{\mathcal{D}\in\mathbf{D}} \mathcal{D} \subset \mathcal{D}'$, $\cap\mathbf{D}$ is a lower bound of $\mathbf{D}$. Now, let $\mathcal{D}'$ be a lower bound of $\mathbf{D}$, *i.e.*, $\mathcal{D}' \subset \mathcal{D}$ for all $\mathcal{D} \in \mathbf{D}$. Thus, $\mathcal{D}' \subset \bigcap_{\mathcal{D}\in\mathbf{D}} \mathbf{D} = \cap\mathbf{D}$. Therefore, $\cap\mathbf{D}$ is the greatest lower bound of $\mathbf{D}$, that is, its infimum. □

1.24. Supremum of a family of diffeologies. Let X be a set and $\mathbf{D}$ be some family of diffeologies of X. The family $\mathbf{D}$ has a supremum for the fineness partial ordering (art. 1.18), this is the infimum of the diffeologies of X containing all the elements of $\mathbf{D}$ (art. 1.23).

$$\sup(\mathbf{D}) = \inf\{\mathcal{D}' \in \text{Diffeologies}(\mathrm{X}) \mid \text{for all } \mathcal{D} \in \mathbf{D}, \mathcal{D} \subset \mathcal{D}'\}.$$

It is the finest diffeology containing every element of the family $\mathbf{D}$,

PROOF. Let us recall that the supremum of $\mathbf{D}$, if it exists, is the smallest upper bound of $\mathbf{D}$. A diffeology $\mathcal{D}'$ is an upper bound of $\mathbf{D}$ if and only if, for all $\mathcal{D} \in \mathbf{D}$, $\mathcal{D} \subset \mathcal{D}'$. The set of the upper bounds of $\mathbf{D}$ is not empty, since it contains the coarsest diffeology (art. 1.21). Let us consider the intersection of all the upper bounds of $\mathbf{D}$. We know that it is a diffeology (art. 1.22). Next, since every $\mathcal{D} \in \mathbf{D}$ is contained in every upper bound, it is also contained in their intersection. Thus, this intersection is an upper bound of $\mathbf{D}$. Then, since the intersection of the upper bounds of the family $\mathbf{D}$ is the infimum of its upper bounds (art. 1.23), this intersection is, by definition, the supremum of the family $\mathbf{D}$. □

1.25. Playing with bounds. It's a nice property that any family of diffeologies of a set X has an infimum and a supremum. This property has even got a name by Bourbaki, the set Diffeologies(X) is said to be *réticulé achevé* [**Bou72**]; in English, it is said to be a *complete lattice*. This property will be heavily used in order to introduce many diffeologies. It is often uncertain to comment upon what is still not defined, but sometimes it can help the reader to guess what is coming. Many of the cases that we shall meet follow the same pattern. We have a set X and some property $\mathcal{P}$ relating to the diffeologies of X — we can understand $\mathcal{P}$ as a function from Diffeologies(X) to the set {true, false}. The diffeologies for which $\mathcal{P}$ is satisfied constitute some set $\mathbf{D} = \mathcal{P}^{-1}(\text{true})$. This set defines two distinguished diffeologies, its infimum $\inf(\mathbf{D})$ and its supremum $\sup(\mathbf{D})$. In most cases, one or the other of these bounds satisfies the property $\mathcal{P}$. Then, we get one (or two) distinguished diffeologies satisfying $\mathcal{P}$, which deserves to be noticed. Translated in terms of diffeology this gives "the finest diffeology such that..." or "the coarsest diffeology such that...". Infimum and supremum change then into minimum and maximum. Here are two examples we shall find again later.

EXAMPLE 1. The diffeology generated by a set $\mathcal{F}$ of parametrizations (art. 1.66) is the minimum of the set of all the diffeologies containing $\mathcal{F}$.

EXAMPLE 2. Let Y be a diffeological space, and let X be a set. Let $f : X \to Y$ be some map. The supremum of the diffeologies of X such that f is smooth is a maximum, and it is called the *pullback* of the diffeology of Y (art. 1.26). In other words, it is the coarsest diffeology such that f is smooth.

These two examples illustrate a frequent phenomenon in diffeology:

CASE 1. If a family of diffeologies clearly contains the coarse diffeology (the right black point of the Figure 1.2), then the distinguished diffeology is the lower bound, the finest diffeology such that.... This is the case of Example 1. The set of diffeologies containing $\mathcal{F}$ clearly contains the coarse diffeology.

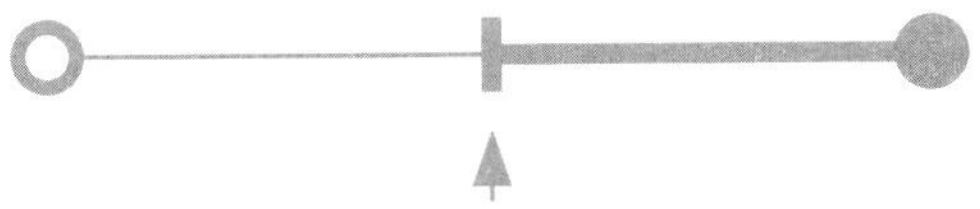

FIGURE 1.2. Case 1 — The finest diffeology such that...

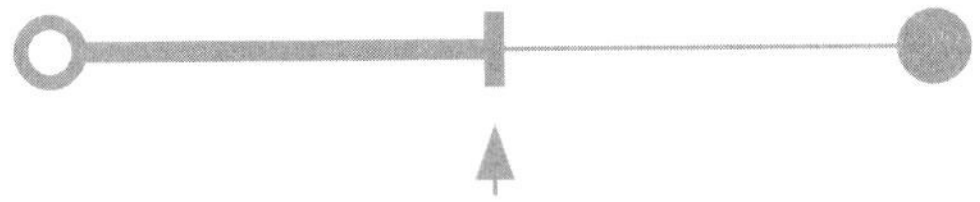

FIGURE 1.3. Case 2 — The coarsest diffeology such that...

CASE 2. If a family of diffeologies clearly contains the discrete diffeology (the left white point of the Figure 1.3), then the distinguished diffeology is the upper bound, the coarsest diffeology such that.... This is the case in the second example. The set of all the diffeologies such that f is smooth clearly contains the discrete diffeology because if the left composite of any map with a locally constant parametrization is locally constant, then a plot of every diffeology.

But Figures 1.2 and 1.3 suggest wrongly that fineness is a total order, which is obviously not the case. Figure 1.4 is more representative of the real situation. In this figure, the gray domains represent families of diffeologies, and the dots their infimum and supremum. The far left dot represents the discrete diffeology and the far right dot, the coarse diffeology. One can argue, of course, that every diffeology is always finer than any other coarser diffeology, or conversely coarser than any other finer. Then, any diffeology is distinguished, and hence, all this discussion does not make sense. But it does not matter, we shall see how this discussion applies in concrete situations.

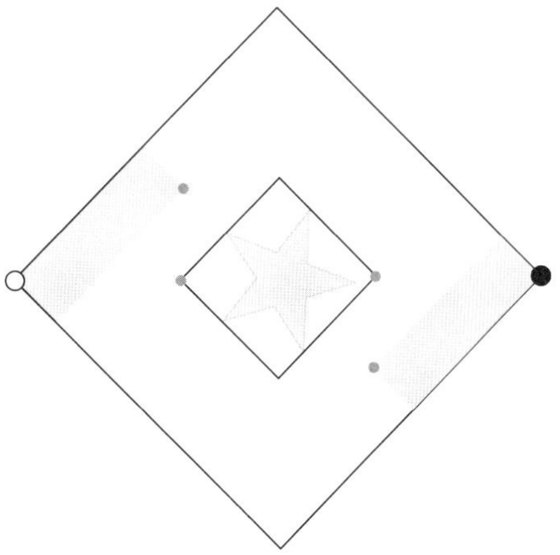

FIGURE 1.4. Diffeologies and bounds of families: inf, sup, min, max.

Exercises

✎ EXERCISE 7 (Locally constant parametrizations). Let $P : U \to X$ be a parametrization in some set X. Show that P is locally constant (art. 1.20) if and only if P is constant on each *pathwise-connected component* of its domain U, that is, if and only if, for every pair of points r and r′ in U, connected by a smooth path $\gamma \in \mathcal{C}^\infty(\mathbf{R}, U)$, $\gamma(0) = r$ and $\gamma(1) = r'$, then $P(r) = P(r')$.

✎ EXERCISE 8 (Diffeology of $\mathbf{Q} \subset \mathbf{R}$). Check that the set of smooth parametrizations in **R** with values in **Q** is a diffeology of **Q**. Show that this diffeology is discrete. Be aware that in the category {Topology} **Q** is not discrete in **R**. More generally, show that every countable subset of every real domain (and by consequence, of every manifold) is diffeologically discrete.

✎ EXERCISE 9 (Smooth maps from discrete spaces). Show that every map defined on a discrete diffeological space to any other diffeological space is smooth, that is, $\mathcal{C}^\infty(X_\circ, X') = \mathrm{Maps}(X, X')$.

✎ EXERCISE 10 (Smooth maps to coarse spaces). Show that every map defined from any diffeological space to a coarse diffeological space is smooth, that is, $\mathcal{C}^\infty(X, X'_\bullet) = \mathrm{Maps}(X, X')$.

Pulling Back Diffeologies

In this section we describe the transfer of diffeology by pullback. *Given any set* X *and any map* f *from* X *to a diffeological space* X′, *the set* X *inherits a natural diffeology by pullback of the diffeology of* X′.

1.26. Pullbacks of diffeologies. Let X be a set. Let X′ be a diffeological space and $\mathcal{D}'$ its diffeology. Let $f : X \to X'$ be some map. There exists a coarsest diffeology of X such that the map f is smooth. This diffeology is called the *pullback* of the diffeology $\mathcal{D}'$ by f, and it is denoted by $f^*(\mathcal{D}')$. A parametrization P in X belongs to $f^*(\mathcal{D}')$ if and only if $f \circ P$ belongs to $\mathcal{D}'$, which writes

$$f^*(\mathcal{D}') = \{P \in \mathrm{Param}(X) \mid f \circ P \in \mathcal{D}'\}.$$

PROOF. Let us consider the family **D** of all the diffeologies of X such that f is smooth. A diffeology of X belongs to **D** if and only if the left composition of each of its elements by f is an element of $\mathcal{D}'$. By the way, note that the discrete diffeology of X belongs to **D** (Exercise 9, p. 14). We know that the family **D** has a supremum (art. 1.24). It is the intersection of all the diffeologies of X containing the set of all the parametrizations P of X such that $f \circ P$ belongs to $\mathcal{D}'$. But this set, let us call it $\mathcal{D}$, is already a diffeology, let us check it.

D1. The composition of every constant map by any function f is constant. Hence, the constant parametrizations belong to $\mathcal{D}$.

D2. Let $P : U \to X$ be a parametrization such that, locally at each point r of U, $f \circ P \in \mathcal{D}'$. Since $\mathcal{D}'$ is a diffeology, $f \circ P$ is a plot for $\mathcal{D}'$. Hence, P belongs to $\mathcal{D}$.

D3. Let P be a parametrization in X. If $f \circ P$ is a plot of X′, for every smooth parametrization F in the domain of P, $f \circ (P \circ F) = (f \circ P) \circ F$ is a plot of X′. Hence, $P \circ F$ belongs to $\mathcal{D}$.

Therefore, $\mathcal{D} = \{P \in \mathrm{Param}(X) \mid f \circ P \in \mathcal{D}'\}$ is the maximum of **D**, that is, the coarsest diffeology such that f is smooth. □

1.27. Smoothness and pullbacks. Let X and X' be two diffeological spaces, whose diffeologies are denoted by $\mathcal{D}$ and $\mathcal{D}'$. The notion of pullback of diffeologies gives a new interpretation of the notion of smooth maps: a map f from X to X' is smooth if and only if $\mathcal{D} \subset f^*(\mathcal{D}')$. In other words,

$$\mathcal{C}^\infty(\mathrm{X}, \mathrm{X}') = \{f \in \mathrm{Maps}(\mathrm{X}, \mathrm{X}') \mid \mathcal{D} \subset f^*(\mathcal{D}')\}.$$

PROOF. If f is smooth, then, for all $P \in \mathcal{D}$, $f \circ P \in \mathcal{D}'$, by definition. But $f^*(\mathcal{D}')$ is the set of parametrizations P in X such that $f \circ P \in \mathcal{D}'$ (art. 1.26). Thus, $P \in f^*(\mathcal{D}')$, that is, $\mathcal{D} \subset f^*(\mathcal{D}')$. Conversely, if $\mathcal{D} \subset f^*(\mathcal{D}')$, then, for all $P \in \mathcal{D}$, $f \circ P \in \mathcal{D}'$, thus f is smooth. Hence, f is smooth if and only if $\mathcal{D} \subset f^*(\mathcal{D}')$. ☐

1.28. Composition of pullbacks. Let X and X' be two sets, and X'' be a diffeological space. Let $f : \mathrm{X} \to \mathrm{X}'$ and $g : \mathrm{X}' \to \mathrm{X}''$ be two maps. The pullback by f of the pullback by g of the diffeology $\mathcal{D}''$ of X'' is equal to the pullback of $\mathcal{D}''$ by $g \circ f$,

$$\mathrm{X} \xrightarrow{f} \mathrm{X}' \xrightarrow{g} (\mathrm{X}'', \mathcal{D}'') \quad \text{and} \quad f^*(g^*(\mathcal{D}'')) = (g \circ f)^*(\mathcal{D}'').$$

In other words, the pullback of diffeologies is contravariant.

PROOF. This is a direct consequence of the associativity of the composition of maps, and of the characterization of the plots of the pullback diffeology (art. 1.26). Let $\mathcal{D}' = g^*(\mathcal{D}'') = \{P \in \mathrm{Param}(\mathrm{X}') \mid g \circ P \in \mathcal{D}''\}$. Then,

$$\begin{aligned} f^*(g^*(\mathcal{D}'')) &= \{P \in \mathrm{Param}(\mathrm{X}) \mid f \circ P \in \mathcal{D}'\} \\ &= \{P \in \mathrm{Param}(\mathrm{X}) \mid g \circ (f \circ P) \in \mathcal{D}''\} \\ &= \{P \in \mathrm{Param}(\mathrm{X}) \mid (g \circ f) \circ P \in \mathcal{D}''\} \\ &= (g \circ f)^*(\mathcal{D}''). \end{aligned}$$

☐

Exercise

✎ EXERCISE 11 (Square root of the smooth diffeology). Let $\mathcal{D}$ be the pullback, by the *square map* $\mathrm{sq} : x \mapsto x^2$, of the smooth diffeology $\mathcal{C}^\infty_\star(\mathbf{R})$ of $\mathbf{R}$ (art. 1.9). Is $\mathcal{D}$ finer or coarser than $\mathcal{C}^\infty_\star(\mathbf{R})$? Check that the map $|\cdot| : x \mapsto |x|$, defined on $\mathbf{R}$, is a plot for the diffeology $\mathcal{D}$. Conclude that all the parametrizations of the kind $P : r \mapsto |Q(r)|$, where $Q \in \mathcal{C}^\infty_\star(\mathbf{R})$, belong to $\mathcal{D}$.

Inductions

Inductions *are injections between diffeological spaces, identifying the source space with the pullback* (art. 1.26) *of the target. They are a key categorical construction, used in particular in the definition of diffeological subspaces.*

1.29. What is an induction? Let X and X' be two diffeological spaces, and let $f : \mathrm{X} \to \mathrm{X}'$ be some map. The map f is said to be *inductive* or to be an *induction* if the following two conditions are satisfied.

1. The map f is injective.
2. The diffeology $\mathcal{D}$ of X is the pullback by f of the diffeology $\mathcal{D}'$ of X', that is, with the notation introduced above, $f^*(\mathcal{D}') = \mathcal{D}$.

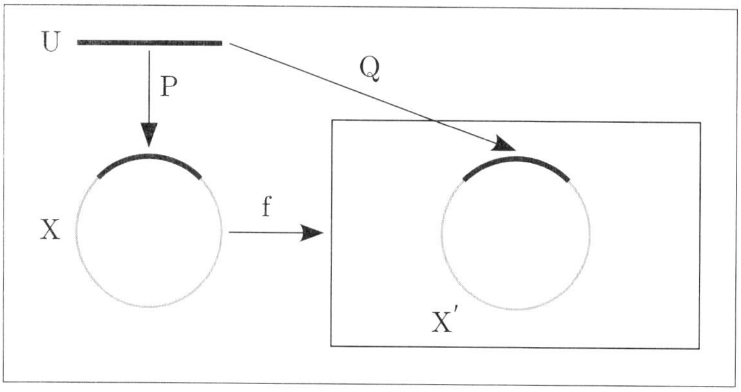

FIGURE 1.5. An induction.

NOTE. The second condition also writes $\mathcal{D} \subset f^*(\mathcal{D}')$ and $f^*(\mathcal{D}') \subset \mathcal{D}$, where the first inclusion $\mathcal{D} \subset f^*(\mathcal{D}')$ just says that f is smooth (art. 1.27).

1.30. Composition of inductions. The composite of two inductions is an induction. Inductions form a subcategory of the category {Diffeology}, which can be naturally denoted by {Inductions}.

PROOF. Let X, X′ and X″ be three diffeological spaces, and let us denote by $\mathcal{D}$, $\mathcal{D}'$ and $\mathcal{D}''$ their diffeologies. Let $f : X \to X'$ and $g : X' \to X''$ be two inductions, that is, $g^*(\mathcal{D}'') = \mathcal{D}'$ and $f^*(\mathcal{D}') = \mathcal{D}$. Since pullbacks are contravariant (art. 1.28), $(g \circ f)^*(\mathcal{D}'') = f^*(g^*(\mathcal{D}'')) = f^*(\mathcal{D}') = \mathcal{D}$. Since the composite of two injections is an injection, $g \circ f$ is injective. Therefore, $g \circ f$ is an induction. □

1.31. Criterion for being an induction. Let X and X′ be two diffeological spaces. A map $f : X \to X'$ is an induction (art. 1.29) if and only if the following two conditions, illustrated by the Figure 1.5, are satisfied:

1. The map f is a smooth injection.
2. For every plot P of X′ with values in f(X), the parametrization $f^{-1} \circ P$ is a plot of X.

PROOF. Let us recall that the map $f : X \to X'$ is an induction if and only if it is injective and $\mathcal{D} = f^*(\mathcal{D}')$. Let us assume first that f is an induction. Hence f is injective and smooth (art. 1.29). Now, let $P : U \to X'$ be a plot with values in f(X). Hence, $Q = f^{-1} \circ P : U \to X$ is a parametrization in X such that $P = f \circ Q$ is a plot of X′. Thus, by the definition of induction, Q is a plot of X (art. 1.29). Therefore, for any plot P of X′ with values in f(X), the map $f^{-1} \circ P$ is a plot of X. Conversely, let f be a smooth injection such that, for any plot P of X′ with values in f(X), the map $f^{-1} \circ P$ is a plot of X. Let P be a parametrization in X such that $f \circ P$ is a plot of X′, that is, $P \in f^*(\mathcal{D}')$. The parametrization $Q = f \circ P$ is a plot of X′ with values in f(X). By hypothesis $f^{-1} \circ Q = f^{-1} \circ f \circ P = P$ is a plot of X, hence $f^*(\mathcal{D}') \subset \mathcal{D}$. Therefore, since f is smooth and injective, f is an induction. □

1.32. Surjective inductions. Let X and X′ be two diffeological spaces. Let $f : X \to X'$ be an induction. If f is surjective, then f is a diffeomorphism. Thus, surjective inductions are diffeomorphisms. Conversely, diffeomorphisms are surjective inductions. This behavior expresses the strong nature of inductions. Compare, for example, with the different behavior of local inductions (art. 2.15).

PROOF. An induction is, by definition, smooth and injective (art. 1.29). If moreover, f is surjective, then f is a bijection, a smooth bijection. By application of the characterization of induction above (art. 1.31), the map f^{-1} also is smooth. Therefore, f is a diffeomorphism. Conversely, let us denote by $\mathcal{D}$ and $\mathcal{D}'$ the diffeologies of X and X'. If f is a diffeomorphism, then it is smooth, thus $f \circ \mathcal{D} \subset \mathcal{D}'$ or $\mathcal{D} \subset f^*(\mathcal{D}')$ (art. 1.27). Now its inverse also is smooth, thus $f^{-1} \circ \mathcal{D}' \subset \mathcal{D}$, that is, $\mathcal{D}' \subset f \circ \mathcal{D}$ or $\mathcal{D} \subset f^*(\mathcal{D}')$. Therefore $f^*(\mathcal{D}') = \mathcal{D}$, f is an induction, and surjective. ☐

Exercises

✎ EXERCISE 12 (Immersions of real domains). Let U be an n-domain, regarded as a diffeological space (art. 1.9). Let $f : U \to \mathbf{R}^m$ be a smooth map, and $r \in U$. Show that, if f is an *immersion* at the point r, that is, if the tangent linear map $D(f)(r) : \mathbf{R}^n \to \mathbf{R}^m$ is injective, then there exists an open neighborhood $\mathcal{O}$ of r such that $f \restriction \mathcal{O}$ is an induction. Apply the implicit function theorem, or the rank theorem, for smooth parametrizations; see, for example, [**Die70a**].

✎ EXERCISE 13 (Flat points of smooth paths). Let γ be a smooth parametrization defined from an interval $]-\varepsilon, +\varepsilon[$ to $\mathbf{R}^n$. Let $\gamma(0) = 0$. We say that γ is *flat* at 0 if all the derivatives of γ vanish at 0, that is, if $\gamma^{(k)}(0) = 0$, for all $k > 0$. Show that, if there exists a sequence of numbers $(t_n)_{n=1}^{\infty}$ converging to 0, such that $t_n < t_{n+1}$ and $\gamma(t_n) = 0$, for all n, then $\gamma(0) = 0$ and γ is flat at 0.

✎ EXERCISE 14 (Induction of intervals into domains). Let $\varepsilon > 0$, let f be a smooth injection from an interval $]-\varepsilon, +\varepsilon[$ to $\mathbf{R}^n$, such that $f(0) = 0$. We say that f is *flat* at the point 0 if, for all positive integers p, the p-th derivative of f at 0 vanishes, which is denoted by $f^{(p)}(0) = 0$ (see Exercise 13, p. 17).

1) Show that, if f is flat at the point 0, then f is not an induction. Thus, an induction from $]-\varepsilon, +\varepsilon[$ to $\mathbf{R}^n$ is nowhere flat.

2) Show that, if f is an induction, there exist a smallest integer $p \geq 0$ and a smooth map $\varphi :]-\varepsilon, +\varepsilon[\to \mathbf{R}^n$, such that

$$\text{for all } t \in]-\varepsilon, +\varepsilon[, \quad f(t) = t^p \times \varphi(t) \quad \text{and} \quad \varphi'(t) \neq 0.$$

NOTE. An injective immersion from an interval to $\mathbf{R}^n$, that is, a smooth injection whose first derivative never vanishes, is not necessarily an induction. A counterexample is treated in the Exercise 59, p. 58. At the moment I write this exercise I still do not know if there exist inductions of domains in $\mathbf{R}^n$ that are not immersions.

✎ EXERCISE 15 (Smooth injection in the corner). Let "the corner" K be the subset of $\mathbf{R}^2$ defined by

$$K = \left\{ \begin{pmatrix} x \\ 0 \end{pmatrix} \text{ with } 0 \leq x < 1 \right\} \cup \left\{ \begin{pmatrix} 0 \\ y \end{pmatrix} \text{ with } 0 \leq y < 1 \right\}.$$

1) Show that if a smooth injection $\gamma :]-\varepsilon, +\varepsilon[\to \mathbf{R}^2$ takes its values in the corner K, and if $\gamma(0) = (0, 0)$, then γ is flat (see the definition in Exercise 13, p. 17) at the

point 0. Exhibit a smooth parametrization γ of the corner, such that $\gamma(0) = (0, 0)$ and γ is not flat at 0.

2) Check that $j : \mathbf{R} \to \mathbf{R}^2$, defined as follows, is a smooth injection whose values are the corner K.

$$j(t) = \begin{pmatrix} e^{\frac{1}{t}} \\ 0 \end{pmatrix} \text{ if } t < 0, \quad j(0) = \begin{pmatrix} 0 \\ 0 \end{pmatrix}, \quad j(t) = \begin{pmatrix} 0 \\ e^{-\frac{1}{t}} \end{pmatrix} \text{ if } t > 0.$$

3) Exhibit a parametrization $c : \mathbf{R} \to \mathbf{R}$ such that $j \circ c$ is smooth, but not c (see Exercise 14, p. 17). Is j an induction?

✎ EXERCISE 16 (Induction into smooth maps). Let n be some integer and equip $\mathcal{C}^\infty(\mathbf{R}, \mathbf{R}^n)$ with the functional diffeology defined in (art. 1.13). Show that the map $f : \mathbf{R}^n \times \mathbf{R}^n \to \mathcal{C}^\infty(\mathbf{R}, \mathbf{R}^n)$ defined by $f(x, v) = [t \mapsto x + tv]$ is an induction.

Subspaces of Diffeological Spaces

Every subset of a diffeological space inherits the subset diffeology *defined as the pullback of the ambient diffeology by the natural inclusion. This construction is related to the notion of induction discussed above* (art. 1.29).

1.33. Subspaces and subset diffeology. Let X be a diffeological space, and let $\mathcal{D}$ be its diffeology. Pick any subset A of X, and let $j_A : A \to X$ be the inclusion. The subset A carries naturally the diffeology $j_A^*(\mathcal{D})$, pullback of the diffeology $\mathcal{D}$ by the inclusion j_A (art. 1.26). This diffeology is said to be *inherited* from X or *induced* by the inclusion, or by the ambient space. It is called the *subset diffeology* of A. The plots of the subset diffeology are just the plots of X with values in A, that is,

$$j_A^*(\mathcal{D}) = \{P \in \mathcal{D} \mid \mathrm{val}(P) \subset A\}.$$

Equipped with the subset diffeology, the subset A is said to be a *diffeological subspace* of X. Thus, a diffeological subspace of X is any subset of X equipped with the subset diffeology.

NOTE. If X' is a diffeological space and $f : X \to X'$ is a smooth map, then just because j_A is smooth, $f \circ j_A$ is smooth, in other words, *the restriction of a smooth map to any subspace is smooth.*

1.34. Smooth maps to subspaces. Let X, X' and X'' be three diffeological spaces. Let $f : X \to X'$ be some map, and let $j : X' \to X''$ be an induction.

$$\begin{array}{ccc} X & & \\ \big\downarrow f & & \\ X' & \xrightarrow{\ j\ } & X'' \end{array}$$

The map f is smooth if and only if $j \circ f$ is smooth. Moreover, the map f is an induction if and only if $j \circ f$ is an induction.

PROOF. Let $\mathcal{D}$, $\mathcal{D}'$ and $\mathcal{D}''$ be the diffeologies of X, X' and X''. The map $j \circ f$ is smooth if and only if $\mathcal{D} \subset (j \circ f)^*(\mathcal{D}'')$ (art. 1.27). But $(j \circ f)^*(\mathcal{D}'') = f^*(j^*(\mathcal{D}''))$ (art. 1.28) and, since j is an induction, $\mathcal{D}' = j^*(\mathcal{D}'')$, thus $f^*(j^*(\mathcal{D}'')) = f^*(\mathcal{D}')$. Hence, $\mathcal{D} \subset (j \circ f)^*(\mathcal{D}'')$ is equivalent to $\mathcal{D} \subset f^*(\mathcal{D}')$, which means that f is smooth.

Note that if $j \circ f$ is injective, then f is injective. Next, replacing the inclusion above by an equality proves the second assertion. □

1.35. Subspaces, subsubspaces, etc. Let X be a diffeological space. Let A and B be two subspaces of X such that $A \subset B \subset X$. It is equivalent to consider A as a subspace of B, regarded as a subspace of X, or to consider A as a subspace of X.

PROOF. The inclusion j_A of A into X is the composite $j_A = j_B \circ j_{AB}$, where j_B is the inclusion of B into X and j_{AB} the inclusion of A into B. Then, it is just a direct consequence of the transitivity of inductions (art. 1.30). □

1.36. Inductions identify source and image. Let X' be a diffeological space and $A \subset X'$. For the subset diffeology (art. 1.33), the inclusion $j_A : A \to X'$ is an induction. More generally, let X and X' be two diffeological spaces. An induction $f : X \to X'$ (art. 1.29) identifies the source X with its image. In other words, f is a diffeomorphism from X onto $f(X)$, where $f(X)$ is equipped with the subset diffeology.

PROOF. The first part of the proposition is clear. Now, if f is an induction, then it is a smooth bijection onto its image $A = f(X)$. Let us consider the image A, equipped with the subset diffeology, and let $f^{-1} : A \to X$ be the inverse map. Let $P : U \to A$ be a plot for the subset diffeology, that is, a plot of X' with values in A (art. 1.33). By criterion (art. 1.31), $f^{-1} \circ P$ is a plot of X. Thus, f^{-1} is smooth and f is a diffeomorphism onto its image, equipped with the subset diffeology. □

1.37. Restricting inductions to subspaces. Let X and X' be two diffeological spaces. Let $F : X \to X'$ be an induction and A be some subspace of X. The restriction f of F to A is a diffeomorphism from A onto its image. In particular restrictions of diffeomorphisms are diffeomorphisms from the source to the image.

PROOF. Let j_A be the inclusion from A into X. Since diffeomorphisms are inductions (art. 1.32), the map $f = F \circ j_A$ is the composite of two inductions, thus an induction (art. 1.30). Now, thanks to (art. 1.36), the map f is a diffeomorphism onto its image, equipped with the subset diffeology. □

1.38. Discrete subsets of a diffeological space. Let X be a diffeological space. A subset $A \subset X$ is said to be *discrete* if its subset diffeology is discrete (art. 1.20), that is, if A, as a subspace of X, is diffeologically discrete. See for example Exercise 8, p. 14 and Exercise 19, p. 20.

Exercises

✎ EXERCISE 17 (Vector subspaces of real vector spaces). Let $\mathbf{R}^n$ be equipped with the standard diffeology. Let $(b_1, \ldots, b_k)$ be k independent vectors of $\mathbf{R}^n$. Let $\mathcal{B}$ be the linear map from $\mathbf{R}^k$ to $\mathbf{R}^n$ defined by $\mathcal{B}(x_1, \ldots, x_k) = \sum_{i=1}^k x_i b_i$. Show that $\mathcal{B}$ is an induction. Deduce that any vector subspace E of $\mathbf{R}^n$, equipped with the subset diffeology, is diffeomorphic to $\mathbf{R}^k$, with $k = \dim(E)$.

✎ EXERCISE 18 (The sphere as diffeological subspace). Let S^n denote the unit n-sphere in $\mathbf{R}^{n+1}$,

$$S^n = \{x \in \mathbf{R}^{n+1} \mid \|x\| = 1\}.$$

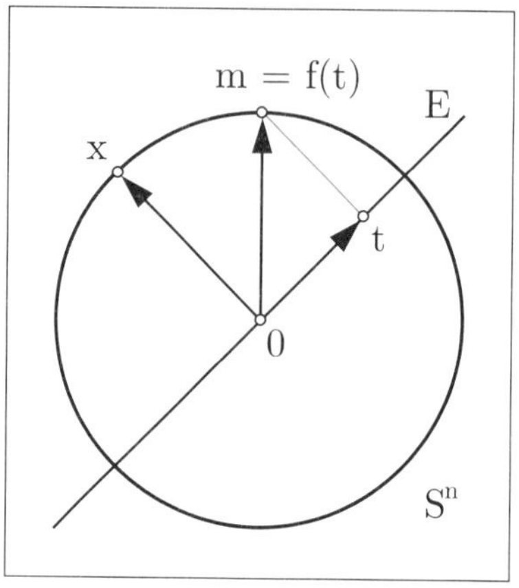

FIGURE 1.6. The projection f.

We equip S^n with the subset diffeology. Let x be a point of the sphere S^n, and $E \subset \mathbf{R}^{n+1}$ be the subspace orthogonal to x. Let f be the map defined on the vectors of E whose norm is strictly less than 1 (Figure 1.6), by

$$f : t \mapsto t + \sqrt{1 - \|t\|^2}\, x, \quad \text{with} \quad t \in E \quad \text{and} \quad \|t\| < 1. \qquad (\diamondsuit)$$

Show that f is an induction from $B = \{t \in E \mid \|t\| < 1\}$ to S^n. Thus, for every point x of S^n, there exists an induction from the open unit ball of $\mathbf{R}^n$ into S^n whose image is the hemisphere made up of the points $x' \in S^n$ such that the scalar product $x \cdot x'$ is strictly positive.

✎ EXERCISE 19 (The pierced irrational torus). Let T_α be the irrational torus, described in Exercise 4, p. 8. Let τ be some element of T_α. Check that T_α is not discrete but show that the subspace $T_\alpha - \{\tau\}$ is discrete.

✎ EXERCISE 20 (A discrete image of $\mathbf{R}$). Let us consider the set $\mathcal{C}^\infty(\mathbf{R}, \mathbf{R})$ of smooth maps from $\mathbf{R}$ to $\mathbf{R}$.

1) Show that the set of parametrizations $P : U \to \mathcal{C}^\infty(\mathbf{R}, \mathbf{R})$ satisfying the following two conditions is a diffeology:

- (♣) The parametrization $\mathbf{P} : (r, s) \mapsto P(r)(s)$ belongs to $\mathcal{C}^\infty(U \times \mathbf{R}, \mathbf{R})$.
- (♠) For all $r_0 \in U$ there exists an open ball B centered at r_0, and for all $r \in B$ there exists a closed interval $[a, b] \subset \mathbf{R}$ such that $P(r)$ and $P(r_0)$ coincide outside of $[a, b]$.

Note that the condition (♣) says that P is a plot for the diffeology defined in (art. 1.13). The second condition (♠) says that the plots are locally compactly supported variations of smooth maps.

2) Show that the image of the injective map $f : \mathbf{R} \to \mathcal{C}^\infty(\mathbf{R}, \mathbf{R})$, which associates with each real a the linear map $[x \mapsto ax]$, is discrete. Is the map f smooth?

3) Generalize this construction to $\mathcal{C}^\infty(\mathbf{R}^n, \mathbf{R}^n)$. What can you say about the injection of $GL(n, \mathbf{R})$ into $\mathcal{C}^\infty(\mathbf{R}^n)$?

Sums of Diffeological Spaces

The category {Diffeology} *is closed for coproducts (also called sums or disjoint unions) of diffeological spaces. There exists a distinguished diffeology, called the* sum diffeology, *on the sum of any family of diffeological spaces.*

Reminder. Let us recall the formal construction of the sum (also called coproduct) of any family of sets [**Bou72**]. Let $\mathcal{E} = \{E_i\}_{i\in\mathcal{J}}$ be an arbitrary family of sets, $\mathcal{J}$ being some set of indices. The set $\mathcal{J}$ can be the set $\mathcal{E}$ itself, in this case the family is said to be self-indexed. The sum of the elements of $\mathcal{E}$ is defined as follows:

$$\coprod_{i\in\mathcal{J}} E_i = \{(i, e) \mid i \in \mathcal{J} \text{ and } e \in E_i\}.$$

The sets E_i are called the *components* of the sum $\coprod_{i\in\mathcal{J}} E_i$. For every index $i \in \mathcal{J}$, we denote by j_i the natural injection

$$j_i : E_i \to \coprod_{i\in\mathcal{J}} E_i \quad \text{defined by} \quad j_i(e) = (i, e).$$

The sum is sometimes simply denoted by $\coprod \mathcal{E}$.

1.39. Building sums with spaces. Let $\{X_i\}_{i\in\mathcal{J}}$ be a family of diffeological spaces for an arbitrary set of indices $\mathcal{J}$. There exists, on the direct sum

$$X = \coprod_{i\in\mathcal{J}} X_i,$$

a finest diffeology such that the natural injections $j_i : X_i \to \coprod_{i\in\mathcal{J}} X_i$ are smooth for each index $i \in \mathcal{J}$. This diffeology is called the *sum diffeology*. The set X, equipped with the sum diffeology, is called the *diffeological sum*, or simply the *sum*, of the spaces X_i. The plots for this diffeology are the parametrizations $\mathbf{P} : U \to \coprod_{i\in\mathcal{J}} X_i$ which are locally plots of one of the components of the sum. Precisely, a parametrization $\mathbf{P} : r \to (i_r, P(r))$ is a plot of the diffeological sum X if and only if it satisfies the following condition.

(♣) For every $r \in U$ there exist an index i and an open neighborhood V of r such that $i_{r'} = i$ for all $r' \in V$, and $P \restriction V$ is a plot of X_i.

NOTE. For the sum diffeology, every natural injection $j_i : X_i \to X$ is an induction (art. 1.29). Each component X_i equipped with the subset diffeology induced by the sum is diffeomorphic to the original X_i.

PROOF. Let us check that (♣) defines a diffeology. By axiom D1, the constant maps take their values in one of the X_i, with $i \in \mathcal{J}$. The locality axiom D2 is satisfied by definition. The smooth compatibility axiom D3 is satisfied thanks to the continuity of smooth parametrizations in real domains, and because each X_i is itself a diffeological space. Therefore, (♣) defines a diffeology of the sum $X = \coprod_{i\in\mathcal{J}} X_i$, for which every injection j_i is obviously smooth.

Conversely, let us consider a diffeology $\mathcal{D}$ on the sum $X = \coprod_{i\in\mathcal{J}} X_i$ such that every injection j_i, $i \in \mathcal{J}$, is smooth. Let $\mathbf{P} : r \mapsto (i_r, P(r)) \in \coprod_{i\in\mathcal{J}} X_i$ be a parametrization satisfying the condition (♣). Thus, for all $r \in U$, there exist an index i and an open neighborhood V of r such that $i_{r'} = i$ for all $r' \in V$, and $P \restriction V$ is a plot of X_i. Since $j_i : X_i \to X$ is smooth for each $i \in \mathcal{J}$, the composite $j_i \circ (P \restriction V) : r \mapsto (i, P(r))$ belongs to $\mathcal{D}$. But $j_i \circ (P \restriction V) = \mathbf{P} \restriction V$, thus $\mathbf{P}$ is the supremum of a family of elements of $\mathcal{D}$, and thanks to the axiom of locality, $\mathbf{P}$ itself belongs to $\mathcal{D}$. Hence, the elements of the diffeology defined by the condition (♣) are elements of every diffeology such that the natural injections j_i, $i \in \mathcal{I}$, are smooth. Therefore the diffeology defined by (♣) is the finest diffeology for which every injection j_i is smooth. The fact that the injections j_i are inductions is obvious. □

1.40. Refining a sum. Let $X = \coprod_{i\in\mathcal{I}} X_i$ be a sum of diffeological spaces. Let, for some $j \in \mathcal{I}$, $X_j = \coprod_{k\in\mathcal{K}} X_j^k$. Let us consider the new family of indices

$$\mathcal{I}' = \{\mathcal{I} - \{j\}\} \coprod \mathcal{K} = \{(\mathcal{I}, i) \mid i \in \mathcal{I} \text{ and } i \neq j\} \cup \{(\mathcal{K}, k) \mid k \in \mathcal{K}\}.$$

For all $i' \in \mathcal{I}'$, let $X_{i'} = X_i$ if $i' = (\mathcal{I}, i)$ and $i \neq j$, and let $X_{i'} = X_j^k$ if $i' = (\mathcal{K}, k)$, with $k \in \mathcal{K}$. Then, there exists a natural equivalence between the two decompositions,

$$X = \coprod_{i\in\mathcal{I}} X_i \simeq \coprod_{i'\in\mathcal{I}'} X_{i'}.$$

This decomposition applies recursively to any subfamily of components which are themselves the sum of diffeological spaces. That leads to the finest decomposition of every diffeological space into the sum of its components (art. 5.8).

1.41. Foliated diffeology. Let X be a diffeological space X, and $\mathcal{R}$ be an equivalence relation defined on X. Let $\mathcal{Q}$ denote the quotient of X by the relation $\mathcal{R}$, that is, $\mathcal{Q}$ is the set of the equivalence classes $q = \mathrm{class}(x) \subset X$, $x \in X$. As a set, X is equivalent to the sum of its classes,

$$X \simeq \coprod_{q\in\mathcal{Q}} q,$$

thanks to the bijection $\sigma : x \mapsto (\mathrm{class}(x), x)$. The pullback of the sum diffeology of the family $\mathcal{Q}$ (art. 1.39), by σ (art. 1.26), where each class is equipped with the subset diffeology of X (art. 1.33), is called the *foliated diffeology* of X (associated with the relation $\mathcal{R}$). The elements of the foliated diffeology are the plots of X, which take locally their values in some equivalence class. More precisely, a parametrization $P : U \to X$ is a plot for the foliated diffeology if and only if the following condition is satisfied.

(♣) For all $r \in U$ there exists an open neighborhood V of r, such that $P \restriction V$ is a plot of X with values in class(P(r)).

The foliated diffeology is obviously finer than the original diffeology.

1.42. Klein structure and singularities of a diffeological space. The notion of singularity takes a special meaning in the diffeological context. For example, the irrational torus $T_\alpha = \mathbf{R}/(\mathbf{Z} + \alpha\mathbf{Z})$ described in Exercise 4, p. 8, often viewed as singular is — as a group — extremely regular. Indeed, the regularity of a diffeological space X lies in its homogeneity, and conversely, its inhomogeneity expresses its singularities. Precisely, the action of the group of diffeomorphisms Diff(X) (art. 1.17) distinguishes between subsets of points of the same kind. A space transitive under the action of its diffeomorphisms — all its points are of the same kind — may be regarded as regular, otherwise it can be split into orbits,

$$X = \bigcup_{x\in X} \mathcal{O}_x \quad \text{with} \quad \mathcal{O}_x = \{g(x) \mid g \in \mathrm{Diff}(X)\}.$$

Each orbit represents a certain degree of singularity. Note that each orbit $\mathcal{O}$ may be equipped with the subset diffeology, and the sum diffeology of the orbits of $\mathrm{Diff}(X)$ may be called the *Klein diffeology* of X. Equipped with the Klein diffeology, the set may be denoted by X_{Klein}, that is,

$$X_{\mathrm{Klein}} = \coprod_{\mathcal{O} \in (X/\mathrm{Diff}(X))} \mathcal{O}.$$

The space X_{Klein} reflects the singular splitting of X. If X is transitive under $\mathrm{Diff}(X)$, then X is said to be a *Klein space*.[2] The orbits of $\mathrm{Diff}(X)$, equipped with the subset diffeology, may be called the *Klein strata* of X. The way the Klein strata are glued together to build the space X is encoded by the quotient $S(X) = X/\mathrm{Diff}(X)$, equipped with the quotient diffeology, as it is described further (art. 1.50). These are perhaps the tools to explore the notion of singularity from the diffeological viewpoint. We could refine this analysis by considering the orbits of the groupoid of germs of local diffeomorphisms (art. 2.5).

EXAMPLE. Let us consider the half-line $\Delta_\infty = [0, \infty[\subset \mathbf{R}$, equipped with the subset diffeology; see Exercise 24, p. 23 and Exercise 64, p. 64. The group $\mathrm{Diff}(\Delta_\infty)$ has two orbits: the point $\bullet = \{0\}$ and the open subset $\circ =]0, \infty[$. Thus, $S(\Delta_\infty)$ is the pair $\{\bullet, \circ\}$. But the diffeology of $S(\Delta_\infty)$ is not the discrete diffeology, not all plots are locally constant (see also Exercise 48, p. 50 for an example of a 2-point nondiscrete space). The diffeology of $S(\Delta_\infty)$ encodes the relationship between the *singularity* $\bullet = \{0\}$ and the *regular stratum* $\circ =]0, \infty[$. Informally, the quotient space $S(\Delta_\infty)$ represents what is called generally the *transverse structure*, here the transverse structure of the partition in Klein strata.

Exercises

✎ EXERCISE 21 (Sum of discrete or coarse spaces). Check that the sum of discrete spaces is discrete. Show that a diffeological space X is discrete if and only if it is the sum of its elements, that is, if and only if $X = \coprod_{x \in X} \{x\}$. What about the sum of coarse spaces?

✎ EXERCISE 22 (Plots of the sum diffeology). Let $X = \coprod_{i \in \mathcal{I}} X_i$ be a sum of diffeological spaces. Show that a parametrization $P : U \to X$ is a plot if and only if there exists a partition $\{U_i\}_{i \in \mathcal{I}}$ of U, that is, $U = \bigcup_{i \in \mathcal{I}} U_i$ and $U_i \cap U_j = \varnothing$ if $i \neq j$, such that, for every nonempty U_i, $P_i = P \restriction U_i$ is a plot of X_i. We say that the plot P is the sum of the P_i.

✎ EXERCISE 23 (Diffeology of $\mathbf{R} - \{0\}$). Check that $\mathbf{R} - \{0\}$, equipped with the subset diffeology of $\mathbf{R}$, is equal to the sum $]-\infty, 0[\coprod]0, \infty[$.

✎ EXERCISE 24 (Klein strata of $[0, \infty[$). Let X be the segment $[0, \infty[$ equipped with the subset diffeology of $\mathbf{R}$. Show that every diffeomorphism φ of X fixes 0, that is, $\varphi(0) = 0$. Justify the fact that the segment has two Klein strata: the singleton $\{0\}$ and the open interval $]0, \infty[$.

[2] In reference to Felix Klein's Erlangen program, where he exposed his ideas about what a geometry is. Here, we describe the geometry defined by a diffeology.

✎ EXERCISE 25 (Compact diffeology). Check that the diffeology of $\mathcal{C}^\infty(\mathbf{R})$ defined in Exercise 20, p. 20 is the foliated diffeology of $\mathcal{C}^\infty(\mathbf{R})$ for the following equivalence relation.

($\diamondsuit$) Two functions f and g are equivalent if there exists a compact K of $\mathbf{R}$ such that f and g coincide outside K.

This diffeology is called the *compact diffeology* of $\mathcal{C}^\infty(\mathbf{R})$ [**Igl87**]. The third part of Exercise 20, p. 20 can be rephrased as follows: the group $\mathrm{GL}(n, \mathbf{R})$ inherits the discrete diffeology from the compact diffeology of $\mathcal{C}^\infty(\mathbf{R}^n)$.

Pushing Forward Diffeologies

In this section we describe the transfer of diffeology by pushforward. *Given a diffeological space* X *and a map* f *from* X *to some set* X', *the set* X' *inherits a natural diffeology by pushing forward the diffeology of* X *to* X'.

1.43. Pushforward of diffeologies. Let X be a diffeological space, and let X' be a set. Let $f : X \to X'$ be a map. There exists a finest diffeology of X' such that the map f is smooth. This diffeology is called the *image*, or the *pushforward*, by f of the diffeology $\mathcal{D}$ of X, and will be denoted by $f_*(\mathcal{D})$. A parametrization $P : U \to X'$ is a plot for $f_*(\mathcal{D})$ if and only if it satisfies the following condition:

($\clubsuit$) For every $r \in U$, there exists an open neighborhood V of r such that, either $P \restriction V$ is a constant parametrization, or there exists a plot $Q : V \to X$ such that $P \restriction V = f \circ Q$.

In the case where $P \restriction V = f \circ Q$, we say that P *lifts locally, at the point* r, *along the map* f. We also say that Q is a *local lift of* P, *over* V, *along* f.

PROOF. The set $\mathbf{D}$ of all the diffeologies of X' such that f is smooth is nonempty, since it contains the coarse diffeology. Now, this family $\mathbf{D}$ has an infimum, let us denote it by $\mathcal{D}'$, it is the intersection of all the elements of $\mathbf{D}$ (art. 1.23). We have now to check that this infimum is actually a minimum, that is, f is smooth for the diffeology $\mathcal{D}'$. But by the very definition of smooth maps, for any plot P belonging to $\mathcal{D}$, $f \circ P$ belongs to every element of $\mathbf{D}$. Thus, $P \circ f$ belongs to their intersection $\mathcal{D}'$. Next, let us check that the set of parametrizations satisfying ($\clubsuit$) is a diffeology. The axioms D1 and D2 are satisfied by definition. Let us consider then a parametrization $P : U \to X'$ satisfying ($\clubsuit$), and let $F : U' \to U$ be a smooth parametrization. Let $P' = P \circ F$, $r' \in U'$, $r = F(r')$, $V' = F^{-1}(V)$. Since F is continuous, V' is a domain, and by construction an open neighborhood of r'. Then, let $Q' = Q \circ F$, then $P \restriction V = f \circ Q$ implies $P' \restriction V' = f \circ Q'$. Therefore, the axiom D3 is satisfied. Now, let $\mathcal{D}'$ be a diffeology such that the map f is smooth, that is, for every plot Q of X, $f \circ Q$ is a plot of X'. Hence, the diffeology $\mathcal{D}'$ contains the set of all the parametrizations of the type $f \circ Q$. By restriction to any subdomain of the plots Q (allowed by the axiom D3), $\mathcal{D}'$ contains the elements of the diffeology described by ($\clubsuit$). It also contains the locally constant parametrizations (art. 1.20). Therefore, this diffeology is the finest diffeology such that f is smooth. □

1.44. Smoothness and pushforwards. Let X and X' be two diffeological spaces, whose diffeologies are denoted by $\mathcal{D}$ and $\mathcal{D}'$. The notion of pushforward of diffeologies gives a new interpretation of the notion of differentiability: a map

$f : X \to X'$ is smooth if and only if $f_*(\mathcal{D}) \subset \mathcal{D}'$. In other words,

$$\mathcal{C}^\infty(X, X') = \{f \in \mathrm{Maps}(X, X') \mid f_*(\mathcal{D}) \subset \mathcal{D}'\}.$$

PROOF. Let us assume first that f is smooth. Said differently, $\mathcal{D}'$ is a diffeology of X' such that f is smooth. But $f_*(\mathcal{D})$ is the intersection of all the diffeologies of X' for which f is smooth (art. 1.43), thus $f_*(\mathcal{D}) \subset \mathcal{D}'$. Now, let us assume that $f_*(\mathcal{D}) \subset \mathcal{D}'$. Let P be a plot of X and $P' = f \circ P$. Thanks to (art. 1.43, (♣)), P' is a plot of $f_*(\mathcal{D})$. Thus, by hypothesis, P' is a plot of $\mathcal{D}'$. Therefore, f is smooth. □

1.45. Composition of pushforwards. Let X be a diffeological space, and let X' and X'' be two sets. Let $f : X \to X'$ and $g : X' \to X''$ be two maps. The pushforward by g of the pushforward by f of the diffeology $\mathcal{D}$ of X is equal to the pushforward of $\mathcal{D}$ by $g \circ f$,

$$(X, \mathcal{D}) \xrightarrow{f} X' \xrightarrow{g} X'' \quad \text{and} \quad g_*(f_*(\mathcal{D})) = (g \circ f)_*(\mathcal{D}).$$

In other words, the pushforward of diffeologies is associative.

PROOF. Let us prove first that $(g \circ f)_*(\mathcal{D}) \subset g_*(f_*(\mathcal{D}))$. Let $P : U \to X''$ be a plot for $(g \circ f)_*(\mathcal{D})$. By (art. 1.43) P is either locally constant or writes locally $P \restriction V = (g \circ f) \circ Q$, where Q is a plot of X. If P is locally constant, every local constant plot is included in every diffeology, *a fortiori* in $g_*(f_*(\mathcal{D}))$. Now, $(g \circ f) \circ Q = g \circ (f \circ Q)$. But $f \circ Q$ belongs to $f_*(\mathcal{D})$, hence $P \restriction V$ belongs to $g_*(f_*(\mathcal{D}))$. Therefore $(g \circ f)_*(\mathcal{D}) \subset g_*(f_*(\mathcal{D}))$. Next, let us check that $g_*(f_*(\mathcal{D})) \subset (g \circ f)_*(\mathcal{D})$. Let $P : U \to X''$ be an element of $g_*(f_*(\mathcal{D}))$. The plot P is either locally constant or writes locally $P \restriction V' = g \circ Q'$, where Q' is a plot for $f_*(\mathcal{D})$. Locally constant parametrizations belong to every diffeology, *a fortiori* to $(g \circ f)_*(\mathcal{D})$. Then, let Q' write locally $Q' \restriction V = f \circ Q$ where Q is a plot of X defined on V. Thus, $P \restriction V = (g \circ f) \circ Q$, that is, P belongs to $(g \circ f)_*(\mathcal{D})$. □

Exercise

✎ EXERCISE 26 (Square of the smooth diffeology). Let $\mathcal{D}$ be the pushforward of the smooth diffeology of $\mathbf{R}$ by the *square map* $\mathrm{sq} = [x \mapsto x^2]$, that is, $\mathcal{D} = \mathrm{sq}_*(\mathcal{C}^\infty_\star(\mathbf{R}))$ (art. 1.9). Is $\mathcal{D}$ finer or coarser than $\mathcal{C}^\infty_\star(\mathbf{R})$? Let $P : U \to \mathbf{R}$ be an n-plot for $\mathcal{D}$, $n > 0$, and let $r \in U$.

1) Check that if $P(r) < 0$, then there exists an open ball B centered at r such that $P \restriction B$ is constant.

2) Check that if $P(r) > 0$, then there exists an open ball B centered at r such that the parametrization $\sqrt{P}$ is smooth.

3) Check that if $P(r_0) = 0$, then the tangent linear map of P at r_0, denoted by $D(P)(r_0)$, vanishes. Check that the Hessian H of P at the point r_0, that is, $H = D^2(P)(r_0) = D(r \mapsto D(P)(r))(r_0)$, regarded as a bilinear map on $\mathbf{R}^n$, is positive, *i.e.*, for all $v \in \mathbf{R}^n$, $H(v)(v) \geq 0$.

4) Is the real function $f : x \mapsto x^2 - x^3$, defined on $]-\infty, 1[$, a plot of $\mathcal{D}$?

Subductions

Subductions are surjections between diffeological spaces, identifying the target with the pushforward (art. 1.43) *of the source. They are a categorical key construction, used in particular in the definition of the diffeological quotients* (art. 1.50).

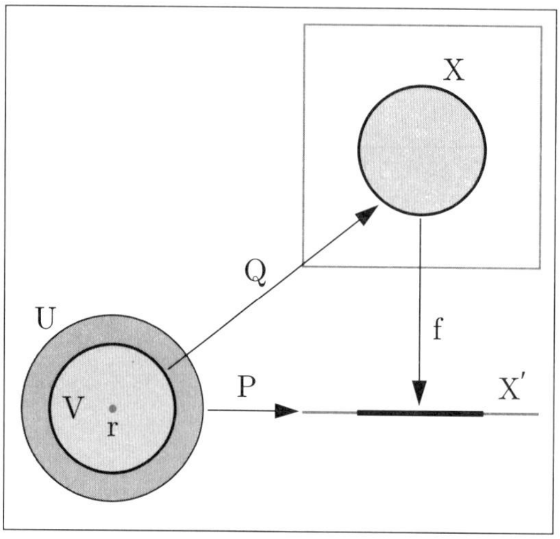

FIGURE 1.7. A subduction.

1.46. What is a subduction? Let X and X' be two diffeological spaces, and let $f : X \to X'$ be some map. The map f is said to be a *subduction* if it satisfies the following conditions:

1. The map f is surjective.
2. The diffeology $\mathcal{D}'$ of X' is the pushforward of the diffeology $\mathcal{D}$ of X, that is, with the notation introduced above, $f_*(\mathcal{D}) = \mathcal{D}'$.

NOTE. The second condition also writes $f_*(\mathcal{D}) \subset \mathcal{D}'$ and $\mathcal{D}' \subset f_*(\mathcal{D})$, where the first inclusion just says that f is smooth (art. 1.44).

1.47. Compositions of subductions. The composite of two subductions is a subduction. Subductions form a subcategory of the category {Diffeology}, which may be naturally denoted by {Subductions}.

PROOF. Let X, X' and X'' be three diffeological spaces, and let $\mathcal{D}$, $\mathcal{D}'$ and $\mathcal{D}''$ be their diffeologies. Let $f : X \to X'$ and $g : X' \to X''$ be two subductions, that is, $f_*(\mathcal{D}) = \mathcal{D}'$ and $g_*(\mathcal{D}') = \mathcal{D}''$. Since the pushforward is associative (art. 1.45), $(g \circ f)_*(\mathcal{D}) = g_*(f_*(\mathcal{D})) = g_*(\mathcal{D}') = \mathcal{D}''$. Therefore, $g \circ f$ is a subduction. □

1.48. Criterion for being a subduction. Let X and X' be two diffeological spaces, and let $f : X \to X'$ be some map. The map f is a subduction if and only the following conditions, illustrated by Figure 1.7, are satisfied:

1. The map f is a smooth surjection.
2. For every plot $P : U \to X'$, for every $r \in U$, there exist an open neighborhood V of r and a plot $Q : V \to X$ such that $P \restriction V = f \circ Q$.

Put differently, using the vocabulary introduced in (art. 1.43), the map f is a subduction if and only if f is smooth and every plot of X' lifts locally along f, at each point of its domain.

PROOF. The first condition is equivalent to $f_*(\mathcal{D}) \subset \mathcal{D}'$ (art. 1.44). The second condition is the reduction of the condition $\mathcal{D}' \subset f_*(\mathcal{D})$ (art. 1.43), in the case of a surjection. □

1.49. Injective subductions. Let X and X' be two diffeological spaces, and let $f : X \to X'$ be a subduction. If f is injective, then f is a diffeomorphism. Conversely, diffeomorphisms are injective subductions.

PROOF. A subduction is, by definition, smooth and surjective (art. 1.46). If moreover, f is injective, then f is a smooth bijection. Now, let $P : U \to X'$ be a plot, for every $r \in U$ there exist an open neighborhood V of r and a plot Q of X such that $f \circ Q = P \restriction V$ (art. 1.48). But since f is bijective, Q is unique and $Q = f^{-1} \circ (P \restriction V)$, that is, $f^{-1} \circ P$ is locally a plot of X at each point r of U. Hence, $f^{-1} \circ P$ is a plot of X (axiom D2) and f^{-1} is smooth. Thus, f is a diffeomorphism. Conversely, let $\mathcal{D}$ and $\mathcal{D}'$ denote the diffeologies of X and X'. If f is a diffeomorphism, then it is smooth and $f_*(\mathcal{D}) \subset \mathcal{D}'$ (art. 1.44). Now, its inverse also is smooth and $f_*^{-1}(\mathcal{D}') \subset \mathcal{D}$. But, thanks to (art. 1.47), this is equivalent to $\mathcal{D}' \subset f_*(\mathcal{D})$. Therefore, $f_*(\mathcal{D}) = \mathcal{D}'$. ☐

Exercises

✎ EXERCISE 27 (Subduction onto the circle). Equip $\mathbf{R}$ with the standard diffeology, and the circle $S^1 \subset \mathbf{R}^2$ with the subset diffeology. Show that the map $\Pi : t \mapsto (\cos(t), \sin(t))$ from $\mathbf{R}$ to S^1 is a subduction.

✎ EXERCISE 28 (Subduction onto diffeomorphisms). Let G be the set of smooth maps defined by

$$G = \{f \in \mathcal{C}^\infty(\mathbf{R}) \mid \text{for all } x \in \mathbf{R},\ f'(x) \neq 0 \text{ and } f(x + 2\pi) = f(x) \pm 2\pi\},$$

where f' denotes the first derivative.

1) Check that G is a group, actually a subgroup of $\mathrm{Diff}(\mathbf{R})$.

2) Let $S^1 \subset \mathbf{R}^2$ be equipped with the subset diffeology, and $\mathrm{Diff}(S^1)$ its group of diffeomorphisms. Show that the set of parametrizations $P : U \to \mathrm{Diff}(S^1)$ satisfying the following condition is a diffeology of $\mathrm{Diff}(S^1)$:

($\diamondsuit$) For all plots $Q : V \to S^1$, the parametrization $(r, s) \mapsto P(r)(Q(s))$, defined on $U \times V$, is a plot of S^1.

3) Show that, for every $f \in G$, there exists a unique diffeomorphism φ of S^1 such that $\Pi \circ f = \varphi \circ \Pi$, where Π is defined in Exercise 27, p. 27. Show that the map $\Phi : f \mapsto \varphi$ is a homomorphism.

4) Let G be equipped with the subset diffeology of the diffeology of $\mathcal{C}^\infty(\mathbf{R})$ defined in Exercise 6, p. 9. We shall admit that any smooth map, from an open ball of a real domain to $S^1 \subset \mathbf{R}^2$, has a global smooth lift to $\mathbf{R}$ along the subduction Π. Show that the map Φ is a subduction from G onto $\mathrm{Diff}(S^1)$. Describe the preimage of any $\varphi \in \mathrm{Diff}(S^1)$.

5) Show that the subgroup of translations $x \mapsto x + a$, where a is any real number, is a subgroup of G, diffeomorphic to $\mathbf{R}$. Show that its image by Φ is a subgroup of $\mathrm{Diff}(S^1)$ diffeomorphic to the circle S^1.

Quotients of Diffeological Spaces

Any quotient of a diffeological space inherits the quotient diffeology *defined as the pushforward of the diffeology of the source space to the quotient. This construction is related to the notion of subduction discussed above in* (art. 1.46). *We say that the category* {Diffeology} *is closed by quotient.*

I remember in high school, I had always been uneasy with the notion of quotient of a set X *by an equivalence relation* $\sim$, *even though I understood pretty well that the class of an element* x *was the set of all its equivalent elements. Only later did I realize my confusion: it was the various representations of the quotients, natural or not, used by my teachers, like magic tricks. I decided then to always regard a class for what it is: a subset of* X, *independently of any representation, and I decided to regard every quotient, the set of all the equivalence classes, as a subset of the powerset* $\mathfrak{P}(X)$. *Embedding the quotient into the powerset comforted me. This is the way I shall understand thereafter the notion of quotient. I shall also often use the generic notation* $\mathrm{class}(x)$ *to denote the class of the element* x, i.e., $\mathrm{class}(x) = \{x' \in X \mid x' \sim x\}$. *So,* $\mathrm{class} : X \to \mathfrak{P}(X)$ *is an application, the* natural projection *or the* canonical projection *from* X *onto its quotient, and by definition* $X/\!\sim\, = \mathrm{class}(X) \subset \mathfrak{P}(X)$.

1.50. Quotient and quotient diffeology. Let X be a diffeological space, and let $\mathcal{D}$ be its diffeology. Pick an equivalence relation $\mathcal{R}$ on X. The quotient set $X/\mathcal{R}$ carries a natural diffeology $\mathrm{class}_*(\mathcal{D})$, the pushforward of the diffeology $\mathcal{D}$ by the natural projection $\mathrm{class} : X \to X/\mathcal{R}$ (art. 1.43). This diffeology is called the *quotient diffeology*. For the quotient diffeology, the projection class is a subduction (art. 1.46). The set $X/\mathcal{R}$ equipped with the quotient diffeology is called the *quotient space* or the *diffeological quotient* of X by the relation $\mathcal{R}$. For example, let Q be a diffeological space. Every surjection $\pi : X \to Q$ defines a natural equivalence relation $\mathcal{R}$, that is, $x \mathcal{R} x'$ if $\pi(x) = \pi(x')$. The quotient space $X/\mathcal{R}$ is often also denoted by X/π. We shall see in (art. 1.52) that, if π is a subduction, then the bijection $\mathrm{class}(x) \mapsto \pi(x)$, from $X/\mathcal{R}$ to Q, is a diffeomorphism.

1.51. Smooth maps from quotients. Let X, X' and X'' be three diffeological spaces. Let $\pi : X \to X'$ be a subduction. A map $f : X' \to X''$ is smooth if and only if $f \circ \pi$ is smooth.

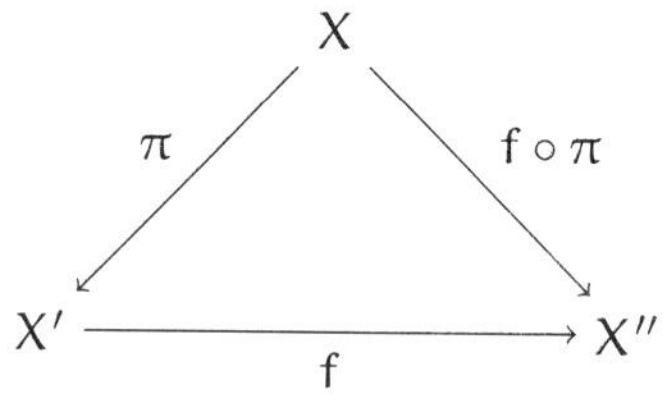

Moreover, the map f is a subduction if and only if $f \circ \pi$ is a subduction.

PROOF. Let $\mathcal{D}$, $\mathcal{D}'$ and $\mathcal{D}''$ be the diffeologies of X, X' and X''. The map $f \circ \pi$ is smooth if and only if $(f \circ \pi)_*(\mathcal{D}) \subset \mathcal{D}''$ (art. 1.44). But $(f \circ \pi)_*(\mathcal{D}) = f_*(\pi_*(\mathcal{D}))$ (art. 1.45). Since π is a subduction, $\pi_*(\mathcal{D}) = \mathcal{D}'$ and $f_*(\pi_*(\mathcal{D})) = f_*(\mathcal{D}')$. Hence, $(f \circ \pi)_*(\mathcal{D}) \subset \mathcal{D}''$ is equivalent to $f_*(\mathcal{D}') \subset \mathcal{D}''$ which means that f is smooth. Note that if $\pi \circ f$ is surjective, then f is surjective. Next, replacing above the inclusion by an equality proves the second assertion. □

1.52. Uniqueness of quotient. Let X, X' and X'' be three diffeological spaces. Let $\pi' : X \to X'$ and $\pi'' : X \to X''$ be two subductions. If $f : X' \to X''$ is a bijection such that $f \circ \pi' = \pi''$, then f is a diffeomorphism. In particular, let X be a diffeological space and $\mathcal{R}$ be an equivalence relation on X. Let $X/\mathcal{R}$ be the diffeological quotient (art. 1.50). Let Q be a set and $p : X \to Q$ be a surjection such that $p(x) = p(x')$ if and only if $x \mathcal{R} x'$. Therefore, the map $f : \mathrm{class}(x) \mapsto p(x)$, is a diffeomorphism from $X/\mathcal{R}$ to Q, where Q is equipped with the pushforward diffeology of X (art. 1.43).

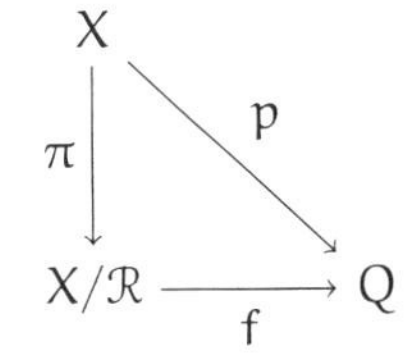

We say that the space Q is a *representation of the quotient* $X/\mathcal{R}$.

PROOF. Since $f \circ \pi' = \pi''$ and π'' is a subduction, f is a subduction (art. 1.51). Since f is bijective, f is a diffeomorphism (art. 1.49). □

1.53. Sections of a quotient. It is sometimes convenient to characterize — or compute — the quotient $X/\mathcal{R}$, of a diffeological space X by an equivalence relation $\mathcal{R}$, by restricting the equivalence relation to a subspace X'. In other words, giving $X' \subset X$ and $\mathcal{R}'$, the restriction of $\mathcal{R}$ to X', what are the conditions for which the natural map

$$\Phi : X'/\mathcal{R}' \to X/\mathcal{R} \quad \text{with} \quad \Phi : \mathrm{class}'(x') \mapsto \mathrm{class}(x')$$

is a diffeomorphism? We remark that since $\mathcal{R}'$ is the restriction of $\mathcal{R}$ to X', then $\mathrm{class}'(x') = \mathrm{class}(x') \cap X' \subset \mathrm{class}(x')$. Note that Φ is necessarily smooth, since class and class$'$ are two subductions, and moreover injective, its inverse is given by

$$\Phi^{-1} : X/\mathcal{R} \to X'/\mathcal{R}' \quad \text{with} \quad \Phi^{-1} : \mathrm{class}(x) \mapsto \mathrm{class}(x) \cap X'.$$

The first condition, that Φ is surjective, writes $X' \cap \mathrm{class}(x) \neq \varnothing$, for all $x \in X$. If this condition is fulfilled, then Φ is a smooth bijection. Now, if Φ is surjective, then Φ^{-1} is smooth if and only if for all plots $P : U \to X$, for all $r \in U$, there exist an open neighborhood V of r and a plot $P' : V \to X'$ such that $\mathrm{class}(P'(r')) = \mathrm{class}(P(r'))$ for all $r' \in V$. If these conditions are fulfilled, then we shall say that $(X', \mathcal{R}')$ is a *reduction* of $(X, \mathcal{R})$. If moreover, X' is such that $\mathrm{class}(x) \cap X'$ is reduced to one point, then we shall say that X' is a *smooth section* of the projection $\mathrm{class} : X \to \mathcal{Q} = X/\mathcal{R}$. Indeed, the map $\sigma : \mathcal{Q} \to X$ defined by $\sigma(q) = q \cap X'$ is a diffeomorphism and satisfies $\mathrm{class} \circ \sigma = \mathbf{1}_{\mathcal{Q}}$. In other words, X' represents $\mathcal{Q}$.

NOTE. If there exists a smooth projector $\rho : X \to X$, that is, $\rho \circ \rho = \rho$, which preserves the equivalence relation $\mathrm{class} \circ \rho = \mathrm{class}$, then $X' = \rho(X)$, together with the restriction $\mathcal{R}'$, is a reduction of $(X, \mathcal{R})$. Moreover, if class$'$ is injective, then X' is a smooth section of class.

PROOF. The map Φ is smooth because for every plot $r \mapsto q'_r$ of $\mathcal{Q}' = X'/\mathcal{R}'$, there exists, locally everywhere, a smooth parametrization $r \mapsto x'_r$ such that $q'_r = \mathrm{class}'(x'_r)$. Then, $\Phi \circ [r \mapsto q'_r] = \mathrm{class} \circ [r \mapsto x'_r]$ is smooth. Next, Φ is clearly injective, let $\Phi(\mathrm{class}'(x')) = \Phi(\mathrm{class}'(x''))$, that is, $\mathrm{class}(x') = \mathrm{class}(x'')$. Thus, $\mathrm{class}'(x') = \mathrm{class}(x') \cap X' = \mathrm{class}(x'') \cap X' = \mathrm{class}'(x'')$. Now, let us assume that Φ is surjective, let $P : r \mapsto q_r$ be a plot of $\mathcal{Q}$, by definition, there exists, locally everywhere, a plot $r \mapsto x_r$ such that $q_r = \mathrm{class}(x_r)$. Let us assume next that, after possibly shrinking the domain of $r \mapsto x_r$, there exists a plot $r \mapsto x'_r$ of X' such that

$\mathrm{class}(x'_r) = \mathrm{class}(x_r)$. Hence, locally,

$$\Phi^{-1}(q_r) = \Phi^{-1}(\mathrm{class}(x_r)) = \mathrm{class}(x_r) \cap X' = \mathrm{class}(x'_r) \cap X' = \mathrm{class}'(x'_r).$$

Therefore, $\Phi^{-1} \circ [r \mapsto q_r]$ is locally smooth, thus Φ^{-1} is smooth, and Φ is a diffeomorphism. Conversely, let us assume that Φ is a diffeomorphism, that is, Φ^{-1} is smooth. Let $P : U \to X$ be a plot, thus $\mathrm{class} \circ P$ is a plot of $\mathcal{Q}$, and $\Phi^{-1} \circ \mathrm{class} \circ P$ is a plot of $\mathcal{Q}'$. Hence, there exists locally everywhere a plot P' of X' such that

$$\mathrm{class}'(P'(r)) = \Phi^{-1}(\mathrm{class}(P(r))), \quad \text{or} \quad \Phi(\mathrm{class}'(P'(r))) = \mathrm{class}(P(r)),$$

that is, $\mathrm{class}(P'(r)) = \mathrm{class}(P(r))$. Now, if $X' \cap \mathrm{class}(x)$ is reduced to one point, then class' is an injective subduction, that is, a diffeomorphism. Let us assume next that there exists a smooth projector ρ satisfying the conditions of the note. The subspace $X' = \rho(X)$ intersects every orbit. Indeed, let $q \in \mathcal{Q}$ and $x \in q$, thus $x' = \rho(x) \in X'$ and $\mathrm{class}(x') = q$. Thus, Φ is surjective. Next, let P be a plot of X, the parametrization $P' = \rho \circ P$ is a plot of X', and since $\mathrm{class} \circ \rho = \mathrm{class}$, the condition of the proposition is satisfied. ☐

1.54. Strict maps, between quotients and subspaces. Let X and X' be two diffeological spaces. We say that a map $f : X \to X'$ is *strict* if f is a subduction onto its image, when its image is equipped with the subset diffeology. We may also use the following criterion:

(♢) A map $f : X \to X'$ is strict if and only if f is smooth and for all plots $P : U \to X$ such that $\mathrm{val}(P) \subset f(X)$, for all $r \in U$, there exist an open neighborhood V of r and a plot $Q : V \to X$ such that $f \circ Q = P \restriction V$.

NOTE. There exists a universal factorization described by the following diagram, where X/f is the quotient by the equivalence relation defined by f (art. 1.50), p is the natural projection, and j is the inclusion $f(X) \subset X'$.

$$\begin{array}{ccc} X & \xrightarrow{\ f\ } & X' \\ {\scriptstyle p}\downarrow & & \uparrow{\scriptstyle j} \\ X/f & \xrightarrow[\ \varphi\ ]{} & f(X) \end{array}$$

The map φ, called the *factorization* of f, is always a smooth bijection from X/f, equipped with the quotient diffeology, to $f(X)$, equipped with the subset diffeology. Then, saying that f is strict is equivalent to say that φ is a diffeomorphism. In other words, a strict smooth map identifies the quotient X/f with $f(X)$.

PROOF. Since f is smooth and p is a subduction, we know that φ is smooth and, by construction, injective. We have just to express that φ^{-1} is smooth. Then, let $P : U \to f(X)$ be a plot, that is, a plot P of X with values in $f(X)$ (art. 1.33). The parametrization $\varphi^{-1} \circ P$ is a plot of X/f if and only if $\varphi^{-1} \circ P$ lifts locally along the projection p, at every point r of U (art. 1.46). Thus, φ^{-1} is smooth if and only if for all r in U there exist an open neighborhood V of r and a plot $Q : V \to X$ such that $p \circ Q = \varphi^{-1} \circ P \restriction V$, *i.e.*, $\varphi \circ p \circ Q = P \restriction V$, that is, $f \circ Q = P \restriction V$. ☐

Exercises

✎ EXERCISE 29 (Quotients of discrete or coarse spaces). Check that any quotient of any discrete diffeological space (art. 1.20) is discrete. As well, any quotient of any coarse space is coarse.

✎ EXERCISE 30 (Examples of quotients). Give a few examples, encountered in the previous exercises, of diffeological quotients.

✎ EXERCISE 31 (The irrational solenoid). The following example of the *irrational solenoid*, or *Kronecker flow*, mixes inductions and subductions. Let T^2 be the 2-torus, defined as the quotient space of $\mathbf{R}^2$ by the subgroup $\mathbf{Z}^2$. In other words, two points (x, y) and (x', y') of $\mathbf{R}^2$ are equivalent if there exist two integers n and m such that $x' = x + n$ and $y' = y + m$.

1) Check that the map $q : (x, y) \mapsto (p(x), p(y))$, with $p(t) = (\cos(2\pi t), \sin(2\pi t))$, from $\mathbf{R}^2$ to $\mathbf{R}^2 \times \mathbf{R}^2$, is strict (art. 1.54), and identifies $\mathbf{R}^2/\mathbf{Z}^2$ with $S^1 \times S^1 \subset \mathbf{R}^2 \times \mathbf{R}^2$.

2) Let $\Delta_\alpha \subset \mathbf{R}^2$ be the line $\{(x, \alpha x) \mid x \in \mathbf{R}\}$, with $\alpha \in \mathbf{R} - \mathbf{Q}$. Let $\mathcal{S}_\alpha = q(\Delta_\alpha) \subset S^1 \times S^1$. The subspace $\mathcal{S}_\alpha$ is the *irrational solenoid* with slope α. Show that $q_\alpha = q \restriction \Delta_\alpha$ is an induction, and thus that $\mathcal{S}_\alpha \subset S^1 \times S^1$ is diffeomorphic to $\mathbf{R}$.

3) Consider the space $S^1 \times S^1$ as a group, after identification with $\mathbf{R}^2/\mathbf{Z}^2$, for the operation inherited from the addition in $\mathbf{R}^2$. Note that $\mathcal{S}_\alpha$ is a subgroup. Show that the quotient $(S^1 \times S^1)/\mathcal{S}_\alpha$ is diffeomorphic to the irrational torus T_α defined in Exercise 4, p. 8.

✎ EXERCISE 32 (A minimal powerset diffeology). Exhibit, for all diffeological spaces X, a diffeology of the powerset $\mathfrak{P}(X)$, that induces for any equivalence relation $\sim$ defined on X, the quotient diffeology on $X/\!\sim$. Let us recall that the canonical projection class : $X \to \mathfrak{P}(X)$ is defined by $\mathrm{class}(x) = \{x' \in X \mid x' \sim x\}$, and that $X/\!\sim\, = \mathrm{class}(X) \subset \mathfrak{P}(X)$.

✎ EXERCISE 33 (Universal construction). Let X be a diffeological space, and let $\mathcal{D}$ be its diffeology. Let $\mathcal{N}$ be the diffeological sum (art. 1.39)

$$\mathcal{N} = \coprod_{P \in \mathcal{D}} \mathrm{def}(P) = \{(P, r) \mid P \in \mathcal{D} \text{ and } r \in \mathrm{def}(P)\},$$

where the domains of the plots are equipped with the standard diffeology.

1) Check that the map $\mathrm{ev} : \mathcal{N} \to X$ defined by $\mathrm{ev}(P, r) = P(r)$ is a subduction. In other words, show that X is equivalent to the quotient $\mathcal{N}/\mathrm{ev}$.

2) Is the subset of $\mathcal{N}$ made up of the 0-plots a smooth section?

NOTE. In other words, every diffeological space is the quotient of the union of some real domains by some equivalence relation. The set $\mathcal{N}$ defined above is called the *Nebula* of the diffeology $\mathcal{D}$; see (art. 1.76). This exercise shows the important role played by the two operations, sum and quotient, in diffeology.

✎ EXERCISE 34 (Strict action of SO(3) on $\mathbf{R}^3$). Let us consider the space $\mathbf{R}^3$, equipped with smooth diffeology. Let SO(3) be the group of all 3×3 real matrices M such that $M^t M = 1$, where M^t is the transposed matrix of M, and $\det(M) = +1$. The group SO(3) is a part of the real space $\mathbf{R}^9$, each matrix M is defined by its nine components M_{ij}, $M = \sum_{i,j=1}^{3} M_{ij} \mathbf{e}_{ij}$, where $\mathbf{e}_{ij}$ is the matrix with a 1 at the line i and the column j and 0 elsewhere. The group SO(3) is equipped with the subset diffeology of $\mathbf{R}^9$. Let X be some point of $\mathbf{R}^3$, show that the *orbit map* $R(X) : SO(3) \to \mathbf{R}^3$, with $R(X)(M) = MX$, is strict (art. 1.54).

Products of Diffeological Spaces

As well as for direct sums of spaces (art. 1.39), *the category* {Diffeology} *is closed for products. In other words, there exists a natural diffeology, called the* product diffeology, *on the product of every family of diffeological spaces. We give here its definition* (art. 1.55), *and then we describe some related constructions.*

Reminder. First of all, let us recall the formal construction of the product for any family of sets [**Bou72**]. Let $\mathcal{E} = \{E_i\}_{i\in\mathcal{J}}$ be an arbitrary family of sets, $\mathcal{J}$ being any set of indices. The set $\mathcal{J}$ can be the set $\mathcal{E}$ itself, in this case one says that the family is self-indexed. The product of the elements of the family $\mathcal{E}$ is the set of all the maps σ which associate with each index $i \in \mathcal{J}$ some element $\sigma(i) \in E_i$. In other words, let $\coprod_{i\in\mathcal{J}} E_i$ be the sum of the members of E, then

$$\coprod_{i\in\mathcal{J}} E_i = \{(i, e) \in \mathcal{J} \times \cup_{i\in\mathcal{J}} E_i \mid e \in E_i\}.$$

Let us denote by pr_1 the projection from $\coprod_{i\in\mathcal{J}} E_i$ onto the first factor:

$$\mathrm{pr}_1 : \coprod_{i\in\mathcal{J}} E_i \to \mathcal{J} \quad \text{with} \quad \mathrm{pr}_1(i, e) = i.$$

The product of the members of the family $\mathcal{E} = \{E_i\}_{i\in\mathcal{J}}$ is the set of all the sections of the projection pr_1, that is,

$$\prod_{i\in\mathcal{J}} E_i = \{s : \mathcal{J} \to \coprod_{i\in\mathcal{J}} E_i \mid \mathrm{pr}_1 \circ s = 1_{\mathcal{J}}\}.$$

For any index $i \in \mathcal{J}$, we denote by π_i the natural projection

$$\pi_i : \prod_{i\in\mathcal{J}} E_i \to E_i \quad \text{with} \quad \pi_i(s) = \mathrm{pr}_2(s(i)),$$

where $\mathrm{pr}_2 : (i, e) \mapsto e$ is the projection from $\coprod_{i\in\mathcal{J}} E_i$ onto the second factor. In other words, $\pi_i(s) = e$ if and only if $s(i) = (i, e)$. For the sake of simplicity, we shall denote sometimes the product $\prod_{i\in\mathcal{J}} E_i$ by $\prod \mathcal{E}$. Note that, if the set of indices $\mathcal{J}$ is finite, $\mathcal{J} = \{1, \ldots, N\}$, then the product $\prod_{i\in\mathcal{J}} E_i$ is denoted by $E_1 \times \cdots \times E_N$.

1.55. Building products with spaces. Let $\{X_i\}_{i\in\mathcal{J}}$ be a family of diffeological spaces, indexed by some set $\mathcal{J}$. Let us denote by $\mathcal{D}_i$, $i \in \mathcal{J}$, their diffeologies. Then, there exists on the product

$$X = \prod_{i\in\mathcal{J}} X_i$$

a coarsest diffeology $\mathcal{D}$ such that, for each index $i \in \mathcal{J}$, the projection $\pi_i : \prod_{i\in\mathcal{J}} X_i \to X_i$ is smooth. This diffeology is called the *product diffeology* of the family $\{\mathcal{D}_i\}_{i\in\mathcal{J}}$. The set X, equipped with the product diffeology, is called the *diffeological product*, or simply the *product*, of the X_i.

NOTE. The product diffeology $\mathcal{D}$ is the intersection, over all indices $i \in \mathcal{J}$, of the pullbacks by the projection π_i of the diffeologies $\mathcal{D}_i$. The plots for the product diffeology are the parametrizations $P : U \to \prod_{i\in\mathcal{J}} X_i$ such that, for each $i \in \mathcal{J}$, the parametrization $\pi_i \circ P$ is a plot of X_i. In other words, a plot $P : U \to \prod_{i\in\mathcal{I}} X_i$ is a family $\{P_i : U \to X_i\}_{i\in\mathcal{I}}$ such that $P_i \in \mathcal{D}_i$, for all $i \in \mathcal{I}$.

$$\mathcal{D} = \bigcap_{i\in\mathcal{I}} \pi_i^*(\mathcal{D}_i).$$

In particular, if $\mathcal{I}$ is a finite set of indices, $\mathcal{I} = \{1,\ldots,N\}$, then the plot P is an N-tuple $(P_1,\ldots,P_N)$, with $P_i \in \mathcal{D}_i$.

PROOF. For each $i \in \mathcal{I}$, there exists a coarsest diffeology of $\prod_{i\in\mathcal{I}} X_i$ such that the projection π_i is smooth, it is the pullback $\pi_i^*(\mathcal{D}_i)$ (art. 1.26). Then, the coarsest diffeology such that π_i is smooth, for any $i \in \mathcal{I}$, contains at least the intersection of all these pullbacks. But the intersection of any family of diffeologies is a diffeology (art. 1.22), and this intersection is the infimum of all the diffeologies containing the $\pi_i^*(\mathcal{D}_i)$. This infimum is clearly a minimum for the property in question. Therefore, the diffeology product is just the intersection $\mathcal{D} = \bigcap_{i\in\mathcal{I}} \pi_i^*(\mathcal{D}_i)$.

Now, the condition for P to be a plot of X, stated in the proposition, is clearly necessary. It just means that each projection is smooth. The three axioms of diffeology (art. 1.5) for this set of parametrizations are inherited from the diffeology of each element of the family $\{X_i\}_{i\in\mathcal{I}}$. Thus, this condition defines a diffeology $\mathcal{D}$ on X such that each projection π_i is smooth. On the other hand, every other diffeology for which each projection π_i is smooth satisfies this condition, so it is contained in $\mathcal{D}$. Therefore, this diffeology $\mathcal{D}$ is indeed the product diffeology. □

1.56. Projections on factors are subductions. Let $X = \prod_{i\in\mathcal{I}} X_i$ be the diffeological product of some family $\{X_i\}_{i\in\mathcal{I}}$ of diffeological spaces. Every projection $\pi_i : X \to X_i$ (art. 1.55) onto each factor is a subduction (art. 1.46).

PROOF. Let $P : U \to X_k$ be a plot. For every index $i \neq k$, let x_i be a point of X_i. The parametrization $\mathbf{P} : U \to X$ defined by $\mathbf{P}(r) = [i \mapsto (i, x_i)$ if $i \neq k$ and $(k, P(r))$ for $i = k]$ is a lift of P, that is, $\pi_k \circ \mathbf{P} = P$. Hence π_k is a subduction, according to the characterization of the plots of a subduction (art. 1.46). Remark that the construction of $\mathbf{P}$ uses the axiom of choice of the theory of sets. But, from the very beginning of the diffeology theory, this axiom has been implicit. □

Exercises

✎ EXERCISE 35 (Products and discrete diffeology). Check that for the discrete diffeology on the product $\prod_{i\in\mathcal{I}} X_i$ (art. 1.55), the projections π_i are smooth. Link that remark to the discussion of (art. 1.25). In the same spirit, comment on the definition of the sum diffeology (art. 1.39).

✎ EXERCISE 36 (Products of coarse or discrete spaces). Check that any product of coarse spaces is coarse. Show directly that finite products of discrete spaces are discrete. Using the result of the Exercise 7, p. 14, show that actually every product of discrete spaces is discrete.

✎ EXERCISE 37 (Infinite product of $\mathbf{R}$ over $\mathbf{R}$). Describe the sum diffeology of an infinite number of copies of $\mathbf{R}$ indexed by $\mathbf{R}$. Describe then the product diffeology of an infinite number of copies of $\mathbf{R}$ indexed by $\mathbf{R}$.

✎ EXERCISE 38 (Graphs of smooth maps). Let X and X' be two diffeological spaces, and let $f : X \to X'$ be any map. Let us denote by $\mathrm{Gr}(f) \subset X \times X'$ the graph of f, that is, the subset of pairs $(x, f(x)) \in X \times X'$, where x runs over X, equipped with the subset diffeology of the product diffeology. Show that f is smooth if and only if the first projection $\mathrm{pr}_X : (x, x') \mapsto x$, restricted to Gr(f), is a subduction (art. 1.46).

✎ EXERCISE 39 (The 2-torus). Check that the diffeology of $S^1 \times S^1$ described in Exercise 31, p. 31 is the product diffeology of S^1 by S^1.

Functional Diffeology

A special and remarkable feature of diffeology is that the set of the smooth maps between two diffeological spaces X and X′ carries a natural diffeology, called the functional diffeology. *This diffeology is used intensively everywhere, in the theory of diffeological spaces: in homotopy theory, for fiber bundles, in Cartan's differential calculus, etc. To be more precise,* $\mathcal{C}^\infty(X, X')$ *carries a whole family of functional diffeologies, all those such that the* evaluation map $\mathrm{ev} : (f, x) \mapsto f(x)$ *is smooth* (art. 1.57). *But if some of them are however interesting, for example the compact controlled diffeology* (art. 1.65), *the supremum of this family, that is, the* standard functional diffeology, *is the most used. Equipped with this functional diffeology, the category of diffeological spaces is Cartesian closed.*

1.57. Functional diffeologies. Let X and X′ be two diffeological spaces, and let $\mathcal{C}^\infty(X, X')$ be the set of smooth maps from X to X′ (art. 1.14). Let ev be the *evaluation map*, defined by

$$\mathrm{ev} : \mathcal{C}^\infty(X, X') \times X \to X' \quad \text{and} \quad \mathrm{ev}(f, x) = f(x).$$

We shall call *functional diffeology* any diffeology of $\mathcal{C}^\infty(X, Y)$ such that the map ev is smooth. Note that the discrete diffeology is a functional diffeology. But there exists a coarsest functional diffeology on $\mathcal{C}^\infty(X, X')$; we shall call it the *standard functional diffeology*, or simply *the functional diffeology*, when there is no risk of confusion. Actually, the plots of this diffeology are explicitly given by the following condition:

(♣) A parametrization $P : U \to \mathcal{C}^\infty(X, X')$ is a plot for the standard functional diffeology if and only if the map $(r, x) \mapsto P(r)(x)$ is smooth. That means that, for every plot $Q : V \to X$, $P \cdot Q : (r, s) \mapsto P(r)(Q(s))$ is a plot of X′.

PROOF. First of all, let us consider the set $\mathcal{C}^\infty(X, X')$ equipped with the discrete diffeology. A plot of the product $\mathcal{C}^\infty(X, X') \times X$ is some parametrization $r \mapsto (f, x)$ such that $r \mapsto f$ is locally constant (art. 1.20) and $r \mapsto x$ is a plot of X. Hence, the parametrization $r \mapsto f(x)$ is a plot of X′ by the very definition of smooth maps (art. 1.14). Therefore, the set of functional diffeologies is not empty, and the discrete diffeology is a functional diffeology.

Let us check the equivalence of conditions in (♣). Let us assume that $(r, x) \mapsto P(r)(x)$ is smooth, for every plot $s \mapsto Q(s)$ in X, then the parametrization $(r, s) \mapsto (r, Q(s)) \mapsto P(r)(Q(s))$ is the composite of smooth maps, thus a plot of X′. Conversely, let us assume $P \times Q$ smooth for every plot Q of X. Let $s \mapsto (r(s), Q(s))$ be a plot of $U \times X$, then $[s \mapsto P(r(s))(Q(s))] = [s \mapsto (r(s), s) \mapsto P(r(s))(Q(s))]$ is the composite of smooth maps, thus $(r, x) \mapsto P(r)(x)$ is smooth.

D1. This is equivalent to the fact that the discrete diffeology of $\mathcal{C}^\infty(X, X')$ is a functional diffeology (art. 1.57).

D2. Let $P : U \to \mathcal{C}^\infty(X, X')$ be a parametrization satisfying locally (♣); that is, for every $r \in U$, there exists an open neighborhood W of r such that for every plot $Q : V \to X$, the parametrization $(P \upharpoonright W) \cdot Q : W \times V \to X'$ is a plot. Hence, the parametrization $P \cdot Q$ is locally a plot of X'. Therefore, $P \cdot Q$ is a plot of X' and the parametrization P satisfies (♣).

D3. Let $P : U \to \mathcal{C}^\infty(X, X')$ be a parametrization satisfying (♣), and $F : W \to U$ be a smooth parametrization. Let $Q : V \to X$ be a plot. The parametrization $(P \circ F) \cdot Q : (t, s) \to P(F(t)(Q(s))$ decomposes into $(t, s) \mapsto (F(t), s) \mapsto P(F(t))(Q(s))$. Now, $(P \circ F) \cdot Q$ is the composite of two smooth maps. Thus, $(P \circ F) \cdot Q$ is smooth. Therefore, $P \circ F$ satisfies (♣).

Hence, the condition (♣) defines a diffeology on $\mathcal{C}^\infty(X, X')$. For this diffeology, the map ev is smooth. Indeed, let $P \times Q : r \mapsto (P(r), Q(r))$ be a plot of $\mathcal{C}^\infty(X, X') \times X$. Then, $\mathrm{ev} \circ (P \times Q) = P \cdot Q = [r \mapsto P(r)(Q(r))]$ is a plot of X', by the very definition of the plot P.

Now, let us show that every diffeology of $\mathcal{C}^\infty(X, X')$ such that ev is smooth satisfies (♣). Since the map ev is smooth, for every plot $P \times Q : r \mapsto (P(r), Q(r))$ of $\mathcal{C}^\infty(X, X') \times X$, the map $P \cdot Q : r \mapsto P(r)(Q(r))$ is a plot of X'. Let $P : U \to \mathcal{C}^\infty(X, X')$ and $Q : V \to X$ be two plots. The parametrizations $P' : (r, s) \mapsto P(r)$ and $Q' : (r, s) \mapsto Q(s)$, defined on $U \times V$, are two plots. Thus, the parametrization $P' \cdot Q' : (r, s) \mapsto P'(r, s)(Q'(r, s))$ is a plot of X'. But, $P'(r, s)(Q'(r, s)) = P(r)(Q(s))$. Hence, P satisfies (♣). Thus, every diffeology of $\mathcal{C}^\infty(X, X')$ such that ev is smooth is contained in the diffeology defined by (♣). Therefore, this is the coarsest diffeology such that ev is smooth. □

1.58. Restriction of the functional diffeology. Let X and X' be two diffeological spaces. Let $\mathcal{C}^\infty(X, X')$ be the set of smooth maps from X to X'. Every subset $\mathcal{M} \subset \mathcal{C}^\infty(X, X')$ inherits the functional diffeology (art. 1.33). The map $\mathrm{ev}_{\mathcal{M}} = \mathrm{ev} \upharpoonright (\mathcal{M} \times X)$ remains smooth, since it is the composition of a smooth map with an induction. The subset functional diffeology of $\mathcal{M}$ is still the coarsest diffeology of $\mathcal{M}$ such that $\mathrm{ev}_{\mathcal{M}}$ is smooth (art. 1.57). This subset diffeology will be called again the *functional diffeology of* $\mathcal{M}$.

NOTE. As an example of such restriction, we can think of the subset of 2π-periodical smooth maps from $\mathbf{R}$ to $\mathbf{R}$, which inherits the functional diffeology.

1.59. The composition is smooth. Let X, X' and X'' be three diffeological spaces. Let $\mathcal{C}^\infty(X, X')$, $\mathcal{C}^\infty(X', X'')$ and $\mathcal{C}^\infty(X, X'')$ be equipped with the standard functional diffeology (art. 1.57) and $\mathcal{C}^\infty(X, X') \times \mathcal{C}^\infty(X', X'')$ with the product diffeology (art. 1.55). Then, the *composition map* is smooth,

$$\circ : \mathcal{C}^\infty(X, X') \times \mathcal{C}^\infty(X', X'') \to \mathcal{C}^\infty(X, X'') \quad \text{with} \quad \circ(f, g) = g \circ f.$$

NOTE. This property, satisfied by the smooth maps between manifolds, was at the origin of Souriau's theory of *groupes différentiels* [**Sou80**], which then led to the introduction of diffeology.

1.60. Functional diffeology and products. Let X, X' and X'' be three diffeological spaces. Let $X \times X'$ be equipped with the product diffeology (art. 1.55). Let $\mathcal{C}^\infty(X', X'')$, $\mathcal{C}^\infty(X \times X', X'')$, and $\mathcal{C}^\infty(X, \mathcal{C}^\infty(X', X''))$ be equipped with the functional diffeology. The spaces $\mathcal{C}^\infty(X, \mathcal{C}^\infty(X', X''))$ and $\mathcal{C}^\infty(X \times X', X'')$ are diffeomorphic. The diffeomorphism ϕ consists in the game of parentheses,

$$\phi(f) : (x, x') \mapsto f(x)(x'), \quad \text{for all} \quad f \in \mathcal{C}^\infty(X, \mathcal{C}^\infty(X', X'')).$$

We say that the category {Diffeology} is *Cartesian closed.*

PROOF. First of all, let us check that $\mathbf{f} = \phi(f) \in \mathcal{C}^\infty(X \times X', X'')$. Since $f \in \mathcal{C}^\infty(X, \mathcal{C}^\infty(X', X''))$, for every plot Q of X, $r \mapsto f(Q(r))$ is a plot of $\mathcal{C}^\infty(X', X'')$, which means that for every plot Q' of X', $(r, s) \mapsto f(Q(r))(Q'(s))$ is a plot of X''. But this implies that, for every plot (Q, Q') of $X \times X'$, $r \mapsto \mathbf{f}(Q(r), Q'(r)) = f(Q(r))(Q'(r))$ is a plot of X''. Then, $\mathbf{f} = \phi(f)$ is smooth, that is, belongs to $\mathcal{C}^\infty(X \times X', X'')$. Now let us remark that ϕ is bijective, since $\mathbf{f}(x, x') = f(x)(x')$ is just a game of parentheses. Now, let us prove that ϕ is smooth. Let P be a plot of $\mathcal{C}^\infty(X, \mathcal{C}^\infty(X', X''))$, that is, a parametrization such that for every plot Q of X, the parametrization $(r, s) \mapsto P(r)(Q(s))$ is a plot of $\mathcal{C}^\infty(X, X')$. This means that for every plot Q' of X' the parametrization $(r, s, s') \mapsto P(r)(Q(s))(Q'(s'))$ is a plot of X''. But this is equivalent to the fact that, for every plot $Q \times Q'$ of $X \times X'$, the parametrization $(r, s, s') \mapsto \phi(P(r))(Q(s), Q'(s'))$ is a plot of X''. Hence, $\phi \circ P$ is a plot of $\mathcal{C}^\infty(X \times X', X'')$. Therefore, ϕ is smooth. Since what we have said is completely reversible, that proves that ϕ is a diffeomorphism. □

1.61. Functional diffeology on diffeomorphisms. Let X be a diffeological space. The group Diff(X), as well as any of its subgroups, inherits the functional diffeology of $\mathcal{C}^\infty(X, X)$. Note that the structure of *diffeological group* (art. 7.1) of Diff(X) is finer. It is the coarsest *group diffeology* such that the evaluation map is smooth (art. 1.57). A parametrization P of Diff(X) is a plot of the *standard diffeology of group of diffeomorphisms* if $r \mapsto P(r)$ and $r \to P(r)^{-1}$ are plots for the functional diffeology.[3]

1.62. Slipping X into $\mathcal{C}^\infty(\mathcal{C}^\infty(X, X), X)$. Let X be a diffeological space. Let us associate with every $x \in X$ the x-*evaluation map*

$$\mathbf{x} : \mathcal{C}^\infty(X, X) \to X \quad \text{with} \quad \mathbf{x} : f \mapsto f(x).$$

By definition of the functional diffeology (art. 1.57), the map $\mathbf{x}$ is smooth, that is, $\mathbf{x} \in \mathcal{C}^\infty(\mathcal{C}^\infty(X, X), X)$. Moreover, the map $j : x \mapsto \mathbf{x}$ is an induction from X into $\mathcal{C}^\infty(\mathcal{C}^\infty(X, X), X)$. We shall see further that j is stronger than an induction, but it is also an embedding (art. 2.13).

PROOF. First of all, applying this definition to the identity $f = \mathbf{1}_X$, we get the injectivity of j, $\mathbf{x}(\mathbf{1}_X) = \mathbf{x}'(\mathbf{1}_X) \Rightarrow x = x'$. Now, let us check that j is smooth. Let Q be a plot of X. The parametrization $j \circ Q$ is a plot of $\mathcal{C}^\infty(\mathcal{C}^\infty(X, X), X)$ if and only if, for every plot P of $\mathcal{C}^\infty(X, X)$, the map $(r, s) \mapsto (j \circ Q(r))(P(s))$ is a plot of X. But, by definition, $(j \circ Q(r))(P(s)) = P(s)(Q(r))$, and to be a plot of $\mathcal{C}^\infty(X, X)$ means precisely that, for every plot Q of X, the parametrization $(r, s) \mapsto P(r)(Q(s))$ is a plot of X (art. 1.57). Thus, j is smooth. Finally, let us check that the pullback of the diffeology of $\mathcal{C}^\infty(\mathcal{C}^\infty(X, X), X)$ by j is finer than the diffeology of X. Let Q be a parametrization in X, such that $j \circ Q$ is a plot of $\mathcal{C}^\infty(\mathcal{C}^\infty(X, X), X)$. Then $(r, f) \mapsto f(Q(r))$ is smooth. Restricting to $f = \mathbf{1}_X$, $r \mapsto (r, \mathbf{1}_X) \mapsto \mathbf{1}_X(Q(r)) = Q(r)$,

[3] For a manifold (art. 4.1), the second condition, $r \to P(r)^{-1}$ smooth, is the consequence of the first one, thanks to the implicit function theorem. For arbitrary diffeological spaces, it is not clear why this condition is needed.

we conclude that Q is a plot of X. Therefore, by application of (art. 1.31), j is an induction. □

1.63. Functional diffeology of a diffeology. Every diffeology, regarded as a set of smooth maps, carries a natural diffeology, which is a specialization of the functional diffeology described above (art. 1.57). Pick a diffeological space X and let $\mathcal{D}$ be its diffeology. The set of parametrizations $P : U \to \mathcal{D}$ satisfying the following condition is a diffeology:

($\diamondsuit$) For all $r_0 \in U$, for all $s_0 \in \operatorname{def}(P(r_0))$, there exist an open neighborhood $V \subset U$ of r_0 and an open neighborhood W of s_0 such that $W \subset \operatorname{def}(P(r))$ for all $r \in V$, and $(r, s) \mapsto P(r)(s)$, defined on $V \times W$, is a plot of X.

Note that a plot P of $\mathcal{D}$ takes locally its values in the set of plots of X with same dimension (art. 1.3). Also note that the last condition writes again $[r \mapsto P(r) \upharpoonright W] \upharpoonright V \in \mathcal{C}^\infty(V, \mathcal{C}^\infty(W, X))$. This is why this diffeology will be called the *standard functional diffeology* of $\mathcal{D}$. It will play afterwards a role in some constructions.

To simplify the vocabulary and to avoid confusing statements, we shall prefer sometimes the wording *smooth family of plots* instead of a *plot of the diffeology*, when the diffeology is regarded itself as a diffeological space.

PROOF. For a constant parametrization $P : s \mapsto Q$, with $Q \in \mathcal{C}^\infty(U, X)$, we can take $W = U$. Hence, the axiom D1 of diffeology is satisfied. The axiom D2 is satisfied by the very definition. Now, let $P : U \to \mathcal{D}$ satisfy ($\diamondsuit$), and let $F : U' \to U$ be a smooth parametrization. Then, let $r'_0 \in U'$ and $r_0 = F(r'_0)$. Let V be an open neighborhood of $r_0 \in U$ satisfying ($\diamondsuit$), and let $V' = F^{-1}(V)$. Then, V' is a domain satisfying ($\diamondsuit$), with the same W. Thus, the axiom D3 is satisfied, and ($\diamondsuit$) defines a diffeology on the set $\mathcal{D}$. □

1.64. Iterating paths. Functional diffeology is heavily used in the theory of homotopy of diffeological spaces (Chapter 5), in particular through the following construction of the iterated spaces of paths. Let X be any diffeological space. The *space of paths* in X, denoted by Paths(X), is defined by

$$\mathrm{Paths}(X) = \mathcal{C}^\infty(\mathbf{R}, X).$$

Now, let us define the following recurrence,

$$\mathrm{Paths}_0(X) = X, \text{ and } \mathrm{Paths}_p(X) = \mathrm{Paths}(\mathrm{Paths}_{p-1}(X)), \text{ for all } p > 0.$$

Said differently,

$$\mathrm{Paths}_1(X) = \mathrm{Paths}(X),\ \mathrm{Paths}_2(X) = \mathrm{Paths}(\mathrm{Paths}(X)), \text{ etc.}$$

The spaces $\mathrm{Paths}_p(X)$ will be called[4] the *iterated spaces of paths* in X. For each integer p, $\mathrm{Paths}_p(X)$ is equipped with the functional diffeology. Let $j_p : \sigma \mapsto \boldsymbol{\sigma}$ be the map defined, for all integers $p \geq 0$, from $\mathrm{Paths}_p(X)$ to $\mathrm{Maps}(\mathbf{R}^p, X)$, by

$$\boldsymbol{\sigma} : (x_1, \ldots, x_p) \mapsto \sigma(x_1)\cdots(x_p), \quad \text{for all } \sigma \in \mathrm{Paths}_p(X). \tag{$\diamondsuit$}$$

Then, the map j_p takes its values in $\mathcal{C}^\infty(\mathbf{R}^p, X)$, and is a diffeomorphism from $\mathrm{Paths}_p(X)$ onto $\mathcal{C}^\infty(\mathbf{R}^p, X)$.

NOTE. The injection $j_X : X \to \mathrm{Paths}(X)$ which associates with each point x the constant path $\mathbf{x} : t \mapsto x$ is not just smooth but it is also an induction. Indeed, for

[4] In reference to Chen's paper *Iterated paths integrals* [**Che77**]; see also the Afterword.

any plot $P : U \to \mathrm{Paths}(X)$ such that $P(U) \subset j_X(X)$, $j_X^{-1} \circ P = [r \mapsto P(r)(0)]$ is smooth. This is also clear for the iterated space of paths.

PROOF. Let us prove the proposition for $p = 0$. In this case, the map j_0 is given by $x \mapsto \mathbf{x} = [0 \mapsto x]$, and maps $\mathrm{Paths}_0(X) = X$ in $\mathrm{Maps}(\mathbf{R}^0, X)$. First of all, let us remark that every map from $\mathbf{R}^0 = \{0\}$ to X is constant, hence smooth. Thus $\mathrm{Maps}(\mathbf{R}^0, X) = \mathcal{C}^\infty(\mathbf{R}^0, X)$, and $j_0 : X \to \mathcal{C}^\infty(\mathbf{R}^0, X)$. Now, j_0 is bijective, let us show that j_0^{-1} is smooth. Let $P : U \to X$ be a plot, $j_0 \circ P : r \mapsto [0 \mapsto P(r)]$ and $Q : V \to \mathbf{R}^0$ be some plot. But Q is constant, $Q(s) = 0$, and $P(r)(Q(s)) = P(r)(0) = P(r)$. Thus, $(r, s) \mapsto P(r)(Q(s)) = P(r)$ is a plot of X and j_0 is smooth (art. 1.57).

Conversely, let $P : U \to \mathcal{C}^\infty(\mathbf{R}^0, X)$ be a plot of the functional diffeology. Since the evaluation map is smooth (art. 1.57), the map $j_p^{-1} \circ P = [r \mapsto P(r)(0)]$ is smooth, and hence is a plot of X. Thus, j_0^{-1} is smooth and j_0 is a diffeomorphism from X to $\mathcal{C}^\infty(\mathbf{R}^0, X)$, equipped with the functional diffeology.

Now, let us complete the proof by a recurrence. Let us assume the proposition is true for some $p \geq 0$. Now $\mathrm{Paths}_{p+1}(X) = \mathrm{Paths}(\mathrm{Paths}_p(X)) = \mathcal{C}^\infty(\mathbf{R}, \mathrm{Paths}_p(X)) \simeq \mathcal{C}^\infty(\mathbf{R}, \mathcal{C}^\infty(\mathbf{R}^p, X))$. But thanks to (art. 1.60), $\mathcal{C}^\infty(\mathbf{R}, \mathcal{C}^\infty(\mathbf{R}^p, X)) \simeq \mathcal{C}^\infty(\mathbf{R} \times \mathbf{R}^p, X)$. Hence, $\mathrm{Paths}_{p+1}(X) \simeq \mathcal{C}^\infty(\mathbf{R}^{p+1}, X)$. Thus, the proposition is still true for $p+1$. □

1.65. Compact controlled diffeology. Functional diffeology can be used as a formal framework for the *variational calculus*. Let us exemplify this claim by the simple classical problem of the extremals of the *energy functional*

$$E(\gamma) = \frac{1}{2}\int \|\dot\gamma(t)\|^2\, dt, \quad \text{with} \quad \gamma \in \mathrm{Paths}(\mathbf{R}^n) \quad \text{and} \quad \dot\gamma(t) = \frac{d\gamma(t)}{dt}.$$

As we know, this integral does not converge in general, but we can avoid this difficulty by just changing the diffeology of the space of paths. Let us begin by computing informally the variation of this integral for a smooth 1-parameter family of paths $s \mapsto \gamma_s$, and let $\gamma = \gamma_0$. The variable s is supposed to belong to some open neighborhood of $0 \in \mathbf{R}$. We get

$$\left.\frac{\partial E(\gamma_s)}{\partial s}\right|_{s=0} = \int \left\langle \dot\gamma(t), \left.\frac{\partial \dot\gamma_s(t)}{\partial s}\right|_{s=0} \right\rangle dt = \int \left\langle \dot\gamma(t), \frac{d}{dt}\left\{\frac{\partial \gamma_s(t)}{\partial s}\right\}_{s=0} \right\rangle dt,$$

which we can also write informally

$$dE(\delta\gamma) = \int \left\langle \dot\gamma(t), \frac{d}{dt}\delta\gamma(t) \right\rangle dt, \quad \text{with} \quad \delta\gamma(t) = \left.\frac{\partial \gamma_s(t)}{\partial s}\right|_{s=0}. \qquad (\bullet)$$

As usual in variational calculus, if we assume that the variation $s \mapsto \gamma_s$ is compactly supported, that is, if γ_s coincides with γ outside an interval $[a, b]$, then the integral will converge since the variation $\delta\gamma$ vanishes outside $[a, b]$. This consideration suggests we equip the space $\mathrm{Paths}(\mathbf{R}^n)$ with the *compact diffeology*, defined in Exercise 25, p. 23. But we need to differentiate under the integral sign, and with boundaries moving with the parameter s, this is a little bit uncomfortable. To secure the place, we introduce a new diffeology, finer than the compact diffeology. Let X and Y be two diffeological spaces. The parametrizations of the functional diffeology $P : U \to \mathcal{C}^\infty(X, Y)$, satisfying the following condition, form the *compact controlled diffeology* [**Igl87**].

(♠) For all $r_0 \in U$ there exist an open neighborhood V of r_0 and a D-compact K of X such that, for all $r \in V$, $P(r)$ and $P(r_0)$ coincide outside K.

A D-compact means a compact for the D-topology (art. 2.8). The difference with Exercise 25, p. 23 is subtle, when r runs around r_0 the compact diffeology implies that there exists a compact K_r, *a priori* depending on r, such that $P(r)$ and $P(r_0)$ coincide outside of K_r. The compact controlled diffeology, as far as it is concerned, implies that we can find one large compact K matching the condition for all r belonging to some small open ball centered at r_0.

Coming back to our problem, these are exactly the conditions of the variational calculus. Let $P : U \to \mathrm{Paths}(\mathbf{R}^n)$ be an m-plot for the compact controlled diffeology. Adjusting the above expression of the variation of the energy, denoted by $dE(P)_r(\delta r)$, where r belongs to U and δr is any vector of $\mathbf{R}^m$, and after a classical integration by parts, we get

$$dE(P)_r(\delta r) = -\int_a^b \left\langle \frac{d^2}{dt^2}\left\{P(r)(t)\right\}, \frac{\partial P(r)(t)}{\partial r}(\delta r)\right\rangle dt,$$

where $[a, b]$ is some interval satisfying the condition ($\spadesuit$) for P in an open neighborhood of r. Now, this last expression of dE, mapping every plot $P : U \to \mathrm{Paths}(\mathbf{R}^n)$ to the 1-form $dE(P)$ of U, is a well defined (closed) differential 1-form of $\mathrm{Paths}(\mathbf{R}^n)$, for the compact controlled diffeology (art. 6.28) and gives to ($\bullet$) its formal status. The critical points of the energy are the zeros of dE. In our very simple case, this 1-form is indeed exact, but the choice of a primitive depends on an arbitrary constant on which dE does not depend.

PROOF. The proof that ($\spadesuit$) defines a diffeology is a simple adaptation of the proof of Exercise 20, p. 20. □

Exercises

✎ EXERCISE 40 (The space of polynomials). Let us consider the space of polynomials with coefficients in $\mathbf{R}^m$, and degree less or equal than n,

$$\mathrm{Pol}_n(\mathbf{R}, \mathbf{R}^m) = \{t \mapsto x_0 + t x_1 + \cdots + t^n x_n \mid t \in \mathbf{R}, \text{ and } x_0, \ldots, x_n \in \mathbf{R}^m\}.$$

1) Show that, the map j_n, defined from $(\mathbf{R}^m)^{n+1}$ to $\mathcal{C}^\infty(\mathbf{R}, \mathbf{R}^m)$ by

$$j_n(x_0, x_1, \ldots, x_n) = [t \mapsto x_0 + t x_1 + \cdots + t^n x_n],$$

where $\mathcal{C}^\infty(\mathbf{R}, \mathbf{R}^m)$ is equipped with the functional diffeology (art. 1.57), and is an induction. Conclude that $\mathrm{Pol}_n(\mathbf{R}, \mathbf{R}^m)$, equipped with the functional diffeology, inherited from $\mathcal{C}^\infty(\mathbf{R}, \mathbf{R}^m)$, is diffeomorphic to $(\mathbf{R}^m)^{n+1}$.

2) Let $\omega \subset (\mathbf{R}^m)^{n+1}$ be any domain. Show that there exists a subset Ω of $\mathcal{C}^\infty(\mathbf{R}, \mathbf{R}^m)$ such that:

($\diamondsuit$) $\Omega \cap \mathrm{Pol}_n(\mathbf{R}, \mathbf{R}^m) = j_n(\omega)$;
($\heartsuit$) for every n-plot P of $\mathcal{C}^\infty(\mathbf{R}, \mathbf{R}^m)$, $P^{-1}(\Omega)$ is an n-domain.

We say that the subset $\mathrm{Pol}_n(\mathbf{R}, \mathbf{R}^m)$ is *embedded* in $\mathcal{C}^\infty(\mathbf{R}, \mathbf{R}^m)$ (art. 2.13).

NOTE. Polynomials are just defined by their coefficients. Since the coefficients belong to some vector space, we decide usually to carry arbitrarily the structure of (a power of) the vector space to the set of polynomials. But polynomials are, above all, maps between vector spaces. This exercise shows how diffeology can give its smooth structure considering only the functional diffeology of polynomials.

✎ Exercise 41 (A diffeology for the space of lines). Let $\mathrm{PL}(\mathbf{R}^n)$ be the set of *parametrized lines* in $\mathbf{R}^n$, defined as the set of all polynomials of $\mathbf{R}^n$ of degree strictly equal to 1 (see Exercise 40, p. 39).

1) Show that two lines f and g have the same images in $\mathbf{R}^n$ if and only if $g(t) = f(at+b)$, where a and b are two real numbers, and $a \neq 0$.

2) Check that 1) defines an action of the affine group of $\mathbf{R}$, denoted by $\mathrm{Aff}(\mathbf{R})$, on $\mathrm{PL}(\mathbf{R}^n)$, that is, $(a,b) : f \mapsto [t \mapsto f(at+b)]$.

3) Let $\mathrm{Aff}_+(\mathbf{R}) \subset \mathrm{Aff}(\mathbf{R})$ be the group of *direct affine transformations*, the ones for which $a > 0$. Let then $\mathrm{UL}_+(\mathbf{R}^n) = \mathrm{PL}(\mathbf{R}^n)/\mathrm{Aff}_+(\mathbf{R})$ be the set of *oriented nonparametrized lines*, equipped with the quotient diffeology. Show that the space $\mathrm{PL}(\mathbf{R}^n)$ is diffeomorphic to $\mathbf{R}^n \times (\mathbf{R}^n - \{0\})$; use the result of Exercise 40, p. 39.

4) Let ρ be the map $(x_0, x_1) \mapsto (r, u)$ defined by

$$r = x_0 - (x_0 \cdot u) \times u \quad \text{with} \quad u = \frac{x_1}{\|x_1\|},$$

where the $\cdot$ denotes the ordinary scalar product in $\mathbf{R}^m$, and $\| * \|$ denotes the associated norm. Show that ρ is a smooth section of the projection $\pi : \mathrm{PL}(\mathbf{R}^n) \to \mathrm{UL}_+(\mathbf{R}^n)$ (art. 1.53). Conclude that the space of oriented nonparametrized lines $\mathrm{UL}_+(\mathbf{R}^n)$, equipped with the diffeology specified in 2), is diffeomorphic to the subspace $\mathrm{TS}^{n-1} \subset \mathbf{R}^n \times \mathbf{R}^n$ defined by

$$\mathrm{TS}^{n-1} = \{(u,r) \in \mathbf{R}^n \times \mathbf{R}^n \mid \|u\| = 1 \text{ and } u \cdot r = 0\}.$$

Deduce that the space $\mathrm{Lines}(\mathbf{R}^2)$ of unparametrized and nonoriented lines of $\mathbf{R}^2$ is diffeomorphic to the Möbius strip, that is, the quotient of $S^1 \times \mathbf{R}$ by the equivalence relation $(u,r) \sim (-u,-r)$.

✎ Exercise 42 (A diffeology for the space of circles). Write down an exercise of the same type as Exercise 41, p. 39, for circles instead of lines.

Generating Families

The diffeologies can be built by generating families. *Any family of parametrizations of a set generates a diffeology. Conversely, any diffeology is generated by some set of parametrizations* (art. 1.66). *This mode of construction of diffeologies is very useful since it can reduce the analysis of the properties of a diffeological space to a subset of its plots, sometimes smaller than the whole diffeology. This definition leads to the definition of the dimension of a diffeological space* (art. 1.78), *which is a first global invariant of the category* {Diffeology}. *But this construction also leads to the introduction of important subcategories of diffeological spaces, for example the category of manifolds* (art. 4.7), *or the category of orbifolds* (art. 4.17), *etc.*

1.66. Generating diffeology. Let X be a set. Pick a set $\mathcal{F}$ of parametrizations of X, that is, $\mathcal{F} \subset \mathrm{Param}(X)$. There exists a finest diffeology containing $\mathcal{F}$, this diffeology will be called the *diffeology generated* by $\mathcal{F}$ and will be denoted by $\langle \mathcal{F} \rangle$. Conversely, let X be a diffeological space and $\mathcal{D}$ be its diffeology. A family $\mathcal{F}$ of plots of X, which generates the diffeology $\mathcal{D}$, will be called a *generating family of the diffeology* $\mathcal{D}$, or a *generating family for the space* X. The set of all the generating families of the diffeology $\mathcal{D}$ will be denoted by $\mathrm{Gen}(\mathcal{D})$, or by $\mathrm{Gen}(X)$.

$$\mathrm{Gen}(\mathcal{D}) = \{\mathcal{F} \subset \mathcal{D} \mid \langle \mathcal{F} \rangle = \mathcal{D}\}, \ \mathrm{Gen}(X) = \mathrm{Gen}(\mathcal{D}).$$

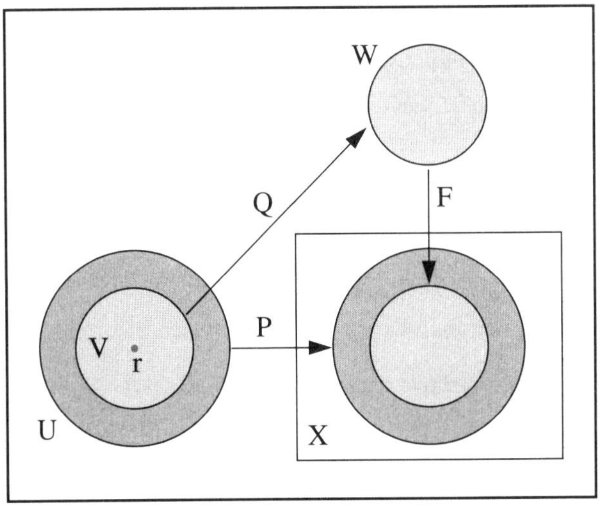

FIGURE 1.8. Smooth local lift of a parametrization.

PROOF. The finest diffeology of a set X containing a family $\mathcal{F}$ of parametrizations is the intersection of all diffeologies containing $\mathcal{F}$ (art. 1.22).

$$\langle \mathcal{F} \rangle = \bigcap_{\mathcal{D} \in \mathbf{D}} \mathcal{D} \quad \text{with} \quad \mathbf{D} = \{\mathcal{D} \in \text{Diffeologies}(X) \mid \mathcal{F} \subset \mathcal{D}\}. \qquad (\diamondsuit)$$

Note that the set $\mathbf{D}$ of all diffeologies containing $\mathcal{F}$ is not empty since it contains, at least, the coarse diffeology. □

1.67. Generated by the empty family. Let X be a set, and let us consider the empty family $\mathcal{F} = \varnothing$ as a subset of Param(X). The diffeology of X generated by the empty family (art. 1.66) is the discrete diffeology (art. 1.20), that is, the intersection of all diffeologies of X. Hence, considering the set X, $\langle \varnothing \rangle = \mathcal{D}_\circ(X)$.

1.68. Criterion of generation. Let X be a set, and $\mathcal{F}$ be a family of parametrizations of X. The plots of the diffeology generated by $\mathcal{F}$ are characterized by the following property (Figure 1.8):

(♣) A parametrization $P : U \to X$ is a plot for the diffeology generated by a family $\mathcal{F}$ if and only if, for all $r \in U$, there exists an open neighborhood V of r such that either $P \restriction V$ is a constant parametrization, or there exist a parametrization $F : W \to X$ belonging to $\mathcal{F}$, and a smooth parametrization $Q : V \to W$ such that $P \restriction V = F \circ Q$.

Now, if the union of the images of the elements F of $\mathcal{F}$ covers X, that is, if $X = \bigcup_{F \in \mathcal{F}} \text{val}(F)$ (we shall say that $\mathcal{F}$ is a *parametrized cover* of X), the criterion above is reduced to the following:

(♠) A parametrization $P : U \to X$ is a plot for the diffeology generated by a parametrized cover $\mathcal{F}$ if and only if, for every point r in U, there exist an open neighborhood V of r, a parametrization $F : W \to X$ belonging to $\mathcal{F}$, and a smooth parametrization $Q : V \to W$ such that $P \restriction V = F \circ Q$.

A generating family $\mathcal{F}$ which is a parametrized cover will be called a *covering generating family*. The criterion above can be rephrased as follows: let $\mathcal{F}$ be a covering generating family for X, a parametrization P is a plot of X if and only if it *lifts locally*, at each point, along an element of $\mathcal{F}$.

PROOF. Let us check first, for the second case (♠), that the defined set of parametrizations of X is a diffeology (art. 1.5). Since $\mathcal{F}$ is a parametrized cover, the plots satisfying (♠) cover X (see Exercise 1, p. 6), and the covering axiom D1 is satisfied. The locality axiom D2 is satisfied by the very definition. Then, let $P : U \to X$ be a parametrization satisfying the condition (♠). Let $\phi : U' \to U$ be a smooth parametrization, where U' is some domain. Let $P' = P \circ \phi$, and $r' \in U'$, let $r = \phi(r')$. Let $F : W \to X$ belonging to $\mathcal{F}$, and $Q : V \to W$ such that $F \circ Q = P \restriction V$, according to (♠). Now, $F \circ Q \circ \phi = (P \restriction V) \circ \phi = (P \circ \phi) \restriction V'$, where $V' = \phi^{-1}(V)$. Thus, $P' \restriction V' = F \circ Q'$, with $Q' = Q \circ \phi$. Therefore, P' satisfies (♠) and the locality axiom D3 is satisfied.

In the general case described by (♣) everything works as well as for (♠), except for D3, where there is something to check. If P is a constant parametrization defined on an open neighborhood V of r, then, ϕ being continuous, $V' = Q^{-1}(V)$ is an open neighborhood of r', and $P \circ \phi \restriction V' = P \restriction V$ is constant. The axiom D3 is checked.

Now, since (♠) is a special case of (♣), we consider only the case (♣). This property defines a diffeology containing $\mathcal{F}$. By application of axioms D1 and D3 of diffeology, any other diffeology containing $\mathcal{F}$ contains either the constant parametrizations or the plots of the form $F \circ Q$ where $F \in \mathcal{F}$ and Q is a smooth parametrization in the domain of F. Thus, the diffeology defined by (♣) is contained in any diffeology containing $\mathcal{F}$. Hence, it is the finest diffeology containing $\mathcal{F}$, that is, the diffeology generated by $\mathcal{F}$. □

1.69. Generating diffeology as projector. Let X be a set, and let $\mathcal{D}$ be a diffeology of X. Since the finest diffeology of X containing $\mathcal{D}$ is $\mathcal{D}$ itself, the diffeology generated by $\mathcal{D}$ (art. 1.66) is $\mathcal{D}$. In other words, for all $\mathcal{D} \in \text{Diffeologies}(X)$, $\langle \mathcal{D} \rangle = \mathcal{D}$. More precisely, the construction of diffeologies by means of generating families can be interpreted as a *projector* from the set of all the subsets of Param(X) onto the set of all the diffeologies of X, which we can write

$$\langle \cdot \rangle : \mathfrak{P}(\text{Param}(X)) \to \text{Diffeologies}(X) \quad \text{and} \quad \langle \langle \cdot \rangle \rangle = \langle \cdot \rangle.$$

1.70. Fineness and generating families. Let X be a set, and let $\mathcal{F}$ and $\mathcal{F}'$ be two families of parametrizations of X such that $\mathcal{F} \subset \mathcal{F}'$. The diffeology generated by $\mathcal{F}$ is finer than the diffeology generated by $\mathcal{F}'$,

$$\mathcal{F} \subset \mathcal{F}' \text{ implies } \langle \mathcal{F} \rangle \subset \langle \mathcal{F}' \rangle.$$

In other words, generating diffeologies by means of families of parametrizations is monotonic, relating to the partial ordering defined by inclusion. In particular, let X be a diffeological space. Let $\mathcal{F}$ be any family of plots of X, that is, a subset of the diffeology $\mathcal{D}$ of X. The diffeology generated by $\mathcal{F}$ is finer than the original $\mathcal{D}$,

$$\mathcal{F} \subset \mathcal{D} \text{ implies } \langle \mathcal{F} \rangle \subset \mathcal{D}.$$

PROOF. The diffeology generated by $\mathcal{F}$ is the intersection of all the diffeologies $\mathcal{D}$ containing $\mathcal{F}$, that is,

$$\langle \mathcal{F} \rangle = \bigcap_{\mathcal{D} \in \mathbf{D}} \mathcal{D} \quad \text{with} \quad \mathbf{D} = \{\mathcal{D} \in \text{Diffeologies}(X) \mid \mathcal{F} \subset \mathcal{D}\}.$$

Let us split the set $\mathbf{D}$ into the following two disjoint sets $\mathcal{A}$ and $\mathcal{B}$:

$$\begin{aligned} \mathcal{A} &= \{\mathcal{D} \in \text{Diffeologies}(X) \mid \mathcal{F} \subset \mathcal{D} \text{ and } \mathcal{F}' \subset \mathcal{D}\}, \\ \mathcal{B} &= \{\mathcal{D} \in \text{Diffeologies}(X) \mid \mathcal{F} \subset \mathcal{D} \text{ and } \mathcal{F}' \not\subset \mathcal{D}\}. \end{aligned}$$

Thus, $\mathbf{D} = \mathcal{A} \cup \mathcal{B}$, and $\mathcal{A} \cap \mathcal{B} = \varnothing$. Then, by associativity of the intersection, $\langle \mathcal{F} \rangle$ is the intersection of the diffeologies belonging to $\mathcal{A}$ and the diffeologies belonging to $\mathcal{B}$, that is,

$$\langle \mathcal{F} \rangle = \bigcap_{\mathcal{D} \in \mathbf{D}} \mathcal{D} = \bigcap_{\mathcal{D} \in \mathcal{A} \cup \mathcal{B}} \mathcal{D} = \Big(\bigcap_{\mathcal{D} \in \mathcal{A}} \mathcal{D} \Big) \bigcap \Big(\bigcap_{\mathcal{D} \in \mathcal{B}} \mathcal{D} \Big).$$

But since $\mathcal{F} \subset \mathcal{F}'$, any diffeology containing $\mathcal{F}'$ contains necessarily $\mathcal{F}$, and the set $\mathcal{A}$ is just the set of all the diffeologies containing $\mathcal{F}'$,

$$\mathcal{A} = \{\mathcal{D} \in \text{Diffeologies}(X) \mid \mathcal{F}' \subset \mathcal{D}\}.$$

Hence,

$$\bigcap_{\mathcal{D} \in \mathcal{A}} \mathcal{D} = \langle \mathcal{F}' \rangle \;\Rightarrow\; \langle \mathcal{F} \rangle = \langle \mathcal{F}' \rangle \bigcap \Big(\bigcap_{\mathcal{D} \in \mathcal{B}} \mathcal{D} \Big).$$

Therefore, $\langle \mathcal{F} \rangle \subset \langle \mathcal{F}' \rangle$, and the first part of the proposition is complete. Now, since $\langle \mathcal{D} \rangle = \mathcal{D}$ (art. 1.69), if $\mathcal{F}' = \mathcal{D}$ is a diffeology, then $\langle \mathcal{F} \rangle \subset \mathcal{D}$. ☐

1.71. Adding and intersecting families. Let X be a set. Let $\mathcal{F}$ and $\mathcal{F}'$ be two families of parametrizations of X, then

$$\langle \mathcal{F} \cup \mathcal{F}' \rangle = \langle \langle \mathcal{F} \rangle \cup \langle \mathcal{F}' \rangle \rangle \text{ and } \langle \mathcal{F} \cap \mathcal{F}' \rangle \subset \langle \mathcal{F} \rangle \cap \langle \mathcal{F}' \rangle.$$

PROOF. Let us prove the first assertion. Let us check first that $\langle \mathcal{F} \cup \mathcal{F}' \rangle \subset \langle \langle \mathcal{F} \rangle \cup \langle \mathcal{F}' \rangle \rangle$. Since $\mathcal{F} \subset \langle \mathcal{F} \rangle$ and $\mathcal{F}' \subset \langle \mathcal{F}' \rangle$, $\mathcal{F} \cup \mathcal{F}' \subset \langle \mathcal{F} \rangle \cup \langle \mathcal{F}' \rangle$. By application of (art. 1.70) we get $\langle \mathcal{F} \cup \mathcal{F}' \rangle \subset \langle \langle \mathcal{F} \rangle \cup \langle \mathcal{F}' \rangle \rangle$. Now, let us check that $\langle \langle \mathcal{F} \rangle \cup \langle \mathcal{F}' \rangle \rangle \subset \langle \mathcal{F} \cup \mathcal{F}' \rangle$. Since $\mathcal{F} \subset \mathcal{F} \cup \mathcal{F}'$, $\langle \mathcal{F} \rangle \subset \langle \mathcal{F} \cup \mathcal{F}' \rangle$ (art. 1.70), as well $\langle \mathcal{F}' \rangle \subset \langle \mathcal{F} \cup \mathcal{F}' \rangle$. Thus, $\langle \mathcal{F} \rangle \cup \langle \mathcal{F}' \rangle \subset \langle \mathcal{F} \cup \mathcal{F}' \rangle$. Hence $\langle \langle \mathcal{F} \rangle \cup \langle \mathcal{F}' \rangle \rangle \subset \langle \langle \mathcal{F} \cup \mathcal{F}' \rangle \rangle = \langle \mathcal{F} \cup \mathcal{F}' \rangle$ (art. 1.69). Therefore, $\langle \mathcal{F} \cup \mathcal{F}' \rangle = \langle \langle \mathcal{F} \rangle \cup \langle \mathcal{F}' \rangle \rangle$. Now concerning the second assertion, since $\mathcal{F} \cap \mathcal{F}' \subset \mathcal{F}$, $\langle \mathcal{F} \cap \mathcal{F}' \rangle \subset \langle \mathcal{F} \rangle$ (art. 1.70). As well, $\mathcal{F} \cap \mathcal{F}' \subset \mathcal{F}$, and then $\langle \mathcal{F} \cap \mathcal{F}' \rangle \subset \langle \mathcal{F}' \rangle$. Hence, $\langle \mathcal{F} \cap \mathcal{F}' \rangle \subset \langle \mathcal{F} \rangle \cap \langle \mathcal{F}' \rangle$. ☐

1.72. Adding constants to generating family. Let X be a set, and let $\mathcal{F}$ be a family of parametrizations of X. Let $\mathbf{X}$ be some family of parametrizations generating the discrete diffeology, that is, $\langle \mathbf{X} \rangle = \mathcal{D}_\circ(X)$, then $\langle \mathcal{F} \rangle = \langle \mathbf{X} \cup \mathcal{F} \rangle$. Thus, it is always possible to add a family of parametrizations, generating the discrete diffeology, to a generating family, without altering the generated diffeology. For example, considering the criterion (art. 1.68), extending the family $\mathcal{F}$ by the constant parametrizations $\mathbf{X} = \mathcal{C}^\infty(\mathbf{R}^0, X)$ reduces the case (♣) to the case (♠). We may denote by $\overline{\mathcal{F}} = \mathcal{C}^\infty(\mathbf{R}^0, X) \cup \mathcal{F}$ the extended family.

PROOF. First of all, by (art. 1.70) we have $\langle \mathcal{F} \rangle \subset \langle \mathbf{X} \cup \mathcal{F} \rangle$. Now, since the diffeology generated by $\mathbf{X}$ is the discrete diffeology, by (art. 1.71) we have $\langle \mathbf{X} \cup \mathcal{F} \rangle = \langle \mathcal{D}_\circ(X) \cup \langle \mathcal{F} \rangle \rangle$. But, since every diffeology contains the discrete diffeology, $\mathcal{D}_\circ(X) \cup \langle \mathcal{F} \rangle = \langle \mathcal{F} \rangle$, and $\langle \mathbf{X} \cup \mathcal{F} \rangle = \langle \mathcal{F} \rangle$. ☐

1.73. Lifting smooth maps along generating families. Let X and and X′ be two diffeological spaces. Let X be generated by a family $\mathcal{F}$, and X′ by a family $\mathcal{F}'$. A map $f : X \to X'$ is smooth if and only if, for every element $F : U \to X$

of $\mathcal{F}$, for every $r \in U$ there exists an open neighborhood $V \subset U$ of r, such that either $f \circ F \upharpoonright V$ is constant, or there exist a parametrization $F' : U' \to X'$ belonging to $\mathcal{F}'$ and a smooth parametrization $\phi : V \to U'$ such that $f \circ F \upharpoonright V = F' \circ \phi$. The following commutative diagram summarizes this criterion of smoothness for diffeologies defined through generating families.

$$
\begin{array}{ccc}
U \supset V & \xrightarrow{\ \ \phi\ \ } & U' \\
{\scriptstyle F}\Big\downarrow & & \Big\downarrow{\scriptstyle F'} \\
X & \xrightarrow[\ \ f\ \ ]{} & X'
\end{array}
$$

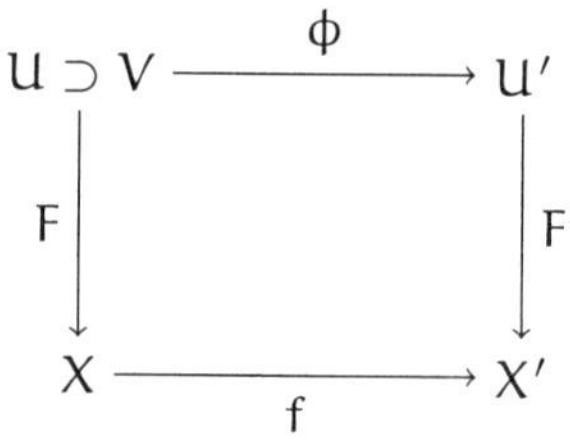

PROOF. Let us assume that f is smooth, for every plot $P : U \to X$ the map $f \circ P$ is a plot of X', in particular for $P = F \in \mathcal{F}$. Then, according to (art. 1.68), since $\mathcal{F}'$ is a generating family of X', for every point r of U there exists an open neighborhood V of r such that either the parametrization $f \circ P \upharpoonright V$ is constant or there exist an element $F' : U' \to X'$ of $\mathcal{F}'$, and a smooth parametrization $\phi : V \to U'$, such that $F' \circ \phi = f \circ F \upharpoonright V$.

Conversely, let $P : \mathcal{O} \to X$ be a plot, and let s be a point in $\mathcal{O}$. Since $\mathcal{F}$ is a generating family for X, there exists an open neighborhood W of r such that either $P \upharpoonright W$ is constant — thus $f \circ P \upharpoonright W$ is a constant parametrization, hence a plot of X' — or there exist a plot $F : U \to X$ belonging to $\mathcal{F}$ and a smooth parametrization $Q : W \to U$ such that $P \upharpoonright W = F \circ Q$. In the second case, since F is an element of $\mathcal{F}$, by assumption there exists an open neighborhood V of $r = F(s)$ such that either $f \circ F \upharpoonright V$ is constant, or there exist an element $F' : U' \to X'$ of $\mathcal{F}'$ and a smooth parametrization $\phi : V \to U'$ such that $f \circ F \upharpoonright V = F' \circ \phi$. Now, let $W' = Q^{-1}(V) \cap W$, the point s belongs to W', and since Q is a smooth parametrization $Q^{-1}(V)$ is a domain, but W is also a domain, hence $V' = Q^{-1}(W) \cap V$ is a domain. Thus, $P \upharpoonright W'$ is a parametrization in X, defined on a neighborhood of s. Now, in the first case, $f \circ P \upharpoonright W' = f \circ F \circ Q \upharpoonright W'$ is a constant parametrization, hence a plot of X'. In the second case, $f \circ P \upharpoonright W' = f \circ F \circ Q \upharpoonright W' = F' \circ \phi \circ Q \upharpoonright W'$. But, since F' is a plot of X' and ϕ and Q are smooth parametrizations, $F' \circ \phi \circ Q$ is a plot of X'. Finally, $f \circ P$ is locally a plot of X' at each point s of $\mathcal{O}$. Therefore, $f \circ P$ is a plot of X' and then f is smooth. $\qquad \square$

1.74. Pushing forward families. Let X and X' be two sets. Let $\mathcal{F}$ be a family of parametrizations of X, and let $f : X \to X'$ be a map. We call *pushforward of the family $\mathcal{F}$ by f*, the family $f_*(\mathcal{F})$ of parametrizations of X', defined by

$$f_*(\mathcal{F}) = \{f \circ F \mid F \in \mathcal{F}\}.$$

Then, the diffeology generated by the pushforward of the family $\mathcal{F}$ by f is the pushforward by f of the diffeology generated by $\mathcal{F}$ (art. 1.43), that is,

$$\langle\, f_*(\mathcal{F})\,\rangle = f_*(\langle\,\mathcal{F}\,\rangle).$$

In particular, let X and X' be two diffeological spaces, and let $f : X \to X'$ be a subduction (art. 1.46). The pushforward $f_*(\mathcal{F})$ of any generating family $\mathcal{F}$ for X is a generating family for X'.

PROOF. Let us denote $\mathcal{D} = \langle \mathcal{F} \rangle$, $\mathcal{F}' = f_*(\mathcal{F})$ and $\mathcal{D}' = \langle \mathcal{F}' \rangle$. We want to show that $f_*(\mathcal{D}) = \mathcal{D}'$. First of all, let us remark that, for X equipped with $\mathcal{D}$ and X' equipped with $\mathcal{D}'$, the map f is smooth. Indeed, let $P : \mathcal{O} \to X$ be a plot of X, and let r be a point of $\mathcal{O}$. Then, according to (art. 1.68), at least one of the two following possibilities occurs. Either there exists an open neighborhood V of r such that if the parametrization $P \restriction V$ is constant, then $f \circ P \restriction V$ is constant, and $f \circ P \restriction V$ belongs to $\mathcal{D}'$. Or there exists an open neighborhood V of r, an element $F : U \to X$ of $\mathcal{F}$ and a smooth parametrization $Q : V \to U$ such that $P \restriction V = F \circ Q$. Thus, $f \circ P \restriction V = f \circ F \circ Q$, but $F' = f \circ F \in \mathcal{F}'$, hence $f \circ P \restriction V = F' \circ Q$ belongs to $\mathcal{D}'$. In the two cases, $f \circ P$ is locally a plot of X' at each point of $\mathcal{O}$, thus $f \circ P$ is a plot of X'. Therefore, f is smooth, and since f is smooth, $f_*(\mathcal{D}) \subset \mathcal{D}'$ (art. 1.44).

Let us show now that $\mathcal{D}' \subset f_*(\mathcal{D})$. Let $P' : \mathcal{O}' \to X'$ be a plot of X', and let r' be a point of $\mathcal{O}'$. Then, according to (art. 1.68), at least one of the two following possibilities occurs: Either there exists an open neighborhood V' of r' such that the parametrization $P' \restriction V'$ is constant, and then $P' \restriction V'$ belongs to $f_*(\mathcal{D})$. Or there exists an open neighborhood V' of r', an element $F' : U' \to X'$ of $\mathcal{F}'$ and a smooth parametrization $Q' : V' \to U'$ such that $P' \restriction V' = F' \circ Q'$. But $F' = f \circ F$, for some $F \in \mathcal{F}$, thus $P' \restriction V' = f \circ F \circ Q' = f \circ Q$, with $Q = F \circ Q'$. Next, since $Q \in \mathcal{D}$ (art. 1.68), $P' \restriction V'$ belongs to $f_*(\mathcal{D})$. In the two cases P' is locally an element of $f_*(\mathcal{D})$ at each point of $\mathcal{O}'$, thus P' belongs to $f_*(\mathcal{D})$. Therefore, $\mathcal{D}' \subset f_*(\mathcal{D})$. ☐

1.75. Pulling back families. Let X and X' be two sets. Let $\mathcal{F}'$ be a family of parametrizations of X', and let $f : X \to X'$ be any map. Let us define the *pullback of the family* $\mathcal{F}'$ *by* f as the family $f^*(\mathcal{F})$ of parametrizations $F : U \to X$ satisfying the following property:

(◇) Either $f \circ F$ is constant or there exist an element $F' : U' \to X'$ of $\mathcal{F}'$ and a smooth parametrization $\phi : U \to U'$ such that $F' \circ \phi = f \circ F$.

Then, the diffeology generated by the pullback $f^*(\mathcal{F}')$ is the pullback by f of the diffeology generated by $\mathcal{F}$, that is,

$$\langle f^*(\mathcal{F}') \rangle = f^*(\langle \mathcal{F}' \rangle).$$

In particular, let X and X' be two diffeological spaces, and let $f : X \to X'$ be an induction (art. 1.29). The pullback $f^*(\mathcal{F}')$ of any generating family $\mathcal{F}'$ for X' is a generating family for X. Unfortunately, compared with the pushforward of a family, pulling back a small generating family may lead to a huge family, almost as big as the diffeology itself; see Exercise 45, p. 47. This is what shows Exercise 51, p. 50.

NOTE. Concerning diffeologies, the choice of a generating family is relatively arbitrary. For example, the empty family is equivalent to the family of constant parametrizations. If the family $\mathcal{F}'$ is empty, its pullback is not empty, but is the set of the parametrizations of X with values in the preimages of points $f^{-1}(x')$, $x' \in X'$. This must not surprise us since the pullback of the discrete diffeology is the sum of the preimages of points, equipped with the coarse diffeology.

PROOF. Let $\mathcal{F} = f^*(\mathcal{F}')$, $\mathcal{D} = \langle \mathcal{F} \rangle$, and $\mathcal{D}' = \langle \mathcal{F}' \rangle$. We want to check that $\mathcal{D} = f^*(\mathcal{D}')$. First of all, let us remark that, for X equipped with $\mathcal{D}$ and X' equipped with $\mathcal{D}'$, the map f is smooth. Indeed, let $P : \mathcal{O} \to X$ be a plot and $r_0 \in \mathcal{O}$. According to (art. 1.68), there exists an open neighborhood V of r_0 such that either $P \restriction V$ is constant or there exist $F \in \mathcal{F}$ and $\psi \in \mathcal{C}^\infty(V, \mathrm{def}(F))$ such that $F \circ \psi = P \restriction V$. In the first case $f \circ (P \restriction V)$ is constant, thus $f \circ (P \restriction V)$ is a plot of X'. in the second case, according to the definition of $\mathcal{F} = f^*(\mathcal{F}')$, either $f \circ F$ is constant or there exist $F' \in \mathcal{F}'$ and $\phi \in \mathcal{C}^\infty(\mathrm{def}(F), \mathrm{def}(F'))$ such that $F' \circ \phi = f \circ F$. If $f \circ F$ is constant, so is $f \circ F \circ \psi = f \circ (P \upharpoonright V)$, then $f \circ (P \upharpoonright V)$ is a plot of X'. If $F' \circ \phi = f \circ F$, then $f \circ (P \upharpoonright V) = f \circ F \circ \psi = F' \circ \phi \circ \psi$, and $f \circ (P \upharpoonright V)$ is a plot of X'. Therefore, $f \circ P$ is a plot of X' (axiom of locality) and f is smooth, that gives $\mathcal{D} \subset f^*(\mathcal{D}')$ (art. 1.27) and thus $\langle f^*(\mathcal{F}') \rangle \subset f^*(\langle \mathcal{F}' \rangle)$.

Let us show next that $f^*(\mathcal{D}') \subset \mathcal{D}$. Let $P : \mathcal{O} \to X$ be an element of $f^*(\mathcal{D}')$, that is, $f \circ P \in \mathcal{D}'$. Let $r_0 \in \mathcal{O}$, according to (art. 1.68) there exists an open neighborhood V of r_0 such that either $f \circ P \upharpoonright V$ is constant or there exist a parametrization $\mathcal{F}' : U' \to X'$ belonging to $\mathcal{F}'$, and a smooth parametrization $\phi : V \to U'$ such that $f \circ P \upharpoonright V = F' \circ \phi$. Let $F = P \upharpoonright V$. We have just said that either $f \circ F$ is constant or $f \circ F = F' \circ \phi$. But this is exactly the definition ($\diamondsuit$) of the elements of the pullback $f^*(\mathcal{F}') = \mathcal{F}$. Hence, P is locally an element of $\mathcal{F}$ at each point of $\mathcal{O}$. Hence, P is an element of $\mathcal{D}$. Therefore $f^*(\mathcal{D}') \subset \mathcal{D}$. $\square$

1.76. Nebula of a generating family. Let X be a diffeological space, and let $\mathcal{F}$ be a generating family for X (art. 1.66). Let us assume that $\mathcal{F}$ is a parametrized cover of X (art. 1.68). Note that if it is not the case, we can always add the constants $\mathcal{C}^\infty(\mathbf{R}^0, X)$ to $\mathcal{F}$ and then consider the extended generating family $\overline{\mathcal{F}} = \mathcal{C}^\infty(\mathbf{R}^0, X) \cup \mathcal{F}$ (art. 1.72). We call *nebula of the generating family* $\mathcal{F}$ the diffeological sum (art. 1.39) of the domains of the elements of $\mathcal{F}$, where the domains of the elements of $\mathcal{F}$ are equipped with the standard diffeology, that is,

$$\mathrm{Nebula}(\mathcal{F}) = \coprod_{F \in \mathcal{F}} \mathrm{def}(F) = \{(F, r) \mid F \in \mathcal{F} \text{ and } r \in \mathrm{def}(F)\}.$$

Then, the natural evaluation map, denoted by $\pi_\mathcal{F}$ and defined by

$$\pi_\mathcal{F} : \mathrm{Nebula}(\mathcal{F}) \to X \quad \text{with} \quad \pi_\mathcal{F}(F, r) = F(r),$$

is a subduction. Therefore, the space X can be regarded as the quotient space $\mathrm{Nebula}(\mathcal{F})/\pi_\mathcal{F}$ (art. 1.50). Now, let us define a *nebula over* X as any diffeological sum $\mathcal{N}$ of real domains together with a subduction $\pi : \mathcal{N} \to X$. Then, the nebula of any generating family $\mathcal{F}$ of X, together with $\pi_\mathcal{F}$, is a nebula over X. Conversely, every nebula over X is the nebula of a generating family of X. The nebula associated with the diffeology $\mathcal{D}$ of X is the maximal nebula of X, every other nebula is some of its subspaces.

NOTE. This construction shows that every finite dimensional diffeological space X (art. 1.78) is equivalent to the quotient of a manifold, maybe with variable dimension; see (art. 4.1) and (art. 4.7, note 2). Indeed, if $\dim(X) = n < \infty$, there exists a generating family $\mathcal{F}$ made up of plots with dimensions less than or equal to n, and $\mathrm{Nebula}(\mathcal{F})$ is clearly a manifold, in a broad sense.

PROOF. Let us show that $\pi_\mathcal{F}$ is smooth. Let $Q : \mathcal{O} \to \coprod_{F \in \mathcal{F}} \mathrm{def}(F)$ be a plot of the nebula. By definition of the sum diffeology (art. 1.39), for every r in $\mathcal{O}$, there exist an open neighborhood V of r and an element $F : U \to X$ of $\mathcal{F}$, such that $Q \upharpoonright V$ is a smooth parametrization in U. Now, $\pi_\mathcal{F} \circ Q \upharpoonright V = F \circ Q \upharpoonright V$, but since F is a plot of X and Q is a smooth parametrization in its domain, $F \circ Q \upharpoonright V$ is a plot of X. Thus, $\pi_\mathcal{F} \circ Q$ is locally, at each point of $\mathcal{O}$, a plot of X, that is, $\pi \circ Q$ is a plot of X, and $\pi_\mathcal{F}$ is smooth. Now, let $P : \mathcal{O} \to X$ be a plot of X and r be a point of $\mathcal{O}$. Since $\mathcal{F}$ is a generating family for X, there exist an open neighborhood V of r, an element $F : U \to X$ of $\mathcal{F}$, and a smooth parametrization $Q : V \to U$ such that $P \upharpoonright V = F \circ Q$

(art. 1.68). But $Q' : r \mapsto (F, Q(r))$, defined on V, is a plot of $\coprod_{F \in \mathcal{F}} \operatorname{def}(F)$, and $P \upharpoonright V = \pi_{\mathcal{F}} \circ Q'$. Thus, the plot P lifts locally at each point of its domain along $\pi_{\mathcal{F}}$. Therefore, $\pi_{\mathcal{F}}$ is a subduction (art. 1.48).

Now, let $\mathcal{N}$ be a nebula over X. By definition $\mathcal{N} = \coprod_{i \in \mathcal{I}} U_i$, where $\{U_i\}_{i \in \mathcal{I}}$ is a family of domains, and $\pi : \mathcal{N} \to X$ is some subduction. Then, let us define the family of parametrizations $\mathcal{F} = \{F_i = \pi \upharpoonright U_i\}_{i \in \mathcal{I}}$. Since π is smooth, the parametrizations are plots of X, and since π is a subduction, $\mathcal{F}$ is a generating family for X (*cf.* the first part of the proof), thus $\mathcal{N} = \operatorname{Nebula}(\mathcal{F})$. Therefore, every nebula over X is the nebula of a generating family. □

Exercises

✎ Exercise 43 (Generating tori). Exhibit a minimal generating family for S^1 defined in (art. 1.11), for the irrational torus T_α defined in Exercise 4, p. 8, and for $\mathbf{R}/\mathbf{Q}$, Exercise 5, p. 9.

✎ Exercise 44 (Global plots as generating families). Let X be a diffeological space. Show that the set of global plots $\mathcal{P} = \bigcup_{n \in \mathbf{N}} \mathcal{C}^\infty(\mathbf{R}^n, X)$ is a generating family for X.

✎ Exercise 45 (Generating the half-line). Consider $[0, \infty[\subset \mathbf{R}$ equipped with the subset diffeology inherited from the standard diffeology of $\mathbf{R}$ (art. 1.33). Let $\mathcal{F}$ be the generating family of $\mathbf{R}$ reduced to the identity, $\mathcal{F} = \{1_{\mathbf{R}}\}$. Show that the pullback of the generating family $\mathcal{F}$ by the inclusion $j : [0, \infty[\to \mathbf{R}$ (art. 1.75) is the whole diffeology of $[0, \infty[$.

✎ Exercise 46 (Generating the sphere). Let us consider the constructions and notations of Exercise 18, p. 19. For each point x of S^n, let us choose, once and for all, an orthonormal basis $(u_1, \ldots, u_n)$ of $E = x^\perp$. Let B be the open unit ball centered at $0 \in \mathbf{R}^n$. Let $F : B \to S^n$ be the parametrization defined by $F(s_1, \ldots, s_n) = f(\sum_{i=1}^n s_i u_i)$, where f is defined in that exercise by (♢). Show that the set of such F, when x runs over S^n, is a generating family for S^n.

✎ Exercise 47 (When the intersection is empty). Let us consider the two families $\mathcal{F} = \{x \mapsto x\}$ and $\mathcal{F}' = \{x \mapsto 2x\}$ of parametrizations in $\mathbf{R}$. Check that $\langle \mathcal{F} \cap \mathcal{F}' \rangle = \mathcal{D}_\circ(\mathbf{R})$ and $\langle \mathcal{F} \rangle \cap \langle \mathcal{F}' \rangle = \mathcal{C}^\infty_\star(\mathbf{R})$. Conclude that if generally $\langle \mathcal{F} \cap \mathcal{F}' \rangle \subset \langle \mathcal{F} \rangle \cap \langle \mathcal{F}' \rangle$, it is not true that $\langle \mathcal{F} \cap \mathcal{F}' \rangle = \langle \mathcal{F} \rangle \cap \langle \mathcal{F}' \rangle$.

Dimension of Diffeological Spaces

The first numerical invariant of a diffeological space is its dimension. The global dimension *of a diffeological space, introduced in this section, has a refinement given further in* (art. 2.22), *the* dimension map *of a diffeological space.*

1.77. Dimension of a family of parametrizations. Let X be a set, and let $\mathcal{F}$ be some nonempty family of parametrizations of X. We define the *dimension* of $\mathcal{F}$ as the supremum of the dimensions of its elements (art. 1.3), that is,

$$\dim(\mathcal{F}) = \sup\{\dim(F) \mid F \in \mathcal{F}\}.$$

By convention, we shall admit that for the empty family $\dim(\varnothing) = 0$. On the other hand, if the dimension of the elements of $\mathcal{F}$ is unbounded, that is, if for all $n \in \mathbf{N}$ there exists an element F of $\mathcal{F}$ such that $\dim(F) = n$, we shall agree that the dimension of $\mathcal{F}$ is infinite and we shall denote $\dim(\mathcal{F}) = \infty$. In other words, $\dim(\mathcal{F}) \in \mathbf{N} \cup \{\infty\}$.

1.78. Dimension of a diffeological space. Let X be a diffeological space, and let $\mathcal{D}$ be its diffeology. We shall call *dimension of* X the infimum of the dimensions of the generating families of $\mathcal{D}$,

$$\dim(X) = \inf_{\langle \mathcal{F} \rangle = \mathcal{D}} \dim(\mathcal{F}) = \inf\{\dim(\mathcal{F}) \mid \mathcal{F} \subset \mathcal{D} \text{ and } \langle \mathcal{F} \rangle = \mathcal{D}\}.$$

If the diffeology $\mathcal{D}$ has no generating family with finite dimension, the dimension of X will be said to be infinite, and we shall denote $\dim(X) = \infty$.

1.79. The dimension is a diffeological invariant. Let X and X′ be two diffeological spaces. If they are diffeomorphic, then they have the same dimension.

PROOF. Let $f : X \to X'$ be a diffeomorphism. Let $\mathcal{F}$ be a generating family for X. Clearly, $f \circ \mathcal{F} = \{f \circ F \mid F \in \mathcal{F}\}$ is a generating family for X′ (art. 1.74). Conversely, let $\mathcal{F}'$ be a generating family of X′, then $f^{-1} \circ \mathcal{F}'$ is a generating family of X. There is a one-to-one correspondence between the generating families of X and X′, therefore $\dim(X) = \dim(X')$. □

1.80. Dimension of real domains. The diffeological dimension of a nonempty n-domain U, equipped with the standard diffeology (art. 1.9), is n. In other words, for real domains, the diffeological dimension coincides with the usual dimension.

$$\text{If } U \in \text{Domains}(\mathbf{R}^n), \text{ then } \dim(U) = n.$$

PROOF. The singleton $\{1_U : U \to U\}$ is a generating family of U, hence $\dim(1_U) = \dim(U) \leq n$. Now, let us assume that there exists a generating family $\mathcal{F}$ of U such that $\dim(\mathcal{F}) < n$. Then, for every $r \in U$ there exist an open neighborhood V of r, an element $F : W \to U$ of $\mathcal{F}$, that is, $F \in \mathcal{C}^\infty(W, U)$ — since F needs to factorize through the identity 1_U — and a smooth map $Q : V \to W$ such that $1_U \upharpoonright V = 1_V = F \circ Q$. But $\dim(\mathcal{F}) < n$ implies that $\dim(F) = \dim(W) < n$. Now, the rank of the linear tangent map $D(F \circ Q)$ is less or equal to $\dim(W) < n$, but $D(F \circ Q) = D(1_V) = 1_{\mathbf{R}^n}$, thus $\text{rank}(D(F \circ Q)) = \text{rank}(1_{\mathbf{R}^n}) = n$. Therefore, there is no generating family $\mathcal{F}$ of U with $\dim(\mathcal{F}) < n$, and $\dim(U) = n$. □

1.81. Dimension zero spaces are discrete. The dimension of a diffeological space is zero if and only if it is discrete. But note that a diffeological space may consist of a finite number of points and have a nonzero dimension; see Exercise 48, p. 50.

PROOF. Let $X_\circ$ be a set X, equipped with the discrete diffeology (art. 1.20). Every plot $P : U \to X$, being locally constant, lifts locally along some 0-plot $\mathbf{x} : 0 \to x$, where $x = P(r)$ (art. 1.3). Hence, $\dim(X_\circ) = 0$. Conversely, let X be a diffeological space such that $\dim(X) = 0$. Then, by application of (art. 1.68), the 0-plots generate the diffeology $\mathcal{D}$ of X. Every plot lifting locally along a 0-plot is locally constant. Therefore, X is discrete. □

1.82. Dimensions and quotients of diffeologies. Let X and X′ be two diffeological spaces. If $\pi : X \to X'$ is a subduction, then $\dim(X') \leq \dim(X)$. Put differently, for any equivalence relation $\mathcal{R}$ defined on X,

$$\dim(X/\mathcal{R}) \leq \dim(X).$$

PROOF. Let $\mathcal{D}$ and $\mathcal{D}'$ be the diffeologies of X and X'. Let $\mathrm{Gen}(\mathcal{D})$ and $\mathrm{Gen}(\mathcal{D}')$ denote the sets of all the generating families of $\mathcal{D}$ and $\mathcal{D}'$. We know that for every generating family $\mathcal{F}$ of $\mathcal{D}$, $\pi \circ \mathcal{F}$ is a generating family of $\mathcal{D}'$, that is, $\pi \circ \mathrm{Gen}(\mathcal{D}) \subset \mathrm{Gen}(\mathcal{D}')$ (art. 1.74). Now, for every plot $P : U \to X$, $\dim(P) = \dim(U) = \dim(\pi \circ P)$. Hence, for every generating family $\mathcal{F} \in \mathrm{Gen}(\mathcal{D})$, $\dim(\mathcal{F}) = \dim(\pi \circ \mathcal{F})$. It follows the series of inequalities,

$$\begin{aligned}
\dim(X') &= \inf_{\mathcal{F}' \in \mathrm{Gen}(\mathcal{D}')} \dim(\mathcal{F}') && \text{by definition} \\
&\leq \inf_{\mathcal{F} \in \mathrm{Gen}(\mathcal{D})} \dim(\pi \circ \mathcal{F}) && \text{since } \pi \circ \mathrm{Gen}(\mathcal{D}) \subset \mathrm{Gen}(\mathcal{D}') \\
&\leq \inf_{\mathcal{F} \in \mathrm{Gen}(\mathcal{D})} \dim(\mathcal{F}) && \text{since } \dim(\pi \circ \mathcal{F}) = \dim(\mathcal{F}) \\
&\leq \dim(X) && \text{by definition.}
\end{aligned}$$

Therefore, $\dim(X') \leq \dim(X)$. □

1.83. Dimensions of a product. Let X and X' be two diffeological spaces. The dimension of the product $X \times X'$ satisfies the following inequality,[5]

$$\max(\dim(X), \dim(X')) \leq \dim(X \times X') \leq \dim(X) + \dim(X').$$

Note that I still do not know, at this moment, if we generally have the equality $\dim(X \times X') = \dim(X) + \dim(X')$ or if there are counterexamples.

PROOF. Let $\mathcal{F}$ and $\mathcal{F}'$ be two generating families, for X and X'. Since every plot of the product $X \times X'$ is a pair $(P, P') \in \mathcal{D} \times \mathcal{D}'$, where $\mathcal{D}$ and $\mathcal{D}'$ are the diffeologies of X and X', the product $\mathcal{F} \times \mathcal{F}'$ is a generating family for $X \times X'$. The elements $(F, F') \in \mathcal{F} \times \mathcal{F}'$ define a parametrization $F \times F' : U \times U' \to X \times X'$ by $F \times F'(r, r') = (F(r), F'(r'))$. Then,

$$\begin{aligned}
\dim(X \times X') &= \inf\{\dim(F'') \mid F'' \in \mathrm{Gen}(X \times X')\} \\
&\leq \inf\{\dim(F \times F') \mid (F, F') \in \mathrm{Gen}(X) \times \mathrm{Gen}(X')\} \\
&\leq \inf\{\dim(F) + \dim(F') \mid (F, F') \in \mathrm{Gen}(X) \times \mathrm{Gen}(X')\} \\
&\leq \inf\{\dim(F) \mid F \in \mathrm{Gen}(X)\} + \inf\{\dim(F') \mid F' \in \mathrm{Gen}(X')\} \\
&\leq \dim(X) + \dim(X').
\end{aligned}$$

Now, let $\mathcal{F}''$ be a generating family for $X \times X'$. The set of parametrizations $\mathcal{F} = \mathrm{pr}_X \circ \mathcal{F}''$, where pr_X is the projection onto the first factor, is a generating family for X. Indeed, let $x'_0 \in X'$, every plot $P : U \to X$ extends into a plot $\bar{P} : r \mapsto (P(r), x'_0)$ of $X \times X'$. Thus, the plot $\bar{P}$ lifts at each point along some element F'' of $\mathcal{F}''$, by composition with pr_X we get a local lift of P along $\mathrm{pr}_X \circ F''$. The same holds for X', and $\mathcal{F}' = \mathrm{pr}_{X'} \circ \mathcal{F}''$ is a generating family for X'. For these generating families we have $\dim(\mathcal{F}) = \dim(\mathcal{F}'')$ and $\dim(\mathcal{F}') = \dim(\mathcal{F}'')$. Thus,

$$\begin{aligned}
\dim(X \times X') &= \inf\{\dim(\mathcal{F}'') \mid \mathcal{F}'' \in \mathrm{Gen}(X \times X')\} \\
&= \inf\{\dim(\mathcal{F}) \mid \mathcal{F} = \mathrm{pr}_X \circ \mathcal{F}'', \mathcal{F}'' \in \mathrm{Gen}(X \times X')\} \\
&\geq \inf\{\dim(\mathcal{F}) \mid \mathcal{F} \in \mathrm{Gen}(X)\} \\
&\geq \dim(X).
\end{aligned}$$

Exchanging X and X' then gives $\dim(X \times X') \geq \max(\dim(X), \dim(X'))$. The inequality is then complete. □

[5]Thanks to the referee of the manuscript who pointed out to me that the proof contained a better lower bound than the one I gave in the statement.

1.84. Dimension of a subspace? There is no simple relation between the dimension of a diffeological space X and the dimension of its subspaces. The dimension of a subspace $A \subset X$ can be less than the dimension of X, equal or even greater. Exercise 51, p. 50 is the simple example of the subspace $[0, \infty[\subset \mathbf{R}$, having infinite dimension, while $\mathbf{R}$ has dimension 1.

Exercises

✎ EXERCISE 48 (Has the set $\{0, 1\}$ dimension 1?). Let $\pi : \mathbf{R} \to \{0, 1\}$ be the parametrization defined by $\pi(x) = 0$ if $x \in \mathbf{Q}$ and $\pi(x) = 1$ otherwise. Let $\{0, 1\}_\pi$ be the set $\{0, 1\}$ representing the quotient $\mathbf{R}/\pi$. Show that $\dim(\{0, 1\}_\pi) = 1$.

✎ EXERCISE 49 (Dimension of tori). Let $\Gamma \subset \mathbf{R}$ be any strict subgroup of $(\mathbf{R}, +)$, and let T_Γ be the quotient $\mathbf{R}/\Gamma$. Show that $\dim(T_\Gamma) = 1$. Note that this applies in particular both to the circles (art. 1.11) and to the irrational tori; see Exercise 4, p. 8, and Exercise 43, p. 47.

✎ EXERCISE 50 (Dimension of $\mathbf{R}^n/O(n, \mathbf{R})$). Let Δ_n be the quotient space of $\mathbf{R}^n$ by the action of the orthogonal group $O(n, \mathbf{R})$ (art. 1.50), that is, two points x and x' of $\mathbf{R}^n$ are equivalent if there exists an element A of $O(n, \mathbf{R})$ such that $x' = Ax$.

1) Show that Δ_n is equivalent to the set $[0, \infty[$ equipped with the pushforward of the standard diffeology of $\mathbf{R}^n$ by the map $\nu_n : x \mapsto \|x\|^2$.

2) Show that the plot ν_n cannot be lifted locally at the point 0 along a p-plot, with $p < n$. Consider the tangent map of ν_n at the point 0.

3) Deduce that $\dim(\Delta_n) = n$ and that $\mathbf{R}^n/O(n, \mathbf{R})$ and $\mathbf{R}^m/O(m, \mathbf{R})$, $n \neq m$, are not diffeomorphic.

NOTE. This exercise shows how a diffeology encodes a smooth structure and not just a set, or even a topology of a set. These quotients $\mathbf{R}^n/O(n, \mathbf{R})$ are homeomorphic two to two, but indeed not diffeomorphic if they are not identical.

✎ EXERCISE 51 (Dimension of the half-line). Let Δ_∞ be $[0, \infty[\subset \mathbf{R}$, equipped with the subset diffeology (art. 1.33). Show that $\dim(\Delta_\infty) = \infty$. Use a similar development to that in Exercise 50, p. 50, and the fact that for any integer n, the map ν_n is a plot of Δ_∞.

NOTE. This exercise and the previous one show in particular that the quotient $\Delta_1 = \mathbf{R}/\{\pm 1\}$, which has a structure of an orbifold (art. 4.17), is not a manifold with boundary (art. 4.15), which is the case of the half-line $\Delta_\infty = [0, \infty[\subset \mathbf{R}$.

CHAPTER 2

Locality and Diffeologies

Diffeology is built on purpose on a dry set, without structure. This may appear unusual, when most of the traditional constructions in differential geometry are built on top of topological spaces. However, that does not mean that topology is completely absent from the theory; it appears actually as a byproduct of local smoothness. This is the way that I chose to introduce these notions, following the logic of diffeology. Indeed, a smooth map f from X to X' transforms, by composition, the plots of X into plots of X'. Thus, a local smooth map f from X to X' will be a map, defined on a subset $A \subset X$, to X', which transforms every plot P of X into a plot of X'. But that raises the question of the domain of definition. Naturally $f \circ P$ is defined on $P^{-1}(A)$, which needs therefore to be a domain (maybe empty). Thus, a local smooth map is defined on a special kind of subset A, the ones for which $P^{-1}(A)$ is open for all plots P of X, and these are the open sets of a special topology, the D-topology of X (art. 2.8). I could have introduced first the D-topology of the diffeological spaces and then the local smooth maps, but I think that it breaks the logic, or self-consistency, of the theory. Moreover, diffeology applies on numerous spaces for which topology is helpless, irrational tori for example. Emphasizing D-topology could have been misinterpreted. However, the reader who is uncomfortable with my choice can begin by the section on D-topology and come back afterwards to local smoothness. That said, locality in diffeology will allow us to refine some diffeological concepts such as subduction or induction. It will suggest the notions of embeddings and embedded subsets. It will give us the notion of local diffeomorphisms which is associated with the definition of manifolds, or more generally, with the concept of modeling diffeological spaces.

To Be Locally Smooth

If we can define local smoothness for diffeological spaces, we must keep in mind that for many singular spaces, like irrational tori, this notion is helpless. For this kind of space, there are no local smooth maps but only global. Nevertheless, local smoothness will be essential to include properly the category of manifolds in the category {Diffeology}, *or the category of orbifolds, or some infinite dimensional spaces, and others.*

2.1. Local smooth maps. Let X and X' be two diffeological spaces. Let $A \subset X$ be some subset. We shall say that a map f from $A \subset X$ to X' is a *local smooth map* if, for every plot P of X, the parametrization $f \circ P$ defined on $P^{-1}(A)$ is a plot of X'. That implies in particular that $P^{-1}(A)$ is a domain. We shall say that a map $f : X \to X'$ is *locally smooth* at the point $x \in X$ if there exists a superset A of x such that the map f, restricted to A, is a local smooth map. We shall admit that the empty map is a local smooth map.

NOTE 1. Since $P^{-1}(A)$ must be open for each plot P of X, the domain of definition $A \subset X$ of the local smooth map f is not any subset, but an open subset for the D-topology of X, see (art. 2.8).

NOTE 2. To clearly express that the property of *local smoothness* of f involves the whole set of plots of X, and not just the ones with values in A, we shall often use this notation $f : X \supset A \to X'$.

NOTE 3. When considering local properties we may use the subscript $\star_{\rm loc}$, for instance, to denote space of local smooth maps or local diffeomorphisms etc.

2.2. Composition of local smooth maps. The composite of two local smooth maps is again a local smooth map. More precisely, let X, Y and Z be three diffeological spaces. Let $f : X \supset A \to Y$ and $g : Y \supset B \to Z$ be two local smooth maps. The map $g \circ f : X \supset f^{-1}(B) \to Z$ is a local smooth map.

PROOF. Let $P : U \to X$ be a plot of X, since f is locally smooth: $P^{-1}(A)$ is a domain, and $Q = f \circ P : P^{-1}(A) \to X$ is a plot of Y. Now, since g is a local smooth map and Q is a plot of Y, $Q^{-1}(B) = (f \circ P)^{-1}(B) = P^{-1}(f^{-1}(B))$ is a domain and $g \circ Q = g \circ (f \circ P) = (g \circ f) \circ P : P^{-1}(f^{-1}(B)) \to Z$ is a plot of Z. Hence, for every plot P of X, the map $(g \circ f) \circ P$ is a plot of Z. Therefore $g \circ f$ is a local smooth map. □

2.3. To be smooth or locally smooth. Let X and X' be two diffeological spaces. A map $f : X \to X'$ is smooth if and only if f is locally smooth at every point x of X.

PROOF. Obviously, if f is smooth it is locally smooth at every point x of X, just choose X as a superset of x (art. 2.1). Conversely, let us assume that f is locally smooth at every point x of X. Let $P : U \to X$ be a plot of X. For all $r \in U$ there exists, by assumption, a superset A of $x = P(r)$ such that $f \restriction A : X \supset A \to X'$ is a local smooth map (art. 2.1). This implies in particular that $(f \restriction A) \circ P$, defined on $V = P^{-1}(A)$, is a plot of X'. Thus, $V \subset U$ is a domain and $(f \circ P) \restriction V$ is a plot of X'. Hence, $f \circ P$ is locally a plot of X' at every point $r \in U$, that is, by axiom D2 of diffeology (art. 1.5), $f \circ P$ is a plot of X'. Therefore, for any plot P of X, $f \circ P$ is a plot of X', and f is smooth. □

2.4. Germs of local smooth maps. Let X and Y be two diffeological spaces, and let x be a point of X. We say that two local smooth maps $f : X \supset A \to Y$ and $f' : X \supset A' \to Y$ *have the same germ at the point* x if and only if there exists a superset $A'' \subset A \cap A'$ of x such that $f \restriction A'' : X \supset A'' \to Y$ and $f' \restriction A'' : X \supset A'' \to Y$ are still two local smooth maps, and $f \restriction A'' = f' \restriction A''$.

NOTE 1. Having the same germ at the point x is an equivalence relation whose classes are called *germs*. The germ of the local smooth map f at the point x will be denoted by $\mathrm{germ}(f)(x)$. If ϕ is the germ of f at the point x, the value $y = f(x)$ is well defined and is called the value of the germ of f at x, and we denote $\phi(x) = y$.

NOTE 2. Let Z be a third diffeological space, and let $f : X \supset A \to Y$ and $g : Y \supset B \to Z$ be two local smooth maps. Let $x \in A$ and $y = f(x) \in B$. The germ of $g \circ f : X \supset f^{-1}(B) \to Z$ at x depends only on the germ of f at x and on the germ of g at y. We shall denote $\mathrm{germ}(g)(y) \cdot \mathrm{germ}(f)(x) = \mathrm{germ}(g \circ f)(x)$. This defines the *composition of germs*. Like composition of functions, composition of germs is associative.

PROOF. For the first point, having the same germ at some point is obviously an equivalence relation. Now, for every local smooth map $f' : X \supset A' \to Y$, having the same germ as f at x implies that the point x belongs to A' and $f'(x) = f(x)$. Hence $y = f(x)$ depends only on the germ of f. For the second point, let $f' : X \supset A' \to Y$ be a local smooth map such that $\mathrm{germ}(f')(x) = \mathrm{germ}(f)(x)$. Then, let $g' : Y \supset B' \to Z$ be a local smooth map such that $\mathrm{germ}(g')(y) = \mathrm{germ}(g)(y)$. Thus, there exists a superset $A'' \subset A \cap A'$ of x such that $f \restriction A''$ and $f' \restriction A''$ are two equal local smooth maps. There exists, as well, a superset $B'' \subset B \cap B'$ of y such that $g \restriction B''$ and $g' \restriction B''$ are two equal local smooth maps. Hence, $g \circ f \restriction f^{-1}(B'') = g' \circ f' \restriction f^{-1}(B'')$, $x \in f^{-1}(B'')$, $g \circ f$ and $g' \circ f'$ are local smooth maps (art. 2.2). Hence, $\mathrm{germ}(g' \circ f')(x) = \mathrm{germ}(g \circ f)(x)$ and $\mathrm{germ}(g \circ f)(x)$ depends only on the germ of f at x and the germ of g at y. Associativity of composition of germs is a direct consequence of the associativity of the composition of maps. □

2.5. Local diffeomorphisms and étale maps. Let X and X' be two diffeological spaces. We say that a map $f : X \supset A \to X'$ is a *local diffeomorphism* if and only if f is injective and f is a local smooth map as well as its inverse $f^{-1} : X' \supset f(A) \to X$. The set of local diffeomorphisms from X to X' will be denoted by $\mathrm{Diff}_{\mathrm{loc}}(X, X')$ and $\mathrm{Diff}_{\mathrm{loc}}(X)$ when $X = X'$.

Let x be a point of X, we say that a map $f : X \to X'$ is a *local diffeomorphism at* x, or is *étale at* x, if there exists a superset A of x such that $f \restriction A : X \supset A \to X'$ is a local diffeomorphism. The map f is said to be *étale* if it is étale at each point.

2.6. Étale maps and diffeomorphisms. An étale map $f : X \to X'$ (art. 2.5), where X and X' are two diffeological spaces, is not necessarily a diffeomorphism, because f is not necessarily bijective. But, if f is bijective and étale, then f is a diffeomorphism. This is a direct consequence of (art. 2.3).

2.7. Germs of local diffeomorphisms as groupoid. The set $\mathrm{Diff}_{\mathrm{loc}}(X)$ of local diffeomorphisms of X is no longer a group, since the domains and the sets of values of local diffeomorphisms do not coincide necessarily. But let ϕ be the germ of a local diffeomorphism f at a point $x \in X$, and let γ be the germ of a local diffeomorphism g at a point $y = f(x)$. Then, the germ $\gamma \cdot \phi$ of $g \circ f$ at the point x is well defined (art. 2.4) and maps the point x to the point $z = g(y)$. This construction gives to the set of germs of local diffeomorphisms of X the structure of a groupoid [**McL71**]; see also (art. 8.3). The objects of this groupoid are the points of X and the morphisms from x to x' are the germs, at the point x, of the local diffeomorphisms of X mapping x to x'. This groupoid may be useful to explore the local structure of a diffeological space and its singularities, in relation to (art. 1.42).

Exercise

✎ EXERCISE 52 (To be a locally constant map). Let X and X' be two diffeological spaces. Let $f : X \to X'$ be a *locally constant map*, that is, for each point $x_0 \in X$ there exists a superset V of x_0 such that $f \restriction V$ is a local smooth map and constant: $f \restriction V : x \to f(x_0)$. Show that f is constant on the *connected components* of X (art. 5.6), that is, if x_0 and x_1 are two points of X such that there exists a 1-plot $\gamma : \mathbf{R} \to X$ connecting x_0 to x_1, *i.e.*, $\gamma(0) = x_0$ and $\gamma(1) = x_1$, then $f(x_0) = f(x_1)$.

D-topology and Local Smoothness

There are several topologies on a diffeological space, compatible with the diffeology, that is, such that smooth maps are continuous, for example, the coarse topology. But one of them plays a particular role, it is the finest topology such that plots are continuous. This topology is called the D-topology *of the space. In particular, local smooth maps are necessarily defined on* D-open *subsets, that is, open subsets for the D-topology. The D-topology defines a faithful functor from the category* {Diffeology} *of diffeological spaces to the category* {Topology} *of topological spaces.*

2.8. The D-Topology of diffeological spaces. There exists, on every diffeological space X, a finest topology such that the plots are continuous. This topology is called the *D-topology* of X [**Igl85**]. The open sets for the D-topology are called *D-open sets*, they are characterized by the following property:

(♣) A subset $A \subset X$ is open for the D-topology if and only if, for every plot P of X, $P^{-1}(A)$ is open.

PROOF. Let us consider the set $\mathcal{T}$ of all the subsets A of X such that for any plot P of X, $P^{-1}(A)$ is open. This set is a topology of X, indeed:

1. The empty set belongs to $\mathcal{T}$, since $P^{-1}(\varnothing) = \varnothing$ is open.

2. The set X belongs to $\mathcal{T}$, since $P^{-1}(X) = \mathrm{def}(P)$ is open by assumption.

3. Let $\{A_i\}_{i \in \mathcal{J}}$ be any finite family, with $A_i \in \mathcal{T}$, and let $A = \bigcap_i A_i$. We have $P^{-1}(A) = P^{-1}(\bigcap_i A_i) = \bigcap_i P^{-1}(A_i)$. But every $P^{-1}(A_i)$ is open. Since the family is finite, the intersection $\bigcap_i P^{-1}(A_i)$ is open. Hence, $P^{-1}(A)$ is open.

4. Let $\{A_i\}_{i \in \mathcal{J}}$ be any family, with $A_i \in \mathcal{T}$, and let $A = \bigcup_i A_i$. We have $P^{-1}(A) = P^{-1}(\bigcup_i A_i) = \bigcup_i P^{-1}(A_i)$. Since every $P^{-1}(A_i)$ is open the union $\bigcup_i P^{-1}(A_i)$ is open. Hence, $P^{-1}(A)$ is open.

Now, let $\mathcal{T}'$ be another topology for which the plots are continuous. Let us recall that $\mathcal{T}$ is finer than $\mathcal{T}'$ if $\mathcal{T}$ contains $\mathcal{T}'$, that is, $\mathcal{T}' \subset \mathcal{T}$. Let $A \subset X$ be open for the topology $\mathcal{T}'$, that is, $A \in \mathcal{T}'$. Because the plots are $\mathcal{T}'$-continuous, $P^{-1}(A)$ is open for all plots P of X, thus A is D-open, that is, $A \in \mathcal{T}$, and thus $\mathcal{T}' \subset \mathcal{T}$. Therefore, the D-topology is finer than $\mathcal{T}'$. Thus, the D-topology is the finest topology such that the plots are continuous. □

2.9. Smooth maps are D-continuous. Let X and X' be two diffeological spaces.

1. Every smooth map f from X to X' is *D-continuous*, that is, continuous for the D-topology.
2. Every diffeomorphism f from X to X' is a *D-homeomorphism*, that is, a homeomorphism for the D-topology.

Associating the underlying D-topological space with a diffeology defines a faithful functor from the category {Diffeology} to the category {Topology}.

PROOF. Let us denote by $\mathcal{D}$ and $\mathcal{D}'$ the diffeologies of X and X'. By the very definition of smooth maps (art. 1.14), f sends every plot $P \in \mathcal{D}$ into a plot $f \circ P \in \mathcal{D}'$. Let $A' \subset X'$ be D-open, and $A = f^{-1}(A')$. For every plot $P \in \mathcal{D}$, $P^{-1}(A) = P^{-1}(f^{-1}(A')) = (f \circ P)^{-1}(A')$. But $f \circ P$ is a plot of $\mathcal{D}'$ and A' is D-open, thus $(f \circ P)^{-1}(A')$ is open. Hence, $P^{-1}(A)$ is open for every plot $P \in \mathcal{D}$.

Therefore $A = f^{-1}(A')$ is D-open and f is continuous. Next, a diffeomorphism is a bijective map, smooth as well as its inverse. Thus, a diffeomorphism is a bijective bicontinuous map, that is, a homeomorphism. □

2.10. Local smooth maps are defined on D-opens. Let f be a map defined on a subset A of a diffeological space X with values in another one X'.

(♣) The map $f : X \supset A \to X'$ is a local smooth map (art. 2.1) if and only if A is D-open and f is smooth as a map from A to X', where A is equipped with the subset diffeology.

In particular, if $f : X \supset A \to X'$ is a local diffeomorphism, then both A and $f(A)$ are D-open, f is a local homeomorphism for the D-topology, and $f \upharpoonright A$ is a diffeomorphism onto $f(A)$.

PROOF. Let us assume first that f is a local smooth map. By definition, for every plot $P : U \to X$, the parametrization $f \circ P : P^{-1}(A) \to X'$ is a plot of X'. Hence, since plots are defined on domains, $P^{-1}(A)$ is a domain. Thus, since for any plot P of X the set $P^{-1}(A)$ is a domain, A is D-open. Now let us consider $A \subset X$ equipped with the subset diffeology (art. 1.33) and $f : A \to X$. Let $P : U \to A$ be a plot, since the injection $j_A : A \to X$ is smooth, P also is a plot of X. Now, since f is a local smooth map, $f \circ P$ is a plot of X'. Therefore, $f : A \to X'$, where A is equipped with the subset diffeology, is smooth.

Conversely, let us assume that A is D-open and that $f : A \to X'$, where A is equipped with the subset diffeology, is smooth. Let $P : U \to X$ be a plot. since A is D-open, $V = P^{-1}(A)$ is a domain. Since the inclusion map $j_V : V \to U$ is a smooth parametrization, $P \upharpoonright V = P \circ j_V$ is a plot of X with values in A, thus $P \upharpoonright V$ is a plot of A for the subset diffeology. Now, since $f : A \to X$ is smooth, $f \circ (P \upharpoonright V)$ is a plot of X'. Then, since $f \circ P = f \circ (P \upharpoonright V)$, the parametrization $f \circ P$ is a plot of X'. Therefore, f is a local smooth map from $A \subset X$ to X'.

Now, if f is a local diffeomorphism, f^{-1}, defined on $f(A)$, is a local smooth map, thus both A and $f(A)$ are D-open. The restrictions $f \upharpoonright A : A \to f(A)$ and $f^{-1} \upharpoonright f(A) : f(A) \to A$, where A and $f(A)$ are equipped with the subset diffeology, are smooth, thus $f \upharpoonright A$ a diffeomorphism onto $f(A)$. Then, $f \upharpoonright A$ is a homeomorphism for the D-topology (art. 2.9), that is, f is a local homeomorphism. □

2.11. D-topology on discrete and coarse spaces. The D-topology of discrete diffeological spaces is discrete. The D-topology of coarse diffeological spaces is coarse. But note that a noncoarse diffeological space can inherit the coarse D-topology (see Exercise 55, p. 56).

PROOF. Let us consider a set X be equipped with the discrete diffeology. Let $x \in X$, and let $P : U \to X$ be a plot of the discrete diffeology, if $x \notin P(U)$, then $P^{-1}(x) = \varnothing$ is open. If $x = P(r)$, $r \in U$, there exists an open neighborhood V of r such that $P \upharpoonright V$ is constant and equal to x. Then, since $P^{-1}(x)$ is a union of open subsets of U, it is open. Thus, every point of X is D-open, and the D-topology of the discrete diffeology is discrete. On the other hand, let X be equipped with the coarse diffeology. Let us assume that X is not reduced to a point. Let $\Omega \subset X$ such that $\Omega \neq \varnothing$ and $\Omega \neq X$. Then, there exist a point $x \in X - \Omega$ and a point $y \in \Omega$. Let us define $P : \mathbf{R} \to X$ by $P(t) = x$ if $t \in \mathbf{R} - \{0\}$ and $P(0) = y$. Since every parametrization in X is a plot, P is a plot. But, $P^{-1}(\Omega) = \{0\}$ is closed in $\mathbf{R}$, then

Ω is not D-open. Therefore, the only D-open subsets of X are the empty set and X, the D-topology associated with the coarse diffeology is coarse. □

2.12. Quotients and D-topology. Let X and X′ be two diffeological spaces. Let $\pi : X \to X'$ be a subduction (art. 1.46). Then, the D-topology of X′ is the quotient of the D-topology of X by π. In other words, a subset $A \subset X'$ is D-open if and only if $\pi^{-1}(A)$ is D-open. This applies in particular to the quotient of X by any equivalence relation (art. 1.50).

PROOF. Let A′ be a subset of X′. Let us assume that $\pi^{-1}(A')$ is D-open. Let $P' : U \to X'$ be a plot. For all $r \in U$, there exist an open neighborhood V of r, and a plot $P : V \to X$, such that $\pi \circ P = P' \upharpoonright V$. Hence, there exists an open covering $\{U_i\}_{i\in\mathcal{J}}$ of U, indexed by some set $\mathcal{J}$ ($\mathcal{J}$ can be U itself), and for each $i \in \mathcal{J}$ there exists a plot $P_i : U_i \to X$, such that $P' = \sup\{\pi \circ P_i\}_{i\in\mathcal{J}}$ (where sup denotes the smallest common extension). Then, $P'^{-1}(A') = (\sup\{\pi\circ P_i\}_{i\in\mathcal{J}})^{-1}(A') = \bigcup_{i\in\mathcal{J}}(\pi\circ P_i)^{-1}(A') = \bigcup_{i\in\mathcal{J}} P_i^{-1}(\pi^{-1}(A'))$. But $\pi^{-1}(A')$ is assumed to be D-open, and the P_i are plots of X. Thus, for all $i \in \mathcal{J}$, $P_i^{-1}(\pi^{-1}(A'))$ are domains, and their union is a domain. Hence, $P'^{-1}(A')$ is a domain for all plots P′ of X′. Therefore, A′ is D-open. Conversely, let us assume that A′ is D-open. Let $P : U \to X$ be a plot. Then, $P^{-1}(\pi^{-1}(A')) = (\pi \circ P)^{-1}(A')$. Since π is smooth, $\pi \circ P$ is a plot of X′, and $(\pi\circ P)^{-1}(A')$ is a domain. Therefore $\pi^{-1}(A')$ is D-open. □

Exercises

✎ EXERCISE 53 (Diffeomorphisms of the square). Let Sq be the square $[0,1] \times \{0,1\} \cup \{0,1\} \times [0,1]$, equipped with the subset diffeology (art. 1.33) of the smooth diffeology of $\mathbf{R}^2$. Justify that the restriction of the diffeomorphisms of Sq to the four corners defines a homomorphism h from Diff(Sq) to the group of permutations $\mathfrak{S}_4$. Describe the image of h.

✎ EXERCISE 54 (Smooth D-topology). Show that the D-topology of a real domain, equipped with the smooth diffeology (art. 1.9), coincides with its standard topology.

✎ EXERCISE 55 (D-topology of irrational tori). Let T_Γ be the quotient of $\mathbf{R}$ by a strict dense subgroup (for example $\mathbf{Q}$ or $\mathbf{Z} + \alpha\mathbf{Z}$ where $\alpha \notin \mathbf{Q}$). Show that the D-topology of T_Γ is coarse. Deduce that the D-topology is not a full functor.

Embeddings and Embedded Subsets

Combining smoothness and D-topology leads to a refinement of the notion of induction and introduces the notion of diffeological embedding. Moreover, since every subset of a diffeological space has a natural subset diffeology, the comparison between the D-topology induced by the ambient space and the D-topology of the induced diffeology distinguishes between embedded subsets of a diffeological space and the others, without referring to anything other than the ambient diffeology.

2.13. Embeddings. Let X and X′ be two diffeological spaces. Let $f : X \to X'$ be an induction (art. 1.29). We shall say that f is an *embedding* if the pullback by f of the D-topology of X′ coincides with the D-topology of X, that is, an induction f is an embedding if and only if, for every D-open subset $A \subset X$, there exists a D-open subset $A' \subset X'$ such that $f^{-1}(A') = A$.

2.14. Embedded subsets of a diffeological space. Let X be a diffeological space and $A \subset X$ be some subset. We shall say that A is *embedded* if the canonical induction $j_A : A \to X$, where A is equipped with the subset diffeology, is an embedding. Every subspace A of X carries two natural topologies: the D-topology given by the induced diffeology of X, and the induced topology of the ambient D-topology of X. The set A is *embedded* if these two topologies coincide. In other words, $A \subset X$ is embedded if every $U \subset A$, open for the D-topology of the induced diffeology, is the imprint of some D-open V of X, that is, $U = A \cap V$.

NOTE. For a subset of a diffeological space, to be embedded depends only on the diffeology of the ambient space X, and does not involve any extra structure.

Exercises

✎ EXERCISE 56 (**Q** is discrete but not embedded in **R**). Show that $\mathbf{Q} \subset \mathbf{R}$, where **R** is equipped with the smooth diffeology, is not embedded.

✎ EXERCISE 57 (Embedding $\mathrm{GL}(n, \mathbf{R})$ in $\mathrm{Diff}(\mathbf{R}^n)$). Let us consider the linear group $\mathrm{GL}(n, \mathbf{R})$ as a group of matrices. The components of a matrix M will be denoted by M_{ij}, $i, j = 1 \cdots n$. The identification between an element of $\mathrm{GL}(n, \mathbf{R})$ and the ordered set of its matrix elements (for any given definite ordering) identifies $\mathrm{GL}(n, \mathbf{R})$ with a domain of $\mathbf{R}^{n \times n}$. By this identification $\mathrm{GL}(n, \mathbf{R})$ inherits the diffeology of $\mathbf{R}^{n \times n}$. This diffeology will be called the *standard diffeology* of $\mathrm{GL}(n, \mathbf{R})$ and will equip $\mathrm{GL}(n, \mathbf{R})$. On the other hand, $\mathrm{GL}(n, \mathbf{R})$ can be regarded as a subgroup of $\mathrm{Diff}(\mathbf{R}^n)$. We equip $\mathrm{Diff}(\mathbf{R}^n)$ with the functional diffeology (art. 1.61).

1) Show that the inclusion $\mathrm{GL}(n, \mathbf{R}) \hookrightarrow \mathrm{Diff}(\mathbf{R}^n)$ is an induction.

2) Show that this induction is actually an embedding.

Hint: Consider the following subsets of $\mathrm{Diff}(\mathbf{R}^n)$:

$$\Omega_\varepsilon = \{f \in \mathrm{Diff}(\mathbf{R}^n) \mid D(f)(0) \in B(\mathbf{1}_n, \varepsilon)\},$$

where $D(f)(0)$ is the linear tangent map of f at the point 0, and $B(\mathbf{1}_n, \varepsilon)$ is the open ball in $\mathrm{GL}(n, \mathbf{R})$, centered at the identity $\mathbf{1}_n$, with radius ε.

✎ EXERCISE 58 (The irrational solenoid is not embedded). Show that the irrational solenoid defined in Exercise 31, p. 31, is not embedded.

Local or Weak Inductions

Some smooth injections look like inductions but are not. The example of the infinite symbol of Exercise 59*, p.* 58*, is a good illustration of this situation. This is why we need to weaken, or refine, the definition of inductions and also consider maps which are only locally inductive. This notion is close to the ordinary concept of immersion (see Exercise* 12*, p.* 17*), but we still do not know if they coincide.*

2.15. Local inductions. Let X and X′ be two diffeological spaces. Let $f : X \to X'$ be a smooth map. We shall say that f is a *local induction at the point* $x \in X$ if there exists an open neighborhood $\mathcal{O}$ of x such that $f \restriction \mathcal{O}$ is injective and for every plot $P : U \to f(\mathcal{O})$ centered at $x' = f(x)$, $P(0) = x'$, $f^{-1} \circ P'$ is a plot of X. Put differently, f is an induction at the level of the germ of the diffeology of X at x (art. 2.19). Note that an induction is a local induction everywhere, but a local induction everywhere

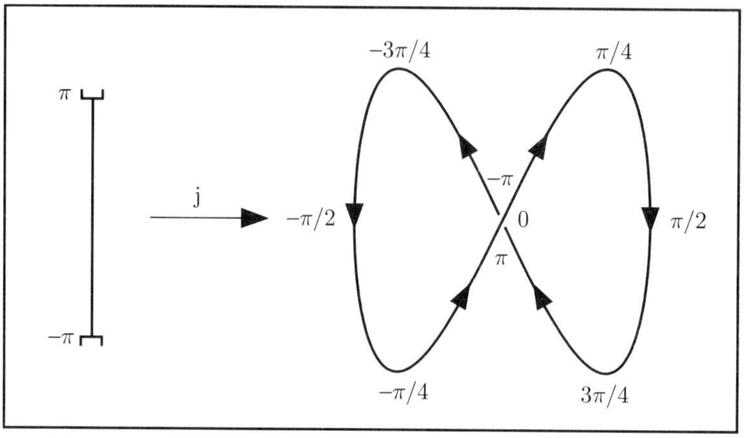

FIGURE 2.1. The infinite symbol.

is not necessarily an induction; see Exercise 59, p. 58. It is not enough for a local induction to be injective for being an induction.

NOTE. For a smooth injection j from an n-domain U into an m-domain V, if j is an immersion at a point x, that is, if its tangent map $D(j)(x)$ is injective, then j is a local induction at x; see Exercise 12, p. 17. I am still not sure whether the converse is true for $n = 1$ and $m > 1$. Actually, it could be weird if local inductions between real domains were not immersions.[1] For $n = m = 1$, j is a local diffeomorphism, thus these notions coincide.

Exercise

✎ EXERCISE 59 (The infinite symbol). Let j be the smooth map described by Figure 2.1 and defined by

$$j :]-\pi, \pi[\to \mathbf{R}^2 \qquad j(t) = \begin{pmatrix} \sin t \\ \sin 2t \end{pmatrix}.$$

1) Check that j is injective and that $\lim_{t \to -\pi} j(t) = \lim_{t \to +\pi} j(t) = (0, 0)$.

2) Let P be the following plot of $\mathbf{R}^2$:

$$P :]-\pi, \pi[\to \mathbf{R}^2 \qquad P(t) = \begin{pmatrix} \sin t \\ -\sin 2t \end{pmatrix}.$$

Check that P and j have the same image. Draw the image of P, similar to the image of j drawn in Figure 2.1.

3) Show that $j^{-1}(P(]-\pi/4, \pi/4[))$ is disconnected. Conclude that the injection j is not an induction.

4) Check that j is everywhere a local induction.

---

[1]When J. M. Souriau wrote one of his first papers on diffeology, he named immersion what is called induction now, but after a remark from J. Pradines he changed his mind. He named also submersions what is now subductions, but for that question things are different; see (art. 2.16).

Local or Strong Subductions

Subductions (art. 1.46) *express a global behavior of some surjections. As we localized the notion of induction and then obtained the notion of local induction* (art. 2.15), *the notion of subduction can be localized, or refined, as well and leads to the notion of* local subduction. *Exercise* 61, *p.* 60, *illustrates the case where a subduction is not everywhere a local subduction.*

2.16. Local subductions. Let X and X' be two diffeological spaces. We shall say that a smooth surjection $f : X \to X'$ is a *local subduction at the point* $x \in X$ if for every plot $P : U \to X'$ pointed at $x' = f(x)$, $P'(0) = x'$, there exist an open neighborhood V of 0 and a plot $Q : V \to X$ such that $Q(0) = x$ and $f \circ Q = P \restriction V$. Said differently, f is a subduction (art. 1.46) at the level of the germ of the diffeology of X at the point x (art. 2.19). We simply shall say local subductions for everywhere local subductions, they are *a fortiori* subductions.

NOTE. For a smooth surjection $\pi : U \to V$ between real domains, if π is a *submersion* at a point $x \in U$, that is, if its tangent map $D(\pi)(x)$ is surjective, then π is a local subduction at x. Conversely, if π is a local subduction at x, then it is a submersion at this point.[2] For real domains these notions are equivalent.

PROOF. Let $\dim(U) = n$ and $\dim(V) = m$. If $\pi : U \to V$ is a submersion at x, thanks to the rank theorem for domains [**Die70a**, 10.3.1], there exists a local diffeomorphism ϕ defined on a neighborhood $\mathcal{O}$ of x to $\mathbf{R}^m \times \mathbf{R}^{n-m}$ such that the projection π is equivalent to the projection onto the first factor. Therefore, every plot centered at x' can be simply locally lifted at x by choosing a constant in the second factor. Conversely, if π is a local subduction at x, for every vector $v \in \mathbf{R}^m$, the path $t \mapsto x' + tv$ has a local lift $c : t \to x_t$, $\pi(x_t) = x' + tv$, centered at x. Thus, $D(\pi)(x)(\dot{c}(0)) = v$, with $\dot{c}(t) = dc(t)/dt$. Therefore, $D(\pi)(x)$ is surjective. □

2.17. Compositions of local subductions. Let X, X' and X'' be three diffeological spaces. Let $f : X \to X'$ and $f' : X' \to X''$ be two local subductions. The composite $f' \circ f$ is still a local subduction.

PROOF. First of all, the composition of two smooth surjections is still a smooth surjection. Let $P : U \to X''$ be a plot centered at x''. Let $x \in X$ such that $f' \circ f(x) = x''$. Let $x' = f(x)$, since f' is a local subduction there exist an open neighborhood V' of 0, and a plot $Q' : V' \to X'$, such that $Q'(0) = x'$ and $P \restriction V' = f' \circ Q'$. Since Q' is a plot of X' and f is a local subduction, there exist an open neighborhood V of 0, and a plot $Q : V \to X$, such that $Q(0) = x$ and $Q' \restriction V = f \circ Q$. Hence, the plot Q satisfies $Q(0) = x$ and $(f' \circ f) \circ Q = P \restriction V$. □

2.18. Local subductions are D-open maps. Let X and X' be two diffeological spaces. Let $f : X \to X'$ be a local subduction. If $A \subset X$ is a D-open set, then $f(A)$ also is D-open. Local subductions are D-open maps.

PROOF. First of all, let us remark that, composing with a translation, we can replace 0 by any r such that $P(r) = x$ in the definition of local subduction (art. 2.16). Now, let $A' = f(A)$ and $P : U \to X'$ be a plot. We want to show that $P^{-1}(A')$ is

[2]It is why we could name local subductions, submersions. But for the balance between local immersion and local subduction, we continue with this vocabulary. If we can prove that a local induction in real domains is always an immersion, then it will be time to adapt the vocabulary.

a domain. For every $r \in P^{-1}(A')$ let us choose $x \in A$ such that $f(x) = P(r)$. Now, since f is a local subduction, there exist an open neighborhood V of r and a plot $Q : V \to X$ such that $Q(r) = x$ and $f \circ Q = P \upharpoonright V$. But since A is open, $Q^{-1}(A)$ is a domain. Let us define $W = Q^{-1}(A) \cap V$, W is still a domain, still containing r. Then, let us define $\bar{Q} = Q \upharpoonright W$, $\bar{Q}$ is still a plot of X. But $\bar{Q}(W) = Q(W) \subset A$, by construction, thus $P(W) = f \circ \bar{Q}(W) \subset f(A) = A'$, that is, $W \subset P^{-1}(A')$. Thus, for every $r \in U$ we found a domain W such that W is an open neighborhood of r and $W \subset P^{-1}(A')$. Hence, since $P^{-1}(A')$ is the union of all the domains W, when r runs over $P^{-1}(A')$, $P^{-1}(A')$ is itself a domain. Hence, $P^{-1}(A')$ is a domain for every plot P of X'. Therefore A' is D-open. □

Exercises

✎ EXERCISE 60 (Quotient by a group of diffeomorphisms). Let X be a diffeological space, and let $G \subset \mathrm{Diff}(X)$ be a group of diffeomorphisms equipped with the functional diffeology (art. 1.61). Choose a point $x \in X$ and denote by $\mathcal{O}$ the orbit $G(x) = \{g(x) \mid g \in G\}$, equipped with the subset diffeology. Let $\mathcal{Q} = G/\pi$ denote the same orbit, but be equipped with the pushforward of the functional diffeology of G by the projection $\pi : g \mapsto g(x)$. Let $j : \mathcal{Q} \to \mathcal{O}$ be the identity map.

1) Check that j is smooth, that is, the quotient diffeology of G by π is finer than the subset diffeology of $G(x)$.

2) Show that π is a local subduction from G to $\mathcal{Q}$.

3) Let us denote by $\mathcal{T}$ the space $\mathbf{R}^2$, equipped with the subdiffeology of the wire diffeology (art. 1.10) generated by the vertical and horizontal lines,[3] that is, the parametrizations $t \mapsto (c, t)$ and $t \mapsto (t, c)$, where c runs over $\mathbf{R}$. Check that for all $u \in \mathbf{R}^2$ the translation $T_u : \mathcal{T} \to \mathcal{T}$, $T_u(\tau) = \tau + u$ is a diffeomorphism. Show that the subgroup $G \subset \mathrm{Diff}(\mathcal{T})$ of these translations, equipped with the functional diffeology, is discrete. For $x = (0,0)$, note that $G(x) = \mathcal{T}$, and that the quotient diffeology of G by π from the first question (which coincides with G), is strictly finer than the diffeology of the orbit $\mathcal{T}$ of x. In other words, the translations, equipped with the functional diffeology, are transitive on $\mathcal{T}$ but do not generate its diffeology.

✎ EXERCISE 61 (A not-so-strong subduction). Let X be the diffeological sum $\mathbf{R} \coprod \mathbf{R}^2$ (art. 1.39), where the real spaces $\mathbf{R}$ and $\mathbf{R}^2$ are equipped with the smooth diffeology. Let $\|\cdot\| : X \to [0, \infty[$ be the modulus $\|x\| = |x|$ if $x \in \mathbf{R}$, and the usual norm when $x \in \mathbf{R}^2$. Let $\mathcal{Q}$ be the quotient $X/\|\cdot\|$ (art. 1.50), that is, the quotient of X by the equivalence relation $x \sim x'$ if $\|x\| = \|x'\|$. Consider the quotient $\mathcal{Q}$ as the half-line $[0, \infty[$, equipped with the pushforward of the diffeology of X by the map $\|\cdot\|^2$. Show that the plot $P : \mathbf{R}^2 \to \mathcal{Q}$, defined by $P(r) = \|r\|^2$, cannot be lifted locally at $(0,0) \in \mathbf{R}^2$ by a plot Q of X, such that $Q(0,0) = 0 \in \mathbf{R} \subset X$. Conclude that $\|\cdot\|$, which is by construction a subduction, is not a local subduction at the point $0 \in \mathbf{R} \subset X$.

✎ EXERCISE 62 (A powerset diffeology). Let X be a diffeological space, and let $\mathcal{D}$ be its diffeology. Let $\mathfrak{P}(X)$ be the powerset of X, that is, $\mathfrak{P}(X) = \{A \mid A \subset X\}$. We have seen in Exercise 32, p. 31, a diffeology on this set, which induces on every quotient of X the quotient diffeology. But this diffeology is too weak to be a good

[3]We denote this space by $\mathcal{T}$ for *Tahar rug*, because it has been imagined by Guillaume Tahar during a private conversation.

diffeology. For example, consider the set of the affine lines in $\mathbf{R}^2$, and choose the parametrization which associates with each rational number the x-axis, and with each irrational number the y-axis. This parametrization is a plot for this weak diffeology, since we can lift it into the point $(0,0)$ which belongs to each of these lines for any number. This is indeed not satisfactory, since the line jumps from the x-axis to the y-axis, when the parameter crosses each number. This is not a behavior we want to regard as smooth. The point that we miss with this diffeology is that we cannot follow smoothly the plots of the line, defined by a value of the parameter, when the parameter moves smoothly in its domain. It is why we shall try to fix this diffeology in the following way. Let us recall, first of all, that the diffeology $\mathcal{D}$ is itself a diffeological space, and thus, we know what a *smooth family of plots of* X is (art. 1.63).

1) Show that the parametrizations $P : U \to \mathfrak{P}(X)$ defined by the following condition form a diffeology. We shall call it the *powerset diffeology.*

(♣) For every $r_0 \in U$ and for every $Q_0 \in \mathcal{D}$ such that $\mathrm{val}(Q_0) \subset P(r_0)$, there exist an open neighborhood V of r_0 and a smooth family of plots $Q : V \to \mathcal{D}$, such that $Q(r_0) = Q_0$ and $\mathrm{val}(Q(r)) \subset P(r)$, for all $r \in V$.

2) Give a more conceptual definition involving local subductions. Compare with being a smooth map between diffeological spaces.

3) Check that the map $j : x \mapsto \{x\}$, from X to $\mathfrak{P}(X)$, is an induction.

4) Show that the *Tzim-Tzum*, the parametrization $\mathcal{T}$ defined by

$$\mathcal{T}(t) = \{x \in \mathbf{R}^2 \mid \|x\| > t\}, \text{ for all } t \in \mathbf{R},$$

is a plot of the powerset diffeology of $\mathfrak{P}(\mathbf{R}^2)$. Notice how the topology of the subset $\mathcal{T}(t)$ changes when t passes through 0.

NOTE. This powerset diffeology avoids the phenomenon described at the beginning. The next exercise shows that it induces, on the set of lines of an affine space, a diffeology of manifold, the one we are used to, which is satisfactory. Also note that this construction is presented as an exercise and not as an independent section because, as it is an attempt to bring out the right concept of a diffeology on $\mathfrak{P}(X)$, it may not be in its final shape.

✎ EXERCISE 63 (The powerset diffeology of the set of lines). Let us denote by $\mathrm{Lines}(\mathbf{R}^n)$ the set of all the (affine) lines of $\mathbf{R}^n$. Let us recall that an affine line $\mathbf{D}$ is a part of $\mathbf{R}^n$ for which there exists a nonzero vector $v \in \mathbf{R}^n$ such that for any two points r and r' of $\mathbf{D}$ there exists $t \in \mathbf{R}$ such that $r' = r + tv$. Let $TS^{n-1} \subset \mathbf{R}^n \times \mathbf{R}^n$ be the *tangent space of the* $(n-1)$*-sphere* defined by

$$TS^{n-1} = \{(u,x) \in \mathbf{R}^n \times \mathbf{R}^n \mid \|u\| = 1 \text{ and } u \cdot x = 0\},$$

where the dot denotes the usual scalar product on $\mathbf{R}^n$. Let $\mathfrak{P}(\mathbf{R}^n)$ be the powerset of $\mathbf{R}^n$. Let

$$j : TS^{n-1} \to \mathrm{Lines}(\mathbf{R}^n) \subset \mathfrak{P}(\mathbf{R}^n) \quad \text{with} \quad j(u,x) = \{x + tu \mid t \in \mathbf{R}\}.$$

Equip TS^{n-1} with the subset diffeology of $\mathbf{R}^n \times \mathbf{R}^n$ and $\mathrm{Lines}(\mathbf{R}^n) \subset \mathfrak{P}(\mathbf{R}^n)$ with the subset diffeology of the powerset diffeology defined in Exercise 62, p. 60. Show that j is a subduction onto its image. Check that the pullback of a line $\mathbf{D} \in \mathrm{Lines}(\mathbf{R}^n)$ by j is just the pair $(\pm u, x)$, such that $\mathbf{D} = \{x + tu \mid t \in \mathbf{R}\}$. Deduce that the set $\mathrm{Lines}(\mathbf{R}^n)$, equipped with the powerset diffeology, is diffeomorphic to the quotient $TS^{n-1}/\{\pm 1\}$.

The Dimension Map of Diffeological Spaces

Because diffeological spaces are neither necessarily homogeneous nor transitive (art. 2.25), *it is necessary to refine the notion of (global) dimension of a diffeological space* (art. 1.78) *by the* dimension map (art. 2.22), *which gives the dimension of a diffeological space at each of its points.*

2.19. Pointed plots and germs of a diffeological space. Let X be a diffeological space, let $\mathcal{D}$ be its diffeology, and let $x \in X$. Let $P : U \to X$ be a plot. We say that P is *centered at* x if $0 \in U$ and $P(0) = x$. We shall agree that the set of germs (art. 2.4) of the centered plots of X at x represents the *germ of the diffeology* at this point, and we shall denote it by $\mathcal{D}_x$.

2.20. Local generating families. Let X be a diffeological space, and let x be some point of X. We shall call *local generating family at* x any family $\mathcal{F}$ of plots of X such that the following conditions are satisfied:

1. Every element P of $\mathcal{F}$ is centered at x, that is, $0 \in \mathrm{def}(P)$ and $P(0) = x$.
2. For every plot $P : U \to X$ centered at x, there exist an open neighborhood V of $0 \in U$, a parametrization $F : W \to X$ belonging to $\mathcal{F}$ and a smooth parametrization $Q : V \to W$, centered at $0 \in W$, such that $F \circ Q = P \restriction V$.

We shall also say that $\mathcal{F}$ *generates the germ* $\mathcal{D}_x$ *of the diffeology* $\mathcal{D}$ *of* X *at the point* x (art. 2.19). We denote, by analogy with (art. 1.66), $\mathcal{D}_x = \langle \mathcal{F} \rangle$.

Note that for all $x \in X$, the set of local generating families at x is not empty, since it contains the set of all the plots centered at x, and this set contains the constant parametrizations with value x (art. 1.5).

2.21. Union of local generating families. Let X be a diffeological space. Let us choose, for every $x \in X$, a local generating family $\mathcal{F}_x$ at x. The union $\mathcal{F}$ of all these local generating families,

$$\mathcal{F} = \bigcup_{x \in X} \mathcal{F}_x,$$

is a generating family of the diffeology of X.

PROOF. Let $P : U \to X$ be a plot, and let $r \in U$ and $x = P(r)$. Let T_r be the translation $T_r(r') = r' + r$. Let $P' = P \circ T_r$ be defined on $U' = T_r^{-1}(U)$. Since the translations are smooth, the parametrization P' is a plot of X. Moreover, P' is centered at x, thus $P'(0) = P \circ T_r(0) = P(r) = x$. By definition of a local generating family (art. 2.20), there exist an element $F : W \to X$ of $\mathcal{F}_x$, an open neighborhood V' of $0 \in U'$ and a smooth parametrization $Q' : V' \to W$, centered at 0, such that $P' \restriction V' = F \circ Q'$. Thus, $P \circ T_r \restriction V' = F \circ Q'$, that is, $P \restriction V = F \circ Q$, where $V = T_r(V')$ and $Q = Q' \circ T_r^{-1}$. Hence, P lifts locally, at every point of its domain, along an element of $\mathcal{F}$. Thus, $\mathcal{F}$ is a generating family (art. 1.66) of the diffeology of X. □

2.22. The dimension map. Let X be a diffeological space and x be a point of X. By analogy with the global dimension of X (art. 1.78), we define the *dimension of* X *at the point* x by

$$\dim_x(X) = \inf\{\dim(\mathcal{F}) \mid \langle \mathcal{F} \rangle = \mathcal{D}_x\},$$

where the dimension of a family of parametrizations has been defined in (art. 1.77). The map $x \mapsto \dim_x(X)$, with values in $\mathbf{N} \cup \{\infty\}$, will be called the *dimension map* of the space X.

2.23. Global dimension and dimension map. Let X be a diffeological space. The global dimension of X (art. 1.78) is the supremum of the dimension map (art. 2.22),

$$\dim(X) = \sup\{\dim_x(X)\}_{x\in X}.$$

PROOF. Let $\mathcal{D}$ be the diffeology of X. Let us prove first that for every $x \in X$, $\dim_x(X) \leq \dim(X)$, which implies $\sup\{\dim_x(X)\}_{x\in X} \leq \dim(X)$. For that we shall prove that for all $x \in X$ and every generating family $\mathcal{F}$ of $\mathcal{D}$, $\dim_x(X) \leq \dim(\mathcal{F})$. Then, since $\dim(X) = \inf\{\dim(\mathcal{F}) \mid \mathcal{F} \in \mathcal{D} \text{ and } \langle\mathcal{F}\rangle = \mathcal{D}\}$, we shall get $\dim_x(X) \leq \dim(X)$. Consider a generating family $\mathcal{F}$ of $\mathcal{D}$. For every plot $P : U \to X$, centered at x, let us choose an element F of $\mathcal{F}$ such that there exist an open neighborhood V of $0 \in U$ and a smooth parametrization $Q : V \to \text{def}(F)$ such that $F \circ Q = P \restriction V$. Then, let $r = Q(0)$ and T_r be the translation $T_r(r') = r' + r$. Let $F' = F \circ T_r$ be defined on $T_r^{-1}(\text{def}(F))$. Thus, $F'(0) = x$, and F' is a plot of X, centered at x, such that $\dim(F') = \dim(F)$. Now, $Q' = T_r^{-1} \circ Q$ is smooth and $P \restriction V = F' \circ Q'$. Hence, the set $\mathcal{F}'_x$ of all these plots F', associated with the plots centered at x, is a generating family of $\mathcal{D}_x$, and for each of them $\dim(F') = \dim(F) \leq \dim(\mathcal{F})$. Therefore, $\dim(\mathcal{F}'_x) \leq \dim(\mathcal{F})$. But $\dim_x(X) \leq \dim(\mathcal{F}'_x)$, thus $\dim_x(X) \leq \dim(\mathcal{F})$. We conclude then that $\dim_x(X) \leq \dim(X)$, for all $x \in X$, and thus $\sup\{\dim_x(X)\}_{x\in X} \leq \dim(X)$. Now, let us prove that $\dim(X) \leq \sup\{\dim_x(X)\}_{x\in X}$. We shall assume that $\sup\{\dim_x(X)\}_{x\in X}$ is finite. Otherwise, according to the previous part, we have $\sup\{\dim_x(X)\}_{x\in X} \leq \dim(X)$, and then $\dim(X)$ is infinite and $\sup\{\dim_x(X)\}_{x\in X} = \dim(X)$. So, since the sequence of the dimensions of the generating families of the $\mathcal{D}_x$ is lower bounded, there exists for every x a generating family $\mathcal{F}_x$ such that $\dim_x(X) = \dim(\mathcal{F}_x)$. Then, for every x in X let us choose one of these families. Now, let us define $\mathcal{F}_m$ as the union of all these families we have chosen. Thanks to (art. 2.21), $\mathcal{F}_m$ is a generating family of $\mathcal{D}$. Hence, $\dim(X) \leq \dim(\mathcal{F}_m)$. But, $\dim(\mathcal{F}_m) = \sup\{\dim(F)\}_{F\in\mathcal{F}_m} = \sup\{\sup\{\dim(F)\}_{F\in\mathcal{F}_x}\}_{x\in X} = \sup\{\dim(\mathcal{F}_x)\}_{x\in X} = \sup\{\dim_x(X)\}_{x\in X}$. Therefore, $\dim(X) \leq \sup\{\dim_x(X)\}_{x\in X}$. We can conclude, from the two parts above, that $\dim(X) = \sup\{\dim_x(X)\}_{x\in X}$. ☐

2.24. The dimension map is a local invariant. Let X and X′ be two diffeological spaces. If $x \in X$ and $x' \in X'$ are two points related by a local diffeomorphism (art. 2.5), then $\dim_x(X) = \dim_{x'}(X')$. In other words, local diffeomorphisms (*a fortiori* global diffeomorphisms) can only exchange points where the spaces have the same dimension. In particular, for $X = X'$, the dimension map is invariant under the local diffeomorphisms of X, that is, constant on the orbits of the germs of local diffeomorphisms (art. 2.5).

PROOF. This proposition is a slight adaptation of (art. 1.79). Let $f : X \supset A \to X'$ be a local diffeomorphism mapping x to x′. Let $\mathcal{F}$ be a local generating family at $x \in X$ (art. 2.20). Clearly, $f \circ \mathcal{F} = \{f \circ F \mid F \in \mathcal{F}\}$ is a generating family at $x' = f(x) \in X'$. Conversely, let $\mathcal{F}'$ be a generating family at $x' \in X'$, then $f^{-1} \circ \mathcal{F}'$ is a generating family at $x \in X$. There is a one-to-one correspondence between the local generating families at $x \in X$ and the local generating families at $x' \in X'$, therefore $\dim_x(X) = \dim_{x'}(X')$. ☐

2.25. Transitive and locally transitive spaces. We shall say that a diffeological space X is *transitive* if for any two points x and x′ there exists a diffeomorphism F (art. 1.17) such that $F(x) = x'$. We shall say that the space is *locally transitive* if for any two points x and x' there exists a local diffeomorphism f (art. 2.5) defined on some superset of x such that $f(x) = x'$. If the space X is transitive, it is *a fortiori* locally transitive. As a direct consequence of (art. 2.24), if a diffeological space X is locally transitive, then the dimension map (art. 2.22) is constant, *a fortiori* if the space X is transitive.

2.26. Local subduction and dimension. Let X and X' be two diffeological spaces, and let $\pi : X \to X'$ be a smooth surjection. Let x be a point of X and $x' = \pi(x)$. If π is a local subduction at the point $x \in X$ (art. 2.16), then $\dim_{x'}(X') \leq \dim_x(X)$.

PROOF. Let $\mathcal{D}_x$ and $\mathcal{D}'_{x'}$ be the germs of the diffeologies of X at the point x and X' at the point $x' = \pi(x)$. Let $\mathrm{Gen}(\mathcal{D}_x)$ and $\mathrm{Gen}(\mathcal{D}'_{x'})$ denote the sets of all the generating families of $\mathcal{D}_x$ and $\mathcal{D}'_{x'}$. We know that for every generating family $\mathcal{F}$ of $\mathcal{D}_x$, $\pi \circ \mathcal{F}$ is a generating family of $\mathcal{D}'_{x'}$, that is, $\pi \circ \mathrm{Gen}(\mathcal{D}_x) \subset \mathrm{Gen}(\mathcal{D}'_{x'})$. Now, for the same reasons as in (art. 1.82), we get $\dim_{x'}(X') \leq \dim_x(X)$. ☐

Exercise

✎ EXERCISE 64 (The diffeomorphisms of the half-line). This exercise illustrates how the invariance of the dimension map (art. 2.22), under diffeomorphisms (art. 2.24), can be used to characterize the diffeomorphisms of the half-line $[0, \infty[\subset \mathbf{R}$. Show that a bijection $f : [0, \infty[\to [0, \infty[$ is a diffeomorphism, for the subset diffeology induced by $\mathbf{R}$, if and only if the three following conditions are fulfilled:

1) The origin is fixed, $f(0) = 0$.
2) The restriction of f to the open half-line is an increasing diffeomorphism of the open half-line, $f \upharpoonright]0, \infty[\in \mathrm{Diff}^+(]0, \infty[)$.
3) The map f is infinitely differentiable at the origin and its first derivative $f'(0)$ does not vanish.

Hint: For the first question use the result of Exercise 51, p. 50, and the invariance of the dimension map by diffeomorphisms (art. 2.24). For the second question use the fact that the open interval $]0, \infty[$ is an orbit of $\mathrm{Diff}([0, \infty[)$. For the third question use the following Whitney theorem:

THEOREM [**Whi43**]. *An even function* $f(x) = f(-x)$, *defined on a neighborhood of the origin, may be written as* $g(x^2)$. *If* f *is smooth,* g *may be made smooth.*

CHAPTER 3

Diffeological Vector Spaces

Some geometrical objects involve infinite vector spaces, for example the construction of the infinite Hilbert sphere (art. 4.9) or the associated infinite projective space (art. 4.11). These constructions are perfect examples of how diffeology handles infinite dimensional objects. But, in order to describe precisely these structures, we need to specialize the notion of vector space according to the diffeological framework. However, this chapter is not devoted to exploring all the connections between the notion of diffeological vector space and the various kinds of vector spaces we find in the literature; this is still not done and may be a program for a future work.

In this chapter, we shall consider the field $\mathbf{R}$ of real numbers equipped with the smooth diffeology (art. 1.4). The field $\mathbf{C}$ of complex numbers is identified, as real vector space, with the product $\mathbf{R} \times \mathbf{R}$ by the real isomorphism fold : $(x, y) \mapsto z = x + iy$. A plot of $\mathbf{C}$ is just a parametrization $r \mapsto Z(r) = P(r) + iQ(r)$ where P and Q are some real smooth parametrizations. The letter $\mathbf{K}$ will denote $\mathbf{R}$ or $\mathbf{C}$.

Basic Constructions and Definitions

3.1. Diffeological vector spaces. Let E be a vector space over $\mathbf{K}$. We shall call *vector space diffeology* on E any diffeology of E such that the addition and the multiplication by a scalar are smooth, that is,

$$[(u, v) \mapsto u + v] \in \mathcal{C}^\infty(E \times E, E) \quad \text{and} \quad [(\lambda, u) \mapsto \lambda u] \in \mathcal{C}^\infty(\mathbf{K} \times E, E),$$

where the spaces $E \times E$ and $\mathbf{K} \times E$ are equipped with the product diffeology (art. 1.55). Every vector space E over $\mathbf{K}$, equipped with a vector space diffeology, is called a *diffeological vector space* over $\mathbf{K}$, or a *diffeological* $\mathbf{K}$*-vector space*.

3.2. Standard vector spaces. Every vector space $\mathbf{K}^n$, equipped with the *standard diffeology*, is a diffeological vector space. But note that every vector space, equipped with the coarse diffeology, is also a diffeological vector space.

3.3. Smooth linear maps. Let E and F be two diffeological $\mathbf{K}$-vector spaces. We denote by $L^\infty(E, F)$ the space of *smooth linear maps* from E to F,

$$L^\infty(E, F) = L(E, F) \cap \mathcal{C}^\infty(E, F).$$

The space $L^\infty(E, F)$ is a $\mathbf{K}$-vector subspace of $L(E, F)$. Since the composite of linear maps is linear and the composite of two smooth maps is smooth, diffeological vector spaces, together with smooth linear maps, form a category which we may denote by {D-Linear}. The isomorphisms of this category are the linear isomorphisms, smooth as well as their inverses.

PROOF. Let A and B belong to $L^\infty(E, F)$. For any plot $P : U \to E$, $(A + B) \circ P = A \circ P + B \circ P$, but $P_A = A \circ P$ and $P_B = B \circ P$ are plots of F. So, $[r \mapsto (A + B) \circ P(r) =$

$P_A(r) + P_B(r)] = [r \mapsto (P_A(r), P_B(r)) \mapsto P_A(r) + P_B(r)]$ is the composite of two smooth maps. Therefore $A + B$ is smooth. Let now $\lambda \in \mathbf{K}$, the map $(\lambda A) \circ P$ splits into $(\lambda A) \circ P = [r \mapsto (\lambda, P_A(r)) \mapsto \lambda \times P_A(r)]$. Thus, the addition of two smooth linear maps, as well as the multiplication of a smooth linear map by a scalar, are smooth linear maps. Therefore, $L^\infty(E, F)$ is a $\mathbf{K}$-vector subspace of $L(E, F)$. □

3.4. Products of diffeological vector spaces. Let $\{E_i\}_{i \in \mathcal{J}}$ be some family of diffeological vector spaces. Let $E = \prod_{i \in \mathcal{J}} E_i$ be the product of the family. The product E is naturally a vector space, for all $v, v' \in E$, for all $\lambda \in \mathbf{K}$,

$$(v + v')(i) = v(i) + v'(i) \quad \text{and} \quad (\lambda v)(i) = \lambda \times v(i).$$

The space E, equipped with the product diffeology (art. 1.55), is a diffeological vector space.

PROOF. Let us check that the addition and the multiplication by a scalar are smooth. Let $P : r \mapsto v_r$ and $P' : r \mapsto v'_r$ be two plots of E. So, for all $i \in \mathcal{J}$ the parametrizations $r \mapsto v_r(i)$ and $r \mapsto v'_r(i)$ are two plots of E_i. Then, $r \mapsto (v_r + v'_r)(i) = v_r(i) + v'_r(i)$ is a plot of E_i, since the addition is smooth in E_i. As well, $(\lambda, r) \mapsto (\lambda v_r)(i) = \lambda \times v_r(i)$ is a plot of E_i. Therefore, E is a diffeological vector space. □

3.5. Diffeological vector subspaces. Any vector subspace F of a diffeological vector space E, equipped with the subset diffeology, is a diffeological vector space.

PROOF. Let $r \mapsto v_r$ and $r \mapsto v'_r$ be two plots of F for the subset diffeology, and defined on the same domain, that is, two plots of E with values in F. So, the sum $r \mapsto v_r + v'_r$ is a plot of E, and since F is a vector subspace of E, $v_r + v'_r$ belongs to F for all r. Thus, $r \mapsto v_r + v'_r$ is a plot of F for the subset diffeology. Thus, the sum is smooth. For analogous reasons, $(\lambda, r) \mapsto \lambda \times v_r$ is a plot of F, and the multiplication by a scalar is smooth. Therefore F, equipped with the subset diffeology, is a diffeological vector space. □

3.6. Quotient of diffeological vector spaces. Let E be a diffeological vector space, and let $F \subset E$ be a vector subspace. The quotient vector space E/F is a diffeological vector space for the quotient diffeology (art. 1.50).

PROOF. Let $E' = E/F$ and $r \mapsto v_r$ be a plot of the quotient diffeology. So, locally, $v_r = [e_r]$, where $r \mapsto e_r$ is a plot of E, and $[e] \in E'$ means the class of $e \in E$. Now, let $r \mapsto v_r$ and $r \mapsto v'_r$ be two plots of E' defined on the same domain U. So, for any $r_0 \in U$ there exist an open neighborhood $\mathcal{O}$ of r_0 and two plots $r \mapsto e_r$ and $r \mapsto e'_r$ defined on $\mathcal{O}$ such that $v_r = [e_r]$ and $e'_r = [v'_r]$. Hence, $v_r + v'_r = [e_r] + [e'_r] = [e_r + e'_r]$, and then $r \mapsto e_r + e'_r$ is a local lift of $r \mapsto v_r + v'_r$. Thus, by definition of the quotient diffeology, the addition on E' is smooth. As well, $(\lambda, r) \mapsto \lambda \times e_r$ lifts $(\lambda, r) \mapsto \lambda \times v_r$, where $\lambda \in \mathbf{K}$. Therefore, E' is a diffeological vector space for the quotient diffeology. □

Exercise

✎ EXERCISE 65 (Vector space of maps into $\mathbf{K}^n$). Let X be a diffeological space. Let $E = \mathcal{C}^\infty(X, \mathbf{K}^n)$ be the space of smooth maps from X to $\mathbf{K}^n$. The set E is a $\mathbf{K}$-vector space for the pointwise addition and multiplication by a scalar. Precisely, for all

f, f' in E and λ in $\mathbf{R}$, $f + f' = [x \mapsto f(x) + f'(x)]$ and $\lambda f = [x \mapsto \lambda \times f(x)]$. Check that E, equipped with the functional diffeology, is a diffeological $\mathbf{K}$-vector space.

Fine Diffeology on Vector Spaces

Every $\mathbf{K}$-vector space equipped with the coarse diffeology is obviously a diffeological vector space, what is not really interesting. But every vector space has a finest vector space diffeology, which we shall call the fine diffeology. *In this section we study some aspects of this fine diffeology.*

3.7. The fine diffeology of vector spaces. There exists, on every $\mathbf{K}$-vector space E, a finest diffeology of vector space. We shall call it *the fine diffeology.* This diffeology is generated (art. 1.66) by the family of parametrizations defined by

$$P : r \mapsto \sum_{\alpha \in A} \lambda_\alpha(r) v^\alpha, \qquad (\heartsuit)$$

where A is a finite set of indices, the λ_α are smooth $\mathbf{K}$-parametrizations defined on the domain of P, and v^α are vectors of E. More precisely, the plots of the fine diffeology are the parametrizations $P : U \to E$ such that, for all $r_0 \in U$ there exist an open neighborhood V of r_0, a family of smooth parametrizations $\lambda_\alpha : V \to \mathbf{K}$, and a family of vectors $v^\alpha \in E$, both indexed by the same finite set of indices A, such that

$$(P \restriction V) : r \mapsto \sum_{\alpha \in A} \lambda_\alpha(r) v^\alpha. \qquad (\diamondsuit)$$

A finite family $(\lambda_\alpha, v^\alpha)_{\alpha \in A}$, defined on some domain V, and satisfying $(\diamondsuit)$ with $\lambda_\alpha \in \mathcal{C}^\infty(V, \mathbf{K})$ and $v^\alpha \in E$, will be called a *local family* for the plot P. We shall agree that plot properties satisfied only locally, over some nonspecified domains, will be denoted with the subscript $_{\mathrm{loc}}$, for example $=_{\mathrm{loc}}$ etc.

PROOF. Let us prove that the set of parametrizations described by $(\diamondsuit)$ is the finest vector space diffeology.

1. (Diffeology) The condition $(\diamondsuit)$ is clearly the specialization, for the family defined in $(\heartsuit)$, of the criterion (art. 1.68) for generating families.

2. (Diffeology of vector space) Let $r \mapsto (P(r), Q(r))$ be a plot of the product $E \times E$. Let $(\lambda_\alpha, u^\alpha)_{\alpha \in A}$ and $(\mu_\beta, v^\beta)_{\beta \in B}$ be two local families such that

$$P(r) =_{\mathrm{loc}} \sum_{\alpha \in A} \lambda_\alpha(r) u^\alpha \quad \text{and} \quad Q(r) =_{\mathrm{loc}} \sum_{\beta \in B} \mu_\beta(r) v^\beta.$$

So, the addition $P + Q$ writes

$$P + Q|_{\mathrm{loc}} : r \mapsto \sum_{\alpha \in A} \lambda_\alpha(r) u^\alpha + \sum_{\beta \in B} \mu_\beta(r) v^\beta = \sum_{\sigma \in C} \nu_\sigma(r) w^\sigma,$$

where C is just the adjunction of the two sets of indices A and B, and the family $(\nu_\sigma, w^\sigma)_{\sigma \in C}$ the adjunction of the local families $(\lambda_\alpha, u^\alpha)_{\alpha \in A}$ and $(\mu_\beta, v^\beta)_{\beta \in B}$. Hence, the addition is smooth. On the other hand, since the multiplication by a scalar is smooth in $\mathbf{K}$, the multiplication by a scalar in E also is smooth. Therefore, this diffeology is a vector space diffeology.

3. (Fineness) Let us consider E, equipped with some other vector space diffeology $\mathcal{D}$. Since the multiplication by a scalar is smooth, for any smooth parametrization λ of $\mathbf{K}$ and any vector $u \in E$, the parametrization $r \mapsto \lambda(r)u$ is smooth. Now, since

the addition is smooth, for any finite local family $(\lambda_\alpha, u^\alpha)_{\alpha\in A}$, the parametrization $r \mapsto \sum_{\alpha\in A} \lambda_\alpha(r)u^\alpha$ is smooth, that is, a plot of the diffeology $\mathcal{D}$. Thus, the diffeology $\mathcal{D}$ is coarser than the fine diffeology defined above. Hence, the fine diffeology is the finest vector space diffeology of E. □

3.8. Generating the fine diffeology. Let E be a vector space on $\mathbf{K}$, and let $L(\mathbf{K}^n, E)$ be the set of all the linear maps from $\mathbf{K}^n$ into E. Let $L^\star(\mathbf{K}^n, E)$ be the set of all the injective linear maps from $\mathbf{K}^n$ into E,

$$L^\star(\mathbf{K}^n, E) = \{j \in L(\mathbf{K}^n, E) \mid \ker(j) = \{0\}\}.$$

The following two families both generate the fine diffeology of E.

$$\mathcal{F} = \bigcup_{n\in\mathbf{N}} L(\mathbf{K}^n, E) \quad \text{and} \quad \mathcal{F}^\star = \bigcup_{n\in\mathbf{N}} L^\star(\mathbf{K}^n, E).$$

NOTE. A parametrization $P : U \to E$ is a plot for the diffeology generated by $\mathcal{F}$ if and only if, for all $r_0 \in U$ there exist an open neighborhood V of r_0 in U, an integer n, a smooth parametrization $\phi : V \to \mathbf{K}^n$, and a linear map $j : \mathbf{K}^n \to E$ such that $P \upharpoonright V = j \circ \phi$. In other words, locally, P takes its values in a constant finite dimensional subspace $F \subset E$, such that the coordinates of P, for some basis of F, are smooth. For the plots generated by $\mathcal{F}^\star$, j is injective.

PROOF. Let us prove that $\mathcal{F}$, as well as $\mathcal{F}^\star$, generates the fine diffeology. Let $P : U \to E$ be a plot of the diffeology generated by $\mathcal{F}$, or by $\mathcal{F}^\star$. Pick a point r_0 in U. By definition, there exist an open neighborhood V of r_0, an integer n, a smooth parametrization $\phi : V \to \mathbf{K}^n$, and a linear map $j : \mathbf{K}^n \to E$ such that $P \upharpoonright V = j \circ \phi$. Thus, for all r in V, $\phi(r) = \sum_{k=1}^n \phi_k(r)\mathbf{e}_k$, where $(\mathbf{e}_1, \ldots, \mathbf{e}_n)$ is the canonical basis of $\mathbf{K}^n$, and $\phi_k \in \mathcal{C}^\infty(V, \mathbf{K})$. Now, $P(r) = j(\sum_{k=1}^n \phi_k(r)\mathbf{e}_k) = \sum_{k=1}^n \phi_k(r)j(\mathbf{e}_k) = \sum_{k=1}^n \phi_k(r)\mathbf{f}_k$, where $\mathbf{f}_k = j(\mathbf{e}_k)$. Therefore, P is a plot of the fine diffeology of E, and $(\phi_k, \mathbf{f}_k)_{k=1}^n$ is a local family of the plot P. Note that j can be chosen to be injective.

Conversely, let $P : U \to E$ be a plot of the fine diffeology, and let r_0 be a point of U. There exist an open neighborhood V of r_0 in U, an integer N, a local family $(\lambda_\alpha, v^\alpha)_{\alpha=1}^N$, with $\lambda_\alpha \in \mathcal{C}^\infty(V, \mathbf{K})$, $v^\alpha \in E$, and such that $P \upharpoonright V = \sum_{\alpha=1}^N \lambda_\alpha(r)v^\alpha$. Let F be the vector space generated by the v^α, and let $\mathbf{f} = (\mathbf{f}_1, \ldots, \mathbf{f}_n)$ be a basis of F. Let us split the vectors v^α on the basis $\mathbf{f}$, $v^\alpha = \sum_{k=1}^n v_k^\alpha \mathbf{f}_k$. Now, $P \upharpoonright V = \sum_{\alpha=1}^N \sum_{k=1}^n \lambda_\alpha(r)v_k^\alpha \mathbf{f}_k = \sum_{k=1}^n \phi_k(r)\mathbf{f}_k$, where $\phi_k(r) = \sum_{\alpha=1}^N \lambda_\alpha(r)v_k^\alpha$. The ϕ_k are smooth maps defined on V with values in $\mathbf{K}$. Now, let $j : \mathbf{K}^n \to E$ be the linear map defined by $j(\mathbf{e}_k) = \mathbf{f}_k$ and $\phi : V \to \mathbf{K}^n$ defined by $\phi = (\phi_1, \ldots, \phi_n)$. Hence, $P \upharpoonright V = j \circ \phi$, where j is an injective linear map from $\mathbf{K}^n$ to E and ϕ belongs to $\mathcal{C}^\infty(V, \mathbf{K}^n)$. Therefore, P is a plot of the diffeology generated by $\mathcal{F}^\star$, *a fortiori* by $\mathcal{F}$. Hence, the fine diffeology of E is generated by the set of linear maps, or injective linear maps, from $\mathbf{K}^n$ into E, when n runs over the integers. □

3.9. Linear maps and fine diffeology. Let E and F be two diffeological vector spaces over $\mathbf{K}$. Let E be equipped with the fine diffeology. Every linear map from E to F is smooth. In other words, if E is fine, $L^\infty(E, F) = L(E, F)$. In particular, if both E to F are fine vector spaces, every linear isomorphism from E to F is a smooth linear isomorphism.

PROOF. Let $(P \upharpoonright V)(r) = \sum_{\alpha=1}^{N} \lambda_\alpha(r)v_\alpha$ be a local expression of some plot P of E. Let $A \in L(E,F)$, thus we have $(A \circ P \upharpoonright V)(r) = \sum_{\alpha=1}^{N} \lambda_\alpha(r)A(v_\alpha)$. Since $A(v_\alpha) \in F$ for each α, P is a plot of the fine diffeology of F, thus a plot of any vector space diffeology. Hence, the linear map A is smooth, and $L(E,F) \subset L^\infty(E,F)$. The converse inclusion is a part of the definition. □

3.10. The fine linear category. Thanks to (art. 3.9) the fine diffeological spaces define a subcategory of the linear diffeological category (art. 3.3), let us denote it by {Fine-Linear}. The objects of this category are all vector spaces, for the field **K**. According to the above proposition (art. 3.9), the morphisms of this category are just the linear maps. Hence, the *fine linear category* coincides with the usual linear category over **K**. In other words, the functor from {Linear} to {Fine-Linear}, which associates with every vector space the same space equipped with the fine diffeology, is a full faithful functor.

3.11. Injections of fine diffeological vector spaces. Let E and E′ be two **K**-vector spaces equipped with the fine diffeology. Every linear injection $f : E \rightarrowtail E'$ is an induction. In particular, every subspace $F \subset E$, where E is a fine diffeological vector space, inherits from E the fine diffeology.

PROOF. The injection $f : E \to E'$ is linear, thus it is smooth (art. 3.9). Let us check now that if a parametrization $P : U \to E$ is such that $f \circ P$ is a plot of E′, then P is a plot of E (art. 1.29). Let $r_0 \in U$ be some point, there exist an open neighborhood V of r_0, an injection $j : \mathbf{K}^n \to E'$, and a smooth parametrization $\phi : V \to \mathbf{K}^n$ such that $f \circ P \upharpoonright V = j \circ \phi$ (art. 3.8). Since $f \circ P \upharpoonright V$ takes its values in f(E), $j(\mathrm{val}(\phi)) \subset f(E)$. Let us denote by $H \subset \mathbf{K}^n$ the vector space spanned by $\mathrm{val}(\phi)$, that is, the smallest vector subspace of $\mathbf{K}^n$ containing $\mathrm{val}(\phi)$. Since j is linear, $j(\mathrm{val}(\phi)) \subset f(E)$ implies $j(H) \subset f(E)$. Now, let $\mathcal{B} : \mathbf{K}^m \to H$ be a linear isomorphism (a basis of H), let $\phi' = \mathcal{B}^{-1} \circ \phi$ and $j' = j \circ \mathcal{B}$. Thus, $f \circ P \upharpoonright V = j' \circ \phi'$, where ϕ' is a smooth parametrization in $\mathbf{K}^m$ and j' is a linear injection from $\mathbf{K}^m$ in f(E). Since f is injective and $\mathrm{val}(j') \subset f(E)$, $j'' = f^{-1} \circ j'$ is an injection from $\mathbf{K}^m$ into E. Thus, $f \circ P \upharpoonright V = j' \circ \phi'$ implies $P \upharpoonright V = j'' \circ \phi'$. Hence, P is a plot of the fine diffeology of E. Therefore, f is an induction. □

3.12. Functional diffeology between fine spaces. Let E and E′ be two fine vector spaces over **K**. The functional diffeology (art. 1.57) of the space of linear maps $L^\infty(E,E') = L^\infty(E,E')$ is characterized as follows.

(♣) If E and F are finite dimensional spaces, $\dim(E) = n$ and $\dim(F) = m$, a parametrization $P : U \to L(E,F)$ is a plot of the functional diffeology if and only if the coefficients $P_{i,j}$ of the matrix associated with P, for some basis $\mathcal{E} = \{e_i\}_{i=1}^n$ of E and $\mathcal{F} = \{f_j\}_{j=1}^m$ of F, are smooth parametrizations of **K**. Briefly, $P_{i,j} \in \mathcal{C}^\infty(U,\mathbf{K})$, for all $i = 1 \cdots n$ and $j = 1 \cdots m$.

(♠) More generally, for two spaces E and F of any dimensions, a parametrization $P : U \to L(E,E')$ is a plot for the functional diffeology if and only if, for all $r_0 \in U$ and for all vector subspaces $F \subset E$ of finite dimension, there exist an open neighborhood V of r_0, and a vector subspace of finite dimension $F' \subset E'$ such that the two following conditions are satisfied:

1. For all $r \in V$, the linear map $P(r) \upharpoonright F$ belongs to $L(F,F')$.
2. The parametrization $r \mapsto P(r) \upharpoonright F$, restricted to V, is a plot of $L(F,F')$.

NOTE. Thanks to (♣), considering the second condition of (♠), the parametrization $r \mapsto P(r) \upharpoonright F$ is a plot of $L(F, F')$ if and only if each coefficient of the associated matrix is a smooth parametrization in $\mathbf{K}$, for some bases of F and F'.

PROOF. (♣) Every basis of a finite dimensional fine vector space is a smooth isomorphism; see Exercise 67, p. 71. Hence, the question is reduced to the functional diffeology of $L(\mathbf{K}^n, \mathbf{K}^m)$, where $\mathbf{K}^n$ and $\mathbf{K}^m$ are equipped with the smooth diffeology. Let $P : U \to L(\mathbf{K}^n, \mathbf{K}^m)$ be a plot of the functional diffeology. By definition, for any vector $\mathbf{e}_i$, $i = 1 \cdots n$, of the canonical basis of $\mathbf{K}^n$, the parametrization $r \mapsto P(r)(\mathbf{e}_i) = \sum_{j=1}^n P_{i,j}(r)\mathbf{f}_j$, where $\mathbf{f}_j$ is the j-th vector of the canonical basis of $\mathbf{K}^m$, is smooth. But the parametrization $P_{i,j}$ is the composite of $r \mapsto P(r)(\mathbf{e}_i)$ with the j-th projection from $\mathbf{K}^m$ to $\mathbf{K}$, which is smooth, by definition of the smooth diffeology of $\mathbf{K}^m$. Hence, the parametrizations $P_{i,j}$ are smooth.

Conversely, let $P : U \to L(\mathbf{K}^n, \mathbf{K}^m)$ be a parametrization such that all the components $P_{i,j}$ of P, for the canonical basis of $\mathbf{K}^n$ and $\mathbf{K}^m$, are smooth. Let $Q : V \to \mathbf{K}^n$ be a plot of $\mathbf{K}^n$. By definition of the smooth diffeology, $Q : s \mapsto \sum_{i=1}^n \phi_i(s)\mathbf{e}_i$, where the ϕ_i are smooth. Thus, for all $(r, s) \in U \times V$:

$$\begin{aligned} P(r)(Q(s)) &= P(r)(\textstyle\sum_{i=1}^n \phi_i(s)\mathbf{e}_i) && = \textstyle\sum_{i=1}^n \phi_i(s)P(r)(\mathbf{e}_i) \\ &= \textstyle\sum_{i=1}^n \phi_i(s)\sum_{j=1}^m P_{i,j}(r)\mathbf{f}_j && = \textstyle\sum_{j=1}^m \sum_{i=1}^n \phi_i(s)P_{i,j}(r)\mathbf{f}_j \\ &= \textstyle\sum_{j=1}^m \psi_j(r,s)\mathbf{f}_j && \& \quad \psi_j(r,s) = \textstyle\sum_{i=1}^n \phi_i(s)P_{i,j}(r). \end{aligned}$$

Since the ψ_j are smooth, the parametrization $(r, s) \mapsto P(r)(Q(s))$ is a plot of $\mathbf{K}^m$. Therefore P is smooth.

Let us now consider (♠). Let $P : U \to L(E, E')$ be a plot for the functional diffeology. Let us show that it satisfies the condition of the proposition. Let $F \subset E$ be a vector subspace of finite dimension. Let $(u_1, \ldots, u_m)$ be a basis of F. Let $r_0 \in U$, by definition of the functional diffeology, for every integer $k = 1 \cdots m$, the map $r \mapsto P(r)(u_k)$ is a plot of E'. So, there exist an open neighborhood V_k of r_0, a finite set of indices A_k, a family $(\lambda_{k,\alpha})_{\alpha \in A_k}$ of smooth parametrizations of $\mathbf{K}$, a family $(w_{k,\alpha})_{\alpha \in A_k}$ of vectors of E, such that, for all $r \in V_k$,

$$P(r)(u_k) = \sum_{\alpha \in A_k} \lambda_{k,\alpha}(r) w_{k,\alpha}.$$

Hence, for all $u = \sum_{k=1}^m c_k u_k$, where $c_k \in \mathbf{K}^m$, for all $r \in V = \bigcap_{k=1}^m V_k$, we have

$$P(r)(u) = P(r)\left(\sum_{k=1}^m c_k u_k\right) = \sum_{k=1}^m c_k P(r)(u_k) = \sum_{k=1}^m \sum_{\alpha \in A_k} c_k \lambda_{k,\alpha}(r) w_{k,\alpha}.$$

Let F' be the subspace of E' spanned by the vectors $\bigcup_{k=1}^m \{w_{k,\alpha}\}_{\alpha \in A_k}$. Hence, for every $u \in F$, $P(r)(u) \in F'$, that is, $P(r)(F) \subset F'$, for all $r \in V$. The first condition of (♠) is checked. Now, let $(v_1, \ldots, v_n)$ be a basis of F', such that for every integer $k = 1 \cdots m$ and every $\alpha \in A_k$, $w_{k,\alpha} = \sum_{j=1}^n w_{k,\alpha}^j v_j$. Replacing $w_{k,\alpha}$ by this expression, we get

$$\begin{aligned} P(r)\left(\sum_{k=1}^{m} c_k u_k\right) &= \sum_{k=1}^{m}\sum_{\alpha\in A_k} c_k\lambda_{k,\alpha}(r)\sum_{j=1}^{n} w^j_{k,\alpha}v_j \\ &= \sum_{j=1}^{n}\left(\sum_{k=1}^{m}\sum_{\alpha\in A_k} c_k\lambda_{k,\alpha}(r)w^j_{k,\alpha}\right)v_j. \end{aligned}$$

Hence, defining

$$\phi_j(r) = \sum_{k=1}^{m}\sum_{\alpha\in A_k} c_k\lambda_{k,\alpha}(r)w^j_{k,\alpha}, \quad \text{we get} \quad P(r)(u) = \sum_{j=1}^{n}\phi_j(r)v_j,$$

where the $(\phi_j)_{j=1}^n$ are a family of smooth parametrizations of $\mathbf{K}$ defined on V. This expression of $P(r)$ clearly shows that $r \mapsto P(r) \restriction F$ is a plot of the functional diffeology of $L(F, F')$. Indeed, choosing for u successively each vector of a basis of F, the last expression shows that the components of $P(r) \restriction F$ are smooth parametrizations of $\mathbf{K}$, which is the condition, for finite dimensional fine spaces, to be a plot of the functional diffeology. Hence, we proved the direct way of the proposition.

Conversely, let us assume that the parametrization P satisfies the conditions 1 and 2 of (♠), and let us show that P is a plot for the functional diffeology. Let us consider a plot $Q : V \to E$. By definition of the fine diffeology, for all $s_0 \in V$ there exist an open neighborhood W of s_0, a finite set of indices A, a family $(\lambda_\alpha)_{\alpha\in A}$ of smooth parametrizations of $\mathbf{K}$ defined on V, and a family $(v_\alpha)_{\alpha\in A}$ of vectors of E, such that for every $s \in V$, $Q(s) = \sum_{\alpha\in A}\lambda_\alpha(s)v_\alpha$. Let $F \subset E$ be the vector subspace spanned by the vectors v_α. Hence, for all $r_0 \in U$, there exist an open neighborhood U' of r_0 and a vector subspace $F' \subset E'$ such that $P(r)(F) \subset F'$, for all $r \in U'$. Thus, for all $(r, s) \in U' \times W$, $P(r)(Q(s)) = P(r)(\sum_{\alpha\in A}\lambda_\alpha(s)v_\alpha) = \sum_{\alpha\in A}\lambda_\alpha(s)P(r)(v_\alpha) \in F'$. Then, since the parametrization $r \mapsto P(r) \restriction F$ is a plot of the functional diffeology, the parametrization $P \cdot Q : (r, s) \mapsto P(r)(Q(s))$ is a smooth parametrization in $F' \subset E'$. Thus, $P \cdot Q$ is a smooth parametrization in E', because any finite subspace is embedded in E'. Therefore, P is a plot of the functional diffeology of $L(E, E')$. This completes the proof of the proposition. □

Exercises

✎ Exercise 66 (Smooth is fine diffeology). Check that the smooth diffeology of $\mathbf{K}^n$ is the fine diffeology.

✎ Exercise 67 (Finite dimensional fine spaces). Let E be a $\mathbf{K}$-vector space of dimension n, equipped with the fine diffeology. Let $\mathcal{B} = \{\mathbf{e}_i\}_{i=1}^n$ be any basis of E. Show that any plot $P : U \to E$ writes $P : r \mapsto \sum_{i=1}^n \phi_i(r)\mathbf{e}_i$, where $\phi_i \in \mathcal{C}^\infty(U, \mathbf{K})$. In other words, a basis $\mathcal{B}$, regarded as the isomorphism $\mathcal{B} : (u_1, \ldots, u_n) \mapsto \sum_{i=1}^n u_i\mathbf{e}_i$, from $\mathbf{K}^n$ to E, where $\mathbf{K}^n$ is equipped with the standard smooth diffeology, is a smooth isomorphism. Any fine vector space of dimension n is isomorphic to the standard $\mathbf{K}^n$.

✎ Exercise 68 (The fine topology). Let E be a fine diffeological vector space. Show that a subset $\Omega \subset E$ is D-open (open for the D-topology) (art. 2.8) if and only if its intersection with any finite dimensional vector space $F \subset E$ is open in F.

This topology is the so-called *finite topology*, introduced in functional analysis by Andrei Tychonoff [**Tyc35**].

Euclidean and Hermitian Diffeological Vector Spaces

The notions of Euclidean or Hermitian structures on diffeological vector spaces are natural extensions of the standard definitions. They are used in particular in the diffeological descriptions of the infinite sphere (art. 4.9) *and the infinite projective space* (art. 4.11).

3.13. Euclidean diffeological vector spaces. Let E be a real diffeological vector space, and let $(X,Y)\mapsto X\cdot Y$ be an *Euclidean product*, that is, a map from $E\times E$ to $\mathbf{R}$ which is

$$\begin{aligned} \text{Symmetric} &: \ X\cdot Y = Y\cdot X,\\ \text{Bilinear} &: \ X\cdot(\alpha Y+\alpha' Y') = \alpha X\cdot Y + \alpha' X\cdot Y',\\ \text{Positive} &: \ X\cdot X \geq 0,\\ \text{Nondegenerate} &: \ X\cdot X = 0 \Leftrightarrow X = 0, \end{aligned}$$

where X, Y and Y' are vectors, and α and α' are real numbers. If the Euclidean product is smooth, the pair $(E,\cdot)$ is an *Euclidean diffeological vector space.*

3.14. Hermitian diffeological vector spaces. Let H be a complex diffeological vector space, and let $(X,Y)\mapsto X\cdot Y$ be a *Hermitian product*, that is, a map from $H\times H$ to $\mathbf{C}$ which is

$$\begin{aligned} \text{Sesquilinear} &: \ X\cdot Y = (Y\cdot X)^*,\\ \text{Bilinear} &: \ X\cdot(\alpha Y+\alpha' Y') = \alpha X\cdot Y + \alpha' X\cdot Y',\\ \text{Positive} &: \ X\cdot X \geq 0,\\ \text{Nondegenerate} &: \ X\cdot X = 0 \Leftrightarrow X = 0, \end{aligned}$$

where X, Y and Y' are vectors, α, α' are complex numbers, and the star $*$ denotes the complex conjugation. The Hermitian product is also denoted sometimes by $\langle X \mid Y\rangle = X\cdot Y$. If the Hermitian product is smooth, then $(H,\cdot)$ is a *Hermitian diffeological vector space.*

3.15. The fine standard Hilbert space. The purpose of this paragraph is to introduce the fine diffeology on the standard Hilbert space, and some related constructions which will be used further for the study of a few diffeological infinite dimensional manifolds. Let $\mathcal{H}_{\mathbf{R}}$ be the real vector space of square-summable real sequences,

$$\mathcal{H}_{\mathbf{R}} = \left\{ X = (X_k)_{k=1}^{\infty} \;\middle|\; X_k \in \mathbf{R},\ k = 1,\ldots,\infty, \text{ and } \sum_{k=1}^{\infty} X_k^2 < \infty \right\}.$$

The space of real numbers is naturally equipped with the smooth diffeology, and the space $\mathcal{H}_{\mathbf{R}}$ is equipped with the fine diffeology (art. 3.7). The usual Euclidean product is defined on $\mathcal{H}_{\mathbf{R}}$ by

$$X\cdot X' = \sum_{k=1}^{\infty} X_k X'_k, \quad \text{for all} \quad X, X' \in \mathcal{H}_{\mathbf{R}}.$$

Hence, the pair $(\mathcal{H}_{\mathbf{R}},\cdot)$ is a fine Euclidean diffeological vector space (see Exercise 69, p. 74). We shall denote as usual, by $\|\cdot\|$, the norm associated with the Euclidean product, that is,

$$\|X\| = \sqrt{X \cdot X}, \quad \text{for all} \quad X \in \mathcal{H}_{\mathbf{R}}.$$

We shall denote by pr_k the k-th projection from $\mathcal{H}_{\mathbf{R}}$ to $\mathbf{R}$,

$$\mathrm{pr}_k(X) = X_k \in \mathbf{R},\ k = 1, \ldots, \infty, \quad \text{for all} \quad X = (X_k)_{k=1}^{\infty} \in \mathcal{H}_{\mathbf{R}}.$$

Since $\mathbf{R}$ and $\mathcal{H}_{\mathbf{R}}$ are fine vector spaces and pr_k is linear, pr_k is smooth (art. 3.9). We shall denote by $\mathbf{e}_k$ the only vector of $\mathcal{H}_{\mathbf{R}}$ defined by

$$\mathrm{pr}_k(\mathbf{e}_k) = 1 \quad \text{and} \quad \mathrm{pr}_j(\mathbf{e}_k) = 0 \quad \text{if} \quad j \neq k.$$

We shall call the family $\{\mathbf{e}_k\}_{k=1}^{\infty}$ the *canonical basis* of $\mathcal{H}_{\mathbf{R}}$. Also note that $\mathrm{pr}_k = \bar{\mathbf{e}}_k : X \mapsto \mathbf{e}_k \cdot X$.

Now, what has been said for $\mathbf{R}$ can be transposed to $\mathbf{C}$. Let $\mathcal{H}_{\mathbf{C}}$ be the set of square-summable complex sequences,

$$\mathcal{H}_{\mathbf{C}} = \left\{ Z = (Z_k)_{k=1}^{\infty} \;\middle|\; Z_k \in \mathbf{C}, k = 1, \ldots, \infty, \text{ and } \sum_{k=1}^{\infty} Z_k^* Z_k < \infty \right\},$$

where z^* denotes the complex conjugate of the complex number z. The space of complex numbers $\mathbf{C}$ is naturally equipped with the standard diffeology, and the space $\mathcal{H}_{\mathbf{C}}$ is equipped with the fine diffeology (art. 3.7). The usual sesquilinear product is defined on $\mathcal{H}_{\mathbf{C}}$ by

$$Z \cdot Z' = \sum_{k=1}^{\infty} Z_k^* Z_k', \quad \text{for all} \quad Z, Z' \in \mathcal{H}_{\mathbf{C}}.$$

The sesquilinear map $(Z, Z') \mapsto Z \cdot Z'$ is a Hermitian product. Thus, the pair $(\mathcal{H}, \cdot)$ is a fine Hermitian diffeological vector space over $\mathbf{C}$ (see Exercise 69, p. 74). The norm associated with the Hermitian product is defined by

$$\|Z\| = \sqrt{Z \cdot Z}, \quad \text{for all} \quad Z \in \mathcal{H}.$$

We preserve the notation pr_k for the k-th projection from $\mathcal{H}_{\mathbf{C}}$ to $\mathbf{C}$,

$$\mathrm{pr}_k(Z) = Z_k \in \mathbf{C},\ k = 1, \ldots, \infty, \quad \text{for all} \quad Z = (Z_k)_{k=1}^{\infty} \in \mathcal{H}.$$

We also preserve the notation $\mathbf{e}_k$ for the only vector of $\mathcal{H}_{\mathbf{C}}$ defined by $\mathrm{pr}_k(\mathbf{e}_k) = 1$ and $\mathrm{pr}_j(\mathbf{e}_k) = 0$ if $j \neq k$. Note that $\mathrm{pr}_k(Z) = \mathbf{e}_k \cdot Z$, but $Z \cdot \mathbf{e}_k = Z_k^*$. These definitions and constructions satisfy the following:

1. The diffeological vector space $\mathcal{H}_{\mathbf{C}}$, regarded as a $\mathbf{R}$ vector space, is still a fine diffeological vector space.

2. The map fold from $\mathcal{H}_{\mathbf{R}}$ to $\mathcal{H}_{\mathbf{C}}$, defined by

$$\mathrm{fold} : (X_k)_{k=1}^{\infty} \mapsto (Z_k)_{k=1}^{\infty}, \text{ with } Z_k = X_{2k-1} + iX_{2k},\ k = 1, \ldots, \infty,$$

is a smooth $\mathbf{R}$-isomorphism, where $\mathcal{H}_{\mathbf{C}}$ is regarded as a $\mathbf{R}$-vector space. Its inverse unfold is given by

$$\mathrm{unfold} : (Z_k)_{k=1}^{\infty} \mapsto (X_k)_{k=1}^{\infty}, \quad \text{with} \quad \begin{cases} X_{2k-1} &= \Re(Z_k), \\ X_{2k} &= \Im(Z_k), \end{cases}$$

where $\Re$ and $\Im$ denote the real and imaginary parts. Moreover, the map fold is an isometry, that is,

$$\|\mathrm{fold}(X)\|^2 = \|X\|^2, \quad \text{for all} \quad X \in \mathcal{H}_{\mathbf{R}}.$$

3. The natural injection $j : \mathcal{H}_{\mathbf{R}} \to \mathcal{H}_{\mathbf{C}}$, defined by the inclusion $X_k \mapsto Z_k = X_k + i \times 0$ of $\mathbf{R}$ into $\mathbf{C}$, is an embedding.

PROOF. 1. Let $P : U \to \mathcal{H}_{\mathbf{C}}$ be a plot of the fine diffeology, that is, $P(r) =_{\mathrm{loc}} \sum_{\alpha \in A} \lambda_\alpha(r) Z^\alpha$, where A is a finite set of indices, the Z^α belong to $\mathcal{H}_{\mathbf{C}}$, and the λ_α are smooth parametrizations in $\mathbf{C}$. Defining a_α and b_α by $\lambda_\alpha(r) = a_\alpha(r) + i b_\alpha(r)$, we get $P(r) =_{\mathrm{loc}} \sum_{\alpha \in A} a_\alpha(r) Z^\alpha + \sum_{\alpha \in A} b_\alpha(r) \times (iZ^\alpha)$. Since the iZ^α are still in $\mathcal{H}_{\mathbf{C}}$, the plot P writes locally $P(r) =_{\mathrm{loc}} \sum_{\beta \in B} \phi_\beta(r) Z'^\beta$, where B is a finite family of indices, the Z'^β belong to $\mathcal{H}_{\mathbf{C}}$ and the ϕ_β are smooth parametrizations in $\mathbf{R}$. Therefore, as a real vector space, $\mathcal{H}_{\mathbf{C}}$ is still a fine diffeological vector space.

2. Since $\mathcal{H}_{\mathbf{C}}$ is fine, regarded as a real vector space, any linear isomorphism is a smooth isomorphism (art. 3.10). Then, it is a simple verification to check that fold and unfold are inverse linear isomorphisms one of each other, and are isometries.

3. Let us check that the injection $j : \mathcal{H}_{\mathbf{R}} \to \mathcal{H}_{\mathbf{C}}$ is an embedding. First of all, since j is a linear injection it is an induction (art. 3.11). Now, let $\mathfrak{R} : \mathcal{H}_{\mathbf{C}} \to \mathcal{H}_{\mathbf{R}}$ be the map $Z \mapsto X = \mathfrak{R}(Z)$ defined by $\mathfrak{R}(Z)_k = \mathfrak{R}(Z_k)$. Note that the subspace $\mathcal{H}_{\mathbf{R}}$ is just the pointwise $\mathfrak{R}$-fixed points set: $\mathcal{H}_{\mathbf{R}} = \{Z \in \mathcal{H}_{\mathbf{C}} \mid \mathfrak{R}(Z) = Z\}$. Now, $\mathfrak{R}$ is $\mathbf{R}$-linear, thus smooth (art. 3.9), thus D-continuous (art. 2.9). Moreover, $\mathfrak{R}$ is surjective over $\mathcal{H}_{\mathbf{R}}$, so for any D-open $A \in \mathcal{H}_{\mathbf{R}}$, $A' = \mathfrak{R}^{-1}(A)$ is D-open and since $Z' \in \mathcal{H}_{\mathbf{R}}$ means $\mathfrak{R}(Z') = Z'$, $A' \cap \mathcal{H} = \{Z' \in \mathcal{H}_{\mathbf{R}} \mid \mathfrak{R}(Z') \in A\} = A$. □

Exercises

✎ EXERCISE 69 (Fine Hermitian vector spaces). Show that every real (or complex) fine diffeological vector space E (art. 3.7) equipped with any Euclidean (or Hermitian) product is an Euclidean (or Hermitian) diffeological vector space.

✎ EXERCISE 70 (Finite dimensional Hermitian spaces). Show that there exists one and only one structure of Euclidean (or Hermitian) diffeological vector space of finite dimension, that is, the fine structure. Compare with the remark of (art. 3.2).

✎ EXERCISE 71 (Topology of the norm and D-topology). The topology of the norm of a Hermitian diffeological space E does not coincide necessarily with the D-topology. Show that the topology of the norm is finer than the D-topology.

✎ EXERCISE 72 (Banach's diffeology). Let E be a Banach space, that is, a complete vector space for a given norm $\|\cdot\|$. Let us recall that a continuous map $\phi : \Omega \to E$, defined on an n-domain, is said to be of *class* $\mathcal{C}^1$ if there exists a continuous linear map, denoted by $D(f) : \Omega \to L(\mathbf{R}^n, E)$, such that,

$$\lim_{t \to 0} \frac{f(x + tu) - f(x)}{t} = D(f)(x)(u), \quad \text{for all } u \in \mathbf{R}^n.$$

The map f is said to be of class $\mathcal{C}^k$, $k > 1$, if f is of class $\mathcal{C}^1$ and if $D(f)$ is of class $\mathcal{C}^{k-1}$. The map f is said to be of class $\mathcal{C}^\infty$ if it is of class $\mathcal{C}^k$ for all $k \in \mathbf{N}$; see [**Die70a**]. Check that the set of parametrizations in E which are of class $\mathcal{C}^\infty$ for the norm is a diffeology. We shall call it the *Banach diffeology* of $(E, \|\cdot\|)$.

1) Show that, for the Banach diffeology, E is a diffeological vector space.

2) Show that the category of Banach spaces is a full subcategory of diffeological vector spaces. Use the Boman theorem: *for any two Banach spaces* E *and* F, *a map*

$f : E \to F$ *is Banach-smooth if and only if it takes smooth curves in* E *to smooth curves in* F [**Bom67**].

✎ EXERCISE 73 ($\mathcal{H}_{\mathbf{C}}$ is isomorphic to $\mathcal{H}_{\mathbf{R}} \times \mathcal{H}_{\mathbf{R}}$). With the notations of (art. 3.15), check that the map $\psi : \mathcal{H}_{\mathbf{R}} \times \mathcal{H}_{\mathbf{R}} \mapsto \mathcal{H}_{\mathbf{C}}$ defined by $\psi(X, Y) = X + iY$ is a smooth isometry, that is, a smooth isomorphism such that $\|\psi(X, Y)\|^2 = \|X\|^2 + \|Y\|^2$.

CHAPTER 4

Modeling Spaces, Manifolds, etc.

Manifolds are the main objects of differential geometry as we know it. It is not necessary to develop the whole theory of diffeological spaces just to introduce or study manifolds. However, manifolds can be found again as a full subcategory of the category {Diffeology}. Roughly speaking, manifolds are diffeological spaces which look like locally real vector spaces. Diffeology gives another insight into this respectable domain: manifolds are not any more regarded as sets equipped with a *manifold structure*, but rather as diffeological spaces whose diffeology satisfies the special property to be generated by local diffeomorphisms with a given vector space.

This rediscovery of manifolds, through diffeology, suggests some generalizations. First of all, we can replace, in the definition of a manifold, real vector spaces by any diffeological vector space (art. 3.1), and we get a larger category of manifolds which contains not only the usual ones, but also infinite dimensional manifolds, which find, that way, a formal framework for their study. If this generalization brings some satisfaction — many infinite dimensional spaces become then manifolds, for example the infinite projective space (art. 4.11), but not only — it also raises new questions. For example, if we have a notion of orientability for diffeological spaces of finite dimension (art. 6.44), what about orientability of infinite dimensional diffeological manifolds? This is not the only question, and it may not be the most relevant. With infinite dimension spaces things are more subtle, and the usual concepts need to be revisited with care if we want to preserve their relevance.

On the other hand, this conception suggests the notion of *locally modeled spaces*, I mean, diffeological spaces which are locally diffeomorphic to some members of a family of *diffeological models*. For manifolds, the family of modeling spaces is usually reduced to a single given diffeological vector space, but diffeological orbifolds (art. 4.17) are defined as diffeological spaces, modeled on the family of the quotients of real spaces by a finite subgroup of its diffeomorphisms.

Standard Manifolds, the Diffeological Way

We should begin with the general definition of diffeological manifolds (art. 4.7), *but, to preserve the intuition of the geometer, I chose to begin with an exposé on usual manifolds — which are regarded as special cases of diffeological spaces — and then to introduce the more general concept of manifolds modeled on general diffeological vector spaces.*

4.1. Manifolds as diffeologies. Let M be a diffeological space, M is said to be a *manifold of dimension* n, or an n-manifold if and only if M is locally diffeomorphic at each point to $\mathbf{R}^n$. This means precisely that, for each point $m \in \mathsf{M}$, there exist a local diffeomorphism $\mathsf{F} : \mathbf{R}^n \supset \mathsf{U} \to \mathsf{M}$ (art. 2.5), and a point $r \in \mathsf{U}$ such that $\mathsf{F}(r) = m$. Such local diffeomorphisms are called *charts* of M. The set of all the charts of M is called the *saturated atlas* of M. Speaking in terms of diffeologies, we shall also say that the diffeology $\mathcal{D}$ of M is a *manifold diffeology*. Manifolds form the subcategory {Manifolds} of the category {Diffeology}.

4.2. Local modeling of manifolds. The previous definition of manifolds (art. 4.1) can be formulated again in terms of generating family (art. 1.66). Let M be a diffeological space.

1. A family $\mathcal{A}$ of local diffeomorphisms (art. 2.5), from $\mathbf{R}^n$ to M, such that

$$\bigcup_{\mathsf{F}\in\mathcal{A}} \operatorname{val}(\mathsf{F}) = \mathsf{M}$$

is a generating family of the diffeology of M.

2. The diffeological space M is an n-manifold (art. 4.1) if and only if there exists a generating family $\mathcal{A}$ of M, made of local diffeomorphisms from $\mathbf{R}^n$ to M.

NOTE 1. If M is a manifold, any generating family $\mathcal{A}$, made of local diffeomorphisms of M, is called an *atlas* of M. The elements of $\mathcal{A}$ are called the *charts of the atlas*. Obviously, if M is a manifold, the set of all local diffeomorphisms from $\mathbf{R}^n$ to M is a generating family of the diffeology of M. This set is the saturated atlas of the manifold M (art. 4.1).

NOTE 2. An n-manifold is a diffeological space of constant dimension n, since local diffeomorphisms preserve the dimension (art. 2.22). Conversely, if the dimension of a manifold is n (art. 1.78), then it is an n-manifold. This is why we sometimes use the wording n-*dimensional manifold* in place of n-manifold.

PROOF. 1. Let us assume that $\mathcal{A}$ is a family of local diffeomorphisms from $\mathbf{R}^n$ to M such that $\bigcup_{\mathsf{F}\in\mathcal{A}} \operatorname{val}(\mathsf{F}) = \mathsf{M}$. Thus, M is locally diffeomorphic to $\mathbf{R}^n$ at each point, since each point m of M is in the set of values of a local diffeomorphism. Let us choose, for each point $m \in \mathsf{M}$, a local diffeomorphism $\mathsf{F} : \mathsf{U} \to \mathsf{M}$ such that $\mathsf{F}(0) = m$, where U is an open neighborhood of $0 \in \mathbf{R}^n$. Such local diffeomorphisms exist since we have just to compose any local diffeomorphism F which maps r to m with the translation mapping 0 to r. Now, let $\mathcal{A}$ be the set of all these chosen local diffeomorphisms, where m runs over M. Let us check that they form a generating family of the diffeology of M. Let $\mathsf{P} : \mathsf{V} \to \mathsf{M}$ be a plot and $r \in \mathsf{V}$, and let $m = \mathsf{P}(r)$ and $\mathsf{F} \in \mathcal{A}$ such that $\mathsf{F} : \mathsf{U} \to \mathsf{M}$, $\mathsf{F}(0) = m$. Let $\mathsf{Q} = \mathsf{F}^{-1} \circ \mathsf{P} \restriction W$, where $W = \mathsf{P}^{-1}(\mathsf{F}(\mathsf{U}))$. Since F is a local diffeomorphism, $\mathsf{F}(\mathsf{U})$ is D-open (art. 2.10), and because P is D-continuous (art. 2.9), W is a domain. Thus, Q is a smooth local lift of P along F. Hence, any plot of M can be lifted along a member of the family $\mathcal{A}$, and this is the criterion for a generating family (art. 1.68). Hence, the diffeology of M is generated by $\mathcal{A}$.

2. Let us assume that the diffeology of M is generated by a family $\mathcal{A}$ of local diffeomorphisms with $\mathbf{R}^n$. Let $m \in \mathsf{M}$ be any point, the constant 0-parametrization $\mathbf{m} : 0 \mapsto m$ is a plot. Thus, by definition of generating families (art. 1.66), there exist an element $\mathsf{F} : \mathsf{U} \to \mathsf{M}$ of $\mathcal{A}$ and a point $r \in \mathsf{U}$ such that $\mathbf{m} = \mathsf{F} \circ \mathbf{r}$, where $\mathbf{r}(0) = r$. Hence, there exist a local diffeomorphism $\mathsf{F} : \mathbf{R}^n \supset \mathsf{U} \to \mathsf{M}$ and a point $r \in \mathsf{U}$ such that $\mathsf{F}(r) = m$. Therefore, M is a manifold (art. 4.1). The converse is a consequence of 1. If M is an n-manifold, it is locally diffeomorphic at each point to $\mathbf{R}^n$. So, there exists a family $\mathcal{A}$ of local diffeomorphisms from $\mathbf{R}^n$ to M such that $\bigcup_{\mathsf{F}\in\mathcal{A}} \operatorname{val}(\mathsf{F}) = \mathsf{M}$. Therefore, $\mathcal{A}$ is a generating family of the diffeology of M.

Now, since n-manifolds are locally diffeomorphic to $\mathbf{R}^n$, their dimension, at every point, is the same dimension as an n-domain (art. 2.24). But the dimension of an n-domain is constant and equal to n (art. 1.80). Hence, the dimension of an n-manifold is constant and equal to n. Conversely, the dimension map of a manifold is constant and equal to the dimension of the domains of its charts, as a direct consequence of the definition. Since $\dim(\mathbf{R}^n) = \dim(\mathbf{R}^m)$ if and only if $n = m$, a manifold has dimension n if and only if it is an n-manifold. □

4.3. Manifolds, the classical way. A full exposé on manifolds, in classical differential geometry context, can be found in every book of differential geometry, for example in [**Bou82**], [**BeGo72**], [**Die70c**], [**Doc76**], [**DNF82**]. Let us just summarize the basic definitions. We choose here the Bourbaki definition [**Bou82**], but we make the inverse convention, made also by some other authors, to regard charts defined from real domains to a manifold M, rather than from subsets of M into real domains.

(♣) Let M be a nonempty set. A *chart* of M is a bijection F defined on an n-domain U to a subset of M. The dimension n is a part of the data. Let $F : U \to M$ and $F' : U' \to M$ be two charts of M. The charts F and F' are said to be *compatible* if and only if the following conditions are fulfilled:

a) The sets $F^{-1}(F'(U'))$ and $F'^{-1}(F(U))$ are open.
b) The two maps, each one the inverse of the other, $F'^{-1} \circ F : F^{-1}(F'(U')) \to F'^{-1}(F(U))$ and $F^{-1} \circ F' : F'^{-1}(F(U)) \to F^{-1}(F'(U'))$, are either empty or smooth. They are called *transition maps.*

An *atlas* is a set of charts, compatible two-by-two, such that the union of the values is the whole M. Two atlases are said to be compatible if their union is still an atlas. This relation is an equivalence relation. A *structure of manifold* on M is the choice of an equivalence class of atlases or, which is equivalent, the choice of a *saturated atlas.* Once a structure of manifold is chosen for M, every compatible chart is called a *chart of the manifold.*

(♡) The simplest example of a classical manifold is a real domain U, with the structure of a manifold given by the atlas reduced to the identity $\{\mathbf{1}_U\}$.

(◇) Let M and M' be two classical manifolds. A map $f : M \to M'$ is said to be infinitely differentiable, or of class $\mathcal{C}^\infty$, or smooth, if and only if, for every pair of charts $F : U \to M$ and $F' : U' \to M'$, the following conditions are fulfilled:

a) The set $(f \circ F)^{-1}(F'(U'))$ is open.
b) The parametrization $F'^{-1} \circ f \circ F : (f \circ F)^{-1}(F'(U')) \to U'$ is either empty or smooth.

The composition of smooth maps between classical manifolds is again a smooth map. Classical manifolds, together with smooth maps, form a category whose isomorphisms are called diffeomorphisms. A diffeomorphism f from a manifold to another is a smooth bijection such that f^{-1} is smooth.

That was for the classical way. Now, manifolds are naturally embedded in the category of diffeological spaces, thanks to the functor defined as follows.

(A) Let M be a classical manifold. Let $\mathcal{D}$ be the set of all the parametrizations $P : U \to M$ which are smooth according to (◇), where U is regarded as the manifold described by (♡). So, $\mathcal{D}$ is a manifold diffeology for which any atlas of M, according to (♣), is a generating family (art. 4.1). This diffeology is the *canonical diffeology* associated with M.

(B) Conversely, let M be a manifold and let $\mathcal{A}$ be any atlas generating the diffeology $\mathcal{D}$ of M, according to the definitions of (art. 4.1). Then, $\mathcal{A}$ is an atlas, according to (♣), for M. The atlas $\mathcal{A}$ gives to M the canonical structure of classical manifold, associated with $\mathcal{D}$.

(C) The constructions (A) and (B) are the inverse of each other. They define a full faithful functor from the classical category of manifolds to the category {Diffeology}. The image of this functor is the category {Manifolds} defined in (art. 4.1).

Therefore, there is no need to distinguish between these two classes of objects. We shall regard finite dimensional real manifolds, always, as defined by (art. 4.1). One can say then, that diffeology is a generalization of the notion of manifolds, but reducing diffeology to be just a generalization of manifolds would be exaggerated.

PROOF. (A) Let M be a classical manifold, and let $\mathcal{A}$ be an atlas of M according to (♣). Let $\mathcal{D}$ be the set of all the smooth parametrizations of M, according to (♢). The domains of these parametrizations are regarded as manifolds, according to (♡). Let us prove the three axioms of diffeology:

D1. Let $m \in M$ and $\mathbf{m} : r \mapsto m$ be a constant parametrization. Let F be a chart of $\mathcal{A}$ such that $F(x) = m$. Since F is injective, $\mathbf{m}$ lifts along F by the constant parametrization $r \mapsto x$. Since constant parametrizations of domains are smooth, $\mathcal{D}$ contains the constant parametrizations of M.

D2. Let $P : V \to M$ be a parametrization which satisfies locally the condition (♢). For all $r \in V$ there exists an open neighborhood W of r such that $P \restriction W$ is a smooth parametrization, according to (♢). In other words, for every chart ϕ of W, for every chart $F : U \to M$, $(P \circ \phi)^{-1}(F(U))$ is open, and $F^{-1} \circ P \circ \phi : (P \circ \phi)^{-1}(F(U)) \to U$ is smooth. But charts of domains are just local diffeomorphisms (art. 2.5), thus ϕ can be simply replaced by the inclusion $j_W : W \hookrightarrow V$. These conditions reduce then to the following one: there exists an open neighborhood W of r such that $(P \restriction W)^{-1}(F(U))$ is open and $F^{-1} \circ (P \restriction W) : (P \restriction W)^{-1}(F(U)) \to U$ is either empty or smooth. But this is clearly a local condition. Therefore, if P belongs locally to $\mathcal{D}$, then it belongs to $\mathcal{D}$.

D3. Let $P : V \to M$ be a parametrization. We have seen just above that P belongs to $\mathcal{D}$ if only if, for any chart $F : U \to M$, $P^{-1}(F(U))$ is open and $F^{-1} \circ P : P^{-1}(F(U)) \to U$ is either empty or smooth. Now, let $Q : W \to V$ be a smooth parametrization. Let $F : U \to M$ be a chart. Since Q is a smooth parametrization and $P^{-1}(F(U))$ is open, $Q^{-1}(P^{-1}(F(U)))$ is open, that is, $(P \circ Q)^{-1}(F(U))$ is open. Then, since Q is smooth and $F^{-1} \circ P : P^{-1}(F(U)) \to U$ is either empty or smooth, $F^{-1} \circ P \circ Q : (P \circ Q)^{-1}(F(U)) \to F(U)$ is either empty or smooth. Therefore, if P belongs to $\mathcal{D}$ and Q is a smooth parametrization in the domain of P, then $P \circ Q$ belongs to $\mathcal{D}$.

In order to complete the first point, it remains to check that any atlas $\mathcal{A}$ of M is a generating family of $\mathcal{D}$. Let $P : U \to M$ be a plot of M, that is, $P \in \mathcal{D}$. Let $r \in U$, there exist a chart F of M and a point $x \in \operatorname{def}(F)$ such that $F(x) = P(r)$, let $U' = P^{-1}(\operatorname{val}(F))$. Since P belongs to $\mathcal{D}$, U' is an open neighborhood of r, and $P' = P \restriction U'$ is still a plot of M. But $\operatorname{val}(P') \subset \operatorname{val}(F)$ implies that $Q = F^{-1} \circ P' \in \mathcal{C}^\infty(U', \operatorname{def}(F))$. Thus, Q is a smooth parametrization in $\operatorname{def}(F)$ and $F \circ Q = P \restriction U'$. This is the condition to be a generating family of the diffeology $\mathcal{D}$.

(B) Let M be a manifold, and let $\mathcal{A}$ be any atlas generating the diffeology $\mathcal{D}$ of M, according to the definitions of (art. 4.1). Since every element F of $\mathcal{A}$ is a local diffeomorphism (art. 2.5), F is an injection defined on a domain. Since compositions of local diffeomorphisms are either empty or local diffeomorphisms, the transition maps associated with the charts of the atlas $\mathcal{A}$ are smooth. Therefore, $\mathcal{A}$ defines on M a structure of classical manifold.

(C) Let M be a set, $\mathcal{A}$ and $\mathcal{A}'$ two maximal atlases according to (♣), defining the same diffeology $\mathcal{D}$. Every chart $F \in \mathcal{A}$ is a local diffeomorphism for $\mathcal{D}$, as well as every chart $F' \in \mathcal{A}'$. Since the composite of two local diffeomorphisms for a given diffeology is either empty or again a local diffeomorphism (art. 2.5), $F'^{-1} \circ F$, defined on $F^{-1}(\mathrm{val}(F'))$, is either empty or a local diffeomorphism between real domains, necessarily of the same dimension. But this is the condition for F and F′ to be compatible. Since $\mathcal{A}$ and $\mathcal{A}'$ are maximal, $\mathcal{A} = \mathcal{A}'$, and the two classical manifold structures are equal. Thus, the map defined in (A) is injective, and clearly surjective, thanks to (B). Therefore, it defines a one-to-one correspondence between the objects of the category of classical manifolds and the objects of the category {Manifolds} defined in (art. 4.1). Then, according to (A) and (B) the saturated atlas of a given classical n-dimensional manifold M is the set of local diffeomorphisms from $\mathbf{R}^n$ to M, regarded as a diffeological space. Therefore, (A) and (B) are the inverse of each other.

Finally, let us check that, given two classical manifolds M and M′, the smooth maps from M to M′, according to (♢), are exactly the smooth maps with M and M′ regarded as diffeological spaces. In fact, since the charts for a given classical n-dimensional manifold are just local diffeomorphisms with $\mathbf{R}^n$, with respect to the associated diffeology, and since the associated diffeology is precisely generated by these local diffeomorphisms, this proposition is a direct application of the criterion (art. 1.73); see also Figure 4.1. Therefore, the functor, mapping a classical manifold to its associated diffeology, is a full faithful functor to {Manifolds}. □

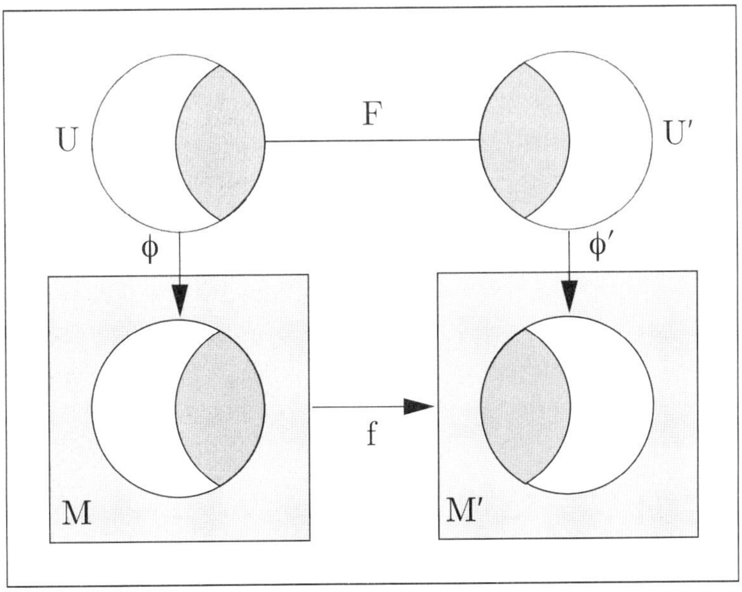

FIGURE 4.1. A smooth map between manifolds.

4.4. Submanifolds of a diffeological space. The definition of *submanifolds* follows immediately the definition of manifolds (art. 4.1). Let X be a diffeological space and $M \subset X$. If M, equipped with the subset diffeology (art. 1.33), is a manifold, then M will be called a *submanifold* of X.

NOTE 1. The vocabulary used here "M is a submanifold of X", can lead to confusion since X itself is not necessarily a manifold (like in the example below). But we have no real choice. We could say "M is a manifold in X" but this is longer, it is not compatible with the standard vocabulary when X is itself a manifold, and it attenuates the fact that M is regarded as equipped with the subset diffeology. So, we choose to say "submanifold" even if the ambient space is not itself a manifold. We have just to be aware of that, and cautious.

NOTE 2. By definition, a submanifold is always *induced* (the injection $M \hookrightarrow X$ is an induction). It can also be *embedded* — if the inclusion of M into X is an embedding (art. 2.13).

EXAMPLE. The sphere (Exercise 18, p. 19) is an example of an embedded submanifold, while the irrational solenoid (Exercise 58, p. 57) is again an example of submanifold, but not embedded. In Exercise 57, p. 57, we have seen that the group $\mathrm{GL}(n, \mathbf{R})$ is embedded in $\mathrm{Diff}(\mathbf{R}^n)$, equipped with the functional diffeology. Since the group $\mathrm{GL}(n, \mathbf{R})$ is a manifold for the subset diffeology, equivalent to the open subset of the $n \times n$ matrices with nonzero determinant, $\mathrm{GL}(n, \mathbf{R})$ is an embedded submanifold of $\mathrm{Diff}(\mathbf{R}^n)$.

4.5. Immersed submanifolds are not submanifolds? Let N and M be two manifolds, and let $j : N \to M$ be an immersion (see Exercise 12, p. 17). Then, the image $j(N)$ is not necessarily a submanifold, because an immersion is not necessarily an induction; see Exercise 59, p. 58. Therefore, if there is a well defined concept of *immersion*, there is no such a concept of "immersed submanifolds".

4.6. Quotients of manifolds. This question is often asked in classical differential geometry: When is the quotient of a manifold — or more generally, the quotient of a diffeological space — by an equivalence relation, a manifold? There is no simple answer to this question because it depends too much on the nature of the diffeological space and of the equivalence relation. But, in some circumstances, the following characterization can be useful. Let X be a diffeological space, and let $\sim$ be an equivalence relation defined on X. Let $M = X/\sim$ be the diffeological quotient (art. 1.50). Let $\pi : X \to M$ be the associated projection. The quotient space M is a manifold of dimension n if and only if there exists a family of n-plots $\{\phi_i : U_i \to X\}_{i \in \mathcal{I}}$, where $\mathcal{I}$ is any set of indices, such that the following conditions are fulfilled:

1. For every $i \in \mathcal{I}$, the map $\phi_i : U_i \to X$ is an induction such that $\pi \circ \phi_i$ is injective. Equivalently, the image of U_i by the induction ϕ_i cuts each class of the relation $\sim$ in one point, at most.

2. The projections of the values of the inductions ϕ_i cover M, that is, $M = \bigcup_{i \in \mathcal{I}} \pi \circ \phi_i(U_i)$.

3. For every $r_i \in U_i$ and $r_j \in U_j$ such that $\pi(\phi_i(r_i)) = \pi(\phi_j(r_j))$, there exists a local diffeomorphism ψ, defined on some open neighborhood V of r_i to U_j, mapping r_i to r_j, and such that $\pi \circ \phi_j \circ \psi = \pi \circ (\phi_i \restriction V)$.

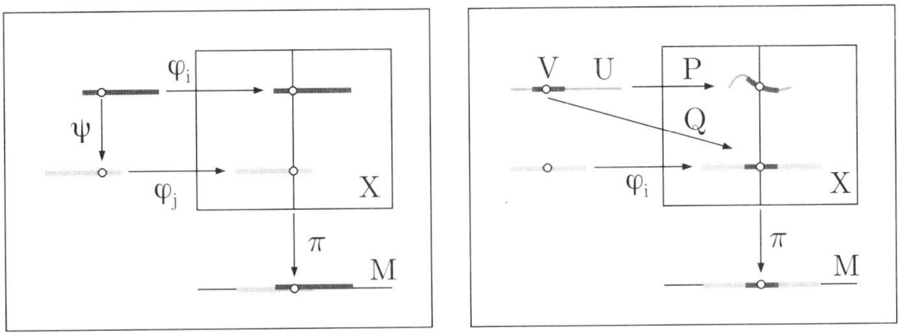

FIGURE 4.2. The quotient is a manifold.

4. For every plot $P : W \to X$, for every $r \in W$, there exist an open neighborhood $V \subset W$ of r, some index $i \in \mathcal{I}$ and a plot $Q : V \to \phi_i(U_i) \subset X$ such that $\pi \circ P \restriction V = \pi \circ Q$.

If these conditions are satisfied, each map $F_i = \pi \circ \phi_i$, with $i \in \mathcal{I}$, is a chart of M, and the set $\mathcal{A} = \{F_i\}_{i \in \mathcal{I}}$ is an atlas of M.

NOTE. The diffeology of the quotient M is well defined as the quotient diffeology of X, there is no choice here. This condition, stated above, answers only this question: Is the quotient diffeology a manifold diffeology? It may be the case, or not.

PROOF. Let us assume that the quotient M is a manifold of dimension n. Then there exists a family of local diffeomorphisms from $\mathbf{R}^n$ to M which generates the diffeology of M (art. 4.2). Let $F : U \to M$ be such a diffeomorphism, and let $r \in U$. Then, since $\pi : X \to M$ is a subduction, there exists a local lift $\phi : V \to X$ such that $\phi \in \mathcal{D}(V)$ and $\pi \circ \phi = F \restriction V$. Let us check that ϕ is an induction, using the criterion (art. 1.31). First of all ϕ is smooth. Since $\pi \circ \phi$ is injective, so is ϕ. Now, let $Q : W \to \phi(V)$ be a plot in X, then $\pi \circ Q \in \mathcal{C}(W)$. Since F is a chart of M, that is, a local diffeomorphism, the composite $F^{-1} \circ \pi \circ Q$, restricted to its domain, is smooth. But, $F \circ \phi = F \restriction V \Rightarrow \phi^{-1} = F^{-1} \circ \pi$, where F^{-1} is restricted to $\pi(\phi(V))$, thus $F^{-1} \circ \pi \circ Q = \phi^{-1} \circ Q$. Hence, $\phi^{-1} \circ Q$ is smooth, and ϕ is an induction. Now, since $\pi \circ \phi = F \restriction V$, $\pi \circ \phi$ is injective, which means that if $\phi(r)$ and $\phi(r')$ belong to the same equivalence class, then $r = r'$. Then, since for every point $m \in M$ there exist a chart $F : U \to M$ and a point $r \in U$ such that $F(r) = m$, there exists a family of such local lifts ϕ whose projections $\pi \circ \phi$ cover M. We proved the first and second points. Let us prove the third one. Let us consider such a family $\{\phi_i : U_i \to X\}_{i \in \mathcal{I}}$ of inductions made of local lifts of charts of M. Let $r_i \in U_i$ and $r_j \in U_j$ such that $\phi_i(r_i) = \phi_j(r_j) = m$. Let F_i and F_j be two charts of M such that $F_i = \pi \circ \phi_i$ and $F_j = \pi \circ \phi_j$. Then, $F_i(r_i) = F_j(r_j) = m$ and there exists a local diffeomorphism ψ defined on an open neighborhood of r_i such that $\psi(r_i) = r_j$ and $F_j \circ \psi = F_i$. Hence, $\pi \circ \phi_j \circ \psi = \pi \circ \phi_i$, and this is the third point. Finally, let us consider a plot $P : W \to X$, thus $\pi \circ P$ is a plot of M. Let $r \in W$ and $m = \pi \circ P(r)$, there exist a chart $F_i : U_i \to M$, an open neighborhood $V \subset W$ of r and a plot $P' : V \to U_i$ such that $F_i \circ P' = \pi \circ P \restriction V$. Now, $F_i = \pi \circ \phi_i$, then $\pi \circ \phi_i \circ P' = \pi \circ P \restriction V$, denoting by $Q = P' \circ \phi_i : V \to \phi_i(U_i)$ we get $\pi \circ Q = \pi \circ P \restriction V$. The fourth point is proved.

Thus, we proved that the four points of the proposition are satisfied if the quotient M is a manifold (see Figure 4.2). Let us prove now that if these four points are satisfied, then M is a manifold. Let us define $F_i = \pi \circ \phi_i$, F_i is injective and smooth. Let us show that F_i is a local diffeomorphism (art. 2.5). Let $P : \mathcal{O} \to M$ be a plot, and let $r \in \mathcal{O}$ such that $m = P(r) \in F_i(U_i)$. Then, there exists a local lift $P' : W \to X$ such that $\pi \circ P' = P \restriction W$ and $r \in W$. Thanks to the fourth point of the proposition, there exist an index $j \in \mathcal{J}$, a subset $V \subset W$, and a plot $Q : V \to \phi_j(U_j)$ of X such that $\pi \circ P' \restriction V = \pi \circ Q$. Hence, $\pi \circ Q = P \restriction V$ and $Q(V) \subset \phi_j(U_j)$. Let $r_j = \phi_j^{-1}(Q(r)) \in U_j$ and $r_i \in U_i$ such that $F_i(r_i) = m$. We have $m = \pi \circ \phi_i(r_i) = \pi \circ \phi_j(r_j)$, and thanks to the third condition of the proposition, there exists a local diffeomorphism ψ defined on an open neighborhood of r_j mapping r_j to r_i and such that $\pi \circ \phi_i \circ \psi = \pi \circ \phi_j$, restricted to this neighborhood. Thus, $F_i^{-1} \circ \pi \restriction Q(V) = \psi \circ \phi_j^{-1} \restriction Q(V)$, hence $F_i^{-1} \circ P \restriction V = F_i^{-1} \circ \pi \circ Q \restriction V = \psi \circ \phi_j^{-1} \circ Q$. But, ϕ_j being an induction, $\phi_j^{-1} \circ Q$ is smooth, and so is $\psi \circ \phi_j^{-1} \circ Q$. Thus, $F_i^{-1} \circ P \restriction V$ is smooth. Therefore, $F_i^{-1} \circ P$ is smooth, that is, F_i^{-1} is smooth. Therefore, F_i is a local diffeomorphism. In conclusion, the set $\mathcal{A} = \{F_i\}_{i \in \mathcal{J}}$ is an atlas of M, where $F_i = \pi \circ \phi_i$, and M is a manifold of dimension n. □

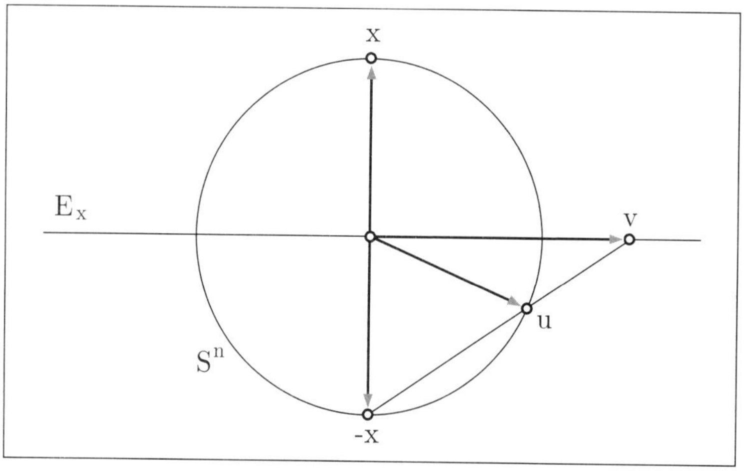

FIGURE 4.3. The stereographic projection.

Exercises

✎ EXERCISE 74 (The irrational torus is not a manifold). Using the criterion (art. 4.6), show that the irrational torus T_α of Exercise 4, p. 8, is not a manifold.

✎ EXERCISE 75 (The sphere as paragon). Let us come back to the sphere S^n of Exercise 18, p. 19. The sphere is an important example which deserves to be watched under several points of view. Let us consider $\mathbf{R}^{n+1}$ equipped with the standard diffeology, the sphere S^n is the subspace defined by

$$S^n = \{x \in \mathbf{R}^{n+1} \mid \|x\| = 1\},$$

where $\|\cdot\|$ denotes the usual Euclidean norm associated with the scalar product, denoted by $(u, v) \mapsto u \cdot v$. Let us recall that a plot of S^n is a plot of $\mathbf{R}^{n+1}$ with values in S^n (art. 1.33). With every $x \in S^n$, let us associate $E_x \subset \mathbf{R}^{n+1}$, defined as the subspace orthogonal to x (see Figure 4.3),

$$E_x = \{v \in \mathbf{R}^{n+1} \mid x \cdot v = 0\}.$$

Let F_x be the map defined by

$$F_x : E_x \to S^n \quad \text{with} \quad v \mapsto u = 2\frac{v + x}{1 + \|v\|^2} - x.$$

The map F_x is called the *stereographic projection* with respect to x.

1) Show that F_x is a local diffeomorphism from E_x to S^n.

2) Deduce that S^n is a manifold of dimension n.

3) Show that the pair $\{F_N, F_{-N}\}$, where $N = (0_n, 1) \in S^n$ is the North Pole, is also a generating family for S^n, that is, an atlas of S^n.

Diffeological Manifolds

The construction of manifolds (art. 4.1) *on the one hand, and the existence of diffeological vector spaces* (art. 3.1) *on the other hand, suggest the introduction of a new diffeological category: the* diffeological manifolds, *which are diffeological spaces* modeled *on diffeological vector spaces. Since finite dimensional vector spaces are diffeological vector spaces, the traditional manifolds then become a special case (or a subcategory) of diffeological manifolds. The goal of this section is to give the general definition of these new objects, and to express some of their properties. We shall illustrate these definitions by two examples: the* infinite sphere *of the fine standard Hilbert space* (art. 3.15) *and the associated* infinite projective space. *We shall have the opportunity, through these examples, to see how diffeology works with infinite dimensional manifolds.*

4.7. Diffeological Manifolds. Let E be a diffeological vector space (art. 3.1). Let X be a diffeological space. We shall say that X is a *diffeological manifold modeled on* E if X is locally diffeomorphic to E at every point, that is, if for every $x \in X$ there exists a local diffeomorphism (art. 2.5) $F : E \supset U \to X$ such that $x \in F(U)$. Such a local diffeomorphism will be called a chart of X. Diffeological manifolds form a subcategory of {Diffeology} still denoted by {Manifolds}. The basic examples of diffeological manifolds are diffeological vector spaces and ordinary manifolds modeled on $\mathbf{R}^n$ (art. 4.1), but also manifolds modeled on a Banach space; see Exercise 72, p. 74.

NOTE 1. In the same spirit of modeling diffeological spaces (see (art. 4.19)), this definition can be altered to be at the same time equivalent and more flexible, practically.[1] Let $\mathcal{E} = \{E_i\}_{i \in \mathcal{I}}$ be a family of isomorphic diffeological vector spaces, a diffeological space X is a manifold if it is locally diffeomorphic, at every point $x \in X$, to one of the element of the family $\mathcal{E}$. Remark that this is exactly the situation we encountered with the sphere in Exercise 75, p. 84.

NOTE 2. Some definitions of manifolds, in classical differential geometry, allow the modeling vector space to change its type from point to point, which gives manifolds with variable dimensions for different connected components. I do not think that it is necessary to introduce this subtlety here, but I do not exclude that it could also be useful. Only experience and time will tell.

[1]This has been suggested to me by one of the referees of the manuscript; it is a good remark.

4.8. Generating diffeological manifolds. The proposition (art. 4.2) adapts itself to the case of diffeological manifolds. Let X be a diffeological space. Let E be a diffeological vector space. Every family $\mathcal{A}$ of local diffeomorphisms (art. 2.5), from E to X, such that

$$\bigcup_{F \in \mathcal{A}} \mathrm{val}(F) = X$$

generates the diffeology of X in the following meaning: for every plot $P : U \to X$, for every point $r \in U$, there exist an open neighborhood V of r, an element F of $\mathcal{A}$, and a plot $Q : V \to E$, such that $P \restriction V = F \circ Q$. If there exists such a family $\mathcal{A}$ of local diffeomorphisms, from E to X, $\mathcal{A}$ will be called an E-*atlas*, or simply an *atlas*, of X, and the elements of A will be called the *charts* of the atlas. Thus, a diffeological space X is a manifold modeled on E (art. 4.7) if and only if there exists an atlas $\mathcal{A}$ of X made of local diffeomorphisms from E to M. Note that, in this case, there exists an atlas made up with all the local diffeomorphisms from E to X, this atlas is called the *saturated* atlas of X.

PROOF. Let $P : U \to X$ be a plot and $r \in U$. Let $x = P(r)$ and $F \in \mathcal{A}$ such that $x \in \mathrm{val}(F)$. Such an F exists by hypothesis. Now, let $V = P^{-1}(\mathrm{val}(F))$ and $Q = F^{-1} \circ (P \restriction V)$, that is, $P \restriction V = F \circ Q$. Now, since F is a local diffeomorphism, val(F) is D-open (art. 2.10), and since P is D-continuous (art. 2.9), V is a domain. Moreover, since F is a local diffeomorphism, Q is a plot of E. Therefore, every plot of X can be smoothly lifted, locally, along some element of the family $\mathcal{A}$, as it is claimed by the proposition. Now, let X be generated by a family $\mathcal{A}$ of local diffeomorphisms from E to X. Pick any point $x \in X$, and let $P : \{0\} \to X$ be the constant plot such that $P(0) = x$. By hypothesis, there exist a chart $F \in \mathcal{A}$ and a lift $Q : \{0\} \to E$ such that $P = F \circ Q$. Hence, F is a local diffeomorphism from E to X, such that $x \in \mathrm{val}(F)$. Therefore, X is a diffeological manifold modeled on E. □

4.9. The infinite sphere. Let $\mathcal{H}_{\mathbf{R}}$ be the real Hilbert space of square summable real sequences, equipped with the fine diffeology (art. 3.15). The unit sphere $\mathcal{S}_{\mathbf{R}}$ of the Hilbert space $\mathcal{H}_{\mathbf{R}}$ is defined as usual by

$$\mathcal{S}_{\mathbf{R}} = \left\{ X \in \mathcal{H}_{\mathbf{R}} \;\middle|\; X \cdot X = \sum_{k=1}^{\infty} X_k^2 = 1 \right\}.$$

The sphere $\mathcal{S}_{\mathbf{R}}$ will be called the *infinite sphere* in the following, and it will be equipped with the subset diffeology (art. 1.33) of the fine diffeology (art. 3.7) of $\mathcal{H}_{\mathbf{R}}$. Let us recall that a plot of $\mathcal{H}_{\mathbf{R}}$ for the fine diffeology is a parametrization such that, for all $r_0 \in U$, there exist an open neighborhood V of r_0 and a finite local family $(\lambda_\alpha, X_\alpha)_{\alpha \in A}$, defined on V, such that

$$P \restriction V : r \mapsto \sum_{\alpha \in A} \lambda_\alpha(r) X_\alpha.$$

Let us recall that to be a *finite local family defined on* V means that the set of indices A is finite, and for every $\alpha \in A$, $\lambda_\alpha \in \mathcal{C}^\infty(V, \mathbf{R})$, and $X_\alpha \in \mathcal{H}_{\mathbf{R}}$. Now, the plot P is a plot of the sphere $\mathcal{S}_{\mathbf{R}}$ if moreover it takes its values in $\mathcal{S}_{\mathbf{R}}$, that is, if $\sum_{\alpha \in A} \sum_{\alpha' \in A} \lambda_\alpha(r) \lambda_{\alpha'}(r)\, X_\alpha \cdot X_{\alpha'} = 1$ for all $r \in V$. Let us then define the following *stereographic maps*:

$$F_+ : \mathcal{H}_{\mathbf{R}} \to \mathcal{S}_{\mathbf{R}} \quad \text{with} \quad F_+ : \xi \mapsto X = \frac{1}{\|\xi\|^2 + 1} \begin{pmatrix} \|\xi\|^2 - 1 \\ 2\xi \end{pmatrix},$$

$$F_- : \mathcal{H}_{\mathbf{R}} \to \mathcal{S}_{\mathbf{R}} \quad \text{with} \quad F_- : \xi \mapsto X = \frac{1}{1+\|\xi\|^2}\begin{pmatrix} 1-\|\xi\|^2 \\ 2\xi \end{pmatrix},$$

where the matrix notation denotes the corresponding sequences, belonging to the vector space $\mathcal{H}_{\mathbf{R}}$ defined by

$$\begin{pmatrix} a \\ \xi \end{pmatrix} = (a, \xi_1, \xi_2, \ldots), \text{ where } a \in \mathbf{R} \text{ and } \xi = (\xi_1, \xi_2, \ldots) \in \mathcal{H}_{\mathbf{R}}.$$

Let us use the notation of (art. 3.15), where $\mathbf{e}_k$ is the vector whose k-th coordinate is 1 and the others are zero. The images of F_+ and F_- are

$$F_+(\mathcal{H}_{\mathbf{R}}) = \mathcal{S}_{\mathbf{R}} - \{\mathbf{e}_1\} \quad \text{and} \quad F_-(\mathcal{H}_{\mathbf{R}}) = \mathcal{S}_{\mathbf{R}} - \{-\mathbf{e}_1\}.$$

Let us denote now, for every $X \in \mathcal{H}_{\mathbf{R}}$

$$X = (X_1, X_+) \quad \text{with} \quad X_1 = \mathrm{pr}_1(X) \quad \text{and} \quad X_+ = (X_2, X_3, \ldots) \in \mathcal{H}_{\mathbf{R}},$$

where pr_k denotes the k-th projection on $\mathbf{R}$, notations of (art. 3.15). The stereographic maps are injective and their inverses are given by

$$F_+^{-1} : \mathcal{S}_{\mathbf{R}} - \{\mathbf{e}_1\} \to \mathcal{H}_{\mathbf{R}} \quad \text{with} \quad F_+^{-1} : X \mapsto \xi = \frac{X_+}{1-X_1},$$

$$F_-^{-1} : \mathcal{S}_{\mathbf{R}} - \{-\mathbf{e}_1\} \to \mathcal{H}_{\mathbf{R}} \quad \text{with} \quad F_-^{-1} : X \mapsto \xi = \frac{X_+}{1+X_1}.$$

Now, the stereographic maps are local diffeomorphisms. Since their images cover the whole sphere $\mathcal{S}_{\mathbf{R}}$, the sphere $\mathcal{S}_{\mathbf{R}}$ is a diffeological manifold modeled on $\mathcal{H}_{\mathbf{R}}$ (art. 4.8). Moreover, the sphere $\mathcal{S}_{\mathbf{R}}$ is embedded in $\mathcal{H}_{\mathbf{R}}$ (art. 2.14).

PROOF. Let us show that the stereographic maps are local diffeomorphisms from $\mathcal{H}_{\mathbf{R}}$ to $\mathcal{S}_{\mathbf{R}}$. We shall just consider F_+, since the case F_- is completely analogous. The fact that the map F_+ is injective is a simple verification. Its domain is $\mathcal{H}_{\mathbf{R}}$ which is, of course, D-open in $\mathcal{H}_{\mathbf{R}}$. We shall prove that $\mathcal{S}_{\mathbf{R}} - \{\mathbf{e}_1\}$, the image of F_+, is D-open in $\mathcal{S}_{\mathbf{R}}$. Then, we shall prove that F_+ is smooth as well as F_+^{-1}, defined on $\mathcal{S} - \{\mathbf{e}_1\}$, equipped with the subset diffeology. Finally, applying the criterion (art. 2.10), it will follow that F_+ is a local diffeomorphism.

a) *The map F_+ is injective.* We already exhibited F_+^{-1}.

b) *The map F_+ is smooth.* Let us consider a parametrization $P : U \to \mathcal{H}_{\mathbf{R}}$. For every $r_0 \in U$ there exist an open neighborhood V of r_0 in U and a local family $(\lambda_\alpha, X_\alpha)_{\alpha \in A}$, defined on V, such that

$$P \upharpoonright V : r \mapsto \sum_{\alpha \in A} \lambda_\alpha(r) X_\alpha.$$

Thus,

$$F_+ \circ (P \upharpoonright V) : r \mapsto \begin{pmatrix} \epsilon(r) \\ \sum_{\alpha \in A} \mu_\alpha(r) X_\alpha \end{pmatrix}$$

with

$$\epsilon(r) = \frac{\|\sum_{\alpha \in A} \lambda_\alpha(r) X_\alpha\|^2 - 1}{\|\sum_{\alpha \in A} \lambda_\alpha(r) X_\alpha\|^2 + 1} \quad \text{and} \quad \mu_\alpha(r) = \frac{2\lambda_\alpha(r)}{\|\sum_{\beta \in A} \lambda_\beta(r) X_\beta\|^2 + 1}.$$

The denominator of ϵ and μ_α never vanishes. Hence, the functions ϵ and μ_α belong to $\mathcal{C}^\infty(V, \mathbf{R})$. Now $F_+ \circ (P \upharpoonright V)$ rewrites

$$F_+ \circ (P \upharpoonright V)(r) = \epsilon(r)\begin{pmatrix} 1 \\ 0 \end{pmatrix} + \sum_{\alpha \in A} \mu_\alpha(r) \begin{pmatrix} 0 \\ X_\alpha \end{pmatrix}.$$

This exhibits the map $F_+ \circ (P \upharpoonright V)$ as a finite linear combination of vectors of $\mathcal{H}_{\mathbf{R}}$ with smooth parametrizations of $\mathbf{R}$ as coefficients. Therefore, $F_+ \circ P$ is a plot of $\mathcal{H}_{\mathbf{R}}$. Since $F_+ \circ P$ takes its values in $\mathcal{S}_{\mathbf{R}}$, it is a plot of the subset diffeology of $\mathcal{S}_{\mathbf{R}} \subset \mathcal{H}_{\mathbf{R}}$, where $\mathcal{H}_{\mathbf{R}}$ is equipped with the fine diffeology. Hence, F_+ is smooth.

c) *The map* F_+^{-1} *is smooth.* Let $P : U \to \mathcal{S}_{\mathbf{R}} - \{\mathbf{e}_1\}$ be a plot. For any $r_0 \in U$, there exist an open neighborhood V of r_0 in U and a local family $(\lambda_\alpha, X_\alpha)_{\alpha \in A}$, defined on V, such that

$$P \upharpoonright V : r \mapsto \sum_{\alpha \in A} \lambda_\alpha(r) X_\alpha,$$

then,

$$F_+^{-1} \circ (P \upharpoonright V) : r \mapsto \sum_{\alpha \in A} \mu_\alpha(r) X_{\alpha,+} \quad \text{with} \quad \mu_\alpha(r) = \frac{\lambda_\alpha(r)}{1 - \sum_{\beta \in A} \lambda_\beta(r) X_{\beta,1}}.$$

Now, the $X_{\beta,1}$ form a finite set of constant numbers, thus the parametrization $r \mapsto \sum_{\beta \in A} \lambda_\beta(r) X_{\beta,1}$ is smooth and is never equal to 1 since P takes its values in $\mathcal{S}_{\mathbf{R}} - \{\mathbf{e}_1\}$. Thus, for each $\alpha \in A$, $\mu_\alpha(r)$ is a smooth parametrization in $\mathbf{R}$. The parametrization $F_+^{-1} \circ (P \upharpoonright V)$ is clearly a finite linear combination of vectors of $\mathcal{H}_{\mathbf{R}}$, with smooth parametrizations of $\mathbf{R}$ as coefficients. Hence, $F_+^{-1} \circ (P \upharpoonright V)$ is a plot for the fine diffeology of $\mathcal{H}_{\mathbf{R}}$. Now, $F_+^{-1} \circ P$ is locally, at each point of U, a plot of $\mathcal{H}_{\mathbf{R}}$, so it is a plot of $\mathcal{H}_{\mathbf{R}}$. Therefore, F_+^{-1} is a smooth map from $\mathcal{S}_{\mathbf{R}} - \{\mathbf{e}_1\}$ to $\mathcal{H}_{\mathbf{R}}$.

d) *The subset* $\mathcal{S}_{\mathbf{R}} - \{\mathbf{e}_1\}$ *is open for the D-topology.* Let us recall that a set is D-open if and only if its preimage by every plot is open (art. 2.8). Let $P : U \to \mathcal{S}_{\mathbf{R}}$ be a plot, for every $r_0 \in U$ there exist an open neighborhood V of r_0 in U and a local family $(\lambda_\alpha, X_\alpha)_{\alpha \in A}$, defined on V, such that

$$P \upharpoonright V : r \mapsto \sum_{\alpha \in A} \lambda_\alpha(r) X_\alpha.$$

Hence,

$$(P \upharpoonright V)^{-1}(\mathcal{S}_{\mathbf{R}} - \{\mathbf{e}_1\}) = \{r \in V \mid \sum_{\alpha \in A} \lambda_\alpha(r) X_{\alpha,1} \neq 1\}.$$

But the real parametrization $P_V : r \mapsto \sum_{\alpha \in A} \lambda_\alpha(r) X_{\alpha,1}$ is smooth, *a fortiori* continuous. So, the preimage of $\mathbf{R} - \{1\}$ is open. Thus, $P^{-1}(\mathcal{S}_{\mathbf{R}} - \{\mathbf{e}_1\})$ is a union of subdomains of U, and therefore a domain. Hence, $P^{-1}(\mathcal{S}_{\mathbf{R}} - \{\mathbf{e}_1\})$ is open for any plot P, that is, $\mathcal{S}_{\mathbf{R}} - \{\mathbf{e}_1\}$ is D-open.

In conclusion, the diffeology of the real infinite sphere $\mathcal{S}_{\mathbf{R}}$ is generated by F_+ and F_-. Therefore, $\mathcal{S}_{\mathbf{R}}$ is a diffeological manifold, modeled on $\mathcal{H}_{\mathbf{R}}$ (art. 4.8). Now, let us prove that $\mathcal{S}_{\mathbf{R}}$ is embedded in $\mathcal{H}_{\mathbf{R}}$ (art. 2.14). Let Ω be D-open in $\mathcal{S}_{\mathbf{R}}$. Let us consider the projection

$$\mathrm{pr}_{\mathcal{S}} : \mathcal{H}_{\mathbf{R}} - \{0\} \to \mathcal{S}_{\mathbf{R}} \quad \text{defined by} \quad \mathrm{pr}_{\mathcal{S}}(X) = \frac{X}{\|X\|}.$$

Since $\|X\|$ never vanishes on $\mathcal{H}_{\mathbf{R}} - \{0\}$, the projection $\mathrm{pr}_{\mathcal{S}}$ is smooth, thus D-continuous (art. 2.9). Hence, the set

$$\tilde{\Omega} = \mathrm{pr}_{\mathcal{S}}^{-1}(\Omega) = \left\{ X \in \mathcal{H} - \{0\} \;\middle|\; \frac{X}{\|X\|} \in \Omega \right\}$$

is D-open in $\mathcal{H}_{\mathbf{R}} - \{0\}$. But $\mathcal{H}_{\mathbf{R}} - \{0\}$ is itself D-open in $\mathcal{H}_{\mathbf{R}}$, since $\mathcal{H}_{\mathbf{R}} - \{0\}$ is the pullback of the domain $]0, \infty[$ by the smooth map $\|\cdot\|$. Hence, $\tilde{\Omega}$ is D-open in $\mathcal{H}_{\mathbf{R}}$. Now, $\Omega = \tilde{\Omega} \cap \mathcal{S}_{\mathbf{R}}$. Therefore, $\mathcal{S}_{\mathbf{R}}$ is embedded. ☐

4.10. The infinite sphere is contractible. The infinite Hilbert sphere $\mathcal{S}_{\mathbf{R}}$, defined in (art. 4.9), is smoothly contractible, that is, there exists

$$\rho \in \mathcal{C}^\infty(\mathbf{R}, \mathcal{C}^\infty(\mathcal{S}_{\mathbf{R}})) \quad \text{such that} \quad \rho(0) = \hat{\mathbf{e}}_1 \quad \text{and} \quad \rho(1) = \mathbf{1}_{\mathcal{S}},$$

where $\hat{\mathbf{e}}_1$ is the constant map $X \mapsto \mathbf{e}_1$, with $\mathbf{e}_1 = (1, 0, 0, \ldots)$, and $\mathbf{1}_{\mathcal{S}}$ is the identity of $\mathcal{S}_{\mathbf{R}}$. The *path* ρ, connecting the identity to the constant map, is called a *smooth retraction* (art. 5.13). Under ρ, the image $\rho(t)(\mathcal{S}_{\mathbf{R}})$ retracts smoothly, when t passes from 1 to 0, from the sphere $\mathcal{S}_{\mathbf{R}}$ to the point $\mathbf{e}_1$.

NOTE. This strange property, for a sphere, has been proved for the topological structure of $\mathcal{S}_{\mathbf{R}}$ by S. Kakutani in 1943 [**Kak43**], and we see that it still persists for the fine diffeology of the sphere.

PROOF. The proof of this proposition uses the following two preliminary constructions:

1. The *shift operator*. It is defined as the linear map

$$\text{Shift} : \mathcal{H}_{\mathbf{R}} \to \mathcal{H}_{\mathbf{R}}, \quad \text{with} \quad \text{Shift}(X) = (0, X) = (0, X_1, X_2, \ldots).$$

Since the shift operator is linear, it is smooth for the fine diffeology (art. 3.9). It is injective and preserves the scalar product. It injects strictly the infinite sphere into an equator.

2. *Connecting points.* Let X be some diffeological space, we say that a path γ connects x_0 and x_1 if $\gamma(0) = x_0$ and $\gamma(1) = x_1$. We also say that x_0 and x_1 are *homotopic*, and that γ is a *homotopy* between x_0 and x_1 (art. 5.6). Now, if there exist a smooth path γ connecting x_0 to x_1 and a smooth path γ' connecting x_1 to x_2, then the path γ'' defined by

$$\gamma'' = \begin{cases} \gamma(\lambda(2t)) & \text{if} \quad t \leq 1/2, \\ \gamma'(\lambda(2t-1)) & \text{if} \quad t \geq 1/2, \end{cases}$$

connects x_0 to x_2, where λ is the smashing function defined in (art. 5.5), and described by Figure 5.1.

Now, we prove the contractibility of the infinite sphere in two steps, first we shall show that the constant map $\hat{\mathbf{e}}_1$ is homotopic to the shift operator, and then, that the shift operator is homotopic to the identity $\mathbf{1}_{\mathcal{S}}$. If we want, we can apply the smashing function to this pair of homotopies to get a smooth path connecting the constant map to the identity.

a) *Homotopy between* $\hat{\mathbf{e}}_1$ *and* Shift. Let us consider the 1-parameter family of deformations,

$$\rho_t(X) = \cos\left(\frac{\pi t}{2}\right) \mathbf{e}_1 + \sin\left(\frac{\pi t}{2}\right) \text{Shift}(X), \quad \text{for all } t \in \mathbf{R}, \text{ and all } X \in \mathcal{S}_{\mathbf{R}}.$$

For all $t \in \mathbf{R}$, $\rho_t(X) \in \mathcal{S}_{\mathbf{R}}$. Since addition and multiplication by a smooth function are smooth, the map $(t, X) \mapsto \rho_t(X)$ is smooth. Thus, the map $t \mapsto \rho_t$ is a path in $\mathcal{C}^\infty(\mathcal{S}_{\mathbf{R}})$ connecting $\hat{\mathbf{e}}_1$ and Shift, precisely,

$$\rho_0 = \hat{\mathbf{e}}_1 \quad \text{and} \quad \rho_1 = \text{Shift}.$$

b) *Homotopy between* Shift *and* $\mathbf{1}_{\mathcal{S}}$. Let us consider the following 1-parameter family of deformations,

$$\sigma_t(X) = tX + (1-t)\text{Shift}(X), \quad \text{for all } t \in \mathbf{R}, \text{ and all } X \in \mathcal{H}_{\mathbf{R}}.$$

Note that $\ker(\sigma_t) = 0$. This is clear for $t = 0$, and for nonzero t it follows inductively by observing that the condition $\sigma_t(X) = 0$ writes

$$(X_1, X_2, X_3, \ldots) = \frac{t-1}{t}(0, X_1, X_2, \ldots).$$

In particular, σ_t is nowhere zero on the sphere, we can define $\rho_t : \mathcal{S}_{\mathbf{R}} \to \mathcal{S}_{\mathbf{R}}$

$$\rho_t(X) = \frac{\sigma_t(X)}{\|\sigma_t(X)\|}.$$

Let us check that $(t, X) \mapsto \rho_t(X)$ is smooth. First of all, $(t, X) \mapsto \sigma_t(X)$ is clearly smooth. Since the scalar product is smooth, it follows that $(t, X) \mapsto \|\sigma_t(X)\|^2$ is smooth, and because this map takes its values in $]0, \infty[$, its square root is smooth. In conclusion, $t \mapsto \rho_t$ is a path in $\mathcal{C}^\infty(\mathcal{S}_{\mathbf{R}})$, and

$$\rho_0 = \text{Shift} \quad \text{and} \quad \rho_1 = \mathbf{1}_{\mathcal{S}}.$$

We proved, with a) and b) above, that $\mathcal{S}_{\mathbf{R}}$ is contractible. □

4.11. The infinite complex projective space. Let us recall some set-theoretic constructions, today classic. Let us introduce

$$\mathbf{C}^\star = \mathbf{C} - \{0\} \quad \text{and} \quad \mathcal{H}_{\mathbf{C}}^\star = \mathcal{H}_{\mathbf{C}} - \{0\},$$

where the Hilbert space $\mathcal{H}_{\mathbf{C}}$ has been described in (art. 3.15). Then, let us consider the multiplicative action of the group $\mathbf{C}^\star$ on $\mathcal{H}_{\mathbf{C}}^\star$, defined by

$$(z, Z) \mapsto zZ \in \mathcal{H}_{\mathbf{C}}^\star, \quad \text{for all } (z, Z) \in \mathbf{C}^\star \times \mathcal{H}_{\mathbf{C}}^\star.$$

The quotient of $\mathcal{H}_{\mathbf{C}}^\star$ by this action of $\mathbf{C}^\star$ is called the *infinite complex projective space*, or simply the *infinite projective space*. We will denote it by

$$\mathcal{P}_{\mathbf{C}} = \mathcal{H}_{\mathbf{C}}^\star / \mathbf{C}^\star.$$

Now, let us consider the space $\mathcal{H}_{\mathbf{C}}$ equipped with the fine diffeology (art. 3.7), the space $\mathcal{H}_{\mathbf{C}}^\star$ equipped with the subset diffeology (art. 1.33) and the infinite projective space $\mathcal{P}_{\mathbf{C}}$ equipped with the quotient diffeology (art. 1.50). Let us denote by $\pi : \mathcal{H}_{\mathbf{C}}^\star \to \mathcal{P}_{\mathbf{C}}$ the canonical projection. Next, for every $k = 1, \ldots, \infty$, let us define the injection $j_k : \mathcal{H}_{\mathbf{C}} \to \mathcal{H}_{\mathbf{C}}^\star$ by

$$j_1(Z) = (1, Z) \quad \text{and} \quad j_k(Z) = (Z_1, \ldots, Z_{k-1}, 1, Z_k, \ldots), \text{ for } k > 1.$$

Let the maps F_k be defined by

$$F_k : \mathcal{H}_{\mathbf{C}} \to \mathcal{P}_{\mathbf{C}} \quad \text{with} \quad F_k = \pi \circ j_k, \quad k = 1, \ldots, \infty.$$

1. For every $k = 1, \ldots, \infty$, j_k is an induction from $\mathcal{H}_{\mathbf{C}}$ into $\mathcal{H}_{\mathbf{C}}^\star$.

2. For every $k = 1, \ldots, \infty$, F_k is a local diffeomorphism from $\mathcal{H}_{\mathbf{C}}$ to $\mathcal{P}_{\mathbf{C}}$. Moreover, their values cover $\mathcal{P}_{\mathbf{C}}$,

$$\bigcup_{k=1}^{\infty} \text{val}(F_k) = \mathcal{P}_{\mathbf{C}}.$$

Thus, $\mathcal{P}_{\mathbf{C}}$ is a diffeological manifold modeled on $\mathcal{H}_{\mathbf{C}}$, for which the family $\{F_k\}_{k=1}^{\infty}$ is an atlas (art. 4.8).

3. The pullback $\pi^{-1}(\text{val}(F_k)) \subset \mathcal{H}_{\mathbf{C}}^\star$ is isomorphic to the product $\mathcal{H}_{\mathbf{C}} \times \mathbf{C}^\star$, where the action of $\mathbf{C}^\star$ on $\mathcal{H}_{\mathbf{C}}^\star$ is transmuted into the trivial action on the factor $\mathcal{H}_{\mathbf{C}}$ and the multiplicative action on the factor $\mathbf{C}^\star$. We say that the projection π is a *locally trivial* $\mathbf{C}^\star$*-principal fibration*; see (art. 8.8) and (art. 8.11).

PROOF. 1. Let us prove first that the maps j_k are inductions from $\mathcal{H}_{\mathbf{C}}$ to $\mathcal{H}^\star_{\mathbf{C}}$. Let $\mathcal{H}_k = j_k(\mathcal{H}_{\mathbf{C}})$, that is,

$$\mathcal{H}_k = \{Z \in \mathcal{H} \mid Z_k = 1\}.$$

Now, let us consider a plot P of $\mathcal{H}_{\mathbf{C}}$ with values in $\mathcal{H}_k$. By definition of the fine diffeology, we have

$$P(r) =_{\text{loc}} \sum_{\alpha \in A} \lambda_\alpha(r) Z_\alpha \text{ and } P_k(r) =_{\text{loc}} \sum_{\alpha \in A} \lambda_\alpha(r) Z_{\alpha,k} = 1, \text{ with } P_k = \text{pr}_k \circ P.$$

The pr_k are the k-projections from $\mathcal{H}_{\mathbf{C}}$ to $\mathbf{C}$, notations of (art. 3.15). Let us now define ζ_α by

$$\zeta_\alpha = (Z_{\alpha,1}, \ldots, Z_{\alpha,k-1}, 1, Z_{\alpha,k+1}, \ldots).$$

For each α in A, ζ_α belongs to $\mathcal{H}_k$. Let $\mathbf{e}_k$ be defined by $\text{pr}_k(\mathbf{e}_k) = 1$ and $\text{pr}_j(\mathbf{e}_k) = 0$ for $j \neq k$, notations of (art. 3.15). From the condition above, we have

$$\begin{aligned} P(r) &=_{\text{loc}} \sum_{\alpha \in A} \lambda_\alpha(r)\zeta_\alpha - \sum_{\alpha \in A} \lambda_\alpha(r)\mathbf{e}_k + \sum_{\alpha \in A} \lambda_\alpha(r) Z_{\alpha,k}\mathbf{e}_k \\ &=_{\text{loc}} \sum_{\alpha \in A} \lambda_\alpha(r)\zeta_\alpha + \Big(1 - \sum_{\alpha \in A} \lambda_\alpha(r)\Big)\mathbf{e}_k. \end{aligned}$$

Now, since the vectors ζ_α and $\mathbf{e}_k$ belong to $\mathcal{H}_k$, the plot $j_k^{-1} \circ P$ writes

$$j_k^{-1} \circ P(r) =_{\text{loc}} \sum_{\alpha \in A} \lambda_\alpha(r) j^{-1}(\zeta_\alpha) + \Big(1 - \sum_{\alpha \in A} \lambda_\alpha(r)\Big) j^{-1}(\mathbf{e}_k),$$

but $j_k(0) = \mathbf{e}_k$ implies $j_k^{-1}(\mathbf{e}_k) = 0$, hence

$$j_k^{-1} \circ P(r) =_{\text{loc}} \sum_{\alpha \in A} \lambda_\alpha(r) j^{-1}(\zeta_\alpha).$$

This exhibits the parametrization $j_k^{-1} \circ P$ as a plot of $\mathcal{H}$, hence j_k is an induction.

2. Let us prove now that the F_k are local diffeomorphisms.

a) F_k *is injective.* First of all, the maps F_k are clearly smooth. Now, since $Z \in \mathcal{H}_k$ implies $Z_k = 1$, the space $\mathcal{H}_k$ intersects the orbits of the group $\mathbf{C}^\star$ in at most one point. The orbits which do not intersect $\mathcal{H}_k$ are those such that $Z_k = 0$. Hence, F_k is injective.

b) *The images of the* F_k *cover* $\mathcal{P}_{\mathbf{C}}$. The orbit of any point $Z \in \mathcal{H}^\star_{\mathbf{C}}$ intersects some $\mathcal{H}_k$, in other words,

$$\bigcup_{k=1}^{\infty} \mathbf{C}^\star \mathcal{H}_k = \mathcal{H}^\star_{\mathbf{C}} \quad \text{or} \quad \pi\Big(\bigcup_{k=1}^{\infty} \mathcal{H}_k\Big) = \mathcal{P} \quad \text{or} \quad \bigcup_{k=1}^{\infty} \text{val}(F_k) = \mathcal{P}_{\mathbf{C}}.$$

c) *The* F_k *are inductions.* Let $Q : U \to \mathcal{P}_{\mathbf{C}}$ be a plot with values in $F_k(\mathcal{H}_{\mathbf{C}})$, and let $r_0 \in U$. By definition of the quotient diffeology, there exist an open neighborhood V of r_0 and a plot $P : V \to \mathcal{H}^\star_{\mathbf{C}}$ such that $Q \upharpoonright V = \pi \circ P$. By hypothesis, for each $r \in V$, $P_k(r) \neq 0$, where $P_k = \text{pr}_k \circ P$, therefore $P' : V \to \mathcal{H}^\star_{\mathbf{C}}$, defined by $P'(r) = P(r)/P_k(r)$, takes its values in $\mathcal{H}_k$. Since P_k is smooth, P' is smooth and $Q \upharpoonright V = \pi \circ P'$. The plot P' takes its values in $\mathcal{H}_k$, and j_k is an induction, thus the composite $j_k^{-1} \circ P'$ is a plot of $\mathcal{H}_{\mathbf{C}}$. But, by construction, $j_k^{-1} \circ P' = F_k^{-1} \circ Q$, thus $F_k^{-1} \circ Q$ is a plot of $\mathcal{H}_{\mathbf{C}}$ and F_k^{-1} is smooth. Therefore, F_k is an induction.

d) *The image of each* F_k *is D-open.* Since the D-topology of the quotient diffeology is the quotient topology of the D-topology (art. 2.12), it is sufficient to prove that the pullback by π of the $F_k(\mathcal{H}_{\mathbf{C}})$ is D-open in $\mathcal{H}_{\mathbf{C}}^\star$. We saw that $\pi^{-1}(F_k(\mathcal{H}_{\mathbf{C}}))$ is the set of all $Z \in \mathcal{H}_{\mathbf{C}}$ such that $Z_k \neq 0$, *i.e.*, $\mathrm{pr}_k^{-1}(\mathbf{C}^\star)$. But pr_k is linear, hence smooth, hence continuous. Since $\mathbf{C}^\star$ is open it follows that $\pi^{-1}(F_k(\mathcal{H}))$ is open.

Thus, by application of (art. 2.9), we proved that the F_k are local diffeomorphisms. Since their images cover $\mathcal{P}_{\mathbf{C}}$, the space $\mathcal{P}_{\mathbf{C}}$ is a diffeological manifold modeled on $\mathcal{H}_{\mathbf{C}}$ (art. 4.8).

3. Let us prove now that π is a locally trivial $\mathbf{C}^\star$-principal fibration. Let

$$\Phi_k : \mathcal{H}_{\mathbf{C}} \times \mathbf{C}^\star \to \mathcal{H}_{\mathbf{C}}^\star \quad \text{defined by} \quad \Phi_k(Z, z) = z j_k(Z).$$

a) The Φ_k are local diffeomorphisms. Indeed, let $\Phi_k(Z, z) = \Phi_k(Z', z')$, that is, $z j_k(Z) = z' j_k(Z')$. So, $\mathrm{pr}_k(z j_k(Z)) = \mathrm{pr}_k(z' j_k(Z'))$, that is, $z = z'$. Thus, $Z = Z'$. So, the Φ_k are injective. They are obviously smooth. Their inverses are

$$\Phi_k^{-1}(Z) = \left(\frac{Z}{Z_k}, Z_k\right), \quad \text{for all } Z \in \mathrm{val}(\Phi_k) = \mathbf{C}^\star \mathcal{H}_k.$$

Since Z_k never vanishes, Φ_k^{-1} also is smooth. Moreover $\mathrm{val}(\Phi_k)$ is D-open since it is the pullback $\pi^{-1}(\mathrm{val}(F_k))$, and we have seen that $\mathrm{val}(F_k)$ is D-open. Therefore, the Φ_k are local diffeomorphisms.

b) The Φ_k commute obviously with the action of $\mathbf{C}^\star$ defined above. Therefore, $\{\Phi_k\}_{k=1}^\infty$ is a family of local diffeomorphisms from $\mathcal{H}_{\mathbf{C}} \times \mathbf{C}^\star$ to $\mathcal{H}_{\mathbf{C}}^\star$ such that

$$\Phi_k(z'(Z, z)) = z' \Phi_k(Z, z) \quad \text{and} \quad \bigcup_{k=1}^{\infty} \mathrm{val}(\Phi_k) = \mathcal{H}_{\mathbf{C}}^\star,$$

where $z'(Z, z) = (Z, z'z)$ is just the action of $\mathbf{C}^\star$ on the second factor. This atlas gives the structure of a locally trivial $\mathbf{C}^\star$-principal bundle to the triple $(\mathcal{H}_{\mathbf{C}}^\star, \mathcal{P}_{\mathbf{C}}, \pi)$ for the given action of $\mathbf{C}^\star$ (art. 8.9, Note 2). ☐

Exercises

✎ EXERCISE 76 (The space of lines in $\mathcal{C}^\infty(\mathbf{R}, \mathbf{R})$). Consider $\mathcal{E} = \mathcal{C}^\infty(\mathbf{R}, \mathbf{R})$, equipped with the functional diffeology. A line of $\mathcal{E}$ is a subset $L \subset \mathcal{E}$ for which there exist $f, g \in \mathcal{E}$, with $g \neq 0$, such that $L = \{f + sg \mid s \in \mathbf{R}\}$. The space of lines, denoted by $\mathrm{Lines}(\mathcal{E})$, will be regarded as the quotient of $\mathcal{E} \times (\mathcal{E} - \{0\})$ by the equivalence relation $(f, g) \sim (f + \lambda g, \mu g)$ where $\lambda \in \mathbf{R}$ and $\mu \in \mathbf{R} - \{0\}$. For all $r \in \mathbf{R}$, let

$$\mathcal{E}_r^0 = \{\alpha \in \mathcal{C}^\infty(\mathbf{R}, \mathbf{R}) \mid \alpha(r) = 0\} \quad \text{and} \quad \mathcal{E}_r^1 = \{\beta \in \mathcal{C}^\infty(\mathbf{R}, \mathbf{R}) \mid \beta(r) = 1\}.$$

Let $F_r : \mathcal{E}_r^0 \times \mathcal{E}_r^1 \to \mathrm{Lines}(\mathcal{E})$ be defined by

$$F_r : (\alpha, \beta) \mapsto \{\alpha + s\beta \mid s \in \mathbf{R}\}.$$

1) Describe the image of F_r and show that F_r is a local diffeomorphism.

2) Show that the images of the F_r, when $r \in \mathbf{R}$, cover $\mathrm{Lines}(\mathcal{E})$. Deduce that $\mathrm{Lines}(\mathcal{E})$ is a manifold modeled on $\mathcal{E}_0^0 \times \mathcal{E}_0^0$.

3) Adapt this construction to $\mathrm{Lines}(\mathbf{R}^2)$, where $\mathbf{R}^2$ is regarded as $\mathcal{C}^\infty(\{1, 2\}, \mathbf{R})$. Deduce that this gives an atlas for the Möbius strip; Exercise 41, p. 39, question 4.

✎ EXERCISE 77 (The Hopf S^1-bundle). Check that the quotient of the unit sphere $\mathcal{S}_{\mathbf{C}} = \{Z \in \mathcal{H}_{\mathbf{C}} \mid Z \cdot Z = 1\}$ by the multiplicative action of the group $U(1)$ of complex numbers with modulus 1,

$$U(1) = \{z \in \mathbf{C} \mid z^* z = 1\},$$

is naturally diffeomorphic to $\mathcal{P}_{\mathbf{C}}$. Transpose this identification to the product $\mathcal{H}_{\mathbf{R}} \times \mathcal{H}_{\mathbf{R}} \simeq \mathcal{H}_{\mathbf{C}}$; see Exercise 73, p. 75.

✎ EXERCISE 78 ($U(1)$ as subgroup of diffeomorphisms). Let $\mathrm{GL}(\mathcal{H}_{\mathbf{C}})$ be the group of $\mathbf{C}$-linear isomorphisms of $\mathcal{H}_{\mathbf{C}}$. Let $U(\mathcal{H}_{\mathbf{C}})$ be the unitary group of $\mathcal{H}_{\mathbf{C}}$, that is, the subgroup of the elements of $\mathrm{GL}(\mathcal{H}_{\mathbf{C}})$ which preserves the Hermitian form,

$$U(\mathcal{H}_{\mathbf{C}}) = \{A \in \mathrm{GL}(\mathcal{H}_{\mathbf{C}}) \mid \text{ for all } Z, Z' \in \mathcal{H}_{\mathbf{C}},\ (AZ) \cdot (AZ') = Z \cdot Z'\}.$$

The group $U(\mathcal{H}_{\mathbf{C}})$ is equipped with the functional diffeology associated with the fine diffeology of $\mathcal{H}_{\mathbf{C}}$ (art. 3.12). Let $U(1)$ be the group of complex numbers with modulus 1; see Exercise 77, p. 93. Let $j : U(1) \to \mathrm{GL}(\mathcal{H}_{\mathbf{C}})$ be the map

$$j(z) : Z \mapsto z \times Z, \text{ with } Z \in \mathcal{H}_{\mathbf{C}}.$$

Show that j is a monomorphic induction from $U(1)$ to $\mathrm{GL}(\mathcal{H}_{\mathbf{C}})$, and thus a monomorphic induction into $U(\mathcal{H}_{\mathbf{C}})$.

Modeling Diffeologies

We have seen, in this chapter, how diffeology offers a new approach for usual objects in classical differential geometry. Manifolds are not regarded anymore as a special class of structure, needing a lot of preparatory material, but as diffeological spaces which satisfy a given property: they are modeled *on finite dimensional real vector spaces. The notion of local diffeomorphism and generating family in diffeology simplify their description. Thanks to this point of view, we have been able to generalize the notion of classical manifolds to diffeological manifolds, simply by changing the* model, *from finite dimensional real vector spaces to general diffeological vector spaces. We have also been able to begin the exploration of this way with two important examples: the Hilbert sphere* (art. 4.9) *and the projective space of the Hilbert space* (art. 4.11).

Modeling diffeological spaces gives us a simple mechanism to understand and present a variety of geometrical objects, which we shall illustrate with two more examples. The first one, manifold with boundary, involves usually a not so simple definition (art. 4.15). *Thanks to the natural subset diffeology of* half-spaces (art. 4.12), *a manifold with boundary becomes simply a diffeological space modeled on a half-space* (art. 4.16); *we include also manifolds with boundary and corners.*

The second example is about orbifolds. *Since the first and original Satake definition* [**Sat56**], [**Sat57**] *they became ordinary objects in mathematics.*[2] *But the concept remained problematic, and different authors give different definitions. Roughly speaking, an orbifold is a topological space which looks like, locally, a quotient* $\mathbf{R}^n/\Gamma$, *where* Γ *is a finite subgroup of linear transformations. Apart from the variety of definitions, what is strange is that orbifolds, or V-manifolds, come to us alone, not included in a specific category, as is usual for mathematical objects. In one of his papers, Satake even quoted "*The notion of $\mathcal{C}^\infty$-map thus defined is inconvenient in

[2]Originally, *orbifolds* were introduced by Ichiro Satake as V-*manifolds*. Later in a series of lectures on the subject, Thurston changed the name from V-*manifolds* to *orbifolds*.

the point that a composite of two $\mathcal{C}^\infty$-maps defined in a different choice of defining families is not always $\mathcal{C}^\infty$-map." *These remarks led us to introduce orbifolds into the category of diffeological spaces, following on the path opened by manifolds* (art. 4.1)*: an orbifold is defined as a diffeological space which is locally diffeomorphic, at each point, to some quotient* $\mathbf{R}^n/\Gamma$, *where* Γ *is a finite subgroup of linear transformations. Then, orbifolds become just a subcategory of diffeological spaces. By this way, orbifolds take advantage of all the theory of diffeologies. In particular, smooth maps between orbifolds are well defined and compose correctly. Fiber bundles will be just a specialization of the notion of diffeological bundles,*[3] *differential forms, covering, etc. We state in* (art. 4.17) *and* (art. 4.18) *two main results without proofs; a complete comparison between this definition and the original Satake definition is published in* [**IKZ05**]. *We shall just read here the main diffeological definitions.*

4.12. Half-spaces. We denote by $\mathbf{H}_n$ the *standard half-space* of $\mathbf{R}^n$, that is, the set of points $x = (r, t) \in \mathbf{R}^n$ such that $r \in \mathbf{R}^{n-1}$ and $t \in [0, +\infty[$, and we denote by $\partial \mathbf{H}_n$ its boundary $\mathbf{R}^{n-1} \times \{0\}$. The subset diffeology of $\mathbf{H}_n$, inherited from $\mathbf{R}^n$, is made of all the smooth parametrizations $P : U \to \mathbf{R}^n$ such that $P_n(r) \geq 0$ for all $r \in U$, $P_n(r)$ being the n-th coordinate of $P(r)$. The D-topology of $\mathbf{H}_n$ is the usual topology defined by its inclusion into $\mathbf{R}^n$.

4.13. Smooth real maps from half-spaces. A map $f : \mathbf{H}_n \to \mathbf{R}^p$ is smooth for the subset diffeology of $\mathbf{H}_n$ if and only if there exists an ordinary smooth map F, defined on an open neighborhood of $\mathbf{H}_n$, such that $f = F \restriction \mathbf{H}_n$. Actually, there exists such an F defined on the whole $\mathbf{R}^n$.

NOTE. As an immediate corollary, any map f defined on $\mathcal{C} \times [0, \varepsilon[$ to $\mathbf{R}^p$, where $\mathcal{C}$ is an open cube of $\partial \mathbf{H}_n$, centered at some point $(r, 0)$, smooth for the subset diffeology, is the restriction of a smooth map $F : \mathcal{C} \times]-\varepsilon, +\varepsilon[\to \mathbf{R}^p$.

PROOF. First of all, if f is the restriction of a smooth map $F : \mathbf{R}^n \to \mathbf{R}^p$, it is obvious that for every smooth parametrization $P : U \to \mathbf{H}_n$, $f \circ P = F \circ P$ is smooth. Conversely, let f_i be a coordinate of f. Let $x = (r, t)$ denote a point of $\mathbf{R}^n$, where $r \in \mathbf{R}^{n-1}$ and $t \in \mathbf{R}$. If f_i is smooth for the subset diffeology, then $\phi_i : (r, t) \mapsto f_i(r, t^2)$, defined on $\mathbf{R}^n$, is smooth. Now, ϕ_i is even in the variable t, $\phi_i(r, t) = \phi_i(r, -t)$, thus, according to Hassler Whitney [**Whi43**, Theorem 1 and final remark] there exists a smooth map $F_i : \mathbf{R}^n \to \mathbf{R}$ such that $\phi_i(r, t) = F_i(r, t^2)$. Hence, $f_i(r, t) = F_i(r, t)$ for all $r \in \mathbf{R}^{n-1}$ and all $t \in [0, +\infty[$. ☐

4.14. Local diffeomorphisms of half-spaces. A map $f : A \to \mathbf{H}_n$, with $A \subset \mathbf{H}_n$, is a local diffeomorphism for the subset diffeology of $\mathbf{R}^n$ if and only if A is open in $\mathbf{H}_n$, f is injective, $f(A \cap \partial \mathbf{H}_n) \subset \partial \mathbf{H}_n$, and for all $x \in A$ there exist an open ball $B \subset \mathbf{R}^n$ centered at x and a local diffeomorphism $F : B \to \mathbf{R}^n$ such that f and F coincide on $B \cap \mathbf{H}_n$.

NOTE. This implies, in particular, that there exist an open neighborhood $\mathcal{U}$ of A and an étale application $g : \mathcal{U} \to \mathbf{R}^n$ such that f and g coincide on A.

---

[3]The notion of *orbibundle* does not seem to coincide with the notion of diffeological bundle (art. 8.8) over an orbifold. It looks like an intertwining of two orbifolds; see for example the definition in [**BoGa07**]. The notion of orbibundle deserves to be included properly later in the diffeological framework.

PROOF. Let us assume that f is a local diffeomorphism for the subset diffeology. Since f is a local diffeomorphism for the D-topology, f is a local homeomorphism, and A is open in $\mathbf{H}_n$. In particular, f maps the boundary $\partial A = A \cap \partial \mathbf{H}_n$ into $\partial \mathbf{H}_n$. As well, f maps the complementary $A - \partial A$ into $\mathbf{H}_n - \partial \mathbf{H}_n$. Now, since A is open in $\mathbf{H}_n$, $A - \partial A$ is open in $\mathbf{H}_n - \partial \mathbf{H}_n$, thus the restriction $f \restriction A - \partial A$ is a local diffeomorphism to $\mathbf{H}_n - \partial \mathbf{H}_n$. Therefore, for every $x \in A - \partial A$ there exists an open ball $B \subset A - \partial A$, centered at x, such that $F = f \restriction B$ is a local diffeomorphism.

Now, let $(r, 0) \in A$. Since A is open in $\mathbf{H}_n$, ∂A is open in $\partial \mathbf{H}_n$. Therefore, there exist an open cube $\mathcal{C} \subset \partial A$ centered at $(r, 0)$, and $\varepsilon > 0$, such that $\mathcal{C} \times [0, +\varepsilon[\subset A$. The restriction of f to $\mathcal{C} \times [0, +\varepsilon[$ is a local diffeomorphism, for the subset diffeology, to $\mathbf{H}_n$. Thanks to (art. 4.13), there exists a smooth map F defined on $\mathcal{C} \times]-\varepsilon, +\varepsilon[$ to $\mathbf{R}^n$, such that f and F coincide on $\mathcal{C} \times [0, +\varepsilon[$. Since f is a diffeomorphism, f maps $\mathcal{C} \times [0, +\varepsilon[$ to some open set $A' \subset \mathbf{H}_n$ and maps $\mathcal{C}$ to some open subset of $\partial \mathbf{H}_n$. We have then $(r', 0) = f(r, 0) \in \partial A' = A' \cap \partial \mathbf{H}_n$. Considering now f^{-1}, for the same reason, there exists an open cube $\mathcal{C}' \subset \partial A'$ centered at $(r', 0)$, there exist $\varepsilon' > 0$ such that $\mathcal{C}' \times [0, +\varepsilon'[\subset A'$, and a smooth map G defined on $\mathcal{C}' \times]-\varepsilon', +\varepsilon'[$ to $\mathbf{R}^n$ such that f^{-1} and G coincide on $\mathcal{C}' \times [0, +\varepsilon'[$. Now, let $\mathcal{O} = F^{-1}(\mathcal{C}' \times]-\varepsilon', +\varepsilon'[)$, and $\mathcal{O}' = G^{-1}(\mathcal{C} \times]-\varepsilon, +\varepsilon[)$, $\mathcal{O}$ and $\mathcal{O}'$ are open subsets of $\mathbf{R}^n$, with $(r, 0) \in \mathcal{O}$ and $(r', 0) \in \mathcal{O}'$. For every $t \geq 0$ such that $(r, t) \in \mathcal{O}$, we have $D(G \circ F)(r, t) = D(f^{-1} \circ f)(r, t) = \mathbf{1}_{n+1}$. Thus, since F and G are smooth parametrizations, we have on the one hand $\lim_{t \to 0^+} D(G \circ F)(r, t) = \mathbf{1}_{n+1}$, and on the other hand $\lim_{t \to 0^+} D(G \circ F)(r, t) = D(G)(r', 0) \circ D(F)(r, 0)$. So, $D(G)(r', 0) \circ D(F)(r, 0) = \mathbf{1}_{n+1}$, and thus $D(F)(r, 0)$ is nondegenerate. Therefore, thanks to the implicit function theorem, there exists an open ball B centered at $x = (r, 0)$ such that $F \restriction B$ is a local diffeomorphism to $\mathbf{R}^n$, and such that f and F coincide on $B \cap \mathbf{H}_n$.

Conversely, let us assume that A is open in $\mathbf{H}_n$, $f : A \to \mathbf{H}_n$ is injective, and for each $x \in A$ there exist an open ball B of $\mathbf{R}^n$ centered at x, and a local diffeomorphism $F : B \to \mathbf{R}^n$ such that f and F coincide on B. Let us prove that f is a local smooth map for the subset diffeology. Let $P : U \to \mathbf{R}^n$ be a smooth parametrization taking its values in $\mathbf{H}_n$. Since A is open in $\mathbf{H}_n$ and P is continuous, for the D-topology, $P^{-1}(A)$ is open. Now let $r \in P^{-1}(A)$ and $x = P(r)$. Since P is continuous, $P^{-1}(B)$ is open, and since F is a local diffeomorphism from B to $\mathbf{R}^n$, $f \circ P \restriction P^{-1}(B) = F \circ P \restriction P^{-1}(B)$ is a smooth parametrization. Thus $f \circ P$ is locally smooth at every point, thus P is smooth, and therefore f is smooth. For the smoothness of f^{-1}, we need only check that $f(A)$ is open, and the rest will follow the same way as for f. So, let $x \in A$. If $x \in A - \partial A$, the ball B can be chosen small enough to fit into $A - \partial A$. Now, by hypothesis ∂A is mapped into $\partial \mathbf{H}_n$ and f is injective, thus f maps the complementary $A - \partial A$ into $\mathbf{H}_n - \partial \mathbf{H}_n$. Then, since $f \restriction B$ is a local diffeomorphism to $\mathbf{R}^n$, $f(B) \subset f(A)$ is an open subset of $\mathbf{H}_n - \partial \mathbf{H}_n$. Next, let $x \in \partial A$, then $f(x) \in \partial \mathbf{H}_n$ and $f(x) \in f(B \cap \mathbf{H}_n) = F(B) \cap \mathbf{H}_n \subset f(A)$. Since F is a local diffeomorphism from B to $\mathbf{R}^n$, $F(B)$ is open in $\mathbf{R}^n$, and $F(B) \cap \mathbf{H}_n$ is an open subset of $\mathbf{H}_n$. Finally, for every $x \in A$, $f(x)$ is contained in some open subset $\mathcal{O}$ of $\mathbf{H}_n$ with $\mathcal{O} \subset f(A)$. Therefore $f(A)$ is a union of open subsets of $\mathbf{H}_n$, that is, an open subset of $\mathbf{H}_n$. ☐

4.15. Classical manifolds with boundary. Smooth manifolds with boundary have been precisely defined in [**GuPo74**]. We use here a more recent definition except that, for our subject, the charts have been inverted.

DEFINITION [**Lee06**]. A *smooth* n*-manifold with boundary* is a topological space M, together with a family of local homeomorphisms F_i defined on some open sets U_i of the half-space $\mathbf{H}_n$ to M, such that the values of the F_i cover M and, for any two elements F_i and F_j of the family, the transition homeomorphism $F_i^{-1} \circ F_j$, defined on $F_i^{-1}(F_i(U_i) \cap F_j(U_j))$ to $F_j^{-1}(F_i(U_i) \cap F_j(U_j))$, is the restriction of some smooth map defined on an open neighborhood of $F_i^{-1}(F_i(U_i) \cap F_j(U_j))$. The boundary ∂M is the union of the $F_i(U_i \cap \partial \mathbf{H}_n)$. Such a family $\mathcal{F}$ of homeomorphisms is called an atlas of M, and its elements are called *charts*. There exists a *maximal atlas* $\mathcal{A}$ containing $\mathcal{F}$, made of all the local homeomorphisms from $\mathbf{H}_n$ to M such that the transition homeomorphisms with every element of $\mathcal{F}$ satisfy the condition given just above. We say that $\mathcal{A}$ gives to M its *structure of manifold with boundary*.

4.16. Diffeology of manifolds with boundary. Let M be a smooth n-manifold with boundary. Let us recall that a parametrization $P : U \to M$ is smooth if for every $r \in U$ there exist an open neighborhood V of r, a chart $F : \Omega \to M$ of M, and a smooth parametrization, $Q : V \to \Omega$ such that $P \restriction V = F \circ Q$. The set of smooth parametrizations of M is then a natural diffeology. Regarded as a diffeological space, M is modeled on $\mathbf{H}_n$ (art. 4.12). Conversely, every diffeological space modeled on $\mathbf{H}_n$ is naturally a manifold with boundary. Moreover, the smooth maps form a manifold with boundary to another, for the category {Smooth Manifolds with Boundary} or for the category {Diffeology} coincide.

NOTE. Nothing now prevents us from defining directly the *manifolds with corners* as the diffeological spaces modeled on corners $\mathbf{K}_n = \{x = (x_1, \ldots, x_n) \in \mathbf{R}^n \mid x_i \geq 0, i = 1, \ldots, n\}$. Indeed, the following proposition, due to Jordan Watts,[4] shows already that $\mathcal{C}^\infty(\mathbf{K}_n, \mathbf{R}^m)$ is made of the restrictions to $\mathbf{K}_n$ of smooth maps defined on an open superset.

PROPOSITION. A map $f : \mathbf{K}_n \to \mathbf{R}$ is smooth for the subset diffeology of $\mathbf{K}_n$ if and only if there exist an open superset W of $\mathbf{K}_n$ and a smooth map $F : W \to \mathbf{R}$, such that $F \restriction \mathbf{K}_n = f$.

This result, adapted to local diffeomorphisms of $\mathbf{K}_n$ as has been done for half-spaces (art. 4.14), will certainly lead to a perfect correspondence between diffeological manifolds with corners and the usual notion that one can find in the literature.

PROOF. We denote by $\mathcal{D}$ the set of smooth parametrizations of M. Now, let us prove that any chart $F : U \to M$ is a local diffeomorphism from $\mathbf{H}_n$ to M, where $\mathbf{H}_n$ is equipped with the subset diffeology, and M is equipped with $\mathcal{D}$. Since for any plot P of $\mathbf{H}_n$, $F \circ P$ — defined on $P^{-1}(U)$ which is open — belongs obviously to $\mathcal{D}$, so F is smooth. Then, let us prove now that F^{-1} is smooth. Let $P : U \to M$ be an element of $\mathcal{D}$, let $r \in U$, let V, Q and F' be as above, such that $P \restriction V = F' \circ Q$. So, $F^{-1} \circ P \restriction V = F^{-1} \circ F' \circ Q$. But since F and F' are charts of M, $F^{-1} \circ F'$ is the restriction of some ordinary smooth map defined on some open neighborhood of $F^{-1}(F(U) \cap F'(U'))$, therefore $F^{-1} \circ F' \circ Q$ is a smooth parametrization in $\mathbf{H}_n$, that is, $F^{-1} \circ P \restriction V$. Now, since $P^{-1}(F(U))$ is open, and since the parametrization $P \circ F^{-1}$ is locally smooth everywhere on $P^{-1}(F(U))$, $F^{-1} \circ P$ is smooth. Thus, F^{-1} is a local smooth map. Therefore, F is a local diffeomorphism. This proves that, for the diffeology $\mathcal{D}$, M is modeled on $\mathbf{H}_n$. Conversely, let us assume that M is a diffeological space modeled on $\mathbf{H}_n$. Let $\mathcal{A}$ be the set of all local diffeomorphisms from $\mathbf{H}_n$ to M, and let us

[4] Private communication.

equip M with the D-topology. So, the elements of $\mathcal{A}$ are local homeomorphisms. Let $F : U \to M$ and $F' : U' \to M$ be two elements of $\mathcal{A}$. Let us assume that $F(U) \cap F'(U')$, which is open, is not empty. Thus $F^{-1}(F(U) \cap F'(U'))$ and $F'^{-1}(F(U) \cap F'(U'))$ are open. But $F'^{-1} \circ F \upharpoonright F^{-1}(F(U) \cap F'(U'))$ and $F^{-1} \circ F' \upharpoonright F'^{-1}(F(U) \cap F'(U'))$ are local diffeomorphisms for the subset diffeology of $\mathbf{H}_n$, and according to (art. 4.14) they are the restrictions of ordinary smooth maps. Therefore, the set $\mathcal{A}$ gives to M a structure of smooth manifold with boundary. It is clear that these two operations just described are inverse one of each other. Now, thanks to (art. 4.14) and (art. 4.13), it is clear that to be a smooth map for the category of smooth manifolds with boundary or to be smooth for the natural diffeology associated is identical.

Let us give now a proof of the proposition in the note. If $f : \mathbf{K}_n \to \mathbf{R}$ is the restriction of a smooth map $F : \mathbf{R}^n \to \mathbf{R}$, then it is clear that f is smooth. Conversely, let $f : \mathbf{K}_n \to \mathbf{R}$ be smooth. Since the map $(x_1, \ldots, x_n) \mapsto (x_1^2, \ldots, x_n^2)$ is smooth and takes its values in the corner $\mathbf{K}_n$, it is a smooth parametrization of $\mathbf{K}_n$, by definition of the subset diffeology. Then, by definition of $\mathcal{C}^\infty(\mathbf{K}_n, \mathbf{R})$, $\phi : (x_1, \ldots, x_n) \mapsto f(x_1^2, \ldots, x_n^2)$ is a smooth real function defined on $\mathbf{R}^n$. Now, ϕ_i is invariant by the action of the discrete group $\{-1, +1\}^n$, acting on $\mathbf{R}^n$ by $(\varepsilon_1, \ldots, \varepsilon_n) \cdot (x_1, \ldots, x_n) = (\varepsilon_1 x_1, \ldots, \varepsilon_n x_n)$, by a result of Gerald Schwarz [**Sch75**], there exists a smooth function $F : \mathbf{R}^n \to \mathbf{R}$ such that $\phi(x_1, \ldots, x_n) = F(x_1^2, \ldots, x_n^2)$. Thus, $f = F \upharpoonright \mathbf{K}_n$ is the restriction of a smooth function defined on an open superset of $\mathbf{K}_n$. Thus smoothness coincides in the two meanings. □

4.17. Orbifolds as diffeologies. A diffeological space M is said to be an *orbifold of dimension* n if M is locally diffeomorphic at every point to $\mathbf{R}^n/\Gamma$, where $\Gamma \subset \mathrm{GL}(\mathbf{R}^n)$ is finite, and $\mathbf{R}^n/\Gamma$ is equipped with the quotient diffeology. The group Γ may change from point to point. Note that n is the dimension, in the diffeological sense (art. 2.22), of the orbifold at each point.

We could say this in other words: an orbifold is a diffeological space *modeled* by the family

$$\mathcal{M} = \{\mathbf{R}^n/\Gamma \mid \Gamma \subset \mathrm{GL}(n, \mathbf{R}) \text{ and } \#\Gamma < \infty\}.$$

The elements of $\mathcal{M}$ can be called the *models* of the category {Orbifolds}, see Figure 4.4.

NOTE. On another side, the relationship between diffeological orbifolds and orbifolds through Lie groupoids is, at this time, a work in progress by Yael Karshon and Masrour Zoghi [**KaZo12**].

4.18. Structure groups of orbifolds. Let M be an orbifold of dimension n and $x \in M$. There exist always a finite linear group $\Gamma \subset \mathrm{GL}(\mathbf{R}^n)$, unique up to conjugation, and a local diffeomorphism $\psi : \mathbf{R}^n/\Gamma \supset \mathcal{O} \to M$ mapping 0 to x, that is, $0 \in \mathcal{O}$ and $\psi(0) = x$. The conjugacy class of the group Γ is called the *structure group* of M at the point x. The point x is said to be *regular* if Γ is trivial, *singular* otherwise.[5] Note that an orbifold with no singular point is just a manifold. For a comprehensive exposé, see [**IKZ05**].

[5]This corresponds to the diffeological notion of singularity, since a local diffeomorphism can only exchange points with the same structure group.

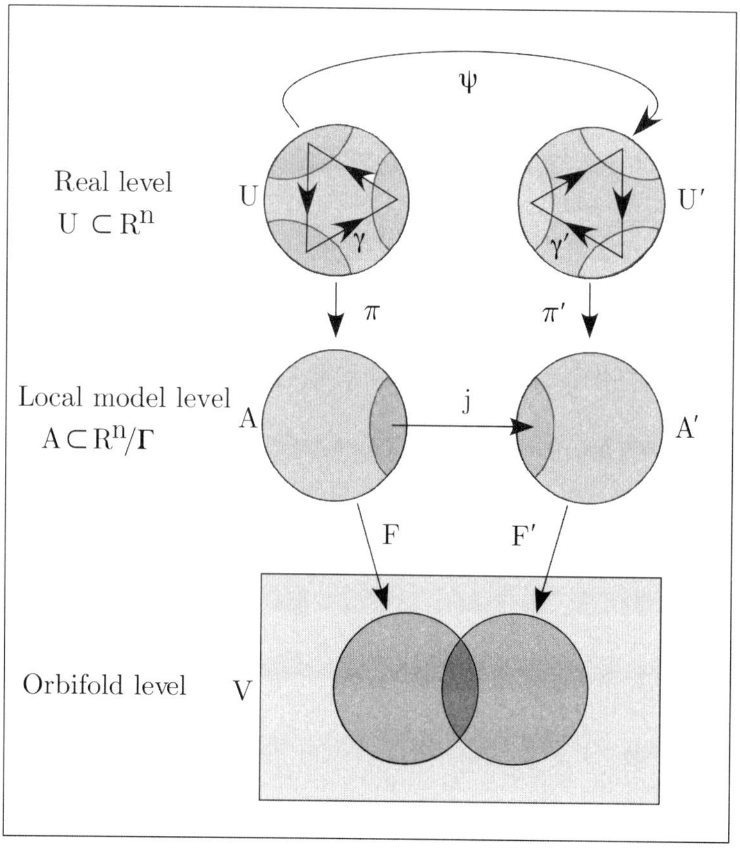

FIGURE 4.4. The three levels of an orbifold structure.

4.19. In conclusion on modeling. These previous examples suggest a precise concept of *modeling* of diffeological spaces, close to the notion of generating family introduced in (art. 1.66), but stronger. We can say formally that a diffeological space X is *modeled* on a family of diffeological spaces $\mathcal{M}$, if X is locally diffeomorphic to some member of $\mathcal{M}$, everywhere. Precisely, for any $x \in X$ there exist a D-open neighborhood $\mathcal{O} \subset X$ of x, a space $A \in \mathcal{M}$, a D-open subset $\mathcal{V} \subset A$ and a local diffeomorphism $\phi : \mathcal{V} \hookrightarrow \mathcal{O}$. So, for every plot $P : U \to X$ and for every $r \in U$, there exist an open neighborhood $V \subset U$ of r, a plot $Q : V \to A$, for some $A \in \mathcal{M}$, and a local diffeomorphism ϕ from A to X, such that $\phi \circ Q = P \restriction V$. This definition covers the various examples we already looked at: the n-dimensional manifolds for which the family is reduced to $\mathcal{M} = \{\mathbf{R}^n\}$, the diffeological manifolds modeled on some diffeological vector space E, or a family $\mathcal{M} = \{E_i\}_{i \in \mathcal{J}}$ of diffeological vector spaces, the manifolds with corners and boundary, modeled on $\mathbf{K}_n$, or the orbifolds where $\mathcal{M}$ is some family of quotients $\{\mathbf{R}^{n_i}/\Gamma_i\}_{i\in\mathcal{J}}$, with Γ_i being some finite subgroup of $\mathrm{GL}(\mathbf{R}^{n_i})$.

Exercises

✎ EXERCISE 79 (Reflexive diffeologies). Let X be any set and $\mathcal{F}$ be any subset of Maps(X, **R**).

1) Describe the coarsest diffeology $\mathcal{D}$ on X such that $\mathcal{F} \subset \mathcal{D}(X, \mathbf{R})$. What about the finest one?

We say that the diffeology $\mathcal{D}$ is *subordinated* to $\mathcal{F}$. Next, we shall say that a diffeology $\mathcal{D}$ of X is *reflexive* if the diffeology $\mathcal{D}'$ subordinated to $\mathcal{F} = \mathcal{D}(X, \mathbf{R})$ coincides with $\mathcal{D}$.

2) Why is the diffeology subordinated to any $\mathcal{F} \subset \mathrm{Maps}(X, R)$ reflexive?

3) Show that the finite dimensional manifolds (art. 4.1) are reflexive.

4) Show that the irrational tori (Exercise 4, p. 8) are not reflexive.

Reflexive spaces form a full subcategory of the category {Diffeology}, extending the ordinary full subcategory {Manifolds}.

✎ EXERCISE 80 (Frölicher spaces). Let X be a set, a Frölicher structure on X is defined by a pair of sets: $\mathcal{F} \subset \mathrm{Maps}(X, \mathbf{R})$ and $\mathcal{C} \subset \mathrm{Maps}(\mathbf{R}, X)$ such that $\mathcal{C} = \mathfrak{C}(\mathcal{F})$ and $\mathcal{F} = \mathfrak{F}(\mathcal{C})$, where

$$\begin{aligned} \mathfrak{C}(\mathcal{F}) &= \{c \in \mathrm{Maps}(\mathbf{R}, X) \mid \mathcal{F} \circ c \subset \mathcal{C}^\infty(\mathbf{R}, \mathbf{R})\}, \\ \mathfrak{F}(\mathcal{C}) &= \{f \in \mathrm{Maps}(X, \mathbf{R}) \mid f \circ \mathcal{C} \subset \mathcal{C}^\infty(\mathbf{R}, \mathbf{R})\}. \end{aligned}$$

Said differently, $\mathcal{C} = \mathfrak{C}(\mathcal{F})$ and $\mathcal{F} = \mathfrak{F}(\mathcal{C})$ are what we call the *Frölicher condition*. Thanks to the second question of Exercise 79, p. 99, we know that the diffeology subordinated to $\mathcal{F}$ is reflexive. We shall admit this version of Boman's theorem.

THEOREM [**Bom67**]. If $F \in \mathrm{Maps}(\mathbf{R}^n, \mathbf{R})$ is such that, for all $\gamma \in \mathcal{C}^\infty(\mathbf{R}, \mathbf{R}^n)$, $F \circ \gamma \in \mathcal{C}^\infty(\mathbf{R}, \mathbf{R})$, then $F \in \mathcal{C}^\infty(\mathbf{R}^n, \mathbf{R})$.

Show that, for every reflexive diffeological space X, $\mathcal{C} = \mathcal{C}^\infty(\mathbf{R}, X)$ and $\mathcal{F} = \mathcal{C}^\infty(X, \mathbf{R})$ satisfy the Frölicher condition.

NOTE. Actually it follows from these statements that there exists an equivalence between the category of Frölicher spaces and the strict subcategory of reflexive diffeological spaces.[6] Note that, as reflexive diffeological spaces, all the diffeological constructions developed in this textbook (homotopy, differential calculus, fiber bundles, etc.) apply in particular to Frölicher spaces. We cannot expect however the same results for Frölicher spaces as for general diffeological spaces. One of many examples: the construction of the integration bundle of a closed 2-form (art. 8.42) does not exist in this subcategory, except for the very special case of an integral form. I will not elaborate on these issues, I leave them open for the reader.

[6]These results have been established collectively in a seminar in Toronto, during the fall of 2010, with the following participants: Augustin Batubenge, Yael Karshon, Jordan Watts, and myself.

CHAPTER 5

Homotopy of Diffeological Spaces

This chapter introduces the elementary constructions and definitions of the theory of homotopy in diffeology. We shall see, in particular, the definitions of connectedness, connected components, homotopic invariants, the construction of the Poincaré groupoid, the fundamental group and the higher homotopy groups, the relative homotopy, and the exact sequence of the homotopy of a pair. Thanks to the functional diffeology on the space of paths of a diffeological space, we define the higher homotopy groups by considering simply the iteration of its space of loops. A loop of a diffeological space being defined as a smooth path having the same ends, that is, a path taking the same values for 0 and 1. This chapter is a rewriting of my thesis *Fibrations difféologiques et homotopie* [**Igl85**], where I first introduced these constructions.

Connectedness and Homotopy Category

In this section we split the diffeological spaces into connected components. This decomposition, applied to the spaces of smooth maps between diffeological spaces, introduces the homotopy category, homotopic equivalences, *and* homotopic invariants *in diffeology.*

5.1. The space of Paths of a diffeological space. Let X be a diffeological space. We call a *smooth path* of X, or simply a *path* of X, any smooth map from $\mathbf{R}$ to X. The set of all the paths in X will be denoted by Paths(X), that is,

$$\mathrm{Paths}(X) = \mathcal{C}^\infty(\mathbf{R}, X).$$

This set will be equipped with the functional diffeology (art. 1.57). For every path γ, we call the *initial point* or *beginning*, or *starting point* or simply *start*, of γ, the point $\gamma(0)$, and we call the *final point*, or *ending point* or simply *end*, of γ, the point $\gamma(1)$. We call the *ends* of γ the pair $(\gamma(0), \gamma(1))$. We denote these *end maps* by

$$\hat{0} : \gamma \mapsto \gamma(0), \quad \hat{1} : \gamma \mapsto \gamma(1) \quad \text{and} \quad \mathrm{ends} : \gamma \mapsto (\gamma(0), \gamma(1)).$$

As a direct consequence of the definition of the functional diffeology, these three maps are smooth:

$$\hat{0}, \hat{1} \in \mathcal{C}^\infty(\mathrm{Paths}(X), X), \ \mathrm{ends} \in \mathcal{C}^\infty(\mathrm{Paths}(X), X \times X).$$

We say that a path γ *connects* x to x′ if $\hat{0}(\gamma) = x$ and $\hat{1}(\gamma) = x'$. Let A and B be two subspaces of X. We say that a path γ of X *starts* in A if $\hat{0}(\gamma) \in A$ and *ends* in B if $\hat{1}(\gamma) \in B$. We also say that the path γ connects A to B. We shall denote by Paths(X, A, B) the subspace of paths $\gamma \in \mathrm{Paths}(X)$ which connect A to B, that is,

$$\mathrm{Paths}(X, A, B) = \mathrm{ends}^{-1}(A \times B).$$

For the sake of simplicity, we may replace A (or B), in $\mathrm{Paths}(X, A, B)$, by a star $\star$ if A (or B) is equal to X. We may also replace A (or B) just by x if A (or B) is the singleton $\{x\}$. For example, for $A = \{x\}$ and $B = X$, $\mathrm{Paths}(X, x, \star)$ denotes the subspace of paths which start at x, that is, $\mathrm{Paths}(X, \{x\}, X)$. For $A = X$ and $B = \{x'\}$, $\mathrm{Paths}(X, \star, x')$ denotes the subspace of paths which end at x', that is, $\mathrm{Paths}(X, X, \{x'\})$. For $A = \{x\}$ and $B = \{x'\}$, $\mathrm{Paths}(X, x, x')$ denotes the subspace of paths which start at x and end at x', that is, $\mathrm{Paths}(X, \{x\}, \{x'\})$. In the special case where $A = B = \{x\}$, an element of $\mathrm{Paths}(X, x, x)$ is called a *loop*, based at x. The space $\mathrm{Paths}(X, x, x)$ of all the loops based at x is denoted by

$$\mathrm{Loops}(X, x) = \mathrm{ends}^{-1}(x, x).$$

The subspace of all the loops in X, independently of the basepoint, is denoted by

$$\mathrm{Loops}(X) = \left\{\ell \in \mathrm{Paths}(X) \mid \hat{0}(\ell) = \hat{1}(\ell)\right\}.$$

In other words, $\mathrm{Loops}(X)$ is the pullback by ends of the diagonal X in $X \times X$.

5.2. Concatenation of paths. Let X be a diffeological space. We say that two paths γ and γ' are *juxtaposable* if $\hat{1}(\gamma) = \hat{0}(\gamma')$ and if there exists a path $\gamma \vee \gamma'$ such that

$$\gamma \vee \gamma'(t) = \begin{cases} \gamma(2t) & \text{if } t \leq \frac{1}{2}, \\ \gamma'(2t-1) & \text{if } \frac{1}{2} \leq t. \end{cases}$$

The path $\gamma \vee \gamma'$ is called the *concatenation* of γ and γ'.

5.3. Reversing paths. Let X be a diffeological space. Let γ be a path in X, let $x = \hat{0}(\gamma)$ and $x' = \hat{1}(\gamma)$. The path

$$\bar{\gamma} : t \mapsto \gamma(1-t)$$

will be called the *reverse path* of γ. It satisfies $\hat{0}(\bar{\gamma}) = \hat{1}(\gamma)$ and $\hat{1}(\bar{\gamma}) = \hat{0}(\gamma)$. The map $\mathrm{rev} : \gamma \mapsto \bar{\gamma}$ is obviously smooth, it is an involution of $\mathrm{Paths}(X)$. If γ and γ' are juxtaposable, then $\mathrm{rev}(\gamma')$ and $\mathrm{rev}(\gamma)$ are juxtaposable, and

$$\mathrm{rev}(\gamma') \vee \mathrm{rev}(\gamma) = \mathrm{rev}(\gamma \vee \gamma').$$

PROOF. First of all, $\hat{1}(\mathrm{rev}(\gamma')) = \hat{0}(\gamma') = \hat{1}(\gamma) = \hat{0}(\mathrm{rev}(\gamma))$. Now let $t \in \mathbf{R}$, and let us express $\mathrm{rev}(\gamma') \vee \mathrm{rev}(\gamma)(t)$,

$$\begin{aligned} \mathrm{rev}(\gamma') \vee \mathrm{rev}(\gamma)(t) &= \begin{cases} \mathrm{rev}(\gamma')(2t) & \text{if } t \leq 1/2, \\ \mathrm{rev}(\gamma)(2t-1) & \text{if } 1/2 \leq t, \end{cases} \\ &= \begin{cases} \gamma'(1-2t) & \text{if } t \leq 1/2, \\ \gamma(1-(2t-1)) = \gamma(2-2t) & \text{if } 1/2 \leq t. \end{cases} \end{aligned}$$

Now let us express $\mathrm{rev}(\gamma \vee \gamma')(t)$,

$$\begin{aligned} \mathrm{rev}(\gamma \vee \gamma')(t) &= \gamma \vee \gamma'(1-t) \\ &= \begin{cases} \gamma(2(1-t)) = \gamma(2-2t) & \text{if } 1-t \leq 1/2, \\ \gamma'(2(1-t)-1) = \gamma'(1-2t) & \text{if } 1/2 \leq 1-t, \end{cases} \\ &= \begin{cases} \gamma'(1-2t) & \text{if } t \leq 1/2, \\ \gamma(2-2t) & \text{if } 1/2 \leq t. \end{cases} \end{aligned}$$

That was what we had to check. ☐

5.4. Stationary paths. Let X be a diffeological space. We say that a path γ is *ends-stationary*, or simply *stationary*, if there exist an open neighborhood of $]-\infty, 0]$ and an open neighborhood of $[+1, +\infty[$, where γ is constant. Formally, the path γ is stationary if there exists $\varepsilon > 0$ such that

$$\gamma \restriction]-\infty, +\varepsilon[= [t \mapsto \gamma(0)] \quad \text{and} \quad \gamma \restriction]1-\varepsilon, +\infty[= [t \mapsto \gamma(1)]. \qquad (\clubsuit)$$

The set of stationary paths in X will be denoted by stPaths(X). The prefix st will be used to denote everything stationary. The paths satisfying ($\clubsuit$) will be also called ε-stationary. Let $\mathrm{stPaths}_\varepsilon(X)$ be the space of ε-stationary paths in X.

1. Two stationary paths γ and γ' are juxtaposable iff $\hat{1}(\gamma) = \hat{0}(\gamma')$.
2. Let $\mathcal{J}_\varepsilon$ be the space of juxtaposable ε-stationary pairs of paths in X,

$$\mathcal{J}_\varepsilon = \left\{(\gamma, \gamma') \in \mathrm{stPaths}_\varepsilon(X) \times \mathrm{stPaths}_\varepsilon(X) \mid \hat{1}(\gamma) = \hat{0}(\gamma')\right\}.$$

The concatenation $\vee : \mathcal{J}_\varepsilon \to \mathrm{stPaths}(X)$, defined by $\vee(\gamma, \gamma') = \gamma \vee \gamma'$, is smooth and takes its values in $\mathrm{stPaths}_{\varepsilon/2}(X)$.

NOTE. The concatenation of stationary paths is not associative, if γ, γ' and γ'' are three stationary paths such that $\hat{1}(\gamma) = \hat{0}(\gamma')$ and $\hat{1}(\gamma') = \hat{0}(\gamma'')$, then $\gamma \vee (\gamma' \vee \gamma'')$ is *a priori* different from $(\gamma \vee \gamma') \vee \gamma''$. For a finite family of stationary paths $(\gamma_k)_{k=1}^n$ such that $\hat{1}(\gamma_k) = \hat{0}(\gamma_{k+1})$, with $1 \leq k < n$, we should prefer, for reason of symmetry, the multiple concatenation defined by

$$\gamma_1 \vee \gamma_2 \vee \cdots \vee \gamma_n : t \mapsto \begin{cases} \gamma_1(nt - 1 + 1) & t \leq \frac{1}{n}, \\ \cdots & \\ \gamma_k(nt - k + 1) & \frac{k-1}{n} \leq t \leq \frac{k}{n}, \\ \cdots & \\ \gamma_n(nt - n + 1) & \frac{n-1}{n} \leq t, \end{cases}$$

which is still a stationary path, connecting $\hat{0}(\gamma_1)$ to $\hat{1}(\gamma_n)$.

PROOF. For the first proposition. Let us assume that γ and γ' are two stationary paths in X such that $\hat{1}(\gamma) = \hat{0}(\gamma')$. We can assume that γ and γ' satisfy ($\clubsuit$) for the same $\varepsilon > 0$. Let γ''_- and γ''_+ be the two 1-parametrizations in X defined by

$$\gamma''_- : \left]-\infty, \frac{1}{2} + \frac{\varepsilon}{2}\right[\to X \quad \text{with} \quad \gamma''_-(t) = \gamma(2t),$$
$$\gamma''_+ : \left]\frac{1}{2} - \frac{\varepsilon}{2}, +\infty\right[\to X \quad \text{with} \quad \gamma''_-(t) = \gamma'(2t - 1).$$

Now, on the intersection of their domains

$$\mathrm{def}(\gamma''_-) \cap \mathrm{def}(\gamma''_+) = \left]\frac{1}{2} - \frac{\varepsilon}{2}, \frac{1}{2} + \frac{\varepsilon}{2}\right[,$$

γ''_- and γ''_+ coincide:

$$\gamma''_- \restriction \mathrm{def}(\gamma''_-) \cap \mathrm{def}(\gamma''_+) = \gamma''_+ \restriction \mathrm{def}(\gamma''_-) \cap \mathrm{def}(\gamma''_+) = [t \mapsto \gamma(1) = \gamma'(0)].$$

By application of the axiom of locality of diffeology (see Exercise 2, p. 6), the supremum of $\{\gamma''_-, \gamma''_+\}$, which is exactly $\gamma \vee \gamma'$, is a plot of X. Thus, γ and γ' are juxtaposable.

Let us now prove the second proposition. Let $r \mapsto (P(r), P'(r)) \in \mathcal{J}_\varepsilon$ be a plot defined on a domain U. Precisely, $P : U \to \mathrm{stPaths}_\varepsilon$ and $P' : U \to \mathrm{stPaths}_\varepsilon$ are two plots such that, for all $r \in U$, $P(r)(1) = P(r)(0)$. Then, let us consider the parametrization $Q : U \times \mathbf{R} \to X$ defined by

$$Q : (r,t) \mapsto (P(r) \vee P'(r))(t) = \begin{cases} P(r)(2t) & \text{if } t \leq \frac{1}{2}, \\ P'(r)(2t-1) & \text{if } \frac{1}{2} \leq t. \end{cases}$$

Let

$$V_- = U \times \left]-\infty, \frac{1}{2} + \frac{\varepsilon}{2}\right[\quad \text{and} \quad V_+ = U \times \left]\frac{1}{2} - \frac{\varepsilon}{2}, \infty\right[.$$

We have then

$$V_- \cup V_+ = U \times \mathbf{R} \quad \text{and} \quad V_- \cap V_+ = U \times \left]\frac{1}{2} - \frac{\varepsilon}{2}, \frac{1}{2} + \frac{\varepsilon}{2}\right[.$$

Let $Q_- : V_- \to X$ be defined by $Q_-(r,t) = P(r)(2t)$, and $Q_+ : V_+ \to X$ be defined by $Q_+(r,t) = P'(r)(2t-1)$. Since $Q_- \upharpoonright V_- \cap V_+ = Q_+ \upharpoonright V_- \cap V_+ = [(r,t) \mapsto P(r)(1) = P'(r)(0)]$, $Q = \sup\{Q_-, Q_+\}$ is a plot of X (see above). Therefore $r \mapsto P(r) \vee P'(r)$ is a plot of Paths(X), and the concatenation is a smooth map from $\mathcal{J}_\varepsilon$ to Paths(X). Then, it is clear that $P(r) \vee P'(r)$ is in $\mathcal{J}_{\varepsilon/2}$, for every $r \in U$. □

5.5. Homotopy of paths. Let X be a diffeological space. Because Paths(X) is itself a diffeological space, it makes sense to say that a path $s \mapsto \gamma_s$ in Paths(X) connects γ and γ', that is,

$$[s \mapsto \gamma_s] \in \text{Paths}(\text{Paths}(X)) = \mathcal{C}^\infty(\mathbf{R}, \text{Paths}(X)),$$

with $\gamma_0 = \gamma$ and $\gamma_1 = \gamma'$.

• *Free-ends homotopy.* Such a path $\gamma \mapsto \gamma_s$ is called a *free-ends homotopy*, connecting γ to γ', or from γ to γ'.

• *Fixed-ends homotopy.* Now, let Paths(X, x, x′) be the set of paths in X connecting x to x′, equipped with the subset diffeology of Paths(X) (art. 1.33). A path $[s \mapsto \gamma_s] \in \text{Paths}(\text{Paths}(X,x,x')) = \mathcal{C}^\infty(\mathbf{R}, \text{Paths}(X,x,x'))$ is called a *fixed-ends homotopy* from γ to γ'. But note that, by definition of the subset diffeology, $[s \mapsto \gamma_s]$ is a fixed-ends homotopy if and only if $[s \mapsto \gamma_s] \in \text{Paths}(\text{Paths}(X))$ and for every $s \in \mathbf{R}$, $\gamma_s(0) = x$ and $\gamma_s(1) = x'$.

A crucial property of homotopy in diffeology is that every path γ is fixed-ends homotopic to a stationary path. Let us consider the *smashing function* λ described by in Figure 5.1, where ε is some strictly positive real number, $0 < \varepsilon \ll 1$. The real function λ satisfies essentially the following conditions, and we can choose it increasing,

$$\lambda \in \mathcal{C}^\infty(\mathbf{R},\mathbf{R}), \ \lambda \upharpoonright]-\infty, \varepsilon[= 0, \ \lambda \upharpoonright]1-\varepsilon, +\infty[= 1.$$

Let $\gamma \in$ Paths(X). We have the following properties:

a) The path $\gamma^\star = \gamma \circ \lambda$ is stationary with the same ends as γ.
b) The path γ is fixed-ends homotopic to $\gamma^\star$.

NOTE. As we know, not any two paths $\gamma, \gamma' \in$ Paths(X) such that $\hat{1}(\gamma) = \hat{0}(\gamma')$ can be juxtaposable, but we can always force the concatenation by smashing them first. Hereafter, we shall use often this *smashed concatenation*, denoted and defined by

$$\gamma \star \gamma' = \gamma^\star \vee \gamma'^\star.$$

As a consequence of the point b), if γ and γ' are juxtaposable, then $\gamma \star \gamma'$ is homotopic to $\gamma \vee \gamma'$.

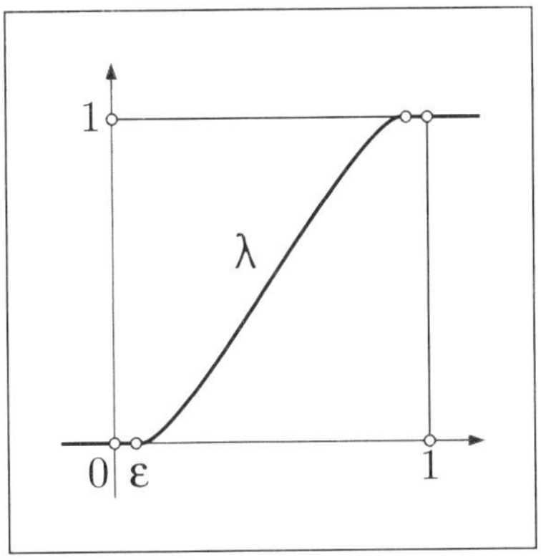

FIGURE 5.1. The smashing function λ.

PROOF. Since $\lambda(0) = 0$ and $\lambda(1) = 1$, $\gamma^\star$ has the same ends as γ. Moreover, $\gamma^\star \restriction]-\infty, \varepsilon[= [t \mapsto \gamma(0)]$ and $\gamma^\star \restriction]1-\varepsilon, +\infty[= [t \mapsto \gamma(1)]$. Therefore, $\gamma^\star$ is stationary, and part a) is proved. Now, consider the following map:

$$[s \mapsto \lambda_s = [t \mapsto s\lambda(t) + (1-s)t]] \in \mathrm{Paths}(\mathcal{C}^\infty(\mathbf{R})).$$

We have then,

$$\lambda_0 = \mathbf{1}_{\mathbf{R}},\ \lambda_1 = \lambda \quad \text{and} \quad \lambda_s(0) = 0,\ \lambda_s(1) = 1 \text{ for all } s \in \mathbf{R}.$$

Thus, $[s \mapsto \gamma_s = \gamma \circ \lambda_s]$ is a path in $\mathrm{Paths}(X, x, x')$, where $x = \gamma(0)$ and $x' = \gamma(1)$, connecting $\gamma = \gamma_0$ to $\gamma^\star = \gamma_1$. □

5.6. Pathwise connectedness. Let X be a diffeological space. We say that two points x and x' of X are *pathwise connected*, or simply *connected*, or *homotopic*, if there exists a path γ *connecting* x to x', that is, $\hat{0}(\gamma) = \gamma(0) = x$ and $\hat{1}(\gamma) = \gamma(1) = x'$. We say that X is *pathwise connected*, or simply *connected*, if any two points x and x' of X can be connected by a path. In other words, X is connected if the map ends (art. 5.1) is surjective. Moreover,

(♢) If X is connected, then ends : $\mathrm{Paths}(X) \to X \times X$ is a subduction, and conversely. This implies, in particular, that the end maps $\hat{1} \restriction \mathrm{Paths}(X, x, \star)$ and $\hat{0} \restriction \mathrm{Paths}(X, \star, x')$ are two subductions onto X.

(♡) If X is connected, then the restrictions ends $\restriction \mathrm{stPaths}(X) \to X \times X$, $\hat{1} \restriction \mathrm{stPaths}(X, x, \star) \to X$ and $\hat{0} \restriction \mathrm{stPaths}(X, \star, x') \to X$ are three subductions, where stPaths(X) is equipped with the subset diffeology.

NOTE. If ends is surjective, we can give a more precise statement: for every plot $P : U \to X \times X$, for every $r_0 \in U$, there exists a smooth lift $Q : V \to \mathrm{Paths}(X)$ of P defined on every star-shaped domain V, centered at r_0. Actually, this lift takes its values in the space stPaths(X).

PROOF. Let us prove that if ends is surjective, then it is a subduction. The converse is just a part of the definition. Let us assume ends : $\mathrm{Paths}(X) \to X \times X$ surjective, let $r \mapsto (P(r), P'(r))$ be a plot of $X \times X$ defined on a domain U. Let $r_0 \in U$, $x_0 = P(r_0)$ and $x'_0 = P'(r_0)$. Let $V \subset U$ be a star-shaped open domain centered at r_0, that is, for every $r \in V$ the segment $\{tr_0 + (1-t)r \mid t \in [0,1]\}$ is contained in V. We can think of V as an open ball, for example. Then, let λ be the smashing function defined in (art. 5.5).

a) According to (art. 5.4), there exists a stationary path $c \in \mathrm{stPaths}_\varepsilon(X)$, connecting x_0 to x'_0, that is, $c(0) = x_0$ and $c(1) = x'_0$.

b) For every point r of V, $\{\lambda(t)r_0 + (1-\lambda(t))r \mid t \in \mathbf{R}\}$ is contained in V, thus $\gamma_r : t \mapsto P(\lambda(t)r_0 + (1-\lambda(t))r)$ is well defined and is an ε-stationary path in X connecting $P(r)$ to x_0. The parametrization $r \mapsto \gamma_r$ is a plot of $\mathrm{stPaths}_\varepsilon(X, \star, x_0)$, defined on V.

c) For every point r of V, $\gamma'_r : t \mapsto P'(\lambda(t)r + (1-\lambda(t))r_0)$ is an ε-stationary path in X connecting x'_0 to $P'(r)$. The parametrization $r \mapsto \gamma'_r$ is a plot of $\mathrm{stPaths}_\varepsilon(X, x'_0, \star)$, defined on V.

Now, according to (art. 5.4), the parametrization $r \mapsto \gamma_r \vee c$, defined on V, is a plot of stPaths(X). More precisely $r \mapsto \gamma_r \vee c$ is a plot of $\mathrm{stPaths}_{\varepsilon/2}(X, \star, x'_0)$. Then, the parametrization $Q : r \mapsto (\gamma_r \vee c) \vee \gamma'_r$, defined on V, is a plot of stPaths(X). More precisely, since γ'_r is ε-stationary, it is $(\varepsilon/2)$-stationary, and since $(\gamma_r \vee c)$ is $(\varepsilon/2)$-stationary, $(\gamma_r \vee c) \vee \gamma'_r$ is $(\varepsilon/4)$-stationary. Thus, Q is a plot of $\mathrm{stPaths}_{\varepsilon/4}(X)$. Finally, Q is a plot of Paths(X) such that, for every $r \in V$, $Q(r)(0) = \gamma_r(0) = P(r)$ and $Q(r)(1) = \gamma'_r(1) = P'(r)$. Hence, Q is a local lift along ends of the plot $r \mapsto (P(r), P'(r))$. Therefore, ends is a subduction (art. 1.48). □

5.7. Connected components. Let X be a diffeological space. To be connected by a path is an equivalence relation on X. The classes of this relation are the *pathwise-connected components* of X. Moreover, the pathwise-connected components of X coincide with the connected components of the D-topology (art. 2.8). There is no ambiguity in simply calling them *connected components*. The component of a point $x \in X$ will be denoted by comp(x),

$$\mathrm{comp}(x) = \{x' \in X \mid \exists \gamma \in \mathrm{Paths}(X) \text{ such that } \mathrm{ends}(\gamma) = (x, x')\}.$$

The set of the connected components of X, the values of the map $\mathrm{comp} : X \to \mathfrak{P}(X)$, is usually denoted by $\pi_0(X)$,

$$\pi_0(X) = \{\mathrm{comp}(x) \mid x \in X\}.$$

PROOF. Let us check the three properties of the equivalence relations for pathwise-connectedness.

Reflexivity. To be connected is of course reflexive: for each $x \in X$, the constant path $[t \mapsto x]$ connects x to x.

Symmetry. For all x and x' in X, if γ connects x to x', then the reverse path $\bar\gamma = [t \mapsto \gamma(1-t)]$ (art. 5.3) connects x' to x.

Transitivity. Let γ be a path connecting x to x', and let γ' be a path connecting x' to x''. We would like to use the concatenation $\gamma \vee \gamma'$ (art. 5.2) to connect x to x'', but γ and γ' are perhaps not juxtaposable. It is why one needs to *smash* γ and γ' first. Let $\gamma^\star = \gamma \circ \lambda$ and $\gamma'^\star = \gamma' \circ \lambda$, where λ is the *smashing function* defined in (art. 5.4); see Figure 5.1. These paths $\gamma^\star$ and $\gamma'^\star$ have, respectively, the same ends as γ and γ', but they are juxtaposable (art. 5.4). Hence, their concatenation $\gamma \star \gamma' = \gamma^\star \vee \gamma'^\star$ connects x to x''.

Now let us consider X, equipped with the D-topology. Let $\mathcal{O}$ be any pathwise-connected component of X. Let us first check that the plots of X take locally their values in pathwise-connected components of X. So, let $P : U \to X$ be a plot. Let

$r_0 \in U$, $x_0 = P(r_0)$, and let X_0 be the pathwise-connected component of x_0. Let $B \in U$ be a ball centered at r_0. For every point $r \in B$, $t \mapsto P(\lambda(t)r + (1-\lambda(t))r_0)$ is a path of X connecting x_0 to $P(r)$. Thus, $P(B) \subset X_0$ and $P \restriction B$ is a plot of X_0. Hence, P takes locally its values in the pathwise-connected components of X. Thus, $P^{-1}(\mathcal{O}) = \{r \in U \mid P(r) \in \mathcal{O}\}$ is also equal to $\bigcup_{r \in P^{-1}(\mathcal{O})} B_r$, where B_r is some ball centered at r and contained in U. Hence, as a union of open balls, $P^{-1}(\mathcal{O})$ is a domain. Therefore $\mathcal{O}$ is D-open. Now, X is the union of its pathwise-connected components, which are all D-open. Since $X - \mathcal{O}$ is a union of pathwise-connected components, $X - \mathcal{O}$ is D-open, and $\mathcal{O}$ is D-closed. Thus, $\mathcal{O}$ is at the same time open and closed for the D-topology. Therefore, $\mathcal{O}$ is a union of D-connected components [**Bou61**, 11-5]. Next, let us show that any pathwise-connected component cannot be the union of two disjoint not empty D-connected components. Let us assume that $\mathcal{O} = \mathcal{A} \cup \mathcal{B}$, where $\mathcal{A}$ and $\mathcal{B}$ are two different D-connected components. Let $x' \in \mathcal{A}$ and $x'' \in \mathcal{B}$. Since $\mathcal{O}$ is a pathwise-connected component, there exists a path γ connecting x' to x''. But, as a smooth map, γ is D-continuous (art. 2.9). Since continuous maps preserve D-connectedness, and since **R** is connected, its image by γ is fully contained in one D-connected component [**Bou61**, 11-2]. But this is in contradiction with the hypothesis. Thus, $\mathcal{O}$ is just one D-connected component. □

5.8. The sum of its components. Let X be a diffeological space. The space X is the sum (art. 1.39) of its connected components (art. 5.7). More precisely, if X is the sum of a family $\{X_i\}_{i \in \mathcal{I}}$, then the connected components of the X_i are the connected components of X. The decomposition of X into the sum of its connected components is the finest decomposition of X into a sum. It follows that the set of components $\pi_0(X)$, equipped with the quotient diffeology of X by the relation *connectedness*, is discrete.

PROOF. We have seen above (art. 5.7) that the plots of X take locally their values in the components of X. Therefore, X is the sum of its connected components (art. 1.39). Now, it is clear that if two points of some component X_i are connected in X_i, they are connected in X. Hence, every connected component of X_i is a subset of a connected component of X. Conversely, let x and x' be two points of X connected by a path γ in X. Now, let $\tau : t \mapsto i$ be the map which associates with every $t \in \mathbf{R}$ the index $i \in \mathcal{I}$ such that $\gamma(t) \in X_i$. By definition of the sum of diffeological spaces, τ is locally constant, and thanks to Exercise 7, p. 14, τ is constant on the connected component of its domain. But τ is defined on **R** which is connected. Thus, τ is constant and x and x' are connected in some X_i. Hence, every connected component of X_i is a connected component of X and conversely every connected component of any X_i is a connected component of X. As a direct consequence of that proposition, the decomposition of X into connected components is the finest decomposition of X into a sum.

Now, let us regard $\pi_0(X)$ as the quotient space X/ comp, notation (art. 1.54). Every plot of $\pi_0(X)$ lifts locally along the projection $X \to \pi_0(X)$ into a plot of X (art. 1.50). But every plot of X takes locally its values into some component of X. Hence, every plot of $\pi_0(X)$ is locally constant. Therefore, $\pi_0(X)$, equipped with the quotient diffeology, is discrete (art. 1.20). □

5.9. Smooth maps and connectedness. Let X be a diffeological space. We say that a subset $A \subset X$ is *connected* if and only if A, equipped with the subset diffeology (art. 1.33), is connected. Note that every connected subset $A \subset X$ is contained in a connected component of X. The connected component of a point $x \in X$ is the greatest connected subset of X containing x, the union of all the connected subsets containing x. Let X' be another diffeological space, and let $f : X \to X'$ be a smooth map. The image by f, of any connected subset $A \subset X$, is connected. In particular, f projects to a map $f_\#$ defined by

$$f_\# : \pi_0(X) \to \pi_0(X'), \quad \text{with} \quad f_\# \circ \mathrm{comp}_X = \mathrm{comp}_{X'} \circ f.$$

Moreover, for any triple of diffeological spaces X, X', X'' and for any pair of smooth maps $f : X \to X'$ and $g : X' \to X''$, we have

$$(g \circ f)_\# = g_\# \circ f_\#.$$

Now, $\mathcal{C}^\infty(X, X')$ being itself a diffeological space, for the functional diffeology, two smooth maps f and f' from X to X' are homotopic if there exists a smooth path $\varphi \in \mathrm{Paths}(\mathcal{C}^\infty(X, X'))$ such that $\varphi(0) = f$ and $\varphi(1) = f'$. If f and f' are homotopic, then they have the same projection in homotopy:

$$f' \in \mathrm{comp}(f) \quad \text{implies} \quad f'_\# = f_\#.$$

We can then denote by $\mathrm{comp}(f)_\#$ the map $f_\#$,

$$\mathrm{comp}(f) \in \pi_0(\mathcal{C}^\infty(X, X')) \quad \text{and} \quad \mathrm{comp}(f)_\# : \pi_0(X) \to \pi_0(X').$$

Therefore, the decomposition of a diffeological space into the sum of its connected components defines π_0 as a functor from the category {Diffeology} to the category {Set} (or to the subcategory of discrete diffeological spaces). It associates with every diffeological space X the set $\pi_0(X)$ of its components, and to every smooth map $f : X \to X'$, the map $\mathrm{comp}(f)_\# : \pi_0(X) \to \pi_0(X')$.

PROOF. Let A be a connected subset of X, and let $f : X \to X'$ be a smooth map. Let x' and y' be two points of $f(A)$, let x and y in X such that $x' = f(x)$ and $y' = f(y)$. Since A is connected, there exists a path $\gamma \in \mathrm{Paths}(A)$ connecting x to y. Thus, the path $f \circ \gamma \in \mathrm{Paths}(f(A))$ connects x' to y'. Therefore $f(A)$ is connected. Now, since connected components of X are connected, it makes sense to define $f_\#(\mathrm{comp}_X(x))$ by $\mathrm{comp}_{X'}(f(x))$, for all $x \in X$. The factorization property $(g \circ f)_\# = g_\# \circ f_\#$ is just obvious. Now, let us consider a homotopy φ connecting f to f', and let $x \in X$. The path $t \mapsto \varphi(t)(x)$ connects $f(x)$ to $f'(x)$. Thus, $\mathrm{comp}(f(x)) = \mathrm{comp}(f(x'))$, and the map $\mathrm{comp}(f)$ is well defined from $\pi_0(X)$ to $\pi_0(X')$. □

5.10. The homotopy category. Let X, X', and X'' be three diffeological spaces. Let $f \in \mathcal{C}^\infty(X, X')$ and $g \in \mathcal{C}^\infty(X', X'')$. If $f' \in \mathcal{C}^\infty(X, X')$ is homotopic to f and $g' \in \mathcal{C}^\infty(X', X'')$ is homotopic to g, then $g' \circ f'$ is homotopic to $g \circ f$. Thus, it makes sense to define a new category {D-Homotopy} whose objects are diffeological spaces and whose morphisms are the connected components of smooth maps:

$$\begin{cases} \mathrm{Obj}\{\text{D-Homotopy}\} = \mathrm{Obj}\{\text{Diffeology}\}, \\ \mathrm{Mor}_{\mathrm{DH}}(X, X') = \pi_0(\mathrm{Mor}_{\mathrm{D}}(X, X')) = \pi_0(\mathcal{C}^\infty(X, X')). \end{cases}$$

Actually the category {D-Homotopy} is a quotient of the category {Diffeology}, where the objects are untouched, but the arrows between two objects are packed by the homotopy relation. We then get a natural functor χ which associates, with every $f \in \mathcal{C}^\infty(X, X')$, its component $\mathrm{comp}(f) \in \pi_0(\mathcal{C}^\infty(X, X'))$. The identity of an object X, in the new category, is the component $\mathrm{comp}(\mathbf{1}_X)$ of the identity map of X. The isomorphisms of this category are called *homotopic equivalences.* Two diffeological spaces X and X′ are homotopy equivalent if and only if there exist a map $f : X \to X'$ and a map $g : X' \to X$ such that $\mathrm{comp}(g \circ f) = \mathrm{comp}(\mathbf{1}_X)$ and $\mathrm{comp}(f \circ g) = \mathrm{comp}(\mathbf{1}_{X'})$.

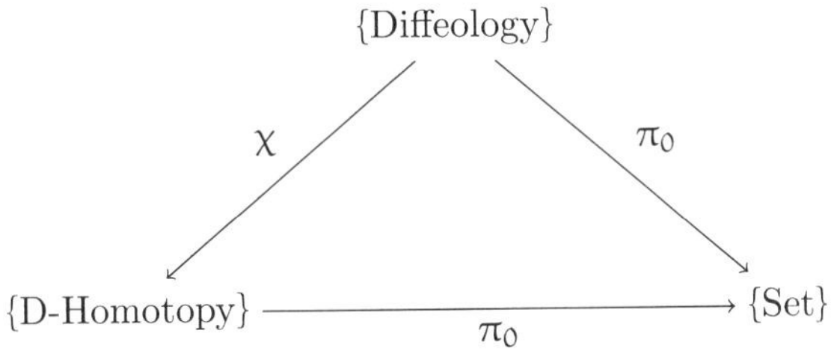

A *homotopic invariant* of the category {Diffeology} is any functor from the category {Diffeology} which factorizes through the functor χ, for example the functor *connected components* π_0, as shown in the above diagram.

PROOF. Let φ be a homotopy connecting f to f′, and let ψ be a homotopy connecting g to g′. The 1-parametrization $t \mapsto \psi(t) \circ \varphi(t)$ splits into $t \mapsto (\psi(t), \varphi(t)) \mapsto \psi(t) \circ \varphi(t)$. But since the composition is smooth (art. 1.59), $t \mapsto \psi(t) \circ \varphi(t)$ is the composite of two smooth maps. Thus, it is a path in $\mathcal{C}^\infty(X, X'')$, connecting $g \circ f$ to $g' \circ f'$. Therefore, the category {D-Homotopy} is well defined. ☐

5.11. Contractible diffeological spaces. Let X be a diffeological space. The final objects of the category {D-Homotopy} are called *contractible diffeological spaces.* Let x_0 be any point of X, we have the following two propositions:

1. The space X is contractible if and only if there exists a smooth section $\sigma : X \to \mathrm{Paths}(X, x_0, \star)$ of the target map $\hat{1}$, that is, $\sigma \in \mathcal{C}^\infty(X, \mathrm{Paths}(X, x_0, \star))$ and $\hat{1} \circ \sigma = \mathbf{1}_X$.

2. The space X is contractible if and only if the constant map $\mathbf{x}_0 = [x \mapsto x_0]$ is homotopic to the identity $\mathbf{1}_X$. In other words, if and only if there exists a smooth path $\rho \in \mathrm{Paths}(\mathcal{C}^\infty(X, X))$ such that $\rho(0) = \mathbf{x}_0$ and $\rho(1) = \mathbf{1}_X$, where $\mathcal{C}^\infty(X, X)$ is equipped with the functional diffeology.

We remark that if these properties are satisfied for some point x_0, they are satisfied for every point of X.

NOTE 1. A diffeological space is contractible if and only if it is homotopy equivalent to a point.

NOTE 2. All the smooth maps from a diffeological space to a contractible diffeological space are homotopic.

NOTE 3. All the smooth maps from a contractible diffeological space to a connected diffeological space are homotopic.

PROOF. Let us prove first that there exists a smooth section σ of $\hat{1}$ if and only if the identity $\mathbf{1}_X$ is homotopic to the constant map $\mathbf{x}_0$. Let $\rho \in \mathrm{Paths}(\mathcal{C}^\infty(X, X))$ be a path such that $\rho(0) = \mathbf{x}_0$ and $\rho(1) = \mathbf{1}_X$. Let $\sigma : x \mapsto [t \mapsto \rho(t)(x)]$. Since the map $(t, x) \mapsto \rho(t)(x)$ is smooth, by the very definition of the functional diffeology, the map σ is smooth. Also, for every $x \in X$, $\sigma(x)(0) = \rho(0)(x) = \mathbf{x}_0(x) = x_0$ and $\sigma(x)(1) = \rho(1)(x) = \mathbf{1}_X(x) = x$. Hence, $\sigma \in \mathcal{C}^\infty(X, \mathrm{Paths}(X, x_0, \star))$ and $\hat{1} \circ \sigma = \mathbf{1}_X$. Thus, σ is a smooth section of the map $\hat{1} : \mathrm{Paths}(X, x_0, \star) \to X$. Now, let σ be a smooth section of the map $\hat{1} : \mathrm{Paths}(X, x_0, \star) \to X$. Then, the map $\rho : t \mapsto$

$[x \mapsto \sigma(x)(t)]$ is a smooth path in $\mathcal{C}^\infty(X,X)$ satisfying $\rho(0)(x) = \sigma(x)(0) = x_0$ and $\rho(1)(x) = \sigma(x)(1) = x$. Thus, ρ is a homotopy from $\mathbf{x}_0$ to $\mathbf{1}_X$.

Let us remark that, if there exists a smooth section σ of $\hat{1}$: Paths$(X, x_0, \star) \to X$, then the map $\sigma^\star : x \mapsto \sigma(x) \circ \lambda$, where λ is the smashing function described in (art. 5.5), is a smooth section of $\hat{1}$ with values in the subspace of stationary paths stPaths$(X, x_0, \star)$. Now, let γ be a stationary path connecting x_1 to x_0, the map $\sigma' : x \mapsto \gamma \vee \sigma^\star(x)$ is smooth (art. 5.4), and it is a section of $\hat{1}$: Paths$(X, x_1, \star) \to X$. Therefore, if there exists a smooth section σ of $\hat{1}$: Paths$(X, x_0, \star) \to X$ for some point x_0, then there exists a smooth section $\hat{1}$: Paths$(X, x, \star) \to X$ for every point x of X. Hence, if there exists a homotopy from $\mathbf{x}_0$ to the identity $\mathbf{1}_X$, for some point x_0, then there exists a homotopy from any constant map $\mathbf{x}$ to $\mathbf{1}_X$.

Now, let X' be another diffeological space. Let ρ be a path, connecting the constant map $\mathbf{x}_0$ to the identity $\mathbf{1}_X$. Let f be a smooth map from X' to X. So, $t \mapsto \rho(t) \circ f$ is a homotopy connecting the constant map $x' \mapsto x_0$ to f. Thus, every smooth map from X' to X are homotopic, and homotopic to a constant map. Therefore, the propositions 1 and 2 are proved.

For Note 1, let us denote by j the injection of $\{x_0\}$ into X. We have, on the one hand, $\mathrm{comp}(\mathbf{x}_0) \circ \mathrm{comp}(j) = \mathrm{comp}(\mathbf{x}_0 \circ j) = \mathrm{comp}(\mathbf{1}_{\{x_0\}})$ and, on the other hand, $\mathrm{comp}(j) \circ \mathrm{comp}(\mathbf{x}_0) = \mathrm{comp}(j \circ \mathbf{x}_0) = \mathrm{comp}(\mathbf{x}_0) = \mathrm{comp}(\mathbf{1}_X)$. Thus, X and $\{x_0\}$ are homotopy equivalent.

Note 2 is just the definition of a final object of the category {D-Homotopy}. Let us recall that a final object of a category is an object X such that, for every other object X', there exists one and only one morphism from X' to X [**McL71**].

Finally, let us consider a connected diffeological space X' and a contractible diffeological space X. Let $f : X \to X'$ be a smooth map. Let $x_0 \in X$ and $x'_0 = f(x_0) \in X'$. The path $[t \mapsto f_t = f \circ \rho(t)]$ connects the constant map $\mathbf{x}'_0 : x \to x'_0$ to f, where ρ is a homotopy from $\mathbf{x}_0$ to $\mathbf{1}_X$. Now, let g be another smooth map from X to X', so g is connected to the constant map $\mathbf{x}''_0 : x \mapsto x''_0 = g(x_0)$. But since X' is connected, there exists a path γ connecting x'_0 to x''_0. Let us denote by $\boldsymbol{\gamma}_t$ the constant map $x \mapsto x_t = \gamma(t)$. Thus, the path $[t \mapsto \boldsymbol{\gamma}_t]$ connects $\mathbf{x}'_0 = \boldsymbol{\gamma}_0$ to $\mathbf{x}''_0 = \boldsymbol{\gamma}_1$. Therefore, f is connected to g, and the third note is proved. ☐

5.12. Local contractibility. We say that a diffeological space X is *locally contractible* if for every point x of X, every D-open (art. 2.8) neighborhood of x contains a contractible D-open neighborhood of x. Note that every diffeological space, modeled on a locally contractible diffeological space (art. 4.19), is obviously locally contractible. The main example of locally contractible diffeological spaces is the category of finite dimensional manifolds.

5.13. Retractions and deformation retracts. Let X be a diffeological space. We call *retraction* any idempotent smooth map $\rho : X \to X$, that is, any $\rho \in \mathcal{C}^\infty(X,X)$ such that $\rho \circ \rho = \rho$. The space $A = \mathrm{val}(\rho)$ is called a *retract* of X, and ρ is said to be a *retraction* from X to A. Note that $\rho \upharpoonright A = \mathbf{1}_A$. We say that ρ is a *deformation retraction* if, moreover, ρ is homotopic to the identity of X. Let ρ be a deformation retraction from X to A. Let $t \mapsto \rho_t$ be a homotopy from ρ to $\mathbf{1}_X$, and let us define σ by

$$\sigma(x) = [t \mapsto \rho_t(x)].$$

Then, σ is a smooth section of $\hat{1}$: Paths$(X, A, \star) \to X$ such that $\hat{0} \circ \sigma \upharpoonright A = \mathbf{1}_A$. Conversely, let $\sigma : X \to$ Paths$(X, A, \star)$ be a smooth section of $\hat{1}$: Paths$(X, A, \star) \to X$ such that $\hat{0} \circ \sigma \upharpoonright A = \mathbf{1}_A$. Then, $\rho = \hat{0} \circ \sigma$ is a deformation retraction from X to A, and $t \mapsto [x \mapsto \sigma(x)(t)]$ is a homotopy from ρ to $\mathbf{1}_X$.

NOTE 1. A diffeological space is homotopy equivalent to any one of its deformation retracts.

NOTE 2. Let $\pi : X \to \bar{X}$ be a subduction, and let ρ be a smooth deformation retraction of X, compatible with π, that is, $\pi(x) = \pi(x')$ implies $\pi(\rho(x)) = \pi(\rho(x'))$. Thus, there exists a smooth deformation retraction $\bar{\rho}$ of $\bar{X}$ such that $\pi \circ \rho = \bar{\rho} \circ \pi$, with retract $\bar{A} = \pi(A)$. Moreover, if $s \mapsto \rho_s$ is a deformation retraction of X such that ρ_s is compatible with π for all s, then $s \mapsto \bar{\rho}_s$ is a deformation retraction of $\bar{X}$.

PROOF. First of all, as an immediate consequence of functional diffeology (art. 1.57), σ is smooth. Now, since $\rho_1 = \mathbf{1}_X$, for all $x \in X$, $\hat{1} \circ \sigma(x) = \sigma(x)(1) = \rho_1(x) = x$. Thus, σ is a section of $\hat{1}$: Paths$(X) \to X$. Moreover, for all $x \in X$, $\hat{0} \circ \sigma(x) = \sigma(x)(0) = \rho_0(x) = \rho(x) \in A$. Thus, σ is a section of $\hat{1}$: Paths$(X, A, \star) \to X$. Now, since $\rho \upharpoonright A = \mathbf{1}_A$, if $x \in A$, then $\hat{0} \circ \sigma(x) = \rho(x) = x$. Thus, $\hat{0} \circ \sigma \upharpoonright A = \mathbf{1}_A$.

Conversely, let σ be a smooth section of $\hat{1}$: Paths$(X, A, \star) \to X$ such that $\hat{0} \circ \sigma \upharpoonright A = \mathbf{1}_A$. Let us define ρ_t by $[x \mapsto \sigma(x)(t)]$, and ρ by $\hat{0} \circ \sigma$, that is, $\rho = \rho_0$. So, $t \mapsto \rho_t$ is smooth, and for every $x \in A$, $\rho(x) = x$. Thus $\rho \circ \rho = \rho$. Now, since σ is a section of $\hat{1}$: Paths$(X, A, \star) \to X$, $\rho_1 = \mathbf{1}_X$. Therefore, ρ is a deformation retraction from X to A.

Now, let us consider the injection $j_A : A \to X$. We have comp$(\rho \circ j_A)$ = comp$(\mathbf{1}_A)$, and comp$(j_A \circ \rho)$ = comp(ρ) = comp$(\mathbf{1}_X)$. Therefore, comp(ρ) and comp(j_A) are the inverse of each other and A is homotopy equivalent to X. For the second note, since π is a subduction, the map $\bar{\rho}$, uniquely defined by $\pi \circ \rho = \bar{\rho} \circ \pi$, is smooth (art. 1.51). Thus, if $s \mapsto \rho_s$ is smooth, that is, if $(s, x) \mapsto \rho_s(x)$ is smooth, then the factorization $(s, \bar{x}) \mapsto \bar{\rho}_s(\bar{x})$ is smooth for the same reason, because $(s, x) \mapsto (s, \bar{x})$ is a subduction. □

Exercises

✎ EXERCISE 81 (Connecting points). Let X be a diffeological space. Show that $\pi_0(X) \simeq \pi_0(\mathcal{C}^\infty(\{0\}, X))$, which justifies the wording "homotopic points" instead of "connected points".

✎ EXERCISE 82 (Connecting segments). Let X be a diffeological space. Let x and x' be any pair of points of X. Show that x and x' are connected if and only if there exists a 1-parametrization $\sigma :]a', b'[\to X$ such that $\sigma(a) = x$, $\sigma(b) = x'$ and $a' < a < b < b'$. Deduce, in particular, that any open ball in $\mathbf{R}^n$ is connected.

✎ EXERCISE 83 (Contractible space of paths). Show that the space of pointed paths Paths$(X, x, \star)$ (art. 5.1) is contractible. More generally, show that Paths$(X, A, \star)$ is homotopy equivalent to A, for every $A \subset X$.

✎ EXERCISE 84 (Deformation onto stationary paths). Let X be a diffeological space. Show that $\gamma \mapsto \gamma^\star$ from Paths(X) to stPaths(X) (art. 5.5, a) is a homotopy equivalence respecting the projection ends, that is, a homotopy equivalence (art. 5.10), and a homotopy equivalence for the diffeology foliated by ends (art. 1.41).

✎ EXERCISE 85 (Contractible quotient). Consider the group of units of the complex plane acting by multiplication $(\zeta_k, z) \mapsto \zeta_k z$, where $(\zeta_k, z) \in \mathbf{Z}_m \times \mathbf{C}$,

$$\mathbf{Z}_m = \left\{ \zeta_k = e^{i\frac{2\pi k}{m}} \mid k = 0, \ldots, m-1 \right\}.$$

Show that $\mathbf{C}/\mathbf{Z}_m$, equipped with the quotient diffeology, is contractible.

✎ EXERCISE 86 (Locally contractible manifolds). Show that every finite dimensional manifold (art. 4.1) is locally contractible (art. 5.12).

Poincaré Groupoid and Homotopy Groups

The main homotopic invariants for a diffeological space are its Poincaré groupoid, and then its various homotopy groups. They are defined for any diffeological space by a recurrence on its spaces of loops. For the diffeological notion of groupoid, see (art. 8.2) *and note that this section is closely related to the construction of the universal covering for diffeological spaces* (art. 8.26).

5.14. Pointed spaces and spaces of components. The category of pointed diffeological spaces is a straightforward specialization of the category of pointed sets. The objects of this category are the pairs (X, x) where X is a diffeological space and x a point of X. A morphism from (X, x) to (X', x') is a smooth map $f : X \to X'$ such that $x' = f(x)$. We shall adopt the following conventions. For a diffeological space X and a point $x \in X$, we shall denote by $\pi_0(X, x)$ the space $\pi_0(X)$ pointed by the component of x,

$$\pi_0(X, x) = (\pi_0(X), \mathrm{comp}(x)). \qquad (\diamondsuit)$$

Groups are naturally pointed by their identity, and there is a natural identification (actually a functor) which associates every group G with the pointed space $(G, 1_G)$. So, we sometimes use one for the other,

$$G \sim (G, 1_G). \qquad (\heartsuit)$$

Morphisms from a group G to a pointed space (X, x) are understood as morphisms of pointed space, that is, maps from G to X mapping 1_G to x.

5.15. The Poincaré groupoid and fundamental group. Let X be a diffeological space. Let Π be the following equivalence relation on Paths(X),

$$\gamma \Pi \gamma' \Leftrightarrow \begin{cases} \text{there exist } x, x' \in X \text{ and } \xi \in \mathrm{Paths}(\mathrm{Paths}(X, x, x')) \\ \text{such that } \xi(0) = \gamma \text{ and } \xi(1) = \gamma'. \end{cases} \qquad (\spadesuit)$$

Said differently, γ and γ' belong to the same component of Paths(X, x, x'), which implies in particular ends(γ) = ends(γ') = (x, x'). We shall denote by $\Pi(X)$ the diffeological quotient (art. 1.50) of Paths(X) by the relation Π, and by

$$\mathrm{class} : \mathrm{Paths}(X) \to \Pi(X) = \mathrm{Paths}(X)/\Pi$$

the canonical projection. We shall denote again by ends the factorization of ends : Paths(X) $\to X \times X$ on $\Pi(X)$. Note that, if X is connected, then ends : $\Pi(X) \to$

$X \times X$ is a subduction, and class $\restriction$ stPaths(X) $\to \Pi(X)$ is also a subduction, where stPaths(X) $\subset$ Paths(X) is the subspace of stationary paths in X (art. 5.4), (art. 5.6). The *Poincaré groupoid* $\mathbf{X}$ (art. 8.2) is then defined by

$$\mathrm{Obj}(\mathbf{X}) = X \quad \text{and} \quad \mathrm{Mor}(\mathbf{X}) = \Pi(X),$$

and for all x and x' in X,

$$\mathrm{Mor}_{\mathbf{X}}(x, x') = \mathrm{Paths}(X, x, x')/\Pi = \pi_0(\mathrm{Paths}(X, x, x'))$$

is the set of fixed-ends homotopy classes of the paths connecting x to x'. The composition in the groupoid is the projection of the concatenation of paths (art. 5.2), For all $\tau \in \mathrm{Mor}_{\mathbf{X}}(x, x')$ and $\tau' \in \mathrm{Mor}_{\mathbf{X}}(x', x'')$,

$$\tau \cdot \tau' = \mathrm{class}(\gamma \vee \gamma'), \quad \text{where} \quad \tau = \mathrm{class}(\gamma) \quad \text{and} \quad \tau' = \mathrm{class}(\gamma').$$

The paths γ and γ' are chosen in stPaths(X), for the concatenation $\gamma \vee \gamma'$ to be well defined; see (art. 5.4).

The construction of the Poincaré groupoid is summarized by the following diagram.

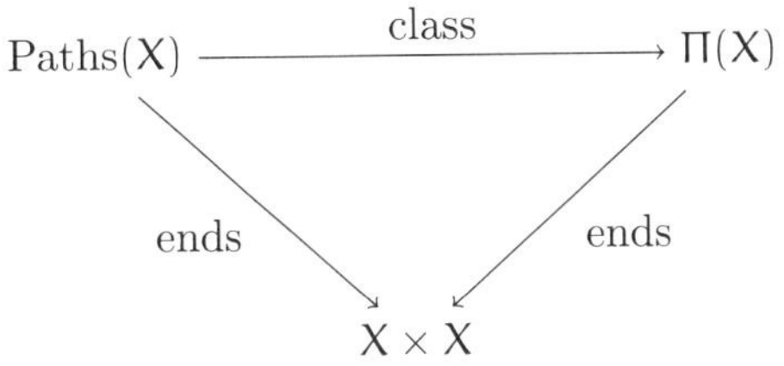

The isotropy groups $\mathbf{X}_x = \mathrm{Mor}_{\mathbf{X}}(x, x)$ and the inverse of the elements of $\mathbf{X}(x, x') = \mathrm{Mor}_{\mathbf{X}}(x, x')$ are described by what follows:

a) For every point x of X, the isotropy group $\mathbf{1}_x \in \mathbf{X}_x$ is the component class($\mathbf{x}$), in Loops(X, x), of the constant path $\mathbf{x} : t \mapsto x$.

b) The inverse τ^{-1} of $\tau = \mathrm{class}(\gamma) \in \mathrm{Mor}_{\mathbf{X}}(x, x')$ is the component class($\bar{\gamma}$) of the reverse path $\bar{\gamma} = \mathrm{rev}(\gamma)$, defined in (art. 5.3).

The structure group $\mathbf{X}_x = \mathrm{Mor}_{\mathbf{X}}(x, x)$, where $x \in X$, is called the *first homotopy group*, or the *fundamental group*, of X at the point x. It is denoted by $\pi_1(X, x)$,

$$\pi_1(X, x) = \pi_0(\mathrm{Loops}(X, x)). \tag{♣}$$

If X is connected, then the fundamental groups are isomorphic. They are precisely conjugate, if $\tau \in \mathrm{Mor}_{\mathbf{X}}(x, x')$, then

$$\pi_1(X, x) = \tau \cdot \pi_1(X, x') \cdot \tau^{-1}.$$

In this case, the type of the homotopy groups $\pi_1(X, x)$ is denoted by $\pi_1(X)$. The space X is said to be *simply connected* if it is connected, $\pi_0(X) = \{X\}$, and if its fundamental group $\pi_1(X)$ is trivial. Or, which is equivalent,

$$X \text{ simply connected} \Leftrightarrow \pi_0(X) = \{X\}, \text{ and for all } x',\ \#\mathrm{Mor}_{\mathbf{X}}(x, x') = 1.$$

In this case, the map ends : $\Pi(X) \to X \times X$ is a diffeomorphism. In other words, X is simply connected if and only if ends : $\Pi(X) \to X \times X$ is a diffeomorphism.

NOTE 1. In the following, $\pi_1(X, x)$ will denote the fundamental group as it is defined above (♣), as well as the pointed space $\pi_1(X, x) \sim (\pi_0(\mathrm{Loops}(X, x)), \mathbf{x})$.

NOTE 2. If X is not connected, then $\pi_1(X, x)$ is also the fundamental group of the connected component of x, that is, $\pi_1(X, x) = \pi_1(\mathrm{comp}(x), x) = \pi_1(\mathrm{comp}(x))$, since there is one type of fundamental group by component.

PROOF. Let us check that we have defined a groupoid, that is, a small category such that every morphism is invertible [**McL75**].

1. The composition of arrows is associative. Let γ, γ' and γ'' be three stationary paths. For s belonging to a small open neighborhood of $[0,1]$, let us define

$$\sigma(s) = t \mapsto \begin{cases} \gamma(s \times 3t + (1-s) \times 4t) & t \leq \frac{1}{4-s}, \\ \gamma'(s \times (3t-1) + (1-s) \times (4t-1)) & \frac{1}{4-s} \leq t \leq \frac{1+s}{4-s}, \\ \gamma''(s \times (3t-2) + (1-s) \times (2t-1)) & \frac{1+s}{4-s} \leq t. \end{cases}$$

Since the paths γ, γ' and γ'' are stationary, the 1-parametrization $\sigma(s)$ is smooth, this is similar to (art. 5.4).

Then, $\sigma(0) = (\gamma \vee \gamma') \vee \gamma''$, and

$$\sigma(1) = \gamma \vee \gamma' \vee \gamma'' : t \mapsto \begin{cases} \gamma(3t) & t \leq \frac{1}{3}, \\ \gamma'(3t-1) & \frac{1}{3} \leq t \leq \frac{2}{3}, \\ \gamma''(3t-2) & \frac{2}{3} \leq t. \end{cases}$$

It is not difficult to find a homotopy of the same kind connecting $\sigma(1)$ to $\gamma \vee (\gamma' \vee \gamma'')$. So, $(\gamma\vee\gamma')\vee\gamma''$ and $\gamma\vee(\gamma'\vee\gamma'')$ are connected in $\mathrm{Paths}(X, \gamma(0), \gamma''(1))$. Now, let $\tau = \mathrm{class}(\gamma)$, $\tau' = \mathrm{class}(\gamma')$ and $\tau'' = \mathrm{class}(\gamma'')$, we have $(\tau\cdot\tau')\cdot\tau'' = \tau\cdot(\tau'\cdot\tau'')$. Therefore, the concatenation of arrows in $\mathbf{X}$ is associative.

2. Let us check now that the component of the constant path $\mathbf{x} : t \mapsto x$ is the identity of the object x. Let $\gamma \in \mathrm{stPaths}(X, x, \star)$. For s belonging to $[-\frac{1}{2}, +\frac{3}{2}]$, for example, let us define

$$\sigma(s) = t \mapsto \begin{cases} x & \text{if} \quad t \leq \frac{1-s}{2-s}, \\ \gamma(2t - 1 + s \times (1-t)) & \text{if} \quad t \geq \frac{1-s}{2-s}. \end{cases} \qquad (\diamondsuit)$$

We can check that $\sigma(s)((1-s)/(2-s)) = \gamma(0) = x$. Since γ is stationary, this concatenation is well defined. Now, $\sigma(0) = \mathbf{x} \vee \gamma$ and $\sigma(1) = \gamma$. Since, $\sigma(s)(0) = x$ and $\sigma(s)(1) = \gamma(1)$, $\mathrm{class}(\mathbf{x} \vee \gamma) = \mathrm{class}(\gamma)$ in $\mathrm{Paths}(X, x, \star)$, that is, $\mathrm{class}(\mathbf{x}) \cdot \mathrm{class}(\gamma) = \mathrm{class}(\gamma)$. It is not difficult to find a homotopy of the same kind, connecting γ' to $\gamma' \vee \mathbf{x}$, with $\gamma' \in \mathrm{stPaths}(X, \star, x)$. Therefore, $\mathrm{class}(\mathbf{x})$ is the identity of the object x.

3. Let us check now that $\mathrm{class}(\bar\gamma)$ and $\mathrm{class}(\gamma)$ are the inverse of each other. Let $\gamma \in \mathrm{stPaths}(X, x, x')$ be a stationary path such that γ and $\bar\gamma$ are juxtaposable. Let us consider

$$\sigma(s) = t \mapsto \begin{cases} \gamma(s \times \lambda(2t)) & \text{if} \quad t \leq \frac{1}{2}, \\ \gamma(s \times \lambda(2(1-t))) & \text{if} \quad \frac{1}{2} \leq t, \end{cases}$$

where λ is the smashing function defined in (art. 5.5). For all s, $\sigma(s)$ is stationary at $t = 1/2$, where $\sigma(s)(1/2) = \gamma(s)$. Thus, $\sigma(s)$ is a well defined path in X. We have, $\sigma(s)(0) = x = \sigma(s)(1)$, and $\sigma(0) = \mathbf{x}$. Now, since $\gamma\circ\lambda$ is fixed-ends homotopic to γ and $\bar\gamma\circ\lambda$ to $\bar\gamma$ (art. 5.5), the path $\sigma(1) = \gamma\circ\lambda\vee\bar\gamma\circ\lambda$ is fixed-ends homotopic to $\gamma\vee\bar\gamma$, that is, $\mathrm{class}(\mathbf{x}) = \mathrm{class}(\gamma\vee\bar\gamma) = \mathrm{class}(\gamma)\cdot\mathrm{class}(\bar\gamma)$. It is not difficult to find a homotopy of the same kind connecting $\mathbf{x}'$ to $\bar\gamma\vee\gamma$. Thus $\mathrm{class}(\gamma)$ and $\mathrm{class}(\bar\gamma)$ are the inverse of each other.

4. Let x and x' be two points of X. Since all the arrows of a groupoid are invertible, if X is connected, then for any $\tau \in \mathrm{Mor}_X(x, x')$ the conjugation by τ is an isomorphism between $\pi_1(X, x) = \mathrm{Mor}_X(x, x)$ and $\pi_1(X, x') = \mathrm{Mor}_X(x', x')$.

5. For every $\tau \in \mathrm{Mor}_X(x, x')$ and every $\tau' \in \mathrm{Mor}_X(x', x)$, $\tau \cdot \tau' \in \mathrm{Mor}_X(x, x)$. If $\pi_1(X, x) = \mathrm{Mor}_X(x, x) = \{\mathbf{1}_x\}$, then $\tau \cdot \tau' = \mathbf{1}_x$, that is, $\tau = \tau'^{-1}$. So, for τ_1 and τ_2 in $\mathrm{Mor}_X(x, x')$, $\tau_1 = \tau'^{-1} = \tau_2$. Thus, there exists only one arrow between x and x', $\#\mathrm{Mor}_X(x, x') = 1$. Conversely, if there exists only one arrow τ from x to x', for every $\nu \in \mathrm{Mor}_X(x, x)$, $\nu \cdot \tau = \tau$, and $\nu = \mathbf{1}_x$, that is, $\pi_1(X, x) = \{\mathbf{1}_x\}$.

6. Let us assume that X is simply connected. Since X is connected, ends is a subduction from $\Pi(X)$ to X. But, $\mathrm{ends}^{-1}(x, x') = \mathrm{Mor}_X(x, x')$, and we know that $\mathrm{Mor}_X(x, x')$ contains only one element. So, ends is injective. On the other hand, we know that injective subductions are diffeomorphisms (art. 1.49). Therefore, X is simply connected if and only if ends : $\Pi(X) \to X \times X$ is a diffeomorphism. □

5.16. Diffeological H-spaces. A *diffeological H-space* is a pointed diffeological space (L, e) (art. 5.14), equipped with a smooth multiplication $(\ell, \ell') \mapsto \ell * \ell'$, which induces a group multiplication on the space of components $\pi_0(L, e)$, that is, $\mathrm{comp}(\ell) \cdot \mathrm{comp}(\ell') = \mathrm{comp}(\ell * \ell')$, with $\mathrm{comp}(e)$ as identity. The main property of an H-space (L, e) is that its first homotopy group $\pi_1(L, e)$ (art. 5.15) is Abelian. The usual examples for H-spaces are the spaces Loops(X, x) of loops in a diffeological space X (art. 5.1), pointed by the constant loop $\mathbf{x} : t \mapsto x$, and equipped with the concatenation (art. 5.2).

Proof. The proof that $\pi_0(L, e)$ is Abelian is a direct adaptation of the classical proof (see **[Hu59]** for example), applied to the subset of stationary paths, in the same vein as the previous paragraph. □

5.17. Higher homotopy groups and functor Π_n. Let X be a diffeological space, and x be a point of X. The *higher homotopy groups* of X, based at x, are defined recursively

$$\pi_{n+1}(X, x) = \pi_n(\mathrm{Loops}(X, x), \mathbf{x}),$$

with $n \in \mathbf{N}$, and where the bold letter $\mathbf{x}$ denotes the constant loop $t \mapsto x$. For $n = 1$ it gives $\pi_1(X, x) = \pi_0(\mathrm{Loops}(X, x), \mathbf{x}) = (\pi_0(\mathrm{Loops}(X, x)), \mathbf{x})$, which is the pointed space associated with the fundamental group; see (art. 5.15, Note 1). Now, let us define the recurrence

$$\mathrm{Loops}_{n+1}(X, x) = \mathrm{Loops}(\mathrm{Loops}_n(X, x), \mathbf{x}_n), \text{ and } \mathbf{x}_{n+1} = [t \mapsto \mathbf{x}_n],$$

initialized by

$$\mathrm{Loops}_0(X, x) = X \quad \text{and} \quad \mathbf{x}_0 = x.$$

We get immediately

$$\pi_n(X, x) = \pi_0(\mathrm{Loops}_n(X, x), \mathbf{x}_n),$$

for all $n \in \mathbf{N}$. For $n \geq 1$, $\pi_n(X, x) = \pi_1(\mathrm{Loops}_{n-1}(X, x), \mathbf{x}_{n-1})$, which shows that the higher homotopy groups of X are the fundamental groups of some loop spaces, and therefore inherit a group operation. For example, $\pi_2(X, x)$ is the fundamental group of the connected component of the constant loop $\mathbf{x}$ in Loops(X, x) etc. Since loop spaces are H-spaces, the groups $\pi_n(X, x)$ are Abelian for $n \geq 2$ (art. 5.16). Now, let X' be another diffeological space, let $f \in \mathcal{C}^\infty(X, X')$, and let $x' = f(x)$. The map f induces naturally a family of smooth maps, on the iterated spaces of loops,

$$f_n : \mathrm{Loops}_n(X, x) \to \mathrm{Loops}_n(X', x'),$$

defined, for all integers n, by recurrence:

$$f_{n+1}(\gamma) = f_n \circ \gamma, \text{ with } f_0 = f.$$

Then, according to (art. 5.9), the map f_n projects naturally on a map $f_{n\#}$, from the space of components $\pi_0(\text{Loops}_n(X,x), x_n) = \pi_n(X,x)$, to the space of components $\pi_0(\text{Loops}_n(X',x'), x'_n) = \pi_n(X',x')$.

(♣) The map $f_{n\#} : \pi_n(X,x) \to \pi_n(X',x')$ is a group morphism for $n > 0$, and a morphism of pointed spaces for $n = 0$.

Now that we have the general context, to make our notations a little bit shorter, let us introduce the following sequence of pointed spaces,

$$L_0(X,x) = (X,x), \quad \text{and} \quad L_n(X,x) = (\text{Loops}_n(X,x), x_n), \text{ for all } n \geq 1.$$

Then, the maps $f_{n\#}$ can be regarded as morphisms of pointed spaces from $L_n(X,x)$ to $L_n(X',x')$ for all $n \geq 0$, with $f_0 = f$. This is represented by the following commutative diagram.

$$\begin{array}{ccc} L_n(X,x) & \xrightarrow{\ f_n\ } & L_n(X',x') \\ \text{comp}\downarrow & & \downarrow\text{comp} \\ \pi_n(X,x) & \xrightarrow[\ f_{n\#}\ ]{} & \pi_n(X',x') \end{array}$$

For all $n \geq 0$, this whole construction defines a functor Π_n, from the category of pointed diffeological spaces to the category of groups, or pointed spaces,

$$\Pi_n(X,x) = \pi_n(X,x) \quad \text{and} \quad \Pi_n(f) = f_{n\#}.$$

The main properties of $f_{n\#}$ are the following:

1. If f is a homotopy equivalence (art. 5.10), then the $f_{n\#}$ are isomorphisms.
2. If $\rho : X \to A$ is a deformation retraction (art. 5.13), then the $\rho_{n\#}$ are isomorphisms from $\pi_n(X,x)$ to $\pi_n(A,x)$, where $x \in A$.
3. If X is contractible (art. 5.11), then $\pi_n(X,x) = \{0\}$ for all $n \geq 1$.

PROOF. According to the recursive definition of the higher homotopy groups, it is enough to check that $f_{n\#}$ is a morphism for $n = 0$ and $n = 1$. For $n = 0$, it is stated in (art. 5.9). For $n = 1$, it is an immediate consequence of the fact that the image by f of the concatenation of two paths γ and γ' is the concatenation of $f \circ \gamma$ and $f \circ \gamma'$. Now, let us prove the following lemma.

LEMMA. If $f : X \to X'$ is a homotopy equivalence, then $f_1 : \text{Loops}(X,x) \to \text{Loops}(X',x')$, where $x' = f(x)$, is a homotopy equivalence.

By definition, the map f is a homotopy equivalence if there exists a smooth map $g : X' \to X$ such that f and g are homotopic inverse one from each other, that is, $g \circ f$ is homotopic to $\mathbf{1}_X$ and $f \circ g$ is homotopic to $\mathbf{1}_{X'}$. So, let $t \mapsto \xi_t$ be the homotopy from $\mathbf{1}_X$ to $g \circ f$, that is,

$$\xi_0 = \mathbf{1}_X \quad \text{and} \quad \xi_1 = g \circ f.$$

Let $y = g(x') = g \circ f(x)$. If $y = x$, then the path $s \mapsto [\ell \mapsto \xi_s \circ \ell]$ is obviously a homotopy from the identity of $\text{Loops}(X,x)$ to $g_1 \circ f_1 = [\ell \mapsto g \circ f \circ \ell]$, and similarly for $f_1 \circ g_1$. We should conclude that f_1 and g_1 are homotopic inverse from each other, but if $y \neq x$, we have to *reconnect* y to x. Now, let us recall that Loops(X, x) and stLoops(X, x) are homotopy equivalent; see Exercise 84, p. 111. So, we shall consider only stationary paths. Let $c = [t \mapsto \xi_{\lambda(t)}(x)]$, where λ is the smashing function defined in (art. 5.5), c is a stationary path in X connecting x to y. Now, let us define, for all $\ell' \in \mathrm{stLoops}(X', x')$,

$$\tilde{g} : \ell' \mapsto c \vee (g \circ \ell') \vee \bar{c}.$$

Since ℓ' is stationary, $g \circ \ell'$ is stationary and $\tilde{g}(\ell')$ is a well defined loop in X, based at x. So, $\tilde{g}$ is a smooth map from stLoops(X′, x′) to stLoops(X, x). Now, for every stationary loop ℓ in X based at x, $\xi_s \circ \ell$ is a stationary loop based at $\xi_s(x)$. Let us define now the stationary path $c_s = [t \mapsto \xi_{s\lambda(t)}(x)]$, connecting x to $\xi_s(x)$. The map $s \mapsto c_s$ is a path in stPaths(X), and satisfies

$$c_s(0) = x, \quad c_s(1) = \xi_s(x), \quad c_0 = \mathbf{x} = [t \mapsto x], \quad \text{and} \quad c_1 = c.$$

Thus, for any s, for any $\ell \in \mathrm{Paths}(X, x)$, c_s, $\xi_s(\ell)$ and $\bar{c}_s$ are juxtaposable, and their concatenation is a stationary path in X based at x. Then, let us define, for all $\ell \in \mathrm{stLoops}(X, x)$,

$$\Xi_s(\ell) = c_s \vee (\xi_s \circ \ell) \vee \bar{c}_s.$$

The map $s \mapsto \Xi_s$ is a path in $\mathcal{C}^\infty(\mathrm{stLoops}(X, x), \mathrm{stLoops}(X, x))$ connecting

$$\Xi_0 = [\ell \mapsto \mathbf{x} \vee \ell \vee \mathbf{x}] \quad \text{to} \quad \Xi_1 = [\ell \mapsto c \vee (g \circ f \circ \ell) \vee \bar{c}] = \tilde{g} \circ f_1.$$

Now, a slight adaptation of part 2 of the proof of (art. 5.15) shows that the map $[\ell \mapsto \mathbf{x} \vee \ell \vee \mathbf{x}]$ is homotopic to the identity of st Loops(X, x). Hence, $\tilde{g} \circ f_1$ is homotopic to the identity of st Loops(X, x). We could prove, in the same way, that $f_1 \circ \tilde{g}$ is homotopic to the identity of stPaths(X′, x′). Therefore, f_1 is a homotopy equivalence and $f_{1\#}$ is an isomorphism. By recurrence, we get that the $f_{n\#}$ are all isomorphisms. Case 1 is proved; cases 2 and 3 are corollaries. □

Relative Homotopy

This section describes the homotopy of a pair (X, A), *where* X *is a diffeological space and* A *is a subspace of* X. *We establish the short and long exact sequences of the homotopy of the pair* (X, A), *pointed at* $a \in A$, *which is a key ingredient of the exact homotopy sequence of the diffeological fiber bundles* (art. 8.21).

5.18. The short homotopy sequence of a pair. Let X be a diffeological space, let A be a subspace of X, and let $a \in A$. Let us recall (art. 5.1) that

$$\mathrm{Paths}(X, A, a) = \left\{ \gamma \in \mathrm{Paths}(X) \mid \hat{0}(\gamma) \in A \text{ and } \hat{1}(\gamma) = a \right\}.$$

Let γ and γ' be two paths belonging to Paths(X, A, a), a *homotopy from* γ *to* γ', *relative to* A, *pointed at* a is a path in Paths(X, A, a), connecting γ to γ'. We shall also say an (A, a)-relative homotopy from γ to γ'. In Figure 5.2 the paths γ and γ' belong to Paths(X, A, a), with $A = A_1 \cup A_2$. The path γ is (A, a)-relatively homotopic to a loop in X, but not γ'. Let us consider the map $\hat{0} : \mathrm{Paths}(X, A, a) \to A$ and the injection $i : A \to X$. They made up a two-terms sequence of smooth maps:

$$\mathrm{Paths}(X, A, a) \xrightarrow{\hat{0}} A \xrightarrow{i} X.$$

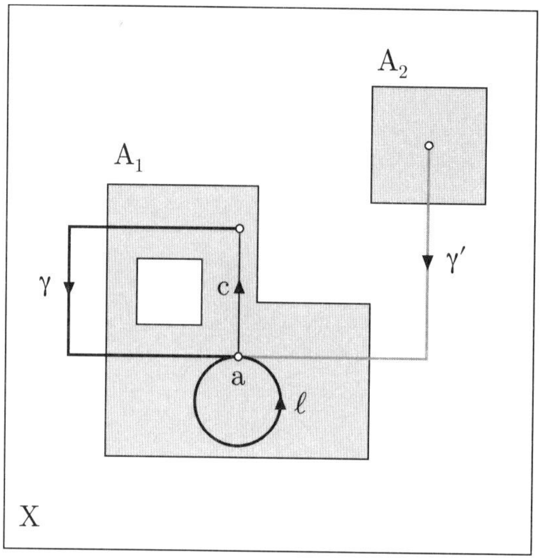

FIGURE 5.2. Relative homotopy of a pair.

This sequence induces naturally the two-terms sequence of morphisms of pointed spaces,

$$(\mathrm{Paths}(X, A, a), \mathbf{a}) \xrightarrow{\hat{0}} (A, a) \xrightarrow{i} (X, a),$$

where $\mathbf{a} = [t \mapsto a]$. Then, this sequence induces a two-terms sequence on the space of components (art. 5.9),

$$\pi_0(\mathrm{Paths}(X, A, a), \mathbf{a}) \xrightarrow{\hat{0}_\#} \pi_0(A, a) \xrightarrow{i_\#} \pi_0(X, a). \qquad (\heartsuit)$$

1. $\ker(i_\#)$ — The *kernel* of $i_\#$ is the subset of components of A, contained in the component of X containing a.

2. $\mathrm{val}(\hat{0}_\#)$ — The *values* of $\hat{0}_\#$ are the components of A, containing the initial points of the paths in X starting in A and ending at a. In other words, the subset of the components of A which can be connected, through X, to a. Now, it is clear that any component of A which can be connected to a by a path in X is included in the component of X containing a. Conversely, every component of A included in the component of X containing a can be connected to a by a path in X, starting in A. So, we get the equality,

$$\ker(i_\#) = \mathrm{val}(\hat{0}_\#).$$

Now, let us consider the inclusion of the triple (X, a, a) into (X, A, a). It induces an injection, denoted by j, on the space of paths,

$$\mathrm{Paths}(X, a, a) = \mathrm{Loops}(X, a) \xrightarrow{j} \mathrm{Paths}(X, A, a).$$

This injection projects, on the space of components, into a morphism of pointed spaces,

$$\pi_0(\mathrm{Loops}(X, a), \mathbf{a}) \xrightarrow{j_\#} \pi_0(\mathrm{Paths}(X, A, a), \mathbf{a}). \qquad (\diamondsuit)$$

Now, the concatenation of (♢) to the two-terms sequence (♡) gives a three-terms sequence of morphisms of pointed spaces,

$$\pi_0(\mathrm{Loops}(X,a),\mathbf{a}) \xrightarrow{j_\#} \pi_0(\mathrm{Paths}(X,A,a),\mathbf{a})) \xrightarrow{\hat{0}_\#} \pi_0(A,a) \xrightarrow{i_\#} \pi_0(X,a). \quad (\spadesuit)$$

Let us call, by abuse of language, *first group of homotopy of* X, *relative to* A, *pointed at* a, the pointed space denoted by $\pi_1(X,A,a)$, and defined by

$$\pi_1(X,A,a) = \pi_0(\mathrm{Paths}(X,A,a),\mathbf{a}).$$

Since, by definition, $\pi_0(\mathrm{Loops}(X,a),\mathbf{a}) = \pi_1(X,a)$ (art. 5.15) — regarded as pointed space — the sequence of morphisms (♠) rewrites,

$$\pi_1(X,a) \xrightarrow{j_\#} \pi_1(X,A,a) \xrightarrow{\hat{0}_\#} \pi_0(A,a) \xrightarrow{i_\#} \pi_0(X,a). \quad (\clubsuit)$$

This sequence is called the *short sequence of the relative homotopy of the pair* (X,A), *at the point* a. We have seen that $\ker(i_\#) = \mathrm{val}(\hat{0}_\#)$; moreover,

$$\ker(\hat{0}_\#) = \mathrm{val}(j_\#).$$

3. $\ker(\hat{0}_\#)$ — The kernel of $\hat{0}_\#$ is the set of the components of $\mathrm{Paths}(X,A,a)$ whose initial point belongs to the component of A containing a.

4. $\mathrm{val}(j_\#)$ — The values of $j_\#$ are the components of the $\gamma \in \mathrm{Paths}(X,A,a)$ which are (A,a)-relatively homotopic to some loops in X, based at a.

In short, the relative homotopy sequence of the pair (X,A), at the point a, is exact.

PROOF. We need only check that $\ker(\hat{0}_\#) = \mathrm{val}(j_\#)$. Let us recall that, on the one hand, $\ker(\hat{0}_\#)$ is made up of the components of $\mathrm{Paths}(X,A,a)$ whose initial point belongs to the same component of A, containing a. On the other hand, $\mathrm{val}(j_\#)$ is the set of the components of paths $\gamma \in \mathrm{Paths}(X,A,a)$ which are (A,a)-relatively homotopic to some loops in X, based at a.

1. $\mathrm{val}(j_\#) \subset \ker(\hat{0}_\#)$. If a path γ is (A,a)-relatively homotopic to some loop in X based at a, its initial point is connected, in A, to a, and belongs to the same component of A containing a.

2. $\ker(\hat{0}_\#) \subset \mathrm{val}(j_\#)$. Let us consider a component of A contained in the same component of X containing a. Let γ be a stationary path in X, beginning in A and ending at a such that its beginning belongs to the component of A containing a. Let $\gamma(0) = x$. Since x and a belong to the same component of A, there exists a stationary path c in A connecting a to x (Figure 5.2). Let $\xi(s) = [t \mapsto c(s + (1-s)\lambda(t))]$, where λ is the smashing function described in Figure 5.1. Thus, ξ belongs to $\mathrm{Paths}(\mathrm{Paths}(X,A,a))$ (art. 5) and $\xi(s)(1) = c(1) = x = \gamma(0)$. So, $\sigma(s) = \xi(s) \vee \gamma$ is a homotopy connecting $(c \circ \lambda) \vee \gamma \in \mathrm{Loops}(X,a)$ to $\mathbf{x} \vee \gamma$, which is homotopic to γ. Therefore γ is (A,a)-relatively homotopic to a loop in X, based at a. □

5.19. The long homotopy sequence of a pair. Let X be a diffeological space, let A be a subspace of X, and let $a \in A$. Let us denote again by i the natural induction $i : \mathrm{Loops}(A,a) \to \mathrm{Loops}(X,a)$. There is no ambiguity with the injection i of (art. 5.18), since the spaces involved are not the same. Then, let us consider the two-terms sequence of smooth maps

$$\mathrm{Paths}(\mathrm{Loops}(X,a),\mathrm{Loops}(A,a),\mathbf{a}) \xrightarrow{\hat{0}} \mathrm{Loops}(A,a) \xrightarrow{i} \mathrm{Loops}(X,a).$$

Or, if we prefer, by denoting

$$X_1 = \mathrm{Loops}(X,a), \quad A_1 = \mathrm{Loops}(A,a) \quad \text{and} \quad a_1 = [t \mapsto a],$$

the above two-terms sequence of smooth maps writes

$$\mathrm{Paths}(X_1, A_1, a_1) \xrightarrow{\hat{0}} A_1 \xrightarrow{i} X_1.$$

We can then apply the construction of (art. 5.18) and get the short sequence of relative homotopy of the pair (X_1, A_1), at the point a_1. Let us denote again by j the natural induction from $\mathrm{Loops}(X_1, a_1)$ to $\mathrm{Paths}(X_1, A_1, a_1)$. Thus, we have,

$$\pi_1(X_1, a_1) \xrightarrow{j_\#} \pi_1(X_1, A_1, a) \xrightarrow{\hat{0}_\#} \pi_0(A_1, a_1) \xrightarrow{i_\#} \pi_0(X_1, a_1). \qquad (\diamondsuit)$$

Let us define the second group of relative homotopy of the pair (X, A), at the point a, by

$$\pi_2(X, A, a) = \pi_1(X_1, A_1, a_1) = \pi_0(\mathrm{Paths}(X_1, A_1, a_1), a_1).$$

So, the short exact sequence $(\diamondsuit)$ writes now

$$\pi_2(X,a) \xrightarrow{j_\#} \pi_2(X,A,a) \xrightarrow{\hat{0}_\#} \pi_1(A,a) \xrightarrow{i_\#} \pi_1(X,a). \qquad (\heartsuit)$$

But the right term $\pi_1(X,a) = \pi_0(X_1, a_1)$ is just $\pi_0(\mathrm{Loops}(X,a),\mathbf{a})$, that is, $\pi_1(X,a)$, regarded as a pointed space. It is also the left term of the relative homotopy sequence of the pair (X, A) at the point a, (art. 5.18). Let us connect the right term of the short homotopy sequences relative to the pair (X_1, A_1), to the left term of the short homotopy sequences relative to the pair (X, A). We get

$$\cdots\ \pi_2(X,A,a) \xrightarrow{\hat{0}_\#} \pi_1(A,a) \xrightarrow{i_\#} \pi_1(X,a) \xrightarrow{j_\#} \pi_1(X,A,a) \xrightarrow{\hat{0}_\#} \pi_0(A,a)\ \cdots.$$

Then, let us describe the connection of the morphisms of these two relative homotopy sequences at the junction $\pi_1(X,a)$.

1. $\ker(j_\# : \pi_1(X,a) \to \pi_1(X,A,a))$ — This kernel is the set of classes of loops of X based at a which can be connected, relatively to (A, a), to the constant loop $\mathbf{a}$.

2. $\mathrm{val}(i_\# : \pi_1(A,a) \to \pi_1(X,a))$ — This is the set of classes of loops in X, based at a, which are fixed-ends homotopic to a loop in A.

Now, if a loop in X, based at a, can be smoothly deformed into a loop contained in A, then it can be retracted relatively to A into the constant loop $\mathbf{a}$. Conversely, if a loop of X, based at a, is connected relatively to (A, a) to the constant loop $\mathbf{a}$, then it is fixed-ends homotopic to a loop in A. In other words,

$$\ker(j_\# : \pi_1(X,a) \to \pi_1(X,A,a)) = \mathrm{val}(i_\# : \pi_1(A,a) \to \pi_1(X,a)).$$

Thus, the connection of the two short exact relative homotopy sequences is exact. Now, let us define the *higher relative homotopy groups* of the pair (X, A) at the point a by recurrence. Let us remark first that the inclusion $i : A \to X$ induces an inclusion $i_n : \mathrm{Loops}_n(A,a) \to \mathrm{Loops}_n(X,a)$ (art. 5.17). Then, we can define

$$\mathrm{Paths}_{n+1}(X,A,a) = \mathrm{Paths}(\mathrm{Loops}_n(X,a), \mathrm{Loops}_n(A,a), \mathbf{a}_n),$$

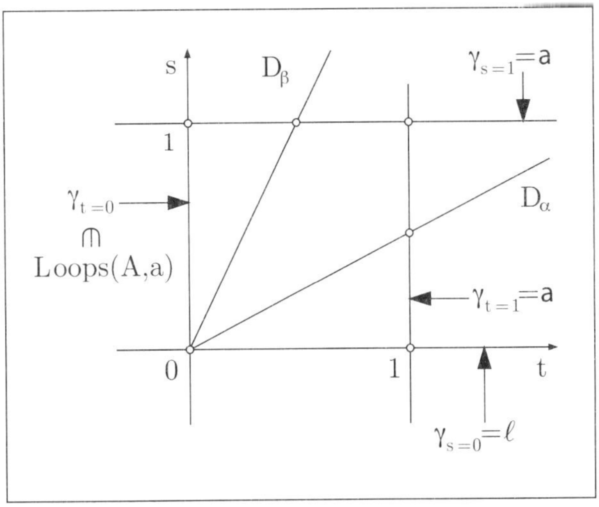

FIGURE 5.3. A relative homotopy of a loop to the constant loop.

for every integer n, and this gives the higher relative homotopy groups

$$\begin{aligned}\pi_{n+1}(X, A, a) &= \pi_0(\mathrm{Paths}_{n+1}(X, A, a), a_n) \\ &= \pi_1(\mathrm{Loops}_n(X, a), \mathrm{Loops}_n(A, a), a_n).\end{aligned}$$

Now, we can iterate the above connection of short relative homotopy sequences for each degree $n+1 \to n$, and we get *the long exact relative homotopy sequence of the pair* (X, A), *at the point* a.

$$\left.\begin{array}{l}\cdots \xrightarrow{i_\#} \pi_n(X, a) \xrightarrow{j_\#} \pi_n(X, A, a) \xrightarrow{\hat{\partial}_\#} \pi_{n-1}(A, a) \xrightarrow{i_\#} \pi_{n-1}(X, a) \cdots \\ \cdots \xrightarrow{i_\#} \pi_1(X, a) \xrightarrow{j_\#} \pi_1(X, A, a) \xrightarrow{\hat{\partial}_\#} \pi_0(A, a) \xrightarrow{i_\#} \pi_0(X, a).\end{array}\right\} \quad (\clubsuit)$$

PROOF. We need only check that $\ker(j_\# : \pi_1(X, a) \to \pi_1(X, A, a))$ is equal to $\mathrm{val}(i_\# : \pi_1(A, a) \to \pi_1(X, a))$. Let us recall that $\ker(j_\#)$ is made up of the loops of X based at a which can be connected, relatively to (A, a), to the constant loop $\mathbf{a}$, and $\mathrm{val}(i_\#)$ is the subset of the classes of loops in X, based at a, which are fixed-ends homotopic to a loop in A.

1. $\ker(j_\#) \subset \mathrm{val}(i_\#)$. Let ℓ be a loop in X, based at a, (A, a)-relatively homotopic to the constant loop $\mathbf{a}$. Let γ be the homotopy. So, for all $s \in \mathbf{R}$,

$$\gamma(0) = \ell, \quad \gamma(1) = \mathbf{a}, \quad \gamma(s)(0) \in A \quad \text{and} \quad \gamma(s)(1) = a.$$

The properties of γ are schematized by Figure 5.3. Let us consider a line of $\mathbf{R}^2$, turning around the origin, its intersection with the cube describes a homotopy connecting ℓ to $\gamma_{t=0} \in \mathrm{Loops}(A, a)$. More precisely, let us consider first the path $\gamma' : s \mapsto [t \mapsto \gamma(t)(st)]$ in $\mathrm{Loops}(X, a)$. The path γ' connects ℓ to $[t \mapsto \gamma(t)(t)]$. Then, let us consider the path $\gamma'' : s \mapsto [t \mapsto \gamma((1-s)t)(t)]$ in $\mathrm{Loops}(X, a)$. The path γ'' connects $[t \mapsto \gamma(t)(t)]$ to $\gamma_{s=0} \in \mathrm{Loops}(A, a)$. Therefore, ℓ is (A, a)-relatively homotopic to a loop in A, based at a.

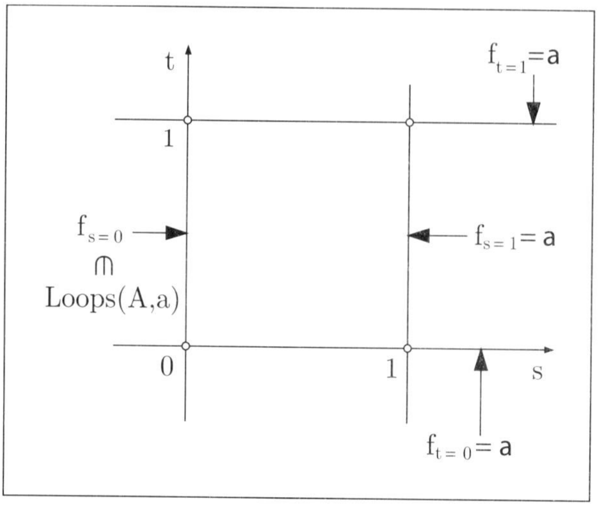

FIGURE 5.4. f belonging to $L_2(X, A, a)$.

2. $\mathrm{val}(i_\#) \subset \ker(j_\#)$. Let $\mathrm{comp}(\gamma) \in \mathrm{val}(i_\#)$. We can choose $\gamma \in \mathrm{Loops}(A, a)$. The path $\xi : s \mapsto [t \mapsto \gamma(s + (1 - s)t)]$ is a (A, a)-relative homotopy connecting γ to the constant loop $\mathbf{a}$. □

5.20. Another look at relative homotopy. Let X be a diffeological space, let A be a subspace of X, and let $a \in A$. We shall give another description of the relative homotopy groups $\pi_n(X, A, a)$. Let us begin by $n = 2$, then $\pi_2(X, A, a) = \pi_1(\mathrm{Loops}(X, a), \mathrm{Loops}(A, a), \mathbf{a})$. Let

$$f \in \mathrm{Paths}(\mathrm{Loops}(X, a), \mathrm{Loops}(A, a), \mathbf{a}).$$

Thus, for all $(s, t) \in \mathbf{R}^2$,

$$f(0) \in \mathrm{Loops}(A, a), \quad f(1)(t) = a, \quad \text{and} \quad f(s)(0) = f(s)(1) = a.$$

We identify f with a smooth plot $f : \mathbf{R}^2 \to X$ by $f(s, t) = f(s)(t)$, the properties of f are described by Figure 5.4. Then, let us permute the variables t and s, and let

$$\tilde{f}(t)(s) = f(s)(t).$$

Now, $\tilde{f}$ satisfies, for all $t \in \mathbf{R}$

$$\tilde{f}(t) \in \mathrm{Paths}(X, A, a), \quad \text{and} \quad \tilde{f}(0) = \tilde{f}(1) = \mathbf{a}.$$

Thus, $\tilde{f} \in \mathrm{Loops}(\mathrm{Paths}(X, A, a), \mathbf{a})$. Now, since $(t, s) \mapsto (s, t)$ is a diffeomorphism, $f \mapsto \tilde{f}$ is a diffeomorphism. More generally,

$$f \in \mathrm{Paths}(\mathrm{Loops}_n(X, a), \mathrm{Loops}_n(A, a), \mathbf{a}_n), \quad \text{for all} \quad n \geq 1.$$

Let $\tilde{f}$ be defined by

$$\tilde{f}(t_0)(t_1)\cdots(t_n) = f(t_1)\cdots(t_n)(t_0).$$

Then, $\tilde{f}$ belongs to $\mathrm{Loops}_n(\mathrm{Paths}(X, A, a), \mathbf{a})$, and the map $f \mapsto \tilde{f}$ is a diffeomorphism. So, thanks to this identification, we get

$$\pi_{n+1}(X, A, a) = \pi_n(\mathrm{Paths}(X, A, a), \mathbf{a}), \quad \text{for all} \quad n \geq 1.$$

Therefore, the relative homotopy groups $\pi_n(X, A, a)$ are pointed spaces for $n = 1$, groups for $n = 2$, and Abelian groups for $n \geq 3$. Moreover, the morphisms of the long exact homotopy sequence of the pair (X, A), at the point a, (art. 5.19) (♣), are group morphisms wherever it makes sense. Indeed, the operator $\hat{0}_\#$ acts on the first variable of the path of loops, while the group acts on the last one.

Exercise

✎ EXERCISE 87 (Homotopy of loops spaces). Let X be a connected diffeological space, $\pi_0(X) = \{X\}$. Show that $\pi_0(\text{Loops}(X))$ is equivalent to the set of conjugacy classes of $\pi_1(X, x)$, where x is any given point in X. Comment on the homotopy of $\text{Loops}(X)$ relatively to $\text{Loops}(X, x)$.

CHAPTER 6

Cartan-De Rham Calculus

The general philosophy of diffeology is to carry, functorially, geometrical objects defined and constructed on real domains (art. 1.1) to diffeological spaces. By nature, since diffeology on a set X is defined by parametrizations in X, geometrical covariant objects are perfectly adapted to this approach. Differential forms are one of the clearest examples. This chapter is devoted to introducing the differential calculus in diffeology.

The first section of this chapter, on *Multilinear Algebra*, puts the notations in place and summarizes the basic notions of linear algebra on which all this building is founded. For the professional mathematician, there is nothing new here, and he can skip it. The reason for this section is to give a self-consistent expository text for the graduate student, or post-graduate, where we regroup what is then used in diffeology. The same holds for the second section, on *Smooth Forms on Real Domains*, where the foundations of differential calculus in $\mathbf{R}^n$ are exposed. It is here for the same reason as the first section. Again, the professional may skip this section.

It is in the third section (art. 6.28), on *Differential Forms on Diffeological Spaces*, where the specificity of differential calculus in diffeology is introduced, and where we begin to establish the main theorems relative to differential forms in diffeology. It continues then along the following sections of this chapter.

It could help, before absorbing all these axiomatics, to first get an idea on what we shall talk about. A *differential form* on a diffeological space is defined as the family of its pullbacks by the plots of the space. In other words, a differential form is defined as a map which associates with each plot a smooth form of its domain, satisfying a compatibility condition when the plots are composed with a smooth parametrization. This definition makes sense, if the diffeological space is a manifold, diffeological forms are just the ones we are used to. Then, exterior *Cartan calculus* and *De Rham cohomological calculus*, the pullbacks by smooth maps, every basic construction associated with differential calculus, pass naturally to diffeology. The Lie derivative of forms extends to diffeological spaces by adapting the ordinary definition to the diffeological context (art. 6.54). The Cartan formula for diffeological spaces is established (art. 6.72). The sets of differential forms on diffeological spaces are naturally diffeological vector spaces.

The definition of *pointed* p*-forms* on a diffeological space X introduces the space $\Lambda^{\mathrm{p}}(X)$ of p-forms over X (art. 6.45). The space $\Lambda^{\mathrm{p}}(X)$ carries a natural functional diffeology such that the projection on X is a subduction, and such that a global form is just one of its smooth sections. This construction gives, in particular, a *cotangent space* T^*X defined over any diffeological space (art. 6.48), without need to refer to some "tangent space". The cotangent space in diffeology is not the dual of a tangent space. The *Liouville forms* (art. 6.49) are introduced and their invariance by the action of the group of diffeomorphisms of the base is established.

Multilinear Algebra

This section establishes some notation and presents a summary of the basic definitions and constructions in real linear and multilinear algebra, used thereafter for Cartan calculus in diffeological spaces. I chose to follow the presentation of Souriau's book "Calcul Linéaire" [**Sou64**]. *I shall mention if a definition or a proposition applies just for finite dimensional real vector spaces. Otherwise, if no such a mention is made, it is assumed to work with any real vector space.*

6.1. Spaces of linear maps. Let E and F be two real vector spaces. We denote by $E * F$ the space $L(E, F)$ of linear maps from E to F.

$$E * F = \{A : E \to F \mid A(ax + by) = aA(x) + bA(y),\ \forall a, b \in \mathbf{R},\ \forall x, y \in E\}.$$

The space $E * F$ is itself a real vector space with

$$(a\,A)(x) = a\,A(x) \text{ and } (A + B)(x) = A(x) + B(x),$$

for all $a \in \mathbf{R}$ and all $A, B \in E * F$.

6.2. Bases and linear groups. Let E be a real vector space of finite dimension, with $\dim(E) = n$. A basis of E is a set $\mathcal{B}$ of n independent vectors $b_1, \ldots, b_n$ of E generating the whole space E.

1. *Independency* — For all $s_i \in \mathbf{R}$, $i = 1, \ldots, n$, $\sum_{i=1}^n s_i b_i = 0$ if and only if $s_i = 0$ for all $i = 1, \ldots, n$.
2. *Generating* — For all $u \in E$, there exist $u_i \in \mathbf{R}$, $i = 1, \ldots, n$, such that $u = \sum_{i=1}^n u_i b_i$.

The numbers $u_i \in \mathbf{R}$ such that $u = \sum_{i=1}^n u_i b_i$ are called the *coordinates* of u in the basis $\mathcal{B}$. The condition to be independent for the b_i implies that the decomposition of u in terms of coordinates is unique. Every basis $\mathcal{B}$ defines a natural isomorphism from $\mathbf{R}^n$ to E, which will be denoted by the same letter,

$$\mathcal{B} \in \mathrm{Isom}(\mathbf{R}^n, E), \quad \mathcal{B}\begin{pmatrix} s_1 \\ \vdots \\ s_n \end{pmatrix} = \sum_{i=1}^n s_i b_i.$$

The two conditions on the b_i, to be independent and generators, is equivalent for $\mathcal{B}$ to be an isomorphism. Conversely, every isomorphism $\mathcal{B} \in \mathrm{Isom}(\mathbf{R}^n, E)$ defines n independent vectors $b_i = \mathcal{B}(\mathbf{e}_i)$, generating E, where the $\mathbf{e}_i$ are the vectors of the canonical basis of $\mathbf{R}^n$. Therefore, we identify a basis with its associated isomorphism, and we consider indifferently the set of vectors of a basis, or the associated isomorphism, as the same thing. We shall denote

$$\mathrm{Bases}(E) = \mathrm{Isom}(\mathbf{R}^n, E).$$

We denote by the indexed bold letters $\mathbf{e}_i$, $i = 1, \ldots, n$, the vectors of the canonical basis of $\mathbf{R}^n$, that is, $\mathbf{e}_i$ is the vector having all its coordinates zero, except at the rank i, where it has the value 1. For example

$$\mathbf{e}_1 = \begin{pmatrix}1\\0\\\downarrow\end{pmatrix}, \quad \mathbf{e}_2 = \begin{pmatrix}0\\1\\0\\\downarrow\end{pmatrix}, \quad \mathbf{e}_3 = \begin{pmatrix}0\\0\\1\\0\\\downarrow\end{pmatrix}, \quad \text{etc.}$$

Regarded as an isomorphism of $\mathbf{R}^n$, the canonical basis $\mathcal{I}_n = (\mathbf{e}_1, \ldots, \mathbf{e}_n)$ of $\mathbf{R}^n$ is just the identity $\mathbf{1}_n$. The set of all the bases of $\mathbf{R}^n$ is just the group of linear transformations,

$$\text{Bases}(\mathbf{R}^n) = \text{GL}(n, \mathbf{R}).$$

As a group, Bases($\mathbf{R}^n$) acts on the bases of E, by right composition,

$$\text{for all } M \in \text{GL}(n, \mathbf{R}), \text{ for all } \mathcal{B} \in \text{Bases}(E), \ \mathcal{B} \circ M^{-1} \in \text{Bases}(E).$$

Note that this action is transitive. For every pair of bases $\mathcal{B}$ and $\mathcal{B}'$ the composition $\mathcal{B}^{-1} \circ \mathcal{B}' \in \text{GL}(n, \mathbf{R})$. This operation is called the *change of basis*. The notion of basis is just a means to reduce vector spaces to standard representatives under the equivalence defined by isomorphisms. Also note that this definition of basis can be extended to any other kind of vector space, finite or infinite dimensional, on any field $\mathbf{K}$, as soon as we have chosen in each class of isomorphisms a favorite representative.

6.3. Dual of a vector space. Let E be a real vector space, the *dual* of E, denoted by E^*, is the space of linear maps from E to $\mathbf{R}$, that is,

$$E^* = E * \mathbf{R} = L(E, \mathbf{R}).$$

An element of E^* is also called a *covector*. If E is finite dimensional, then E and E^* are isomorphic,

$$\dim(E) < \infty \quad \Rightarrow \quad E \sim E^* \quad \text{and} \quad \dim(E^*) = \dim(E).$$

PROOF. Let $\mathcal{B} = (\mathbf{e}_1, \mathbf{e}_2, \ldots, \mathbf{e}_n)$, where $\dim(E) = n$. Let us define the *dual basis* $\mathcal{B}^* = (\mathbf{e}^1, \mathbf{e}^2, \ldots, \mathbf{e}^n)$ by

$$\mathbf{e}^i(\mathbf{e}_j) = \delta^i_j,$$

where δ^i_j is the Kronecker symbol: $\delta^i_j = 1$ if $i = j$ and 0 otherwise. Then, for every $w \in E^*$ and for every vector $x \in E$, we have

$$w(x) = w\left(\sum_{i=1}^n x^i \mathbf{e}_i\right) = \sum_{i=1}^n x^i w(\mathbf{e}_i) = \sum_{i=1}^n x^i w_i = \sum_{i=1}^n w_i \mathbf{e}^i(x),$$

where $w_i = w(\mathbf{e}_i)$. Then, $w = \sum_{i=1}^n w_i \mathbf{e}^i$, and B^* is a generator system of E^*. It is independent, since $\sum_{i=1}^n s_i \mathbf{e}^i = 0$ implies $(\sum_{i=1}^n s_i \mathbf{e}^i)(\mathbf{e}_j) = s_j = 0$. Therefore, $\mathcal{B}^*$ is a basis of E^*. □

6.4. Bilinear maps. Let E, F and G be three real vector spaces. Let us interpret the space $E * (F * G)$. By definition, $E * (F * G)$ is the space of linear maps from E to the space of linear maps from F to G. Let us consider a linear map $A : E \to F * G$. For any $x \in E$, $A(x) \in F * G$, then, for any $y \in F$, $[A(x)](y) \in G$. We can summarize this double mapping by

$$A = [x \mapsto [y \mapsto z]],$$

that is, for all $x \in E$ and all $y \in F$ there exists $z \in G$, such that $z = [A(x)](y)$. We can forget the superfluous brackets and denote $z = A(x)(y)$. So, the map A can be identified with some map from $E \times F$ to G, and $A(x)(y)$ could also be be denoted by $A(x,y)$. From the linearity of the operator A, and from the linearity of the operator $A(x)$, for all $x \in E$, we get the following characterization of $A(x)(y)$:

$$\left.\begin{array}{rcl} A(x+x')(y) & = & A(x)(y) + A(x')(y) \\ A(sx)(y) & = & sA(x)(y) \end{array}\right\} \text{ linearity of } A,$$

$$\left.\begin{array}{rcl} A(x)(y+y') & = & A(x)(y) + A(x)(y') \\ A(x)(sy) & = & sA(x)(y) \end{array}\right\} \text{ linearity of } A(x),$$

where $x, x' \in E$, $y, y' \in F$, and $s \in \mathbf{R}$. This is equivalent to

$$A(\sum_i t_i x_i)(\sum_j s_j y_j) = \sum_{i,j} t_i s_j A(x_i)(y_j),$$

where $s_i, t_j \in \mathbf{R}$, $x_i \in E$ and $y_j \in F$ are finite families. Such operators are called *bilinear maps*.

6.5. Symmetric and antisymmetric bilinear maps. Let E and F be two real vector spaces. The *symmetric* of a bilinear map $A \in E * (E * F)$ is the map $B \in E*(E*F)$ defined by $B(y)(x) = A(x)(y)$. The operator A is said to be *symmetric* if $B = A$, and to be *antisymmetric* if $B = -A$.

Any bilinear map $A \in E * (E * F)$ clearly decomposes, in a unique way, into the sum of a symmetric and an antisymmetric component,

$$A(x)(y) = \frac{1}{2}\,[A(x)(y) + A(y)(x)] + \frac{1}{2}\,[A(x)(y) - A(y)(x)].$$

We deduce, from this decomposition, that an operator is antisymmetric if and only if $A(x)(x) = 0$.

PROOF. If A is antisymmetric, clearly $A(x)(x) = -A(x)(x) = 0$. Now, if $A(x)(x) = 0$, then $0 = A(x+y)(x+y) = A(x)(x) + A(x)(y) + A(y)(x) + A(y)(y) = A(x)(y) + A(y)(x)$. Therefore, $A(x)(y) = -A(y)(x)$. □

6.6. Multilinear operators. Let $E_1, \ldots, E_N, E_{N+1}$ be a family of real vector spaces. Extending the previous construction (art. 6.4), we can interpret the space $E_1 * (E_2 * (E_3 * (\cdots * E_{N+1})))$ as an operator which associates with N vectors $(x, y, \ldots, z) \in E_1 \times E_2 \times \cdots \times E_N$, some element of E_{N+1}, such that

$$A(\sum_i s_i x_i)(\sum_j t_j y_j) \cdots (\sum_k r_k z_k) = \sum_{i,j,\ldots,k} s_i t_j \cdots r_k A(x_i)(y_j) \cdots (z_k),$$

where $s_i, t_j, \ldots r_k \in \mathbf{R}$, $x_i \in E_1$, $y_j \in E_2$, ..., $z_k \in E_N$ are finite families. This last condition expresses the linearity of all the spaces involved in the definition of A. Such operators are called *multilinear operators*. The sum of two elements of $E_1 * (E_2 * (E_3 * (\cdots * E_{N+1})))$, and the multiplication of an element of $E_1 * (E_2 * (E_3 * (\cdots * E_{N+1})))$ by a scalar, are given by

$$\begin{array}{rcl} [A+B](x)(y)\cdots(z) & = & A(x)(y)\cdots(z) + B(x)(y)\cdots(z), \\ [sA](x)(y)\cdots(z) & = & s[A(x)(y)\cdots(z)]. \end{array}$$

For a multilinear operator A, we use indifferently the multiparentheses notation $A(x)(y)\cdots(z)$ or the simple parentheses notation $A(x,y,\ldots,z)$, depending upon which is more suitable.

6.7. Symmetric and antisymmetric multilinear operators. Let E and F be two real vector spaces. Let $A \in E*(E*(E*(\cdots*F)))$ be an n-multilinear operator.

1. The operator A is said to be *totally symmetric*, or simply *symmetric*, if $A(x)(y)\cdots(z)$ does not change when one exchanges any two vectors.

2. The operator A is said to be *totally antisymmetric*, or simply *antisymmetric*, if $A(x)(y)\cdots(z)$ changes sign when one exchanges any two vectors.

Let us consider the group of *permutations* $\mathfrak{S}_n$ of the set of indices $\{1,\ldots,n\}$, acting on the n-tuple $(x_1,\ldots,x_n) \in E^n$ by

$$\sigma(x_1,\ldots,x_n) = (x_{\sigma(1)},\ldots,x_{\sigma(n)}),$$

for all $\sigma \in \mathfrak{S}_n$ and all $(x_1,\ldots,x_n) \in E^n$. Totally symmetric or antisymmetric operators are also characterized by

$$A(x_{\sigma(1)})\cdots(x_{\sigma(n)}) = \begin{cases} A(x_1)\cdots(x_n) & \text{if } A \text{ is symmetric,} \\ \operatorname{sgn}(\sigma)A(x_1)\cdots(x_n) & \text{if } A \text{ is antisymmetric,} \end{cases}$$

where $\sigma \in \mathfrak{S}_n$, and $\operatorname{sgn}(\sigma)$ is the *signature* of the permutation σ, that is, $\operatorname{sgn}(\sigma)$ is equal to 1 if σ decomposes into an even number of *transpositions* (permutations of only two distinct elements) and $\operatorname{sgn}(\sigma) = -1$ otherwise. The number n is called the *order* or *degree* of the operator A, and it is usually denoted by $\deg(A)$. Note that A is antisymmetric if and only if, evaluated on any family of vectors with two repeated vectors, A vanishes:

$$A \text{ antisymmetric} \Leftrightarrow A\cdots(x)\cdots(x)\cdots = 0.$$

PROOF. The first part of the proposition, that is, the formula with the permutations, results immediately from a decomposition of a permutation in a finite product of transpositions. Now, it is clear that if A is antisymmetric, then

$$A\cdots(x)\cdots(x)\cdots = -A\cdots(x)\cdots(x)\cdots = 0.$$

Conversely, if $A\cdots(x)\cdots(x)\cdots = 0$, we have

$$\begin{aligned} A\cdots(x+x')\cdots(x+x')\cdots &= A\cdots(x)\cdots(x)\cdots + A\cdots(x)\cdots(x')\cdots \\ &+ A\cdots(x')\cdots(x)\cdots + A\cdots(x')\cdots(x')\cdots \\ &= A\cdots(x)\cdots(x')\cdots + A\cdots(x')\cdots(x)\cdots. \end{aligned}$$

Therefore, $A\cdots(x)\cdots(x')\cdots = -A\cdots(x')\cdots(x)\cdots$. □

6.8. Tensors. Let E be a real vector space, and let $E^* = E*\mathbf{R} = L(E,\mathbf{R})$ be the dual space (art. 6.3).

1. A *covariant* p-*tensor* of E is a p-multilinear operator A defined p-times on E with values in $\mathbf{R}$, that is, an element of $E*(E*(\cdots*E)*\mathbf{R})$, where E is repeated p-times. A covariant 1-tensor is just an element of E^*.

$$A(x_1)(x_2)\cdots(x_p) \in \mathbf{R} \quad \text{where} \quad x_1,x_2,\ldots,x_p \in E.$$

2. A *contravariant* p-*tensor* of E is a p-multilinear map B defined p-times on E^* with values in $\mathbf{R}$, that is, an element of $E^**(E^**(\cdots*E^*)*\mathbf{R})$, where E^* is repeated p-times

$$B(w_1)(w_2)\cdots(w_p) \in \mathbf{R} \quad \text{where} \quad w_1,w_2,\ldots,w_p \in E^*.$$

3. A *mixed tensor*, p-covariant and q-contravariant, is a $(p+q)$-multilinear operator C defined p-times on E and q-times on E^* with values in $\mathbf{R}$.

$$C(x_1)(x_2)\cdots(x_p)(w_1)(w_2)\cdots(w_q) \in \mathbf{R} \quad \text{where} \quad \begin{cases} x_1, x_2, \ldots, x_p \in E, \\ w_1, w_2, \ldots, w_q \in E^*. \end{cases}$$

Note that a 0-tensor is any map from $E^0 = \{0\}$ to $\mathbf{R}$. Thus, it is just a number. Therefore, the space of 0-tensors of any vector space is $\mathbf{R}$.

6.9. Vector space and bidual. Every element of a vector space E defines an element of its *bidual* $(E^*)^* = (E * \mathbf{R}) * \mathbf{R}$, as follows:

$$\text{for all } x \in E, \text{ for all } w \in E^*, \quad \hat{x}(w) = w(x) \quad \text{and} \quad \hat{x} \in (E^*)^* = E^* * \mathbf{R}.$$

If $\dim(E) < \infty$, then the map $x \mapsto \hat{x}$ is a canonical isomorphism which identifies the space E with its bidual $(E^*)^*$.

PROOF. The map $\hat{x}$ is obviously linear. Let $x \in E$ such that $\hat{x} = 0$, so $\hat{x}(w) = w(x) = 0$ for all $w \in E^*$. If $\dim(E) < \infty$, then $x = 0$, and $x \mapsto \hat{x}$ is injective. Thus, $\dim(E) = \dim(E^*) = \dim((E^*)^*)$ (art. 6.3), and $x \mapsto \hat{x}$ injective implies $x \mapsto \hat{x}$ is bijective. ☐

6.10. Tensor products. Let E be a real vector space. Let A and B be any two tensors of E, covariant, contravariant or mixed. Let A be of order p and B be of order q. We define the tensor product $A \otimes B$ by

$$A \otimes B(a_1)\cdots(a_p)(b_1)\cdots(b_q) = A(a_1)\cdots(a_p) \times B(b_1)\cdots(b_q),$$

where the (a_i) are in the domain of A and the (b_j) in the domain of B. The sign $\times$ denotes the ordinary multiplication of real numbers. The tensor product $A \otimes B$ is a $(p+q)$-tensor. It can be covariant if A and B are covariant, contravariant if A and B are contravariant, mixed otherwise. Moreover, the tensor product is associative, $(A \otimes B) \otimes C = A \otimes (B \otimes C)$. If A or B is a 0-tensor, that is, a real s, then $s \otimes A = A \otimes s = s \times A$.

6.11. Components of a tensor. Let E be a real vector space of finite dimension, and let $\dim(E) = n$. Let A be a tensor of E. Let $\mathcal{B} = (\mathbf{e}_1, \ldots, \mathbf{e}_n)$ be a basis of E, and let $\mathcal{B}^* = (\mathbf{e}^1, \ldots, \mathbf{e}^n)$ be the dual basis of $\mathcal{B}$ (art. 6.3). Let us consider, for example, a tensor A, 2-times covariant and 1-time contravariant. Then, for any $x, y \in E$ and any $w \in E^*$,

$$\begin{aligned} A(x)(y)(w) &= A\Big(\sum_i x^i \mathbf{e}_i\Big)\Big(\sum_j y^j \mathbf{e}_j\Big)\Big(\sum_k w_k \mathbf{e}^k\Big) \\ &= \sum_{i,j,k} x^i y^j w_k A_{ij}^k, \text{ where } A_{ij}^k = A(\mathbf{e}_i)(\mathbf{e}_j)(\mathbf{e}^k). \end{aligned}$$

The numbers A_{ij}^k are called the *components of the tensor* A for the basis $\mathcal{B}$. In fact, the space of these tensors — 2-times covariant and 1-time contravariant — has a natural basis associated with $\mathcal{B}$, the set of all the tensors $\mathbf{e}^i \otimes \mathbf{e}^j \otimes \mathbf{e}_k$, where $i, j, k = 1 \cdots n$. In other words,

$$A = \sum_{i,j,k} A_{ij}^k \, \mathbf{e}^i \otimes \mathbf{e}^j \otimes \mathbf{e}_k.$$

Let us consider the product of two tensors A and B. Let us assume that A^k_{ij} are the components of A for the basis $\mathcal{B}$ and B^{mn}_{pqr} are the components of B, then

$$[A \otimes B]^{kmn}_{ijpqr} = A^k_{ij} \times B^{mn}_{pqr}.$$

Note that the symmetry or the antisymmetry of a tensor — covariant or contravariant — can be checked on its components for any basis. Let us consider, for example, a covariant tensor A, let $\mathcal{B}$ be a basis of E,

$$A_{\sigma(i)\sigma(j)\cdots\sigma(k)} = \begin{cases} A_{ij\ldots k} & \text{if } A \text{ is symmetric,} \\ \operatorname{sgn}(\sigma) \times A_{ij\ldots k} & \text{if } A \text{ is antisymmetric.} \end{cases}$$

For example, for covariant 2-tensors the condition of antisymmetry writes

$$A_{ij} + A_{ji} = 0.$$

6.12. Symmetrization and antisymmetrization of tensors. Let E be a real vector space, and let T be a covariant p-tensor of E.

1. The *symmetrization* of the tensor T is the tensor $\mathrm{Sym}(T)$ defined by

$$\mathrm{Sym}(T)(x_1)\cdots(x_p) = \frac{1}{p!} \sum_{\sigma \in \mathfrak{S}_p} T(x_{\sigma(1)})\cdots(x_{\sigma(p)}).$$

The tensor T is symmetric if and only if $\mathrm{Sym}(T) = T$. The map Sym is a projector from the space of all covariant p-tensors onto the subspace of symmetric covariant p-tensors, $\mathrm{Sym} \circ \mathrm{Sym} = \mathrm{Sym}$.

2. The *antisymmetrization* of T is the tensor $\mathrm{Alt}(T)$ defined by

$$\mathrm{Alt}(T)(x_1)\cdots(x_p) = \frac{1}{p!} \sum_{\sigma \in \mathfrak{S}_p} \operatorname{sgn}(\sigma) \times T(x_{\sigma(1)})\cdots(x_{\sigma(p)}).$$

The tensor T is antisymmetric if and only if $\mathrm{Alt}(T) = T$. The map Alt is a projector from the space of all covariant p-tensors onto the subspace of antisymmetric covariant p-tensors, $\mathrm{Alt} \circ \mathrm{Alt} = \mathrm{Alt}$.

What has been said above, for covariant tensors, can be transposed to contravariant tensors and, then, give projectors to the spaces of symmetric or antisymmetric contravariant tensors.

PROOF. 1. Let us check that $\mathrm{Sym}(T)$ is symmetric. Let $\epsilon \in \mathfrak{S}_p$. We have

$$\begin{aligned} \mathrm{Sym}(T)(x_{\epsilon(1)})\cdots(x_{\epsilon(p)}) &= \frac{1}{p!} \sum_{\sigma \in \mathfrak{S}_p} T(x_{\sigma\circ\epsilon(1)})\cdots(x_{\sigma\circ\epsilon(p)}) \\ &= \frac{1}{p!} \sum_{\sigma' \in \mathfrak{S}_p} T(x_{\sigma'(1)})\cdots(x_{\sigma'(p)}), \quad \text{with } \sigma' = \sigma \circ \epsilon \\ &= \mathrm{Sym}(T)(x_1)\cdots(x_p). \end{aligned}$$

If T is already symmetric, then for all $\sigma \in \mathfrak{S}_p$ $T(x_{\sigma(1)})\cdots(x_{\sigma(p)}) = T(x_1)\cdots(x_p)$, and since there are $p!$ permutations in $\mathfrak{S}_p$, $\mathrm{Sym}(T)(x_1)\cdots(x_p) = T(x_1)\cdots(x_p)$, that is, $\mathrm{Sym}(T) = T$.

2. Let us check that $\mathrm{Alt}(T)$ is antisymmetric. Let $\epsilon \in \mathfrak{S}_p$, we have

$$\begin{aligned}\mathrm{Alt}(T)(x_{\epsilon(1)})\cdots(x_{\epsilon(p)}) &= \frac{1}{p!}\sum_{\sigma\in\mathfrak{S}_p} \mathrm{sgn}(\sigma)\times T(x_{\sigma\circ\epsilon(1)})\cdots(x_{\sigma\circ\epsilon(p)}) \\ &= \frac{1}{p!}\sum_{\sigma'\in\mathfrak{S}_p} \mathrm{sgn}(\sigma'\circ\epsilon^{-1})\times T(x_{\sigma'(1)})\cdots(x_{\sigma'(p)}) \\ &= \mathrm{sgn}(\epsilon)\times \mathrm{Alt}(T)(x_1)\cdots(x_p),\end{aligned}$$

where we denoted $\sigma' = \sigma\circ\epsilon$, and used $\mathrm{sgn}(\epsilon^{-1}) = \mathrm{sgn}(\epsilon)$. Now, if T is already antisymmetric, then for all $\sigma \in \mathfrak{S}_p$, $T(x_{\sigma(1)})\cdots(x_{\sigma(p)}) = \mathrm{sgn}(\sigma)\times T(x_1)\cdots(x_p)$. Thus,

$$\begin{aligned}\mathrm{Alt}(T)(x_1)\cdots(x_p) &= \frac{1}{p!}\sum_{\sigma\in\mathfrak{S}_p} \mathrm{sgn}(\sigma)\times T(x_{\sigma(1)})\cdots(x_{\sigma(p)}) \\ &= \frac{1}{p!}\sum_{\sigma\in\mathfrak{S}_p} \mathrm{sgn}(\sigma)^2\times T(x_1)\cdots(x_p) \\ &= \frac{1}{p!}\sum_{\sigma\in\mathfrak{S}_p} T(x_1)\cdots(x_p) \\ &= T(x_1)\cdots(x_p).\end{aligned}$$

Therefore, $\mathrm{Alt}(T) = T$. □

6.13. Linear p-forms. The antisymmetric covariant tensors play an important role in differential geometry; they are at the basis of the Cartan calculus. It is why they got a special name and notation. Let E be a real vector space, a covariant antisymmetric p-tensor of E is called a *linear* p*-form* of E. The vector space of linear p-forms of E is denoted by $\Lambda^p(E)$. Note that $\Lambda^0(E) = \mathbf{R}$ and $\Lambda^1(E) = E^*$.

NOTE. On the other hand, the space of symmetric covariant p-tensors of E is sometimes denoted by $\mathrm{V}^p(E)$. But symmetric covariant p-tensors did not get a special name, except in special cases, as for Euclidean scalar product for instance.

6.14. Inner product. Let E be a real vector space. Let A be a covariant p-tensor, $p \geq 1$. For any $x \in E$, we denote by $A(x)$ the covariant $(p-1)$-tensor defined by

$$[A(x)](x_1)\cdots(x_{p-1}) = A(x)(x_1)\cdots(x_{p-1}).$$

The tensor $A(x)$ is called the *inner* (or *interior*) *product* of A with x. The map $A \mapsto A(x)$, from $\Lambda^p(E)$ to $\Lambda^{p-1}(E)$, is a linear operator denoted by $\mathrm{Int}(x)$, that is, $\mathrm{Int}(x)(A) = A(x)$. Note that if A is antisymmetric or symmetric, that is, it changes or not its sign under a permutation of two vectors, then $A(x)$ is still antisymmetric or symmetric. Then, the interior product is also an inner product inside the subspace of antisymmetric or symmetric multilinear covariant tensors.

6.15. Exterior product. Let E be a real vector space. Let $A \in \Lambda^p(E)$ and $B \in \Lambda^q(E)$ (art. 6.13), the *exterior product* $A \wedge B \in \Lambda^{p+q}(E)$ is defined by

$$A\wedge B = \frac{(p+q)!}{p!\,q!}\,\mathrm{Alt}(A\otimes B),$$

where Alt has been defined in (art. 6.12). The exterior product with a 0-form $s \in \mathbf{R}$ is simply given by $s\wedge A = A\wedge s = s\times A$.

1. The exterior product is associative, for any three linear forms A, B, C,

$$A \wedge (B \wedge C) = (A \wedge B) \wedge C.$$

2. The commutation of the exterior product is given by the rule,

$$A \wedge B = (-1)^{pq}\, B \wedge A.$$

3. For all $a \in E^* = \Lambda^1(E)$, the map $B \mapsto a \wedge B$, from $\Lambda^p(E)$ to $\Lambda^{p+1}(E)$, is a linear operator denoted by $\mathrm{Ext}(a)$, that is, $\mathrm{Ext}(a)(B) = a \wedge B$. The exterior product $a \wedge B$ takes a special expression,

$$\begin{aligned}(a \wedge B)(x)(x_1)\cdots(x_q) &= a(x) \times B(x_1)(x_2)\cdots(x_q)\\ &- a(x_1) \times B(x)(x_2)\cdots(x_q)\\ &- a(x_2) \times B(x_1)(x)(x_3)\cdots(x_q)\\ &- \cdots\\ &- a(x_q) \times B(x_1)\cdots(x_{q-1})(x).\end{aligned}$$

EXAMPLE. Let $a, b \in E^*$, be two 1-forms, then $a \wedge b = a \otimes b - b \otimes a$. Note that this is different from the antisymmetrization $\mathrm{Alt}(a \otimes b) = (1/2)\; a \wedge b$.

PROOF. 1. Let C be of order r, and let $x_1, \ldots, x_{p+q+r} \in E$. Let $a.bc = A \wedge (B \wedge C)(x_1)\cdots(x_{p+q+r})$ and $ab.c = (A \wedge B) \wedge C(x_1)\cdots(x_{p+q+r})$. We have,

$$\begin{aligned}
a.bc &= \frac{1}{p!\,(q+r)!} \sum_{\sigma \in \mathfrak{S}_{p+q+r}} \mathrm{sgn}(\sigma) \times A(x_{\sigma(1)})\cdots(x_{\sigma(p)})\\
&\times \frac{1}{q!\,r!} \sum_{\epsilon \in \{\sigma(p+1)\cdots\sigma(p+q+r)\}!} \mathrm{sgn}(\epsilon) \times (B \otimes C)(x_{\epsilon\circ\sigma(p+1)})\cdots(x_{\epsilon\circ\sigma(p+q+r)})\\
&= \frac{1}{p!\,q!\,r!} \sum_{\sigma \in \mathfrak{S}_{p+q+r}} \frac{1}{(q+r)!} \sum_{\epsilon \in \{\sigma(p+1)\cdots\sigma(p+q+r)\}!} \mathrm{sgn}(\sigma) \times \mathrm{sgn}(\epsilon)\\
&\times (A \otimes B \otimes C)(x_{\sigma(1)})\cdots(x_{\sigma(p)})(x_{\epsilon\circ\sigma(p+1)})\cdots(x_{\epsilon\circ\sigma(p+q+r)})\\
&= \frac{1}{p!\,q!\,r!} \sum_{\bar\sigma \in \mathfrak{S}_{p+q+r}} \frac{1}{(q+r)!} \sum_{\epsilon \in \{\sigma(p+1)\cdots\sigma(p+q+r)\}!} \mathrm{sgn}(\bar\sigma)\\
&\times (A \otimes B \otimes C)(x_{\bar\sigma(1)})\cdots(x_{\bar\sigma(p+q+r)})\\
&= \frac{1}{p!\,q!\,r!} \sum_{\bar\sigma \in \mathfrak{S}_{p+q+r}} \mathrm{sgn}(\bar\sigma) \times (A \otimes B \otimes C)(x_{\bar\sigma(1)})\cdots(x_{\bar\sigma(p+q+r)}),
\end{aligned}$$

where $\{\sigma(p+1)\cdots\sigma(p+q+r)\}!$ denotes the set of permutations of the indices $\sigma(p+1)\cdots\sigma(p+q+r)$, and where $\bar\sigma = \sigma\circ\bar\epsilon$, with $\bar\epsilon$ defined by $\bar\epsilon \restriction \{\sigma(1)\cdots\sigma(p)\} = \mathbf{1}$ and $\bar\epsilon \restriction \{\sigma(p+1)\cdots\sigma(p+q+r)\} = \epsilon$. So, except the fact that this last expression of $a.bc$ is clearly associative, a simple analogous computation gives the same expression for $ab.c$, thus the exterior product is associative.

2. Let $x_1, \ldots, x_{p+q} \in E$, let us denote $ab = (A \wedge B)(x_1)\cdots(x_{p+q})$, and let $ba = (B \wedge A)(x_1)\cdots(x_{p+q})$. We have

$$\begin{aligned}
ab &= k \sum_{\sigma \in \mathfrak{S}_{p+q}} \mathrm{sgn}(\sigma) \times A(x_{\sigma(1)})\cdots(x_{\sigma(p)}) \times B(x_{\sigma(p+1)})\cdots(x_{\sigma(p+q)})\\
&= k \sum_{\sigma \in \mathfrak{S}_{p+q}} \mathrm{sgn}(\sigma) \times B(x_{\sigma(p+1)})\cdots(x_{\sigma(p+q)}) \times A(x_{\sigma(1)})\cdots(x_{\sigma(p)})
\end{aligned}$$

$$= k \sum_{\sigma \in \mathfrak{S}_{p+q}} \operatorname{sgn}(\sigma) \times B(x_{\sigma\circ\epsilon(1)}) \cdots (x_{\sigma\circ\epsilon(q)}) \times A(x_{\sigma\circ\epsilon(q+1)}) \cdots (x_{\sigma\circ\epsilon(q+p)}),$$

where ϵ is the permutation

$$\epsilon = \left\{ \begin{array}{cccccccc} 1 & 2 & \cdots & q & q+1 & q+2 & \cdots & q+p \\ \downarrow & \downarrow & & \downarrow & \downarrow & \downarrow & & \downarrow \\ p+1 & p+2 & \cdots & p+q & 1 & 2 & \cdots & p \end{array} \right. .$$

Thus,

$$\begin{aligned} ab &= k \sum_{\sigma' \in \mathfrak{S}_{p+q}} \frac{\operatorname{sgn}(\sigma')}{\operatorname{sgn}(\epsilon)} \times B(x_{\sigma'(1)}) \cdots (x_{\sigma'(q)}) \times A(x_{\sigma'(q+1)}) \cdots (x_{\sigma'(q+p)}) \\ &= \frac{k}{(-1)^{p+q}} \sum_{\sigma' \in \mathfrak{S}_{p+q}} \operatorname{sgn}(\sigma') \times B(x_{\sigma'(1)}) \cdots (x_{\sigma'(q)}) \times A(x_{\sigma'(q+1)}) \cdots (x_{\sigma'(q+p)}) \\ &= (-1)^{p+q} \times ba. \end{aligned}$$

Then, since the permutation ϵ is made of pq successive transpositions, $\operatorname{sgn}(\epsilon) = (-1)^{pq}$. Hence, $A \wedge B = (-1)^{pq} B \wedge A$.

3. To compute $a \wedge B$, let $x_0, x_1, \ldots, x_q$ be $q+1$ vectors of E, and let $R = (a \wedge B)(x_0)(x_1) \cdots (x_q)$. Then, by definition,

$$\begin{aligned} R &= \frac{(p+q)!}{1!\,q!} \frac{1}{(p+q)!} \sum_{\sigma \in \mathfrak{S}_{1+q}} \operatorname{sgn}(\sigma) \times a(x_{\sigma(0)}) \times B(x_{\sigma(1)}) \cdots (x_{\sigma(q)}) \\ &= \frac{1}{q!} \sum_{\sigma \mid \sigma(0)=0} \operatorname{sgn}(\sigma) \times a(x_0) \times B(x_{\sigma(1)}) \cdots (x_{\sigma(q)}) \\ &+ \frac{1}{q!} \sum_{\sigma \mid \sigma(0)\neq 0} \operatorname{sgn}(\sigma) \times a(x_{\sigma(0)}) \times B(x_{\sigma(1)}) \cdots (x_{\sigma(q)}) \\ &= \frac{1}{q!} \sum_{\sigma \mid \sigma(0)=0} \operatorname{sgn}(\sigma) \times a(x_0) \times \operatorname{sgn}(\sigma \restriction \{1 \cdots q\}) \times B(x_1) \cdots (x_q) \\ &+ \frac{1}{q!} \sum_{i=1}^{q} \sum_{\sigma \mid \sigma(0)=i} \operatorname{sgn}(\sigma) \times a(x_i) \times B(x_{\sigma(1)}) \cdots (x_{\sigma(q)}). \end{aligned}$$

But, on the one hand, we have $\operatorname{sgn}(\sigma) = \operatorname{sgn}(\sigma \restriction \{1 \cdots q\})$. On the other hand we can decompose every permutation σ such that $\sigma(0) = i$ into the product of the following two permutations.

$$\sigma = \left\{ \begin{array}{ccccccc} 0 & 1 & \cdots & i & \cdots & q \\ \downarrow & \downarrow & & \downarrow & & \downarrow \\ i & 1 & \cdots & 0 & \cdots & q \\ \downarrow & \downarrow & & \downarrow & & \downarrow \\ i & \sigma(1) & \cdots & \sigma(0) & \cdots & \sigma(q) \end{array} \right. \begin{array}{l} \left.\vphantom{\begin{array}{c}a\\a\end{array}}\right\} \tau \\ \left.\vphantom{\begin{array}{c}a\\a\end{array}}\right\} \epsilon \end{array}$$

Thus, for each σ and i such that $\sigma(0) = i$, we have $B(x_{\sigma(1)}) \cdots (x_{\sigma(q)}) = \operatorname{sgn}(\epsilon) \times B(x_1) \cdots (x_0) \cdots (x_q)$. Therefore, because $\operatorname{sgn}(\tau) = -1$, we get

$$\begin{aligned}
R &= \frac{1}{q!} \sum_{\sigma \in \mathfrak{S}_q} \operatorname{sgn}(\sigma) \times a(x_0) \times \operatorname{sgn}(\sigma) \times B(x_1) \cdots (x_q) \\
&+ \frac{1}{q!} \sum_{i=1}^{q} \sum_{\epsilon \in \mathfrak{S}_q} \operatorname{sgn}(\epsilon) \times \operatorname{sgn}(\tau) \times a(x_i) \times \operatorname{sgn}(\epsilon) \times B(x_1) \cdots (x_0) \cdots (x_q) \\
&= a(x_0) \times B(x_1) \cdots (x_q) \\
&- \sum_{i=1}^{q} a(x_i) \times B(x_1) \cdots (x_0) \cdots (x_q),
\end{aligned}$$

where, in the second term of the equalities, x_0 is at the rank i. □

6.16. Exterior monomials and basis of $\Lambda^p(E)$. Let E be a real vector space. Let $a_1, \ldots, a_p$ be p covectors of E. Let us define

$$a_1 \wedge a_2 \wedge \cdots \wedge a_p = a_1 \wedge (a_2 \wedge (\cdots \wedge a_p) \cdots),$$

where the exterior product has been defined in (art. 6.15). Such linear p-forms $a_1 \wedge a_2 \wedge \cdots \wedge a_p$ are called *exterior monomials* of degree p.

1. The exterior monomial $a_1 \wedge \cdots \wedge a_p$ is explicitly given by

$$a_1 \wedge \cdots \wedge a_p = \sum_{\sigma \in \mathfrak{S}_p} \operatorname{sgn}(\sigma) \times a_{\sigma(1)} \otimes \cdots \otimes a_{\sigma(p)}.$$

2. The value of $a_1 \wedge \cdots \wedge a_p$, applied to $v_1, \ldots, v_p \in E$, is given by

$$(a_1 \wedge \cdots \wedge a_p)(v_1) \cdots (v_p) = \sum_{\sigma \in \mathfrak{S}_p} \operatorname{sgn}(\sigma) \times a_1(v_{\sigma(1)}) \cdots a_p(v_{\sigma(p)}).$$

3. If one of the covectors of the family is a linear combination of the others, the exterior monomial $a_1 \wedge \cdots \wedge a_p$ is zero.

4. Let us assume that E is finite dimensional, with $\dim(E) = n$, and let $\mathcal{B} = (\mathbf{e}_1, \cdots, \mathbf{e}_n)$ be a basis. The set of monomials of degree p

$$\{\mathbf{e}^i \wedge \mathbf{e}^j \wedge \cdots \wedge \mathbf{e}^k\}, \quad \text{where} \quad i < j < \cdots < k \in \{1, \cdots n\},$$

is a basis of the vector space $\Lambda^p(E)$. In other words, for every linear p-form A, there exists a family of numbers $A_{ij\cdots k}$ such that

$$A = \sum_{i<j<\cdots<k} A_{ij\cdots k}\, \mathbf{e}^i \wedge \mathbf{e}^j \wedge \cdots \wedge \mathbf{e}^k, \text{ with } A_{ij\cdots k} = A(\mathbf{e}_i)(\mathbf{e}_j) \cdots (\mathbf{e}_k).$$

The numbers $A_{ij\cdots k}$, for all increasing sequences $i < j < \cdots < k$, where the indices are running from 1 to n, are called the *components*, or the *coordinates*, of the linear p-form A in the basis $\mathcal{B}$.

5. Let $\dim(E) = n$. Since every element of a basis of $\Lambda^p(E)$ is defined by a choice of p different indices out of n, the dimension of $\Lambda^p(E)$ is the Pascal binomial coefficient C_n^p, that is,

$$\dim(\Lambda^p(E)) = \frac{n!}{p!\,(n-p)!} = C_n^p.$$

Note that, thanks to this decomposition and to antisymmetry,

$$\text{if } p > \dim(E), \text{ then } \Lambda^p(E) = \{0\}.$$

PROOF. 1. We prove the first assertion by recurrence. Let us assume that the proposition is true for $a_1 \wedge \cdots \wedge a_p$, then,

$$\begin{aligned}
a_0 \wedge (a_1 \wedge \cdots \wedge a_p) &= a_0 \wedge \sum_{\sigma \in \mathfrak{S}_p} \operatorname{sgn}(\sigma) \times a_{\sigma(1)} \otimes \cdots \otimes a_{\sigma(p)} \\
&= \sum_{\sigma \in \mathfrak{S}_p} \operatorname{sgn}(\sigma) \times a_0 \wedge (a_{\sigma(1)} \otimes \cdots \otimes a_{\sigma(p)}) \\
&= \sum_{\sigma \in \mathfrak{S}_p} \operatorname{sgn}(\sigma) \times a_0 \otimes a_{\sigma(1)} \otimes \cdots \otimes a_{\sigma(p)} \\
&- \sum_{\sigma \in \mathfrak{S}_p} \operatorname{sgn}(\sigma) \times a_{\sigma(1)} \otimes a_0 \otimes \cdots \otimes a_{\sigma(p)} \\
&- \cdots \\
&- \sum_{\sigma \in \mathfrak{S}_p} \operatorname{sgn}(\sigma) \times a_{\sigma(p)} \otimes a_{\sigma(1)} \otimes \cdots \otimes a_0 \\
&= \sum_{\sigma' \in \mathfrak{S}_{p+1}} \operatorname{sgn}(\sigma') \times a_{\sigma'(0)} \otimes a_{\sigma'(1)} \otimes \cdots \otimes a_{\sigma'(p)} \\
&= a_0 \wedge a_1 \wedge \cdots \wedge a_p.
\end{aligned}$$

Now, the proposition being true for $p = 1$, it is true for any p.

2. Let us make the change of variable, σ into σ^{-1}, which does not change the sum, and since the products involved in the sum are independent of the ordering, we get the result:

$$\begin{aligned}
(a_1 \wedge \cdots \wedge a_p)(v_1) \cdots (v_p) &= \sum_{\sigma \in \mathfrak{S}_p} \operatorname{sgn}(\sigma) \times a_{\sigma(1)}(v_1) \cdots a_{\sigma(p)}(v_p) \\
&= \sum_{\sigma \in \mathfrak{S}_p} \operatorname{sgn}(\sigma^{-1}) \times a_1(v_{\sigma(1)}) \cdots a_p(v_{\sigma(p)}) \\
&= \sum_{\sigma \in \mathfrak{S}_p} \operatorname{sgn}(\sigma) \times a_1(v_{\sigma(1)}) \cdots a_p(v_{\sigma(p)}).
\end{aligned}$$

3. Let us first assume that $a_i = a_j = a$. By 2 we have

$$\begin{aligned}
(a_1 \wedge \cdots \wedge a_p)(v_1) \cdots (v_p) &= \sum_{\sigma \in \mathfrak{S}_p} \operatorname{sgn}(\sigma) \times a_1(v_{\sigma(1)}) \cdots a_p(v_{\sigma(p)}) \\
&= \sum_{\sigma \in \mathfrak{S}_p} \operatorname{sgn}(\sigma) \times a(v_{\sigma(i)}) a(v_{\sigma(j)}) \times (\star),
\end{aligned}$$

where $(\star)$ contains terms involving all the indices different from i and j. Thus, denoting by ε the transposition $i \leftrightarrow j$, we have

$$\begin{aligned}
(a_1 \wedge \cdots \wedge a_p)(v_1) \cdots (v_p) &= \sum_{\sigma \in \mathfrak{S}_p} \operatorname{sgn}(\sigma) \times a(v_{\sigma(i)}) a(v_{\sigma(j)}) \times (\star) \\
&= \frac{1}{2}\Bigg[\sum_{\sigma \in \mathfrak{S}_p} \operatorname{sgn}(\sigma) \times a(v_{\sigma(i)}) a(v_{\sigma(j)}) \times (\star) \\
&+ \sum_{\sigma \in \mathfrak{S}_p} \operatorname{sgn}(\sigma) \times a(v_{\sigma(j)}) a(v_{\sigma(i)}) \times (\star) \Bigg]
\end{aligned}$$

$$= \frac{1}{2}\Big[\sum_{\sigma\in\mathfrak{S}_p} \operatorname{sgn}(\sigma)\times a(v_{\sigma(i)})a(v_{\sigma(j)}\times(\star)$$
$$+ \sum_{\sigma\in\mathfrak{S}_p} \operatorname{sgn}(\sigma)\times a(v_{\sigma\varepsilon(i)})a(v_{\sigma\varepsilon(j)})\times(\star)\Big].$$

Let $\sigma' = \sigma\varepsilon$. Since ε acts trivially on $(\star)$, and since $\operatorname{sgn}(\sigma') = \operatorname{sgn}(\sigma)\times\operatorname{sgn}(\varepsilon) = \operatorname{sgn}(\sigma)\times(-1) = -\operatorname{sgn}(\sigma)$, we get

$$\begin{aligned}(a_1\wedge\cdots\wedge a_p)(v_1)\cdots(v_p) &= \frac{1}{2}\Big[\sum_{\sigma\in\mathfrak{S}_p} \operatorname{sgn}(\sigma)\times a(v_{\sigma(i)})a(v_{\sigma(j)}\times(\star)\\ &- \sum_{\sigma'\in\mathfrak{S}_p} \operatorname{sgn}(\sigma')\times a(v_{\sigma'(i)})a(v_{\sigma'(j)})\times(\star)\Big]\\ &= 0.\end{aligned}$$

We then get the result by linearity.

4. Let us consider a tensor A. In the basis $\mathcal{B}$, we get from (art. 6.11)

$$A = \sum_{i,j,\cdots,k=1}^{n} A_{ij\cdots k}\,\mathbf{e}^i\otimes\mathbf{e}^j\cdots\otimes\mathbf{e}^k \quad\text{where}\quad A_{ij\cdots k} = A(\mathbf{e}_i)(\mathbf{e}_j)\cdots(\mathbf{e}_k).$$

Moreover, the components of the tensor A in the basis $\mathcal{B}$ are totally antisymmetric (art. 6.11), that is,

$$\text{for all } \sigma\in\mathfrak{S}_p,\ A_{\sigma(i)\sigma(j)\cdots\sigma(k)} = \operatorname{sgn}(\sigma)\times A_{ij\cdots k}.$$

Thus, the form A writes

$$\begin{aligned}A &= \sum_{i<j<\cdots<k=1}^{n} A_{ij\cdots k}\sum_{\sigma\in\mathfrak{S}_p}\operatorname{sgn}(\sigma)\times\mathbf{e}^{\sigma(i)}\otimes\mathbf{e}^{\sigma(j)}\cdots\otimes\mathbf{e}^{\sigma(k)}\\ &= \sum_{i<j<\cdots<k=1}^{n} A_{ij\cdots k}\,\mathbf{e}^i\wedge\mathbf{e}^j\wedge\cdots\wedge\mathbf{e}^k.\end{aligned}$$

Hence, the set of monomials $\mathbf{e}^i\wedge\mathbf{e}^j\wedge\cdots\wedge\mathbf{e}^k$, where $i<j<\cdots<k$ run from 1 to n, is a basis of the vector space $\Lambda^p(E)$. The numbers $A_{ij\cdots k}$, where $i<j<\cdots<k$, are the *components*, or the *coordinates*, of the form A in the basis $\mathcal{B}$.

5. If $p > \dim(E)$, then each monomial of the basis will contain necessarily two repeated terms, thus all the monomials of the basis are zero, thanks to 3. Hence, $\Lambda^p(E) = 0$. □

6.17. Pullbacks of tensors and forms. Let E and F be two real vector spaces, let $M : E \to F$ be a linear map, $M \in E * F = L(E,F)$. Let A be a covariant p-tensor of F (art. 6.8). The *pullback* of A by M is the covariant p-tensor of E defined by

$$M^*(A)(v_1)\cdots(v_p) = A(M(v_1))\cdots(M(v_p)),\quad\text{where } v_1,\ldots,v_p\in E.$$

Here are some of the most important properties of the pullback operation. The pullback is a linear operator, for any pair of covariant p-tensors A and B, and for any real λ

$$M^*(A+B) = M^*(A) + M^*(B) \quad\text{and}\quad M^*(\lambda\times A) = \lambda\times M^*(A).$$

The pullback is contravariant, let G be a third vector space, $N : F \to G$ be a linear map, A a p-tensor on G, then

$$(N \circ M)^* = M^* \circ N^*, \text{ that is, } (N \circ M)^*(A) = M^*(N^*(A)).$$

The pullback clearly respects the tensor product of covariant tensors (art. 6.10). Now, if A is a p-form, that is, $A \in \Lambda^p(F)$ (art. 6.13), then its pullback by M also is a p-form. The pullback respects the exterior product too, that is, for any pair of linear forms A and B, we have

$$M^*(A \wedge B) = M^*(A) \wedge M^*(B).$$

In other words, the pullback is a morphism for the exterior algebra of linear forms.

6.18. Coordinates of the pullback of linear forms. Let E and F be two real vector spaces of dimensions n and m. Let A be a linear p-form of F. Let $\mathcal{E} = (\mathbf{e}_1, \ldots, \mathbf{e}_n)$ be a basis of E and $\mathcal{F} = (\mathbf{f}_1, \ldots, \mathbf{f}_m)$ be a basis of F. Let $A_{i\cdots k}$ be the coordinates of A in the basis $\mathcal{F}$ (art. 6.16),

$$A = \sum_{i<\cdots<j} A_{i\cdots j} \times \mathbf{f}^i \wedge \cdots \wedge \mathbf{f}^j.$$

Then, the coordinates of $M^*(A)$ in the basis $\mathcal{E}$ are given by

$$[M^*(A)]_{k\cdots\ell} = \sum_{i<\cdots<j} \sum_{\sigma\in\{i\cdots j\}!} \operatorname{sgn}(\sigma) \times M_k^{\sigma(i)} \cdots M_\ell^{\sigma(j)} A_{i\cdots j}, \text{ with } M_k^i = \mathbf{f}^i(M\mathbf{e}_k),$$

where $\{i \cdots j\}!$ denotes the group of permutations of the set of indices $\{i \cdots j\}$.

PROOF. By definition $[M^*(A)]_{k\cdots\ell} = M^*(A)(\mathbf{e}_k)\cdots(\mathbf{e}_\ell) = A(M\mathbf{e}_k)\cdots(M\mathbf{e}_\ell)$, but $M\mathbf{e}_k = \sum_i M_k^i \mathbf{f}_i$. Thus,

$$\begin{aligned}
[M^*(A)]_{k\cdots\ell} &= A\Big(\sum_i M_k^i \mathbf{f}_i\Big)\cdots\Big(\sum_j M_\ell^i \mathbf{f}_j\Big) \\
&= \sum_i \cdots \sum_j M_k^i \cdots M_\ell^i A(\mathbf{f}_i)\cdots A(\mathbf{f}_j) \\
&= \sum_i \cdots \sum_j M_k^i \cdots M_\ell^i A_{i\cdots j} \\
&= \sum_{i<\cdots<j} \sum_{\sigma\in\{i\cdots j\}!} \operatorname{sgn}(\sigma) \times M_k^{\sigma(i)} \cdots M_\ell^{\sigma(j)} A_{i\cdots j},
\end{aligned}$$

since $A_{i\cdots j} = \operatorname{sgn}(\sigma) \times A_{\sigma(i)\cdots\sigma(j)}$ (art. 6.16). □

6.19. Volumes and determinants. Let E be a real vector space of dimension n. The space $\Lambda^n(E)$ of linear n-forms of E has dimension 1. Thanks to the decomposition of forms on any basis $\mathcal{B} = (\mathbf{e}_1, \ldots, \mathbf{e}_n)$ (art. 6.16), every linear n-form of E is proportional to $\mathrm{vol}_\mathcal{B}$, defined by

$$\mathrm{vol}_\mathcal{B} = \mathbf{e}^1 \wedge \mathbf{e}^2 \cdots \wedge \mathbf{e}^n, \quad \text{and} \quad \Lambda^n(E) = \mathbf{R} \times \mathrm{vol}_\mathcal{B}. \qquad (\diamondsuit)$$

We call *volume* any nonzero form belonging to $\Lambda^n(E)$. We shall denote the set of all the volumes of E by Volumes(E),

$$\mathrm{Volumes}(E) = \{\mathrm{vol} \in \Lambda^n(E) \mid \mathrm{vol} \neq 0\}.$$

For $E = \mathbf{R}^n$, the canonical basis (art. 6.2) defines by ($\diamondsuit$) a *canonical volume* denoted by vol_n. Now, let M be a linear map from E to E, and let $M^*(\mathrm{vol})$ be the pullback (art. 6.17) of some volume $\mathrm{vol} \in \Lambda^n(E)$ by M. Let us recall that, by definition, for all $v_1, \ldots, v_n \in E$, $M^*(\mathrm{vol})(v_1)\cdots(v_n) = \mathrm{vol}(Mv_1)\cdots(Mv_n)$. Since the space $\Lambda^n(E)$ is 1-dimensional and $\mathrm{vol} \neq 0$, the n-form $M^*(\mathrm{vol})$ is proportional to vol. The coefficient of proportionality is independent on the choice of the volume, it is called the *determinant* of the linear map M, and it is denoted by $\det(M)$,

$$M^*(\mathrm{vol}) = \det(M) \times \mathrm{vol} \quad \text{or} \quad \det(M) = \frac{M^*(\mathrm{vol})}{\mathrm{vol}}. \qquad (\clubsuit)$$

So, for every family of vectors $v_1, \ldots, v_n \in E$,

$$\mathrm{vol}(Mv_1)\cdots(Mv_n) = \det(M) \times \mathrm{vol}(v_1)\cdots(v_n).$$

Actually, the computation of $\det(M)$ can be done by evaluating the previous identity on the vectors of some basis $\mathcal{B} = (\mathbf{e}_1, \ldots, \mathbf{e}_n)$ of E, that is,

$$\det(M) = \mathrm{vol}_{\mathcal{B}}(M\mathbf{e}_1)\cdots(M\mathbf{e}_n) \quad \text{with} \quad \mathrm{vol}_{\mathcal{B}} = \mathbf{e}^1 \wedge \mathbf{e}^2 \cdots \wedge \mathbf{e}^n. \qquad (\heartsuit)$$

Now, let $M_1 = M(\mathbf{e}_1), \ldots, M_n = M(\mathbf{e}_n)$ be the images of the vectors of the basis $\mathcal{B}$ by the linear map M. The determinant of M is given, in terms of the matrix coefficients M_i^j of M in the basis $\mathcal{B}$, by

$$\det(M) = \sum_{\sigma \in \mathfrak{S}_n} \mathrm{sgn}(\sigma) \times M_1^{\sigma(1)} \times \cdots \times M_n^{\sigma(n)} \quad \text{with} \quad M_i^j = \mathbf{e}^j(M\mathbf{e}_i). \qquad (\spadesuit)$$

The choice of a basis $\mathcal{B}$ of E defines an *orientation*, that is, the set of the bases $\mathcal{B}'$ such that $\det(\mathcal{B}'^{-1}\mathcal{B}) = 1$. With what precedes, it is equivalent to choose an orientation or to choose a volume for the vector space E. Finally, by direct application of the definition (Exercise 92, p. 140) we can establish the usual representation formulas:

$$\det(MN) = \det(M) \times \det(N) \quad \text{and} \quad \det(s \times M) = s^n \times \det(M),$$

where M, N are two linear maps from $\mathbf{R}^n$ to $\mathbf{R}^n$, and $s \in \mathbf{R}$.

PROOF. Let us check first that the determinant of a linear map M does not depend on the choice of the volume. Let vol' be another volume. Thus, $\mathrm{vol}' = c \times \mathrm{vol}$, where $c \neq 0$. Now, $M^*(\mathrm{vol}') = M^*(c \times \mathrm{vol}) = c \times M^*(\mathrm{vol}) = c \times \det(M) \times \mathrm{vol} = \det(M) \times \mathrm{vol}'$, that is, $M^*(\mathrm{vol}') = \det(M) \times \mathrm{vol}'$. Now, let us prove the formula ($\spadesuit$). On the one hand, by application of the definition of the exterior product (art. 6.16), we get $\mathrm{vol}_{\mathcal{B}}(M_1)\cdots(M_n) = \sum_{\sigma \in \mathfrak{S}_n} \mathrm{sgn}(\sigma) \times M_1^{\sigma(1)} \times \cdots \times M_n^{\sigma(n)}$, where the $M_i^j = \mathbf{e}^j(M_i) = \mathbf{e}^j(M\mathbf{e}_i)$, $j = 1 \cdots n$, are the coordinates of the vector M_i in the basis $\mathcal{B}$. On the other hand, $\mathrm{vol}_{\mathcal{B}}(M\mathbf{e}_1)\cdots(M\mathbf{e}_n) = \det(M) \times \mathrm{vol}_{\mathcal{B}}(\mathbf{e}_1)\cdots(\mathbf{e}_n)$. But $\mathrm{vol}_{\mathcal{B}}(\mathbf{e}_1)\cdots(\mathbf{e}_n) = (\mathbf{e}^1 \wedge \mathbf{e}^2 \cdots \wedge \mathbf{e}^n)(\mathbf{e}_1)\cdots(\mathbf{e}_n) = 1$. So, $\det(M) = \sum_{\sigma \in \mathfrak{S}_n} \mathrm{sgn}(\sigma) \times M_1^{\sigma(1)} \times \cdots \times M_n^{\sigma(n)}$. ☐

Exercises

✎ EXERCISE 88 (Antisymmetric 3-form). Let A be a 3-form on a finite dimensional vector space. Show that an antisymmetric tensor A is zero if and only if, for some basis, its coordinates satisfy $A_{ijk} + A_{jki} + A_{kij} = 0$, for every triple of indices i, j, k.

✎ EXERCISE 89 (Expanding the exterior product). Let E be a vector space. Let a, b and c be three 1-forms of E. Expand the evaluation of the triple product $\mathrm{Ext}(a)(\mathrm{Ext}(b)(c))$ on the triple of vectors $(x_1)(x_2)(x_3)$, where Ext is defined in (art. 6.15). Check that

$$[\mathrm{Ext}(a)(\mathrm{Ext}(b)(c))](x_1)(x_2)(x_3) = \sum_{\sigma \in \mathfrak{S}_3} \mathrm{sgn}(\sigma) a(x_{\sigma(1)}) b(x_{\sigma(2)}) c(x_{\sigma(3)}).$$

✎ EXERCISE 90 (Determinant and isomorphisms). Let E be an n-dimensional vector space, and let vol be a volume. Check that, for n vectors $v_i \in E$, $\mathrm{vol}(v_1)\cdots(v_n) = 0$ if and only if the v_i are not linearly independent. Deduce that the determinant of $M \in L(E)$ is not zero if and only if $\ker(M) = \{0\}$, that is, if and only if M is an isomorphism.

✎ EXERCISE 91 (Determinant is smooth). Check that the determinant is a smooth map from $L(E)$ to $\mathbf{R}$. Let $P : U \to \mathrm{GL}(E)$ be any m-plot. Let $r \in U$, $M = P(r)$ and δr be any vector of $\mathbf{R}^m$. Let δM be the variation $\delta M = D(P)(r)(\delta r)$, and let $\delta[\det(M)] = D(r \mapsto \det(P(r))(r)(\delta r)$ be the variation of the determinant. Show that

$$\delta[\det(M)] = \det(M) \times \mathrm{Tr}(M^{-1}\delta M),$$

where Tr is the trace operator on matrices.

✎ EXERCISE 92 (Determinant of a product). Show that for every pair M, N, of elements of $L(E)$, where E is a finite dimensional vector space, we have $\det(MN) = \det(M) \times \det(N)$. Deduce that $\det(s \times M) = s^n \times \det(M)$, with $s \in \mathbf{N}$.

Smooth Forms on Real Domains

This section introduces the notion of smooth forms defined on real domains, and most of the related material. Then, this definition and associated constructions will be naturally extended to diffeological spaces, in order to define the Cartan-De Rham calculus in the larger context of diffeology, (art. 6.28) *and after.*

6.20. Smooth forms on real domains. Let $U \subset \mathbf{R}^n$ be a real domain (art. 1.1). Equip $\Lambda^p(\mathbf{R}^n)$ with its smooth structure of finite dimensional vector space (art. 6.16). A *smooth* p*-form* of (or on) U is any smooth map $\omega : U \to \Lambda^p(\mathbf{R}^n)$. The set of all the smooth p-forms of U,

$$\mathcal{C}^\infty(U, \Lambda^p(\mathbf{R}^n)),$$

is naturally a real vector space. Let $\omega, \omega' \in \mathcal{C}^\infty(U, \Lambda^p(\mathbf{R}^n))$ and $s \in \mathbf{R}$. The sum $\omega + \omega'$ and the product $s\omega$ are given by

$$\text{for all } x \in U : \quad \left\{ \begin{array}{rcl} (\omega + \omega')(x) & = & \omega(x) + \omega'(x), \\ (s \times \omega)(x) & = & s \times (\omega(x)). \end{array} \right.$$

For example, any form $A \in \Lambda^p(\mathbf{R}^n)$ defines the constant smooth p-form $x \mapsto A$ on U. Thus, $\Lambda^p(\mathbf{R}^n)$ is naturally a subspace of the space $\mathcal{C}^\infty(U, \Lambda^p(\mathbf{R}^n))$ of smooth p-forms of U.

NOTE. More generally, *smooth tensors* on U are defined the same way, as smooth maps defined on U with values in some fixed space of linear tensors (art. 6.8). But the smooth tensors which may have a role in diffeology are the *covariant smooth tensors*, since they can be transported by pullback; for example (art. 6.28).

6.21. Components of smooth forms. Let $\mathcal{B} = (\mathbf{e}_1, \ldots, \mathbf{e}_n)$ be the canonical basis of $\mathbf{R}^n$ (art. 6.2). So, for every $x \in U$, for every smooth p-form ω, the p-form $\omega(x)$ breaks down into components over the basis $\{\mathbf{e}^i \wedge \mathbf{e}^j \wedge \cdots \wedge \mathbf{e}^k\}_{i<j<\cdots<k}$ (art. 6.16),

$$\omega(x) = \sum_{i<j<\cdots<k} \omega_{ij\cdots k}(x) \times \mathbf{e}^i \wedge \mathbf{e}^j \wedge \cdots \wedge \mathbf{e}^k.$$

Since ω is a smooth map, the C^p_n parametrizations $\omega_{ij\cdots k} : U \to \mathbf{R}$ are smooth. They are still called the *components of the smooth form* ω *in the basis* $\mathcal{B}$, even if they are not anymore constant but smooth real parametrizations. Conversely, C^p_n smooth real parametrizations $\{\omega_{ij\cdots k}\}_{i<j<\cdots<k}$, defined on U, define by the expression above a unique smooth p-form ω of U.

NOTATION. We shall sometimes denote by ω_x or by $\omega(x)$ the value of the smooth form ω, at the point x.

6.22. Pullbacks of smooth forms. Let $U \subset \mathbf{R}^m$ and $V \subset \mathbf{R}^n$ be two real domains. Let $\omega \in \mathcal{C}^\infty(V, \Lambda^p(\mathbf{R}^n))$, and let $f : U \to V$ be a smooth parametrization. The *pullback* of ω by f, denoted by $f^*(\omega)$, is defined by

$$f^*(\omega)(u)(x_1)\cdots(x_p) = \omega(f(u))(D(f)(u)(x_1))\cdots(D(f)(u)(x_p)),$$

for every $u \in U$, and every $x_1, \ldots, x_p \in \mathbf{R}^n$. We have denoted by $D(f)(u)$ the tangent map of f at the point u (art. 1.4). Now, since components of the pullback $f^*(\omega)$ are linear combinations of the components of ω with the components of the linear map $D(f)$, which are also smooth functions, the pullback $f^*(\omega)$ is a smooth p-form of U, that is, $f^*(\omega) \in \mathcal{C}^\infty(U, \Lambda^p(\mathbf{R}^m))$. Note that, by definition (art. 6.17),

$$f^*(\omega)(u) = (D(f)(u))^*(\omega(f(u))).$$

Next, since tangent maps satisfy the chain-rule property,

$$D(g \circ f)(u) = D(g)(f(u)) \circ D(f)(u),$$

and thanks to the composition of pullbacks of linear forms (art. 6.17), we get the contravariant property of the pullback. Let U, V and W be three real domains. Let $f : U \to V$ and $g : V \to W$ be two smooth maps, for any smooth p-form ω on W,

$$(g \circ f)^*(\omega) = f^*(g^*(\omega)).$$

NOTE. The definition of pullback of smooth forms, and its contravariant property, applies to general covariant smooth tensors, and not just to forms.

PROOF. We need only check the last sentence. Let $u \in U$ and $v = f(u)$. We have

$$\begin{aligned} (g \circ f)^*(\omega)(u) &= (D(g \circ f)(u))^*(\omega(g \circ f(u))) \\ &= [D(g)(f(u)) \circ D(f)(u)]^*[\omega(g(f(u)))] \\ &= D(f)(u)^*[D(g)(v)^*(\omega(g(v))] \\ &= D(f)(u)^*[g^*(\omega)(v)] \\ &= f^*(g^*(\omega))(u). \end{aligned}$$

Thus, $(g \circ f)^*(\omega) = f^*(g^*(\omega))$. ☐

6.23. The differential of a function as pullback. As an example of pullback of form (art. 6.22), the pullback of the volume form of $\mathbf{R}$ is interesting. Let us consider the canonical constant 1-form θ on $\mathbf{R}$:

$$\Lambda^1(\mathbf{R}) = \mathbf{R}\,\theta \quad \text{with} \quad \theta(t) = \mathbf{1}_{\mathbf{R}}, \ \textit{i.e.}, \quad \theta_t(s) = s, \quad \forall s \in \mathbf{R}.$$

Let $f : U \to \mathbf{R}$ be an smooth n-parametrization. Let $u \in U$ and $x \in \mathbf{R}^n$. Then,

$$\begin{aligned} f^*(\theta)_u(x) &= \theta_{f(x)}(D(f)(u)(x)) \\ &= D(f)(u)(x) \\ &= df_u(x). \end{aligned}$$

Hence,

$$f^*(\theta) = df$$

is the *differential* of the function f. Applied to $\mathbf{R}$ itself, the form θ is the differential of the identity map $\mathbf{1}_{\mathbf{R}} = [t \mapsto t]$, and the form θ is also denoted by $\mathbf{dt}$, which is summarized by

$$f^*(dt) = df, \quad \text{where} \quad \Lambda^1(\mathbf{R}) = \mathbf{R}\,dt.$$

Note that this notation, which is the standard notation, is not very coherent. The differential $\mathbf{dt}$ should be denoted by $d[t \mapsto t]$ or $d\mathbf{1}_{\mathbf{R}}$, and then $df = f^*(d\mathbf{1}_{\mathbf{R}})$, which is the coherent notation. Otherwise, we have to interpret differently the symbolic construction $\mathbf{dt}$. In the same way, let us consider $\mathbf{R}^n$, and the projections *coordinates*:

$$\mathbf{x}^k : \mathbf{R}^n \to \mathbf{R} \quad \text{such that} \quad \mathbf{x}^k : x \mapsto x^k.$$

For all points $x \in \mathbf{R}^n$ and all vectors $u \in \mathbf{R}^n$, the differential of the function $\mathbf{x}^k$, that is, $\mathbf{dx}^k = (\mathbf{x}^k)^*(\theta)$, satisfies

$$\mathbf{dx}^k_x(u) = \mathbf{dx}^k_x\left(\sum_{i=1}^n u^i \mathbf{e}^i\right) = u^k = \mathbf{e}^k(u), \quad \text{which implies} \quad \mathbf{dx}^k = \mathbf{e}^k.$$

Thus, any exterior monomial writes also

$$\mathbf{e}^i \wedge \mathbf{e}^j \wedge \cdots \wedge \mathbf{e}^k = \mathbf{dx}^i \wedge \mathbf{dx}^j \wedge \cdots \wedge \mathbf{dx}^k.$$

By abuse of notation, or for some better reason, the monomial $\mathbf{dx}^i \wedge \mathbf{dx}^j \wedge \cdots \wedge \mathbf{dx}^k$ is also simply written $dx^i \wedge dx^j \wedge \cdots \wedge dx^k$.

6.24. Exterior derivative of smooth forms. There exists an operation called *exterior differentiation*, denoted by the letter d and defined by the following properties:

1. For all integers n and all n-domains U,

$$d : \mathcal{C}^\infty(U, \Lambda^p(\mathbf{R}^n)) \to \mathcal{C}^\infty(U, \Lambda^{p+1}(\mathbf{R}^n)).$$

2. If $p = 0$, for every smooth real parametrization $f : x \mapsto a$,

$$df(x) = \sum_{\ell=1}^n \frac{\partial a}{\partial x^\ell}\,\mathbf{e}^\ell = \sum_{\ell=1}^n \frac{\partial a}{\partial x^\ell}\,dx^\ell.$$

3. If α is monomial, $\alpha : x \mapsto a\,\mathbf{e}^i \wedge \cdots \wedge \mathbf{e}^k$ with $[x \mapsto a] \in \mathcal{C}^\infty(U, \mathbf{R})$, then

$$(d\alpha)(x) = \sum_{\ell=1}^n \frac{\partial a}{\partial x^\ell}\,\mathbf{e}^\ell \wedge \mathbf{e}^i \wedge \cdots \wedge \mathbf{e}^k = \sum_{\ell=1}^n \frac{\partial a}{\partial x^\ell}\,dx^\ell \wedge dx^i \wedge \cdots \wedge dx^k.$$

4. The operator d is extended by linearity to any smooth p-form ω,

$$\omega = \sum_{i<\cdots<k} \omega_{i\cdots k}\, \mathbf{e}^i \wedge \cdots \wedge \mathbf{e}^k \quad \Rightarrow \quad d\omega = \sum_{i<\cdots<k} d[\omega_{i\cdots k}\, \mathbf{e}^i \wedge \cdots \wedge \mathbf{e}^k].$$

The smooth form $d\omega$ is called the *exterior derivative* of ω. For example, the exterior derivative of a 1-form $\alpha = \sum_{i=1}^n a_i\, dx^i$, defined on some domain U, is given by

$$d\alpha = \sum_{k=1}^n \sum_{i=1}^n \frac{\partial a_k}{\partial x^i}\, dx^i \wedge dx^k = \sum_{1\le i<k\le n} \left(\frac{\partial a_k}{\partial x^i} - \frac{\partial a_i}{\partial x^k}\right) dx^i \wedge dx^k.$$

6.25. Exterior derivative commutes with pullback. Let $U \subset \mathbf{R}^n$ and $U' \subset \mathbf{R}^{n'}$ be two real domains. Let $f : U' \to U$ be a smooth map, and let $\alpha \in \mathcal{C}^\infty(U, \Lambda^p(\mathbf{R}^n))$. We have, $f^*(d\alpha) = d(f^*\alpha)$. In other words, the exterior derivative and pullback commutes, $f^* \circ d = d \circ f^*$.

$$\begin{array}{ccc}
\mathcal{C}^\infty(U, \Lambda^p(\mathbf{R}^n)) & \xrightarrow{\ d\ } & \mathcal{C}^\infty(U, \Lambda^{p+1}(\mathbf{R}^n)) \\
\Big\downarrow f^* & & \Big\downarrow f^* \\
\mathcal{C}^\infty(U', \Lambda^p(\mathbf{R}^{n'})) & \xrightarrow[\ d\ ]{} & \mathcal{C}^\infty(U', \Lambda^{p+1}(\mathbf{R}^{n'}))
\end{array}$$

PROOF. Let x' be a point in U', and $x = f(x')$. Let M be the linear tangent map $M = D(f)(x')$, and $M_i = M\mathbf{e}_i$, where the $\mathbf{e}_i$ are the vectors of the canonical basis. We shall use the expression of the exterior derivative given in Exercise 93, p. 146, $(d\omega)_{ijk\cdots\ell} = \partial_i \omega_{jk\cdots\ell} - \partial_j \omega_{ik\cdots\ell} - \partial_k \omega_{ji\cdots\ell} - \cdots - \partial_\ell \omega_{jk\cdots i}$, to show that, for every family of indices, we have $[d(f^*(\alpha))]_{ijk\cdots\ell} = [f^*(d\alpha)]_{ijk\cdots\ell}$. First of all, we have $[f^*(\alpha)]_{j\cdots\ell} = \sum_{r\cdots t} M_j^r \cdots M_\ell^t\, \alpha_{r\cdots t}$. Hence, the coordinates of the differential $d(f^*(\alpha))$ are given by

$$\begin{aligned}
[d(f^*(\alpha))]_{ij\cdots\ell} &= \partial_i[\sum_{r\cdots t} M_j^r \cdots M_\ell^t\, \alpha_{r\cdots t}] \\
&- \partial_j[\sum_{r\cdots t} M_i^r \cdots M_\ell^t\, \alpha_{rs\cdots t}] \\
&- \cdots \\
&- \partial_\ell[\sum_{rs\cdots t} M_j^r \cdots M_i^t\, \alpha_{r\cdots t}] \\
&= \sum_{r\cdots t} \partial_i[M_j^r \cdots M_\ell^t]\, \alpha_{r\cdots t} + \sum_{r\cdots t} M_j^r \cdots M_\ell^t\, \partial_i \alpha_{r\cdots t} \\
&- \sum_{r\cdots t} \partial_j[M_i^r \cdots M_\ell^t]\, \alpha_{r\cdots t} - \sum_{r\cdots t} M_i^r \cdots M_\ell^t\, \partial_j \alpha_{r\cdots t} \\
&- \cdots \\
&- \sum_{r\cdots t} \partial_\ell[M_j^r \cdots M_i^t]\, \alpha_{r\cdots t} - \sum_{r\cdots t} M_j^r \cdots M_i^t\, \partial_\ell \alpha_{r\cdots t}.
\end{aligned}$$

Let $T^{r\cdots t}_{j\cdots\ell} = M^r_j \cdots M^t_\ell$, and let us develop $\partial_i \alpha_{s\cdots t} = \sum_r M^r_i \partial_r \alpha_{s\cdots t}$. After reordering some of the indices, we get

$$\begin{aligned}[d(f^*(\alpha))]_{ij\cdots\ell} &= \sum_{rs\cdots t} M^r_i M^s_j \cdots M^t_\ell (\partial_r \alpha_{s\cdots t} - \partial_s \alpha_{r\cdots t} - \cdots - \partial_t \alpha_{s\cdots r}) \\ &- \sum_{r\cdots t} [\partial_j T^{r\cdots t}_{i\cdots\ell} + \cdots + \partial_\ell T^{r\cdots t}_{j\cdots i} - \partial_i T^{r\cdots t}_{j\cdots\ell}]\, \alpha_{r\cdots t}.\end{aligned}$$

But

$$\begin{aligned}[f^*(d\alpha)]_{ij\cdots\ell} &= \sum_{rs\cdots t} M^r_i M^s_j \cdots M^t_\ell\, [d(\alpha)]_{rs\cdots t} \\ &= \sum_{rs\cdots t} M^r_i M^s_j \cdots M^t_\ell (\partial_r \alpha_{s\cdots t} - \partial_s \alpha_{r\cdots t} - \cdots - \partial_t \alpha_{s\cdots r}).\end{aligned}$$

Hence,

$$[d(f^*(\alpha))]_{ij\cdots\ell} = [f^*(d\alpha)]_{ij\cdots\ell} - \sum_{r\cdots t} [\partial_j T^{r\cdots t}_{i\cdots\ell} + \cdots + \partial_\ell T^{r\cdots t}_{j\cdots i} - \partial_i T^{r\cdots t}_{j\cdots\ell}]\, \alpha_{r\cdots t}.$$

Now, let us consider the coefficients of the summands involved in the right-hand side,

$$S^{r\cdots t}_{i,j\cdots\ell} = \partial_j T^{r\cdots t}_{i\cdots\ell} + \cdots + \partial_\ell T^{r\cdots t}_{j\cdots i} - \partial_i T^{r\cdots t}_{j\cdots\ell},$$

that is,

$$\begin{aligned}S^{r\cdots t}_{i,j\cdots\ell} &= \partial_j (M^r_i \cdots M^t_\ell) + \cdots + \partial_\ell (M^r_j \cdots M^t_i) - \partial_i (M^r_j \cdots M^t_\ell) \\ &= \partial_j M^r_i (\cdots M^t_\ell) + M^r_i \partial_j (\cdots M^t_\ell) \\ &+ \cdots \\ &+ (M^r_j \cdots) \partial_\ell M^t_i + M^t_i \partial_\ell (M^r_j \cdots) \\ &- \partial_i M^r_j (\cdots M^t_\ell) - \cdots - (M^r_j \cdots) \partial_i M^t_\ell.\end{aligned}$$

But $\partial_i M^r_j = \partial_i \partial_j x^r = \partial_j \partial_i x^r = \partial_j M^r_i$, thus each term of the last line of the second equality cancels with a corresponding term in the previous lines. So, the sum rewrites

$$\begin{aligned}S^{rs\cdots t}_{i,jk\cdots\ell} &= M^r_i \partial_j (M^s_k \cdots M^t_\ell) + M^s_i \partial_k (M^r_j \cdots M^t_l) + \cdots + M^t_i \partial_\ell (M^r_j M^s_k \cdots) \\ &= M^r_i \partial_j M^s_k (\cdots) M^t_\ell + M^r_i M^s_k \partial_j (\cdots) M^t_\ell + M^r_i M^s_k (\cdots) \partial_j M^t_\ell \\ &+ M^s_i \partial_k M^r_j (\cdots) M^t_\ell + M^s_i M^r_j \partial_j (\cdots) M^t_\ell + M^s_i M^r_j (\cdots) \partial_j M^t_\ell \\ &+ \cdots \\ &+ M^t_i \partial_\ell M^r_j M^s_k (\cdots) + M^t_i M^r_j \partial_\ell M^s_k (\cdots) + M^t_i M^r_j M^s_k \partial_j (\cdots).\end{aligned}$$

Each term looks like $M^r_i \partial_j M^s_k [\cdots]$, but each such term has its counterpart in the sum, which writes $M^s_i \partial_k M^r_j [\cdots]$. Replacing the letter M by what it represents, the term $S^{rs\cdots t}_{i,jk\cdots\ell}$ is the sum of pairs of the following kind:

$$\partial_i x^r \partial_j \partial_k x^s [\cdots] + \partial_i x^s \partial_k \partial_j x^r [\cdots] = (\partial_i x^r \partial_j \partial_k x^s + \partial_i x^s \partial_j \partial_k x^r)[\cdots].$$

The term between parentheses is clearly symmetric in (r, s). Hence, summed with the coefficient $\alpha_{rs\cdots t}$, which is antisymmetric in (r, s), it vanishes. Thus, all these terms vanish and $[d(f^*(\alpha))]_{ij\cdots\ell} = [f^*(d\alpha)]_{ij\cdots\ell}$, for every family of indices. Therefore, $d(f^*(\alpha)) = f^*(d\alpha)$. $\square$

6.26. Exterior product of smooth forms. Let $U \subset \mathbf{R}^n$ be some real domain. Let $\alpha \in \mathcal{C}^\infty(U, \Lambda^p(\mathbf{R}^n))$ and $\beta \in \mathcal{C}^\infty(U, \Lambda^q(\mathbf{R}^n))$. The *exterior product* $\alpha \wedge \beta$ is defined by

$$\text{for all } x \in U, \quad (\alpha \wedge \beta)(x) = \alpha(x) \wedge \beta(x),$$

where $\alpha(x) \wedge \beta(x)$ is given in (art. 6.15). So, $\alpha \wedge \beta$ is a smooth $(p+q)$-form of U. The exterior product is a bilinear map,

$$\wedge : \mathcal{C}^\infty(U, \Lambda^p(\mathbf{R}^n)) \times \mathcal{C}^\infty(U, \Lambda^q(\mathbf{R}^n)) \to \mathcal{C}^\infty(U, \Lambda^{p+q}(\mathbf{R}^n)).$$

PROOF. The components of $\alpha \wedge \beta$ are linear combinations of products of the components of α and β. Since the components of α and β are smooth, the components of $\alpha \wedge \beta$ are smooth. □

6.27. Integration of smooth p-forms on p-cubes. Let ω be a smooth p-form defined on a domain $U \subset \mathbf{R}^p$, that is, $\omega \in \mathcal{C}^\infty(U, \Lambda^p(\mathbf{R}^p))$. Let $\mathcal{B} = (\mathbf{e}_1, \ldots, \mathbf{e}_p)$ be the canonical basis of $\mathbf{R}^p$. The basis $\mathcal{B}$ defines the *positive orientation* of $\mathbf{R}^p$, and let vol_p be the associated volume $\mathrm{vol}_p = \mathbf{e}^1 \wedge \cdots \wedge \mathbf{e}^p$ (art. 6.19). At each point $x \in U$, the form ω is proportional to vol_p. Let us define $f : U \to \mathbf{R}$ by $f(x) = \omega(x)(\mathbf{e}_1)\cdots(\mathbf{e}_p)$. Thus, the smooth form ω writes unambiguously

$$\omega(x) = f(x) \times \mathrm{vol}_p \quad \text{with} \quad f \in \mathcal{C}^\infty(U, \mathbf{R}).$$

Now, let us define a p-cube in U as a product of closed intervals,

$$C = [a_1, b_1] \times \cdots \times [a_p, b_p] \subset U.$$

The *integral of the smooth form* ω on the cube C is defined as the real number

$$\int_C \omega = \int_C f \times \mathrm{vol}_p = \int_{a_1}^{b_1} dx_1 \cdots \int_{a_p}^{b_p} dx_p \; f(x_1, \ldots, x_p).$$

The ordinary *multiple integral* of the right-hand side is defined as usual. For each given value $(x_1, \ldots, x_{p-1})$, the function $x \mapsto f(x_1, \ldots, x_{p-1}, x)$, defined on a small open neighborhood of $[a_1, b_1]$, is smooth. Thus, this function has primitives, let $F[x_1 \cdots x_{p-1}]$ be one of them,

$$F[x_1 \cdots x_{p-1}]'(x) = f(x_1, \ldots, x_{p-1}, x).$$

Now, the integral of $x \mapsto f(x_1, \ldots, x_{p-1}, x)$ on $[a_p, b_p]$,

$$\int_{a_p}^{b_p} f(x_1, \ldots, x_{p-1}, x)dx = F[x_1 \cdots x_{p-1}](b_p) - F[x_1 \cdots x_{p-1}](a_p),$$

is a smooth real function of the variables $(x_1, \ldots, x_{p-1})$. Then, the integral of ω on the cube C is given by

$$\begin{aligned} \int_C \omega &= \int_{a_1}^{b_1} dx_1 \cdots \int_{a_p}^{b_p} dx_p \; f(x_1, \ldots, x_p) \\ &= \int_{a_1}^{b_1} dx_1 \cdots \int_{a_{p-1}}^{b_{p-1}} dx_{p-1} \{F[x_1 \cdots x_{p-1}](b_p) - F[x_1 \cdots x_{p-1}](a_p)\}. \end{aligned}$$

$$\begin{aligned}
&= \int_{a_1}^{b_1} dx_1 \cdots \int_{a_{p-2}}^{b_{p-2}} dx_{p-2}\{F[x_1 \cdots x_{p-2}](b_{p-1})(b_p) \\
&- \int_{a_1}^{b_1} dx_1 \cdots \int_{a_{p-2}}^{b_{p-2}} dx_{p-2} F[x_1 \cdots x_{p-2}](a_{p-1})(b_p)\} \\
&- \int_{a_1}^{b_1} dx_1 \cdots \int_{a_{p-2}}^{b_{p-2}} dx_{p-2}\{F[x_1 \cdots x_{p-2}](b_{p-1})(a_p) \\
&+ \int_{a_1}^{b_1} dx_1 \cdots \int_{a_{p-2}}^{b_{p-2} dx_{p-2}} F[x_1 \cdots x_{p-2}](a_{p-1})(a_p)\},
\end{aligned}$$

where $x \mapsto F[x_1 \cdots x_{p-2}](x)(x_p)$ is a primitive of $x \mapsto F[x_1 \cdots x_{p-2}\ x](x_p)$, etc. Finally, the integration of ω over C is obtained after p iterations of this process. Note that the result does not depend on the choice of the primitives. Let us illustrate this construction by integrating the first forms of low degree.

1-*form.* If ω is a smooth 1-form on $U \subset \mathbf{R}$, that is, $\omega(t) = f(t) \times dt$, then

$$\int_{[a,b]} \omega = \int_a^b f(t)\, dt = F(b) - F(a),$$

where F is a primitive of f, $F' = f$.

2-*form.* If ω is a smooth 2-form on $U \subset \mathbf{R}^2$, and $C = [a_1, b_1] \times [a_2, b_2]$, then

$$\int_C \omega = [F(b_1)(b_2) - F(a_1)(b_2)] - [F(b_1)(a_2) - F(a_1)(a_2)],$$

where $x \mapsto F(x)(x_2)$ is a primitive of $x \mapsto F[x](x_2)$ and $x \mapsto F[x_1](x)$ is a primitive of $x \mapsto f(x_1, x)$.

NOTE. Since, by convention, $\mathrm{vol}_p = \mathbf{e}^1 \wedge \cdots \wedge \mathbf{e}^p = dx^1 \wedge \cdots \wedge dx^p$ (art. 6.23), we have the equivalence of notation

$$\int_C f \times dx^1 \wedge \cdots \wedge dx^p = \int_{a_1}^{b_1} dx_1 \cdots \int_{a_p}^{b_p} dx_p\ f(x_1, \ldots, x_p).$$

Exercises

✎ EXERCISE 93 (Coordinates of the exterior derivative). Let $U \subset \mathbf{R}^n$ be a domain. Let ω be a smooth 1-form defined on U, and let x denote a generic point of U. Let

$$\omega = \sum_{i=1}^n \omega_i \mathbf{e}^i \quad \text{and} \quad x = \sum_{i=1}^n x^i \mathbf{e}_i.$$

Let ∂_i denote the partial derivative with respect to x^i. Using antisymmetry and reordering indices, we have

$$\begin{aligned}
d\omega &= \sum_{j=1}^n \sum_{i=1}^n \partial_i \omega_j\, \mathbf{e}^i \wedge \mathbf{e}^j \\
&= \sum_{1 \le i < j \le n} \partial_i \omega_j\, \mathbf{e}^i \wedge \mathbf{e}^j + \sum_{1 \le j < i \le n} \partial_i \omega_j\, \mathbf{e}^i \wedge \mathbf{e}^j \\
&= \sum_{1 \le i < j \le n} \partial_i \omega_j\, \mathbf{e}^i \wedge \mathbf{e}^j + \sum_{1 \le i < j \le n} \partial_j \omega_i\, \mathbf{e}^j \wedge \mathbf{e}^i
\end{aligned}$$

$$= \sum_{1\leq i<j\leq n} \partial_i\omega_j\, \mathbf{e}^i \wedge \mathbf{e}^j - \sum_{1\leq i<j\leq n} \partial_j\omega_i\, \mathbf{e}^i \wedge \mathbf{e}^j$$

$$= \sum_{1\leq i<j\leq n} (\partial_i\omega_j - \partial_j\omega_i)\mathbf{e}^i \wedge \mathbf{e}^j.$$

Show that, for a general smooth p-form ω, the coordinates of the exterior derivative $d\omega$ are given by

$$(d\omega)_{ijk\cdots\ell} = \partial_i\omega_{jk\cdots\ell} - \partial_j\omega_{ik\cdots\ell} - \partial_k\omega_{ji\cdots\ell} - \cdots - \partial_\ell\omega_{jk\cdots i}.$$

✎ EXERCISE 94 (Integral of a 3-form on a 3-cube). Make explicit the integral of a smooth 3-form on a 3-cube using iterated primitives; see (art. 6.27).

Differential Forms on Diffeological Spaces

Differential forms on diffeological spaces are defined by their evaluations on the plots, which are regarded as the pullbacks of the forms by the plots. These pullbacks are ordinary smooth forms (art. 6.20). *Hence, a differential form of a diffeological space is known as soon as we know all its pullbacks by all the plots. This is the idea behind the definition of differential forms on diffeological spaces. The condition on the pullbacks, to represent a differential form of a diffeological space, expresses just the condition of compatibility under composition.*

6.28. Differential forms on diffeological spaces. Let X be a diffeological space. A *differential* k*-form* on X, or of X, is a map α which associates, with every plot P of X, a smooth k-form $\alpha(P)$ defined on the domain of P (see (art. 6.20)) such that, for every smooth parametrization F in the domain of the plot P,

$$\alpha(P \circ F) = F^*(\alpha(P)). \qquad (\clubsuit)$$

To be more precise, α is a k-form of X if and only if the following two conditions are fulfilled:

1. For all integers n, for all n-plots $P : U \to X$, $\alpha(P) \in \mathcal{C}^\infty(U, \Lambda^k(\mathbf{R}^n))$.
2. For all m-domains V, for all smooth parametrizations $F : V \to U$, for all $\nu \in V$ and for all k vectors $\xi_1 \cdots \xi_k \in \mathbf{R}^m$,

$$\alpha(P \circ F)(\nu)(\xi_1)\cdots(\xi_k) = \alpha(P)(F(\nu))(D(F)(\nu)(\xi_1))\cdots(D(F)(\nu)(\xi_k)).$$

The condition $\alpha(P \circ F) = F^*(\alpha(P))$ is the *smooth compatibility condition*, and we shall say that $\alpha(P)$ *represents* the differential form α in the plot P. The set of differential k-forms of X is clearly a real vector space; it will be denoted by $\Omega^k(X)$.

NOTE. Since smooth forms are a special case of covariant smooth tensors, which satisfy the same contravariant pullback property, there is no reason *a priori* to not define a larger class of objects, the *differential covariant tensors*, on any diffeological space X. By definition, a differential covariant p-tensor τ on X is a map $P \mapsto \tau(P)$ such that $\tau(P)$ is a smooth covariant p-tensor (maybe with a given type of symmetry) satisfying the compatibility condition ($\clubsuit$) above.

6.29. Functional diffeology of the space of forms. The set $\Omega^k(X)$ of all differential k-forms of a diffeological space X is a real vector space. For all $\alpha, \alpha' \in \Omega^k(X)$, for all real numbers s, for all plots P of X:

$$\begin{cases} (\alpha + \alpha')(P) = \alpha(P) + \alpha'(P), \\ (s \times \alpha)(P) = s \times \alpha(P). \end{cases}$$

The sum $\alpha(P) + \alpha'(P)$ and the product by a scalar $s \times \alpha(P)$, of smooth differential forms, have been defined in (art. 6.20). The set of parametrizations $\phi : V \mapsto \Omega^k(X)$, defined by the following condition, is a diffeology of vector space (art. 3.1).

(♣) For every plot $P : U \to X$, the map $(s, r) \mapsto \phi(s)(P)(r)$ defined from $V \times U$ to $\Lambda^k(\mathbf{R}^n)$ is smooth. Briefly, $[(s, r) \mapsto \phi(s)(P)(r)] \in \mathcal{C}^\infty(V \times U, \Lambda^k(\mathbf{R}^n))$.

We shall call this diffeology the *standard functional diffeology*, or simply the *functional diffeology*, of $\Omega^k(X)$.

PROOF. Let us check the axioms of the diffeology.

D1. By the very definition of differential forms, the constant m-plots $s \mapsto \omega$ satisfy the condition $[(s, r) \mapsto \omega(P)(r)] \in \mathcal{C}^\infty(\mathbf{R}^m \times U, \Lambda^k(\mathbf{R}^n))$.

D2'. Let $\phi : s \mapsto \omega_s$ be a parametrization in $\Omega^k(X)$, defined on V such that, for all $s_0 \in V$ there exists an open neighborhood $W \subset V$ of s_0 such that $[(s, r) \mapsto \omega_s(P)(r)] \upharpoonright W \times U \in \mathcal{C}^\infty(W \times U, \Lambda^k(\mathbf{R}^n))$. Then, the map $(s, r) \mapsto \omega_s(P)(r)$ is a parametrization in $\Lambda^k(\mathbf{R}^n)$, locally smooth at each point, thus smooth.

D3. Let $F \in \mathcal{C}^\infty(W, V)$ and $\phi : s \mapsto \omega_s$, defined on V, satisfying (♣). Then, $\phi \circ F = [(t, r) \mapsto (s = F(t), r) \mapsto \omega_s(P)(r)]$ is smooth, since it is the composite of two smooth parametrizations. Thus, $\phi \circ F$ satisfies (♣).

Therefore, the condition (♣) defines a diffeology of $\Omega^k(X)$. Moreover, since for smooth forms, defined on domains, addition and multiplication by a scalar are smooth operations, these operations are also smooth for differential forms defined on diffeological spaces. In conclusion, the diffeology of $\Omega^k(X)$ defined by (♣) is a vector space diffeology. ☐

6.30. Smooth forms and differential forms. Let $U \subset \mathbf{R}^n$ be a real domain, regarded as a diffeological space (art. 1.9). Let a be a smooth k-form on U (art. 6.20), that is, $a \in \mathcal{C}^\infty(U, \Lambda^k(U))$. Let us define, for every plot in U, $P \in \mathcal{C}^\infty(V, U)$, the smooth k-form on V,

$$\alpha(P) = P^*(a).$$

Thanks to the chain-rule property of the pullback (art. 6.22), for any smooth parametrization $F \in \mathcal{C}^\infty(W, V)$, we have

$$\alpha(P \circ F) = (P \circ F)^*(a) = F^*(P^*(a)) = F^*(\alpha(P)).$$

Thus, α is a differential k-form of U (art. 6.28). Now, the map defined by $a \mapsto \alpha$ is clearly a linear isomorphism from $\mathcal{C}^\infty(U, \Lambda^k(U))$ to $\Omega^k(U)$. Its inverse is given by $a = \alpha(\mathbf{1}_U)$, where $\mathbf{1}_U$ is the identity plot of U. Therefore, for real domains, this map is a natural identification between the smooth forms defined in (art. 6.20) and the differential forms defined in (art. 6.28), $\Omega^k(U) \simeq \mathcal{C}^\infty(U, \Lambda^k(\mathbf{R}^n))$. We may sometimes identify a differential form α, defined on a domain U, with its value on the identity $a = \alpha(\mathbf{1}_U)$.

6.31. Zero-forms are smooth functions. Let X be a diffeological space. Let us associate with each 0-form $\varphi \in \Omega^0(X)$ (art. 6.28) the map $f : X \to \mathbf{R}$ defined by

$$f(x) = \varphi([0 \mapsto x]),$$

where $[0 \mapsto x]$ is the 0-plot with value x. Then, f is a smooth real map from X to $\mathbf{R}$, $f \in \mathcal{C}^\infty(X, \mathbf{R})$. This construction is a natural identification

$$\Omega^0(X) \simeq \mathcal{C}^\infty(X, \mathbf{R}).$$

PROOF. Let $P : U \to X$ be a plot, for every $r \in U$, $f \circ P(r) = \varphi([0 \mapsto P(r)] = \varphi(P \circ [0 \mapsto r]) = [0 \mapsto r]^*(\varphi(P))$, but $[0 \mapsto r]^*(\varphi(P)) = \varphi(P)(r)$, by definition. Then, $f \circ P$ is smooth. Therefore f is a smooth real function of X. Conversely, every smooth function f from X to $\mathbf{R}$ defines a 0-form φ by $\varphi(P) = f \circ P$. Therefore, $\Omega^0(X) \simeq \mathcal{C}^\infty(X, \mathbf{R})$. ☐

6.32. Pullbacks of differential forms. Let X and X' be two diffeological spaces. Let $\alpha' \in \Omega^k(X')$ and $f : X \to X'$ be a smooth map. There exists a differential k-form on X, denoted by $f^*(\alpha')$ and defined by

$$(f^*(\alpha'))(P) = \alpha'(f \circ P), \text{ for all plots } P \text{ of } X.$$

The k-form $f^*(\alpha')$ is called *pullback* of α' by f. The pullback of differential forms is contravariant. Let X, X' and X'' be three diffeological spaces. Let $f : X \to X'$ and $g : X' \to X''$ be two smooth maps, let $\alpha'' \in \Omega^k(X'')$, then

$$(g \circ f)^*(\alpha'') = f^*(g^*(\alpha'')).$$

Moreover, the pullback operation

$$f^* : \Omega^k(X') \to \Omega^k(X')$$

is a smooth linear map for the functional diffeology of the spaces of forms defined in (art. 6.28).

PROOF. Let us check that $f^*(\alpha')$ is a form on X. Let $P : U \to X$ be a plot of X and $F \in \mathcal{C}^\infty(V, U)$, where V is some real domain. Then, $(f^*(\alpha))(P \circ F) = \alpha(f \circ P \circ F) = F^*(\alpha(f \circ P)) = F^*((f^*(\alpha))(P))$. Now, let us prove that $f^* : \Omega^k(X') \to \Omega^k(X')$ is smooth. Let $\phi : V \to \Omega^k(X')$ be a plot, we want to check that $f^* \circ \phi$ is smooth. Let $P : U \to X$ be a plot, then for every $(s, r) \in V \times U$, $(f^* \circ \phi)(s)(P)(r) = f^*((\phi)(s))(P)(r) = \phi(s)(f \circ P)(r)$. But, since f is smooth, $P' = f \circ P$ is a plot of X' and since ϕ is smooth, $(s, r) \mapsto \phi(s)(P')(r)$ is smooth. Therefore, f^* is smooth. ☐

6.33. Pullbacks by the plots. Let X be a diffeological space and $\alpha \in \Omega^k(X)$. Let $P : U \to X$ be a plot and U regarded as a smooth diffeological space (art. 1.9). The plot P is a smooth map from U to X (art. 1.16), so $P^*(\alpha) \in \Omega^k(U)$. The k-form $P^*(\alpha)$ is characterized by its values on the plot 1_U (art. 6.30), that is, $P^*(\alpha)(1_U) = \alpha(P \circ 1_U) = \alpha(P)$. Therefore, the values of α on the plots, which define the form α, represent just the pullbacks of α by the plots. In other words, the differential form α is just defined by its pullbacks by the plots of X, and the condition of compatibility just writes $(P \circ F)^*(\alpha) = F^*(P^*(\alpha))$, for any plot P of X and any smooth parametrization F in the domain of P.

6.34. Exterior derivative of forms. Let X be a diffeological space. Let α be a p-form of X. The exterior derivative of α is the differential $(p+1)$-form defined by

$$(d\alpha)(P) = d(\alpha(P)),$$

for all plots P of X. The exterior derivative d is a smooth linear operator,

$$d : \Omega^p(X) \to \Omega^{p+1}(X),$$

with $\Omega^p(X)$ and $\Omega^{p+1}(X)$ equipped with the functional diffeology (art. 6.29).

NOTE. Thanks to the commutativity between pullback and exterior derivative of smooth forms (art. 6.17), the pullback of differential forms on diffeological spaces commutes with the exterior derivative. Let X' be another diffeological space, and let $f: X \to X'$ be a smooth map. Then, for all differential forms α' of X' we have

$$d(f^*(\alpha')) = f^*(d\alpha').$$

PROOF. First of all, $d\alpha$ is well defined. Indeed, since, for differential forms, the pullback commutes with the exterior derivative (art. 6.24), the definition of $d\alpha$ gives a well defined form of X,

$$(d\alpha)(P \circ F) = d(\alpha(P \circ F)) = d(F^*(\alpha(P))) = F^*(d(\alpha(P))) = F^*((d\alpha)(P)).$$

Next, the differential d is smooth. Indeed, for every plot P of X, the components of $(d\alpha)(P)$ are linear combinations of partial derivatives of smooth real functions, thus they are smooth real functions. □

6.35. Exterior product of differential forms. Let X be a diffeological space, let $\alpha \in \Omega^k(X)$ and $\beta \in \Omega^\ell(X)$. The *exterior product* $\alpha \wedge \beta$ is the differential $(k+\ell)$-form defined on X by

$$(\alpha \wedge \beta)(P) = \alpha(P) \wedge \beta(P),$$

for all plots P of X. Regarded as the map

$$\wedge : \Omega^k(X) \times \Omega^\ell(X) \to \Omega^{k+\ell}(X) \quad \text{with} \quad \wedge(\alpha, \beta) = \alpha \wedge \beta,$$

the exterior product is smooth and bilinear, for the functional diffeology of the spaces of forms defined in (art. 6.28). The following properties of the exterior product of smooth forms pass naturally to the exterior product of differential forms on diffeological spaces,

$$\alpha \wedge \beta = (-1)^{k\ell} \beta \wedge \alpha \text{ and } f^*(\alpha \wedge \beta) = f^*(\alpha) \wedge f^*(\beta),$$

where $f: X' \to X$ is a smooth map.

PROOF. First of all, the exterior product is well defined. Indeed, let $P: U \to X$ be a plot, and $F \in \mathcal{C}^\infty(V, U)$ be a smooth parametrization, then

$$\begin{aligned} (\alpha \wedge \beta)(P \circ F) &= \alpha(P \circ F) \wedge \beta(P \circ F) \\ &= F^*(\alpha(P)) \wedge F^*(\beta(P)) \\ &= F^*(P^*(\alpha) \wedge P^*(\beta)) \quad \text{(art. 6.17)} \\ &= F^*((\alpha \wedge \beta)(P)). \end{aligned}$$

Hence, the map $P \mapsto P^*(\alpha \wedge \beta)$ defines a differential form $\alpha \wedge \beta \in \Omega^{k+\ell}(X)$. Next, the exterior product is smooth. Indeed, let $r \mapsto (\phi(r), \psi(r))$ be a plot of $\Omega^k(X) \times \Omega^\ell(X)$. Thus, $\phi : V \to \Omega^k(X)$ and $\psi : V \to \Omega^\ell(X)$ are two plots. Let $P: U \to X$ be a plot. Then, $(s, r) \mapsto [\phi(s) \wedge \psi(s)](P)(s) = [\phi(s)(P)(r)] \wedge [\psi(s)(P)(r)]$. But the product of smooth plots is smooth (art. 6.26), thus the map $(s, r) \mapsto [\phi(s)(P)(r)] \wedge [\psi(s)(P)(r)]$ is smooth. Therefore, the exterior product is a smooth operation. What remains of the proposition is a direct consequence of the properties of linear forms; see (art. 6.15) and (art. 6.17). □

6.36. Differential forms are local. Let X be a diffeological space. Let α and β be two differential p-forms of X. Let x be a point of X. We shall say that α and β *have the same germ* at x if for every plot $P: U \to X$, such that $0 \in U$ and $P(0) = x$ (that is, P is centered at x), there exists an open neighborhood V of $0 \in U$ such that $\alpha(P) \restriction V = \beta(P) \restriction V$.

1. The form α is zero if and only if its germ vanishes at each point.

The following propositions are equivalent to the first one:

2. The form α is zero if and only if for every $x \in X$ there exists a D-open neighborhood V of x (art. 2.8) such that $\alpha \upharpoonright V = 0$.

3. Two differential forms α and β of X coincide if and only if they have the same germ at every point.

4. Two forms $\alpha = \beta$ of X coincide if and only if there exists a D-open covering $\mathcal{U} = \{U\}_{i \in \mathcal{I}}$ of X such that $\alpha \upharpoonright U_i = \beta \upharpoonright U_i$.

NOTE. This is not in contradiction with the fact that the D-topology of diffeological spaces can be trivial, as it happens, for example, with the irrational torus; see Exercise 4, p. 8, and Exercise 74, p. 84. It just says that, in this case, a differential form is only defined globally; see Exercise 105, p. 170.

PROOF. 1. It is obvious that the zero form, defined by $\alpha(P) = 0$ for all plots P of X, has a zero germ at every point of X. Conversely, if the germ of α vanishes at each point, for every plot $P : U \to X$, for every $r \in U$ such that $P(r) = x$, there exists an open neighborhood V of r such that $\alpha(P) \upharpoonright W = 0$. Indeed, we just have to compose the plot P with the translation mapping 0 to r to get a centered plot at x. Thus, $\alpha(P)$ vanishes locally everywhere. Since smooth forms on real domains are local, $\alpha(P) = 0$, that is, $\alpha = 0$.

2. It is obvious that if $\alpha = 0$, its restriction on every D-open vanishes. Conversely, let us assume that α vanishes D-locally everywhere on X. Let $P : U \to X$ be a plot, let $r \in U$ and $x = P(r)$. Let V be a D-open neighborhood of x such that $\alpha \upharpoonright V = 0$. Since, by definition of the D-topology, plots are continuous, $W = P^{-1}(V)$ is an open neighborhood of r and $\alpha(P \upharpoonright W) = (\alpha \upharpoonright V)(P \upharpoonright W) = 0$. Hence, $\alpha(P)$ vanishes locally at each point $r \in U$, thus $\alpha(P) = 0$, for all plots P of X, that is, $\alpha = 0$.

3. We apply item 2 to $\alpha - \beta$.

4. From item 2 we have immediately that a form α of X is zero if and only if there exists a D-open covering of X such that α vanishes on each of its elements. Then, we apply that to $\alpha - \beta$. □

6.37. The k-forms are defined by the k-plots. A k-form α, on a diffeological space X, is zero if and only if $\alpha(P) = 0$ for every k-plot P of X. Formally, denoting by $\mathcal{D}_k(X)$ the set of the k-plots of X,

$$\alpha = 0 \quad \Leftrightarrow \quad \alpha(P) = 0, \quad \text{for all } P \in \mathcal{D}_k(X).$$

Then, for α and β, two k-forms on X, $\alpha = \beta$ if and only if $\alpha(P) = \beta(P)$ for every k-plot of X.

$$\alpha = \beta \quad \Leftrightarrow \quad \alpha(P) = \beta(P), \quad \text{for all } P \in \mathcal{D}_k(X).$$

In other words, differential k-forms of X are defined uniquely by their values on the k-plots. But the smoothness, defined by the compatibility axiom, still needs to be checked against all the plots of the space.

PROOF. Only one way of this proposition needs to be proved. Let us assume that $\alpha(P) = 0$ for all k-plots. Let $Q : U \to X$ be any plot, let $n = \dim(U)$, let $r \in U$ and $v_1, \ldots, v_k$ be k vectors of $\mathbf{R}^n$. Let us define the following smooth parametrization in $\mathbf{R}^n$:

$$T^r_{v_1 \cdots v_k} : (s_1, \ldots, s_k) \mapsto r + s_1 \times v_1 + \cdots + s_k \times v_k.$$

The parametrization $T^r_{v_1\cdots v_k}$ maps $0 \in \mathbf{R}^k$ to r and, restricted to some open neighborhood of 0, $T^r_{v_1\cdots v_k}$ is smooth. Then, $Q\circ T^r_{v_1\cdots v_k}$ is a k-plot of X and, by hypothesis, $\alpha(Q \circ T^r_{v_1\cdots v_k}) = 0$. But

$$\begin{aligned}\alpha(Q \circ T^r_{v_1\cdots v_k})(0)(\mathbf{e}_1)\cdots(\mathbf{e}_k) &= (T^r_{v_1\cdots v_k})^*(\alpha(Q))(0)(\mathbf{e}_1)\cdots(\mathbf{e}_k) \\ &= \alpha(Q)(r)(v_1)\cdots(v_k).\end{aligned}$$

Hence, $\alpha(Q)(r)(v_1)\cdots(v_k) = 0$, for any $r \in U$ and any $v_1, \ldots, v_k \in \mathbf{R}^n$. Therefore $\alpha(Q) = 0$. Now, if $\alpha(P) = \beta(P)$ for any k-plot P, then $(\alpha - \beta)(P) = 0$, so $\alpha - \beta = 0$ and $\alpha = \beta$. □

6.38. Pushing forms onto quotients. Let X and X′ be two diffeological spaces. Let $\pi : X \to X'$ be a subduction (art. 1.46), and let α be a differential k-form on X. The k-form α is the pullback of a k-form β defined on X′, $\alpha = \pi^*(\beta)$, if and only if, for any two plots P and Q of X such that $\pi \circ P = \pi \circ Q$, $\alpha(P) = \alpha(Q)$. We also say that β is the *pushforward* of α by π. The differential forms of X satisfying this property may be called *basic forms*, with respect to π.

NOTE 1. For every integer k, the pullback $\pi^* : \Omega^k(X') \to \Omega^k(X)$ is always a smooth linear map as soon as π is smooth (art. 6.32). The previous proposition is a characterization of the image of π^*, when π is a subduction.

NOTE 2. This property can be expressed with the help of a diagram and will be used further this way (art. 8.42). Consider the pullback of π by itself

$$\pi^*(X) = \{(x_1, x_2) \in X \times X \mid \pi(x_1) = \pi(x_2)\},$$

with projections pr_1 and pr_2.

$$\begin{array}{ccc} \pi^*(X) & \xrightarrow{\ \mathrm{pr}_1\ } & X \\ {\scriptstyle \mathrm{pr}_2}\downarrow & & \downarrow{\scriptstyle \pi} \\ X & \xrightarrow[\ \pi\ ]{} & X' \end{array}$$

Then, α is basic with respect to π if and only if $\mathrm{pr}_1^*(\alpha) - \mathrm{pr}_2^*(\alpha) = 0$, in other words, if and only if $\mathrm{pr}_1^*(\alpha) = \mathrm{pr}_2^*(\alpha)$.

PROOF. One way is clear, if $\pi^*(\beta) = \alpha$, then for every pair of plots such that $\pi\circ P = \pi\circ Q$, we have $P^*(\alpha) = P^*(\pi^*(\beta)) = (\pi\circ P)^*(\beta) = (\pi\circ Q)^*(\beta) = Q^*(\pi^*(\beta)) = Q^*(\alpha)$. Conversely, let α be a k-form of X such that for every pair of plots P and Q of X, $\pi \circ P = \pi \circ Q$ implies $P^*(\alpha) = Q^*(\alpha)$. Let $F : W \to X'$ be a plot. Since π is a subduction, there exists a family $\mathcal{P} = \{P_i : W_i \to X\}_{i\in\mathcal{I}}$ of plots, such that $\bigcup_{i\in\mathcal{I}} W_i = W$, and for every $i \in \mathcal{I}$, $\pi \circ P_i = F \restriction W_i$. We shall say that $\mathcal{P}$ is a covering of F. So, let us define $\beta_i = P_i^*(\alpha)$, β_i is a form defined on W_i. Now, $\beta_i \restriction W_i \cap W_j = P_i^*(\alpha) \restriction W_i \cap W_j$, but $\pi \circ P_i \restriction W_i \cap W_j = \pi \circ P_j \restriction W_i \cap W_j = F \restriction W_i \cap W_j$, then $\beta_i \restriction W_i \cap W_j = \beta_j \restriction W_i \cap W_j$. Hence, there exists a form $\beta(F) \in \Omega^k(W)$, such that

$$\beta(F) = \sup_{i\in\mathcal{I}} P_i^*(\alpha), \quad \text{with} \quad \beta(F) \restriction W_i = \beta_i.$$

The $\sup_{i\in\mathcal{I}}$ denotes, as usual, the smallest common extension. But $\beta(F)$ could depend *a priori* on the choice of the covering $\mathcal{P}$. Let us prove that it does not. Let $\mathcal{P}' = \{P'_{i'} : W'_{i'} \to X\}_{i'\in\mathcal{I}'}$ be another covering of F, and $\beta'(F)$ be the form associated with $\mathcal{P}'$. Then, the union $\mathcal{P}'' = \{P''_{i''} : W''_{i''} \to X\}_{i''\in\mathcal{I}''}$ of the coverings $\mathcal{P}$

and $\mathcal{P}'$ is another covering of F. But the form $\beta''(F)$, associated with $\mathcal{P}''$, coincides by construction with $\beta(F)$, by choosing just the members of $\mathcal{P}$ for building it, and coincides with $\beta'(F)$, by choosing just the members of $\mathcal{P}'$. Hence, $\beta(F)$ does not depend on the covering.

Now, we have to prove that the map β, we have just defined, is a k-form of X'. Let $\phi : T \to W$ be a smooth parametrization. Let $\mathcal{P} = \{P_i : W_i \to X\}_{i\in\mathcal{J}}$ be a covering of F, let $\mathcal{Q} = \{Q_i : T_i \to X\}_{i\in\mathcal{J}}$ be the pullback of the covering $\mathcal{P}$, that is, $T_i = \phi^{-1}(W_i)$ and $Q_i = P_i \circ \phi$. By the previous construction we have $\beta(F \circ \phi) = \sup_{i\in\mathcal{J}} Q_i^*(\alpha) = \sup_{i\in\mathcal{J}}(P_i \circ \phi)^*(\alpha) = \sup_{i\in\mathcal{J}} \phi^*(P_i^*(\alpha)) = \phi^*(\sup_{i\in\mathcal{J}} P_i^*(\alpha)) = \phi^*(\beta(F))$. Therefore, β is a k-form of X'. We still need to check that $\pi^*(\beta) = \alpha$. By definition (art. 6.32), α is the pullback of β by π if and only if, for every plot P of X, $\alpha(P) = \beta(\pi\circ P)$. But, by construction, $\beta(\pi\circ P) = \alpha(P)$, thus $\alpha = \pi^*(\beta)$.

For Note 2, we remark that a plot of $\pi^*(X)$ is just a pair (P, Q) of plots of X, such that $\pi \circ P = \pi \circ Q$, and that $[\mathrm{pr}_1^*(\alpha) - \mathrm{pr}_2^*(\alpha)](P, Q) = \alpha(P) - \alpha(Q)$. ☐

6.39. Vanishing forms on quotients. Let X and X' be two diffeological spaces and $f : X \to X'$ be a subduction. Let α be a p-form on X', $\alpha \in \Omega^p(X')$, $p \in \mathbf{N}$. Then, $f^*(\alpha) = 0$ if and only if $\alpha = 0$. Equivalently, for any two p-forms α and β in X' and for every subduction f from X to X',

$$f^*(\alpha) = f^*(\beta) \quad \Rightarrow \quad \alpha = \beta. \qquad (\diamondsuit)$$

In other words, for every subduction $f : X \to X'$, the pullback by $f^* : \Omega^p(X') \to \Omega^p(X)$ is injective.

NOTE. This implies in particular that if a diffeological space X has a finite dimension $n \in \mathbf{N}$ (art. 1.78), then every $n+k$ differential form, with $k > 0$, is zero. Formally,

$$\dim(X) = n < \infty \quad \text{implies} \quad \Omega^{n+k}(X) = \{0\}, \text{ for all } k > 0. \qquad (\heartsuit)$$

PROOF. Let us consider a plot $P : U \to X'$, $r \in U$. Since f is a subduction, there exist an open neighborhood V of r and a plot $Q : V \to X$ such that $f\circ Q = P \upharpoonright V$, then $\alpha(P \upharpoonright V) = \alpha(f \circ Q) = (f \circ Q)^*(\alpha) = Q^*(f^*(\alpha)) = 0$. So, since $\alpha(P \upharpoonright V) = \alpha(P) \upharpoonright V$, the form $\alpha(P)$ vanishes locally at each point $r \in U$, then $\alpha(P) = 0$. Since $\alpha(P) = 0$ for all plots P, $\alpha = 0$. Now, let us assume that $\dim(X) = n < \infty$. Thus, there exists a generating family $\mathcal{F}$ such that the dimension of all its elements is less than or equal to n. But the space X is the quotient of $\mathrm{Nebula}(\mathcal{F})$ by the natural projection $\pi_{\mathcal{F}}$, from $\mathrm{Nebula}(\mathcal{F})$ onto X (art. 1.76). Now, the nebula is the sum of the domains of the elements of $\mathcal{F}$, whose dimensions are less or equal than n. Hence, for $k > 0$, every $(n + k)$-form on the nebula is zero, since every linear $(n + k)$-form on $\mathbf{R}^m$ is zero, for $m \leq n$ (art. 6.16). Thus, for every $(n + k)$-form α of X, with $k > 0$, $\pi_{\mathcal{F}}^*(\alpha) = 0$. By application of the first proposition, $\alpha = 0$. ☐

6.40. The values of a differential form. Let X be a diffeological space. Let us recall that a plot *centered* at $x \in X$ is a plot $P : U \to X$ such that $0 \in U$ and $P(0) = x$. Let p be any integer, and let us consider the following relation:

($\diamondsuit$) We say that two p-forms $\alpha, \beta \in \Omega^p(X)$ *have the same value in* x if and only if, for every plot P centered at x, we have $\alpha(P)(0) = \beta(P)(0)$, or, which is equivalent, $(\alpha - \beta)(P)(0) = 0$.

Having the same value at the point x is an equivalence relation, we shall denote it by $\sim_x$. The class of α for this relation will be called the *value of* α *at the point*

x and will be denoted by α_x. The set of all the values at the point x, of all the p-forms of X, will be denoted by $\Lambda^p_x(X)$,

$$\Lambda^p_x(X) = \Omega^p(X)/\sim_x = \{\alpha_x \mid \alpha \in \Omega^p(X)\}.$$

An element $a \in \Lambda^p_x(X)$ will be called a p-*form of* X *at the point* x, the *basepoint* of a, and the space $\Lambda^p_x(X)$ will be called the *space of* p-*forms of* X *at the point* x. We shall say that a form α *vanishes at the point* x if and only if, for every plot P centered at x, $\alpha(P)(0) = 0$, that is, if α is equivalent to the zero form in this point, $\alpha \sim_x 0$. We shall denote that by $\alpha_x = 0_x$. Then, two p-forms α and β have the same value at the point x if and only if their difference vanishes at this point,

$$\alpha_x = \beta_x \quad \Leftrightarrow \quad (\alpha - \beta)_x = 0_x.$$

Now, it is clear that the set $\{\alpha \in \Omega^p(X) \mid \alpha_x = 0_x\}$ of the p-forms of X vanishing at the point x is a vector subspace of $\Omega^p(X)$, and it is also clear that

$$\Lambda^p_x(X) = \Omega^p(X)/\{\alpha \in \Omega^p(X) \mid \alpha_x = 0_x\}.$$

Thus, as a quotient of a diffeological vector space by a vector subspace, the space $\Lambda^p_x(X)$ is naturally a diffeological vector space (art. 3.6). The addition and the multiplication by a scalar on $\Lambda^p_x(X)$ are given by

$$\alpha_x + \beta_x = (\alpha + \beta)_x \quad \text{and} \quad s(\alpha_x) = (s\alpha)_x,$$

where $\alpha, \beta \in \Omega^p(X)$ and $s \in \mathbf{R}$. Naturally, the zero form 0_x is the zero of the vector space $\Lambda^p_x(X)$.

6.41. Differential forms through generating families. Let X be a diffeological space, and let $\mathcal{F}$ be a generating family of its diffeology $\mathcal{D}$ (art. 1.66). A collection $\{\alpha_F\}_{F \in \mathcal{F}}$ of smooth k-forms is made of the values of a differential form $\alpha \in \Omega^k(X)$, that is, $\alpha_F = \alpha(F)$ — or equivalently $\alpha_F = F^*(\alpha)$ (art. 6.33) — if and only if the following two conditions are fulfilled:

1. For all $F \in \mathcal{F}$, α_F is a smooth k-form defined on $\mathrm{def}(F)$.

2. For all $F, F' \in \mathcal{F}$, for every smooth parametrization P of the domain of F, and for every smooth parametrization P' of the domain of F',

$$F \circ P = F' \circ P' \Rightarrow P^*(\alpha_F) = P'^*(\alpha_{F'}).$$

NOTE 1. Every differential k-form α of X, pulled back by the elements of the family $\mathcal{F}$, satisfies these properties. It seems that there is no better characterization of α using only its values on the elements of the family $\mathcal{F}$. We cannot hope for anything really better since all the plots of X form a generating family (see Note 2). However, in some cases, in particular for manifolds but not only, this condition is useful and used (art. 6.42).

NOTE 2. Applied to the whole diffeology $\mathcal{D}$, which is obviously a generating family, this condition is reduced to the compatibility condition (art. 6.28),

$$P' = P \circ F \quad \Rightarrow \quad \alpha_{P'} = F^*(\alpha_P),$$

where P is a plot of X and F is a smooth parametrization in $\mathrm{def}(P)$. This is another way to talk about a differential k-form α, as the family of smooth k-forms $\{\alpha_P\}_{P \in \mathcal{D}}$ such that α_P is a smooth k-form defined on $\mathrm{def}(P)$, satisfying the compatibility condition above.

PROOF. Let $\mathrm{Nebula}(\mathcal{F}) = \{(F,r) \mid F \in \mathcal{F} \text{ and } r \in \mathrm{def}(F)\}$ (art. 1.76). We know that the projection $\pi : \mathrm{Nebula}(\mathcal{F}) \to X$, defined by $\pi(F,r) = F(r)$, is a subduction (art. 1.76). Let $\alpha \in \Omega^k(X)$, the pullback of α by π is exactly given by $\pi^*(\alpha) = \{F^*(\alpha)\}_{F\in\mathcal{F}}$. Conversely, let $a \in \Omega^k(\mathrm{Nebula}(\mathcal{F}))$, that is, $a = \{a_F\}_{F\in\mathcal{F}}$. The form a is a pullback of $\alpha \in \Omega^k(X)$ if and only if, for any pair of plots P and P' of Nebula(X), $\pi \circ P = \pi \circ P'$ implies $P^*(a) = P'^*(a)$ (art. 6.38). Now, a plot P of $\mathrm{Nebula}(\mathcal{F})$ is just, locally at every point, a smooth parametrization in some domain $U = \mathrm{def}(F)$, with $F \in \mathcal{F}$. Then $P^*(a) = P^*(a \restriction F) = P^*(a_F)$, at least locally. On the other hand, $\pi \circ P(s) = F(P(s))$, that is, $\pi \circ P = P \circ F$. Therefore, the last condition writes $F \circ P = F' \circ P'$, which implies $P^*(a_F) = P'^*(a_{F'})$. Now, let us consider the special case where $\mathcal{F} = \mathcal{D}$. Let us assume that the condition above is satisfied, let $F, F' \in \mathcal{D}$, let $P : \mathrm{def}(F') \to \mathrm{def}(F)$ be a smooth parametrization, let $P' = \mathbf{1}_{\mathrm{def}(F')}$, and let us assume that $F \circ P = F' \circ P'$, that is, $F' = F \circ P$. Then, condition 2 above gives $\alpha_{F'} = P^*(\alpha_F)$. This is the compatibility condition of (art. 6.28), where the role of P and F have been inverted. Conversely, let us assume that the compatibility condition of (art. 6.28) is satisfied and let P, P', F, F' as above, such that $F \circ P = F' \circ P'$. So, $\alpha(F \circ P) = \alpha(F' \circ P')$, thus $P^*(\alpha(F)) = P'^*(\alpha(F'))$, that is, $P^*(\alpha_F) = P'^*(\alpha_{F'})$. ☐

6.42. Differential forms on manifolds. Let M be a manifold modeled on some diffeological vector space E (art. 4.1). Let $\mathcal{A}$ be an atlas, that is, a generating family of M whose elements are local diffeomorphisms from E to X. A family of smooth k-forms $\{\alpha_F\}_{F\in\mathcal{A}}$ are the values of a differential form $\alpha \in \Omega^k(M)$ if and only if, for every nonempty transition function $\phi = F^{-1} \circ F'$, defined on $F'^{-1}(\mathrm{val}(F))$, we have

$$\alpha'_F \restriction F'^{-1}(\mathrm{val}(F)) = \phi^*(\alpha_F).$$

This condition is the specialization of the characterization of forms on manifolds, through generating families (art. 6.41). This is actually the usual way of defining differential forms in classical differential geometry, in the case of finite dimensional manifolds.

6.43. Linear differential forms on diffeological vector spaces. Let E be a diffeological vector space on $\mathbf{R}$ (art. 3.1). Let $E^* = E * \mathbf{R} = L(E,\mathbf{R})$, and let E^*_∞ be the space of smooth linear maps from E to $\mathbf{R}$, $E^*_\infty = L^\infty(E,\mathbf{R}) = L(E,\mathbf{R}) \cap \mathcal{C}^\infty(E,\mathbf{R})$. Every element $\alpha \in E^*_\infty$ defines naturally a differential 1-form $\underline{\alpha} \in \Omega^1(E)$. Indeed, let $P : U \to E$ be an n-plot, so $\alpha \circ P \in \mathcal{C}^\infty(U,\mathbf{R})$. We define, for all $r \in U$ and $\delta r \in \mathbf{R}^n$,

$$\underline{\alpha}(P)_r(\delta r) = d[\alpha \circ P]_r(\delta r) = \frac{\partial[\alpha \circ P(r)]}{\partial r}(\delta r).$$

This makes sense because $\alpha \circ P$ is a smooth real function on U, and then $\underline{\alpha}(P)$ is a smooth 1-form on U. Moreover, the map $\alpha \mapsto \underline{\alpha}$ is linear and injective, indeed, for all $u \in E$, $\alpha(u) = \underline{\alpha}(s \mapsto su)_{s=0}(1)$. Thus, we can identify α with $\underline{\alpha}$. In other words, E^*_∞ can be regarded as a subset of $\Omega^1(E)$, which is satisfactory. This construction is a particular case of the following general construction, which appears in some examples from mathematical physics; see (art. 9.27) and after. Let $\Lambda^k(E)$ be the vector space of k-linear forms of E and $\omega \in \Lambda^k(E)$. We shall say that ω is a *smooth linear* k-*form of* E if, for all plots $Q : V \to E^k$, $\omega \circ Q \in \mathcal{C}^\infty(V,\mathbf{R})$. Note that $\omega \circ Q(s) = \omega(Q_1(s))\cdots(Q_k(s))$, with $Q = (Q_1,\ldots,Q_k)$. We shall denote by $\Lambda^k_\infty(E)$ the space of smooth linear k-forms of E, it is a vector subspace of $\Lambda^k(E)$.

Let $P : U \to E$ be an n-parametrization and $P^k : U^k \to E^k$ be the k-th power of P,

$$\text{for all } (r_1, \ldots, r_k) \in U^k, \quad P^k(r_1, \ldots, r_k) = (P(r_1), \ldots, P(r_k)).$$

Then, let us define $\underline{\omega}$ by

$$\underline{\omega}(P)_r(\delta_1 r, \ldots, \delta_k r) = \left.\frac{\partial^k[\omega \circ P^k(r_1, \ldots, r_k)]}{\partial r_1 \cdots \partial r_k}\right|_{r_1 = \cdots = r_k = r} (\delta_1 r) \cdots (\delta_k r), \qquad (\diamondsuit)$$

where $r \in U$ and the $\delta_i r$ belong to $\mathbf{R}^n$. For $k = 1$, we find again the above definition of a smooth linear 1-form. For $k = 2$, $r \in U$ and $u_1, u_2 \in \mathbf{R}^n$, we have

$$\begin{aligned} \underline{\omega}(P)_r(u_1, u_2) &= \left.\frac{\partial^2[\omega(P(r_1), P(r_2))]}{\partial r_1 \partial r_2}\right|_{r_1 = r_2 = r} (u_1, u_2) \\ &= D\{r_1 \mapsto D[r_2 \mapsto \omega(P(r_1), P(r_2))](r_2 = r)(u_2)\}(r_1 = r)(u_1), \end{aligned}$$

and so on. Now, $\underline{\omega}$ defined by $(\diamondsuit)$ is a smooth k-form of E. The map $\omega \mapsto \underline{\omega}$ is linear and injective. Indeed, for any k vectors $u_1, \ldots, u_k$ of E, we have

$$\omega(u_1, \ldots, u_k) = \underline{\omega}(P)_0(\mathbf{e}_1, \ldots, \mathbf{e}_k) \quad \text{with} \quad P : (t_1, \ldots, t_k) \mapsto \sum_{i=1}^{k} t_i u_i,$$

where the t_i belong to $\mathbf{R}$ and the $\mathbf{e}_i$ are the vectors of the canonical basis of $\mathbf{R}^k$. Therefore, we can regard $\Lambda^k_\infty(E)$ as a vector subspace of $\Omega^k(E)$.

Proof. Since $\alpha \in \mathcal{C}^\infty(E, \mathbf{R})$ and P is an n-plot of E, $\alpha \circ P \in \mathcal{C}^\infty(U, \mathbf{R})$ and the notation $d[\alpha \circ P]$ makes sense, it is a smooth 1-form on U, $d[\alpha \circ P] \in \mathcal{C}^\infty(U, L(\mathbf{R}^n, \mathbf{R})) = \mathcal{C}^\infty(U, \Lambda^1(\mathbf{R}^n))$. Thus, the first condition for $\underline{\alpha}$ to be a differential 1-form of E is satisfied. Next, let $F : V \to U$ be a smooth parametrization, then $\underline{\alpha}(P \circ F) = d[\alpha \circ (P \circ F)] = d[(\alpha \circ P) \circ F] = F^*[d(\alpha \circ P)] = F^*(\underline{\alpha}(P))$. Hence, the second condition is satisfied and $\underline{\alpha}$ is a differential 1-form of E. The linearity of $\alpha \mapsto \underline{\alpha}$ is clear. Let us consider its kernel: let α be such that $\underline{\alpha} = 0$, that is, for all plots P, $d[\alpha \circ P] = 0$. Let $u \in E$ be any vector and let $P : s \mapsto su$, with $s \in \mathbf{R}$, P is a 1-plot by the very definition of diffeological vector spaces. Then, $d[\alpha \circ P] = 0$, and $\alpha \circ P \in \mathcal{C}^\infty(\mathbf{R}, \mathbf{R})$ implies that $\alpha \circ P$ is a constant map, thus $\alpha \circ P(1) = \alpha \circ P(0)$, that is, $\alpha(u) = \alpha(0) = 0$. Hence, $\alpha = 0$ and $\alpha \mapsto \underline{\alpha}$ is injective. Now, since $P : s \mapsto su$ is a plot of E, we have $\underline{\alpha}(P)_{s=0}(1) = D(\alpha \circ P)(0)(1) = D(s \mapsto \alpha(su))(0)(1) = D(s \mapsto s\alpha(u))(0)(1) = D(s \mapsto [\alpha(u)] \times s)(0)(1) = \alpha(u)$.

Let us consider the general case. First of all, the partial derivative is linear in each of its arguments, thus $\underline{\omega}(P)_r$ is multilinear. Since $\underline{\omega} \circ P^k$ is smooth, all of its partial derivatives, of any order, are smooth, hence $r \mapsto \underline{\omega}(P)_r$ is smooth. Now, the partial derivatives are totally symmetric, thus, for any couple of indices j and ℓ, we have

$$\begin{aligned} &\left.\frac{\partial^k[\omega \circ P^k(r_1, \ldots, r_j, \ldots, r_\ell, \ldots, r_k)]}{\partial r_1 \cdots \partial r_j \cdots \partial r_\ell \cdots \partial r_k}\right|_{r_i = r} (u_1, \ldots, u_j, \ldots, u_\ell, \ldots, u_k) \\ = &\left.\frac{\partial^k[\omega \circ P^k(r_1, \ldots, r_j, \ldots, r_\ell, \ldots, r_k)]}{\partial r_1 \cdots \partial r_\ell \cdots \partial r_j \cdots \partial r_k}\right|_{r_i = r} (u_1, \ldots, u_\ell, \ldots, u_j, \ldots, u_k). \end{aligned}$$

But $\omega \circ P^k(r_1, \ldots, r_j, \ldots, r_\ell, \ldots, r_k) = -\omega \circ P^k(r_1, \ldots, r_\ell, \ldots, r_j, \ldots, r_k)$, hence

$$\begin{aligned} &\left.\frac{\partial^k[\omega \circ P^k(r_1, \ldots, r_j, \ldots, r_\ell, \ldots, r_k)]}{\partial r_1 \cdots \partial r_\ell \cdots \partial r_j \cdots \partial r_k}\right|_{r_i=r} (u_1, \ldots, u_\ell, \ldots, u_j, \ldots, u_k) \\ = &-\left.\frac{\partial^k[\omega \circ P^k(r_1, \ldots, r_\ell, \ldots, r_j, \ldots, r_k)]}{\partial r_1 \cdots \partial r_\ell \cdots \partial r_j \cdots \partial r_k}\right|_{r_i=r} (u_1, \ldots, u_\ell, \ldots, u_j, \ldots, u_k). \end{aligned}$$

Hence, $\underline{\omega}(P)_r(u_1, \ldots, u_j, \ldots, u_\ell, \ldots, u_k) = -\underline{\omega}(P)_r(u_1, \ldots, u_\ell, \ldots, u_j, \ldots, u_k)$, and $\underline{\omega}(P)_r$ is totally antisymmetric. Therefore, $\underline{\omega}(P) \in \mathcal{C}^\infty(U, \Lambda^p(\mathbf{R}^n))$. Now, let $F : V \mapsto U$ be some smooth m-parametrization. Let s_i denote a point in V and $r_i = F(s_i)$. Let $s \in V$, $r = F(s)$, $M = D(F)(s)$ and let $v_1, \ldots, v_k \in \mathbf{R}^m$. We have,

$$\begin{aligned} \underline{\omega}(P \circ F)_s(v_1, \ldots, v_k) &= \left.\frac{\partial^k[\omega(P \circ F(s_1), \ldots, P \circ F(s_k))]}{\partial s_1 \cdots \partial s_k}\right|_{s_i=s} (v_1, \ldots, v_k) \\ &= \left.\frac{\partial^k[\omega(P(r_1), \ldots, P(r_k))]}{\partial r_1 \cdots \partial r_k}\right|_{r_i=r} (Mv_1, \ldots, Mv_k) \\ &= \underline{\omega}(P)_{F(s)}(D(F)(s)(v_1), \ldots, D(F)(s)(v_k)) \\ &= F^*(\underline{\omega}(P))_s(v_1, \ldots, v_k). \end{aligned}$$

Therefore, $\underline{\omega}(P \circ F) = F^*(\underline{\omega}(P))$, and $\underline{\omega}$ is a differential k-form of E. Let us consider now $\underline{\omega}(P)_0(\mathbf{e}_1, \ldots, \mathbf{e}_k)$ with $P(r) = \sum_{i=1}^k t_i u_i$ and $r = (t_1, \ldots, t_k)$, as defined above. The plot $P : \mathbf{R}^k \to E$ is a linear map, thus $\omega \circ P^k(r_1, \ldots, r_k) = \omega(P(r_1), \ldots, P(r_k)) = P^*(\omega)(r_1, \ldots, r_k)$, by definition of the pullback of a linear form by a linear map. But since $P^*(\omega)$ is a linear k-form of $\mathbf{R}^k$, $P^*(\omega)$ is proportional to the canonical volume, that is, the determinant. Let $P^*(\omega) = c \times \det$, the coefficient c is given by $c = c \times \det(\mathbf{e}_1, \ldots, \mathbf{e}_k) = P^*(\omega)(\mathbf{e}_1, \ldots, \mathbf{e}_k) = \omega(P(\mathbf{e}_1), \ldots, P(\mathbf{e}_k)) = \omega(u_1, \ldots, u_k)$. Thus, $\omega \circ P^k(r_1, \ldots, r_k) = \omega(u_1, \ldots, u_k) \times \det(r_1, \ldots, r_k)$. Now, since the determinant is multilinear, we get

$$\left.\frac{\partial^k \det(r_1, \ldots, r_k)}{\partial r_1 \cdots \partial r_k}\right|_{r_i=0} (\mathbf{e}_1, \ldots, \mathbf{e}_k) = \det(\mathbf{e}_1, \ldots, \mathbf{e}_k) = 1.$$

Therefore, $\underline{\omega}(P)_0(\mathbf{e}_1, \ldots, \mathbf{e}_k) = \omega(u_1, \ldots, u_k)$. The map $\omega \mapsto \underline{\omega}$, from $\Lambda^k_\infty(E)$ to $\Omega^k(E)$, is clearly linear. If $\underline{\omega} = 0$, then, applied to any linear plot P as defined just above, we get $\omega(u_1, \ldots, u_k) = 0$ for all k vectors $u_1, \ldots, u_k$ of E, and $\omega = 0$. The map $\omega \mapsto \underline{\omega}$ is thus injective. ☐

6.44. Volumes on manifolds and diffeological spaces. Let M be a finite dimensional manifold (art. 4.1), and let $n = \dim(M)$. Recall that a *volume* of M is any n-form ω of M nowhere vanishing. The set of volumes of M will be denoted by

$$\mathrm{Volumes}(M) = \{\omega \in \Omega^n(M) \mid \forall x \in M, \omega_x \neq 0\}.$$

The manifold M is said to be *orientable* if there exists a volume, *i.e.*, if $\mathrm{Volumes}(M)$ is not empty. A pair (M, vol), where $\mathrm{vol} \in \mathrm{Volumes}(M)$, is called an *oriented manifold*, and the manifold M is said to be *oriented* by vol. The following proposition is a classic result.

1. If M is oriented by vol, then, for every n-form $\omega \in \Omega^n(M)$, there exists a function $f \in \mathcal{C}^\infty(M, \mathbf{R})$ such that

$$\omega = f \times \mathrm{vol}, \quad \text{and we denote} \quad f = \frac{\omega}{\mathrm{vol}}.$$

More generally, let X be a diffeological space of dimension $n < \infty$. By analogy with the definition above, we could call *volume* of X any differential n-form ω of X nowhere vanishing. We denote the same way, by Volumes(X), the set of all the volumes of X. In these conditions the following proposition gives more precision.

2. If ω is a volume of X and $\dim(X) = n$, then the dimension of X at each point $x \in X$ is n. Moreover, there exists $\mathcal{F}$, a parametrized covering family of plots, such that for all $F \in \mathcal{F}$, $\omega(F)$ is a volume on $\mathrm{def}(F)$.

NOTE 1. In the case of an arbitrary diffeological space of finite dimension, the property of proportionality between volumes is uncertain. There exist indeed diffeological spaces which are not manifolds, with volumes; see for example Exercise 101, p. 160. But it is not clear in general why, except for particular situations, two volumes should be proportional.

NOTE 2. In the second statement, if the diffeology $\langle \mathcal{F} \rangle$ generated by $\mathcal{F}$ (art. 1.66) is obviously finer than the diffeology $\mathcal{D}$ of X, nothing proves *a priori* that $\langle \mathcal{F} \rangle$ coincides with $\mathcal{D}$.

PROOF. 1. For the space M, to be a manifold of dimension n means that its diffeology is generated by a family $\mathcal{A}$ of local diffeomorphisms from $\mathbf{R}^n$ to M (art. 4.1). Let us assume that there exists a volume vol $\in$ Volumes(M), and let $\omega \in \Omega^n(M)$. For any chart $[F : U \to M] \in \mathcal{A}$, $\omega(F)$ and $\mathrm{vol}(F)$ are n-forms. But since F is a local diffeomorphism, $\mathrm{vol}(F)$ is a volume of U. Hence, there exists a function $f(F) \in \mathcal{C}^\infty(U, \mathbf{R})$ such that $\omega(F) = f(F) \times \mathrm{vol}(F)$. Now, let $\phi \in \mathrm{Diff}_{\mathrm{loc}}(\mathbf{R}^n)$ be a transition function between two charts, that is, $F' \restriction W = F \circ \phi$, where $W = \mathrm{def}(\phi)$. Then, on the one hand $\omega(F' \restriction W) = (f(F') \times \mathrm{vol}(F')) \restriction W = f(F' \restriction W) \times \mathrm{vol}(F \circ \phi) = f(F' \restriction W) \times \phi^*(\mathrm{vol}(F))$, and on the other hand $\omega(F' \restriction W) = \omega(F \circ \phi) = \phi^*(\omega(F)) = \phi^*(f(F) \times \mathrm{vol}(F)) = \phi^*(f(F)) \times \phi^*(\mathrm{vol}(F))$. Thus, $f(F' \restriction W) \times \phi^*(\mathrm{vol}(F)) = \phi^*(f(F)) \times \phi^*(\mathrm{vol}(F))$, which implies $f(F' \restriction W) = f(\phi^*(f(F))$, that is, $f(F \circ \phi) = f(F) \circ \phi$. Hence, $F \mapsto f(F)$ is the expression of a smooth real function defined on M in the atlas $\mathcal{A}$ (art. 6.42). Therefore, there exists a function $f \in \mathcal{C}^\infty(M, \mathbf{R})$ such that $\omega = f \times \mathrm{vol}$.

2. Let us prove first that if there exists a volume on X and if $\dim(X) = n$, then the dimension at each point of X is n (art. 2.22). We know that, for all $x \in X$, $\dim_x(X) \leq n$ (art. 2.23). Let us assume that $x \in X$ and $\dim_x(X) = p < n$. Since $\omega_x \neq 0$, there exists a plot $P : U \to X$ such that $0 \in U$, $P(0) = x$ and $\omega(P)(0) \neq 0$ (art. 6.40). But $\dim_x(X) = p$ implies that there exists a family $\mathcal{F}_x$ of centered plots, generating the germ of the diffeology at the point x, with $\dim(\mathcal{F}_x) = p$. Thus, there exist an open neighborhood V of $0 \in U$, a plot $F \in \mathcal{F}_x$ and a smooth parametrization Q in $\mathrm{def}(F)$ such that $P \restriction V = F \circ Q$. Hence, $\omega(P \restriction V) = \omega(F \circ Q) = Q^*(\omega(F))$. But, $\dim(F) \leq p < n$ implies $\omega(F) = 0$ (art. 6.16). Thus, $Q^*(\omega(F)) = 0$, which implies $\omega(P)(0) = 0$. But $\omega(P)(0) \neq 0$, therefore $\dim_x(X) = n$. Moreover, what we proved there is that for each point $x \in X$ there exists a plot $\phi = F \restriction V$ such that $\omega(\phi)$ is a volume. ☐

Exercises

✎ EXERCISE 95 (Functional diffeology of 0-forms). Let X be a diffeological space. Show that the functional diffeology of $\Omega^0(X)$ defined in (art. 6.29) coincides with the functional diffeology of $\mathcal{C}^\infty(X, \mathbf{R})$ defined in (art. 1.57).

✎ EXERCISE 96 (Differential forms against constant plots). Check that for any differential form α on a diffeological space X, if $P : U \to X$ is a locally constant plot, then $\alpha(P) = 0$.

✎ EXERCISE 97 (The equi-affine plane). Let us consider the *wire plane*, that is, $\mathbf{R}^2$ equipped with the wire diffeology (art. 1.10). Recall that the 1-plots of $\mathbf{R}^2$ generate the wire diffeology. Let γ be a 1-plot of $\mathbf{R}^2$, let $\dot\gamma$ and $\ddot\gamma$ be the first and second derivatives of γ, and let ω be the canonical volume on $\mathbf{R}^2$, that is, $\omega(V, W) = \det[V\ W]$. Let $\alpha(\gamma)$ be the covariant 3-tensor defined on $\mathrm{def}(\gamma)$ by

$$\alpha(\gamma)_t(\delta t, \delta' t, \delta'' t) = \omega(\dot\gamma(t)(\delta t), \ddot\gamma(t)(\delta' t)(\delta'' t)),$$

where $\delta t, \delta' t, \delta'' t \in \mathbf{R}$. Use the criterion of (art. 6.41) to prove that α defines, through the generating family of 1-plots, a differential covariant 3-tensor on the wire plane. The cubic root $\sqrt[3]{\alpha}$ appears in the geometric analysis of human arm movements, and it is called the *equi-affine arc length*; see for example [**BFBF09**].

✎ EXERCISE 98 (Liouville 1-form of the Hilbert space). Let $\mathcal{H}_\mathbf{R}$ be the Hilbert space equipped with the fine diffeology (art. 3.15). Let us recall that a parametrization $P : U \to \mathcal{H}_\mathbf{R} \times \mathcal{H}_\mathbf{R}$ is a plot if for all $r \in U$ there exist an open neighborhood V of r and a finite local family $(\lambda_\alpha, (X_\alpha, Y_\alpha))$, where the λ_α are smooth real functions defined on V and the (X_α, Y_α) are vectors of $\mathcal{H}_\mathbf{R} \times \mathcal{H}_\mathbf{R}$, such that

$$P \restriction V : r \mapsto \sum_{\alpha \in A} \lambda_\alpha(r)(X_\alpha, Y_\alpha), \quad \#A < \infty.$$

Let $\Lambda(P \restriction V)$ be the following 1-form, defined on V:

$$\Lambda(P \restriction V) = \frac{1}{2} \sum_{\alpha,\beta \in A} (X_\alpha \cdot Y_\beta - Y_\alpha \cdot X_\beta)(\lambda_\alpha d\lambda_\beta - \lambda_\beta d\lambda_\alpha).$$

1) Show that if P' is a plot of $\mathcal{H}_\mathbf{R} \times \mathcal{H}_\mathbf{R}$ such that $P \restriction V = P' \restriction V$, then $\Lambda(P \restriction V) = \Lambda(P' \restriction V)$.

2) Show that there exists a 1-form $\Lambda(P)$ on U such that, for every open subset $V \subset U$, $\Lambda(P \restriction V) = \Lambda(P) \restriction V$.

3) Show that the map $\Lambda : P \mapsto \Lambda(P)$ is a 1-form of $\mathcal{H}_\mathbf{R} \times \mathcal{H}_\mathbf{R}$.

✎ EXERCISE 99 (The complex picture of the Liouville form). Let us identify $\mathcal{H}_\mathbf{C}$ with $\mathcal{H}_\mathbf{R} \times \mathcal{H}_\mathbf{R}$, defined by the unique decomposition $Z = X + iY$, with $(X, Y) \in \mathcal{H}_\mathbf{R} \times \mathcal{H}_\mathbf{R}$; see Exercise 73, p. 75. Let $P : r \mapsto \sum_{\alpha \in A} \lambda_\alpha(r) Z_\alpha$ be a plot of $\mathcal{H}_\mathbf{C}$, where $(\lambda_\alpha, Z_\alpha)_{\alpha \in A}$ is a local family. The λ_α are complex valued functions and the Z_α are vectors of $\mathcal{H}_\mathbf{C}$. Let us define the symbol dZ by

$$dZ(P) : r \mapsto \sum_{\alpha \in A} d\lambda_\alpha(r) Z_\alpha \quad \text{with} \quad P : r \mapsto \sum_{\alpha \in A} \lambda_\alpha(r) Z_\alpha.$$

In this formula, $d\lambda_\alpha$ needs to be understood as

$$d\lambda_\alpha = da_\alpha + i db_\alpha, \text{ where } \lambda_\alpha = a_\alpha + i b_\alpha.$$

Show that the pullback of the Liouville form Λ by the isomorphism $\Phi : Z \mapsto (X, Y)$ writes

$$\Phi^*(\Lambda) = \frac{1}{2i}[Z \cdot dZ - dZ \cdot Z].$$

✎ EXERCISE 100 (The Fubini-Study 2-form). Let $\mathcal{S}_{\mathbf{C}}$ be the Hilbert sphere $\mathcal{S}_{\mathbf{C}} = \{Z \in \mathcal{H}_{\mathbf{C}} \mid Z \cdot Z = 1\}$, equipped with the fine diffeology (art. 3.15). Let $U(1)$ be the group of complex numbers with modulus 1, acting on $\mathcal{S}_{\mathbf{C}}$ by multiplication (Exercise 78, p. 93). Let $\mathcal{P}_{\mathbf{C}} = \mathcal{S}_{\mathbf{C}}/U(1)$ be the Hilbert projective space (Exercise 77, p. 93), and $\pi : \mathcal{S}_{\mathbf{C}} \to \mathcal{P}_{\mathbf{C}}$ be the projection. Let ϖ be the 1-form of $\mathcal{S}_{\mathbf{C}}$ defined by restriction of the Liouville 1-form of $\mathcal{H}_{\mathbf{C}}$ (Exercise 99, p. 159).

1) Show that the 1-form ϖ is invariant under the action of $U(1)$, that is, for all $z \in U(1)$, $z^*(\varpi) = \varpi$.

2) Show that if $P : \mathcal{O} \to \mathcal{S}_{\mathbf{C}}$ and $P' : \mathcal{O} \to \mathcal{S}_{\mathbf{C}}$ are two plots such that $\pi \circ P = \pi \circ P'$. Then there exists a smooth parametrization $\zeta : \mathcal{O} \to U(1)$ such that $P'(r) = \zeta(r) \times P(r)$, for all $r \in \mathcal{O}$.

3) With P and P' of question 2, show that $\varpi(P') = \varpi(P) + \zeta^*(\theta)$, where θ is the 1-form of $U(1)$ defined by $[t \mapsto e^{it}]^*(\theta) = dt$.

4) Deduce that there exists a closed 2-form ω on $\mathcal{P}_{\mathbf{C}}$ such that $\pi^*(\omega) = d\varpi$.

NOTE. The 2-form ω is called the Fubini-Study form of the projective space $\mathcal{P}_{\mathbf{C}}$.

✎ EXERCISE 101 (Irrational tori are orientable). Let $\Gamma \subset \mathbf{R}^n$ be a generating discrete subgroup of $(\mathbf{R}^n, +)$. Let us recall that *diffeologically discrete* does not mean *topologically discrete*, but that the plots of the subset diffeology of Γ are locally constant (art. 1.33). Generating $\mathbf{R}^n$ means here that the vector space $\mathbf{R}^n$ is spanned by Γ. Let $T_\Gamma = \mathbf{R}^n/\Gamma$, T_Γ is a *torus*, *rational* or *irrational*, depending if Γ is embedded or not in $\mathbf{R}^n$ (art. 2.14). Let us remark that this implies $\dim(T_\Gamma) = n$ (art. 49). Let π_Γ be the canonical projection from $\mathbf{R}^n$ onto its quotient T_Γ. The projection π_Γ is a generating family of T_Γ. The canonical volume form $\mathrm{vol}_n = \mathbf{e}^1 \wedge \cdots \wedge \mathbf{e}^n \in \Lambda^n(\mathbf{R}^n)$ is invariant by translation, *a fortiori* invariant by Γ.

1) Show that there exists, on T_Γ, a nowhere vanishing n-form $\mathrm{vol}_\Gamma \in \Omega^n(T_\Gamma)$ such that $\pi_\Gamma^*(\mathrm{vol}_\Gamma) = \mathrm{vol}_n$.

2) Conclude that, even if the group Γ is not embedded, that is, not topologically discrete — which implies that the quotient T_Γ is not a manifold (art. 74) — the torus T_Γ is orientable, and oriented by vol_Γ.

3) Show that, if Γ is dense in $\mathbf{R}^n$, then all the volumes of T_Γ are proportional, by a constant, to vol_Γ.

Forms Bundles on Diffeological Spaces

In this section we bundle *all the vector spaces* $\Lambda^p_x(X)$ *of values of* p*-forms of* X (art. 6.40), *when* x *runs over* X, *into one geometrical object: the* p-form bundle $\Lambda^p(X)$. *This* bundle *is equipped with the pushforward diffeology of the product* $X \times \Omega^p(X)$ *by the map associating with the pair* (x, α) *the value* α_x. *Then, every differential* p*-form of* X *becomes a smooth section of the projection* $\pi : \Lambda^p_x(X) \to X$, *which maps each value of a* p*-form to its basepoint, and vice versa.*

Then, we define on the product $X \times \Omega^p(X)$ *a tautological* p*-form denoted by* Taut, *satisfying* $\mathrm{Taut} \restriction X \times \{\alpha\} = \alpha$. *We show that this tautological form passes to the quotient* $\Lambda^p(X)$ *into a form called Liouville form, and denoted by* Liouv. *The form* Liouv *is characterized by the following property: for every* p*-form* $\alpha \in \Omega^p(X)$, $\boldsymbol{\alpha}^*(\mathrm{Liouv}) = \alpha$, *where* $\boldsymbol{\alpha}$ *is the section* $x \mapsto \alpha_x$ *of the bundle* $\Lambda^p(X)$.

We suggest in the last paragraph a construction for the p*-vector bundles of a diffeological space. We give, in particular, a definition for the tangent bundle. This* *question, of a good definition of a tangent bundle in diffeology, has been recurrently discussed, and different answers have been advanced by various authors. The reason for which there is no unique answer to the question of the tangent bundle in diffeology lies in the covariant nature of diffeology. There are indeed many ways to think about tangent spaces, which are equivalent for manifolds but not when applied to diffeological spaces. We discuss this question in (art. 6.53, Note 5) and we give a construction which tries to fit the general philosophy of the theory: since the nature of diffeology is to be covariant, contravariant objects need to be defined by duality with their covariant natural counterparts, and not the opposite. But, actually, different kind of questions may need different versions of tangent bundles.*

6.45. The p-form bundle of diffeological spaces. Let X be a diffeological space. Let us recall that the vector space $\Lambda^p_x(X)$, of values of p-forms at the point $x \in X$, is the quotient of the space of differential p-forms $\Omega^p(X)$ by the subspace of all p-forms of X vanishing in x (art. 6.40). We define the *bundle of p-forms of X* (or *over X*) as the union of all these spaces $\Lambda^p_x(X)$, and we denote it by

$$\Lambda^p(X) = \coprod_{x \in X} \Lambda^p_x(X) = \{(x, a) \mid a \in \Lambda^p_x(X)\}.$$

This bundle will be equipped with the pushforward diffeology of $X \times \Omega(X)$ by the projection

$$\mathbf{p} : X \times \Omega^p(X) \to \Lambda^p(X) \quad \text{such that} \quad \mathbf{p}(x, \alpha) = (x, \alpha_x).$$

We shall denote by $\pi : (x, a) \mapsto x$ the natural projection from the bundle $\Lambda^p(X)$ to X. The construction is illustrated by the following diagram.

$$
\begin{array}{ccc}
X \times \Omega^p(X) & \xrightarrow{\ \ \mathbf{p}\ \ } & \Lambda^p(X) \\
& \searrow_{\mathrm{pr}_1} \quad \swarrow_{\pi} & \\
& X &
\end{array}
$$

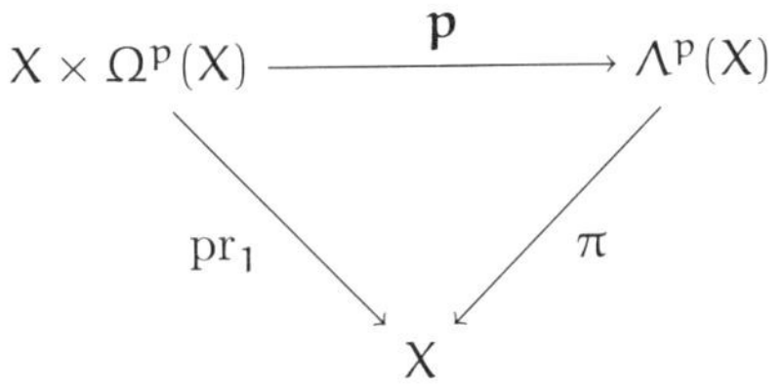

NOTE. The space $\Lambda^p(X)$ is not the diffeological sum of the subsets $\Lambda^p_x(X)$, $x \in X$. Actually, $\Lambda^p(X)$ is the diffeological quotient of $X \times \Omega^p(X)$ by the equivalence relation $(x, \alpha) \sim (x', \alpha')$ if and only if $x = x'$ and $\alpha_x = \alpha'_{x'}$.

6.46. Plots of the bundle $\Lambda^p(X)$. A parametrization $\Pi : r \mapsto (Q(r), P(r))$ in $\Lambda^p(X)$, defined on some domain U, is a plot if and only if the following two conditions are fulfilled:

    1. The parametrization Q is a plot of X.
    2. For all $r_0 \in U$ there exist an open neighborhood V of r_0 and a plot $A : V \to \Omega^p(X)$ such that for all $r \in V$, $P(r) = A(r)_{Q(r)}$, *i.e.*, $P(r)$ is the value of $A(r)$ at the point $Q(r)$.

Note that the standard diffeology of $\Lambda^p(X)$, illustrated by Figure 6.1, satisfies the following properties:

    a) The projection $\pi : \Lambda^k(X) \to X$ is a local subduction (art. 2.16).
    b) Each subspace $\pi^{-1}(x)$ is smoothly isomorphic to $\Lambda^p_x(X)$.

PROOF. 1. Let $\Pi : U \to \Lambda^p(X)$ be a plot, thanks to the diagram (art. 6.45), and let $Q = \pi \circ \Pi$ be a plot of X.

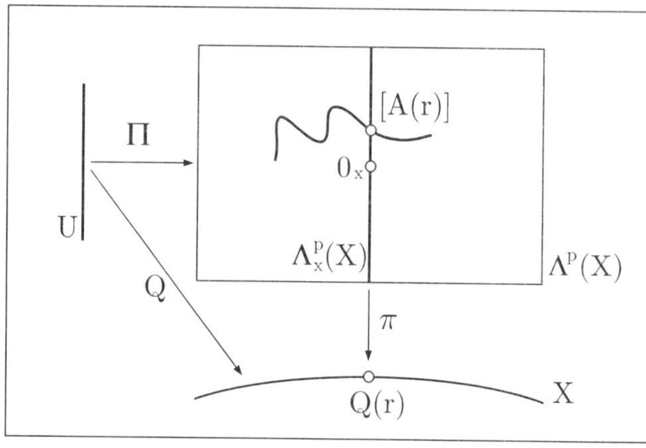

FIGURE 6.1. The p-form bundle.

2. By definition of the pushforward diffeology (art. 1.43), for every $r_0 \in U$ there exist an open neighborhood V of r_0 and a plot $A : V \to \Omega^p(X)$, such that $\mathbf{p} \circ A = \Pi \restriction V$, that is, for any $r \in V$, $\mathbf{p}(Q(r), A(r)) = \Pi(r) = (Q(r), P(r))$. But $\mathbf{p}(Q(r), A(r)) = (Q(r), A(r)_{Q(r)})$, hence $P(r) = A(r)_{Q(r)}$.

a) Let $Q : U \to X$ be a plot, $x = Q(r)$ and $(x, a) \in \Lambda^p(X)$. By construction, there exists a p-form $\alpha \in \Omega^p(X)$ such that $a = \alpha_x$. The parametrization $Q \times \hat{\alpha} : r \mapsto (Q(r), \alpha)$ is a plot of the product $X \times \Omega^p(X)$. Hence, $\mathbf{p} \circ (Q \times \hat{\alpha})$ is a plot of $\Lambda^p(X)$, lifting Q and passing through (x, a). Therefore, $\pi : \Lambda^k(X) \to X$ is a local subduction.

b) Now, for every $x \in X$ the injection $\Lambda^p_x(X) \to \Lambda^p(X)$, defined by $a \mapsto (x, a)$, is an induction. Indeed, every plot of $\Lambda^p(X)$ with values in $\{x\} \times \Lambda^p_x(X)$ is just a parametrization a of $\Lambda^p_x(X)$ such that there exists a plot A of $\Omega^p(X)$ with $a(r) = A(r)_x$. This is exactly the condition to be an element of the diffeology of $\Lambda^p_x(X)$. □

6.47. The zero-forms bundle. The bundle of 0-forms $\Lambda^0(X)$ of a diffeological space X is just the product $X \times \mathbf{R}$. Indeed, the value of a 0-form $f \in \Omega^0(X) = \mathcal{C}^\infty(X, \mathbf{R})$, at any point $x \in X$, is just a number, and any number is the value of a 0-form, at least the value of a constant function,

$$\Lambda^0(X) \simeq X \times \mathbf{R}.$$

The projection $(x, f) \mapsto (x, f(x))$, from $X \times \mathcal{C}^\infty(X, \mathbf{R})$ to $X \times \mathbf{R}$, is equivalent to the projection from $X \times \mathcal{C}^\infty(X, \mathbf{R})$ to $\Lambda^0(X)$.

6.48. The cotangent space. The space $\Lambda^1(X)$ will be called the *cotangent space* of X, even if it is not defined *a priori* by duality with some *tangent space*; see the discussion in (art. 6.53). Since in classical differential geometry, for a finite dimensional manifold M, the cotangent space is denoted by $T^*(M)$, we should also denote sometimes $\Lambda^1(X)$ by $T^*(X)$.

6.49. The Liouville forms on the spaces of p-forms. Let X be a diffeological space. For every plot $Q \times A : U \to X \times \Omega^p(X)$, let us define

$$\mathrm{Taut}(Q \times A) = [r \mapsto A(r)(Q)(r)] \in \Omega^p(U). \qquad (\diamondsuit)$$

Taut is a differential p-form of $X \times \Omega^p(X)$ such that

$$\text{for all } \alpha \in \Omega^p(X), \quad \mathrm{Taut} \restriction X \times \{\alpha\} = \alpha. \qquad (\heartsuit)$$

We shall call this p-form the *tautological* p-*form* of $X \times \Omega^p(X)$. Moreover, there exists a differential p-form of $\Lambda^p(X)$, denoted by Liouv, such that

$$\mathrm{Taut} = \mathbf{p}^*(\mathrm{Liouv}) \quad \text{with} \quad \mathrm{Liouv} \in \Omega^p(\Lambda^p(X)),$$

where $\mathbf{p}$ is the projection from $X \times \Omega^p(X)$ onto its quotient $\Lambda^p(X)$ (art. 6.45). We shall call this form the *Liouville* p-*form* of $\Lambda^p(X)$.

PROOF. 1. Let us prove that the condition $(\diamondsuit)$ above defines a differential form on $X \times \Omega^p(X)$ satisfying $(\heartsuit)$. Let (Q, A) be a plot of $X \times \Omega^p(X)$, defined on an n-domain U. Let $F : V \to U$ be a smooth parametrization with $\dim(V) = m$. Now, let $s \in V$, $v_1, \ldots v_p \in \mathbf{R}^m$, $r = F(s)$ and $u_i = D(F)(s)(v_i)$, $i = 1 \cdots p$. Then,

$$\begin{aligned}
\mathrm{Taut}((Q,A)\circ F)(s)(v_1)\cdots(v_p) &= \mathrm{Taut}((Q\circ F)\times(A\circ F))(s)(v_1)\cdots(v_p)\\
&= [(A\circ F)(s)](Q\circ F)(s)(v_1)\cdots(v_p)\\
&= A(r)(Q\circ F)(s)(v_1)\cdots(v_p)\\
&= F^*[A(r)(Q)](s)(v_1)\cdots(v_p)\\
&= A(r)(Q)(r)(D(F)_s(v_1))\cdots(D(F)_s(v_p))\\
&= A(r)(Q)(r)(u_1)\cdots(u_p)\\
&= \mathrm{Taut}(Q,A)(r)(u_1)\cdots(u_p)\\
&= F^*[\mathrm{Taut}(Q,A)](s)(v_1)\cdots(v_p).
\end{aligned}$$

Hence, $\mathrm{Taut}((Q,A)\circ F) = F^*(\mathrm{Taut}(Q,A))$, and Taut is a well defined p-form on $X \times \Omega^p(X)$. Now, let us consider the restriction of Taut to a subspace $X \times \{\alpha\}$. A plot of $X \times \{\alpha\}$ is just a parametrization $(Q, \hat{\alpha}) : r \mapsto (Q(r), \alpha)$, where Q is a plot of X. Then, by definition, $[\mathrm{Taut} \restriction X \times \{\alpha\}](Q, \hat{\alpha})(r) = \alpha(Q)(r)$. Hence $\mathrm{Taut} \restriction X \times \{\alpha\} = \alpha$.

2. Now, let us prove that there exists a differential p-form Liouv on $\Lambda^p(X)$ such that $\mathbf{p}^*(\mathrm{Liouv}) = \mathrm{Taut}$. We shall use the criterion (art. 6.38). Let (Q, A) and (Q', A') be two plots of $X \times \Omega^p(X)$, defined on a domain U, such that $\mathbf{p} \circ (Q, A) = \mathbf{p} \circ (Q', A')$, that is,

$$Q = Q' \quad \text{and for every } r \in U, \quad A(r)_{Q(r)} = A'(r)_{Q'(r)}.$$

Thus, the value of $A(r)$ and $A'(r)$ coincide at the point $Q(r)$, for all $r \in U$. That means that for any plot P centered at $Q(r)$, $A(r)(P)(0) = A'(r)(P)(0)$. Let T_r be the translation $T_r(r') = r + r'$, and $P = Q \circ T_r$. Then, P is a plot centered at $Q(r)$, since $Q \circ T_r(0) = Q(r)$. Hence, $A(r)(Q \circ T_r)(0) = A'(r)(Q \circ T_r)(0)$, that is, $T_r^*(A(r)(Q))(0) = T_r^*(A'(r)(Q))(0)$. But since T_r is a translation, $T_r^*(A(r)(Q))(0) = A'(r)(Q)(r)$ and $T_r^*(A(r)(Q))(0) = A'(r)(Q)(r)$. Thus, for every $r \in U$, $A(r)(Q)(r) = A'(r)(Q)(r)$. But $A(r)(Q)(r) = \mathrm{Taut}(Q,A)(r)$ and $A'(r)(Q)(r) = \mathrm{Taut}((Q, A')(r)$, so $\mathrm{Taut}(Q, A) = \mathrm{Taut}(Q, A')$. Therefore, for any two plots (Q, A) and (Q', A') of $X \times \Omega^p(X)$ such that $\mathbf{p} \circ (Q, A) = \mathbf{p} \circ (Q', A')$, $\mathrm{Taut}(Q, A) = \mathrm{Taut}(Q', A')$. By application of the criterion (art. 6.38), there exists a p-form $\mathrm{Liouv} \in \Omega^p(\Lambda^p(X))$ such that $\mathbf{p}^*(\mathrm{Liouv}) = \mathrm{Taut}$. ☐

6.50. Differential forms are sections of bundles. Let X be a diffeological space. Let $\Lambda^p(X)$ be the bundle of p-forms over X, and let $\pi : \Lambda^p(X) \to X$ be the canonical projection (art. 6.45). A *smooth section* of π is a smooth map $\sigma : X \to \Lambda^p(X)$ such that $\pi \circ \sigma = \mathbf{1}_X$. Let

$$\mathrm{Sec}(\pi) = \{\sigma \in \mathcal{C}^\infty(X, \Lambda^p(X)) \mid \pi \circ \sigma = \mathbf{1}_X\}.$$

Let χ be the map

$$\chi : \mathrm{Sec}(\pi) \to \Omega^p(X) \quad \text{defined by} \quad \chi(\sigma) = \sigma^*(\mathrm{Liouv}).$$

This map is smooth and surjective, and for every $\alpha \in \Omega^p(X)$,

$$\alpha = [x \mapsto \alpha_x]^*(\mathrm{Liouv}), \ \textit{i.e.}, \ \alpha = \chi([x \mapsto \alpha_x]).$$

Hence, every differential form can be regarded as the pullback of the Liouville form by its associated section $x \mapsto \alpha_x$.

NOTE. For the 0-forms, $\Lambda^0(X) = X \times \mathbf{R}$ (art. 6.47), a section $\mathbf{f} : X \to \Lambda^0(X)$ is just a map $\mathbf{f} : x \mapsto (x, f(x))$, where $f : X \to \mathbf{R}$, and the section $\mathbf{f}$ is smooth if and only if f is smooth, which is satisfactory. Now, the Liouville form on $X \times \Omega^0(X)$ is a 0-form, that is, a real function, and it is given by $\mathrm{Taut}(Q \times \varphi)(r) = f_r(Q(r))$, where $f_r = \varphi(r)$, and $Q \times \varphi$ is any plot of $X \times \Omega^0(X)$. Hence, the Liouville form on $\Lambda^0(X)$ is just the second projection $\mathrm{Liouv}(x, t) = \mathrm{pr}_2(x, t) = t$, and we can check that $\mathbf{f}^*(\mathrm{Liouv})(x) = \mathrm{Liouv}(x, f(x)) = f(x)$.

PROOF. Since the pullback is a smooth map (art. 6.22), the map χ is smooth. Let us check then that the map $\alpha \mapsto \boldsymbol{\alpha} = [x \mapsto \alpha_x]$, from $\Omega^p(X)$ to $\mathrm{Sec}(\pi)$, where α_x is the value of α at x, is smooth. Note that we commit an abuse with the notation $[x \mapsto \alpha_x]$, since α_x does not belong to $\Lambda^p(X)$, actually $\boldsymbol{\alpha}(x) = (x, \alpha_x)$. Next, $\boldsymbol{\alpha}$ is smooth because $\boldsymbol{\alpha} = [x \mapsto \mathbf{p}(x, \alpha)]$ is the composite of two smooth maps $\boldsymbol{\alpha} = \mathbf{p} \circ \hat{\alpha}$, where $\hat{\alpha} = [x \mapsto (x, \alpha)]$, and $\mathbf{p}$ has been defined in (art. 6.45). Now, let α be a differential p-form, and let $\boldsymbol{\alpha} : x \mapsto (x, \alpha_x)$ be the associated section. Thus, $\boldsymbol{\alpha}^*(\mathrm{Liouv}) = (\mathbf{p} \circ \hat{\alpha})^*(\mathrm{Liouv}) = \hat{\alpha}^*(\mathbf{p}^*\mathrm{Liouv}) = \hat{\alpha}^*(\mathrm{Taut})$. But for any plot Q of X, $\hat{\alpha}^*(\mathrm{Taut})(Q) = [r \mapsto \alpha(Q)(r)]$, hence $\boldsymbol{\alpha}^*(\mathrm{Liouv}) = \alpha$, and the map χ is surjective. ☐

6.51. Pointwise pullback or pushforward of forms. Let X and X' be two diffeological spaces. Let $f : X \to X'$ be a smooth map.

1. Let $x \in X$ and $x' = f(x)$. For every $a' \in \Lambda^p_{x'}(X')$, the *pointwise pullback* $f^*_x(a') \in \Lambda^p_x(X)$ is defined by

$$f^*_x(a') = (f^*\alpha')_x,$$

for any $\alpha' \in \Omega^p(X')$ such that $a' = \alpha'_{x'}$.

$$\begin{array}{ccc} \Lambda^p_x(X) & \xrightarrow{\ f^*_x\ } & \Lambda^p_{x'}(X') \\ \Big\downarrow{\scriptstyle \pi_X} & & \Big\downarrow{\scriptstyle \pi_{X'}} \\ \{x\} & \xrightarrow[\ f\ ]{} & \{x'\} \end{array}$$

The composition of pointwise pullbacks is contravariant,

$$(g \circ f)^*_x = f^*_x \circ g^*_{x'},$$

where X, X′ and X″ are three diffeological spaces, $f : X \to X'$ and $g : X' \to X''$ are smooth maps, and $x' = f(x)$, with $x \in X$.

2. Let f be a diffeomorphism from X to X′, and let $a \in \Lambda^p_x(X)$. The *pointwise pushforward* $f^x_*(a)$ is the p-form at the point $x' = f(x)$, defined by

$$f^x_*(a) = (f^{-1}_{x'})^*(a) \in \Lambda^p_{x'}(X').$$

As a consequence of the chain rule of pointwise pullbacks, the pointwise pushforward is covariant, $(g \circ f)^x_* = g^{x'}_* \circ f^x_*$, where $f : X \to X'$ and $g : X' \to X''$ are diffeomorphisms. The pushforward of forms, by diffeomorphisms, defines an *action* of the group Diff(X) on the bundle $\Lambda^p(X)$. The map f_* is a diffeomorphism of $\Lambda^p(X)$ for each $f \in \mathrm{Diff}(X)$.

PROOF. What we have to check is only that if two forms α' and β' have the same value at the point $x' = f(x)$, their pullbacks $\alpha = f^*(\alpha')$ and $\beta = f^*(\beta')$ have the same value in x. But this is a consequence of linearity of pullback of linear forms. □

6.52. Diffeomorphisms invariance of the Liouville form. Let X be a diffeological space. The group of diffeomorphisms of X acts naturally on the product $X \times \Omega^p(X)$ by

$$\varphi(x, \alpha) = (\varphi(x), \varphi_*(\alpha)),$$

for all $\varphi \in \mathrm{Diff}(X)$ and all $(x, \alpha) \in X \times \Omega^p(X)$, where $\varphi_*(\alpha) = (\varphi^{-1})^*(\alpha)$. The Liouville form Taut (art. 6.49) is invariant by this action, that is,

$$\varphi^*(\mathrm{Taut}) = \mathrm{Taut}, \text{ for all } \varphi \in \mathrm{Diff}(X).$$

The action of Diff(X) on $X \times \Omega^p(X)$ projects to a pointwise action (art. 6.51) on the bundle of p-forms $\Lambda^p(X)$,

$$\varphi(x, a) = (\varphi(x), (\varphi^x_*(a)),$$

for all $\varphi \in \mathrm{Diff}(X)$ and all $(x, a) \in \Lambda^p(X)$. Moreover, the Liouville form Liouv of the p-form bundle (art. 6.49) is invariant by this projected action, that is,

$$\varphi^*(\mathrm{Liouv}) = \mathrm{Liouv}, \text{ for all } \varphi \in \mathrm{Diff}(X).$$

PROOF. The pullback of the Liouville form Taut on $X \times \Omega^p(X)$ is given, for every plot $Q \times A$, by

$$\varphi^*(\mathrm{Taut})(Q \times A) = \mathrm{Taut}(\varphi \circ (Q \times A)) = \mathrm{Taut}((\varphi \circ Q) \times (\varphi_* \circ A)).$$

But $\mathrm{Taut}((\varphi \circ Q) \times (\varphi_* \circ A))(r) = \varphi_*(A(r))(\varphi \circ Q(r)) = A(r)(\varphi^{-1} \circ \varphi \circ Q)(r) = A(r)(Q)(r) = \mathrm{Taut}(Q \times A)(r)$. Therefore, $\varphi^*(\mathrm{Taut}) = \mathrm{Taut}$. For equivariance reasons, by projection, $\varphi^*(\mathrm{Liouv}) = \mathrm{Liouv}$. □

6.53. The p-vector bundle of diffeological spaces. Let X be a diffeological space. Let $\mathrm{Plots}_p(X)$ be the set of p-plots of X, for some integer p. Let $P \in \mathrm{Plots}_p(X)$ and $r \in \mathrm{def}(P)$, for every $\alpha \in \Omega^p(X)$, $\alpha(P)(r) \in \Lambda^p(\mathbf{R}^p)$, thus $\alpha(P)(r)$ is proportional to the standard volume vol_p of $\mathbf{R}^p$ (art. 6.19). Let us denote by $\dot{P}(r)(\alpha)$ the coefficient of proportionality. The real map

$$\dot{P}(r) : \Omega^p(X) \to \mathbf{R} \quad \text{defined by} \quad \dot{P}(r) : \alpha \mapsto \frac{\alpha(P)(r)}{\mathrm{vol}_p}$$

belongs to the dual vector space $\Omega^p(X)^*$, and moreover to the smooth dual.

1. Let $P : U \to X$ be a p-plot, and let $r_0 \in U$. There exist an open neighborhood $V \subset U$ of r_0 and a plot $r \mapsto Q_r$ of $\mathrm{Paths}_p(X)$ (art. 1.64), defined on V, such that, for all $r \in V$, $P(r) = Q_r(0)$ and $\dot{P}(r) = \dot{Q}_r(0)$.

2. Thanks to the previous proposition, we redirect our attention to the set of global p-plots $\mathrm{Paths}_p(X)$. For every $x \in X$, we define $S^p_x(X) \subset \Omega^p(X)^*$ by

$$S^p_x(X) = \{\dot{Q}(0) \in \Omega^p(X)^* \mid Q \in \mathrm{Paths}_p(X) \text{ and } Q(0) = x\}.$$

For all $x \in X$, $0 \in S^p_x(X)$. For all $x \in X$, for all $v \in S^p_x(X)$ and for all $s \in \mathbf{R}$, $s \times v \in S^p_x(X)$. Thus, $S^p_x(X)$ is a star-shaped subset of $\Omega^p(X)^*$ with origin 0.

3. However, if $S^p_x(X)$ is not *a priori* a vector subspace of $\Omega^p(X)^*$, then we shall use the vector space structure of the dual space $\Omega^p(X)^*$ to extend, by linearity, the star $S^p_x(X)$ into a vector subspace. Let $T^p_x(X) = \mathrm{Span}(S^p_x(X))$, $T^p_x(X) \subset \Omega^p(X)^*$, be the smallest vector subspace containing $S^p_x(X)$. The elements of $T^p_x(X)$ may be called the *tangent* p*-vectors* of X at the point x. The space $T^p_x(X)$ is made of the finite sums of elements of $S^p_x(X)$,

$$T^p_x(X) = \left\{ v = \sum_{i \in \mathcal{I}} v_i \;\middle|\; \#\mathcal{I} < \infty \text{ and for all } i \in \mathcal{I},\ v_i \in S^p_x(X) \right\}.$$

The bundle of all the p-vector spaces $T^p_x(X)$, when x runs over X, defines the p*-vector bundle* of X, denoted by $T^p(X)$, that is,

$$T^p(X) = \{(x, v) \mid x \in X \text{ and } v \in T^p_x(X)\}.$$

4. The space $T^p(X)$ may be equipped with a *standard diffeology*. A parametrization $P : U \to T^p(X)$, with $P(r) = (x_r, v_r)$, is a plot for the standard diffeology if for all $r_0 \in U$ there exist an open neighborhood V of r_0, a finite family of indices $\mathcal{I}$, a family $\{[r \mapsto Q_{i,r}]\}_{i \in \mathcal{I}}$ of plots of $\mathrm{Paths}_p(X)$ defined on V, such that $Q_{i,r}(0) = x_r$ for all $i \in \mathcal{I}$, and $v_r = \sum_{i \in \mathcal{I}} \dot{Q}_{i,r}(0)$.

5. The restriction of the standard diffeology to each $T^p_x(X)$ is a diffeology of vector space (art. 3.1). The zero section $x \mapsto (x, 0)$ of $T^p(X)$ is smooth. The first projection $\pi : (x, v) \mapsto x$ is obviously a subduction.

NOTE 1. By analogy with classical differential geometry, for $p = 1$, the space $T^1_x(X)$ will simply be denoted by $T_x(X)$, and may be interpreted as the *tangent space* to X at the point x. The space $T(X) = T^1(X)$ becomes then the *tangent bundle* of X. Moreover, for $\gamma \in \mathrm{Plots}_1(X)$, the vector $\dot{\gamma}(t) \in T_{\gamma(t)}(X)$ is the *speed* of γ at *time* t.

NOTE 2. By definition each vector space $T^p_x(X)$ is a subspace of the smooth dual of the space of p-forms $\Lambda^p_x(X)$. Conversely, $\Lambda^p_x(X)$ is a subspace of the smooth dual of $T^p_x(X)$, thanks to the pairing from $\Lambda^p_x(X) \times T^p_x(X)$ to $\mathbf{R}$, $(a, v) \mapsto \alpha(Q)(0)$ where $v = \dot{Q}(0)$ and $a = \alpha_x$ is the value of α at the point x (art. 6.40).

NOTE 3. The map $\dot{Q}(0)$ is not just local but depends only on the 1*-jet* of Q at 0. Precisely, let $Q = Q' \circ F$, with $F \in \mathcal{C}^\infty(\mathbf{R}^p, \mathbf{R}^p)$ and $F(0) = 0$. If $D(F)(0) = 1_{\mathbf{R}^p}$ then $\dot{Q} = \dot{Q}'$. Actually, since the plot Q is evaluated on the p-forms, $\dot{Q}$ is "antisymmetric" in the sense that it depends only on the determinant of $D(F)(0)$, $\dot{Q}(0) = \det(D(F)(0)) \times \dot{Q}'(0)$.

NOTE 4. Since $T^p(X)$ is a subset of the product $X \times L^\infty(\Omega^p(X), \mathbf{R})$, where the smooth dual $L^\infty(\Omega^p(X), \mathbf{R}) = \Omega^p(X)^* \cap \mathcal{C}^\infty(\Omega^p(X))$ of $\Omega^p(X)$ is equipped with the functional diffeology, $T^p(X)$ can also be equipped with the subset diffeology of this product. We shall call this subset diffeology the *functional diffeology* of $T^p(X)$. The standard diffeology of $T^p(X)$ is finer than its functional diffeology.

NOTE 5. There are a few ways to look at vectors in differential geometry. We can see a vector v at a some point x as the class of a path γ. We can see this vector as a derivation $f \mapsto df_x(v)$ on the space of smooth functions. We can see also this vector v as the value of a vector field F, $v = F(x)$. In classical differential geometry, for manifolds, all these approaches lead to the same objects. It is no more the case in diffeology. Essentially, a vector v at x is the class of a path γ pointed at x, two paths γ and γ' define the same tangent vector at x if $\alpha(\gamma)_0 = \alpha(\gamma')_0$, for all 1-forms α. Considering a vector as a derivation consists of testing the paths γ on exact 1-forms only, that is, $\alpha = df$, and that shrinks *a priori* the tangent space. For instance, regarded this way, the tangent space of an n-dimensional irrational torus is reduced to $\{0\}$, while the construction above gives $\mathbf{R}^n$, which coincides with what we expect; see Exercise 105, p. 170. In relation with vector fields in ordinary differential geometry, and thanks to the Cauchy-Lipschitz theorem, a vector field F generates a local flow φ_t, and $v = F(x)$ is just the vector associated with the path $t \mapsto \varphi_t(x)$, which brings us back to the construction above. As we can see, this is the key concept, which drags everything else. But every new concept is justified by its achievements, until now there is no real deep result involving or needing tangent vectors or tangent spaces in diffeology. In general, using directly smooth paths, for instance here (art. 8.29) or there (art. 8.42), or the local flows, for example in (art. 6.72), to prove or construct what we need is sufficient.

PROOF. 1. First we claim that for any real $\varepsilon > 0$ there exists a diffeomorphism $\lambda : \mathbf{R} \to]-\varepsilon, +\varepsilon[$ such that $\lambda \upharpoonright]-\varepsilon/2, +\varepsilon/2[$ is the identity; see Figure 6.2. Now, let $P : U \to X$ be a p-plot and $r_0 \in \mathrm{def}(P)$, there exists $\varepsilon > 0$ such that the open ball $\mathcal{B}(r_0, 2\varepsilon) \subset U$. For all $r \in \mathcal{B}(r_0, \varepsilon)$ let $\phi_r : \mathbf{R}^p \to \mathbf{R}^p$ be defined by

$$\phi_r(t) = r + \frac{\lambda(\|t\|)}{\|t\|} \times t.$$

The map ϕ_r is well defined since on an open neighborhood of $0 \in \mathbf{R}^p$, $\lambda(\|t\|) = \|t\|$ so $\phi_r(t) = r + t$. Now,

$$\begin{aligned} \|\phi_r(t) - r_0\| &= \|r - r_0 + \frac{\lambda(\|t\|)}{\|t\|} \times t\| \\ &< \|r - r_0\| + |\lambda(\|t\|)| \\ &< \varepsilon + \varepsilon. \end{aligned}$$

Thus, $\phi_r(t) \in \mathcal{B}(r_0, 2\varepsilon)$ for all $t \in \mathbf{R}^p$. Then, we can define

$$Q_r = P \circ \phi_r, \text{ that is, } Q_r(t) = P\left(r + \frac{\lambda(\|t\|)}{\|t\|} \times t\right).$$

For all $r \in \mathcal{B}(r_0, \varepsilon)$, Q_r is a global p-plot of X, and since the map $r \mapsto \phi_r$ is clearly smooth for the functional diffeology, $r \mapsto Q_r$ is a plot of $\mathrm{Paths}_p(X)$. We have obviously $Q_r(0) = P(r)$. Now, for all $\alpha \in \Omega^p(X)$, for computing $\alpha(Q_r)(0)$, it is sufficient to restrict Q_r to an open neighborhood of $0 \in \mathbf{R}^p$. Let us choose the ball $\mathcal{B}(0, \varepsilon/2)$, on which $Q_r(t) = P(r+t)$. We have $\alpha(Q_r)(0) = \alpha(Q_r \upharpoonright \mathcal{B}(0, \varepsilon/2))(0) = \alpha(t \mapsto P(r+t))(0) = \alpha(r \mapsto r+t \mapsto P(r+t))(0) = [t \mapsto r+t]^*(\alpha(P))(0) = \alpha(P)(r)$. Thus, $\dot{Q}_r(0) = \dot{P}(r)$.

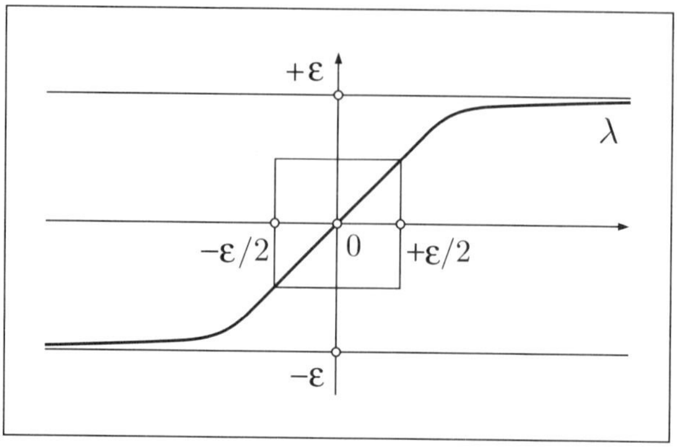

FIGURE 6.2. Contraction of $\mathbf{R}$ to an interval.

2. Every differential form evaluated on a constant plot vanishes (see Exercise 96, p. 158). Thus, for all $x \in X$, $\alpha(\mathbf{x}) = 0$ for all $\alpha \in \Omega^p(X)$, with $\mathbf{x} = [r \mapsto x]$, $r \in \mathbf{R}^p$, that is, $\dot{\mathbf{x}}(0) = 0 \in \Omega^p(X)^*$, hence $0 \in S^p_x(X)$. Let us show now that if $v = \dot{Q}(0)$, then $s \times v = \dot{Q}_s(0)$ with $Q_s(t) = Q(st)$. For all $\alpha \in \Omega^p(X)$ we have $\dot{Q}_s(0)(\alpha) \times \mathrm{vol}_p = \alpha(Q_s)(0) = \alpha(t \mapsto Q(st))(0) = \alpha(t \mapsto st \mapsto Q(st))(0) = \{[s \mapsto st]^*(\alpha(Q))\}(0) = s \times \alpha(Q)(0) = s \times \dot{Q}(0)(\alpha) \times \mathrm{vol}_p$. So, $s \times v = \dot{Q}_s(0) \in S^p_x(X)$.

3. Let us check now that $\mathrm{Span}(S^p_x(X))$ is made of the finite sums of elements of $S^p_x(X)$. First of all, the set $T^p_x(X)$ of these finite sums is a vector subspace of $\Omega^p(X)^*$. It is obviously closed by addition and by scalar multiplication because $S^p_x(X)$ is already closed by scalar multiplication. Now, by definition, $\mathrm{Span}(S^p_x(X))$ is the intersection of all the vector subspaces of $\Omega^p(X)^*$ containing $S^p_x(X)$. But every vector subspace of $\Omega^p(X)^*$ containing $S^p_x(X)$ will contain every finite sum of elements of $S^p_x(X)$, so will contain $T^p_x(X)$, thus $T^p_x(X) \subset \mathrm{Span}(S^p_x(X))$. Conversely, every vector of $\mathrm{Span}(S^p_x(X))$ is by definition contained in every vector subspace of $\Omega^p(X)^*$ containing $S^p_x(X)$, but $T^p_x(X)$ is one of them, thus $\mathrm{Span}(S^p_x(X)) \subset T^p_x(X)$. Therefore, $\mathrm{Span}(S^p_x(X)) = T^p_x(X)$.

4. The proof that the property described in the fourth point is a diffeology is left as an exercise. Let us just remark that this diffeology is the image of the sum $\coprod_{k\in\mathbf{N}} \mathrm{Paths}^k_p(X, \bullet)$ by the map $(k, Q_1, \ldots, Q_k) \mapsto (Q_1(0), \sum_{i=1}^k \dot{Q}_i(0))$, where $\mathrm{Paths}^k_p(X, \bullet)$ is the subspace of $[\mathrm{Paths}_p(X)]^k$ of elements $(Q_1, \ldots, Q_k)$ such that $Q_1(0) = \cdots = Q_k(0)$. The fact that the restriction of this diffeology to the subspaces $T^p_x(X)$ is a diffeology of vector space is a simple verification, and the fact that the zero section is smooth follows that $0 = \dot{\mathbf{x}}(0)$ (see point 2 above). □

Exercises

✎ EXERCISE 102 (The k-forms bundle on a real domain). Let U be an n-domain. For $\alpha \in \Omega^k(U)$, let $a = \alpha(\mathbf{1}_U) \in \mathcal{C}^\infty(U, \Lambda^k(\mathbf{R}^n))$.

1) Let $P : V \to \Omega^k(U)$ be a parametrization, $\alpha_r = P(r)$ and $a_r = \alpha_r(\mathbf{1}_U)$. Show that P is a plot for the functional diffeology (art. 6.29) if and only if $(r, x) \mapsto a_r(x)$, defined on $V \times U$ with values in $\Lambda^k(\mathbf{R}^n)$, is smooth.

2) Let $\alpha, \beta \in \Omega^k(U)$ and $x \in U$. Show that $\alpha_x = \beta_x$ if and only if $a(x) = b(x)$, where $b = \beta(\mathbf{1}_U)$. Deduce that $\Lambda^k_x(U) \simeq \Lambda^k(\mathbf{R}^n)$.

3) Show that the map $\phi : (x, \alpha_x) \mapsto (x, a(x) = \alpha(1_U)(x))$, from $\Lambda^k(U)$ to $U \times \Lambda^k(\mathbf{R}^n)$, is a diffeomorphism.

✎ EXERCISE 103 (The p-form bundle on a manifold). Let M be a finite dimensional manifold, $\dim(M) = n$ (art. 4.1). Let $m \in M$, $F : U \to M$ and $F \in \mathcal{A}$ such that $F(x) = m$, let $a \in \Lambda^p(\mathbf{R}^n)$.

1) Show that, for $\varepsilon > 0$ small enough, there exists a p-form $\bar{a}$ defined on U such that $\bar{a}$ is zero outside a ball $B(x, \varepsilon)$, centered at x with radius ε, and equal to a at the point x, $\bar{a}(x) = a$.

2) Show that there exists a p-form α, defined on M, such that $F^*(\alpha) = \bar{a}$.

3) Deduce that $\Lambda^p_m(M) \simeq \Lambda^p(\mathbf{R}^n)$.

4) For every chart $F : U \to M$, define

$$\mathcal{F} : U \times \Lambda^p(\mathbf{R}^n) \to \Lambda^p(M) \quad \text{by} \quad \mathcal{F}(r, a) = (F(r), F_*(a)),$$

where $F_*(a) \in \Lambda^p_{F(r)}(M)$ is the pushforward of the form $a \in \Lambda^p(\mathbf{R}^n)$ by F, according to (art. 6.51). Check that $\mathcal{F}$ is a chart of $\Lambda^p(M)$.

$$\begin{array}{ccc} U \times \Lambda^p(\mathbf{R}^n) & \xrightarrow{\ \mathcal{F}\ } & \Lambda^p(M) \\ {\scriptstyle \mathrm{pr}_1}\big\downarrow & & \big\downarrow{\scriptstyle \pi} \\ U & \xrightarrow[\ F\ ]{} & M \end{array}$$

Conclude that the bundle of p-forms $\Lambda^p(M)$ is a manifold of dimension $n + C^p_n$, modeled on $\mathbf{R}^n \times \Lambda^p(\mathbf{R}^n)$, and that the set of $\mathcal{F}$, when F runs over an atlas of M, is an atlas of $\Lambda^p(M)$.

✎ EXERCISE 104 (Smooth forms on diffeological vector spaces). Let E be a diffeological vector space and $\mathcal{O} \subset E$ be a D-open subset.

1) Let α be a differential 1-form on $\mathcal{O}$ and $A(x)(u) = \alpha(t \mapsto x + tu)_0(1)$, for all $x \in \mathcal{O}$ and all $u \in E$. Show that $A \in \mathcal{C}^\infty(\mathcal{O}, E^*_\infty)$.

2) Conversely, let $A : \mathcal{O} \to E^*_\infty$ be a smooth map. Let $P : U \to \mathcal{O}$ be a plot. For all $r \in U$, let $\alpha(P)_r = d[A(x) \circ P]_r$, where $x = P(r)$. Show that α is a differential 1-form on $\mathcal{O}$.

3) Let $\pi : \Omega^1(\mathcal{O}) \to \mathcal{C}^\infty(\mathcal{O}, E^*_\infty)$ and $\sigma : \mathcal{C}^\infty(\mathcal{O}, E^*_\infty) \to \Omega^1(\mathcal{O})$ be the maps $\pi : \alpha \mapsto A$ and $\sigma : A \mapsto \alpha$. Show that $\pi \circ \sigma = 1$.

4) Assume that every arc in E at the origin is tangential to some ray, that is, for every pointed 1-plot c, $c(0) = 0_E$, there exists $u \in E$ such that for every differential 1-form α, $\alpha(c)_0(1) = \alpha(t \mapsto tu)_0(1)$. Observe that $A(x)$ represents the value of α at x (art. 6.40) and show that π is an isomorphism.

NOTE 1. For the last question, we cannot exclude *a priori* the case for which an arc could be not tangential to some vector, which does not happen in finite dimensions, for example.

NOTE 2. This construction is related to the characterization of the cotangent space $T^*(M)$ (art. 6.48) of a manifold M modeled on E (art. 4.7). It is possible to show that, if for every $m \in M$, every element $a \in E^*_\infty$ represents, in some chart F of M, the value of a differential 1-form α at m, that is, if a is the value of $F^*(\alpha)$ at x, with

$m = F(x)$, then the cotangent space is a (locally trivial) fiber bundle (art. 8.8) with fiber E^*_∞. The same kinds of consideration can help to deconstruct the bundles of differential forms in any degree over diffeological manifolds.

✎ EXERCISE 105 (Forms bundles of irrational tori). Let T_Γ be the irrational torus $\mathbf{R}^n/\Gamma$ where Γ is a dense generating subgroup of $\mathbf{R}^n$. Let $\pi : \mathbf{R}^n \to T_\Gamma$ be the canonical projection.

1) Let α be a p-form of T_Γ. Let $a = \pi^*(\alpha)$. Show that a is invariant under the group Γ, that is, for all $\gamma \in \Gamma$ $\gamma^*(a) = a$. Deduce that every component of the form a is invariant under Γ. Deduce that a is constant and $\Omega^p(T_\Gamma) \simeq \Lambda^p(\mathbf{R}^n)$.

2) Deduce that $\Lambda^p(T_\Gamma) \simeq T_\Gamma \times \Lambda^p(\mathbf{R}^n)$.

✎ EXERCISE 106 (Vector bundles of irrational tori). Let T_Γ be the irrational torus $\mathbf{R}^n/\Gamma$ where Γ is a dense generating subgroup of $\mathbf{R}^n$; see Exercise 105, p. 170. Show that, for the p-vector bundles (art. 6.53) of T_Γ,

$$T^p(T_\Gamma) \simeq T_\Gamma \times [\Lambda^p(\mathbf{R}^n)]^* \simeq T_\Gamma \times \mathbf{R}^{\frac{n!}{p!(n-p)!}}.$$

In particular, for the tangent space of the irrational torus,

$$T_x(T_\Gamma) \simeq \mathbf{R}^n \quad \text{and} \quad T(T_\Gamma) \simeq T_\Gamma \times \mathbf{R}^n.$$

✎ EXERCISE 107 (Differential 1-forms on $\mathbf{R}/\{\pm 1\}$). Let $\Delta = \mathbf{R}/\{\pm 1\}$ be the quotient of the real line by the action of the multiplicative group $\{\pm 1\}$. Identify Δ with the half-line $[0, \infty[$, equipped with the pushforward of the smooth diffeology of $\mathbf{R}$ by the square map $\mathrm{sq}(x) = x^2$.

1) Show that every differential 1-form on Δ writes $\alpha = f \times \theta$, where θ is the 1-form defined by $\mathrm{sq}^*(\theta) = d[t^2] = 2t \times dt$, and $f \in \mathcal{C}^\infty(\Delta, \mathbf{R})$.

2) Deduce that every differential 1-form on Δ vanishes at the origin, as well as its tangent space at this point.

Lie Derivative of Differential Forms

In this section, we introduce the notion of Lie derivative *of differential forms, on a diffeological space, along* slidings, *that is, germs of paths of diffeomorphisms. We introduce the notion of* contraction of forms *by* arcs of plots *which generalizes the notion of contraction of smooth forms by a vectors on real domains. We generalize then the famous Cartan-Lie formula for differential forms in the context of diffeology.*

6.54. The Lie derivative of differential forms. Let X be a diffeological space, let $\alpha \in \Omega^k(X)$, and let $h : \mathbf{R} \to \mathrm{Diff}(X)$ be a 1-plot for the functional diffeology (art. 1.61) centered at the identity $h(0) = \mathbf{1}_X$. The *Lie derivative* $\pounds_h(\alpha)$, of α by h, is the k-form defined by

$$\pounds_h(\alpha) = \frac{\partial}{\partial t}\left\{h(t)^*\alpha\right\}_{t=0}.$$

This means precisely that, for every n-plot $P : U \to X$, for all $r \in U$ and all $v_1, \ldots, v_k \in \mathbf{R}^n$,

$$\pounds_h(\alpha)(P)(r)(v_1)\cdots(v_k) = \frac{\partial}{\partial t}\left\{\alpha(h(t) \circ P)(r)(v_1)\cdots(v_k)\right\}_{t=0}.$$

NOTE 1. The Lie derivative $\pounds_h(\alpha)$ depends only on the 1-jet of the plot h at 0. Precisely, if h and h' are two 1-plots of $\mathrm{Diff}(X)$ centered at the identity such that $h' = h \circ f$, with $f(0) = 0$ and $D(f)(0) = 1$, then $\pounds_h(\alpha) = \pounds_{h'}(\alpha)$. This is the case in particular when h and h' coincide on an open neighborhood of 0 (they have the same germ at 0).

NOTE 2. The Lie derivative $\pounds_h : \Omega^k(X) \to \Omega^k(X)$ is a smooth linear map for the functional diffeology of forms (art. 6.29).

PROOF. First of all, let us show that $\pounds_h(\alpha)$ is a well defined differential form of X. Let $P : U \to X$ be an n-plot of X, let $h \cdot P$ be the $(p+1)$-plot defined on $\mathbf{R} \times U$ by

$$h \cdot P(t, r) = (h(t) \circ P)(r) = h(t)(P(r)), \quad \text{where } (t, r) \in \mathbf{R} \times U.$$

By definition, $\alpha(h \cdot P)$ is a differential form on $\mathbf{R} \times U$. But $\alpha(h(t) \circ P)$ is the pullback of $\alpha(h \cdot P)$ by the injection $j_t : r \mapsto (t, r)$ from U into $\mathbf{R} \times U$, i.e., $\alpha(h(t) \circ P)$ is the restriction of $\alpha(h \cdot P)$ to $\{t\} \times U$. Since $\alpha(h(t) \circ P)$ is a smooth form, the map $t \mapsto \alpha(h(t) \circ P) \restriction \{t\} \times U$ is a smooth parametrization in $\Lambda^k(\mathbf{R}^n)$. Thus, its derivative is still a smooth parametrization of $\Lambda^k(\mathbf{R}^n)$. Therefore, the definition $\pounds_h(\alpha)$ makes sense. Let us check now the fundamental property of differential forms (art. 6.28). Let $F : V \to U$ be a smooth parametrization in U:

$$
\begin{aligned}
\pounds_h(\alpha)(P \circ F) &= \frac{\partial}{\partial t}\left\{\alpha(h(t) \circ (P \circ F))\right\}_{t=0} = \frac{\partial}{\partial t}\left\{\alpha((h(t) \circ P) \circ F)\right\}_{t=0} \\
&= \frac{\partial}{\partial t}\left\{F^*(\alpha(h(t) \circ P))\right\}_{t=0} = F^*\left(\frac{\partial}{\partial t}\left\{\alpha(h(t) \circ P)\right\}_{t=0}\right) \\
&= F^*\left(\pounds_h(\alpha)(P)\right).
\end{aligned}
$$

Then $\pounds_h(\alpha)$ is a differential k-form of X. For the first note, since the derivative is local, the Lie derivative is local and $\pounds_h(\alpha)$ does not depend on more than the germ of h. Now, if $h' = h \circ f$ with $f(0) = 0$, then

$$
\begin{aligned}
\pounds_{h'}(\alpha)(P)(r)[v] &= D(s \mapsto \alpha(h(f(s)) \circ P)(r)[v])(s = 0)(1) \\
&= D(s \mapsto t = f(s) \mapsto \alpha(h(t) \circ P)(r)[v])(s = 0)(1) \\
&= D(t \mapsto \alpha(h(t) \circ P)(r)[v])(0)(D(f)(0)(1)) \\
&= D(f)(0)(1) \times \pounds_h(\alpha)(P)(r)[v],
\end{aligned}
$$

where we denoted a k-uple of vectors $(v_1) \cdots (v_k)$ by $[v]$. Thus, if $D(f)(0)(1) = 1$, then $\pounds_h(\alpha) = \pounds_{h'}(\alpha)$. For the second note, it is clear that the Lie derivative is linear. Let us prove that $\pounds_h$ is smooth. Let $\phi : V \to \Omega^k(X)$ and $P : U \to X$ be two plots. We want to check that $(s, r) \mapsto \pounds_h(\phi(s))(P)(r)$ is smooth. But

$$
\begin{aligned}
\pounds_h(\phi(s))(P)(r) &= \frac{\partial}{\partial t}\left(h(t)^*(\phi(s))(P)(r)\right)_{t=0} \\
&= \frac{\partial}{\partial t}\left(\phi(s)(h(t) \circ P)(r)\right)_{t=0} \\
&= \frac{\partial}{\partial t}\left(j_t^*(\phi(s)(h \cdot P))(r)\right)_{t=0}.
\end{aligned}
$$

Now, $(s, t, r) \mapsto \phi(s)(h \cdot P)(t, r)$ is a smooth map, and $j_t^*(\phi(s)(h \cdot P))(r)$ is just its restriction to $V \times \{t\} \times U$. Thus, $\pounds_h(\phi(s))(P)(r)$ is a partial derivative of a smooth map, hence it is smooth. Therefore, the Lie derivative $\pounds_h$ is a smooth endomorphism of $\Omega^k(X)$. □

6.55. Lie derivative along homomorphisms. Let X be a diffeological space. We call a *ray* of Diff(X) every smooth homomorphism from $\mathbf{R}$ to Diff(X), and we call a *flow* every 1-plot h of Diff(X), defined on an interval I and centered at the identity, such that for all $t, t' \in I$, $t + t' \in I$ implies $h(t+t') = h(t) \circ h(t')$.

1. Let h be a ray of Diff(X). Then

$$\left.\frac{\partial h(t)^*(\alpha)}{\partial t}\right|_{t=t_0} = h(t_0)^*(\pounds_h(\alpha)), \text{ for all } t_0 \in \mathbf{R}.$$

2. Let α be a k-form on X and h be a flow defined on I. If $\pounds_h(\alpha) = 0$, then $h(t)^*(\alpha) = \alpha$, for all $t \in I$.

3. Let $\mathrm{Diff}(X)^\star \subset \mathrm{Diff}(X)$ be the subgroup generated by the elements of $h(t) \in \mathrm{Diff}(X)$ where h is a flow. The subgroup $\mathrm{Diff}(X)^\star$ is made of finite products $h_1(t_1) \circ \cdots \circ h_N(t_N)$, where N runs over $\mathbf{N}$, and the h_i are flows.[1] The subgroup $\mathrm{Diff}(X)^\star$ is normal in Diff(X), that is, for all $g \in \mathrm{Diff}(X)$, and all $h \in \mathrm{Diff}(X)^\star$, $g \circ h \circ g^{-1} \in \mathrm{Diff}(X)^\star$. Now, let $H \subset \mathrm{Diff}(X)^\star$ be a subgroup and let α be a k-form of X. If $\pounds_F(\alpha) = 0$ for every 1-plot F of H, centered at the identity, then, for every $h \in H$, $h^*(\alpha) = \alpha$.

PROOF. 1. Let us compute the derivative of $[t \mapsto h(t)^*(\alpha)]$ at the point t_0:

$$\begin{aligned}\left.\frac{\partial h(t)^*(\alpha)}{\partial t}\right|_{t=t_0} &= \lim_{\epsilon \to 0} \frac{h(t_0+\epsilon)^*(\alpha) - h(t_0)^*(\alpha)}{\epsilon} \\ &= \lim_{\epsilon \to 0} h(t_0)^* \left[\frac{h(\epsilon)^*(\alpha) - \alpha}{\epsilon}\right].\end{aligned}$$

Let $\beta_\epsilon = (h(\epsilon)^*(\alpha) - \alpha)/\epsilon$. Then for every n-plot $P : U \to X$, for every point $r \in U$, and for any k vectors $u_1, \ldots, u_k \in \mathbf{R}^n$,

$$\begin{aligned}\left.\frac{\partial h(t)^*(\alpha)}{\partial t}\right|_{t=t_0} (P)(r)(u_1)\cdots(u_k) &= \lim_{\epsilon \to 0} h(t_0)^*(\beta_\epsilon)(P)(r)(u_1)\cdots(u_k) \\ &= \lim_{\epsilon \to 0} \beta_\epsilon(h(t_0) \circ P)(r)(u_1)\cdots(u_k) \\ &= \pounds_h(\alpha)(h(t_0) \circ P)(r)(u_1)\cdots(u_k) \\ &= h(t_0)^*(\pounds_h(\alpha))(P)(r)(u_1)\cdots(u_k).\end{aligned}$$

Hence,

$$\left.\frac{\partial h(t)^*(\alpha)}{\partial t}\right|_{t=t_0} = h(t_0)^*(\pounds_h(\alpha)).$$

2. Now, if $\pounds_h(\alpha) = 0$, then $[\partial(h(t_0)^*(\alpha))/\partial t]_{t=t_0} = 0$ for all $t_0 \in I$, and $h(t)^*(\alpha)$ is constant on I. Hence, $h(t)^*(\alpha)$ is equal to $h(0)^*(\alpha) = 1_X^*(\alpha) = \alpha$.

3. This is a direct application of 2. □

6.56. Contraction of differential forms. Let X be a diffeological space, and let $\mathcal{D}$ be its diffeology. Let $P : U \to X$ be an n-plot. We shall call an *arc of plots* or

[1]The group generated by the flows is clearly a subgroup of the identity component of the group of diffeomorphisms. For a compact second countable manifold they coincide, but I do not think that they coincide in general, for diffeological spaces.

an *arc in* $\mathcal{D}$, any 1-parameter family $F : s \mapsto P_s$, centered at some $P \in \mathcal{D}$, defined on an interval $]-\varepsilon, +\varepsilon[$ with values in $\mathcal{D}$ such that

a) $P_0 = P$ and $\mathrm{def}(P_s) = U$, for all s,
b) $\bar{P} : (s, r) \mapsto P_s(r)$, defined on $]-\varepsilon, \varepsilon[\times U$, is a plot.

In particular, F is a 1-plot of $\mathcal{D}$, equipped with the functional diffeology (art. 1.63). The plot $P = P_0$ will be called the *target* of F. Let F be an arc in $\mathcal{D}$, centered at P, and let α be a p-form of X, $p \geq 1$. For all $r \in U$, $\alpha(\bar{P})$ is a smooth p-form of $]-\varepsilon, +\varepsilon[\times U$. The restriction to $\{0\} \times U$ of the contraction of $\alpha(\bar{P})$ by the vector $(1, 0) \in \mathbf{R} \times \mathbf{R}^n$ (art. 6.14), is the following smooth $(p-1)$-form of U

$$r \mapsto \left[(v_2, \ldots, v_p) \mapsto \alpha(\bar{P})_{\binom{0}{r}} \begin{pmatrix}1\\0\end{pmatrix} \begin{pmatrix}0\\v_2\end{pmatrix} \cdots \begin{pmatrix}0\\v_p\end{pmatrix}\right] \in \mathcal{C}^\infty(U, \Lambda^{p-1}(\mathbf{R}^n)),$$

with r in U, and $v_2, \ldots, v_p$ being $(p-1)$ vectors of $\mathbf{R}^n$. Let $F : s \mapsto P_s$ and $F' : s \mapsto P'_s$ be two arcs of plots centered at the same plot $P = P_0 = P'_0$. We shall say that F and F′ define the same *variation* δP of the n-plot P if, for every p-form α of X, $p \geq 1$, for all r in U and for all $v_2, \ldots, v_p \in \mathbf{R}^n$,

$$\alpha(\bar{P}')_{\binom{0}{r}} \begin{pmatrix}1\\0\end{pmatrix} \begin{pmatrix}0\\v_2\end{pmatrix} \cdots \begin{pmatrix}0\\v_p\end{pmatrix} = \alpha(\bar{P})_{\binom{0}{r}} \begin{pmatrix}1\\0\end{pmatrix} \begin{pmatrix}0\\v_2\end{pmatrix} \cdots \begin{pmatrix}0\\v_p\end{pmatrix}. \qquad (\heartsuit)$$

Thus, the variation δP of the plot P can be regarded as the class of the arc of plots F, for the equivalence relation defined by $(\heartsuit)$. Let $\alpha \in \Omega^p(X)$ with $p \geq 1$, and $P : U \to X$ be an n-plot, let δP be a variation of the plot P. We shall call a *contraction* of α by δP the smooth $(p-1)$-form defined on U by

$$\alpha(\delta P)(r)(v_1) \cdots (v_{p-1}) = \alpha(\bar{P})_{\binom{0}{r}} \begin{pmatrix}1\\0\end{pmatrix} \begin{pmatrix}0\\v_1\end{pmatrix} \cdots \begin{pmatrix}0\\v_{p-1}\end{pmatrix}, \qquad (\diamondsuit)$$

where $F : s \mapsto P_s$ represents δP, $r \in U$, and $v_1, \ldots v_{p-1} \in \mathbf{R}^n$. Be aware that

$$\alpha(\delta P) \in \mathcal{C}^\infty(U, \Lambda^{p-1}(\mathbf{R}^n)).$$

NOTE 1. To be coherent and to avoid possible confusion, we shall denote formally by $\alpha \lrcorner \delta P \in \Omega^{p-1}(U)$ the differential form, where U is regarded as a diffeological space, defined by $(\alpha \lrcorner \delta P)(\mathbf{1}_U) = \alpha(\delta P)$.

NOTE 2. The contraction of a p-form of X, by some arc of plots F, is not a $(p-1)$ form of X, since it is not defined on the plots of X but on the domain of the target $P = F(0)$. However, in some particular situations, for example in (art. 6.57), this definition may lead to a true $(p-1)$-form of X.

NOTE 3. In the definition of the variation δP, the value $s = 0$, where the variation is computed, does not really matter. We can define as well the variation denoted by δP_s, as the variation for $s = 0$ of the translated arc $F_s : s' \mapsto F(s' + s)$, that is,

$$\alpha(\delta P_s)(r)(v_2) \cdots (v_p) = \alpha(\bar{P})_{\binom{s}{r}} \begin{pmatrix}1\\0\end{pmatrix} \begin{pmatrix}0\\v_2\end{pmatrix} \cdots \begin{pmatrix}0\\v_p\end{pmatrix}.$$

6.57. Contracting differential forms on slidings. Let X be a diffeological space and Diff(X) be its group of diffeomorphisms, equipped with the functional diffeology (art. 1.61). Let $\alpha \in \Omega^p(X)$ be any differential p-form with $p \geq 1$. Let $F :]-\epsilon, +\epsilon[\to \mathrm{Diff}(X)$, $\epsilon > 0$, be a *sliding* of X, that is, any 1-plot centered at the identity, $F(0) = \mathbf{1}_X$. Let $P : U \to X$ be any n-plot of X. We denote by $F \cdot P$ the $(1+n)$-plot of X defined by

$$F \cdot P :]-\epsilon, +\epsilon[\times U \to X \quad \text{and} \quad F \cdot P : (t, r) \mapsto F(t)(P(r)).$$

1. The contraction of α by the arc of plots $t \mapsto [r \mapsto F \cdot P(t,r)]$ (art. 6.56) will be denoted by $i_F(\alpha)(P)$. It is defined, for every $r \in U$ and for any $(p-1)$ vectors $v_1, \ldots, v_{p-1} \in \mathbf{R}^n$, by

$$i_F(\alpha)(P)_r(v_1)\cdots(v_{p-1}) = \alpha(F \cdot P)\begin{pmatrix}0\\r\end{pmatrix}\begin{pmatrix}1\\0\end{pmatrix}\begin{pmatrix}0\\v_1\end{pmatrix}\cdots\begin{pmatrix}0\\v_{p-1}\end{pmatrix}. \qquad (\diamondsuit)$$

The mapping $i_F(\alpha)$ defined by $(\diamondsuit)$ is a differential $(p-1)$-form of X. It will be called the *contraction of* α *by the sliding* F.

2. The contraction operation $i_F : \Omega^p(X) \to \Omega^{p-1}(X)$, with $p \geq 1$, is a smooth linear map, where the spaces of differential forms are equipped with the functional diffeology (art. 6.28).

NOTE 1. Diffeomorphisms are automorphisms, they preserve the *morphology* of the space. A 1-plot F of diffeomorphisms of X, centered at the origin represents some sliding from X onto itself, and this explains the vocabulary.

NOTE 2. Actually, the map $(F, \alpha) \mapsto i_F(\alpha)$, where F belongs to the space of slidings of X and α belongs to $\Omega^p(X)$, is smooth.

NOTE 3. To use the notation of (art. 6.56), $i_F(\alpha)(P) = \alpha(\delta P)$, where δP is the variation of the arc of plots $[t \mapsto [r \mapsto F(t)(P(r))]]$.

PROOF. 1. The condition for $i_F(\alpha)$ to be a differential form on U is to satisfy the compatibility relation

$$i_F(\alpha)(P \circ \psi) = \psi^*(i_F(\alpha)(P)),$$

where $\psi : V \to U$ is some smooth parametrization in U. Let ψ be a m-plot of U, let $s \in V$ and let $u_1, \ldots, u_{p-1} \in \mathbf{R}^m$. We have then

$$i_F(\alpha)(P \cdot \psi)_s(u_1, \ldots, u_{p-1}) = \alpha(F \cdot (P \circ \psi))\begin{pmatrix}0\\s\end{pmatrix}\begin{pmatrix}1\\0\end{pmatrix}\begin{pmatrix}0\\u_1\end{pmatrix}\cdots\begin{pmatrix}0\\u_{p-1}\end{pmatrix}.$$

But

$$\begin{aligned} F \cdot (P \circ \psi)(t,s) &= F(t)(P \circ \psi(s)) \\ &= F(t)(P(\psi(s))) \\ &= F \cdot P(t, \psi(s)) \\ &= (F \cdot P) \circ (\mathbf{1} \times \psi)(t,s), \end{aligned}$$

where $\mathbf{1} \times \psi(t,v) = (t, \psi(v))$. Let us use the more compact notation

$$\begin{bmatrix}0\\u\end{bmatrix} = \begin{pmatrix}0\\u_1\end{pmatrix}\cdots\begin{pmatrix}0\\u_{p-1}\end{pmatrix}.$$

We have, from above,

$$\begin{aligned} \alpha(F \cdot (P \circ \psi))\begin{pmatrix}0\\s\end{pmatrix}\begin{pmatrix}1\\0\end{pmatrix}\begin{bmatrix}0\\u\end{bmatrix} &= \alpha((F \cdot P) \circ (\mathbf{1} \times \psi))\begin{pmatrix}0\\s\end{pmatrix}\begin{pmatrix}1\\0\end{pmatrix}\begin{bmatrix}0\\u\end{bmatrix} \\ &= (\mathbf{1} \times \psi)^*(\alpha(F \cdot P))\begin{pmatrix}0\\s\end{pmatrix}\begin{pmatrix}1\\0\end{pmatrix}\begin{bmatrix}0\\u\end{bmatrix} \\ &= \alpha(F \cdot P)\begin{pmatrix}0\\\psi(s)\end{pmatrix}\begin{pmatrix}1\\0\end{pmatrix}\begin{bmatrix}0\\v\end{bmatrix}, \\ &\qquad \text{with} \quad v_i = D(\psi)_s(u_i). \end{aligned}$$

Hence, we get finally, with $D(\psi)_s[u]$ for $(D(\psi)_s(u_1))\cdots(D(\psi)_s(u_{p-1}))$,

$$\begin{aligned}\alpha(F\cdot(P\circ\psi))\begin{pmatrix}0\\s\end{pmatrix}\begin{pmatrix}1\\0\end{pmatrix}\begin{bmatrix}0\\u\end{bmatrix} &= i_F(\alpha)(P)_{\psi(s)}(D(\psi)_s[u]) \\ &= \psi^*(i_F(\alpha)(P))_s[u].\end{aligned}$$

Therefore, $i_F(\alpha)(P\circ\psi)=\psi^*(i_F(\alpha)(P))$, and $i_F(\alpha)$ is a $(p-1)$-form of X.

2. The contraction i_F is obviously linear. Let us prove that it is smooth. Let $s\mapsto\beta_s$ be a plot of $\Omega^p(X)$. We have to check that $s\mapsto i_F(\beta_s)$ is a plot of $\Omega^{p-1}(X)$. Let $P:U\to X$ be a plot, we have

$$i_F(\beta_s)(P)_r[v]=\beta_s(F\cdot P)\begin{pmatrix}0\\r\end{pmatrix}\begin{pmatrix}1\\0\end{pmatrix}\begin{bmatrix}0\\v\end{bmatrix},$$

with the same notation as above for v. But $F\cdot P:(t,r)\mapsto F(t)(P(r))$ is a plot of X, thus $(t,r,s)\mapsto\beta_s(F\cdot P)_{(t,r)}$ is smooth. Then,

$$(r,s)\mapsto\beta_s(F\cdot P)\begin{pmatrix}0\\r\end{pmatrix}\begin{pmatrix}1\\0\end{pmatrix}$$

is smooth. Hence, $(r,s)\mapsto i_F(\beta_s)(P)_r$ is a plot of $\Omega^{p-1}(X)$ and the contraction $i_F:\Omega^p(X)\to\Omega^{p-1}(X)$ is a smooth linear map. □

Exercises

✎ EXERCISE 108 (Anti-Lie derivative). Let X be a diffeological space, let α be a p-form of X, and let F be a 1-plot of Diff(X) centered at the identity. For all $A\in\mathrm{Diff}(X)$, let A_* denote the pushforward $A_*(\alpha)=(A^{-1})^*(\alpha)$. Use the identity $1_X=F(t)\circ F(t)^{-1}$, and that $F(0)=1_X$, to show that

$$\frac{\partial}{\partial t}\Big\{F(t)_*(\alpha)\Big\}_{t=0}=-\pounds_F(\alpha).$$

✎ EXERCISE 109 (Multi-Lie derivative). The notion of Lie derivative can be extended to any plot of Diff(X). Let us define, with the kinds of notation and conventions of (art. 6.54), for every q-plot h of Diff(X) centered at the identity and for every vector $v\in\mathbf{R}^q$,

$$\pounds_h(\alpha)(v)=\frac{\partial}{\partial s}\Big\{h(s)^*\alpha\Big\}_{s=0}(v).$$

1) Check that $\pounds_h(\alpha)(v)$ is a k-form of X. Actually, check that $\pounds_h(\alpha)$ is a smooth linear map from $\mathbf{R}^q$ to $\Omega^k(X)$.

2) Let $v=\sum_i v_i\,\mathbf{e}_i$, where the $\mathbf{e}_i$ are the vectors of the canonical basis of $\mathbf{R}^q$. And let $h_i:t\mapsto h(t\mathbf{e}_i)$ be defined on some open neighborhood of $0\in\mathbf{R}$. Show that

$$\pounds_h(\alpha)(v)=\sum_{i=1}^q v^i\,\pounds_{h_i}(\alpha).$$

As a corollary of this linearity, $\pounds_h(\alpha)(P)=0$ for every q-plot h of Diff(X) if and only if $\pounds_h(\alpha)(P)=0$ for every 1-plot h of Diff(X).

✎ EXERCISE 110 (Variations of points of domains). Let U be an n-domain and $x\in U$. Let $\mathbf{x}=[0\mapsto x]$ be the 0-plot with value x. Describe the variations $\delta\mathbf{x}$ of $\mathbf{x}$.

✎ EXERCISE 111 (Liouville rays). Let ω be a p-form on a diffeological space X, assume $\omega \neq 0$. Let $h \in \mathrm{Hom}^\infty(\mathbf{R}, \mathrm{Diff}(X))$ be a ray such that, for all $t \in \mathbf{R}$, $h(t)^*(\omega) = \lambda(t) \times \omega$. Show that $\lambda(t) = e^t$; see Exercise 114, p. 198.

Cubes and Homology of Diffeological Spaces

We shall use the singular cubic homology and cohomology because they are naturally adapted, in the diffeological framework, to the integration of forms (art. 6.27) *and to the variational calculus* (art. 6.70). *We shall see further the main theorems relative to the De Rham theory in this context. For a short description of cubic homology in classical differential geometry, see for example* [**HoYo61**].

6.58. Cubes and cubic chains on diffeological spaces. We call *standard* p*-cube* the subset $[0,1]^p$ of $\mathbf{R}^p$, and we denote it by I^p,

$$I^p = [0,1]^p \subset \mathbf{R}^p.$$

Let X be a diffeological space. We call a *smooth* p*-cube* in X any smooth map from $\mathbf{R}^p$ to X. And we denote by $\mathrm{Cub}_p(X)$ the set of all the smooth p-cubes in X,

$$\mathrm{Cub}_p(X) = \mathcal{C}^\infty(\mathbf{R}^p, X).$$

The set $\mathrm{Cub}_p(X)$ will be equipped with the functional diffeology; see (art. 1.57) and (art. 1.60). Note that since 0-cubes are any maps from $I^0 = \mathbf{R}^0 = \{0\}$ to X, then $\mathrm{Cub}_0(X)$ is naturally equivalent to X, thanks to the diffeomorphism $x \mapsto \mathbf{x} = [0 \mapsto x]$. Hence,

$$\mathrm{Cub}_0(X) \simeq X.$$

Then, we define the *smooth cubic* p*-chains* in X, with coefficients in $\mathbf{Z}$, as the free Abelian group generated by $\mathrm{Cub}_p(X)$, and we denote it by $C_p(X)$. Thus, a (smooth) cubic p-chain c, in X, is any finite $\mathbf{Z}$-linear combination of p-cubes, that is,

$$c = \sum_\sigma n_\sigma\, \sigma, \ \text{ with } \sigma \in \mathrm{Cub}_p(X), \ \text{ and } n_\sigma \in \mathbf{Z},$$

where the sum is performed over a finite set of p-cubes called the *support* of c, and denoted by

$$\mathrm{Supp}(c) = \{\sigma \in \mathrm{Cub}_p(X) \mid n_\sigma \neq 0\}.$$

The group of cubic p-chains $C_p(X)$ can be represented by

$$C_p(X) \simeq \{c \in \mathrm{Maps}(\mathrm{Cub}_p(X), \mathbf{Z}) \mid \#\mathrm{Supp}(c) < \infty\}.$$

Note that in the writing $\sum_\sigma n_\sigma \sigma$ of the chain c, $n_\sigma = c(\sigma)$. Then, the sum of two cubic p-chains c and c', and the multiplication of a cubic p-chain c by an integer m, are defined as usual:

$$(c + c')(\sigma) = c(\sigma) + c'(\sigma) \quad \text{and} \quad (mc)(\sigma) = m \times c(\sigma).$$

NOTE 1. A cubic chain can also be regarded as any finite family $\{(n_i, \sigma_i)\}_{i \in \mathcal{J}}$ and can be written $\sum_{i\in\mathcal{J}} n_i\sigma_i$, with the convention that if $\sigma_i = \sigma_j$, then $\sum_{i\in\mathcal{J}} n_i\sigma_i = \sum_{i'\in\mathcal{J}'} n_{i'}\sigma_{i'} + (n_i + n_j)\sigma_i$, where $\mathcal{J}' = \mathcal{J} - \{i,j\}$. Since the family is finite, the sum of the coeffcients of a same cube is finite and both aspects are equivalent.

NOTE 2. With smooth homology or cohomology in mind, there is no contradiction in defining smooth p-cubes in X as smooth maps from $\mathbf{R}^p$ to X, as we do here, or as the maps from I^p to X which are the restrictions of smooth maps defined on an

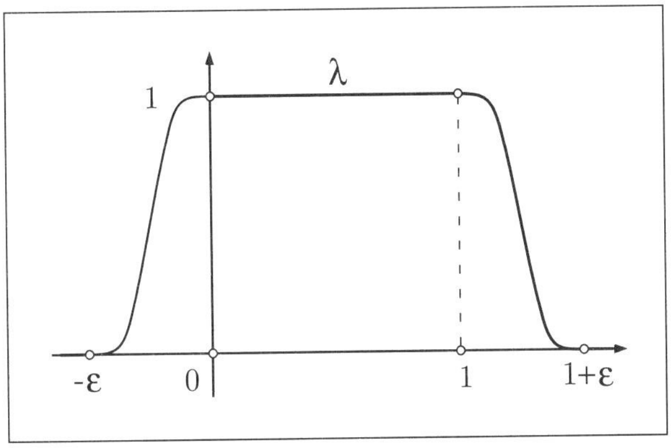

FIGURE 6.3. The cube's smashing function.

open neighborhood of I^p, as we could have also chosen to do. Indeed the following proposition addresses this issue.

($\diamondsuit$) Every p-plot of X defined on a small open neighborhood of I^p coincides, on I^p, with some global p-plot of X.

This is why, for sake of simplicity and without loss of generality, the smooth p-cubes in X have been defined as the global p-plots of X. But to focus our attention on $I^p \in \mathbf{R}^p$ we have introduced a special name, p-cube instead of global p-plot, and a special notation $\mathrm{Cub}_p(X)$ for $\mathcal{C}^\infty(\mathbf{R}^p, X)$).

PROOF. For the second note, let P be a p-plot of X, defined on some open neighborhood of I^p. Since I^p is compact there exists a real $\epsilon > 0$, small enough, such that $]-\epsilon, 1+\epsilon[^p \subset \mathrm{def}(P)$. Let us restrict P to this small open p-cube. Now, let $\lambda \in \mathcal{C}^\infty(\mathbf{R}, \mathbf{R})$ be a smooth function, as shown in Figure 6.3, equal to 1 on a superset of $[0,1]$, equal to 0 on a superset of $]-\infty, \epsilon] \cup [1+\epsilon, +\infty[$, increasing between $-\epsilon$ and 0, decreasing between 1 and $1+\epsilon$. Then, let us define

$$P'(t_1, \ldots, t_p) = P(\lambda(t_1) \times t_1, \ldots, \lambda(t_p) \times t_p) \text{ if } (t_1, \ldots, t_p) \in]-\epsilon, 1+\epsilon[^p,$$

thus $P'(t_1, \ldots, t_p) = 0$ outside $]-\epsilon, 1+\epsilon[^p$. The parametrization P' is a global p-plot of X, equal to P on I^p and 0 outside of a small open neighborhood of I^p. □

6.59. Boundary of cubes and chains. Let X be a diffeological space. The integration of a smooth p-form on a p-cube (art. 6.27) suggests the introduction of a *boundary* operation on cubic chains, denoted by ∂, and satisfying the *homological condition* $\partial \circ \partial = 0$. Let us consider the simplest example: Let f be a smooth function defined on the cube $c = [a, b]$. Let F be a primitive of f, $F' = f$ or $dF = f$. We can write

$$\int_c f = \int_c dF = F(b) - F(a) = [F]_a^b = \int_{\partial c} F \quad \text{with} \quad \partial c = b - a.$$

Here, the string $\partial c = b - a$ is interpreted as a 0-chain $\partial c = 1 \times \mathbf{b} + (-1) \times \mathbf{a}$, where $\mathbf{a} : 0 \mapsto a$ and $\mathbf{b} : 0 \mapsto b$. We shall extend this particular case to the general case on X as follows. Let us describe the cubic chain-complex.

0-Chains. For every $x \in X$, let $\mathbf{x} : 0 \mapsto x$ be the 0-cube with value x. Let

$$\partial \mathbf{x} \text{ (or } \partial x) = 0.$$

Now, ∂ is extended by linearity on every 0-chain $c = \sum_x n_x x \in C_0(X)$, and

$$\partial\left(\sum_x n_x x\right) = \sum_x n_x \partial x = 0.$$

In other words, $\partial(C_0(X)) = \{0\}$, or $\partial : C_0(X) \to \{0\}$.

1-Chains. Let $\sigma \in \mathrm{Cub}_1(X)$ be a smooth 1-cube, and we define

$$\partial\sigma = \sigma(1) - \sigma(0).$$

Extended by linearity on any 1-chain $c = \sum_\sigma n_\sigma \sigma \in C_1(X)$, we have $\partial : C_1(X) \to C_0(X)$ with

$$\partial\left(\sum_\sigma n_\sigma \sigma\right) = \sum_\sigma n_\sigma \partial\sigma = \sum_\sigma n_\sigma(\sigma(1) - \sigma(0)).$$

Now, since for all 1-chains c, $\partial(c)$ is a 0-chain, and since ∂ vanishes on the 0-chain, we have $\partial(\partial(c)) = 0$, for all $c \in C_1(X)$.

2-Chains. Let $\sigma \in \mathrm{Cub}_2(X)$ be any smooth 2-cube, and we define

$$\partial\sigma(t) = [\sigma(1)(t) - \sigma(0)(t)] - [\sigma(t)(1) - \sigma(t)(0)].$$

Extended by linearity on any 2-chain $c = \sum_\sigma n_\sigma \sigma \in C_2(X)$, we have $\partial : C_2(X) \to C_1(X)$ with

$$\partial\left(\sum_\sigma n_\sigma \sigma\right) = \sum_\sigma n_\sigma \partial\sigma = \sum_\sigma n_\sigma\{[\sigma(1)(t) - \sigma(0)(t)] - [\sigma(t)(1) - \sigma(t)(0)]\}.$$

Now, for any 2-cube σ we have

$$\begin{aligned} \partial(\partial\sigma) &= \partial([\sigma(1)(t) - \sigma(0)(t)] - [\sigma(t)(1) - \sigma(t)(0)]) \\ &= \{[\sigma(1)(1) - \sigma(1)(0)] - [\sigma(0)(1) - \sigma(0)(0)]\} \\ &- \{[\sigma(1)(1) - \sigma(0)(1)] - [\sigma(1)(0) - \sigma(0)(0)]\} \\ &= +\sigma(1)(1) - \sigma(1)(0) - \sigma(0)(1) + \sigma(0)(0) \\ &\quad -\sigma(1)(1) + \sigma(0)(1) + \sigma(1)(0) - \sigma(0)(0) \\ &= 0. \end{aligned}$$

Hence, by linearity, for all 2-chains $c \in C_2(X)$, we have also $\partial(\partial c) = 0$.

3-Chains. Let us continue once more, to see clearly the pattern, before giving the general definition. Let $\sigma \in \mathrm{Cub}_3(X)$ be a smooth 3-cube, and we define its boundary

$$\begin{aligned} \partial\sigma(t_1)(t_2) &= [\sigma(1)(t_1)(t_2) - \sigma(0)(t_1)(t_2)] \\ &- [\sigma(t_1)(1)(t_2) - \sigma(t_1)(0)(t_2)] \\ &+ [\sigma(t_1)(t_2)(1) - \sigma(t_1)(t_2)(0)]. \end{aligned}$$

As usual the boundary operator ∂ is extended by linearity on any 3-chain $c \in C_3(X)$, and $\partial : C_3(X) \to C_2(X)$. The verification that $\partial(\partial\sigma) = 0$, for any $\sigma \in \mathrm{Cub}_3(X)$, is left as an exercise.

p-Chains. We are now ready to give the general definition for the boundary of a cubic $(p+1)$-chain of a diffeological space X. Let σ be a $(p+1)$-cube of X, that is, $\sigma \in \mathrm{Cub}_{p+1}(X) = \mathcal{C}^\infty(\mathbf{R}^{p+1}, X)$, with $p \geq 0$. We define the boundary $\partial\sigma$ of the $(p+1)$-cube σ as the following cubic p-chain in X:

$$\begin{aligned}
\partial\sigma(t_1)(t_2)\cdots(t_p) &= (-1)^1 \times [\sigma(0)(t_1)(t_2)\cdots(t_p) - \sigma(1)(t_1)(t_2)\cdots(t_p)] \\
&+ (-1)^2 \times [\sigma(t_1)(0)(t_2)\cdots(t_p) - \sigma(t_1)(1)(t_2)\cdots(t_p)] \\
&+ (-1)^3 \times [\sigma(t_1)(t_2)(0)\cdots(t_p) - \sigma(t_1)(t_2)(1)\cdots(t_p)] \\
&\cdots \\
&+ (-1)^{p+1} \times [\sigma(t_1)(t_2)\cdots(t_p)(0) - \sigma(t_1)(t_2)\cdots(t_p)(1)].
\end{aligned}$$

To give a more compact expression of the boundary operator ∂, we introduce the following family of injections $j_k(a) : \mathbf{R}^p \to \mathbf{R}^{p+1}$, $k = 1, \ldots, p+1$ and $a \in \mathbf{R}$:

$$\begin{array}{lrcl}
k = 1 & j_1(a) : (t_1)\cdots(t_p) & \mapsto & (a)(t_1)\cdots(t_p), \\
1 < k \leq p & j_k(a) : (t_1)\cdots(t_p) & \mapsto & (t_1)\cdots(t_{k-1})(a)(t_k)\cdots(t_p), \\
k = p+1 & j_{p+1}(a) : (t_1)\cdots(t_p) & \mapsto & (t_1)\cdots(t_p)(a).
\end{array}$$

Given a p-tuple of numbers, $j_k(a)$ puts a at the place number k, preserving the numbers before and shifting the numbers after. Therefore, the boundary operator ∂ defined above writes again, for $p \geq 1$,

$$\text{for all } \sigma \in \mathrm{Cub}_p(X), \quad \partial\sigma = \sum_{k=1}^{p} (-1)^k [\sigma \circ j_k(0) - \sigma \circ j_k(1)]. \qquad (\diamondsuit)$$

The operator ∂ defined by $(\diamondsuit)$ is naturally extended by linearity on all cubic p-chains, with $p \geq 1$, and

$$\text{for all } c = \sum_\sigma n_\sigma \sigma \in C_p(X), \quad \partial c = \sum_\sigma n_\sigma \sum_{k=1}^{p} (-1)^k [\sigma \circ j_k(0) - \sigma \circ j_k(1)]. \quad (\heartsuit)$$

The operator ∂ defined by $(\heartsuit)$ is a boundary operator, that is, $\partial \circ \partial = 0$, and

$$\cdots \xrightarrow{\partial} C_p(X) \xrightarrow{\partial} C_{p-1}(X) \xrightarrow{\partial} \cdots \xrightarrow{\partial} C_0(X) \xrightarrow{\partial} \{0\}. \qquad (\clubsuit)$$

PROOF. What we have to prove is that the operator ∂ defined by $(\heartsuit)$ satisfies $\partial \circ \partial = 0$. According to the definition $(\diamondsuit)$, it is sufficient to check it on cubes. Let $p \geq 0$ and σ be a $(p+2)$-cube, thus $\partial\sigma$ is a cubic $(p+1)$-chain, and $\partial^2\sigma = \partial\partial\sigma$ is a cubic p-chain. We have

$$\partial\sigma = \sum_{\ell=1}^{p+2} (-1)^\ell [\sigma \circ j_\ell(0) - \sigma \circ j_\ell(1)].$$

Then,

$$\begin{aligned}
\partial^2\sigma &= \sum_{\ell=1}^{p+2} (-1)^\ell \{\partial[\sigma \circ j_\ell(0)] - \partial[\sigma \circ j_\ell(1)]\} \\
&= \sum_{\ell=1}^{p+2} (-1)^\ell \partial[\sigma \circ j_\ell(0)] - \sum_{\ell=1}^{p+2} (-1)^\ell \partial[\sigma \circ j_\ell(1)].
\end{aligned}$$

But $\sigma \circ j_\ell(a)$ is a $(p+1)$-cube, with here $a = 0$ or 1, hence

$$\begin{aligned}\partial[\sigma \circ j_\ell(a)] &= \sum_{k=1}^{p+1} (-1)^k [(\sigma \circ j_\ell(a)) \circ j_k(0) - (\sigma \circ j_\ell(a)) \circ j_k(1)] \\ &= \sum_{k=1}^{p+1} (-1)^k [\sigma \circ j_\ell(a) \circ j_k(0) - \sigma \circ j_\ell(a) \circ j_k(1)].\end{aligned}$$

Thus,

$$\begin{aligned}\partial^2\sigma &= \sum_{\ell=1}^{p+2} (-1)^\ell \sum_{k=1}^{p+1} (-1)^k \sigma \circ j_\ell(0) \circ j_k(0) - \sum_{\ell=1}^{p+2} (-1)^\ell \sum_{k=1}^{p+1} (-1)^k \sigma \circ j_\ell(0) \circ j_k(1) \\ &- \sum_{\ell=1}^{p+2} (-1)^\ell \sum_{k=1}^{p+1} (-1)^k \sigma \circ j_\ell(1) \circ j_k(0) + \sum_{\ell=1}^{p+2} (-1)^\ell \sum_{k=1}^{p+1} (-1)^k \sigma \circ j_\ell(1) \circ j_k(1) \\ &= \sum_{\ell=1}^{p+2} \sum_{k=1}^{p+1} (-1)^{k+l} \sigma \circ j_\ell(0) \circ j_k(0) \\ &- \sum_{\ell=1}^{p+2} \sum_{k=1}^{p+1} (-1)^{k+l} [\sigma \circ j_\ell(0) \circ j_k(1) + \sigma \circ j_\ell(1) \circ j_k(0)] \\ &+ \sum_{\ell=1}^{p+2} \sum_{k=1}^{p+1} (-1)^{k+l} \sigma \circ j_\ell(1) \circ j_k(1).\end{aligned}$$

Let us denote

$$\begin{aligned}(A) &= \sum_{\ell=1}^{p+2} \sum_{k=1}^{p+1} (-1)^{k+l} \sigma \circ j_\ell(0) \circ j_k(0), \\ (B) &= \sum_{\ell=1}^{p+2} \sum_{k=1}^{p+1} (-1)^{k+l} [\sigma \circ j_\ell(0) \circ j_k(1) + \sigma \circ j_\ell(1) \circ j_k(0)], \\ (C) &= \sum_{\ell=1}^{p+2} \sum_{k=1}^{p+1} (-1)^{k+l} \sigma \circ j_\ell(1) \circ j_k(1).\end{aligned}$$

Now, let us prove that (A), (B), and (C) are zero. Let us define first, for any pair $(a, b) \in \{0, 1\}^2$, the map

$$j_{\ell,k}(a)(b) : \mathbf{R}^p \to \mathbf{R}^{p+2},$$

which puts a at the place ℓ and b at the place k, distributing the other variables into the free places, respecting their initial distribution. This map is defined for any pair of indices $k \neq \ell$ which makes sense.

$$\begin{aligned}\text{For example} \quad & j_{1,2}(a)(b) : \mathbf{R}^0 \to \mathbf{R}^2 \quad j_{1,2}(a)(b)(0) = (a)(b), \\ \text{or} \quad & j_{1,3}(a)(b)(t) = (a)(t)(b), \\ \text{etc.} \quad & j_{4,2}(a)(b)(r)(t)(s) = (r)(b)(t)(a)(s).\end{aligned}$$

Note the variance of $j_{\ell,k}$,

$$j_{\ell,k}(a)(b) = j_{k,\ell}(b)(a).$$

Now, a simple verification shows that

$$(*) \quad \begin{cases} j_\ell \circ j_k = j_{\ell,k+1} & \text{if} \quad \ell \leq k, \\ j_\ell \circ j_k = j_{\ell,k} & \text{if} \quad \ell > k. \end{cases}$$

Let us denote, for the sake of convenience,

$$S_p(a)(b) = \sum_{\ell=1}^{p+2} \sum_{k=1}^{p+1} (-1)^{k+l} \sigma \circ j_\ell(a) \circ j_k(b),$$

such that $(A) = S_p(0)(0)$, $(B) = S_p(1)(0) + S_p(0)(1)$, $(C) = S_p(1)(1)$. Let us show that S_p is antisymmetric, $S_p(a)(b) = -S_p(b)(a)$, and we shall conclude that (A), (B), and (C) are zero. Using the identity $(*)$ above, we have

$$\begin{aligned} \sum_{\ell=1}^{p+2} \sum_{k=1}^{p+1} (-1)^{k+l} \sigma \circ j_\ell(a) \circ j_k(b) &= \sum_{\substack{1 \leq \ell \leq k \\ 1 \leq k < p+2}} (-1)^{k+\ell} \sigma \circ j_\ell(a) \circ j_k(b) \\ &+ \sum_{\substack{k < \ell \leq p+2 \\ 1 \leq k < p+2}} (-1)^{k+\ell} \sigma \circ j_\ell(a) \circ j_k(b) \\ &= \sum_{\substack{1 \leq \ell \leq k \\ 1 \leq k < p+2}} (-1)^{k+\ell} \sigma \circ j_{\ell,k+1}(a)(b) \\ &+ \sum_{\substack{k < \ell \leq p+2 \\ 1 \leq k < p+2}} (-1)^{k+\ell} \sigma \circ j_{\ell,k}(a)(b). \end{aligned}$$

Now, the first term of the right-hand side of the last equality becomes

$$\begin{aligned} \sum_{\substack{1 \leq \ell \leq k \\ 1 \leq k < p+2}} (-1)^{k+\ell} \sigma \circ j_{\ell,k+1}(a)(b) &= \sum_{\substack{1 \leq \ell < k \\ 2 \leq k \leq p+2}} (-1)^{k+\ell-1} \sigma \circ j_{\ell,k}(a)(b) \\ &= \sum_{1 \leq \ell < k \leq p+2} (-1)^{k+\ell-1} \sigma \circ j_{\ell,k}(a)(b) \\ &= -\sum_{1 \leq \ell < k \leq p+2} (-1)^{k+\ell} \sigma \circ j_{\ell,k}(a)(b). \end{aligned}$$

Then, using the variance of $j_{\ell,k}$, the second term writes

$$\begin{aligned} \sum_{\substack{k < \ell \leq p+2 \\ 1 \leq k < p+2}} (-1)^{k+\ell} \sigma \circ j_{\ell,k}(a)(b) &= \sum_{\substack{\ell < k \leq p+2 \\ 1 \leq \ell < p+2}} (-1)^{\ell+k} \sigma \circ j_{k,\ell}(a)(b) \\ &= \sum_{\substack{\ell < k \leq p+2 \\ 1 \leq \ell < p+2}} (-1)^{\ell+k} \sigma \circ j_{\ell,k}(b)(a) \\ &= \sum_{1 \leq \ell < k \leq p+2} (-1)^{k+\ell} \sigma \circ j_{\ell,k}(b)(a). \end{aligned}$$

Hence, joining the last two expressions, we get

$$\begin{aligned} \sum_{\ell=1}^{p+2} \sum_{k=1}^{p+1} (-1)^{k+l} \sigma \circ j_\ell(a) \circ j_k(b) &= \sum_{1 \leq \ell < k \leq p+2} (-1)^{k+\ell} \sigma \circ j_{\ell,k}(b)(a) \\ &- \sum_{1 \leq \ell < k \leq p+2} (-1)^{k+\ell} \sigma \circ j_{\ell,k}(a)(b). \end{aligned}$$

And, finally

$$S_p(a)(b) = \sum_{1 \leq \ell < k \leq p+2} (-1)^{k+\ell} [\sigma \circ j_{\ell,k}(b)(a) - \sigma \circ j_{\ell,k}(a)(b)].$$

Thus, $S_p(a)(b)$ is clearly antisymmetric, hence (A), (B), and (C) are zero. Therefore, $\partial(\partial\sigma) = (A) - (B) + (C) = 0$, and the operator ∂ is a boundary operator. □

6.60. Degenerate cubes and chains. Let p and q be two integers such that $0 \leq q < p$. We call a *reduction* from $\mathbf{R}^p$ to $\mathbf{R}^q$ any projection $\mathrm{pr} : \mathbf{R}^p \to \mathbf{R}^q$ such that $\mathrm{pr}(t_1, \ldots, t_p) = (t_{i_1}, \ldots, t_{i_q})$, where $\{i_1, \ldots, i_q\} \subset \{1, \ldots, p\}$ is a subset of indices, and $i_1 < \cdots < i_q$. For $q = 0$, there is only one reduction: the constant map $\hat{0} : (t_1, \ldots, t_p) \mapsto 0$. So, a reduction from $\mathbf{R}^p$ to $\mathbf{R}^q$ consists of just "forgeting" some, or all, of the components of $t = (t_1, \ldots, t_p) \in \mathbf{R}^p$. Now, let X be a diffeological space. Let $p > 0$ be an integer, we say that a p-cube $\sigma \in \mathrm{Cub}_p(X)$ is *degenerate* if there exist an integer q such that $0 \leq q < p$, a reduction pr from $\mathbf{R}^p$ to $\mathbf{R}^q$ and a q-cube $\sigma' \in \mathrm{Cub}_q(X)$ such that $\sigma = \sigma' \circ \mathrm{pr}$. In other words, a p-cube is degenerate if it does not depend on some coordinates of $\mathbf{R}^p$. Let us denote by $\mathrm{Cub}^\bullet_p(X)$ the set of degenerate p-cubes of X, and let us denote by $C^\bullet_p(X)$ the free Abelian group generated by $\mathrm{Cub}^\bullet_p(X)$. The elements of $C^\bullet_p(X)$ will be called the *degenerate cubic* p-*chains* of X. For $p = 0$, we agree that

$$\mathrm{Cub}^\bullet_0(X) = \varnothing \quad \text{and} \quad C^\bullet_0(X) = \{0\}.$$

We define the *reduced group of cubic* p-*chains of* X, denoted by $\mathbf{C}_p(X)$, as the quotient of the group of cubic p-chains of X by its subgroup of degenerate cubic p-chains, that is,

$$\mathbf{C}_p(X) = C_p(X)/C^\bullet_p(X).$$

Note that $\mathbf{C}_0(X) = C_0(X)/C^\bullet_0(X) = C_0(X)/\{0\} = C_0(X)$. Now, for any integer $p > 0$, the boundary (art. 6.59) of any degenerate p-cube is a degenerate cubic p-chain, that is,

$$\text{for all } \sigma \in \mathrm{Cub}^\bullet_p(X), \quad \partial\sigma \in C^\bullet_{p-1}(X).$$

Then by linearity we get immediately that

$$\text{for all } c \in C^\bullet_p(X), \quad \partial c \in C^\bullet_{p-1}(X) \quad \text{or} \quad \partial[C^\bullet_p(X)] \subset C^\bullet_{p-1}(X).$$

Thus, there exists an operator, denoted again by ∂, from $\mathbf{C}_p(X)$ to $\mathbf{C}_{p-1}(X)$, such that the following diagram commutes

$$\begin{array}{ccc}
C_p(X) & \xrightarrow{\ \partial\ } & C_{p-1}(X) \\
\pi_p \big\downarrow & & \big\downarrow \pi_{p-1} \\
\mathbf{C}_p(X) & \xrightarrow[\ \partial\ ]{} & \mathbf{C}_{p-1}(X)
\end{array}$$

where π_p is the natural projection from $C_p(X)$ onto its quotient $\mathbf{C}_p(X)$. Moreover, the operator $\partial : \mathbf{C}_p(X) \to \mathbf{C}_{p-1}(X)$ again satisfies the boundary property $\partial \circ \partial = 0$.

PROOF. Let us begin with the case $p = 1$. Let $\sigma \in \mathrm{Cub}^\bullet_1(X)$, for all $t \in \mathbf{R}$, $\sigma(t) = \sigma(0) = \sigma(1)$, since the reduction can be only $\hat{0} : t \mapsto 0$. Now, $\partial\sigma = \sigma(1) - \sigma(0) = 0$, that is, $\partial\sigma \in C^\bullet_0(X) = \{0\}$. Now, let us consider the case $p > 1$, and let σ be a degenerate p-cube, that is, $\sigma = \sigma' \circ \mathrm{pr}$ with $\mathrm{pr}(t_1, \ldots, t_p) = (t_{i_1}, \ldots, t_{i_q})$,

$\{i_1, \ldots, i_q\} \subset \{1, \ldots, p\}$, $0 \leq q < p$, and $i_1 < \cdots < i_q$. Applying the definition (art. 6.59) ($\diamondsuit$), we have

$$\begin{aligned}\partial\sigma &= \sum_{k=1}^{p} (-1)^k [\sigma \circ j_k(0) - \sigma \circ j_k(1)] \\ &= \sum_{k=1}^{p} (-1)^k [\sigma' \circ \mathrm{pr} \circ j_k(0) - \sigma' \circ \mathrm{pr} \circ j_k(1)].\end{aligned}$$

Let us consider the term $[\sigma' \circ \mathrm{pr} \circ j_k(0) - \sigma' \circ \mathrm{pr} \circ j_k(1)]$, for $1 \leq k \leq p$. There are two possibilities:

a) The index k belongs to the complement of $\{i_1, \ldots, i_q\}$ in $\{1, \ldots, p\}$. Thus, $\mathrm{pr} \circ j_k(0)(t_1, \ldots, t_{p-1}) = \mathrm{pr} \circ j_k(1)(t_1, \ldots, t_{p-1})$ and the term $[\sigma' \circ (\mathrm{pr} \circ j_k(0)) - \sigma' \circ (\mathrm{pr} \circ j_k(1))]$ disappears from the sum.

b) The index k belongs to $\{i_1, \ldots, i_q\}$. Let us assume that $q > 0$. Now, $\sigma' \circ \mathrm{pr} \circ j_k(a)(t_1, \ldots, t_{p-1}) = \sigma' \circ \mathrm{pr}(\ldots, a, \ldots)$, where the dots represent the $(p-1)$ numbers $t_1, \ldots, t_{p-1}$, distributed around a, and $a = 0$ or 1. Thus, $\sigma' \circ \mathrm{pr} \circ j_k(a)(t_1, \ldots, t_{p-1}) = \sigma'(\ldots, a, \ldots)$, where the dots represent $q-1$ of the numbers $\{t_{i_1}, \ldots, t_{i_q}\} \subset \{t_1, \ldots, t_p\}$, distributed around a. Thus, $\sigma' \circ \mathrm{pr} \circ j_k(a)$ depends only on $q-1$ variables, and, as a $(p-1)$-cube, is degenerate. If $q = 0$, then σ' is a 0-cube and $\partial\sigma = 0$.

Therefore, as a linear combination of degenerate $(p-1)$-cubes, $\partial\sigma$ is a degenerate cubic $(p-1)$-chain. Now, let us define ∂ on $\mathbf{C}_p(X)$ naturally by $\partial\bar{c} = \pi_{p-1}(\partial c)$, with $\bar{c} \in \mathbf{C}_p(X)$ and $\pi_p(c) = \bar{c}$. Since $\partial[C^\bullet_p(X)] \subset C^\bullet_{p-1}(X)$, $\partial\bar{c}$ does not depend on the choice of the representative c and ∂ is well defined. The fact that ∂ on $\mathbf{C}_p(X)$ is a boundary operator is then obvious. □

6.61. Cubic homology. Let X be a diffeological space. As usual in homology theory [**McL75**], when we have a chain complex, here $\mathbf{C}_\star(X)$ with $\star = 0, 1, \ldots \infty$, and the boundary operator

$$\partial : \mathbf{C}_\star(X) \to \mathbf{C}_{\star-1}(X) \quad \text{with} \quad \partial \circ \partial = 0,$$

the space of p-*cycles* is defined as the kernel in $\mathbf{C}_p(X)$ of the operator ∂, and the space of p-*boundary* as the image, in $\mathbf{C}_p(X)$, of the operator ∂ defined on $\mathbf{C}_{p+1}(X)$. These spaces will be denoted by

$$\begin{cases} \mathbf{Z}_p(X) = \ker[\partial : \mathbf{C}_p(X) \to \mathbf{C}_{p-1}(X)] & \text{with } p \geq 1, \\ \mathbf{B}_p(X) = \partial(\mathbf{C}_{p+1}(X)) \subset \mathbf{Z}_p(X) \subset \mathbf{C}_p(X) & \text{with } p \geq 0. \end{cases}$$

Then, the homology groups are defined as the quotients of the spaces of cycles by the spaces of boundaries

$$\mathbf{H}_p(X) = \mathbf{Z}_p(X)/\mathbf{B}_p(X).$$

Let us recall that for $p = 0$, $\partial : \mathbf{C}_0(X) \to \{0\}$, thus $\mathbf{Z}_0(X) = \mathbf{C}_0(X)$, and in this case $\mathbf{H}_0(X) = \mathbf{C}_0(X)/\partial\,\mathbf{C}_1(X) = C_0(X)/\partial\,\mathbf{C}_1(X)$. We call this homology $\mathbf{H}_*(X)$, the *cubic homology*[2] of the space X.

NOTE. It is not clear if cubic (or singular) homology will play, in diffeology, the crucial role it plays in the theory of manifolds. But since it is a traditional tool, and since it is still a smooth invariant, it was worth extending it to the general case.

[2]In topology the cubic homology and the singular homology coincide [**HoYo61**, Ex. 8.1]. For a natural singular homology in diffeology, these two homologies will coincide also.

6.62. Interpreting $\mathbf{H}_0$. Let X be a diffeological space. The group $\mathbf{H}_0(X)$ is the free Abelian group generated by the connected components $\pi_0(X)$ (art. 5.7) of X,

$$\mathbf{H}_0(X) = \mathrm{Maps}(\pi_0(X), \mathbf{Z}).$$

In particular if X is connected (art. 5.6), $\mathbf{H}_0(X) = \mathbf{Z}$.

PROOF. On $\mathbf{C}_0(X)$, the boundary operator is the zero homomorphism, $\partial : \mathbf{C}_0(X) \to \{0\}$. Then, $\mathbf{H}_0(X) = \mathbf{C}_0(X)/\partial(\mathbf{C}_1(X))$. Let us consider the homomorphism $\nu : \mathbf{C}_0(X) \to \mathbf{Z}$ defined by $\nu(\sum_x n_x x) = \sum_x n_x$. On the one hand, $\partial(\mathbf{C}_1(X)) \subset \ker(\nu)$. On the other hand, $\sum_x n_x = 0$ means that, by decomposing $\sum_x n_x$ as an indexed sum of $+1$ and -1, there are as many terms associated with $+1$ as terms associated with -1.

1) *If X is connected.* For any pair $(x, x') \in X \times X$ there exists a 1-cube connecting x to x', that is, a path γ such that $\gamma(0) = x$ and $\gamma(1) = x'$. So, for any pair of $(+1, x)$ and $(-1, x')$ appearing in the sum, we can find a 1-cube connecting x to x'. In other words, if $\sum_x n_x = 0$ there exists a 1-chain c such that $\partial c = \sum_x n_x$. Hence, $\ker(\nu) \subset \partial(\mathbf{C}_1(X))$ and $\partial(\mathbf{C}_1(X)) = \ker(\nu)$. Therefore, $\mathbf{H}_0(X) = \mathbf{C}_0(X)/\ker(\nu) = \nu(\mathbf{C}_0(X)) = \mathbf{Z}$.

2) *If X is not connected.* First of all, we decompose every chain c into a sum of chains $c = \sum_{a \in A} c_a$, where A is a finite set of components of X and c_a takes its values in a. Then, we define a set of homomorphisms ν_a, for each index a, as above, such that $c \in \partial(\mathbf{C}_1(X))$ if and only if each c_a belongs to the kernel of ν_a. Therefore, $\mathbf{H}_0(X)$ is the free Abelian group generated by the connected components. □

6.63. Cubic cohomology. Let X be a diffeological space. Let $C_p(X)$ be the space of cubic p-chains of X (art. 6.58). A cubic p-cochain of X, with coefficients in $\mathbf{R}$, is any homomorphism from $C_p(X)$ to $\mathbf{R}$. But $C_p(X)$ is the free Abelian group generated by $\mathrm{Cub}_p(X)$, thus every cubic p-cochain is defined by its values on the set of generators $\mathrm{Cub}_p(X)$. Since $\mathrm{Cub}_p(X) = \mathcal{C}^\infty(\mathbf{R}^p, X)$ is naturally a diffeological space, we shall define the *smooth cubic p-cochains* as the homomorphisms from $C_p(X)$ to $\mathbf{R}$ generated by $\mathcal{C}^\infty(\mathrm{Cub}_p(X), \mathbf{R})$. Now, since we are only interested in smooth cubic p-cochains, we shall omit the adjective "smooth". In other words, a *cubic p-cochain of X* is a linear map $f : C_p(X) \to \mathbf{R}$ such that

$$f : \sum_\sigma n_\sigma \sigma \mapsto \sum_\sigma n_\sigma f(\sigma) \quad \text{and} \quad f \restriction \mathrm{Cub}_p(X) \in \mathcal{C}^\infty(\mathrm{Cub}_p(X), \mathbf{R}),$$

with the notations of (art. 6.58). We shall denote the spaces of cubic p-cochains by $C^p(X)$, that is,

$$C^p(X) = \mathrm{Hom}^\infty(C_p(X), \mathbf{R}) \simeq \mathcal{C}^\infty(\mathrm{Cub}_p(X), \mathbf{R}).$$

Note that, as spaces of smooth functions in $\mathbf{R}$, the spaces of cubic p-cochains $C^p(X)$ are naturally real diffeological vector spaces (art. 3.1). Now, the boundary ∂ defined from $C_{p+1}(X)$ to $C_p(X)$ (art. 6.61) induces, by duality, a *coboundary operator* d such that

$$d : C^p(X) \to C^{p+1}(X) \quad \text{with} \quad df(c) = f(\partial c),$$

for all $f \in C^p(X)$, and all $c \in C_{p+1}(X)$. But the value of the cubic $(p+1)$-cochain df is characterized by its values on the $(p+1)$-cubes, by definition of the operator

$\partial\sigma$ (art. 6.61) applied on cubes,

$$\mathrm{d}f(\sigma) = \sum_{k=1}^{p} (-1)^k [f(\sigma \circ j_k(0)) - f(\sigma \circ j_k(1))]$$

for all $\sigma \in \mathrm{Cub}_{p+1}(X)$. The injections j_k have been defined above (art. 6.61). Then, by transfer of property, the coboundary d satisfies

$$\mathrm{d} \in \mathrm{Hom}(C^p(X), C^{p+1}(X)) \quad \text{and} \quad \mathrm{d} \circ \mathrm{d} = 0.$$

Now, a *reduced cubic* p*-cochain* is a cubic p-cochain f which vanishes on the degenerate p-cubes (art. 6.58), or equivalently, which is defined on the set of reduced cubic p-chains $\mathbf{C}_p(X)$. Let us denote the space of reduced cubic p-cochains by $\mathbf{C}^p(X)$,

$$\mathbf{C}^p(X) = \{f \in C^p(X) \mid f \upharpoonright \mathrm{Cub}^\bullet_p(X) = 0\}.$$

Since $\partial[C^\bullet_{p+1}(X)] \subset C^\bullet_p(X)$, if $f \in \mathbf{C}^p(X)$, then $\mathrm{d}f \in \mathbf{C}^{p+1}(X)$. Thus, the coboundary of a reduced cubic cochain is a reduced cubic cochain, and we can define, by duality with the cubic homology, a *cubic cohomology*, with coefficients in $\mathbf{R}$. The spaces of *cocycles* and *coboundaries* of this cohomology are defined, and denoted, by

$$\begin{cases} \mathbf{Z}^p(X) = \ker[\mathrm{d} : \mathbf{C}^p(X) \to \mathbf{C}^{p+1}(X)] & \text{for } p \geq 0, \\ \mathbf{B}^p(X) = \mathrm{d}(\mathbf{C}^{p-1}(X)) \subset \mathbf{Z}^p(X) \subset \mathbf{C}^p(X) & \text{for } p \geq 1. \end{cases}$$

The elements of $\mathbf{Z}^p(X)$ are called *cubic* p*-cocycles of* X and the elements of $\mathbf{B}^p(X)$ are called *cubic* p*-coboundaries of* X, we omit the word "reduced". The *cubic cohomology spaces* are defined as the quotient of the spaces of cocycles by the spaces of coboundaries,

$$\begin{cases} \mathbf{H}^p(X) = \mathbf{Z}^p(X)/\mathbf{B}^p(X) & \text{for } p > 0, \\ \mathbf{H}^0(X) = \mathbf{Z}^0(X) & \text{for } p = 0. \end{cases}$$

The Abelian group $\mathbf{H}^p(X)$ is called the p*-th space of cubic cohomology of* X*, with coefficients in* $\mathbf{R}$.

NOTE. We may be interested in a cubic homology with coefficients in a group other than $\mathbf{R}$, for example a cohomology with coefficients in some Abelian diffeological group A (an irrational torus for example). In this case the morphisms are generated by the maps from the space of p-cubes to A, and the cohomology groups are denoted by $\mathbf{H}^p(X, A)$, and the spaces $\mathbf{H}^p(X)$ also are denoted by $\mathbf{H}^p(X, \mathbf{R})$. We use A*-cubic cochains* or *real cubic cochain* if we want to specify.

6.64. Interpreting $\mathbf{H}^0(X, \mathbf{R})$. Let X be a diffeological space, and let us consider the group $\mathbf{H}^0(X, \mathbf{R})$. By definition (art. 6.63) it is the set of functions $f \in \mathcal{C}^\infty(\mathrm{Cub}_0(X), \mathbf{R})$ such that $\mathrm{d}f = 0$ (there are no degenerate 0-cubes). But $\mathrm{Cub}_0(X) = X$ (art. 6.58). Hence $f \in \mathcal{C}^\infty(X, \mathbf{R})$, and if f is a cocycle, that is, if $\mathrm{d}f = 0$, then $f(\partial\gamma) = f(\gamma(1)) - f(\gamma(0)) = 0$, for all $\gamma \in \mathrm{Cub}_1(X) = \mathrm{Paths}(X)$. But this just means that f is constant on the connected components of X (art. 5.6). Hence, $\mathbf{H}^0(X, \mathbf{R})$ is the group of real maps defined on the space $\pi_0(X)$ (art. 5.7),

$$\mathbf{H}^0(X, \mathbf{R}) = \mathrm{Maps}(\pi_0(X), \mathbf{R}).$$

NOTE. This is also the group $H^0_{\mathrm{dR}}(X)$ of De Rham cohomology (art. 6.75), and also the group of homomorphisms from $\mathbf{H}_0(X)$ to $\mathbf{R}$,

$$\mathbf{H}^0(X, \mathbf{R}) = \mathrm{Hom}(\mathbf{H}_0(X), \mathbf{R}) = H^0_{\mathrm{dR}}(X).$$

Exercises

✎ EXERCISE 112 (The boundary of a 3-cube). For a 3-cube, $\sigma \in \mathrm{Cub}_3(X)$, check explicitly that $\partial\partial\sigma = 0$.

✎ EXERCISE 113 (Cubic homology of a point). Let $H_*(X)$ be the homology, defined by the boundary operator ∂ on the space of p-chains, before the reduction by the degenerate chains (art. 6.61). Let us call it the *nonreduced cubic homology* of X. Let $\star = \{0\}$ be the singleton. Show that for the nonreduced homology $H_p(\star)$ is equal to $\mathbf{Z}$ for all p. And check that $\mathbf{H}_0(\star) = \mathbf{Z}$ and $\mathbf{H}_p(\star) = 0$ for all $p > 0$, for the reduced homology.

Integration of Differential Forms

This section describes the integration of forms on chains, and the essential properties related to this construction. I chose, for the sake of simplicity, to integrate differential forms on cubic chains, *which suggested the cubic homology* (art. 6.61), *because it is closely related to the computation of multiple integrals* (art. 6.27). *This is a simple way to introduce such further important formulas as, for example, the Cartan formula relating the Lie derivative to contraction and the exterior derivative of differential forms.*

6.65. Integrating forms on chains. Let us consider the real vector space $\mathbf{R}^p$, *oriented* by its canonical basis $\mathcal{B} = (\mathbf{e}_1, \dots, \mathbf{e}_p)$, that is, oriented by the canonical volume vol_p associated with $\mathcal{B}$ (art. 6.19) (art. 6.23),

$$\mathrm{vol}_p = \mathbf{e}^1 \wedge \cdots \wedge \mathbf{e}^p = dx^1 \wedge \cdots \wedge dx^p.$$

Every smooth p-form ω, on a real domain $U \subset \mathbf{R}^p$, is proportional to vol_p (art. 6.19). For every $\omega \in \mathcal{C}^\infty(U, \Lambda^p(\mathbf{R}^p))$, there exists a unique $f \in \mathcal{C}^\infty(\mathbf{R}^p, \mathbf{R})$, such that

$$\omega = f \times \mathrm{vol}_p, \quad \text{and} \quad f(x) = \omega(x)(\mathbf{e}_1, \dots, \mathbf{e}_p).$$

Let α be a p-form on a diffeological space X. And let $\sigma \in \mathrm{Cub}_p(X)$ be a smooth p-cube (art. 6.58). The *integral* of the p-form α on the p-cube σ is defined by

$$\int_\sigma \alpha = \int_{I^p} \alpha(\sigma), \qquad (\diamondsuit)$$

where $I^p = [0,1]^p$, and the integration of a smooth p-form on a p-cube has been defined in (art. 6.27). Since $\alpha(\sigma)$ is a smooth p-form of $\mathbf{R}^p$, there exists a smooth function f_σ such that $\alpha(\sigma) = f_\sigma \times \mathrm{vol}_p$. Hence, the integral of α over σ writes also

$$\int_\sigma \alpha = \int_{I^p} f_\sigma \times dx^1 \wedge \cdots \wedge dx^p \quad \text{with} \quad \alpha(\sigma) = f_\sigma \times \mathrm{vol},$$

or, in terms of multiple integrals,

$$\int_\sigma \alpha = \int_0^1 dx_1 \cdots \int_0^1 dx_p \; f_\sigma(x) \quad \text{with} \quad x = (x_1, \dots, x_p),$$

where

$$f_\sigma(x) = \alpha(\sigma)(x)(\mathbf{e}_1, \dots, \mathbf{e}_p).$$

Note that for the space $\mathbf{R}^p$, having been oriented once and for all, there is no ambiguity on the value, or the sign, of the function f_σ. The integral of p-forms on cubic p-chains is defined by linear extension of the integral of p-forms on p-cubes,

$$\int_c \alpha = \sum_\sigma n_\sigma \int_\sigma \alpha \quad \text{with} \quad c = \sum_\sigma n_\sigma \sigma.$$

The way the sum is written, it appears to be infinite but it is actually performed only on the support of c which is a finite set of cubes.

NOTE. For $p = 0$, a 0-form on X is any smooth function $f \in \mathcal{C}^\infty(X, \mathbf{R})$, and a 0-chain is any finite sum $c = \sum_x n_x x$. The integral of f on c is then defined as the sum

$$\int_c f = \sum_x f(x) \quad \text{with} \quad c = \sum_x n_x x.$$

6.66. Pairing chains and forms. The map defined above (art. 6.65) by integrating differential p-forms on cubic p-chains is known as the *pairing operation*,

$$(c, \alpha) \mapsto \int_c \alpha, \text{ for all } (c, \alpha) \in C_p(X) \times \Omega^p(X).$$

This pairing of forms and chains satisfies, in particular, the following properties:

1. The pairing is a *bilinear* operation,

$$\int_{nc+n'c'} (s\alpha + s'\alpha') = ns \int_c \alpha + ns' \int_c \alpha' + n's \int_{c'} \alpha + n's' \int_{c'} \alpha'.$$

2. On cubes, the pairing is smooth,

$$\left[(\sigma, \alpha) \mapsto \int_\sigma \alpha\right] \in \mathcal{C}^\infty(\mathrm{Cub}_p(X) \times \Omega^p(X), \mathbf{R}),$$

where $\mathrm{Cub}_p(X)$ is equipped with its functional diffeology of space of smooth maps from $\mathbf{R}^p$ to X (art. 6.58), and $\Omega^p(X)$ is equipped with its functional diffeology of space of forms (art. 6.28).

3. A p-form α is zero if and only if its integral, on every smooth p-cube, vanishes. Formally,

$$\alpha = 0 \quad \Leftrightarrow \quad \int_\sigma \alpha = 0, \quad \text{for all} \quad \sigma \in \mathrm{Cub}_p(X).$$

Equivalently, two p-forms coincide if and only if their integrals, on every smooth p-cube, coincide.

PROOF. 1. The bilinearity of the pairing is a direct consequence of the definitions of sums of chains and sums of forms.

2. Let $r \mapsto (\sigma_r, \alpha_r)$ be a plot of $\mathcal{C}^\infty(\mathrm{Cub}_p(X) \times \Omega^p(X))$. By definition,

$$\int_{\sigma_r} \alpha_r = \int_{I^p} \alpha_r(\sigma_r).$$

On the one hand $r \mapsto \alpha_r$ is a plot of $\Omega^p(X)$, which means that for every plot P of X the map $(r, r') \mapsto \alpha_r(P)(r')$ is smooth (art. 6.28). On the other hand, $r \mapsto \sigma_r$ is a plot of $\mathrm{Cub}_p(X)$, which means that $(r, t) \mapsto \sigma_r(t)$ is a plot of X. Combined, we get that for all $v_1, \ldots, v_p \in \mathbf{R}^p$ the map

$$(r, t) \mapsto \alpha_r\left(\begin{pmatrix} r' \\ s \end{pmatrix} \mapsto \sigma_{r'}(s)\right)_{\binom{r'=r}{s=t}} \begin{pmatrix} 0 \\ v_1 \end{pmatrix} \cdots \begin{pmatrix} 0 \\ v_p \end{pmatrix} = \alpha_r(\sigma_r)_t(v_1) \cdots (v_p)$$

is smooth. Thus, $(r,t) \mapsto \alpha_r(\sigma_r)_t$ is smooth, therefore there exists a smooth function $(r,t) \mapsto f_r(t)$ such that $\alpha_r(\sigma_r)_t = f_r(t) \times \mathrm{vol}_p$. Hence,

$$\int_{\sigma_r} \alpha_r = \int_{I^p} f_r \times \mathrm{vol}_p,$$

and, since the integration over the cube I^p is a smooth operation, $r \mapsto \int_{\sigma_r} \alpha_r$ is smooth.

3. Let us assume that $\int_\sigma \alpha = 0$ for all p-cubes σ, and that $\alpha \neq 0$. Then, there exists a p-plot $P : U \to X$ such that $\alpha(P) \neq 0$, that is, there exists $r \in U$ such that $\alpha(P)(r) \neq 0$. But α is a p-form on a p-domain, thus there exists $f \in \mathcal{C}^\infty(U, \mathbf{R})$ such that $\alpha(P) = f \times \mathrm{vol}$ (art. 6.19), and $\alpha(P)(r) \neq 0$ means that $f(r) \neq 0$. Let us assume that $f(r) > 0$ (it would be equivalent to assume that $f(r) < 0$). Since f is smooth, there exists a small cube C, centered at the point r, such that $f(r') > 0$ for all $r' \in C$. Since $f \restriction C$ is positive, the integral of f over the cube C is positive:

$$\int_C f \times \mathrm{vol}_p > 0.$$

But there exists a positive diffeomorphism φ from $\mathbf{R}^p$ onto an open neighborhood of r, mapping the standard cube I^p to C. Then, $\sigma = P \circ \varphi$ is a smooth p-cube of X and

$$\int_\sigma \alpha = \int_{I^p} \alpha(P \circ \varphi) = \int_{I^p} \varphi^*(\alpha(P)) = \int_{I^p} \varphi^*(f \times \mathrm{vol}_p).$$

But since φ is a positive diffeomorphism, by a change of coordinates we get

$$\int_{I^p} \varphi^*(f \times \mathrm{vol}_p) = \int_{I^p} f \circ \varphi \times \det(D(\varphi)) \times \mathrm{vol}_p = \int_C f \times \mathrm{vol}_p .$$

Hence, $\int_\sigma \alpha > 0$ and we have a smooth p-cube σ on which the integral of α does not vanish. This is in contradiction with the hypothesis. Therefore, for each plot P of X, $P^*(\alpha) = 0$, that is, $\alpha = 0$. ☐

6.67. Pulling back and forth forms and chains. Let X and X′ be two diffeological spaces and $f \in \mathcal{C}^\infty(X, X')$. For all p-cubes $\sigma \in \mathrm{Cub}_p(X)$, the *pushforward* of σ by f is denoted and defined by

$$f_* : \mathrm{Cub}_p(X) \to \mathrm{Cub}_p(X') \quad \text{with} \quad f_*(\sigma) = f \circ \sigma, \text{ for all } \sigma \in \mathrm{Cub}_p(X).$$

The pushforward of cubic p-chains by f is defined by linear extension from the pushforward of p-cubes, briefly written as

$$f_*\Big(\sum_\sigma n_\sigma \, \sigma\Big) = \sum_\sigma n_\sigma \, f_*(\sigma) = \sum_{\sigma'} n_{\sigma'} \, \sigma' \quad \text{with} \quad n_{\sigma'} = \sum_{\substack{\sigma \\ f_*(\sigma) = \sigma'}} n_\sigma.$$

More precisely, let $c = \sum_\sigma n_\sigma \, \sigma$, that is, $c = \sum_{\sigma \in \mathrm{Supp}(c)} n_\sigma \sigma$. For σ and τ in $\mathrm{Supp}(c)$, let $\sigma \sim \tau$ if $f_*(\sigma) = f_*(\tau)$. Let $\mathcal{I} = \mathrm{Supp}(c)/\sim$, then choose, for each $i \in \mathcal{I}$, one $\sigma_i \in i$, and let $\sigma'_i = f_*(\sigma_i)$. Therefore, $f_*(c) = \sum_{i \in \mathcal{I}} n'_i \, \sigma'_i$, with $n'_i = \sum_{\sigma \in i} n_\sigma$. The support of $f_*(c)$ is the subset of these σ'_i, such that $n'_i \neq 0$, which is of course finite. All that being specified, we have now, for all $\alpha' \in \Omega^p(X')$,

$$\int_{f_*(c)} \alpha' = \int_c f^*(\alpha').$$

NOTE. Since for every p-cube $\sigma = \mathbf{1}_{p*}(\sigma)$, where $\mathbf{1}_p : \mathbf{R}^p \to \mathbf{R}^p$ is the identity, we have an equivalent writing for the integral of a p-form on a p-cube,

$$\int_\sigma \alpha = \int_{\mathbf{1}_p} \sigma^*(\alpha).$$

This formulation is sometimes useful, to avoid ambiguity, in particular in the section about variation of integral of forms on chains.

PROOF. By definition (art. 6.65), for every smooth p-cube σ, we have

$$\int_{f_*(\sigma)} \alpha' = \int_{\mathbf{1}_p} [f_*(\sigma)]^*(\alpha') = \int_{\mathbf{1}_p} (f \circ \sigma)^*(\alpha') = \int_{\mathbf{1}_p} \sigma^*(f^*(\alpha')) = \int_\sigma f^*(\alpha').$$

Now, let $c = \sum_\sigma n_\sigma \sigma$, and let $c' = f_*(c)$. On the one hand, we have

$$\int_{f_*(c)} \alpha' = \sum_{\sigma'} n_{\sigma'} \int_{\sigma'} \alpha' = \sum_{\sigma'} \Bigg[\sum_{\substack{\sigma \\ f_*(\sigma)=\sigma'}} n_\sigma \Bigg] \int_{f_*(\sigma)} \alpha',$$

and on the other hand,

$$\int_c f^*(\alpha') = \sum_\sigma n_\sigma \int_\sigma f^*(\alpha') = \sum_\sigma n_\sigma \int_{f_*(\sigma)} \alpha' = \sum_{\sigma'} \Bigg[\sum_{\substack{\sigma \\ \sigma'=f_*(\sigma)}} n_\sigma \Bigg] \int_{f_*(\sigma)} \alpha'.$$

Therefore, $\int_{f_*(c)} \alpha' = \int_c f^*(\alpha')$. ☐

6.68. Changing the coordinates of a cube. Let X be a diffeological space. Let $\sigma \in \mathrm{Cub}_p(X)$ and $\alpha \in \Omega^p(X)$. Let φ be a *positive diffeomorphism* of I^p, that is, $\varphi \in \mathrm{Diff}(\mathbf{R}^p)$, $\varphi(I^p) = I^p$, and for all $x \in I^p$, $\det[D(\varphi)(x)] > 0$. Then,

$$\int_{\sigma_*(\varphi)} \alpha = \int_\sigma \alpha.$$

Note that in the notation $\sigma_*(\varphi)$, φ is regarded as a smooth cube and σ as a smooth map. Also note that φ does not need to be defined on the whole $\mathbf{R}^p$, but only on some open superset of the cube $I^p \subset \mathbf{R}^p$.

PROOF. Let f be such that $\alpha(\sigma) = f \times \mathrm{vol}_p$, then

$$\int_{\sigma_*(\varphi)} \alpha = \int_{I^p} \alpha(\sigma \circ \varphi) = \int_{I^p} \varphi^*(\alpha(\sigma)) = \int_{I^p} \varphi^*(f \times \mathrm{vol}_p).$$

But $\varphi^*(f \times \mathrm{vol}_p) = (f \circ \varphi) \times \varphi^*(\mathrm{vol}_p)$, thus

$$\int_{I^p} \varphi^*(f \times \mathrm{vol}_p) = \int_{I^p} (f \circ \varphi) \times \varphi^*(\mathrm{vol}_p) = \int_{I^p} (f \circ \varphi) \times \det(D(\varphi)) \times \mathrm{vol}_p\,.$$

Next, since $\det(D(\varphi)) > 0$,

$$\int_{I^p} (f \circ \varphi) \times \det(D(\varphi)) \times \mathrm{vol}_p = \int_{I^p} (f \circ \varphi) \times |\det(D(\varphi))| \times \mathrm{vol}_p\,.$$

By application of the change of variables $x \mapsto \varphi(x)$ in a multiple integral,

$$\int_{I^p} (f \circ \varphi) \times |\det(D(\varphi))| \times \mathrm{vol}_p = \int_{\varphi(I^p)} f \times \mathrm{vol}_p = \int_{I^p} f \times \mathrm{vol}_p = \int_\sigma \alpha.$$

Therefore, $\int_{\sigma_*(\varphi)} \alpha = \int_\sigma \alpha$. ☐

Variation of the Integrals of Forms on Chains

In this section we shall establish some theorems relative to the variation of integrals of forms on chains. First of all, we give the diffeological version of the Stokes theorem. Then, we give a formula for any variation of the integral of a p*-form on a* p*-chain. This formula mixes the form, its exterior derivative and the contractions with the variation of the chain. We deduce then the homotopic invariance of the De Rham cohomology* (art. 6.88) *and the diffeological version of the classical Cartan formula* (art. 6.72)*, relating the Lie derivative, contraction of forms and the exterior derivative.*

6.69. The Stokes theorem. Let X be a diffeological space, and let α be a $(p-1)$-form on X (art. 6.28), with $p \geq 1$, let c be a cubic p-chain in X (art. 6.58). With the integration of differential forms on chains defined in (art. 6.65), we have

$$\int_c d\alpha = \int_{\partial c} \alpha.$$

This is the diffeological version of classical Stokes' theorem.

NOTE. A p-form α is zero if and only if its integral vanishes on every p-cube (art. 6.66). As an immediate corollary of Stokes' theorem, α is closed if and only if its integral vanishes on every boundary ∂c, where c is a $(p+1)$-cube.

$$d\alpha = 0 \quad \text{if and only if} \quad \int_{\partial c} \alpha = 0.$$

PROOF. Let us prove first the theorem for a p-cube $\sigma \in \mathrm{Cub}_p(X)$. We have

$$\int_\sigma d\alpha = \int_{I^p} d\alpha(\sigma) = \int_{I^p} d[\alpha(\sigma)].$$

Let $a = \alpha(\sigma)$, a is a smooth $(p-1)$-form of $\mathbf{R}^p$, let

$$a = \sum_{k=1}^{p} a_k\, \mathbf{e}^1 \wedge \cdots [\mathbf{e}^k] \cdots \wedge \mathbf{e}^p, \qquad (\spadesuit)$$

where the brackets $[\mathbf{e}^k]$ mean that $\mathbf{e}^k$ is omitted. This is the general expression of a smooth $(p-1)$-form on $\mathbf{R}^p$. The a_k are smooth real functions defined on $\mathbf{R}^p$. Now, the exterior derivative da is given by

$$\begin{aligned} da(x) &= \left\{ \frac{\partial a_1}{\partial x_1} - \cdots + (-1)^{p-1} \frac{\partial a_p}{\partial x_p} \right\} \mathbf{e}^1 \wedge \cdots \wedge \mathbf{e}^p \\ &= -\left\{ \sum_{k=1}^{p} (-1)^k \frac{\partial a_k}{\partial x_k} \right\} \mathbf{e}^1 \wedge \cdots \wedge \mathbf{e}^p. \end{aligned}$$

Then, the integral of $d\alpha$ over σ splits into

$$\int_\sigma d\alpha = \int_{I^p} da = -\sum_{k=1}^{p} (-1)^k \int_0^1 dx_1 \cdots \int_0^1 \frac{\partial a_k}{\partial x_k} dx_x \cdots \int_0^1 dx_p.$$

Next, after integration by parts, we get

$$\begin{aligned}\int_\sigma d\alpha &= -\sum_{k=1}^{p}(-1)^k\int_0^1 dx_1\cdots\Big[a_k\Big]_{x_k=0}^{x_k=1}\cdots\int_0^1 dx_p\\ &= +\sum_{k=1}^{p}(-1)^k\int_0^1 dx_1\cdots\Big[a_k\Big]_{x_k=1}^{x_k=0}\cdots\int_0^1 dx_p\\ &= +\sum_{k=1}^{p}(-1)^k\int_{I^{p-1}}\Big[a_k\Big]_{x_k=1}^{x_k=0}\mathbf{e}^1\wedge\cdots[\mathbf{e}^k]\cdots\wedge\mathbf{e}^p. \qquad (\heartsuit)\end{aligned}$$

Since we have the choice for the name of the coordinates in $\mathbf{R}^{p-1}$, for each $k = 1,\ldots,p$, we denote by $(x_1\cdots[x_k]\cdots x_p)$ a current point in $\mathbf{R}^{p-1}$, where the brackets $[x_k]$ mean that there is no coordinate with index k, and the $\mathbf{e}_i$, in the last right-hand term above, are the vectors of the canonical basis of $\mathbf{R}^{p-1}$ according to the new indexation. For example,

$$k=1 \quad p=3 \quad x\in\mathbf{R}^2 \quad : \quad x=\begin{pmatrix}x_2\\x_3\end{pmatrix} \quad \text{and} \quad \mathbf{e}_1=\begin{pmatrix}1\\0\end{pmatrix} \quad \mathbf{e}_2=\begin{pmatrix}0\\1\end{pmatrix}.$$

Now, to compute the integral of α on the boundary $\partial\sigma$, with

$$\partial\sigma=\sum_{k=1}^{p}(-1)^k[\sigma\circ j_k(0)-\sigma\circ j_k(1)],$$

we need a workable expression of $j_k(t)^*(a)$. But $j_k(t)^*(a)$ is a $(p-1)$-form on $\mathbf{R}^{p-1}$, then

$$j_k(t)^*(a)_x = c(x)\,\mathbf{e}^1\wedge\cdots[\mathbf{e}^k]\cdots\wedge\mathbf{e}^p,$$

with

$$c(x)=j_k(t)^*(a)_x(\mathbf{e}_1)\cdots[\mathbf{e}_k]\cdots(\mathbf{e}_p).$$

Thus,

$$\begin{aligned}c(x) &= j_k(t)^*(a)_x(\mathbf{e}_1)\cdots[\mathbf{e}_k]\cdots(\mathbf{e}_p)\\ &= a_{j_k(t)(x)}(M\mathbf{e}_1)\cdots[M\mathbf{e}_k]\cdots(M\mathbf{e}_p),\end{aligned}$$

with $M = D(j_k(t))(x)$. But

$$D(j_k(t))(x)(\mathbf{e}_i)=\frac{\partial}{\partial s}\Big\{j_k(t)(x+s\mathbf{e}_i)\Big\}_{s=0}=j_k(0)(\mathbf{e}_i)=\mathbf{e}_i\in\mathbf{R}^p.$$

Now, the $\mathbf{e}_i$ are the vectors of the canonical basis of $\mathbf{R}^p$. Thus,

$$\begin{aligned}c(x) &= a_{j_k(t)(x)}(M\mathbf{e}_1)\cdots[M\mathbf{e}_k]\cdots(M\mathbf{e}_p)\\ &= a(x_k=t)(\mathbf{e}_1)\cdots[\mathbf{e}_k]\cdots(\mathbf{e}_p)\\ &= \sum_{j=1}^{p}a_j(x_k=t)\mathbf{e}^1\wedge\cdots[\mathbf{e}^j]\cdots\wedge\mathbf{e}^p(\mathbf{e}_1)\cdots[\mathbf{e}_k]\cdots(\mathbf{e}_p)\\ &= a_k(x_k=t).\end{aligned}$$

Hence, $j_k(t)^*(a)_x = \sum_{k=1}^p a_k(x_k = t)\, \mathbf{e}^1 \cdots \wedge [\mathbf{e}^k] \cdots \wedge \mathbf{e}^p$, we have then

$$\begin{aligned}
\int_{\partial\sigma} \alpha &= \sum_{k=1}^p (-1)^k \left[\int_{\sigma \circ j_k(0)} \alpha - \int_{\sigma \circ j_k(1)} \alpha \right] \\
&= \sum_{k=1}^p (-1)^k \int_{I^{p-1}} j_k(0)^*(\alpha(\sigma)) - \int_{I^{p-1}} j_k(1)^*(\alpha(\sigma)) \\
&= \sum_{k=1}^p (-1)^k \int_{I^{p-1}} j_k(0)^*(a) - \int_{I^{p-1}} j_k(1)^*(a) \\
&= \sum_{k=1}^p (-1)^k \int_{I^{p-1}} a_k(x_k = 0)\, \mathbf{e}^1 \wedge \cdots [\mathbf{e}^k] \cdots \wedge \mathbf{e}^p \\
&- \sum_{k=1}^p (-1)^k \int_{I^{p-1}} a_k(x_k = 1)\, \mathbf{e}^1 \wedge \cdots [\mathbf{e}^k] \cdots \wedge \mathbf{e}^p \\
&= \sum_{k=1}^p (-1)^k \int_{I^{p-1}} [a_k(x_k = 0) - a_k(x_k = 1)]\, \mathbf{e}^1 \wedge \cdots [\mathbf{e}^k] \cdots \wedge \mathbf{e}^p \\
&= \sum_{k=1}^p (-1)^k \int_{I^{p-1}} \Big[a_k \Big]_{x_k=1}^{x_k=0} \mathbf{e}^1 \wedge \cdots [\mathbf{e}^k] \cdots \wedge \mathbf{e}^p. \qquad (\diamondsuit)
\end{aligned}$$

Thus, $(\heartsuit) = (\diamondsuit)$, and the Stokes theorem is proved for the integral of forms on cubes. It extends by linearity on every chain:

$$\int_c d\alpha = \sum_\sigma n_\sigma \int_\sigma d\alpha = \sum_\sigma n_\sigma \int_{\partial\sigma} \alpha = \int_{\partial c} \alpha, \text{ for all } c = \sum_\sigma n_\sigma\, \sigma.$$

Note that the covariant nature of diffeology reduces the Stokes theorem to the simplest case of smooth $(p-1)$-forms on standard p-cubes. ☐

6.70. Variation of the integral of a form on a cube. Let X be a diffeological space and $r \mapsto (\sigma_r, \alpha_r) \in \mathcal{C}^\infty(U, \mathrm{Cub}_p(X) \times \Omega^p(X))$ be a plot of the product, defined on some real domain U of $\mathbf{R}^m$, $m \in \mathbf{N}$. Let us denote simply α for α_0 and σ for σ_0. Then, since the pairing of chains and forms is a smooth map (art. 6.66), we have

$$\left[r \mapsto \int_{\sigma_r} \alpha_r \right] \in \mathcal{C}^\infty(U, \mathbf{R}).$$

The *variation of the integral* of α on σ, at some point $r \in U$, applied to a vector $\delta r \in \mathbf{R}^m$, is the number denoted and defined by

$$\delta \int_\sigma \alpha = \frac{\partial}{\partial r} \left\{ \int_{\sigma_r} \alpha_r \right\}_r (\delta r).$$

The partial derivative $\partial/\partial r$ denotes the tangent linear map (art. 1.4). Let us give an equivalent formulation of the variation of the integral. Let us consider the following function, defined on a small real interval $]-\epsilon, \epsilon[$, with $\epsilon > 0$,

$$s \mapsto \int_{\sigma_s} \alpha_s \quad \text{where} \quad \sigma_s = \sigma_{r+s\delta r} \text{ and } \alpha_s = \alpha_{r+s\delta r}.$$

Then,

$$\delta \int_\sigma \alpha = \frac{\partial}{\partial s} \left\{ \int_{\sigma_s} \alpha_s \right\}_{s=0}.$$

Thus, the variation of the integral involves only 1-plot of cubes and forms. For this reason we shall continue only with 1-*plot variations* of $\int_\sigma \alpha$. Now, for any arc of a p-cube $s \mapsto \sigma_s$ of X centered at σ (art. 6.56), for any arc of a p-form $s \mapsto \alpha_s$ of X centered at α, we have the following identity:

$$\delta \int_\sigma \alpha = \int_{1_p} d\alpha \rfloor \delta\sigma + \int_{1_p} d[\alpha \rfloor \delta\sigma] + \int_{1_p} \sigma^*(\delta\alpha). \qquad (\diamondsuit)$$

a) $d\alpha$ is the exterior derivative of α defined in (art. 6.34).

b) $\delta\alpha$ is the p-form of X denoted, for every n-plot P of X, by

$$\delta\alpha = P \mapsto \frac{\partial}{\partial s}\left\{\alpha_s(P)\right\}_{s=0},$$

and defined by

$$\frac{\partial}{\partial s}\left\{\alpha_s(P)\right\}_{s=0}(r)(v_1)\cdots(v_p) = \frac{\partial}{\partial s}\left\{\alpha_s(P)(r)(v_1)\cdots(v_p)\right\}_{s=0},$$

for all $r \in U$ and $v_1, \ldots, v_p \in \mathbf{R}^n$.

c) $\alpha \rfloor \delta\sigma$ and $d\alpha \rfloor \delta\sigma$ are the contractions of the forms α and $d\alpha$ with the arc of plots $s \mapsto \sigma_s$ (art. 6.56).

NOTE 1. Thanks to the Stokes theorem (art. 6.69), the variation of the integral α on the cube σ writes also

$$\delta \int_\sigma \alpha = \int_{1_p} d\alpha \rfloor \delta\sigma + \int_{\partial 1_p} \alpha \rfloor \delta\sigma + \int_{1_p} \sigma^*(\delta\alpha).$$

NOTE 2. The variation formula $(\diamondsuit)$ still applies *mutatis mutandis* for the variation $\delta\sigma_s$ (art. 6.56, Note 3), for any $s \in\,]-\varepsilon, +\varepsilon[$.

PROOF. Let us consider the decomposition of the pairing

$$s \mapsto \begin{pmatrix} s \\ s \end{pmatrix} \mapsto \int_{\sigma_s} \alpha_s \quad \text{with} \quad \begin{pmatrix} s \\ t \end{pmatrix} \mapsto \int_{\sigma_s} \alpha_t,$$

such that,

$$\frac{\partial}{\partial s}\left\{\int_{\sigma_s} \alpha_s\right\}_{s=0} = \frac{\partial}{\partial s}\left\{\int_{\sigma_s} \alpha\right\}_{s=0} + \frac{\partial}{\partial t}\left\{\int_{\sigma} \alpha_t\right\}_{t=0}.$$

Let us use the variable $r \in \mathbf{R}^p = \mathrm{def}(\sigma)$, and let $r = \sum_{k=1}^p r^k \mathbf{e}_k$. The second term of the right-hand sum of this identity gives immediately,

$$\begin{aligned}
\frac{\partial}{\partial t}\left\{\int_\sigma \alpha_t\right\}_{t=0} &= \frac{\partial}{\partial t}\left\{\int_{I^p} \alpha_t(\sigma)\right\}_{t=0} \\
&= \frac{\partial}{\partial t}\left\{\int_{I^p} \alpha_t(\sigma)(r)(\mathbf{e}_1)\cdots(\mathbf{e}_p)\, dr^1 \wedge \cdots \wedge dr^p\right\}_{t=0} \\
&= \int_{I^p} \frac{\partial}{\partial t}\left\{\alpha_t(\sigma)(r)(\mathbf{e}_1)\cdots(\mathbf{e}_p)\right\}_{t=0} dr^1 \wedge \cdots \wedge dr^p \\
&= \int_{I^p} (\delta\alpha)(\sigma)(r)(\mathbf{e}_1)\cdots(\mathbf{e}_p)\Big\}_{t=0} dr^1 \wedge \cdots \wedge dr^p \\
&= \int_\sigma \delta\alpha.
\end{aligned}$$

Hence, the variation of the pairing splits into two parts,

$$\delta \int_\sigma \alpha = \delta \int_\sigma \alpha \Big|_{\delta\alpha=0} + \int_\sigma \delta\alpha. \qquad (\clubsuit)$$

Let us focus on the first integral of the right-hand side of ($\clubsuit$), which corresponds to $\delta\alpha = 0$, and let us define

$$\sigma(s,r) = \sigma_s(r) \quad \text{and} \quad j_s : r \mapsto (s,r).$$

Thus, $\boldsymbol{\sigma}$ is a $(p+1)$-plot defined on $]-\epsilon,\epsilon[\times I^p$ and j_s is the injection from I^p to $]-\epsilon,\epsilon[\times I^p$ at the height s. Now, using $\sigma_s = \boldsymbol{\sigma} \circ j_s$, we have

$$\begin{aligned}
\int_{\sigma_s} \alpha &= \int_{I^p} \alpha(\sigma_s)(r)(\mathbf{e}_1)\cdots(\mathbf{e}_p)\, dr^1 \wedge \cdots \wedge dr^p \\
&= \int_{I^p} \alpha(\boldsymbol{\sigma} \circ j_s)(r)(\mathbf{e}_1)\cdots(\mathbf{e}_p)\, dr^1 \wedge \cdots \wedge dr^p \\
&= \int_{I^p} j_s^*(\alpha(\boldsymbol{\sigma}))(r)(\mathbf{e}_1)\cdots(\mathbf{e}_p)\, dr^1 \wedge \cdots \wedge dr^p \\
&= \int_{I^p} \alpha(\boldsymbol{\sigma}) \begin{pmatrix} s \\ r \end{pmatrix} \begin{pmatrix} 0 \\ \mathbf{e}_1 \end{pmatrix} \cdots \begin{pmatrix} 0 \\ \mathbf{e}_p \end{pmatrix} dr^1 \wedge \cdots \wedge dr^p.
\end{aligned}$$

Note that $\alpha(\boldsymbol{\sigma})$ is a p-form on the $(p+1)$-domain $]-\epsilon,\epsilon[\times \mathbf{R}^p$. Let us introduce then the $(p+1)$ coordinates $\mathbf{a}_i = [(s,r) \mapsto a_i]$, with $i = 0, 1, \ldots, p$, in this way

$$\begin{aligned}
\alpha(\boldsymbol{\sigma})_{\binom{s}{r}} &= a_0\, dr^1 \wedge dr^2 \wedge dr^3 \wedge \cdots \wedge dr^p \\
&+ a_1\, ds \wedge dr^2 \wedge dr^3 \wedge \cdots \wedge dr^p \\
&+ a_2\, ds \wedge dr^1 \wedge dr^3 \wedge \cdots \wedge dr^p \\
&+ \cdots \\
&+ a_p\, ds \wedge dr^1 \wedge dr^2 \wedge \cdots \wedge dr^{p-1},
\end{aligned}$$

or, equivalently

$$\alpha(\boldsymbol{\sigma})_{\binom{s}{r}} = \sum_{k=0}^{p} a_k\, dr^0 \wedge \cdots [dr^k] \cdots \wedge dr^p = \sum_{k=0}^{p} a_k\, \mathbf{e}^0 \wedge \cdots [\mathbf{e}^k] \cdots \wedge \mathbf{e}^p,$$

where $r^0 = s$, and the brackets $[dr^k]$ mean that dr^k is omitted. Thus,

$$\int_{\sigma_s} \alpha = \int_{I^p} a_0\, dr^1 \wedge \cdots \wedge dr^p \quad \text{with} \quad a_0 = \alpha(\boldsymbol{\sigma}) \begin{pmatrix} s \\ r \end{pmatrix} \begin{pmatrix} 0 \\ \mathbf{e}_1 \end{pmatrix} \cdots \begin{pmatrix} 0 \\ \mathbf{e}_p \end{pmatrix}.$$

Now, since everything is smooth, integration and derivation commute, and the derivative with respect to the variable s becomes

$$\frac{\partial}{\partial s} \left\{ \int_{\sigma_s} \alpha \right\}_{s=0} = \int_{I^p} \frac{\partial a_0}{\partial s} \Big|_{s=0} dr^1 \wedge \cdots \wedge dr^p.$$

Next, let us introduce the exterior derivative of $\alpha(\boldsymbol{\sigma})$,

$$d[\alpha(\boldsymbol{\sigma})]_{\binom{s}{r}} = d\alpha(\boldsymbol{\sigma})_{\binom{s}{r}} = \left\{ \frac{\partial a_0}{\partial s} - \frac{\partial a_1}{\partial r^1} + \cdots + (-1)^p \frac{\partial a_p}{\partial r^p} \right\} ds \wedge dr^1 \wedge \cdots \wedge dr^p.$$

After contracting the two terms of this identity by the vector of coordinates $(1,0) \in \mathbf{R} \times \mathbf{R}^p$, we get

$$\frac{\partial a_0}{\partial s} dr^1 \wedge \cdots \wedge dr^p = d\alpha(\sigma)_{\binom{s}{r}} \begin{pmatrix} 1 \\ 0 \end{pmatrix} - \left\{ \sum_{k=1}^{p} (-1)^k \frac{\partial a_k}{\partial r^k} \right\} dr^1 \wedge \cdots \wedge dr^p.$$

The first term of the right-hand side is the inner product (art. 6.14) of the value of the $(p+1)$-form $d\alpha(\sigma)$ at the point (s,r), by the vector of coordinates $(1,0) \in \mathbf{R} \times \mathbf{R}^p$. Evaluated at the point $(0,r)$, and restricted to $\mathbf{R}^p$, it is exactly the contraction $d\alpha$ by the arc of p-cubes $s \mapsto \sigma_s$ (art. 6.56, ($\diamondsuit$)). Hence,

$$\left.\frac{\partial a_0}{\partial s}\right|_{s=0} dr^1 \wedge \cdots \wedge dr^p = d\alpha(\delta\sigma)_r - \left\{ \sum_{k=1}^{p} (-1)^k \frac{\partial a_k(0,r)}{\partial r^k} \right\} dr^1 \wedge \cdots \wedge dr^p.$$

The second term of the right-hand side of this identity is just the exterior derivative of the contraction $\alpha(\delta\sigma)$. Indeed, let $v_2, \ldots, v_p$ be $(p-1)$ vectors of $\mathbf{R}^p$. Let us abridge the notation,

$$[v] = (v_2) \cdots (v_p), \quad \text{and} \quad \begin{bmatrix} 0 \\ v \end{bmatrix} = \begin{pmatrix} 0 \\ v_2 \end{pmatrix} \cdots \begin{pmatrix} 0 \\ v_p \end{pmatrix}.$$

Using the above expression of $\alpha(\sigma)$, we get

$$\begin{aligned} \alpha(\delta\sigma)_r[v] &= \alpha(\sigma)_{\binom{0}{r}} \begin{pmatrix} 1 \\ 0 \end{pmatrix} \begin{bmatrix} 0 \\ v \end{bmatrix} \\ &= \left[\sum_{k=0}^{p} a_k(0,r)\, dr^0 \wedge \cdots [dr^k] \cdots \wedge dr^p \right] \begin{pmatrix} 1 \\ 0 \end{pmatrix} \begin{bmatrix} 0 \\ v \end{bmatrix} \\ &= \left[\sum_{k=1}^{p} a_k(0,r)\, dr^1 \wedge \cdots [dr^k] \cdots \wedge dr^p \right] [v]. \end{aligned}$$

Thus,

$$\alpha(\delta\sigma)_r = \sum_{k=1}^{p} a_k(0,r)\, dr^1 \wedge \cdots [dr^k] \cdots \wedge dr^p.$$

Then,

$$\begin{aligned} d[\alpha(\delta\sigma)]_r &= \sum_{k=1}^{p} \frac{\partial a_k(0,r)}{\partial r^k} dr^k \wedge dr^1 \wedge \cdots [dr^k] \cdots \wedge dr^p \\ &= \sum_{k=1}^{p} (-1)^{p-1} \frac{\partial a_k(0,r)}{\partial r^k} dr^1 \wedge \cdots \wedge dr^k \wedge \cdots \wedge dr^p \\ &= -\left\{ \sum_{k=1}^{p} (-1)^k \frac{\partial a_k(0,r)}{\partial r^k} \right\} dr^1 \wedge \cdots \wedge dr^p. \end{aligned}$$

Hence,

$$\left.\frac{\partial a_0}{\partial s}\right|_{s=0} dr^1 \wedge \cdots \wedge dr^p = d\alpha(\delta\sigma)_r + d[\alpha(\delta\sigma)]_r,$$

and

$$\begin{aligned}\frac{\partial}{\partial s}\left\{\int_{\sigma_s}\alpha\right\}_{s=0} &= \left.\frac{\partial a_0}{\partial s}\right|_{s=0} dr^1\wedge\cdots\wedge dr^p \\ &= \int_{I^p} d\alpha(\delta\sigma) + d[\alpha(\delta\sigma)] \\ &= \int_{1_p} d\alpha\rfloor\delta\sigma + \int_{1_p} d[\alpha\rfloor\delta\sigma]. \qquad (\spadesuit)\end{aligned}$$

Finally, combining ($\clubsuit$) and ($\spadesuit$), we obtain the formula ($\diamondsuit$) of the variation of the integral of a p-form on a p-cube. ☐

6.71. Variation of the integral of forms on chains. We defined the integral of a p-form on a cubic p-chain by $\int_c \alpha = \sum_\sigma n_\sigma \int_\sigma \alpha$, where $c = \sum_\sigma n_\sigma \sigma$ (art. 6.65). To extend the variation of the integral of a p-form, from a p-cube (art. 6.70) to a cubic p-chain, we need a suitable diffeology on the set of cubic p-chains. Let $r \mapsto c_r$ be a parametrization from U in $C_p(X)$, the following property defines a diffeology, the proof is left as an exercise.

($\heartsuit$) For all $r_0 \in U$, there exist an open neighborhood V of r_0, a finite family of indices $\mathcal{J}$, together with a family of integers $\{n_i\}_{i\in\mathcal{J}}$ and a family $\{\sigma_i\}_{i\in\mathcal{J}}$ of elements of $\mathcal{C}^\infty(V, \mathrm{Cub}_p(X))$ such that $c_r = \sum_{i\in\mathcal{J}} n_i\,\sigma_{i,r}$ for all $r \in V$.

Now, let $s \mapsto c_s$ be an arc of cubic p-chain centered at c and $s \mapsto \alpha_s$ be an arc of p-form centered at α, the variation of the integral of α on c is given by

$$\delta\int_c \alpha = \int_{1_p} d\alpha\rfloor\delta c + \int_{\partial 1_p} \alpha\rfloor\delta c + \int_{1_p} c^*(\delta\alpha).$$

We need to explain the terms involved in the formula. Every cubic p-chain c decomposes into a finite sum $c = \sum_{i\in\mathcal{J}} n_i\,\sigma_i$. The pullback of a p-form α by c is defined by linearity, $c^*(\alpha) = \sum_{i\in\mathcal{J}} n_i\,\sigma_i^*(\alpha)$, as well $\alpha(c) = \sum_{i\in\mathcal{J}} n_i\,\alpha(\sigma_i)$. The contraction $d\alpha\rfloor\delta c$ and $\alpha\rfloor\delta c$ also are given by linearity, $d\alpha\rfloor\delta c = \sum_{i\in\mathcal{J}} n_i\, d\alpha\rfloor\delta\sigma_i$ and $\alpha\rfloor\delta c = \sum_{i\in\mathcal{J}} n_i\,\alpha\rfloor\delta\sigma_i$. They are p-forms on $\mathbf{R}^p$, or a $(p-1)$-form on $\mathbf{R}^{p-1}$, and do not depend on the decomposition of c.

PROOF. Consider a small interval around $0 \in \mathbf{R}$ on which $c_s = \sum_{i\in\mathcal{J}} n_i\,\sigma_{i,s}$, then

$$\delta\int_c \alpha = \frac{\partial}{\partial s}\left\{\sum_{i\in\mathcal{J}} n_i \int_{\sigma_{i,s}} \alpha_s\right\}_{s=0} = \sum_{i\in\mathcal{J}} n_i \frac{\partial}{\partial s}\left\{\int_{\sigma_{i,s}} \alpha_s\right\}_{s=0} = \sum_{i\in\mathcal{J}} n_i\,\delta\int_{\sigma_i}\alpha.$$

The formula of the variation of the integral of a p-form on a p-cube (art. 6.70) gives

$$\begin{aligned}\delta\int_c \alpha &= \sum_{i\in\mathcal{J}} n_i \int_{I^p} d\alpha(\delta\sigma_i) + \sum_{i\in\mathcal{J}} n_i \int_{I^p} d[\alpha(\delta\sigma_i)] + \sum_{i\in\mathcal{J}} n_i \int_{I^p} \delta\alpha(\sigma_i) \\ &= \int_{I^p} d\alpha\Big(\sum_{i\in\mathcal{J}} n_i\,\delta\sigma_i\Big) + \int_{I^p} d\Big[\alpha\Big(\sum_{i\in\mathcal{J}} n_i\,\delta\sigma_i\Big)\Big] + \int_{I^p} \delta\alpha\Big(\sum_{i\in\mathcal{J}} n_i\,\sigma_i\Big) \\ &= \int_{I^p} d\alpha(\delta c) + \int_{I^p} d[\alpha(\delta c)] + \int_{I^p} \delta\alpha(c).\end{aligned}$$

This makes sense according to the definition (art. 6.56). Let $\mathbf{c} : (s,t) \mapsto c_s(t)$ and $\boldsymbol{\sigma}_i : (s,t) \mapsto \sigma_{i,s}(t)$ such that $\mathbf{c} = \sum_{i\in\mathcal{J}} \boldsymbol{\sigma}_i$. We have for all $t \in \mathbf{R}^p$ and all

$v_2, \ldots, v_p \in \mathbf{R}^p$,

$$\begin{aligned}\alpha(\delta c)(t)(v_2)\cdots(v_p) &= \alpha(\mathbf{c})_{\binom{0}{t}}\begin{pmatrix}1\\0\end{pmatrix}\begin{pmatrix}0\\v_2\end{pmatrix}\cdots\begin{pmatrix}0\\v_p\end{pmatrix} \\ &= \sum_{i\in\mathcal{J}} n_i\,\alpha(\sigma_i)_{\binom{0}{t}}\begin{pmatrix}1\\0\end{pmatrix}\begin{pmatrix}0\\v_2\end{pmatrix}\cdots\begin{pmatrix}0\\v_p\end{pmatrix}.\end{aligned}$$

The same holds for $d\alpha(\delta c)$. We have still to be sure that the computation does not depend on the choice of the decomposition of the arc $s \mapsto c_s$. Let us check generally that the evaluation of a p-form α on a cubic p-chain c does not depend on a decomposition $c = \sum_{i\in\mathcal{J}} n_i\,\sigma_i$. Equivalently, if $\sum_{i\in\mathcal{J}} n_i\,\sigma_i = 0$, then $\sum_{i\in\mathcal{J}} n_i\,\alpha(\sigma_i) = 0$. Let $i \sim j$ if $\sigma_i = \sigma_j$, let $\mathcal{A} = \mathcal{J}/{\sim}$, and let us denote $\sigma_a = \sigma_i$ for every $a \in \mathcal{A}$, where $i \in a$, then $\sum_{i\in\mathcal{J}} n_i\,\sigma_i = \sum_{a\in\mathcal{A}}(\sum_{i\in a} n_i)\,\sigma_a$. The same factorization gives $\sum_{i\in\mathcal{J}} n_i\,\alpha(\sigma_i) = \sum_{a\in\mathcal{A}}(\sum_{i\in a} n_i)\,\alpha(\sigma_a)$. But since $\sum_{a\in\mathcal{A}}(\sum_{i\in a} n_i)\sigma_a = 0$ and since the σ_a are all different, $\sum_{i\in a} n_i = 0$ for all $a \in \mathcal{A}$. Therefore, $\sum_{a\in\mathcal{A}}(\sum_{i\in a} n_i)\,\alpha(\sigma_a) = 0$, that is, $\sum_{i\in\mathcal{J}} n_i\,\alpha(\sigma_i) = 0$. Now, if for all s, $c_s = \sum_{i\in\mathcal{J}} n_i\,\sigma_{i,s} = \sum_{i'\in\mathcal{J}'} n_{i'}\,\sigma_{i',s}$, then, from what precedes, $\alpha_s(\sum_{i\in\mathcal{J}} n_i\,\sigma_{i,s}) = \alpha_s(\sum_{i'\in\mathcal{J}'} n_{i'}\,\sigma_{i',s})$. Thus, the evaluation of the variation does not depend on the choice of the decomposition. □

6.72. The Cartan-Lie formula. Let X be a diffeological space, and let Diff(X) be its group of diffeomorphisms equipped with the functional diffeology (art. 1.61). Let $F : \mathbf{R} \to \mathrm{Diff}(X)$ be a sliding, that is, a 1-plot centered at the identity (art. 6.57). Let α be any differential k-form on X, with $k \geq 1$. Then,

$$\pounds_F(\alpha) = i_F(d\alpha) + d(i_F(\alpha)),$$

where $\pounds_F(\alpha)$ is the Lie derivative of α by F (art. 6.54), and i_F denotes the contraction by F (art. 6.57). For a 0-form f, that is, a smooth function from X to $\mathbf{R}$, the Cartan formula is reduced to $\pounds_F(f) = i_F(df)$. The identity above extends, to the diffeological spaces, the *Cartan-Lie formula* of classical differential geometry.

PROOF. Let $\sigma \in \mathrm{Cub}_p(X)$ (art. 6.58) and

$$\begin{cases} \alpha_t = F(t)_*(\alpha) = (F(t)^{-1})^*(\alpha), \\ \sigma_t = (F(t))_*(\sigma) = F(t)\circ\sigma. \end{cases}$$

Thanks to (art. 6.67), for all $t \in \mathbf{R}$,

$$\int_{\sigma_t} \alpha_t = \int_{F(t)_*(\sigma)} F(t)_*(\alpha) = \int_\sigma F(t)^* \circ F(t)_*(\alpha) = \int_\sigma \alpha.$$

Now, by differentiation with respect to the parameter t, we get on the one hand

$$\delta\int_{\sigma_t} \alpha_t = \delta\int_\sigma \alpha = 0, \quad \text{with} \quad \delta = \left.\frac{\partial}{\partial t}\right|_{t=0},$$

and on the other hand, by the formula of the variation of integral of differential forms (art. 6.70),

$$\delta\int_{\sigma_t} \alpha_t = \int_{1_p} d\alpha\rfloor\delta\sigma + \int_{1_p} d[\alpha\rfloor\delta\sigma] + \int_{1_p} \delta\sigma^*(\alpha). \qquad (\diamondsuit)$$

But

a) $\delta\alpha = \left.\frac{\partial \alpha_t}{\partial t}\right|_{t=0} = \left.\frac{\partial}{\partial t}F(t)_*(\alpha)\right|_{t=0} = \left.\frac{\partial}{\partial t}(F(t)^{-1})^*(\alpha)\right|_{t=0} = -\pounds_F(\alpha).$

b) $\sigma_t = F(t) \circ \sigma$ and $\delta = \left.\dfrac{\partial}{\partial t}\right|_{t=0}$ which implies $\alpha \rfloor \delta\sigma = \sigma^*(i_F(\alpha))$.

The point a) is Exercise 108, p. 175. Then, the identity ($\diamondsuit$) above becomes

$$0 = \int_\sigma i_F[d\,\alpha] + \int_\sigma d\,[i_F(\alpha)] - \int_\sigma \pounds_F(\alpha) = \int_\sigma i_F[d\,\alpha] + d\,[i_F(\alpha)] - \pounds_F(\alpha).$$

Since this is satisfied for any p-cube σ, we get $i_F[d\,\alpha)] + d\,[i_F(\alpha)] - \pounds_F(\alpha) = 0$ (art. 6.66). Thus, $\pounds_F(\alpha) = i_F[d\,\alpha)] + d\,[i_F(\alpha)]$. □

Exercises

✎ EXERCISE 114 (Liouville rays and closed forms). Consider the notations and hypothesis of Exercise 111, p. 176. Show that, for the p-form ω, with $p \geq 1$, if $d\omega = 0$, then ω is exact and $\varpi = i_h(\omega)$ is a primitive, $d\varpi = \omega$.

✎ EXERCISE 115 (Integrals on homotopic cubes). Let X be a diffeological space. Let α be a closed p-form, $\alpha \in \Omega^p(X)$ and $d\alpha = 0$. Let $s \mapsto \sigma_s$ be an arc of p-cubes of X centered at σ such that $\sigma_s \restriction \partial I^p = \sigma \restriction \partial I^p$ for all s, where $\partial I^p = \bigcup_{k=1}^{p} \bigcup_{a=0,1} j_p(a)(I^{p-1})$ and $I = [0,1]$. Use the formula of the variation of the integral of a p-form on a p-cube (art. 6.70) to show that $\delta \int_\sigma \alpha = 0$.

✎ EXERCISE 116 (Closed 1-forms on connected spaces). Let X be a connected diffeological space, and α be a closed 1-form of X. Show that the integral of α on any $\ell \in \mathrm{Loops}(X, x)$ depends only on the fixed-ends homotopy classes of ℓ (art. 5.5). Let P_α be the set of all the numbers $\int_\ell \alpha \in \mathbf{R}$ where $\ell \in \mathrm{Loops}(X, x)$, show that P_α does not depend on the basepoint x. Conclude that P_α is a homomorphic image of $\pi_1(X)$ (art. 5.15).

✎ EXERCISE 117 (Closed 1-forms on simply connected spaces). Let X be a diffeological space and α be a closed 1-form of X. Show that if there exists a loop ℓ based at a point x (art. 5.1) such that $\int_\ell \alpha \neq 0$, then the connected component of x (art. 5.6) is not simply connected (art. 5.15).

De Rham Cohomology

In this section we introduce the De Rham cohomology of diffeological spaces, according to the definition of differential forms (art. 6.28) *and the exterior derivative* (art. 6.34). *We give some examples to show how homotopy and De Rham cohomology can be related. A precise statement will be given later* (art. 6.88).

6.73. The De Rham cohomology. Let X be a diffeological space. The exterior derivative defined above (art. 6.34) satisfies the *coboundary condition*

$$d : \Omega^p(X) \to \Omega^{p+1}(X), \ p \geq 0 \quad \text{and} \quad d \circ d = 0.$$

As is usual in cohomology theories [**McL75**], when we have a chain complex — here the chain complex of real vector spaces $\Omega^\star(X) = \{\Omega^p(X)\}_{p=0}^\infty$ with a coboundary operator d — the space of p-*cocycles* is defined as the kernel in $\Omega^p(X)$ of the operator d, and the space of p-*coboundary* is defined as the image, in $\Omega^p(X)$, of the operator d. They will be denoted by

$$\begin{cases} Z^p_{\mathrm{dR}}(X) &= \ker\left[d : \Omega^p(X) \to \Omega^{p+1}(X)\right], \\ B^p_{\mathrm{dR}}(X) &= d(\Omega^{p-1}(X)) \subset Z^p_{\mathrm{dR}}(X) \quad \text{with} \quad B^0_{\mathrm{dR}}(X) = \{0\}. \end{cases}$$

The *De Rham cohomology groups* of X are then defined as the quotients of the spaces of cocycles by the spaces of coboundaries, we denote them by

$$H^p_{\rm dR}(X) = Z^p_{\rm dR}(X)/B^p_{\rm dR}(X).$$

Since the operator d is linear, and since the space of differential p-forms $\Omega^p(X)$, equipped with the functional diffeology (art. 6.29), is a diffeological vector space (art. 3.1), the De Rham cohomology group $H^p_{\rm dR}(X)$, equipped with the quotient diffeology (art. 3.6), is a diffeological vector space.

6.74. The De Rham homomorphism. Let X be a diffeological space, let p be any positive integer, and let $\alpha \in \Omega^p(X)$. The integration of α on the cubic p-chains (art. 6.65) defines a cubic p-cochain f_α for the reduced cubic cohomology (art. 6.63), for all $c \in C_p(X)$,

$$f_\alpha(c) = \int_c \alpha, \quad \text{and} \quad f_\alpha \in \mathbf{C}^p(X, \mathbf{R}).$$

Then, thanks to Stokes' theorem (art. 6.69), if α is closed, $d\alpha = 0$, then f_α is closed as cochain, $df_\alpha = 0$. Thus, the integration of α on chains defines a morphism from $Z^p_{\rm dR}(X)$ to $\mathbf{Z}^p(X)$. If α is exact, $\alpha = d\beta$, then f_α is exact as cochain, $f_\alpha = df_\beta$. Hence, the integration on a chain defines a linear map from $H^p_{\rm dR}(X)$ to $\mathbf{H}^p(X)$. This morphism is called the *De Rham homomorphism*, and we shall denote it by $h^p_{\rm dR}$,

$$h^p_{\rm dR} \in L(H^p_{\rm dR}(X), \mathbf{H}^p(X)) \quad \text{with} \quad h^p_{\rm dR} : \text{class}(\alpha) \mapsto \text{class}\left(c \mapsto \int_c \alpha\right).$$

NOTE 1. For a closed p-form α, representing some class in $H^p_{\rm dR}(X)$, the set of values of the p-cochain f_α on the p-cycles is a homomorphic image of the group $\mathbf{H}_p(X)$ (art. 6.61) in $\mathbf{R}$, it is generally called the *group of periods* of α, and denoted by

$$P_\alpha = \left\{ \int_c \alpha \in \mathbf{R} \;\middle|\; c \in C_p(X) \text{ and } \partial c = 0 \right\}. \qquad (\clubsuit)$$

It appears that in diffeology the groups of periods of a closed p-form have some refinements and may better be defined by iterations through the space of loops; see for instance (art. 8.42) and Exercise 141, p. 297.

NOTE 2. The De Rham homomorphism is generally not an isomorphism in diffeology; see Exercise 134, p. 272. There exists however a spectral sequence describing the relationship between the two cohomologies [**Igl87b**]. The first obstruction is interpreted in (art. 8.30), but very little, perhaps nothing, is known about a geometric interpretation of the other terms.

PROOF. Since the integration of a p-form on a p-cube is smooth (art. 6.66), the De Rham homomorphism is smooth. Now, if σ is a degenerate p-cube, that is, $\sigma = \sigma' \circ \text{pr}$, where $\text{pr} : \mathbf{R}^p \to \mathbf{R}^q$ and $q < p$, then

$$\int_\sigma \alpha = \int_{\sigma'_*(\text{pr})} \alpha = \int_{\text{pr}} \sigma'^*(\alpha).$$

But $\sigma'^*(\alpha)$ is a p-form on $\mathbf{R}^q$ with $q < p$, thus $\sigma'^*(\alpha) = 0$ and the integral vanishes. Hence, the morphism $c \mapsto \int_c \alpha$ vanishes on degenerate cubes, and then on degenerate cubic chains. Therefore, this is a cochain for the reduced cubic cohomology. ☐

6.75. The functions whose differential is zero. Let X be a diffeological space. Let us recall that two points x_0 and x_1 are said to be connected if there exists a path $\gamma \in \mathrm{Paths}(X) = \mathcal{C}^\infty(\mathbf{R}, X)$ such that $\gamma(0) = x_0$ and $\gamma(1) = x_1$ (art. 5.6). Connectedness is an equivalence relation on X, the classes of this relation are the components of X and the set of components of X has been denoted by $\pi_0(X)$ (art. 5.7). Next, according to the definition above (art. 6.73) and given $\Omega^0(X) = \mathcal{C}^\infty(X, \mathbf{R})$ (art. 6.31), we have

$$H^0_{\mathrm{dR}}(X) = \ker\left[d : \Omega^0(X) \to \Omega^1(X)\right] = \{f \in \mathcal{C}^\infty(X, \mathbf{R}) \mid df = 0\}.$$

Now, every function whose differential is zero is constant on the connected components of X, and is therefore characterized by the set of values it takes on each connected component, thus

$$H^0_{\mathrm{dR}}(X) = \mathrm{Maps}(\pi_0(X), \mathbf{R}).$$

NOTE 1. The group of periods P_f of f (art. 6.74, (♣)) is generated by the following set of periods

$$\mathrm{Periods}(f) = \{f(X_i) \mid X_i \in \pi_0(X)\} \subset \mathbf{R}.$$

NOTE 2. The group $H^0_{\mathrm{dR}}(X)$ also coincides with the cubic cohomology group $\mathbf{H}^0(X)$ (art. 6.64). The De Rham homomorphism h^0_{dR} (art. 6.74) is an isomorphism.

PROOF. Let x and x′ be two connected points, let $\gamma \in \mathcal{C}^\infty(\mathbf{R}, X)$ such that $x = \gamma(0)$, and let $x' = \gamma(1)$. Then $df = 0$ implies $df(\gamma) = \gamma^*(df) = d[\gamma^*(f)] = 0$. Hence, $\gamma^*(f) = f \circ \gamma = \mathrm{cst}$. Therefore, the function f is constant on any connected component of X. Conversely, the differential of any smooth function, constant on each connected component of X, is zero. Therefore, $H^0_{\mathrm{dR}}(X)$ is the Abelian group generated by the set of components of X, that is, $H^0_{\mathrm{dR}}(X) = \mathrm{Maps}(\pi_0(X), \mathbf{R})$. □

6.76. Vanishing De Rham 1-cohomology. Let us recall that a diffeological space X is said to be *connected* if for any two points $x, x' \in X$ there exists a path $\gamma \in \mathrm{Paths}(X) = \mathcal{C}^\infty(\mathbf{R}, X)$ such that $\gamma(0) = x$ and $\gamma(1) = x'$ (art. 5.6). This is denoted by $\pi_0(X) = \{X\}$. Then, the space X is said to be *simply connected* if X is connected, and if for any two points $x, x' \in X$, and for any two paths γ and γ' connecting x to x′, there exists a path $[s \mapsto \gamma_s] \in \mathrm{Paths}(\mathrm{Paths}(X)) = \mathcal{C}^\infty(\mathbf{R}, \mathrm{Paths}(X))$ such that $\gamma_0 = \gamma$, $\gamma_1 = \gamma'$, $\gamma_s(0) = x$ and $\gamma_s(1) = x'$, for all s (art. 5.15). This is denoted by $\pi_1(X) = \{0\}$. Now, for every simply connected diffeological space X, the first De Rham cohomology group is trivial, $H^1_{\mathrm{dR}}(X) = \{0\}$, that is, every closed 1-form is exact. Precisely, for all $\alpha \in \Omega^1(X)$ such that $d\alpha = 0$ the functions

$$f : x \mapsto \int_{x_0}^{x} \alpha + \mathrm{cst}$$

are smooth and are the primitives of α, $\alpha = df$. The integral is taken on any path in X, connecting an arbitrary chosen basepoint x_0 to x.

PROOF. Let $x_0 \in X$ be a point, chosen as origin. Let $\mathrm{Paths}(X, x_0, \star)$ be the subspace of paths in X having x_0 as origin (art. 5.1), that is,

$$\mathrm{Paths}(X, x_0, \star) = \{\gamma \in \mathcal{C}^\infty(\mathbf{R}, X) \mid \gamma(0) = x_0\}.$$

Then, let

$$F : \mathrm{Paths}(X, x_0, \star) \to \mathbf{R} \quad \text{with} \quad F(\gamma) = \int_\gamma \alpha.$$

The value $F(\gamma)$ is just the pairing of α with γ. Moreover, thanks to (art. 6.66), F belongs to $\mathcal{C}^\infty(\mathrm{Paths}(X, x_0, \star), \mathbf{R})$. Since the space X is connected, the projection $\hat{1} : \mathrm{Paths}(X, x_0, \star) \to X$, defined by $\hat{1}(\gamma) = \gamma(1)$, is a subduction (art. 5.1). Let us show the following propositions.

1. There exists $f : X \to \mathbf{R}$ such that $F = f \circ \hat{1}$. Then, since $\hat{1}$ is a subduction and F is smooth, f is smooth (art. 1.51).
2. $dF = \hat{1}^*(\alpha)$. Then, since $dF = d(f \circ \hat{1}) = \hat{1}^*(df)$, $\hat{1}^*(\alpha) = \hat{1}^*(df)$. Next, $\hat{1}$ being a subduction, $\alpha = df$ (art. 6.38).

Let us establish first that $dF = \hat{1}^*(\alpha)$. Let $P : U \to \mathrm{Paths}(X, x_0, \star)$ be an n-plot, and let us denote $P(r) = \gamma_r$ for $r \in U$. Let $\delta r \in \mathbf{R}^n$ be any vector, then

$$dF(P)_r(\delta r) = \frac{\partial F(P)(r)}{\partial r}(\delta r) = \frac{\partial}{\partial r}\left\{\int_{\gamma_r} \alpha\right\}(\delta r) = \delta \int_\gamma \alpha.$$

But thanks to the formula of the variation of the integral (art. 6.70),

$$\delta \int_\gamma \alpha = \int_0^1 d\alpha(\delta\gamma) + \int_0^1 d[\alpha(\delta\gamma)] = 0 + \Big[\alpha(\delta\gamma)\Big]_{t=0}^{t=1},$$

that is,

$$\delta \int_\gamma \alpha = \left[\alpha(\gamma)_{\binom{r}{t}} \begin{pmatrix} \delta r \\ 0 \end{pmatrix}\right]_{t=0}^{t=1},$$

where $\gamma(r, t) = \gamma_r(t) = P(r)(t)$. Hence,

$$\begin{aligned}
dF(P)_r(\delta r) &= \alpha[(r,t) \mapsto P(r)(t)]_{\binom{r}{1}} \begin{pmatrix} \delta r \\ 0 \end{pmatrix} - \alpha[(r,t) \mapsto P(r)(t)]_{\binom{r}{0}} \begin{pmatrix} \delta r \\ 0 \end{pmatrix} \\
&= \alpha[r \mapsto P(r)(1)]_r(\delta r) - \alpha[r \mapsto P(r)(0)]_r(\delta r) \\
&= (\hat{1}^*(\alpha))(P)_r(\delta r) - \alpha[r \mapsto x_0]_r(\delta r) \\
&= (\hat{1}^*(\alpha))(P)_r(\delta r) - 0.
\end{aligned}$$

Therefore, $dF = \hat{1}^*(\alpha)$. Now, for all $x \in X$, the restriction of dF to any pullback $\hat{1}^{-1}(x)$ vanishes, that is, $dF \restriction \hat{1}^{-1}(x) = 0$. Thus, dF is locally constant on the subspace $\hat{1}^{-1}(x)$. But $\hat{1}^{-1}(x)$ is the subspace of paths in X with origin x_0 and end x. By hypothesis X is simply connected, hence $\hat{1}^{-1}(x)$ is connected. Therefore, F is constant on $\hat{1}^{-1}(x)$ Exercise 52, p. 53, and there exists $f : X \to \mathbf{R}$ such that $F = f \circ \hat{1}$. □

6.77. Closed 1-forms on locally simply connected spaces. Let X be a diffeological space, X is said to be *locally simply connected* if every D-open (art. 2.8) neighborhood of every point $x \in X$ contains a simply connected (art. 5.15) D-open neighborhood of x. In particular, if X is locally simply connected, there exists a D-open covering $\mathcal{U} = \{U_i\}_{i \in \mathcal{J}}$ of X, such that each element U_i of $\mathcal{U}$ is simply connected. As a consequence of the previous proposition (art. 6.76), every closed 1-form α on a locally simply connected diffeological space X is locally exact, that is, for every $x \in X$ there exist a D-open neighborhood U of x and a function $f \in \mathcal{C}^\infty(U, \mathbf{R})$ such that $\alpha \restriction U = df$. Finite dimensional manifolds (art. 4.1), for example, are locally simply connected. Indeed, manifolds are locally diffeomorphic to real domains, and real domains are locally simply connected. Thus, any closed form defined on a finite dimensional manifold is locally exact.

Exercises

✎ EXERCISE 118 (1-forms vanishing on loops). Let X be a diffeological space. Show that if the integral of a differential 1-form α vanishes on every loop, then the form is closed. Use the Stokes theorem (art. 6.69) and the fact that a 2-form is characterized by its values on the 2-plots (art. 6.37). Note that the form α is actually exact thanks to (art. 6.89).

✎ EXERCISE 119 (Forms on irrational tori are closed). Show that every differential form on an irrational torus $T_\Gamma = \mathbf{R}^n/\Gamma$, where Γ is a dense discrete generating subgroup of $\mathbf{R}^n$, $n \geq 1$, is closed. Deduce that $H^p_{dR}(T_\Gamma) \simeq \Lambda^p(\mathbf{R}^n)$.

✎ EXERCISE 120 (Is the group Diff(S^1) simply connected?). Let $S^1 \subset \mathbf{R}^2$ be the subspace of unit vectors and consider the group Diff(S^1) equipped with the functional diffeology. Let $X \in \mathcal{C}^\infty(\mathbf{R}, S^1)$ and $J \in GL(\mathbf{R}^2)$, defined by

$$\theta \mapsto X(\theta) = \begin{pmatrix} \cos(\theta) \\ \sin(\theta) \end{pmatrix} \quad \text{and} \quad J = \begin{pmatrix} 0 & -1 \\ 1 & 0 \end{pmatrix}.$$

For every n-plot $P : U \to \mathrm{Diff}(S^1)$, for every $r \in U$ and for every $\delta r \in \mathbf{R}^n$, let

$$\alpha(P)_r(\delta r) = \int_0^{2\pi} \left\langle J[P(r)(X(\theta))], \frac{\partial P(r)(X(\theta))}{\partial r}(\delta r) \right\rangle d\theta,$$

where

$$\begin{aligned} \frac{\partial P(r)(X(\theta))}{\partial r}(\delta r) &= D(r \mapsto P(r)(X(\theta)))(r)(\delta r) \\ &= \lim_{t \to 0} \frac{P(r + t\delta r)(X(\theta)) - P(r)(X(\theta))}{t}, \end{aligned}$$

and the difference is computed in $\mathbf{R}^2$. The brackets denote the standard scalar product on $\mathbf{R}^2$.

1) Check that α is a differential 1-form on Diff(S^1), and compute $d\alpha$.

2) Compute the integral of α on the following loop in Diff(S^1)

$$\sigma : t \mapsto \begin{pmatrix} \cos(2\pi t) & -\sin(2\pi t) \\ \sin(2\pi t) & \cos(2\pi t) \end{pmatrix}.$$

3) Deduce that the identity component of Diff(S^1) is not simply connected.

Chain-Homotopy Operator

The Chain-Homotopy *operator on a diffeological space* X *is a smooth linear map* $\mathcal{K} : \Omega^p(X) \to \Omega^{p-1}(\mathrm{Paths}(X))$ *which satisfies* $\mathcal{K} \circ d + d \circ \mathcal{K} = \hat{1}^* - \hat{0}^*$ (art. 6.83), *with* $\mathrm{Paths}(X) = \mathcal{C}^\infty(\mathbf{R}, X)$, $\hat{0}$ *and* $\hat{1}$ *map a path* γ *to its source* $\hat{0}(\gamma) = \gamma(0)$ *and its target* $\hat{1}(\gamma) = \gamma(1)$. *Since the space of paths in* X *is naturally a diffeological space, it is legitimate to consider differential forms on* Paths(X) *and its subspaces, it is the ability of the diffeological approach to stay in the same category and avoid parallel and tedious constructions. The Chain-Homotopy operator has multiple applications, for instance, it leads to the homotopic invariance of the De Rham cohomology* (art. 6.88), *and it is crucial for the construction of the moment map (Chapter 9).*

6.78. Integration operator of forms along paths. Let X be a diffeological space and let $\mathrm{Paths}(X) = \mathcal{C}^\infty(\mathbf{R}, X)$ be the space of smooth paths in X, equipped with the functional diffeology (art. 1.57). Let us consider, for each $t \in \mathbf{R}$, the *evaluation map* of paths at the point t, that is,

$$\mathbf{t} : \text{Paths}(X) \to X \quad \text{with} \quad \mathbf{t}(\gamma) = \gamma(t).$$

The map $\mathbf{t}$ is a smooth map, this is an immediate consequence of the definition of the functional diffeology.

1. We call the *integration operator* the map $\Phi : \Omega^p(X) \to \Omega^p(\text{Paths}(X))$ defined, for all integers $p > 0$, by

$$\text{for all } \alpha \in \Omega^p(X), \quad \Phi(\alpha) = \int_0^1 \mathbf{t}^*(\alpha)\, dt.$$

It maps any differential p-form α on X to the p-form on Paths(X) obtained by integrating α along the paths. Precisely, let $P : U \to \text{Paths}(X)$ be an n-plot, let $r \in U$, and let $v = (v_1 \cdots v_p)$ denote p vectors of $\mathbf{R}^n$. Then,

$$\Phi(\alpha)(P)(r)(v) = \int_0^1 \alpha(\mathbf{t} \circ P)(r)(v)\, dt, \quad \mathbf{t} \circ P = [r \mapsto P(r)(t)].$$

2. The integration operator Φ is linear. For any two p-forms α and α' on X, and for all $s \in \mathbf{R}$,

$$\Phi(\alpha + \alpha') = \Phi(\alpha) + \Phi(\alpha') \quad \text{and} \quad \Phi(s\alpha) = s \times \Phi(\alpha).$$

3. The integration operator Φ is smooth, $\Phi \in \mathcal{C}^\infty(\Omega^p(X), \Omega^p(\text{Paths}(X)))$.

PROOF. 1. Let us check first that $\Phi(\alpha)$ is a well defined p-form on Paths(X). Let $P : U \to X$ be a plot and let $F \in \mathcal{C}^\infty(V, U)$, where V is some real domain. Then,

$$\begin{aligned} \Phi(\alpha)(P \circ F) &= \int_0^1 \alpha(\mathbf{t} \circ P \circ F)\, dt \\ &= \int_0^1 F^*(\alpha(\mathbf{t} \circ P))\, dt \\ &= F^*\left(\int_0^1 \alpha(\mathbf{t} \circ P)\, dt\right) \\ &= F^*(\Phi(\alpha)(P)). \end{aligned}$$

2. Now, let us check that the integration operator Φ is linear. Let α and α' be any two p-forms of X. Let $P_t = \mathbf{t} \circ P$, we have

$$\begin{aligned} \Phi(\alpha + \alpha') &= [P \mapsto \int_0^1 (\alpha + \alpha')(P_t)\, dt] \\ &= \int_0^1 \alpha(P_t)\, dt + \int_0^1 \alpha'(P_t)\, dt] \\ &= \Phi(\alpha) + \Phi(\alpha'). \end{aligned}$$

And, for all $s \in \mathbf{R}$,

$$\Phi(s\alpha) = [P \mapsto \int_0^1 s\alpha(P_t)\, dt] = [P \mapsto s \int_0^1 \alpha(P_t)\, dt] = s \times \Phi(\alpha).$$

3. Since the map $\mathbf{t} : \text{Paths}(X) \to X$ is smooth and since integration preserves smoothness, the integration operator Φ is a smooth linear map from $\Omega^p(X)$ to $\Omega^p(\text{Paths}(X))$. □

6.79. The operator Φ is a morphism of De Rham complex. The integration operator Φ, of a diffeological space X (art. 6.78), is a morphism from the De Rham complex of X to the De Rham complex of Paths(X), that is,

$$d \circ \Phi = \Phi \circ d.$$

This is summarized by the following commutative diagram, with $p > 0$.

$$\begin{array}{ccc} \Omega^p(X) & \xrightarrow{\Phi} & \Omega^p(\mathrm{Paths}(X)) \\ \Big\downarrow d & & \Big\downarrow d \\ \Omega^{p+1}(X) & \xrightarrow[\Phi]{} & \Omega^{p+1}(\mathrm{Paths}(X)) \end{array}$$

PROOF. Let α be a p-form on X, $p > 0$. Then,

$$\begin{aligned} \Phi(d\alpha)(P) &= \int_0^1 (d\alpha)(P_t)\, dt = \int_0^1 d[\alpha(P_t)]\, dt = d\left(\int_0^1 \alpha(P_t)\, dt\right) \\ &= d(\Phi(\alpha)(P)). \end{aligned}$$

Thus, $\Phi(d\alpha) = d(\Phi(\alpha))$. ☐

6.80. Variance of the integration operator Φ. Let X and X′ be two diffeological spaces, and let $f : X \to X'$ be a smooth map. The map f induces a smooth map on the spaces of paths,

$$f_{\mathcal{P}} : \mathrm{Paths}(X) \to \mathrm{Paths}(X') \quad \text{with} \quad f_{\mathcal{P}}(\gamma) = f \circ \gamma.$$

The map f also induces the two pullbacks

$$f^* : \Omega^\star(X') \to \Omega^\star(X) \quad \text{and} \quad f^*_{\mathcal{P}} : \Omega^\star(\mathrm{Paths}(X')) \to \Omega^\star(\mathrm{Paths}(X)).$$

Let Φ_X and $\Phi_{X'}$ be the two associated integration operators (art. 6.78), then

$$\Phi_X \circ f^* = f^*_{\mathcal{P}} \circ \Phi_{X'}.$$

This is summarized by the following commutative diagram.

$$\begin{array}{ccc} \Omega^p(X') & \xrightarrow{\Phi_{X'}} & \Omega^p(\mathrm{Paths}(X')) \\ \Big\downarrow f^* & & \Big\downarrow f^*_{\mathcal{P}} \\ \Omega^p(X) & \xrightarrow[\Phi_X]{} & \Omega^p(\mathrm{Paths}(X)) \end{array}$$

PROOF. Let α be a differential p-form on X′ and $P : U \to X$ be a plot. Let us denote $P_t = \mathbf{t} \circ P$. On the one hand we have

$$[(\Phi_X \circ f^*)(\alpha)](P) = [\Phi_X(f^*(\alpha))](P) = \int_0^1 f^*(\alpha)(P_t)\, dt = \int_0^1 \alpha(f \circ P_t)\, dt,$$

and on the other hand

$$[(f^*_{\mathcal{P}} \circ \Phi_{X'})(\alpha)](P) = [f^*_{\mathcal{P}}(\Phi_{X'}(\alpha))](P) = [\Phi_{X'}(\alpha)](f_{\mathcal{P}} \circ P) = \int_0^1 \alpha[(f_{\mathcal{P}} \circ P)_t]\, dt.$$

But, for every $r \in U$, $(f \circ P_t)(r) = f(P_t(r)) = f(P(r)(t))$, and $(f_{\mathcal{P}} \circ P)_t(r) = (f_{\mathcal{P}} \circ P)_t(r) = f(P(r)(t))$. Hence, $f \circ P_t = (f_{\mathcal{P}} \circ P)_t$, thus $\Phi_X \circ f^* = f^*_{\mathcal{P}} \circ \Phi_{X'}$. ☐

6.81. Derivation along time reparametrization. Let X be a diffeological space and Paths(X) be the space of its smooth paths, equipped with the functional diffeology. The group of translations $(\mathbf{R}, +)$ acts by reparametrization on Paths(X), as a 1-parameter group of diffeomorphisms. Let us denote by τ this action. For all $\gamma \in$ Paths(X) and for all $e \in \mathbf{R}$,

$$\tau(e) : \gamma \mapsto \gamma \circ T_e \quad \text{with} \quad T_e : t \mapsto t + e.$$

Let α be a p-form of X, the Lie derivative (art. 6.54) of the p-form $\Phi(\alpha)$ by the 1-parameter group τ satisfies the identity

$$\pounds_\tau(\Phi(\alpha)) = \hat{1}^*\alpha - \hat{0}^*\alpha.$$

PROOF. Let us check first that $\tau : e \mapsto [\gamma \mapsto \gamma \circ T_e]$ is a smooth homomorphism from $(\mathbf{R}, +)$ into Diff(Paths(X)). It takes its values in Diff(X). Indeed, for all $e \in \mathbf{R}$, $\gamma' = \tau(e)(\gamma)$ implies $\gamma = \tau(-e)(\gamma')$, $\tau(e)^{-1} = \tau(-e)$. For all $e \in \mathbf{R}$, $\tau(e)$ (and thus $\tau(e)^{-1}$) is smooth by the very definition of the functional diffeology of Diff(X), and τ is clearly a homomorphism, $\tau(e+e') = \tau(e) \circ \tau(e')$, thus τ is a 1-parameter group of diffeomorphisms of Paths(X).

Now, let us denote $\boldsymbol{\alpha} = \Phi(\alpha)$, for every plot P de Paths(X),

$$[\pounds_\tau \boldsymbol{\alpha}](P) = \frac{\partial}{\partial t}\Big\{[\tau(t)^* \boldsymbol{\alpha}](P)\Big\}_{t=0} = \frac{\partial}{\partial t}\Big\{\boldsymbol{\alpha}(\tau(t) \circ P)\Big\}_{t=0}.$$

But $\tau(t) \circ P : r \mapsto P(r) \circ T_t$, then

$$\boldsymbol{\alpha}[\tau(t) \circ P] = \int_0^1 \alpha[(\tau(t) \circ P)_s]\, ds = \int_0^1 \alpha[r \mapsto P(r)(t+s)]\, ds.$$

Let $u = t + s$, we have

$$\boldsymbol{\alpha}[\tau(t) \circ P] = \int_t^{1+t} \alpha[r \mapsto P(r)(u)]\, du.$$

After derivation we get, for all plots P of X, $[\pounds_\tau \boldsymbol{\alpha}](P) = \alpha[r \mapsto P(r)(1)] - \alpha[r \mapsto P(r)(0)] = [\hat{1}^*\alpha - \hat{0}^*\alpha](P)$, that is, $\pounds_\tau(\Phi(\alpha)) = \hat{1}^*\alpha - \hat{0}^*\alpha$. ☐

6.82. Variance of the time reparametrization. Let X and X′ be two diffeological spaces and $f : X \to X'$ be a smooth map. Let $f_{\mathcal{P}} :$ Paths(X) $\to$ Paths(X′) be the action of f, on the paths, defined in (art. 6.80). Let $f^*_{\mathcal{P}} : \Omega^p(\text{Paths}(X')) \to \Omega^p(\text{Paths}(X))$ be the induced action of $f_{\mathcal{P}}$ at the level of p-forms. Let τ and τ' denote the action $(\mathbf{R}, +)$ on Paths(X) and Paths(X′), as defined in (art. 6.81). Let i_τ and $i_{\tau'}$ be the contractions associated with these 1-parameter groups of diffeomorphisms (art. 6.57). Then, $i_\tau \circ f^*_{\mathcal{P}} = f^*_{\mathcal{P}} \circ i_{\tau'}$.

$$\begin{array}{ccc}
\Omega^p(\text{Paths}(X')) & \xrightarrow{\;f^*_{\mathcal{P}}\;} & \Omega^p(\text{Paths}(X)) \\
\Big\downarrow i_{\tau'} & & \Big\downarrow i_\tau \\
\Omega^{p-1}(\text{Paths}(X')) & \xrightarrow[\;f^*_{\mathcal{P}}\;]{} & \Omega^{p-1}(\text{Paths}(X))
\end{array}$$

PROOF. Let β be a p-form on Paths(X′), $P : U \to \text{Paths}(X)$ be an n-plot, $r \in U$ and ν represents $(p-1)$ vectors of $\mathbf{R}^n$. By definition of the contraction of a p-form by a 1-parameter group of diffeomorphisms (art. 6.57), we have

$$\begin{aligned}[(i_\tau \circ f_{\mathcal{P}}^*)(\beta)](P)_r(\nu) &= [i_\tau(f_{\mathcal{P}}^*(\beta))](P)_r(\nu) \\ &= f_{\mathcal{P}}^*(\beta)(\tau \cdot P)_{\binom{0}{r}} \begin{pmatrix}1\\0\end{pmatrix} \begin{pmatrix}0\\\nu\end{pmatrix} \\ &= \beta(f_{\mathcal{P}} \circ (\tau \cdot P))_{\binom{0}{r}} \begin{pmatrix}1\\0\end{pmatrix} \begin{pmatrix}0\\\nu\end{pmatrix},\end{aligned}$$

where $\tau \cdot P(t, r) = \tau(t)(P(r))$. But

$$\begin{aligned}[f_{\mathcal{P}} \circ (\tau \cdot P)](t, r) &= f_{\mathcal{P}}((\tau \cdot P)(t, r)) \\ &= f_{\mathcal{P}}(\tau(t)(P(r)) \\ &= f_{\mathcal{P}}[s \mapsto \tau(t)(P(r))(s)] \\ &= f_{\mathcal{P}}[s \mapsto P(r)(s+t)] \\ &= [s \mapsto f(P(r)(s+t))]. \end{aligned} \qquad (\clubsuit)$$

On the other hand,

$$\begin{aligned}[(f_{\mathcal{P}}^* \circ i_{\tau'})(\beta)](P)_r(\nu) &= [f_{\mathcal{P}}^*(i_{\tau'}(\beta))](P)_r(\nu) \\ &= [(i_{\tau'}(\beta)(f_{\mathcal{P}} \circ P)]_r(\nu) \\ &= \beta(\tau' \cdot (f_{\mathcal{P}} \circ P))_{(0,r)} \begin{pmatrix}1\\0\end{pmatrix} \begin{pmatrix}0\\\nu\end{pmatrix}.\end{aligned}$$

But

$$\begin{aligned}[\tau' \cdot (f_{\mathcal{P}} \circ P)](t, r) &= \tau'(t)(f_{\mathcal{P}} \circ P(r)) \\ &= [s \mapsto f_{\mathcal{P}} \circ P(r))(s+t)], \\ [\tau' \cdot (f_{\mathcal{P}} \circ P)] &= [s \mapsto f(P(r)(s+t))]. \end{aligned} \qquad (\spadesuit)$$

Now, comparing $(\clubsuit)$ and $(\spadesuit)$, we get

$$[(i_\tau \circ f_{\mathcal{P}}^*)(\beta)](P)_r(\nu) = [(f_{\mathcal{P}}^* \circ i_{\tau'})(\beta)](P)_r(\nu),$$

that is, $i_\tau \circ f_{\mathcal{P}}^* = f_{\mathcal{P}}^* \circ i_{\tau'}$. ☐

6.83. The Chain-Homotopy operator $\mathcal{K}$. Let X be a diffeological space, let Φ be the integration operator defined in (art. 6.78), let τ be the time-reparametrization defined in (art. 6.81), and let i_τ be the contraction by the sliding τ (art. 6.57). The operator $\mathcal{K} : \Omega^p(X) \to \Omega^{p-1}(\text{Paths}(X))$ defined, for all $p > 0$, by

$$\mathcal{K} = i_\tau \circ \Phi, \qquad (\spadesuit)$$

satisfies

$$\mathcal{K} \circ d + d \circ \mathcal{K} = \hat{1}^* - \hat{0}^*. \qquad (\clubsuit)$$

The operator $\mathcal{K}$ will be called the *Chain-Homotopy operator*, it is smooth and linear,

$$\mathcal{K} \in L^\infty(\Omega^p(X), \Omega^{p-1}(\text{Paths}(X))).$$

Let α be a p-form of X, with $p > 1$, and $P : U \to \text{Paths}(X)$ be an n-plot. The value of $\mathcal{K}\alpha$ on the plot P, at the point $r \in U$, evaluated on $(p-1)$ vectors $v_2, \ldots, v_p$ of

$\mathbf{R}^n$, is explicitly given by

$$(\mathcal{K}\alpha)(P)_r(v_2)\cdots(v_p) = \int_0^1 \alpha\left(\begin{pmatrix} t \\ r \end{pmatrix} \mapsto P(r)(t)\right)_{\binom{t}{r}} \begin{pmatrix} 1 \\ 0 \end{pmatrix}\begin{pmatrix} 0 \\ v_2 \end{pmatrix}\cdots\begin{pmatrix} 0 \\ v_p \end{pmatrix} dt. \qquad (\diamondsuit)$$

For $p = 1$, see (art. 6.86).

PROOF. Let α be a p-form of X, $p > 0$. On the one hand (art. 6.81) we have

$$\pounds_\tau(\Phi(\alpha)) = \hat{1}^*\alpha - \hat{0}^*\alpha,$$

and on the other hand, applying the Cartan formula (art. 6.72) and the commutation $d \circ \Phi = \Phi \circ d$ (art. 6.79), we have

$$\begin{aligned} \pounds_\tau(\Phi(\alpha)) &= d[i_\tau(\Phi(\alpha))] + i_\tau(d[\Phi(\alpha)]) \\ &= d[i_\tau\Phi(\alpha)] + i_\tau\Phi[d\alpha] \\ &= d[\mathcal{K}(\alpha)] + \mathcal{K}(d\alpha). \end{aligned}$$

Hence, $d[\mathcal{K}(\alpha)] + \mathcal{K}(d\alpha) = \hat{1}^*\alpha - \hat{0}^*\alpha$, that is, $\mathcal{K} \circ d + d \circ \mathcal{K} = \hat{1}^* - \hat{0}^*$. Now, because the contraction operation and the map Φ are smooth (art. 6.56), (art. 6.78), the Chain-Homotopy operator $\mathcal{K}$ is a smooth linear map. A direct application of all these successive definitions gives

$$(\mathcal{K}\alpha)(P)_r(v_2)\cdot(v_p) = \int_0^1 \alpha\left(\begin{pmatrix} s \\ r \end{pmatrix} \mapsto P(r)(s+t)\right)_{\binom{0}{r}} \begin{pmatrix} 1 \\ 0 \end{pmatrix}\begin{pmatrix} 0 \\ v_2 \end{pmatrix}\cdot\begin{pmatrix} 0 \\ v_p \end{pmatrix} dt,$$

which becomes $(\diamondsuit)$ after the change of variable $s \mapsto s + t$. ☐

6.84. Variance of the Chain-Homotopy operator. Let X et X' be two diffeological spaces. Let $f : X \to X'$ be a smooth map. Let us use the notations of the proposition (art. 6.80) and let $\mathcal{K}_X$ et $\mathcal{K}_{X'}$ be the two Chain-Homotopy operators of X and X'. The variance of the Chain-Homotopy operators is given by

$$\mathcal{K}_X \circ f^* = f^*_{\mathcal{P}} \circ \mathcal{K}_{X'},$$

where $f_{\mathcal{P}}$ has been defined in (art. 6.80), as the action of the function f on the paths. This is summarized by the following commutative diagram.

$$\begin{array}{ccc} \Omega^p(X') & \xrightarrow{\mathcal{K}_{X'}} & \Omega^{p-1}(\mathrm{Paths}(X')) \\ {\scriptstyle f^*}\big\downarrow & & \big\downarrow{\scriptstyle f^*_{\mathcal{P}}} \\ \Omega^p(X) & \xrightarrow[\mathcal{K}_X]{} & \Omega^{p-1}(\mathrm{Paths}(X)) \end{array}$$

NOTE. Let $\mathrm{Diff}(X, \alpha)$ be the *group of automorphisms* of α, that is the group of diffeomorphisms of X preserving α,

$$\mathrm{Diff}(X, \alpha) = \{f \in \mathrm{Diff}(X) \mid f^*(\alpha) = \alpha\}.$$

As an application of the above proposition we get immediately that

$$f \in \mathrm{Diff}(X, \alpha) \quad \Rightarrow \quad f_{\mathcal{P}} \in \mathrm{Diff}(\mathrm{Paths}(X), \mathcal{K}\alpha).$$

In other words, if a diffeomorphism f of X preserves α, then the action $f_{\mathcal{P}}$, of f on $\mathrm{Paths}(X)$, preserves $\mathcal{K}\alpha$.

PROOF. Let τ and τ' be the action of $(\mathbf{R},+)$ on Paths(X) and Paths(X'), defined in (art. 6.81). Let $\alpha \in \Omega^p(X')$, by definition $\mathcal{K}_X(f^*(\alpha)) = (i_\tau \circ \Phi_X)(f^*(\alpha)) = i_\tau(\Phi_X(f^*(\alpha)))$, but $\Phi_X \circ f^* = f^*_{\mathcal{P}} \circ \Phi_{X'}$ (art. 6.80), thus $\mathcal{K}_X(f^*(\alpha)) = i_\tau(\Phi_X \circ f^*(\alpha)) = i_\tau(f^*_{\mathcal{P}} \circ \Phi_{X'}(\alpha)) = (i_\tau \circ f^*_{\mathcal{P}})(\Phi_{X'}(\alpha))$. Now, thanks to (art. 6.82), $i_\tau \circ f^*_{\mathcal{P}} = f^*_{\mathcal{P}} \circ i_{\tau'}$, hence $\mathcal{K}_X(f^*(\alpha)) = (f^*_{\mathcal{P}} \circ i_{\tau'})(\Phi_{X'}(\alpha)) = f^*_{\mathcal{P}}(i_{\tau'} \circ \Phi_{X'}(\alpha)) = f^*_{\mathcal{P}}(\mathcal{K}_{X'}(\alpha))$. Therefore, $\mathcal{K}_X \circ f^* = f^*_{\mathcal{P}} \circ \mathcal{K}_{X'}$. ☐

6.85. Chain-Homotopy and paths concatenation. Let X be a diffeological space and α be a differential k-form on X, $k \geq 1$. Let $\gamma \in \text{stPaths}(X, x, x')$, the space of stationary paths, we consider the *preconcatenation* $L(\gamma)$ defined on $\text{stPaths}(X, x', \star)$ and the *postconcatenation* $R(\gamma)$ defined on $\text{stPaths}(X, \star, x)$, that is,

$$L(\gamma)(\gamma') = \gamma \vee \gamma' \quad \text{and} \quad R(\gamma)(\gamma') = \gamma' \vee \gamma.$$

These operations preserve the $(k-1)$-form $\mathcal{K}\alpha$, where $\mathcal{K}$ is the Chain-Homotopy operator (art. 6.83), precisely,

$$\begin{cases} L(\gamma)^*(\mathcal{K}\alpha \restriction \text{stPaths}(X, x, \star)) & = \mathcal{K}\alpha \restriction \text{stPaths}(X, x', \star)), \\ R(\gamma)^*(\mathcal{K}\alpha \restriction \text{stPaths}(X, \star, x')) & = \mathcal{K}\alpha \restriction \text{stPaths}(X, \star, x)). \end{cases}$$

PROOF. Let $P : U \to \text{stPaths}(X, x', \star)$ be an n-plot. Let r be a generic point in U and, in the following, as is usual now, let the surrounding square brackets represent $(k-1)$ vectors, for example $[\delta r] = (\delta r_2) \cdots (\delta r_k)$ with $\delta r_i \in \mathbf{R}^n$, etc. We have According to (art. 6.83, $(\diamondsuit)$),

$$\begin{aligned} L(\gamma)^*(\mathcal{K}\alpha)(P)_r[\delta r] &= \mathcal{K}\alpha(L(\gamma) \circ P)_r[\delta r] \\ &= \int_0^1 \alpha\left(\begin{pmatrix} t \\ r \end{pmatrix} \mapsto L(\gamma) \circ P(r)(t)\right)_{\binom{t}{r}} \begin{pmatrix} 1 \\ 0 \end{pmatrix} \begin{bmatrix} 0 \\ \delta r \end{bmatrix} dt \\ &= \int_0^1 \alpha\left(\begin{pmatrix} t \\ r \end{pmatrix} \mapsto (\gamma \vee P(r))(t)\right)_{\binom{t}{r}} \begin{pmatrix} 1 \\ 0 \end{pmatrix} \begin{bmatrix} 0 \\ \delta r \end{bmatrix} dt \\ &= \int_0^1 \alpha\left(\begin{pmatrix} t \\ r \end{pmatrix} \mapsto \begin{Bmatrix} \gamma(2t) & t \leq 1/2 \\ P(r)(2t-1) & t \geq 1/2 \end{Bmatrix}\right)_{\binom{t}{r}} \begin{pmatrix} 1 \\ 0 \end{pmatrix} \begin{bmatrix} 0 \\ \delta r \end{bmatrix} dt \\ &= \int_0^{1/2} \alpha\left(\begin{pmatrix} t \\ r \end{pmatrix} \mapsto \gamma(2t)\right)_{\binom{t}{r}} \begin{pmatrix} 1 \\ 0 \end{pmatrix} \begin{bmatrix} 0 \\ \delta r \end{bmatrix} dt \\ &+ \int_{1/2}^1 \alpha\left(\begin{pmatrix} t \\ r \end{pmatrix} \mapsto P(r)(2t-1)\right)_{\binom{t}{r}} \begin{pmatrix} 1 \\ 0 \end{pmatrix} \begin{bmatrix} 0 \\ \delta r \end{bmatrix} dt. \end{aligned}$$

The first term of the right-hand side of the last identity vanishes, since $\gamma(2t)$ does not depend on r, and after a change of variable $t' = 2t - 1$, the second term gives

$$\begin{aligned} L(\gamma)^*(\mathcal{K}\alpha)(P)_r[\delta r] &= \int_0^1 \alpha\left(\begin{pmatrix} t' \\ r \end{pmatrix} \mapsto P(r)(t')\right)_{\binom{t'}{r}} \begin{pmatrix} 1 \\ 0 \end{pmatrix} \begin{bmatrix} 0 \\ \delta r \end{bmatrix} dt' \\ &= \mathcal{K}\alpha(P)_r[\delta r]. \end{aligned}$$

Thus, $L(\gamma)^*(\mathcal{K}\alpha)(P)_r[\delta r] = \mathcal{K}\alpha(P)_r[\delta r]$, and this proof applies *mutatis mutandis* to the postconcatenation operator $R(\gamma)$. ☐

6.86. The Chain-Homotopy operator for p = 1. Let X be a diffeological space. Let $\text{Paths}(X) = \mathcal{C}^\infty(\mathbf{R}, X)$ be the space of paths in X, equipped with the functional diffeology. For every 1-form $\alpha \in \Omega^1(X)$, $F_\alpha = \mathcal{K}(\alpha)$ belongs to $\Omega^0(\mathrm{Paths}(X)) = \mathcal{C}^\infty(\mathrm{Paths}(X), \mathbf{R})$,

$$\mathcal{K} : \Omega^1(X) \to \mathcal{C}^\infty(\mathrm{Paths}(X), \mathbf{R}) \quad \text{and} \quad \mathcal{K}(\alpha) = F_\alpha = \left[\gamma \mapsto \int_\gamma \alpha\right].$$

The function $F_\alpha = \mathcal{K}(\alpha)$ can be extended, by linearity, over the whole space $C_1(X)$ of 1-chains of X (art. 6.58),

$$\text{for all } \sum_\gamma n_\gamma \gamma \in C_1(X), \quad F_\alpha\Big(\sum_\gamma n_\gamma \gamma\Big) = \sum_\gamma n_\gamma F_\alpha(\gamma) = \sum_\gamma n_\gamma \int_\gamma \alpha.$$

Hence, for $p = 1$, the Chain-Homotopy operator is just the pairing of 1-forms with 1-chains (art. 6.66). Moreover, if γ and γ' are two paths such that the concatenation $\gamma \vee \gamma'$ is defined (art. 5.6), then

$$F_\alpha(\gamma \vee \gamma') = F_\alpha(\gamma + \gamma') = F_\alpha(\gamma) + F_\alpha(\gamma'). \qquad (\diamondsuit)$$

NOTE. If $\gamma(1) = \gamma'(0)$ but the concatenation of γ and γ' is not smooth, then the smashed concatenation $\gamma \star \gamma' = \gamma^\star \vee \gamma'^\star$ is smooth (art. 5.5, Note), and also satisfies

$$F_\alpha(\gamma \star \gamma') = F_\alpha(\gamma + \gamma') = F_\alpha(\gamma) + F_\alpha(\gamma'). \qquad (\heartsuit)$$

In other words, $F_\alpha = \mathcal{K}(\alpha)$ is a morphism from the magma $(\mathrm{Paths}(X), \star)$ to the Abelian group $(\mathbf{R}, +)$.

PROOF. If $\gamma \vee \gamma'$ is a path in X, then the identity $(\diamondsuit)$ is just the additivity of the integral,

$$\begin{aligned} F_\alpha(\gamma \vee \gamma') &= \int_0^1 \alpha(\gamma \vee \gamma')(t)\, dt \\ &= \int_0^{1/2} \alpha(s \mapsto \gamma(2s))(t)\, dt + \int_{1/2}^1 \alpha(s \mapsto \gamma'(2s-1))(t)\, dt. \end{aligned}$$

After a suitable change of variable, we get

$$F_\alpha(\gamma \vee \gamma') = \int_0^1 \alpha(\gamma)(t)\, dt + \int_0^1 \alpha(\gamma')(t)\, dt = F_\alpha(\gamma) + F_\alpha(\gamma').$$

Now, if we need to smash the paths γ and γ', we just remark that according to (art. 6.68) we have

$$F_\alpha(\gamma^\star) = F_\alpha(\gamma \circ \lambda) = \int_I \alpha(\gamma \circ \lambda) = \int_I \lambda^*(\alpha(\gamma)) = \int_{\lambda(I)} \alpha(\gamma) = \int_I \alpha(\gamma) = F_\alpha(\gamma),$$

where λ is the smashing function described in (art. 5.6), satisfying $\lambda(I) = I$. Therefore, $F_\alpha(\gamma \star \gamma') = F_\alpha(\gamma) + F_\alpha(\gamma')$. ☐

6.87. The Chain-Homotopy operator for manifolds. Let M be a manifold of finite dimension (art. 4.1). The Chain-Homotopy operator $\mathcal{K}$ (art. 6.83) of M can be expressed in terms of path integral. Let $P : U \to \mathrm{Paths}(M)$ be an n-plot. Let us denote, for any $r \in U$ and for any vector $\delta_i r \in \mathbf{R}^n$,

$$\gamma_r = P(r) : t \mapsto \gamma_r(t) \quad \text{and} \quad \delta\gamma_r : t \mapsto D[r \to \gamma_r(t)](r)(\delta r).$$

Let $\alpha \in \Omega^p(M)$, the real $\mathcal{K}\alpha(P)(r)(\delta_2 r)\cdots(\delta_p r)$ can be interpreted as $\mathcal{K}\alpha$, computed at the point γ_r and applied to the $(p-1)$ *variations* $\delta_i\gamma_r$ associated with the $(p-1)$ vectors $\delta_i r$. It can be written as follows:

$$\mathcal{K}\alpha_{\gamma_r}(\delta_2\gamma_r)\cdots(\delta_p\gamma_r) = \int_0^1 \alpha_{\gamma_r(t)}\left(\frac{d\gamma_r(t)}{dt}\right)(\delta_2\gamma_r(t))\cdots(\delta_p\gamma_r(t))\,dt.$$

Homotopy and Differential Forms, Poincaré's Lemma

One of the crucial properties of De Rham cohomology is its homotopic invariance: the mapping induced in De Rham cohomology by the pullback of a smooth map depends only on the homotopy class of the map. The equivalent theorem in classical differential geometry, for the De Rham cohomology of finite dimensional manifolds, can be regarded as a specialization of general diffeological theory. The transition through the space of paths, which is impossible in the restricted category of manifolds, offers a great simplification of this theorem, even for the sole case of manifolds. The main tool used in this section is the Chain-Homotopy operator defined in the previous section (art. 6.83). *This proves that the nature of this invariance dwells deeply in the diffeological structure.*

6.88. Homotopic invariance of the De Rham cohomology. Let X and X' be two diffeological spaces and $f \in \mathcal{C}^\infty(X, X')$. Since the exterior derivative commutes with the pullback (art. 6.34), the action of f passes from the differential forms to the De Rham cohomology, as a linear action denoted by f^*_{dR}. For all $\alpha' \in \Omega^\star(X')$,

$$f^*_{dR}(\mathrm{class}(\alpha')) = \mathrm{class}(f^*(\alpha')), \quad f^*_{dR} \in L^\infty(H^\star_{dR}(X'), H^\star_{dR}(X)).$$

Let $s \mapsto f_s$ be a *homotopy of smooth maps* from X to X' and let α' be a closed differential form on X', that is,

$$[s \mapsto f_s] \in \mathrm{Paths}(\mathcal{C}^\infty(X, X')) \text{ and } \alpha' \in \Omega^p(X'), \text{ with } d\alpha' = 0.$$

Let f_0 and f_1 be the ends of this path. Then, there exists a differential form $\beta \in \Omega^{p-1}(X)$ such that

$$f_1^*(\alpha') = f_0^*(\alpha') + d\beta \quad \text{and,} \quad f^*_{0\,dR} = f^*_{1\,dR}.$$

In other words, the pullback $f^*_{dR} : H^\star_{dR}(X') \to H^\star_{dR}(X)$ by smooth maps depends only on the homotopy class of f (art. 5.9). We say that f_0 *and* f_1 *are equivalent in De Rham cohomology.*

PROOF. Let us consider the following map φ from X to $\mathrm{Paths}(X')$,

$$\begin{array}{ccl} X & \xrightarrow{\ \varphi\ } & \mathrm{Paths}(X') \\ & & \Big\downarrow \text{ends} \qquad \text{defined by} \quad \varphi : x \mapsto [s \mapsto f_s(x)], \\ & & X' \times X' \end{array}$$

where $\mathrm{ends}(\gamma') = (\gamma'(0), \gamma'(1))$. The map φ is clearly smooth for the functional diffeology, $\varphi \in \mathcal{C}^\infty(X, \mathrm{Paths}(X'))$. Let us compose φ^* with the identity satisfied by the Chain-Homotopy operator $\mathcal{K} \circ d + d \circ \mathcal{K} = \hat{1}^* - \hat{0}^*$ (art. 6.83). Using the hypothesis $d\alpha' = 0$ and the commutativity between the exterior derivative and pullback (art. 6.34), we have on the one hand

$$\varphi^*(\mathcal{K}d\alpha' + d\mathcal{K}\alpha') = \varphi^*(d\mathcal{K}\alpha') = d(\varphi^*(\mathcal{K}\alpha')),$$

and on the other hand

$$\begin{aligned} \varphi^* \circ (\hat{1}^* - \hat{0}^*)(\alpha) &= \varphi^* \circ \hat{1}^*(\alpha) - \varphi^* \circ \hat{0}^*(\alpha) \\ &= (\hat{1} \circ \varphi)^*(\alpha) - (\hat{0} \circ \varphi)^*(\alpha) \\ &= f_1^*(\alpha) - f_0^*(\alpha). \end{aligned}$$

Thus, $f_1^*(\alpha) = f_0^*(\alpha) + d\,\beta$, with $\beta = \varphi^*(\mathcal{K}\alpha')$. □

6.89. Closed 1-forms vanishing on loops. Let X be a connected diffeological space (art. 5.6). A closed 1-form α, $\alpha \in \Omega^1(X)$ and $d\alpha = 0$, is exact, $\alpha = df$, if and only if its integral vanishes on every loop, that is, $\int_\ell \alpha = 0$ for all $\ell \in \mathrm{Loops}(X)$. The primitive f can be chosen as the integral of α along any path connecting a chosen origin $x_0 \in X$ to the current point, which is summarized by

$$\alpha = df \quad \text{with} \quad f(x) = \int_{x_0}^{x} \alpha + \mathrm{cst}.$$

NOTE 1. This proposition looks like (art. 6.76) for which the hypothesis to be simply connected — which means that every loop is homotopic to a constant loop — implies, by homotopic invariance, that the integral of the 1-form vanishes on every loop, and thus satisfies the above condition.

NOTE 2. Let η be the *first De Rham homomorphism*,

$$\eta : H^1_{\mathrm{dR}}(X) \to \mathrm{Hom}(\pi_1(X), \mathbf{R}) \quad \text{with} \quad \eta(\mathrm{class}(\alpha)) = \left[\mathrm{class}(\ell) \mapsto \int_\ell \alpha\right].$$

The map η is well defined since the integral of a closed 1-form on a loop depends only on the homotopy class of the loop and the cohomology class of the form. The proposition above can be reformulated as follows: "The first De Rham homomorphism is injective", that is,

$$\ker(\eta) = \{0\}.$$

PROOF. One way is a direct consequence of the Stokes theorem (art. 6.69). If $\alpha = df$, then

$$\int_\ell \alpha = \int_\ell df = \int_{\partial\ell} f = f(\ell(1)) - f(\ell(0)) = 0.$$

Conversely, let us assume that the integral of α vanishes on any loop in X,

$$\text{for all } \ell \in \mathrm{Loops}(X), \quad \int_\ell \alpha = \int_0^1 \alpha(\ell)_t(1)\,dt = 0.$$

Let γ and γ' be two paths in X such that $\gamma(0) = \gamma'(0)$ and $\gamma(1) = \gamma'(1)$. Let $\bar{\gamma}$ defined by $\bar{\gamma}(t) = \gamma'(1-t)$, then the path $\ell = \gamma' \star \bar{\gamma} = \gamma'^{\star} \vee \bar{\gamma}^{\star}$ (art. 5.5, note) is a loop in X based in $x_0 = \gamma(0)$. Hence, thanks to (art. 6.86), we have

$$F_\alpha(\ell) = F_\alpha(\gamma') - F_\alpha(\gamma), \text{ that is, } F_\alpha(\gamma') = F_\alpha(\gamma) + F_\alpha(\ell).$$

Since, by hypothesis, $F_\alpha(\ell) = 0$, $F_\alpha(\gamma') = F_\alpha(\gamma)$. Hence, there exists a real function $f : X \times X \to \mathbf{R}$ such that $F(\gamma) = f(x, x')$, where $x = \gamma(0)$ and $x' = \gamma(1)$. Then, since $\hat{0} \times \hat{1} : \gamma \mapsto (\gamma(0), \gamma(1))$ is a subduction (art. 5.1), the map f is smooth. The application of the Chain-Homotopy operator gives

$$dF_\alpha = d[\mathcal{K}(\alpha)] = \hat{1}^*(\alpha) - \hat{0}^*(\alpha) - \mathcal{K}[d(\alpha)],$$

but $d\alpha = 0$, thus $dF_\alpha = \hat{1}^*(\alpha) - \hat{0}^*(\alpha)$. Restricting this identity to the subspace $\text{Paths}(X, x_0, \star)$ of paths with origin x_0, we get

$$dF_\alpha \restriction \text{Paths}(X, x_0, \star) = \hat{1}^*(\alpha), \text{ that is, } d(f_{x_0} \circ \hat{1}) = \hat{1}^*(df_{x_0}) = \hat{1}^*(\alpha),$$

where $f_{x_0}(x) = f(x_0, x)$. Then, since $\hat{1}$ is a subduction, $\alpha = df_{x_0}$ (art. 6.38). This is summarized by

$$f : x \mapsto \int_{x_0}^{x} \alpha + \text{cst},$$

as the general solution of the equation $\alpha = df$. □

6.90. Closed forms on contractible spaces are exact. As a corollary of the above proposition (art. 6.88), we get that every closed form on a contractible diffeological space X (art. 5.11) is exact. For finite dimensional manifolds, this theorem is a variant of the *Poincaré lemma.*

PROOF. Let ρ be a *deformation retraction* to the point $x_0 \in X$ (art. 5.13), that is, $\rho \in \text{Paths}(\mathcal{C}^\infty(X, X))$ such that $\rho(0) = [x \mapsto x_0]$ and $\rho(1) = \mathbf{1}_X$. With the same notation as above (art. 6.88), for every closed p-form α, $\rho(1)^*(\alpha) = \rho(0)^*(\alpha) + d\beta$, but $\rho(1)^*(\alpha) = \alpha$ and $\rho(0)^*(\alpha) = [x \mapsto x_0]^*(\alpha) = 0$. Therefore $\alpha = d\,\beta$. □

6.91. Closed forms on centered paths spaces. Let X be a diffeological space and $x_0 \in X$ be some point. The space $\text{Paths}(X, x_0, \star)$ of pointed paths (art. 5.1) is a particular example of contractible space; see Exercise 83, p. 111. Therefore, any closed form on X $\text{Paths}(X, x_0, \star)$ is exact (art. 6.90). It is, in particular, the role of the Chain-Homotopy operator (art. 6.83) to give a primitive for the pullback, by the end map, of a closed form on X.

6.92. The Poincaré lemma. Let X be a diffeological space. As a corollary of (art. 6.90), if X is locally contractible (art. 5.12), then every closed p-form α, $p > 0$, is locally exact. For each point $x \in X$ there exist a D-open neighborhood U of x (art. 2.8) and a $(p-1)$-form β, defined on U, such that $\alpha \restriction U = d\beta$. This proposition extends the Poincaré lemma about integration of closed forms on star-shaped domains.

6.93. The Poincaré lemma for manifold. As a corollary of (art. 6.92), since manifolds are locally diffeomorphic to $\mathbf{R}^n$ and since $\mathbf{R}^n$ is locally contractible, every closed form of a finite dimensional manifold is locally exact. This is not the case with diffeological spaces which are not locally contractible, for example the irrational tori, see Exercise 134, p. 272.

Exercises

✎ EXERCISE 121 (The Fubini-Study form is locally exact). Show that the Fubini-Study form ω, defined on the infinite projective space in Exercise 100, p. 160, is locally exact.

✎ EXERCISE 122 (Closed but not locally exact). Check that all forms on the irrational tori are closed but not locally exact; see Exercise 119, p. 202.

✎ EXERCISE 123 (A morphism from $H^\star_{dR}(X)$ to $H^\star_{dR}(\mathrm{Diff}(X))$). Let X be a diffeological space, and let $\mathrm{Diff}(X)$ be its group of diffeomorphisms equipped with the functional diffeology (art. 1.61). Let $\hat{x}$ be the orbit map of the point $x \in X$, that is, $\hat{x} : \mathrm{Diff}(X) \to X$ with $\hat{x}(\varphi) = \varphi(x)$. Check that the morphism $\hat{x}^*_{dR}$, from $H^\star_{dR}(X)$ to $H^\star_{dR}(\mathrm{Diff}(X))$, induced by $\hat{x}$ (art. 6.88), depends only on the connected component of x. For $X = S^1$, use the result of Exercise 133, p. 266, to show that this morphism is injective. What does this example make you think of?

CHAPTER 7

Diffeological Groups

The notion of *diffeological group* is the natural adaptation of the classical notion of Lie group to diffeology. Diffeological groups are groups equipped with a diffeology compatible with the group multiplication and the inversion (art. 7.1). This definition has been introduced originally by J.-M. Souriau [**Sou80**] as *groupes différentiels*. They appeared with their own axiomatics, and the main examples were the groups of diffeomorphisms of manifolds. Later they found their place in the general diffeological framework. However, the groups of diffeomorphisms of diffeological spaces, equipped with their functional diffeology (art. 1.61), are still the main examples of diffeological groups since every diffeological group is a subgroup of its group of diffeomorphisms (art. 7.7). The introduction of diffeological groups leads naturally to the notion of smooth actions on diffeological spaces as smooth homomorphisms (art. 7.4).

After the basic definitions, we focus our attention, in this chapter, on the momenta of diffeological groups, that is, the space of their left-invariant 1-forms (art. 7.12). The classical way for studying Lie groups uses the notion of Lie algebra. But, because diffeology is adapted better to covariant objects than to contravariant ones, the usual point of view is skipped. Lie algebra could be possibly defined afterwards, by duality with the space of momenta, or as the space of 1-parameter subgroups, or whatever else if necessary.

The notion of coadjoint orbits persists in the diffeological framework and plays an important role. They are introduced directly as the orbits of the group acting, by pushforward of the adjoint action, on its space of momenta. Because there is a natural equivalence between left-invariant and right-invariant 1-forms of diffeological groups (art. 7.19), which intertwines the coadjoint action, the choice of right- or left-invariant 1-forms is not relevant. The left momenta come naturally in the picture with the study of invariant differential forms, especially in symplectic diffeology (art. 9.9).

Basics of Diffeological Groups

After their introduction by J.-M. Souriau in 1979 as groupes différentiels [**Sou80**], *diffeological groups, and their relations with diffeological spaces, have been a little bit developed, in particular in* [**Sou84**], [**Don84**], [**DoIg85**], [**Igl85**], *but overall they remain* terra incognita. *For example, is a finite-dimensional diffeological group always the quotient of a Lie group?*

7.1. Diffeological groups. A *diffeological group* is a group G equipped with a compatible diffeology, that is, such that the multiplication and the inversion are smooth:

$$\mathrm{mul} = [(\mathsf{g}, \mathsf{g}') \mapsto \mathsf{g}\mathsf{g}'] \in \mathcal{C}^\infty(\mathsf{G} \times \mathsf{G}, \mathsf{G}) \text{ and } \mathrm{inv} = [\mathsf{g} \mapsto \mathsf{g}^{-1}] \in \mathcal{C}^\infty(\mathsf{G}, \mathsf{G}).$$

A diffeology $\mathcal{D}$ on a group G which satisfies these two conditions will be called a *group diffeology*. Let G and G′ be two diffeological groups, $\mathrm{Hom}^\infty(G, G')$ will denote the space of smooth homomorphisms from G to G′, and by $\mathrm{Isom}^\infty(G, G')$ the space of smooth isomorphisms from G to G′.

$$\begin{cases} \mathrm{Hom}^\infty(G, G') & = & \mathrm{Hom}(G, G') \cap \mathcal{C}^\infty(G, G'), \\ \mathrm{Isom}^\infty(G, G') & = & \mathrm{Isom}(G, G') \cap \mathrm{Diff}(G, G'). \end{cases}$$

Diffeological groups, together with smooth homomorphisms is a category we denoted by {D-Groups}. There is a natural forgetful functor from {D-Groups} to {Diffeology}.

Traditional *Lie groups* are diffeological groups G which are manifolds, modeled on a finite-dimensional real vector spaces (art. 4.1). A large literature exists on Lie groups, which occupy a big part of modern mathematics. A diffeological group which is a diffeological manifold modeled on a diffeological vector space can be regarded as *diffeological Lie group*. Very little has been written on diffeological Lie groups [**Les03**].

7.2. Subgroups of diffeological groups. Every subgroup H of a diffeological group G is canonically a diffeological group for the subset diffeology (art. 1.33). When we refer in the following to a subgroup of a diffeological group, it will be always equipped with the subset diffeology.

PROOF. Let $j : H \to G$ be the inclusion. The inclusion j is smooth, it is even an induction. Let mul_H and inv_H be the multiplication and the inversion of H. Since $\mathrm{mul}_H = \mathrm{mul} \circ (j \times j)$ and $\mathrm{inv}_H = \mathrm{inv} \circ j$ are composites of smooth maps, they are smooth. □

7.3. Quotients and extensions of diffeological groups. Let G be a diffeological group. Let $N \subset G$ be a normal subgroup, that is, $gNg^{-1} = N$ for all $g \in G$. Considering indifferently the right or the left coset,

(♣) the quotient group $H = G/N$ is canonically a diffeological group for the quotient diffeology (art. 1.50).

We represent this situation by the short exact sequence of homomorphisms

$$1_N \longrightarrow N \xrightarrow{\ j\ } G \xrightarrow{\ \pi\ } H = G/N \longrightarrow 1_H,$$

where $j : N \to G$ is the inclusion and $\pi : G \to H$ is the projection. By construction, j is an induction and π is a subduction. Actually, π is more than a subduction, it is a principal fibration, see (art. 8.15). Conversely, let H and N be just two groups. An *extension* of H by N is any group G with two morphisms $j : N \to G$ and $\pi : G \to H$ satisfying the diagram above; see for example [**Kir76**] for a full discussion. Now, if N and H are diffeological groups, a *diffeological* or *smooth extension* of H by N is an extension such that j is an induction and π is a subduction. As a special case of diffeological fibrations, *diffeological group extensions* form a subcategory of the category of principal fiber bundles (art. 8.12). A morphism of this category is a morphism of the category of principal diffeological fiber bundles which is also a group morphism (or conversely). The simplest case of such extensions is the direct product $G = H \times N$, with $(h, n) \cdot (h', n') = (hh', nn')$, $j : n \mapsto (1_H, n)$ and $\pi = \mathrm{pr}_1 : (h, n) \mapsto h$. Such an extension is said to be the *trivial extension* of H by N. An extension which is equivalent to a direct product is said to be *trivial*. When it comes to this question of trivial extension, the following two different points must be inspected.

a) The projection $\pi : G \to H = G/N$, as a principal diffeological fiber bundle, must be trivial, which is equivalent to the existence of a smooth section $\sigma : H \to G$ (art. 8.12, Note 2).

b) This equivalence must be an equivalence of extension, that is, σ is a group homomorphism from H to G, and $\sigma(H)$ is a subgroup of the centralizer of N, that is, $n\sigma(h) = \sigma(h)n$ for all $(h, n) \in H \times N$.

NOTE. There exist smooth extensions of H by N, on the product $H \times N$, which are not trivial as group extensions. The first example which comes in mind is the Heisenberg group, in the special case of central extensions. These extensions are given by smooth cocycles, the condition a) is fulfilled but not the condition b). It is why it is worth insisting on these two aspects.

PROOF. Let $\mathrm{mul}_{G/N}$ and $\mathrm{inv}_{G/N}$ be the multiplication and the inversion of G/N. Since $\mathrm{mul}_{G/N} \circ (\pi \times \pi) = \pi \circ \mathrm{mul}$ and since $\mathrm{inv}_{G/N} \circ \pi = \pi \circ \mathrm{inv}$, and π and $(\pi \times \pi)$ are subductions, $\mathrm{mul}_{G/N}$ and $\mathrm{inv}_{G/N}$ are smooth (art. 1.51). The other parts of the proposition are simple adaptations of classical results. □

7.4. Smooth action of a diffeological group. Let X be a diffeological space, and let G be a diffeological group. A *smooth action* of G on X is any smooth homomorphism $\rho : G \to \mathrm{Diff}(X)$, where Diff(X) is equipped with its functional diffeology of group of diffeomorphisms (art. 1.61). To simplify the notations, and where there is no risk of confusion, we shall sometimes denote g_X for $\rho(g)$.

NOTE 1. A homomorphism $\rho : G \to \mathrm{Diff}(X)$ is smooth if and only if the evaluation map $\mathrm{ev}_\rho : (x, g) \mapsto \rho(g)(x)$ defined on $X \times G$ into X is smooth.

NOTE 2. Let us recall that the kernel of a group action is always a normal subgroup, and that the action is said to be *effective* if its kernel is trivial, ρ is then a monomorphism. For every smooth action ρ of G on X, there exists a natural effective smooth action of $K = G/\ker(\rho)$ on X, where $\ker(\rho)$ is equipped with the subset diffeology and K with the quotient diffeology.

NOTE 3. In the paper [**IZK10**] we have shown that any effective smooth action $\rho : G \to \mathrm{Diff}(M)$, where G is a Lie group and M is a manifold — effective Hausdorff and second countable — is actually an induction, that is, a diffeomorphism onto its image. In other words, an effective smooth action, in the category of Hausdorff and second countable finite-dimensional manifolds, is equivalent to giving a subgroup of Diff(M) which, equipped with the induced functional diffeology, is a manifold.

PROOF. The second note is a direct application of (art. 7.3). So, let us prove the first note. Let us assume that ρ is smooth. The map ev_ρ splits into $(x, g) \mapsto (x, \rho(g)) \mapsto \rho(g)(x)$ from $X \times G$ to $X \times \mathrm{Diff}(G)$ to X. But, since ρ is smooth, $\mathrm{Ev}_\rho : (x, g) \mapsto (x, \rho(g))$ is smooth, and the second factor $(x, \phi) \mapsto \phi(x)$ from $X \times \mathrm{Diff}(G)$ to X is the evaluation map, which is smooth by the very definition of the functional diffeology (art. 1.57). Therefore, the map ev_ρ is smooth. Conversely, let us assume that the map ev_ρ is smooth. Now, ρ is smooth if and only if for every plot $P' : U' \to G$ the parametrization $r \mapsto \rho(P'(r'))$ is a plot of Diff(X), that is, if and only if for every plot $P'' : U'' \to X$ the parametrizations $(r', r'') \mapsto \rho(P'(r'))(P''(r''))$ and $(r', r'') \mapsto \rho(P'(r'))^{-1}(P''(r''))$ are plots of X. By hypothesis, for every plot $P : U \to X \times G$, the parametrization $r \mapsto \rho(P(r))(Q(r))$ is a plot of X. This, applied to $U = U' \times U''$ and to the plot $P : r = (r', r'') \mapsto P(r) = (P'(r'), P''(r''))$ gives exactly that $(r', r'') \mapsto \rho(P'(r'))(P''(r''))$ is a plot of X. Applied to the plot $P : r = (r', r'') \mapsto P(r) = (P'(r')^{-1}, P''(r''))$ (because the inversion in G is smooth), we get that $(r', r'') \mapsto \rho(P'(r')^{-1})(P''(r'')) = \rho(P'(r'))^{-1}(P''(r''))$ is a plot of X. Therefore, ρ is smooth. ☐

7.5. Orbit map. Let X be a diffeological space. Let G be a diffeological group, and let $g \mapsto g_X$ be a smooth action (art. 7.4) of G on X. For all points $x \in X$, we denote by $R(x)$ the *orbit map*

$$R(x) : G \to X \quad \text{with} \quad R(x)(g) = g_X(x).$$

As an immediate consequence of the definition of smooth actions, the orbit maps are smooth maps from G to X.

NOTE. Let M be a manifold and G be a Lie group, both Hausdorff and second countable. Let $\rho : G \to \mathrm{Diff}(M)$ be a smooth action and $m \in M$. Then, the orbit map $R(m) : G \to M$ is a strict map (art. 1.54) [**IZK10**], that is, the projection of $R(m)$, from G/G_m to M, where G_m is the stabilizer of m, is an induction. Consequently, every orbit of G in M is a submanifold, that is, for all $m \in M$, the orbit $G \cdot m = \{\rho(g)(m) \mid g \in G\}$, equipped with the subset diffeology, is a manifold (art. 4.4).

7.6. Left and right multiplications. Let G be a diffeological group, and let $g \in G$. We denote by $L(g)$ the left multiplication by g,

$$L(g) : k \mapsto gk, \; L(g) \in \mathcal{C}^\infty(G).$$

Left multiplication is a smooth effective action of G onto itself, that is, $L \in \mathrm{Hom}^\infty(G, \mathrm{Diff}(G))$. Then, we denote by $R(g)$, the right multiplication by $g \in G$,

$$R(g) : k \mapsto kg, \; R(g) \in \mathcal{C}^\infty(G).$$

The right multiplication is a smooth effective *antiaction* of G onto itself, that is, $R(gg') = R(g') \circ R(g)$. The map R is a smooth antihomomorphism from G to Diff(G), which can also be expressed as $R \circ \mathrm{inv} \in \mathrm{Hom}^\infty(G, \mathrm{Diff}(G))$. We denote by $\mathrm{Ad}(g)$ the *adjoint action* of G onto itself,

$$\mathrm{Ad}(g) : k \mapsto gkg^{-1}, \; \mathrm{Ad}(g) \in \mathcal{C}^\infty(G).$$

The adjoint action is smooth, $\mathrm{Ad} \in \mathrm{Hom}^\infty(G, \mathrm{Diff}(G))$. Actually

$$\mathrm{Ad}(g) = L(g) \circ R(g^{-1}) = R(g^{-1}) \circ L(g).$$

Note that the adjoint action is not necessarily effective. The kernel of the adjoint action is called the *center* of G, denoted sometimes by Z(G). It is the subgroup of G whose elements commute with every element of G.

7.7. Left and right multiplications are embeddings. Let G be a diffeological group. Left and right multiplications (art. 7.6) are embeddings from G to Diff(G), where Diff(G) is equipped with the functional diffeology (art. 1.57). Explicitly, left and right multiplications are inductions such that the pullback of the D-topology of the functional diffeology of Diff(G) coincides with the D-topology of G (art. 2.14).

NOTE. In particular, this shows that every diffeological group is (equivalent to) a subgroup of a group of diffeomorphisms, for the functional diffeology.

PROOF. Let us prove this proposition for the left multiplication. And first, let us prove that L is an induction. We know already that $L : G \to \mathrm{Diff}(G)$ is injective and smooth (art. 7.6). So, let $P : U \to G$ be a parametrization such that $L \circ P$ is a plot of Diff(G). Thus, for every plot $Q : V \to G$, the parametrization $(r, s) \mapsto L(P(r))(Q(s)) = P(r)Q(s)$ is a plot of G. But since the multiplication and the inversion are smooth, the parametrization $(r, s) \mapsto (P(r)Q(s), Q(s)^{-1}) \mapsto [P(r)Q(s)]Q(s)^{-1} = P(r)$ is a plot of G, that is, P is a plot of G. Therefore, L is an induction. Next, let $\mathcal{O}$ be D-open in G, the subset

$$\Omega = \{f \in \mathrm{Diff}(G) \mid f(1_G) \in \mathcal{O}\}$$

is D-open. Indeed, let P be a plot of Diff(G), $P^{-1}(\Omega)$ is the subset of $r \in \mathrm{def}(P)$ such that $P(r)(1_G) \in \mathcal{O}$, but $P(r)(1_G) = (R(1_G) \circ P)(r)$, where $R(1_G)$ is the orbit map of the point 1_G for the left multiplication (art. 7.5), thus $P^{-1}(\Omega) = (R(1_G) \circ P)^{-1}(\mathcal{O})$. Since $R(1_G)$ and P are smooth, thus D-continuous, and since $\mathcal{O}$ is D-open, $P^{-1}(\Omega)$ is open and thus Ω is D-open in Diff(G). Now, $L^{-1}(\Omega)$ is the set of $g \in G$ such that $L(g)(1_G) \in \mathcal{O}$, that is, $L^{-1}(\Omega) = \mathcal{O}$. □

7.8. Transitivity and homogeneity. Let X be a diffeological space, and let G be a diffeological group. Let ρ be smooth action of G on X, that is, a homomorphism from G to Diff(X). The action of G is said to be *transitive* on X if the *evaluation map*

$$\mathrm{Ev}_\rho : X \times G \to X \times X \quad \text{defined by} \quad \mathrm{Ev}_\rho(x, g) = (x, \rho(g)(x))$$

is surjective, that is, if for any $x, x' \in X$ there exists $g \in G$ such that $x' = \rho(g)(x)$. We say that X is *homogeneous* for the action ρ of G if the evaluation map Ev_ρ is a subduction. Now, the space X is a homogeneous space of G (for ρ) if and only if the action ρ is smooth and for some $x_0 \in X$ the *orbit map*

$$R(x_0) : G \to X \quad \text{defined by} \quad R(x_0)(g) = \rho(g)(x_0)$$

is a subduction (if it is true for some point, then it will be true for every point). Moreover, the subduction $R(x_0)$ is a local subduction (art. 2.16). In this case we also say that G *generates* X, or is a *generating group* for X.

PROOF. First of all, if Ev_ρ is a subduction, then ρ is smooth, since Ev_ρ is smooth and $\mathrm{ev}_\rho = \mathrm{pr}_2 \circ \mathrm{Ev}_\rho$ (art. 7.4, Note 1). Then, $R(x_0)$ is the restriction of Ev_ρ to $\{x_0\} \times G$ with values $\{x_0\} \times X$. Thus, every plot $r \mapsto (x_0, Q(r))$ lifts locally into a plot $r \mapsto (x_0(r) = x_0, \gamma(r))$ such that $\rho(\gamma(r)(x_0)) = Q(r)$. Hence, $R(x_0)$ is a subduction. Now, let us assume that ρ is smooth and $R(x_0)$ is a subduction. First of all, if ev_ρ is smooth so is Ev_ρ. Next, let $P : r \mapsto (P'(r), P''(r))$ be a plot of $X \times X$, defined on a domain U. By hypothesis, for every r_0 in U, there exist two open neighborhoods V' and V'' of r_0, there exist two plots $\gamma' : V' \to G$ and $\gamma'' : V'' \to G$ such that $P'(r) = \rho(\gamma'(r))(x_0)$ and $P''(r) = \rho(\gamma''(r))(x_0)$. Thus, on $V = V' \cap V''$, we have $P''(r) = \rho(\gamma''(r))[\rho(\gamma(r')^{-1})(P'(r)]$, that is, $P''(r) = \rho(\gamma(r))(P'(r))$, with $\gamma(r) = \gamma''(r)\gamma'(r)^{-1}$. Hence, there exists a plot $Q : r \mapsto (P'(r), \gamma(r))$ of $X \times G$, defined on V, such that $P \restriction V = \mathrm{Ev}_\rho \circ Q$. Therefore, Ev_ρ is a subduction. Then, since G is transitive, if this is true for one point x_0, by composition with some element of G, it will be true for every point. Finally, pick a plot $P : U \to X$, let $r_0 \in U$, and let $g_0 \in G$ such that $R(x_0)(g_0) = P(r_0)$, that is, $\rho(g_0)(x_0) = P(r_0)$. The plot P lifts locally, on an open neighborhood of r_0, into a plot Q of G, $R(x_0) \circ Q =_{\mathrm{loc}} P$. Let $g_0' = Q(r_0)$, the parametrization $r \mapsto g_0(g_0')^{-1}Q(r)$ is a local lift of P passing through g_0 at r_0. Therefore, $R(x_0)$ is a local subduction. □

7.9. Connected homogeneous spaces. Let X be a connected diffeological space. If X is homogeneous for an action ρ of a diffeological group G, then it is homogeneous for its identity component $G^\circ = \mathrm{comp}(\mathbf{1}_G)$.

PROOF. Let $x_0, x_1 \in X$ and $t \mapsto x_t$ be a smooth path connecting x_0 to x_1. Since $\mathrm{R}(x_0) : g \mapsto \rho(g)(x_0)$ is a subduction from G onto X, there exist a family of intervals $\{]a_i, b_i[\}_{i\in I}$ covering $[0,1]$ and a family of plots $t \mapsto g_i(t)$ in G, $t \in I$, such that $\rho(g_i(t))(x_0) = x_t$. Since $[0,1]$ is compact, we can find a family $\{]a'_n, b'_n[\}_{n=1}^N$ such that each $]a'_n, b'_n[$ is contained in some $]a_i, b_i[$ (chosen once and for all), and such that $]a'_n, b'_n[\cap]a'_m, b'_m[\neq \varnothing$, with $1 \leq n \leq N-1$ and $n > m$, if and only if $m = n+1$. We denote by g_n the restriction of g_i to $]a'_n, b'_n[$. Let $t_0 = 0$, $t_N = 1$, and for each n such that $1 \leq n \leq N-1$, we chose $t_n \in]a'_n, b'_n[\cap]a'_{n+1}, b'_{n+1}[$. So, we have $\rho(g_n(t_n))(x_0) = \rho(g_{n+1}(t_n))(x_0) = x_{t_n}$, that is, $g_{n+1}(t_n)^{-1} g_n(t_n) \in \mathrm{St}_G(x_0)$. Now, let us define $g'_n(t) = g_n(t)k_n$ with $k_n = g_n(t_{n-1})^{-1} g_{n-1}(t_{n-1})k_{n-1}$, for $2 \leq n \leq N$ and $k_1 = g_1(t_0)^{-1}$. So, all the k_n belong to $\mathrm{St}_G(x_0)$, and therefore $\rho(g'_n(t))(x_0) = \rho(g_n(t))(x_0) = x_t$. Moreover, $g'_n(t_{n-1}) = g_n(t_{n-1})k_n = g_n(t_{n-1})g_n(t_{n-1})^{-1}g_{n-1}(t_{n-1})k_{n-1} = g_{n-1}(t_{n-1})k_{n-1} = g'_{n-1}(t_{n-1})$. Thus, the family $\{g'_i\}_{i=1}^N$ connects $\mathbf{1}_G = g'_1(0)$ to a diffeomorphism $g = g'_N(1)$ which satisfies $\rho(g)(x_0) = x_1$. Therefore, the connected component G° is transitive on X, and since the projection $\mathrm{R}(x_0)$ is a local subduction, its restriction to G° is a local subduction too. ☐

7.10. Covering diffeological groups. Let $\mathrm{pr} : \hat{G} \to G$ be a subduction, where $\hat{G}$ and G are two diffeological groups. We say that pr is a *group covering* if pr is a homomorphism and the kernel $K = \mathrm{pr}^{-1}(\mathbf{1}_G)$ is discrete. Let us recall that *discrete* means that the plots — here the plots for the subset diffeology — are locally constant (art. 1.20). Let G be a connected diffeological group, its universal covering $\tilde{G}$ (art. 8.26) has a natural structure of diffeological group such that the subduction $\pi : \tilde{G} \to G$ is a homomorphism. The first homotopy group $\pi_1(G) = \ker(\pi)$ (art. 5.15) is a discrete normal subgroup of $\tilde{G}$, and π is a group covering. Any connected covering $\mathrm{pr} : \hat{G} \to G$ (art. 8.24) is the quotient of the universal covering by a subgroup K of $\pi_1(G)$. If the subgroup K is normal then pr is also a group covering.

PROOF. This property has been established originally in [**Sou84**], [**Don84**], inside the strict framework of diffeological groups. But let us show how the general construction given in [**Igl85**] or in (art. 8.24) applies to the special case of diffeological groups. Let X be a connected diffeological space, let x_0 be a point of X, chosen as a basepoint. Let $\mathrm{Paths}(X, x_0, \star)$ be the space of paths starting at x_0. First of all, the end map $\hat{1} : p \mapsto p(1)$, defined on $\mathrm{Paths}(X, x_0, \star)$, is a subduction (art. 5.6). The quotient of $\mathrm{Paths}(X, x_0, \star)$ by the fixed ends homotopy equivalence relation is exactly the universal covering, pointed by the constant map $\hat{x}_0 : t \mapsto x_0$, over the pointed space (X, x_0) (art. 8.26). The fiber over x_0 is the homotopy group $\pi_1(X, x_0)$. Now, if $X = G$ we choose the identity $\mathbf{1}_G$ as base point. Thus, the multiplication of paths $(p, p') \mapsto [t \mapsto p(t) \cdot p'(t)]$ defines on $\tilde{G}$ a group multiplication such that the projection $\pi : \tilde{G} \to G$, defined by $\pi(\mathrm{class}(p)) = \hat{1}(p)$, is a homomorphism. The kernel of this morphism is clearly the fiber over $\mathbf{1}_G$, that is, $\pi_1(G)$. The kernel of a homomorphism is always a normal subgroup. Then, since π is a covering, $\pi^{-1}(\mathbf{1}_G)$ is discrete (art. 8.24). ☐

7.11. Covering smooth actions. Let X be a connected diffeological space. Let G be a connected diffeological group. Let $\rho : G \to \mathrm{Diff}(X)$ be a smooth action of G on X, hence ρ takes its values in the identity component of Diff(X). Then, there exists a unique smooth action $\tilde{\rho}$, of the universal covering $\tilde{G}$ of G on the universal covering $\tilde{X}$ of X, covering ρ. Precisely, let us denote briefly by $\mathbf{D}_X$ the group Diff(X), by $\mathbf{D}_X^\bullet$ the identity component and by $\tilde{\mathbf{D}}_X^\bullet$ its universal covering. Let $\pi_G : \tilde{G} \to G$ and $\pi_D : \tilde{\mathbf{D}}_X^\bullet \to \mathbf{D}_X^\bullet$ be the projections, $\tilde{\rho} : \tilde{G} \to \tilde{\mathbf{D}}_X^\bullet$ is a smooth homomorphism such that the following diagram commutes.

$$\begin{array}{ccc} \tilde{G} & \xrightarrow{\ \tilde{\rho}\ } & \tilde{\mathbf{D}}_X^\bullet \\ {\scriptstyle \pi_G}\downarrow & & \downarrow{\scriptstyle \pi_D} \\ G & \xrightarrow[\ \rho\ ]{} & \mathbf{D}_X^\bullet \end{array}$$

PROOF. The map $\rho \circ \pi_G$ is smooth and $\tilde{G}$ is simply connected. Thanks to the monodromy theorem (art. 8.25), there exists a unique lift $\tilde{\rho}$ of $\rho \circ \pi$ mapping the identity of $\tilde{G}$ to the identity of $\tilde{\mathbf{D}}_X^\bullet$. This lift is a homomorphism because its restriction to $\ker(\pi_G)$ and its projection ρ are both homomorphisms. □

Exercises

✎ EXERCISE 124 (Subgroups of $\mathbf{R}$). Show that every strict subgroup K of $\mathbf{R}$, that is, $K \subset \mathbf{R}$ and $K \neq \mathbf{R}$, is diffeologically discrete and that the only embedded strict subgroups are the groups $a\mathbf{Z} \subset \mathbf{R}$, where $a \in \mathbf{R}$.

✎ EXERCISE 125 (Diagonal diffeomorphisms). Let X be a diffeological space. Let $N > 0$ be any integer. Let Diff(X) and $\mathrm{Diff}(X^N)$ be equipped with the functional diffeology (art. 1.61). Show that the diagonal injection $\Delta : \mathrm{Diff}(X) \to \mathrm{Diff}(X^N)$, defined by $\Delta(\varphi)(x_1, \ldots, x_N) = (\varphi(x_1), \ldots, \varphi(x_N))$, is an induction.

✎ EXERCISE 126 (The Hilbert sphere is homogeneous). Let $\mathcal{H}_\mathbf{C}$ be the standard Hilbert space, described in (art. 3.15). Let $\mathcal{S}_\mathbf{C}$ be the Hilbert unit sphere (art. 4.9) and $\mathcal{P}_\mathbf{C}$ be the Hilbert projective space; see (art. 4.11) and Exercise 77, p. 93. Let $\mathbf{U}(\mathcal{H}_\mathbf{C})$ be the unitary group of $\mathcal{H}_\mathbf{C}$, equipped with the functional diffeology (art. 3.12). Using the fact that a subspace $E \subset \mathcal{H}_\mathbf{C}$ and its orthogonal $E^\perp$ are supplementary [**Bou55**], show that $\mathcal{S}_\mathbf{C}$ is homogeneous under $\mathbf{U}(\mathcal{H}_\mathbf{C})$. Deduce that $\mathcal{P}_\mathbf{C}$ also is homogeneous.

The Spaces of Momenta

Diffeological groups are a huge category of groups, not very well known and with much still to be explored. Their left (or right) invariant differential 1*-forms, called* momenta, *and their orbits under the coadjoint action are key concepts, at a classical level in symplectic mechanics, but also for the development of* symplectic diffeology. *In this section we introduce these objects and we give some of their properties. They will be used further, in Chapter 9. In particular, the notion of coadjoint orbit, more precisely* (Γ, θ)*-coadjoint orbit, plays a crucial role in the construction of the moment maps, for an arbitrary action of a diffeological group* *preserving a closed* 2-*form* (art. 9.9), (art. 9.10). *It is in this sense, for example, that "every symplectic manifold is a coadjoint orbit"* (art. 9.23).

7.12. Momenta of a diffeological group. We call *left momentum* — or simply *momentum* — of a diffeological group G any 1-form on G, invariant by the left action of G onto itself. We denote by $\mathcal{G}^*$ the *space of momenta* of G. The space of momenta of a diffeological group is naturally a diffeological vector space, equipped with the functional diffeology,

$$\mathcal{G}^* = \{\alpha \in \Omega^1(G) \mid \text{for all } g \in G, \ L(g)^*(\alpha) = \alpha\}.$$

Note that, in spite of what the notation $\mathcal{G}^*$ suggests, the space of momenta of a diffeological group is not defined by some duality. This notation is chosen here just to remind ourselves of the connection with the dual of the Lie algebra in the classical case of Lie groups.

7.13. Momenta and connectedness. Let G be a diffeological group. Let G° be its identity component, that is, $G^\circ = \mathrm{comp}(\mathbf{1}_G) \subset G$. Then, the pullback $j^* : \mathcal{G}^* \to \mathcal{G}^{\circ *}$ of the injection $j : G^\circ \to G$ is an isomorphism. This property is quite natural but deserves to be checked in the context of diffeological groups.

NOTE. Said differently, the space of momenta of a connected diffeological group, or any of its extensions by a discrete group, coincide. In particular, the only momentum of a discrete group is the momentum zero.

PROOF. Let us check first the injectivity. Let $\alpha \in \mathcal{G}^*$ such that $j^*(\alpha) = 0$, and let $P : U \to G$ be a plot. Let $r_0 \in U$, and let $B \subset U$ be a small open ball centered at r_0, and $g_0 = P(r_0)$. Since B is connected, since $L(g_0^{-1}) \circ P(r_0) = \mathbf{1}_G$, and thanks to the smoothness of group operations, the parametrization $Q = [L(g_0^{-1}) \circ P] \upharpoonright B$ is a plot of G°. Since $j^*(\alpha) = 0$, $\alpha(Q) = 0$. But, $\alpha(Q) = \alpha(L(g_0^{-1}) \circ (P \upharpoonright B)) = L(g_0^{-1})^*(\alpha)(P \upharpoonright B) = \alpha(P \upharpoonright B)$. Thus, $\alpha(P \upharpoonright B) = 0$. Since α vanishes locally at each point of U, $\alpha = 0$, and j^* is injective. Now, let us prove the surjectivity. Let $\alpha \in \mathcal{G}^{\circ *}$. For any component G_i of G, let us choose an element $g_i \in G_i$, and the identity for the identity component. Let $P : U \to G$ be a plot, and let us assume that U is connected. So, $P(U)$ is contained in one connected component of G, let us say the component G_i. Let us define then, $\bar{\alpha}(P) = \alpha(R(g_i^{-1}) \circ P)$. Since $R(g_i^{-1}) \circ P(r) \in G^\circ$ for all $r \in U$, this is well defined. Now, since any plot is the sum of its restrictions on the components of its domain, the map $\bar{\alpha}$ extends naturally to every plot of G. Now, let $P : U \to G$ be a plot, let V be a domain, and let $F \in \mathcal{C}^\infty(V, U)$. Let $s_0 \in V$, let V_0 be the component of s_0 in V, let $r_0 = F(s_0)$ and U_0 be the component of r_0 in U. Let G_i be the component of $P \circ F(s_0) = P(r_0)$ in G. We have $\bar{\alpha}((P \circ F) \upharpoonright V_0) = \bar{\alpha}((P \upharpoonright U_0) \circ (F \upharpoonright V_0)) = \alpha(R(g_i^{-1}) \circ (P \upharpoonright U_0) \circ (F \upharpoonright V_0)) = \alpha([R(g_i^{-1}) \circ (P \upharpoonright U_0)] \circ (F \upharpoonright V_0)) = (F \upharpoonright V_0)^*[\alpha(R(g_i^{-1}) \circ (P \upharpoonright U_0)] = (F \upharpoonright V_0)^*[\bar{\alpha}(P \upharpoonright U_0)]$. Hence, $\bar{\alpha}(F \circ P) =_{\mathrm{loc}} F^*(\bar{\alpha}(P))$, and if it is satisfied locally, then it is satisfied globally, thus $\bar{\alpha}(F \circ P) = F^*(\bar{\alpha}(P))$. The map $\bar{\alpha}$ is a well defined differential 1-form on G. Let us check now that $\bar{\alpha}$ is invariant by left multiplication. Let $g \in G$, let $P : U \to G$ be a plot, let $r_0 \in U$, let U_0 be the component of r_0 in U, let G_i be the component of $P(r_0)$ in G, so $P(U_0) \subset G_i$. We have $L(g)^*(\bar{\alpha}(P \upharpoonright U_0)) = \bar{\alpha}(L(g) \circ (P \upharpoonright U_0)) = \alpha(R(g_i^{-1}) \circ L(g) \circ (P \upharpoonright U_0)) = \alpha(L(g) \circ R(g_i^{-1}) \circ (P \upharpoonright U_0)) = [L(g)^*(\alpha)](R(g_i^{-1}) \circ (P \upharpoonright U_0)) = \alpha(R(g_i^{-1}) \circ (P \upharpoonright U_0)) = \bar{\alpha}(P \upharpoonright U_0)$. So, $L(g)^*(\bar{\alpha})(P) =_{\mathrm{loc}} \bar{\alpha}(P)$, and hence globally. Thus, $L(g)^*(\bar{\alpha}) = \bar{\alpha}$. Therefore, $\bar{\alpha}$ is an element of $\mathcal{G}^*$ which coincides with α on G°. Since the pullback is a smooth operation, j^* is a smooth linear bijection; on the other hand, since the multiplication by the basepoints $g_i \in G_i$ is smooth, the process of extension $\alpha \mapsto \bar{\alpha}$ is smooth and j^* is an isomorphism of diffeological vector spaces. □

7.14. Momenta of coverings of diffeological groups. Let G be a diffeological group, let $\mathrm{pr} : \widehat{G} \to G$ be some group covering, see (art. 7.10). Let $\mathcal{G}^*$ and $\widehat{\mathcal{G}}^*$ be the spaces of momenta of G and $\widehat{G}$. Then, the pullback $\mathrm{pr}^* : \mathcal{G}^* \to \widehat{\mathcal{G}}^*$ is an isomorphism.

PROOF. Thanks to (art. 7.13), it is sufficient to assume that $\widehat{G}$ and G are connected, and thanks to (art. 7.10), it is sufficient to check it for the universal covering $\pi : \tilde{G} \to G$. Now, π^* is obviously linear, let us show that π^* is surjective. Let $\tilde{\alpha} \in \tilde{\mathcal{G}}^*$. The group G is isomorphic to $\tilde{G}/\pi_1(G)$, with respect to the left action of $\pi_1(G)$, that is, $\tilde{g} \sim k\tilde{g}$, for all $k \in \pi_1(G)$. Now, let $\tilde{\alpha} \in \tilde{\mathcal{G}}^*$, $\tilde{\alpha}$ is left invariant by $\tilde{G}$, thus by $\pi_1(G)$, that is, for all $k \in \pi_1(G)$, $L(k)^*(\tilde{\alpha}) = \tilde{\alpha}$. But, since $\pi_1(G) = \ker(\pi)$ is discrete, this is sufficient for the existence of a 1-form α on G such that $\tilde{\alpha} = \pi^*(\alpha)$ (art. 6.38), and the map $\tilde{\alpha} = \alpha$ is smooth. Now, let $\tilde{g} \in \tilde{G}$ and $g = \pi(\tilde{g})$. Since π is a homomorphism, $\pi \circ L(\tilde{g}) = L(g) \circ \pi$. Thus, on the one hand we have $L(\tilde{g})^*(\tilde{\alpha}) = L(\tilde{g})^*(\pi^*(\alpha)) = (\pi \circ L(\tilde{g}))^*(\alpha) = (L(g) \circ \pi)^*(\alpha) = \pi^*(L(g)^*(\alpha))$, and on the other hand $L(\tilde{g})^*(\tilde{\alpha}) = \tilde{\alpha} = \pi^*(\alpha)$. Hence, $\pi^*(L(g)^*(\alpha)) = \pi^*(\alpha)$. But since π is a subduction, $L(g)^*(\alpha) = \alpha$, $\alpha \in \mathcal{G}^*$ and the map π^* is surjective. Next, let $\tilde{\alpha}$ and $\tilde{\beta}$ be such that $\pi^*(\tilde{\alpha}) = \pi^*(\tilde{\beta})$. Since π is a subduction, $\tilde{\alpha} = \tilde{\beta}$, π^* is injective and thus bijective. Finally, $\pi^* : \mathcal{G}^* \to \tilde{\mathcal{G}}^*$ is a smooth linear bijection, its inverse is smooth, therefore it is an isomorphism of diffeological vector spaces. □

7.15. Linear coadjoint action and coadjoint orbits. Let G be a diffeological group, and let $\mathcal{G}^*$ be the space of its momenta. The pushforward $\mathrm{Ad}(g)_*(\alpha)$ of a momentum $\alpha \in \mathcal{G}^*$, by the adjoint action of any element g of G, is again a momentum of G, that is, again a left-invariant 1-form. This defines a linear smooth action of G on $\mathcal{G}^*$ called *coadjoint action*, and denoted by Ad_*,

$$\mathrm{Ad}_* : (g, \alpha) \mapsto \mathrm{Ad}(g)_*(\alpha) = \mathrm{Ad}(g^{-1})^*(\alpha).$$

We check immediately that for all g, g' in G, $\mathrm{Ad}_*(gg') = \mathrm{Ad}_*(g) \circ \mathrm{Ad}_*(g')$, and that $\mathrm{Ad}_*(g)$ is linear. Note that, since α is left-invariant, $\mathrm{Ad}_*(g)(\alpha) = R(g)^*(\alpha)$. The orbit of α by G is by definition a *coadjoint orbit* of G, it will be denoted by

$$\mathcal{O}_\alpha \text{ or } \mathrm{Ad}_*(G)(\alpha) = \{\mathrm{Ad}_*(g)(\alpha) \mid g \in G\}.$$

The orbit $\mathcal{O}_\alpha$ can be regarded as a subset of $\mathcal{G}^*$, but also as the quotient of the group G by the stabilizer of the momentum α,

$$\mathcal{O}_\alpha \simeq G/\mathrm{St}_G(\alpha), \text{ with } \mathrm{St}_G(\alpha) = \{g \in G \mid \mathrm{Ad}(g)_*(\alpha) = \alpha\}.$$

NOTE. The orbit $\mathcal{O}_\alpha$ can be equipped with the subset diffeology of the functional diffeology of $\mathcal{G}^*$ or with the quotient diffeology of G. For a Hausdorff second countable Lie group it has been proved that these diffeologies coincide [**IZK10**], but there is no reason *a priori* that these two diffeologies coincide in general. It could be interesting, however, to understand in which conditions they do.

7.16. Affine coadjoint actions and (Γ,θ)-coadjoint orbits. Let G be a diffeological group, and let $\mathcal{G}^*$ be the space of its momenta. Let $\Gamma \subset \mathcal{G}^*$ be a subgroup of $(\mathcal{G}^*,+)$, invariant by the coadjoint action Ad_*, that is,

$$\mathrm{Ad}_*(g)(\Gamma) \subset \Gamma,$$

for all $g \in G$. Then, the coadjoint action of G on $\mathcal{G}^*$ projects to the quotient $\mathcal{G}^*/\Gamma$, regarded as an Abelian group, as a smooth action denoted by Ad_*^Γ,

$$\mathrm{Ad}_*^\Gamma(g)(\tau) = \mathrm{class}(\mathrm{Ad}_*(g)(\mu)) \quad \text{with} \quad \tau = \mathrm{class}(\mu) \in \mathcal{G}^*/\Gamma,$$

for all $g \in G$ and all $\tau \in \mathcal{G}^*/\Gamma$. Now, let θ be a smooth map, from G to the space $\mathcal{G}^*/\Gamma$, such that

$$\theta(gg') = \mathrm{Ad}_*^\Gamma(g)(\theta(g')) + \theta(g),$$

for all $g, g' \in G$. Such maps are formally known, in the literature, as twisted 1-*cocycles* of G with values in $\mathcal{G}^*/\Gamma$ [**Kir76**]. We shall call them cocycles of G, with values in $\mathcal{G}^*/\Gamma$, or simply $(\mathcal{G}^*/\Gamma)$-cocycles. A cocycle θ is a *coboundary* if there exists a constant $c \in \mathcal{G}^*/\Gamma$, such that $\theta = \Delta c$, with

$$\Delta c : g \mapsto \mathrm{Ad}_*^\Gamma(g)(c) - c.$$

Cocycles modulo coboundaries define a cohomology group denoted by $H^1(G, \mathcal{G}^*/\Gamma)$. Every such cocycle θ defines a new action of G on $\mathcal{G}^*/\Gamma$ by

$$\mathrm{Ad}_*^{\Gamma,\theta} : (g, \tau) \mapsto \mathrm{Ad}_*^\Gamma(g)(\tau) + \theta(g).$$

The orbits for these actions will be called the (Γ,θ)-*coadjoint orbits* of G. If $\Gamma = \{0\}$, we shall call them simply θ-coadjoint orbits. If $\theta = 0$, we shall call them simply Γ-coadjoint orbits. And, if $\Gamma = \{0\}$ and $\theta = 0$, we find again the ordinary coadjoint orbits defined in (art. 7.15).

7.17. Closed momenta of a diffeological group. Let G be a diffeological group, and let $\mathcal{G}^*$ be its space of momenta. Let us denote by Z the subset of closed momenta of G, and by B the subset of exact momenta of G, that is,

$$Z = Z^1_{\mathrm{dR}}(G) \cap \mathcal{G}^* \quad \text{and} \quad B = B^1_{\mathrm{dR}}(G) \cap \mathcal{G}^*.$$

1. Let us assume that G is connected, and let $\tilde{G}$ be its universal covering. By factorization, the Chain-Homotopy operator defines a canonical De Rham isomorphism k, from the space of closed momenta Z to the vector space $\mathrm{Hom}^\infty(\tilde{G}, \mathbf{R})$, that is, for all $\zeta \in Z$,

$$k(\zeta) = [\tilde{g} \mapsto \mathcal{K}\zeta(p)], \quad \text{where} \quad \mathcal{K}\zeta(p) = \int_p \zeta \quad \text{and} \quad \tilde{g} = \mathrm{class}(p).$$

Here, we have denoted by $\mathrm{class}(p)$ the fixed-ends homotopy class of the path $p \in \mathrm{Paths}(G, 1_G)$. The subspace of exact momenta B identifies, through the isomorphism k, with the subspace $\mathrm{Hom}^\infty(G, \mathbf{R})$,

$$Z \simeq \mathrm{Hom}^\infty(\tilde{G}, \mathbf{R}) \quad \text{and} \quad B \simeq \mathrm{Hom}^\infty(G, \mathbf{R}).$$

2. Let G be a diffeological group, connected or not. Let $\zeta \in \mathcal{G}^*$. If ζ is closed, then ζ is Ad_* invariant. For all $\zeta \in \mathcal{G}^*$,

$$d\zeta = 0 \quad \text{implies} \quad \mathrm{Ad}_*(g)(\zeta) = \zeta, \text{ for all } g \in G.$$

NOTE. Every homomorphism from a group G to an Abelian group factorizes through the *Abelianized group* $G^{Ab} = G/[G, G]$, where $[G, G]$ is the normal subgroup of the commutators of G. Thus actually, $Z \simeq \mathrm{Hom}^\infty(\tilde{G}^{Ab}, \mathbf{R})$ and $B \simeq \mathrm{Hom}^\infty(G^{Ab}, \mathbf{R})$.

PROOF. 1. Let $\pi : \tilde{G} \to G$ be the universal covering defined in (art. 7.10). Since $\tilde{G}$ is simply connected, every closed 1-form is exact (art. 6.89). Thus, for every $\zeta \in Z$, the pullback $\pi^*(\zeta)$ is exact. Let F be a primitive of $\pi^*(\alpha)$, that is, $dF = \pi^*(\alpha)$. We can even set F uniquely by choosing $F(1_{\tilde{G}}) = 0$. Actually F is defined by integrating the form ζ along the paths starting at the identity, that is, $F = k(\zeta)$. Since α is left invariant and since the projection π commutes with the left actions, on G and $\tilde{G}$, $\pi^*(\alpha)$ is left invariant. So, for every $\tilde{g} \in \tilde{G}$, $d[F \circ L(\tilde{g})] = dF$. Since $\tilde{G}$ is connected, for every $\tilde{g}$, $\tilde{g}'$ in $\tilde{G}$, $F(\tilde{g}\tilde{g}') = F(\tilde{g}') + f(\tilde{g})$, where f is a smooth real function. But since $F(1_G) = 0$, $f(\tilde{g}) = F(\tilde{g})$, and F is a smooth homomorphism from $\tilde{G}$ to $\mathbf{R}$. So, for every closed momentum $\zeta \in Z$, there exists a unique homomorphism $F \in \mathrm{Hom}^\infty(\tilde{G}, \mathbf{R})$ such that $\zeta = \pi_*(dF)$. The homomorphism k is thus injective, and it is obviously surjective. Now, if ζ is exact, that is, if $\zeta = df$, then $F = \pi^*(f)$. So, $k(B) = \pi^*(\mathrm{Hom}^\infty(G, \mathbf{R})) \simeq \mathrm{Hom}^\infty(G, \mathbf{R})$.

2. Thanks to (art. 7.13) we can assume that G is connected. Now, for every $\tilde{g}$, $\tilde{g}'$ in $\tilde{G}$, $F(\tilde{g}\tilde{g}'\tilde{g}^{-1}) = F(\tilde{g}')$, that is, $F \circ \mathrm{Ad}(\tilde{g}) = \mathrm{Ad}(\tilde{g})^*(F) = F$, for all $\tilde{g} \in \tilde{G}$. Thus, $d[\mathrm{Ad}(\tilde{g})^*(F)] = dF$, or $\mathrm{Ad}^*(\tilde{g})(\pi^*(\zeta)) = \pi^*(\zeta)$, or $(\pi \circ \mathrm{Ad}(\tilde{g}))^*(\zeta) = \pi^*(\zeta)$. But $\pi \circ \mathrm{Ad}(\tilde{g}) = \mathrm{Ad}(g) \circ \pi$, where $g = \pi(\tilde{g})$. Hence, $\pi^*(\mathrm{Ad}(g)^*(\zeta)) = \pi^*(\zeta)$, and since π is a subduction, $\mathrm{Ad}(g)^*(\zeta) = \zeta$, that is, $\mathrm{Ad}_*(g)(\zeta) = \zeta$. ☐

7.18. The value of a momentum. Let G be a diffeological group. Every momentum α of G is characterized by its value at the identity (art. 6.40). In other words, if the value α_1 of α at the identity $1_G \in G$ vanishes, then α is zero. Equivalently, $\alpha = 0$ if and only if for any 1-plot F of G, centered at the identity, $\alpha(F)(0) = 0$. Equivalently, two momenta of G, α and β, coincide if and only if, for any 1-plot F of G centered at the identity, $\alpha(F)(0) = \beta(F)(0)$.

NOTE. It is tempting to reduce this proposition for F being a germ of a 1-parameter subgroup of G. But, if it is clear that this is possible for Lie groups, it is not clear that it is still the case for any diffeological group.

PROOF. Let us consider an n-plot $P : U \to G$ and $r \in U$. Let T_r be the translation $T_r : s \mapsto s + r$, and let us denote $g = P(r)$. The parametrization $\bar{P} = L(g^{-1}) \circ P \circ T_r$, defined on $\bar{U} = T_r^{-1}(U)$, is a plot of G, centered at the identity: $0 \in \bar{U}$ and $\bar{P}(0) = 1_G$. So, $P = L(g) \circ \bar{P} \circ T_{-r}$, and $\alpha(P)_r = \alpha(L(g) \circ \bar{P} \circ T_{-r})_r = (L(g) \circ \bar{P} \circ T_{-r})^*(\alpha)_r = T_{-r}^* \circ \bar{P}^* \circ L(g)^*(\alpha)_r$. But $L(g)^*(\alpha) = \alpha$, thus $\alpha(P)_r = T_{-r}^* \circ \bar{P}^*(\alpha)_r = T_{-r}^*[\bar{P}^*(\alpha)]_r = \bar{P}^*(\alpha)_0 = 0$. Hence, if $\alpha(F)(0) = 0$ for every 1-plot F centered at 1_G, then $\alpha(P)(r) = 0$, for every plot P and every $r \in \mathrm{def}(P)$, that is, $\alpha = 0$. Therefore, the momenta of G are characterized by their values at the identity. Now, we know that every 1-form is characterized by its values on the 1-plot (art. 6.37), which completes the proposition. ☐

7.19. Equivalence between right and left momenta. Let G be a diffeological group, and let $\mathcal{G}^\star$ denote the space of *right momenta* of the group G, that is, the space of 1-forms of G invariant by the right multiplication,

$$\mathcal{G}^\star = \{\alpha \in \Omega^1(G) \mid \text{for all } g \in G,\ R(g)^*(\alpha) = \alpha\}.$$

There exists a natural linear isomorphism flip : $\mathcal{G}^* \to \mathcal{G}^\star$ equivariant with respect to the coadjoint action, $\mathrm{Ad}_*(g) \circ \mathrm{flip} = \mathrm{flip} \circ \mathrm{Ad}_*(g)$, for all $g \in G$. In other words, there is no reason to prefer left or right momenta of a diffeological group. The particular role of left momenta exists because we are dealing with actions of groups and not antiactions. This situation is summarized by the following commutative diagram.

$$\begin{array}{ccc} \mathcal{G}^* & \xrightarrow{\ \mathrm{flip}\ } & \mathcal{G}^\star \\ {\scriptstyle \mathrm{Ad}_*(g)}\Big\downarrow & & \Big\downarrow{\scriptstyle \mathrm{Ad}_*(g)} \\ \mathcal{G}^* & \xrightarrow[\ \mathrm{flip}\ ]{} & \mathcal{G}^\star \end{array}$$

PROOF. Let us denote by a dot the multiplication in G. Let α be any left p-momentum of G. Let $P : U \to G$ be an n-plot. Let $\bar{\alpha}(P)$ be defined by

$$\bar{\alpha}(P)_r = \alpha\left[s \mapsto P(s) \cdot P(r)^{-1}\right]_{s=r},$$

where r belongs to U. Let us show that $\bar{\alpha}$ defines a p-form of G. First of all let us remark that $\bar{\alpha}(P)$ is the restriction of the 1-form $\alpha((s,r) \mapsto P(s) \cdot P(r)^{-1})$ to the diagonal $s = r$. Thus, $\bar{\alpha}(P)$ is a smooth 1-form of U. Now, let us check that $\bar{\alpha}$ is a well defined differential 1-form on G. let $F : V \to U$ be a smooth m-parametrization, ν be a point of V and $\delta\nu$ be a vector of $\mathbf{R}^m$. We have

$$\begin{aligned}
\bar{\alpha}(P \circ F)_\nu(\delta\nu) &= \alpha\left[s \mapsto (P \circ F)(s) \cdot (P \circ F)(\nu)^{-1}\right]_\nu (\delta\nu) \\
&= \alpha\left[s \mapsto F(s) \mapsto (P \circ F)(s) \cdot (P \circ F)(\nu)^{-1}\right]_\nu (\delta\nu) \\
&= \alpha\left[s \mapsto (r = F(s)) \mapsto P(r) \cdot P(F(\nu))^{-1}\right]_\nu (\delta\nu) \\
&= \alpha\left(\left[r \mapsto P(r) \cdot P(F(\nu))^{-1}\right] \circ F\right)_\nu (\delta\nu) \\
&= F^*\left[\alpha\left(r \mapsto P(r) \cdot P(F(\nu))^{-1}\right)\right]_\nu (\delta\nu) \\
&= \alpha\left[r \mapsto P(r) \cdot P(F(\nu))^{-1}\right]_{F(\nu)} (D(F)(\nu)(\delta\nu)) \\
&= \bar{\alpha}(P)_{F(\nu)}(D(F)(\nu)(\delta\nu)) \\
&= F^*\left[\bar{\alpha}(P)\right]_\nu (\delta\nu).
\end{aligned}$$

Next, let us check that the 1-form $\bar{\alpha}$ is right-invariant, that is, $\bar{\alpha} \in \mathcal{G}^\star$. For all $g \in G$, we have

$$\begin{aligned}
R(g)^*(\bar{\alpha})(P)_r &= \bar{\alpha}(R(g) \circ P)_r \\
&= \alpha\left[s \mapsto (R(g) \circ P)(s) \cdot (R(g) \circ P)(r)^{-1}\right]_{s=r} \\
&= \alpha\left[s \mapsto P(s) \cdot g \cdot (P(r) \cdot g)^{-1}\right]_{s=r} \\
&= \alpha\left[s \mapsto P(s) \cdot g \cdot g^{-1} \cdot P(r)^{-1}\right]_{s=r} \\
&= \alpha\left[s \mapsto P(s) \cdot P(r)^{-1}\right]_{s=r} \\
&= \bar{\alpha}(P)_r.
\end{aligned}$$

Thus, we just defined a map flip : $\alpha \mapsto \bar{\alpha}$, from $\mathcal{G}^*$ to $\mathcal{G}^\star$. Let us check now that flip is bijective. Let $\beta = \bar{\alpha}$. Let $P : U \to G$ be a plot, and let us define $\bar{\beta}$ by

$\bar{\beta}(P)(r) = \beta[s \mapsto P(r)^{-1} \cdot P(s)](s = r)$, for all $r \in U$. We have then,

$$\begin{aligned} \bar{\beta}(P)_r &= \beta\left[s \mapsto P(r)^{-1} \cdot P(s)\right]_{s=r} \\ &= \bar{\alpha}\left[s \mapsto P(r)^{-1} \cdot P(s)\right]_{s=r} \\ &= \alpha\left[s \mapsto P(r)^{-1} \cdot P(s) \cdot P(r)^{-1} \cdot P(r)\right]_{s=r} \end{aligned}$$

$$\begin{aligned} &= \alpha\left[s \mapsto P(r)^{-1} \cdot P(s)\right]_{s=r} \\ &= L(P(r)^{-1})^*(\alpha)\,[s \mapsto P(s)]_{s=r} \\ &= \alpha(P)_r. \end{aligned}$$

Hence, $\bar{\beta} = \alpha$. Thus, flip is bijective and is clearly linear. Therefore, flip is a linear isomorphism from $\mathcal{G}^*$ to $\mathcal{G}^*$. It is easy to check then that it is a smooth isomorphism. Let us check now that flip is equivariant under the coadjoint action. Let $\alpha \in \mathcal{G}^*$, let $P : U \to G$ be a plot and $r \in U$. On the one hand we have

$$\begin{aligned} \mathrm{flip}[\mathrm{Ad}(g)^*(\alpha)](P)_r &= \mathrm{flip}[R(g)^*(\alpha)](P)_r \\ &= R(g)^*(\alpha)[s \mapsto P(s) \cdot P(r)^{-1}]_{s=r} \\ &= \alpha[s \mapsto P(s) \cdot P(r)^{-1} \cdot g]_{s=r}, \end{aligned}$$

and on the other hand,

$$\begin{aligned} [\mathrm{Ad}(g)^*(\mathrm{flip}(\alpha))](P)_r &= [L(g)_*(\mathrm{flip}(\alpha))](P)_r \\ &= \mathrm{flip}(\alpha)(L(g^{-1}) \circ P)_r \\ &= \alpha[s \mapsto (L(g^{-1}) \circ P)(s) \cdot (L(g^{-1}) \circ P)(r))^{-1}]_{s=r} \\ &= \alpha[s \mapsto g^{-1} \cdot P(s) \cdot P(r)^{-1} \cdot g]_{s=r} \\ &= L(g^{-1})^*(\alpha)[s \mapsto P(s) \cdot P(r)^{-1} \cdot g]_{s=r} \\ &= \alpha[s \mapsto P(s) \cdot P(r)^{-1} \cdot g]_{s=r}. \end{aligned}$$

Therefore, $\mathrm{flip} \circ \mathrm{Ad}(g)^* = \mathrm{Ad}(g)^* \circ \mathrm{flip}$ for all $g \in G$. □

Exercises

✎ EXERCISE 127 (Pullback of 1-forms by multiplication). Let G be a diffeological group, and let $\alpha \in \Omega^1(G)$. Let $\mathrm{mul} : G \times G \to G$ be the multiplication map, $\mathrm{mul}(a, b) = ab$. Let $P : U \to G$ be an n-plot, and $Q : V \to G$ be a m-plot. Let $P \times Q$ be the $(n+m)$-plot of $G \times G$ defined by $(r, s) \mapsto (P(r), Q(s))$. Let $r \in U$ and $s \in V$. Let $\delta r \in \mathbf{R}^n$, and $\delta s \in \mathbf{R}^m$. Show that

$$\mathrm{mul}^*(\alpha)(P \times Q)_{\binom{r}{s}} \begin{pmatrix} \delta r \\ \delta s \end{pmatrix} = L(P(r))^*(\alpha)(Q)_s(\delta s) + R(Q(s))^*(\alpha)(P)_r(\delta r),$$

and, in particular, $\alpha[r \mapsto P(r) \cdot Q(r)]_r = L(P(r))^*(\alpha)(Q)_r + R(Q(r))^*(\alpha)(P)_r$.

✎ EXERCISE 128 (Liouville form on groups). Let G be a diffeological group, and let $\mathcal{G}^*$ be its space of momenta. Let $G \times \mathcal{G}^*$ be the diffeological product where $\mathcal{G}^*$ is equipped with its functional diffeology. Let λ be the map defined for every n-plot $Q : U \to G \times \mathcal{G}^*$ by

$$\lambda(Q)_r(\delta r) = A(r)(P)_r(\delta r),$$

where $r \in U$, $Q(r) = (P(r), A(r))$ and $\delta r \in \mathbf{R}^n$.

1) Show that λ is a be differential 1-form on the product $G \times \mathcal{G}^*$.

2) Let $\alpha \in \mathcal{G}^*$, and let $j_\alpha : G \to G \times \mathcal{G}^*$ be defined by $j_\alpha : g \mapsto (g, \alpha)$. Show that $j_\alpha^*(\lambda) = \alpha$.

3) Show that the form λ is invariant by the following action of G on $G \times \mathcal{G}^*$

$$\text{for all } g' \in G, \quad g'_{G \times \mathcal{G}^*} : (g, \alpha) \mapsto (\mathrm{Ad}(g')(g), \mathrm{Ad}_*(g')(\alpha)).$$

Note the similitude of this construction with the general construction of the Liouville form for any diffeological space (art. 6.49).

CHAPTER 8

Diffeological Fiber Bundles

Finding the right notion of fiber bundle for diffeology **[Igl85]** has been a question raised by the study of the irrational torus T_α **[DoIg85]**. The direct computation of homotopy groups showed that the projection $\mathrm{T}^2 \to \mathrm{T}_\alpha$ behaves like a fibration but without being locally trivial, since T_α inherits the coarse topology. So, it was necessary to adapt the notion of fiber bundle from classical differential geometry to diffeology in order to include such objects, regarded at this time as *singular*, without losing the main properties of this theory. We shall see, thereafter, two equivalent definitions of *diffeological bundles*. One is pedestrian and operative, involving *local triviality along the plots* (art. 8.9), the other one involves *groupoid* (art. 8.8), is more sophisticated but seems also to be deeper. And indeed, according to this definition, diffeological fiber bundles satisfy the *exact homotopy sequence* (art. 8.21), one of their major properties. The category of diffeological fibrations also includes all the quotients of diffeological groups by any subgroup, and this is what explains *a posteriori* the homotopy of the irrational torus, computed previously by another method. In parallel, many classical constructions can be extended to this category: *pullbacks bundles*, *products bundles*, *principal bundles*, *associated bundles*, etc. Because of the wide category of diffeological groups and of the flexibility of diffeological fibrations, the notion of principal bundle in diffeology plays an even more important role than in classical differential geometry, if only for the fact that every diffeological fiber bundle is associated with a principal fiber bundle. To reconnect with homotopy in diffeology, *coverings* are defined as fiber bundles with (diffeologically) discrete fiber (art. 8.22), and the main property is that every diffeological space has a unique simply connected covering up to equivalence (art. 8.26), which is a principal bundle with structure group the fundamental group (art. 8.24). Any other covering is a quotient of this universal covering by a subgroup of the fundamental group. Another important property is the *monodromy theorem* of lifting maps which applies in this context of diffeological coverings (art. 8.25). These constructions are then illustrated by the 1-dimensional irrational tori (art. 8.38).

Building Bundles with Fibers

Intuitively, a fiber bundle is a kind of projection such that the preimage of any value is equivalent to any other one, and equivalent to a given fiber. *This is the minimal condition required for using the word "fiber bundle". The notion of equivalence depends uniquely and naturally on the category we are working in. Here the equivalence will be understood obviously as diffeomorphic. But what is then of major importance is how these fibers are* glued together *to make a bundle. This is the critical point which can make the difference between one choice and another, and this is the point we describe and discuss in this section.*

8.1. The category of smooth surjections. We shall consider the category whose objects are *smooth surjections* $\pi : \mathrm{T} \to \mathrm{B}$, T and B are diffeological spaces, π is surjective and smooth. The space T will be called the *total space* of the surjection and B its *base space*. Let $\pi' : \mathrm{T}' \to \mathrm{B}'$ be another surjection, a *morphism* from π to π' is a couple of smooth maps $\Phi : \mathrm{T} \to \mathrm{T}'$ and $\phi : \mathrm{B} \to \mathrm{B}'$ such that the following diagram commutes.

$$\begin{array}{ccc} \mathrm{T} & \xrightarrow{\ \Phi\ } & \mathrm{T}' \\ {\scriptstyle\pi}\downarrow & & \downarrow{\scriptstyle\pi} \\ \mathrm{B} & \xrightarrow[\ \phi = \mathrm{pr}(\Phi)\ ]{} & \mathrm{B}' \end{array}$$

Note that, for any Φ, if there exists a map ϕ such that $\pi' \circ \Phi = \phi \circ \pi$, then ϕ is unique, it is called the *projection* of Φ on B, and denoted by $\mathrm{pr}(\Phi)$. The identity morphism of the object $\pi : \mathrm{T} \to \mathrm{B}$ is the pair $(\mathbf{1}_\mathrm{T}, \mathbf{1}_\mathrm{B})$. A morphism (Φ, ϕ) from π to π' is an isomorphism if and only if $\Phi \in \mathrm{Diff}(\mathrm{T}, \mathrm{T}')$ and $\phi \in \mathrm{Diff}(\mathrm{B}, \mathrm{B}')$, in this case we say that π and π' are equivalent. We say particularly that π and π' are B-equivalent if $\mathrm{B}' = \mathrm{B}$ and $\phi = \mathbf{1}_\mathrm{B}$.

1. *Trivial projections.* We say that a projection $\pi : \mathrm{T} \to \mathrm{B}$ is *trivial* with *fiber* F if π is equivalent to the first factor projection $\mathrm{pr}_1 : \mathrm{B} \times \mathrm{F} \to \mathrm{B}$, where F is some diffeological space. In this case π is B-equivalent to pr_1, that is, there exists $\Phi \in \mathrm{Diff}(\mathrm{T}, \mathrm{B} \times \mathrm{F})$ such that $\mathrm{pr}_1 \circ \Phi = \pi$.

2. *Pullbacks.* Let $\pi' : \mathrm{T}' \to \mathrm{B}'$ be a smooth surjection. Let $\phi : \mathrm{B} \to \mathrm{B}'$ be a smooth map. The *pullback* of T' by ϕ is denoted by $\phi^*(\mathrm{T}')$, and defined by

$$\phi^*(\mathrm{T}') = \{(b, t') \in \mathrm{B} \times \mathrm{T}' \mid \phi(b) = \pi'(t')\}.$$

It is equipped with the subset diffeology of the product diffeology. Now, let $\pi = \mathrm{pr}_1 \restriction \phi^*(\mathrm{T}')$, $\Phi = \mathrm{pr}_2 \restriction \phi^*(\mathrm{T}')$ and $\mathrm{T} = \phi^*(\mathrm{T}')$, so $\pi : \mathrm{T} \to \mathrm{B}$ is a smooth surjection, called *pullback* of π by ϕ, and (Φ, ϕ) is the natural morphism from π to π' associated with this construction.

$$\begin{array}{ccc} \mathrm{T} = \phi^*(\mathrm{T}') \subset \mathrm{B} \times \mathrm{T}' & \xrightarrow{\ \Phi = \mathrm{pr}_2 \restriction \phi^*(\mathrm{T}')\ } & \mathrm{T}' \\ {\scriptstyle \pi = \mathrm{pr}_1 \restriction \phi^*(\mathrm{T}')}\downarrow & & \downarrow{\scriptstyle\pi'} \\ \mathrm{B} & \xrightarrow[\ \phi\ ]{} & \mathrm{B}' \end{array}$$

The pullback is associative up to equivalence. Let us denote by π'_ϕ the pullback of π' by ϕ. If $\pi'' : \mathrm{T}'' \to \mathrm{B}''$ is a smooth projection, if $\phi' : \mathrm{B}' \to \mathrm{B}''$ and $\phi : \mathrm{B} \to \mathrm{B}'$ are two smooth maps, then $(\pi''_{\phi'})_\phi$ is naturally equivalent to $\pi''_{\phi' \circ \phi}$. Indeed, on the one hand we have $(\phi' \circ \phi)^*(\mathrm{T}'') = \{(b, t'') \in \mathrm{B} \times \mathrm{T}'' \mid \pi''(t'') = \phi' \circ \phi(b)\}$, and on the other hand, $\phi^*(\phi'^*(\mathrm{T}'')) = \{(b, (b', t'')) \in \mathrm{B} \times \mathrm{B}' \times \mathrm{T}'' \mid b' = \phi(b) \text{ and } \phi'(b') = \pi''(t'')\}$. Thus, the map $(b, t'') \mapsto (b, (\phi(b), t''))$ defined on $(\phi' \circ \phi)^*(\mathrm{T}'')$ takes its values in $\phi^*(\phi'^*(\mathrm{T}''))$ and is clearly an isomorphism. Also note that, if π' is a subduction, then π is also a subduction, and symmetrically, if ϕ is a subduction, then Φ is also a subduction.

3. *Restriction.* Let $\pi : T \to B$ be a smooth surjection, and let $A \subset B$ be any subset. We call the *restriction of* π *over* A the restriction $\pi \restriction \pi^{-1}(A) : \pi^{-1}(A) \to A$, denoted sometimes by $\pi^{\restriction A}$, where $\pi^{-1}(A)$ and A are equipped with the subset diffeology. It is equivalent to the pullback of π by the inclusion $j_A : A \hookrightarrow B$. When $A = \{b\}$ is a single point, the restriction $\pi^{-1}(b)$ is denoted by T_b and is called the *fiber* of π (or T) over b.

4. *Reductions and sections.* Let $\pi : T \to B$ be a smooth surjection, and let $\Sigma \subset T$. If $\pi \restriction \Sigma$ is surjective, then it is a smooth surjection called the *reduction of* π *to* Σ. If $\Sigma \cap T_b$ is reduced to a single point for all $b \in B$, we say that Σ is a *section* of π (over B). In this case, we also call section the map σ which associates with every $b \in B$ the unique element contained in $\Sigma \cap T_b$. Conversely, any map $\sigma : B \to T$ satisfying $\pi \circ \sigma = \mathbf{1}_B$ defines a section $\Sigma = \sigma(B)$. We say that the section Σ is smooth when σ is smooth. Note that, if σ is smooth, then the projection π is not just a smooth projection but also a subduction.

5. *Local triviality.* Let $\pi : T \to B$ be a smooth surjection and F be some diffeological space. We say that π is *locally trivial* with *fiber* F if there exists a cover of B by a family of D-open sets $\{U_i\}_{i \in \mathcal{J}}$ such that the restriction of π over each U_i is trivial with fiber F.

PROOF. There is not much to prove here. Let us consider the first point. Let us check that if $\pi : T \to B$ is trivial with fiber F, then π is B-equivalent to $\mathrm{pr}_1 : B \times F \to B$. Let (Ψ, ψ) be an equivalence from π to pr_1. From $\mathrm{pr}_1 \circ \Psi = \psi \circ \pi$ we get $\Psi(x) = (\psi(\pi(x)), f(x))$, where $f \in \mathcal{C}^\infty(X, F)$. Now, since ψ is an automorphism of B, $\psi \times \mathbf{1}_F : (b, y) \mapsto (\psi(b), z)$ is an automorphism of $B \times F$, thus $\Phi = (\psi \times \mathbf{1}_F)^{-1} \circ \Psi : X \to B \times F$ is a diffeomorphism satisfying $\mathrm{pr}_1 \circ \Phi = \pi$. Let us now consider the second point. On the one hand, $(\phi' \circ \phi)^*(T'')$ is made of the elements $(b, t'') \in B \times T''$ such that $\phi' \circ \phi(b) = \pi''(t'')$, on the other hand $\phi^*(\phi'^*(T''))$ is made of the elements $(b, (b', t'')) \in B \times \phi'^*(T'')$ such that $\phi(b) = b'$, that is, $(b, (b', t'')) \in B \times (B' \times T'')$ with $\phi(b) = b'$ and $\phi'(b') = \pi''(t'')$. Thus, the diffeomorphism $(b, t'') \mapsto (b, (\phi(b), t''))$ from $(\phi' \circ \phi)^*(T'')$ to $\phi^*(\phi'^*(T''))$ is a natural B-equivalence from $\pi''_{\phi' \circ \phi}$ to $(\pi''_{\phi'})_\phi$. Now, if π' is a subduction, let $P : U \to B$ be a plot, and let $\phi \circ P$ be a plot of B' which can be lifted locally in a plot Q of T' (art. 1.48). Hence, $\pi' \circ Q = \phi \circ P$, everywhere Q is defined, thus $r \mapsto (P(r), Q(r))$ is a plot of $\phi^*(T')$, and π is a subduction. A symmetrical reasoning applies for ϕ and Φ. For the third point, let us notice that the map $t \mapsto (b = \pi(t), t)$, defined from $\pi^{-1}(A)$ to $j_A^*(T)$ is an A-equivalence from $\pi \restriction \pi^{-1}(A)$ to $\mathrm{pr}_1 \restriction j_A^*(T)$. For the fourth point, note that if $\sigma : B \to T$ is a smooth section, then every plot $P : U \to B$ lifts on T by $Q = \sigma \circ P$, and thus π is a subduction. □

8.2. The category of groupoids. Let us recall some elements of the theory of groupoids, and let us set some notation; see [**McL75**] and [**McK87**] for a comprehensive exposé. A groupoid $\mathbf{K}$ is a category such that the objects constitute a set (and so do the morphisms) and such that every morphism (also called an *arrow*) is invertible, that is, every morphism is an isomorphism.

1. *Objects and Morphisms.* We denote by $\mathrm{Obj}(\mathbf{K})$ and by $\mathrm{Mor}(\mathbf{K})$ respectively, the set of objects and morphisms of $\mathbf{K}$. For all $x, x' \in \mathrm{Obj}(\mathbf{K})$, we denote by $\mathrm{Mor}_{\mathbf{K}}(x, x')$ the set of morphisms from x to x'. This set is also denoted sometimes by $\mathbf{K}(x, y)$, or even by $\mathbf{K}_{x,y}$.

2. *Source and Target.* Let $f \in \mathrm{Mor}(\mathbf{K})$, $\mathrm{src}(f)$ and $\mathrm{trg}(f)$ denote the *source* and the *target* of f, that is, if $f \in \mathrm{Mor}_{\mathbf{K}}(x, x')$, then $x = \mathrm{src}(f)$ and $x' = \mathrm{trg}(f)$.

3. *Groupoid Operation.* We denote by $f \cdot g$ (multiplication) or by $g \circ f$ (composition) the groupoid composite of two morphisms f and g, defined when $\mathrm{trg}(f) = \mathrm{src}(g)$.

4. *Characteristic Map.* We call *characteristic map* the map defined by

$$\chi : \mathrm{Mor}(\mathbf{K}) \to \mathrm{Obj}(\mathbf{K}) \times \mathrm{Obj}(\mathbf{K}) \quad \text{with} \quad \chi(f) = (\mathrm{src}(f), \mathrm{trg}(f)).$$

And thus, $\mathrm{Mor}_{\mathbf{K}}(x, x') = \chi^{-1}(x, x')$.

5. *Subgroupoids.* A *subgroupoid* $\mathbf{H}$ is a subcategory which is a groupoid, that is, a subcategory such that if $f \in \mathrm{Obj}(\mathbf{K})$, then $f^{-1} \in \mathrm{Obj}(\mathbf{K})$. A subgroupoid $\mathbf{H}$ of a groupoid $\mathbf{K}$ is said to be *wide* if $\mathrm{Obj}(\mathbf{H}) = \mathrm{Obj}(\mathbf{K})$; we shall also say that $\mathbf{H}$ is a *reduction* of $\mathbf{K}$.

6. *Structure Groups and Units.* For all $x \in \mathrm{Obj}(\mathbf{K})$,

$$\mathbf{K}_x = \mathrm{Mor}_{\mathbf{K}}(x, x)$$

is a group called the *isotropy* of x, or the *structure group* of the groupoid $\mathbf{K}$ at x. The identity of $\mathbf{K}_x$ is denoted by $\mathbf{1}_x$ and is a *unit* of $\mathbf{K}$. The units of $\mathbf{K}$ form a (wide) subgroupoid defined by

$$\mathrm{Obj}(\mathrm{U}(\mathbf{K})) = \mathrm{Obj}(\mathbf{K}) \quad \text{and} \quad \mathrm{Mor}(\mathrm{U}(\mathbf{K})) = \{\mathbf{1}_x \mid x \in \mathrm{Obj}(\mathbf{K})\}.$$

Said differently, $\mathrm{Mor}_{\mathrm{U}(\mathbf{K})}(x, x') = \varnothing$ if $x \neq x'$, and $\mathrm{Mor}_{\mathrm{U}(\mathbf{K})}(x, x) = \{\mathbf{1}_x\}$. We denote by $\mathfrak{i}_{\mathrm{Obj}(\mathbf{K})}$ the injection from the objects to the morphisms,

$$\mathfrak{i}_{\mathrm{Obj}(\mathbf{K})} : \mathrm{Obj}(\mathbf{K}) \to \mathrm{Mor}(\mathbf{K}) \quad \text{with} \quad \mathfrak{i}_{\mathrm{Obj}(\mathbf{K})}(x) = \mathbf{1}_x.$$

7. *Components.* There exists a natural equivalence relation on the objects of a groupoid $\mathbf{K}$. Two elements x and x' of $\mathrm{Obj}(\mathbf{K})$ are equivalent if there exists an arrow f connecting x to x', that is, $\mathrm{src}(f) = x$ and $\mathrm{trg}(f) = x'$. This relation decomposes $\mathrm{Obj}(\mathbf{K})$ into classes called *components*. Each component defines naturally a subgroupoid, called again a component of $\mathbf{K}$. The groupoid is said to be connected if it has only one component, that is, if the characteristic map is surjective.

8. *Morphisms of Groupoids.* A *morphism* from a groupoid $\mathbf{K}$ to a groupoid $\mathbf{K}'$ is a covariant functor Φ, which can be represented by a couple of maps $(\Phi_{\mathrm{O}}, \Phi_{\mathrm{M}})$ with

$$\Phi_{\mathrm{O}} : \mathrm{Obj}(\mathbf{K}) \to \mathrm{Obj}(\mathbf{K}') \quad \text{and} \quad \Phi_{\mathrm{M}} : \mathrm{Mor}(\mathbf{K}) \to \mathrm{Mor}(\mathbf{K}'),$$

such that the following diagram commutes.

$$\begin{array}{ccc} \mathrm{Mor}(\mathbf{K}) & \xrightarrow{\quad \Phi_{\mathrm{M}} \quad} & \mathrm{Mor}(\mathbf{K}') \\ \chi \downarrow & & \downarrow \chi' \\ \mathrm{Obj}(\mathbf{K}) \times \mathrm{Obj}(\mathbf{K}) & \xrightarrow[\Phi_{\mathrm{O}} \times \Phi_{\mathrm{O}}]{} & \mathrm{Obj}(\mathbf{K}') \times \mathrm{Obj}(\mathbf{K}') \end{array}$$

Also

$$\Phi_{\mathrm{M}}(f \cdot g) = \Phi_{\mathrm{M}}(f) \cdot \Phi_{\mathrm{M}}(g).$$

The groupoids and their morphisms form a category we shall denote by {Groupoids}. The commutativity of the diagram $\chi' \circ \Phi_{\mathrm{M}} = (\Phi_{\mathrm{O}} \times \Phi_{\mathrm{O}}) \circ \chi$ also writes

$$\mathrm{src} \circ \Phi_{\mathrm{M}} = \Phi_{\mathrm{O}} \circ \mathrm{src} \quad \text{and} \quad \mathrm{trg} \circ \Phi_{\mathrm{M}} = \Phi_{\mathrm{O}} \circ \mathrm{trg}.$$

The *kernel* of Φ, denote by $\ker(\Phi)$, is the subgroupoid of $\mathbf{K}$ defined by

$$\mathrm{Obj}(\ker(\Phi)) = \mathrm{Obj}(\mathbf{K}) \quad \text{and} \quad \mathrm{Mor}(\ker(\Phi)) = \Phi_M^{-1}(\mathrm{Mor}(U(\mathbf{K}'))).$$

Note that the restrictions of Φ_M to the isotropy groups $\mathbf{K}_x$ are group homomorphisms, precisely

$$\Phi_M \restriction \mathbf{K}_x \in \mathrm{Hom}(\mathbf{K}_x, \mathbf{K}'_{\Phi_O(x)}).$$

Note that since $U(\mathbf{K}')$ is totally disconnected, $\ker(\Phi)$ also is totally disconnected, if $x \neq x'$, then $\mathrm{Mor}_{\ker(\Phi)}(x, x') = \varnothing$, and $\nu \in \mathrm{Mor}(\ker(\Phi))$ means that, for some $x \in \mathrm{Obj}(\mathbf{K})$, $\nu \in \mathbf{K}_x$ and $\Phi_M(\nu) = \mathbf{1}_{x'}$ with $x' = \Phi_O(x)$. Thus,

$$\mathrm{Mor}(\ker(\Phi)) = \bigcup_{x \in \mathrm{Obj}(\mathbf{K})} \ker(\Phi_M \restriction \mathbf{K}_x).$$

Remark that $U(\mathbf{K})$ is always a subgroupoid of $\ker(\Phi)$, we say that Φ is faithful if $\ker(\Phi)$ is reduced to $U(\mathbf{K})$. The *quotient groupoid*, denoted by $\mathbf{K}/\ker(\Phi)$, is defined in any case by

$$\mathrm{Obj}(\mathbf{K}/\ker(\Phi)) = \mathrm{Obj}(\mathbf{K}) \text{ and } \mathrm{Mor}(\mathbf{K}/\ker(\Phi)) = \mathrm{Mor}(\mathbf{K})/\mathrm{Mor}(\ker(\Phi)),$$

where the quotient of the set of morphisms is defined by the equivalence relation: for all f and f' in $\mathrm{Mor}(\mathbf{K})$, $f \sim f'$ if there exists $\nu \in \mathrm{Mor}(\ker(\Phi))$ such that $f' = \nu \cdot f$. Note that $\nu \in \ker(\Phi_M \restriction \mathbf{K}_x)$, where $x = \mathrm{src}(f)$, and $\chi(f) = \chi(f')$. The class of $f \in \mathrm{Mor}(\mathbf{K})$ is the subset of morphisms

$$\mathrm{class}(f) = \{\nu \cdot f \mid \nu \in \ker(\Phi_M \restriction \mathbf{K}_x) \text{ with } x = \mathrm{src}(f)\}.$$

The composition of classes on the quotient $\mathbf{K}/\ker(\Phi)$ is naturally defined by

$$\mathrm{class}(f) \cdot \mathrm{class}(f') = \mathrm{class}(f \cdot f').$$

Then, $\mathrm{class}(f)^{-1} = \mathrm{class}(f^{-1})$ and the structure groups $[\mathbf{K}/\ker(\Phi)]_x$ of the quotient are the quotient groups $\mathbf{K}_x/\ker(\Phi_M \restriction \mathbf{K}_x)$. There exist two natural functors, the *projection functors* π from $\mathbf{K}$ to $\mathbf{K}/\ker(\Phi)$ and the *quotient functor* $\mathrm{pr}(\Phi)$ from $\mathbf{K}/\ker(\Phi)$ to $\mathbf{K}'$ such that the following diagram commutes.

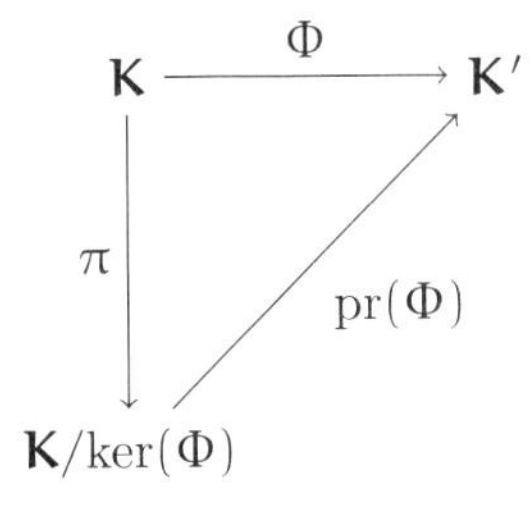

The projection functor π is naturally defined by

$$\pi_O = \mathbf{1}_{\mathrm{Obj}(\mathbf{K})} \quad \text{and} \quad \pi_M(f) = \mathrm{class}(f),$$

for all $f \in \mathrm{Mor}(\mathbf{K})$, where $\mathrm{class}(f) = \ker(\Phi_M \restriction \mathbf{K}_x) \cdot f$, with $x = \mathrm{src}(f)$. The quotient functor is naturally defined by

$$\mathrm{pr}(\Phi)_O = \Phi_O \quad \text{and} \quad \mathrm{pr}(\Phi)_M(\mathrm{class}(f)) = \Phi_M(f).$$

NOTE. Let $\mathbf{K}$ be a connected groupoid, let us choose a point $x \in \mathrm{Obj}(\mathbf{K})$. The data of the groupoid is just the choice of an arrow $a_{x,x'} \in \mathrm{Mor}_{\mathbf{K}}(x, x')$ by element

$x' \in \mathrm{Obj}(\mathbf{K})$ and the group $\mathbf{K}_x = \mathrm{Mor}(x,x)$. Indeed, for any pair $x', x'' \in \mathrm{Mor}_{\mathbf{K}}(x', x'')$ we have

$$\mathrm{Mor}_{\mathbf{K}}(x', x'') = a_{x,x''} \circ \mathbf{K}_x \circ a_{x,x'}^{-1}.$$

Now, if the groupoid is not connected, it splits into components for which the previous construction applies. And therefore, the structure group of every component are isomorphic.

PROOF. For item 8, let us check that $f \sim f'$ is a well defined equivalence relation. It is reflexive, $f = \mathbf{1}_x \cdot f$. It is is symmetric, if $f' = \nu \cdot f$, then $f = \nu^{-1} \cdot f'$. It is transitive, if $f' = \nu \cdot f$ and $f'' = \nu' \cdot f'$, then $f'' = \nu' \cdot \nu \cdot f$, with $\nu' \cdot \nu \in \ker(\Phi)_x$ and $x = \mathrm{src}(f)$. Next, let us check that the multiplication of classes is well defined. Let f and f' be two composable arrows, let $\nu \in \ker(\Phi)_x$ and $\nu' \in \ker(\Phi)_{x'}$ with $x = \mathrm{src}(f)$ and $x' = \mathrm{src}(f')$. Thus, $\nu \cdot f \cdot \nu' \cdot f' = \nu \cdot f \cdot \nu' \cdot f^{-1} \cdot f \cdot f'$, but $\Phi_M(\nu \cdot f \cdot \nu' \cdot f^{-1}) = \Phi_M(\nu) \cdot \Phi_M(f) \cdot \Phi_M(\nu') \cdot \Phi_M(f^{-1}) = \mathbf{1}_{\Phi_O(x)} \cdot \Phi_M(f) \cdot \mathbf{1}_{\Phi_O(x')} \cdot \Phi_M(f^{-1}) = \Phi_M(f) \cdot \Phi_M(f^{-1}) = \Phi_M(f \cdot f^{-1}) = \mathbf{1}_{\Phi_O(x)}$. Hence $\nu \cdot f \cdot \nu' \cdot f^{-1} \in \ker(\Phi_M \restriction \mathbf{K}_x)$. Therefore, $\mathrm{class}(\nu \cdot f \cdot \nu' \cdot f') = \mathrm{class}(f \cdot f')$. Let us check also that $\mathrm{pr}(\Phi)_M$ is well defined, let $\mathrm{class}(f) = \mathrm{class}(f')$, then $f' = \nu \cdot f$, with $\nu \in \ker(\Phi_M \restriction \mathbf{K}_x)$ and $x = \mathrm{src}(f)$. Therefore, $\Phi_M(f') = \Phi_M(\nu \cdot f) = \Phi_M(\nu) \cdot \Phi_M(f) = \mathbf{1}_{\Phi_O(x)} \cdot \Phi_M(f) = \Phi_M(f)$. For the note, every morphism $b \in \mathrm{Mor}_{\mathbf{K}}(x, x')$ writes in a unique way $b = a_{x,x''} \circ a \circ a_{x,x'}^{-1}$, where $a = a_{x,x''}^{-1} \circ b \circ a_{x,x'} \in \mathbf{K}_x$. ☐

8.3. Diffeological groupoids. Let $\mathbf{K}$ be a groupoid, we shall use the notations introduced previously (art. 8.2). We call *groupoid diffeology* on the groupoid $\mathbf{K}$ a diffeology defined on $\mathrm{Mor}(\mathbf{K})$ and on $\mathrm{Obj}(\mathbf{K})$ such that the following conditions are satisfied.[1]

a) The composition $(f, g) \mapsto f \cdot g$, defined on the set $\{(f,g) \in \mathrm{Mor}(\mathbf{K}) \times \mathrm{Mor}(\mathbf{K}) \mid \mathrm{trg}(f) = \mathrm{src}(g)\}$ equipped with the subset diffeology of the product diffeology, with values in $\mathrm{Mor}(\mathbf{K})$, is smooth.
b) The inversion $f \mapsto f^{-1}$, from $\mathrm{Mor}(\mathbf{K})$ to itself, is smooth.
c) The two maps src and trg, from $\mathrm{Mor}(\mathbf{K})$ to $\mathrm{Obj}(\mathbf{K})$, are smooth. Or, which is equivalent, the characteristic map χ is smooth.
d) The identity injection $i_{\mathrm{Obj}(\mathbf{K})} : x \mapsto \mathbf{1}_x$, from $\mathrm{Obj}(\mathbf{K})$ into $\mathrm{Mor}(\mathbf{K})$, is smooth.

A groupoid equipped with a groupoid diffeology is a *diffeological groupoid.*

1. *Morphisms.* A morphism from a diffeological groupoid $\mathbf{K}$ to another one $\mathbf{K}'$ is a morphism of groupoid Φ such that $\Phi_M : \mathrm{Mor}(\mathbf{K}) \to \mathrm{Mor}(\mathbf{K}')$ is smooth, which implies that $\Phi_O : \mathrm{Obj}(\mathbf{K}) \to \mathrm{Obj}(\mathbf{K}')$ is also smooth. This defines the category {D-Groupoids} of diffeological groupoids. An isomorphism Φ is a morphism such that Φ_M and Φ_O are diffeomorphisms.

2. *Subgroupoids.* Any subgroupoid $\mathbf{H}$ of a diffeological group $\mathbf{K}$ is a diffeological groupoid when $\mathrm{Mor}(\mathbf{H})$ and $\mathrm{Obj}(\mathbf{H})$ are equipped with the subset diffeology. In this case we say that $\mathbf{H}$ is a *diffeological subgroupoid* of $\mathbf{K}$.

3. *Morphisms and quotients.* Let $\mathbf{K}$ and $\mathbf{K}'$ be two diffeological groupoids. Let Φ be a morphism from $\mathbf{K}$ to $\mathbf{K}'$. Equipped with the quotient diffeology, the groupoid $\mathbf{K}/\ker(\Phi)$ is a diffeological groupoid and the natural associated morphisms π, from

[1] This notion is close to the concept of Lie-groupoid but differs in that we just require that source and target are smooth maps and not necessarily stronger; see also (art. 8.4) and (art. 8.6).

$\mathbf{K}$ to $\mathbf{K}/\ker(\Phi)$, and $\mathrm{pr}(\Phi)$, from $\mathbf{K}/\ker(\Phi)$ to $\mathbf{K}'$ (art. 8.2), are smooth, that is, morphisms of diffeological groupoids.

NOTE 1. The inclusion $i_{\mathrm{Obj}(\mathbf{K})}$ of a diffeological groupoid is an induction (art. 1.29). The space of objects $\mathrm{Obj}(\mathbf{K})$ identifies naturally, in the category {Diffeology}, with the subspace of identities

$$\mathrm{Obj}(\mathbf{K}) \simeq \{\mathbf{1}_x \mid x \in \mathrm{Obj}(\mathbf{K})\} \subset \mathrm{Mor}(\mathbf{K}).$$

Therefore, a diffeological groupoid is uniquely characterized by $\mathrm{Mor}(\mathbf{K})$, its groupoid operation and its diffeology.

NOTE 2. The structure groups of a component of a diffeological groupoid $\mathbf{K}$, equipped with the subset diffeology of $\mathrm{Mor}(\mathbf{K})$, are isomorphic in the category {D-Groups}, which is a full subcategory of the category {D-Groupoids}.

PROOF. For point 1: Note that $\Phi_O = \mathrm{src} \circ \Phi_M \circ i_{\mathrm{Obj}(\mathbf{K})}$, indeed we have $x \mapsto \mathbf{1}_x = i_{\mathrm{Obj}(\mathbf{K})}(x) \mapsto \mathbf{1}_{\Phi_O(x)} = \Phi_M(\mathbf{1}_x) \mapsto \Phi_O(x) = \mathrm{src}(\mathbf{1}_{\Phi_O(x)})$. Thus, since src and $i_{\mathrm{Obj}(\mathbf{K})}$ are smooth, if Φ_M is smooth, then Φ_O is smooth. For Note 1, the inverse map $i^{-1}_{\mathrm{Obj}(\mathbf{K})} : \mathbf{1}_x \mapsto x$ defined on $\{\mathbf{1}_x \mid x \in \mathrm{Obj}(\mathbf{K})\} \subset \mathrm{Mor}(\mathbf{K})$ coincides with the restriction of the source map, $\mathrm{src}(\mathbf{1}_x) = x$ (and coincides with the restriction of the target map too). Thus, $i^{-1}_{\mathrm{Obj}(\mathbf{K})}$ is smooth when $\{\mathbf{1}_x \mid x \in \mathrm{Obj}(\mathbf{K})\}$ is equipped with the subset diffeology. Then, since $i_{\mathrm{Obj}(\mathbf{K})}$ and $i^{-1}_{\mathrm{Obj}(\mathbf{K})}$ are smooth, $i_{\mathrm{Obj}(\mathbf{K})}$ is an induction. Point 3 is a consequence of the game of subductions. For Note 2, the conjugation given in (art. 8.2, Note) is clearly smooth and thus the structure groups of every component are isomorphic in the category of diffeological groups. □

8.4. Fibrating groupoids. Let $\mathbf{K}$ be a diffeological groupoid, we shall say that $\mathbf{K}$ is *fibrating* if the characteristic map $\chi : \mathrm{Mor}(\mathbf{K}) \to \mathrm{Obj}(\mathbf{K}) \times \mathrm{Obj}(\mathbf{K})$ defined by $\chi(f) = (\mathrm{src}(f), \mathrm{trg}(f))$ is a subduction.[2] Note that this implies in particular that χ is surjective and thus the groupoid is connected (art. 8.2, Note). As well, all the isotropy groups are isomorphic in the category of diffeological groups (art. 8.3, Note 2). This definition is the central point of the theory of diffeological fiber bundles. We shall see in the following that the theory of diffeological fiber bundles is reduced to the theory of fibrating diffeological groupoids.

8.5. Trivial diffeological groupoids. Let us describe a simple but important example of fibrating diffeological groupoid. Let X be a diffeological space and G be a diffeological group. Let us define Γ by

$$\mathrm{Obj}(\Gamma) = X \quad \text{and} \quad \mathrm{Mor}(\Gamma) = X \times G \times X,$$

with the multiplication defined on

$$\mathrm{def}(\cdot) = \{((x, g, y), (y, k, z)) \mid x, y, z \in X \text{ and } g, k \in G\}$$

by

$$(x, g, y) \cdot (y, k, z) = (x, gk, z).$$

We clearly defined, this way, a groupoid. The source and the target are obviously

$$\mathrm{src}(x, g, y) = x \quad \text{and} \quad \mathrm{trg}(x, g, y) = y.$$

[2] In [**Igl85**] these groupoids were called "parfaits" but this terminology is confusing since to be perfect for a group means something else.

The inverses, the identities, and the isotropy groups are given by

$$(x, g, y)^{-1} = (y, g^{-1}, x), \quad \mathrm{id}_x = (x, \mathbf{1}_G, x), \quad \text{and} \quad \Gamma_x = \{(x, g, x) \mid g \in G\}.$$

Note that def$(\cdot)$ is naturally diffeomorphic to $X \times G \times X \times G \times X$ and the multiplication as well as the inversion are clearly smooth. The identity injection from X to $\mathrm{Mor}(\Gamma)$ and the characteristic map from $\mathrm{Mor}(\Gamma)$ to $X \times X$

$$i_{\mathrm{Obj}(\Gamma)} : x \mapsto (x, \mathbf{1}_G, x) \quad \text{and} \quad \chi : (x, g, y) \mapsto (x, y)$$

are again smooth. Therefore Γ is a diffeological groupoid. We shall name this groupoid *the trivial groupoid* with *base* X and *structure group* G, and we may denote it sometimes by $\mathrm{Triv}(X, G)$. Every diffeological groupoid K isomorphic to $\mathrm{Triv}(X, G)$, for some X and G, will be said *trivial* (with base X and structure group G). Moreover, the map χ being clearly a subduction, thanks to the section $(x, y) \mapsto (x, \mathbf{1}_G, y)$, the trivial groupoid $\mathrm{Triv}(X, G)$ is fibrating.

NOTE. A diffeological groupoid $\mathbf{K}$ is trivial if and only there exists a map $\sigma : \mathrm{Obj}(K)^2 \to \mathrm{Mor}(\mathbf{K})$ such that the following conditions are satisfied.

A. σ is a smooth section of the characteristic map χ, $\chi \circ \sigma = \mathbf{1}_{\mathrm{Obj}(\mathbf{K})^2}$.

B. σ satisfies the cocycle relation $\sigma(x, y) \cdot \sigma(x, z) = \sigma(x, z)$, for all triples of points x, y, z in $\mathrm{Obj}(\mathbf{K})$.

Picking a basepoint $o \in \mathrm{Obj}(\mathbf{K})$, and denoting by $\mathrm{Mor}(\mathbf{K}, o)$ the subspace of arrows of $\mathbf{K}$ with origin o, it is equivalent to say that

A′. There exists a smooth section s of $\mathrm{trg} \upharpoonright \mathrm{Mor}(\mathbf{K}, o)$, that is, a smooth map $s : \mathrm{Obj}(\mathbf{K}) \to \mathrm{Mor}(\mathbf{K}, o)$ such that $\mathrm{trg} \circ s = \mathbf{1}_{\mathrm{Obj}(\mathbf{K})}$.

PROOF. Let us assume first that $\mathbf{K}$ is trivial with structure group G, let us denote by X the space of objects. Let Φ be an isomorphism from $\mathbf{K}$ to $\mathrm{Triv}(X, G)$. Thus, $\Phi_M : \mathrm{Mor}(\mathbf{K}) \to X \times G \times X$ is a diffeomorphism satisfying $\Phi_M(f \cdot g) = \Phi_M(f) \cdot \Phi_M(g)$. So, σ defined by $\sigma(x, y) = \Phi_M^{-1}(x, \mathbf{1}_G, y)$ satisfies the conditions above. Now, let us assume that there exists such a section σ. Since σ is a section, we get χ surjective, the groupoid $\mathbf{K}$ is connected. Thus choosing a basepoint $o \in X$, we can define $s(x) = \sigma(o, x)$, for all $x \in X$. The map s is a smooth section of $\mathrm{trg} \upharpoonright \mathrm{Mor}(\mathbf{K}, o)$. Let us denote by G the isotropy group $\mathbf{K}_o$, for each $f \in \mathrm{Mor}(\mathbf{K})$, $s(x) \cdot f \cdot s(y)^{-1}$ belongs to G, where $x = \mathrm{src}(f)$ and $y = \mathrm{trg}(f)$. The map $\Phi_M : f \mapsto (\mathrm{src}(f), s(x) \cdot f \cdot s(y)^{-1}, \mathrm{trg}(f))$, with values in $X \times G \times X$, is smooth and satisfies: 1) $\Phi_M(f \cdot g) = \Phi_M(f) \cdot \Phi_M(g)$ for all f and g in $\mathrm{Mor}(\mathbf{K})$ such that $\mathrm{trg}(f) = \mathrm{src}(g)$, and 2) $\chi \circ \Phi_M = \chi$, where χ denotes generically the characteristic maps. Moreover Φ_M is bijective, its inverse is given by $\Phi_M^{-1}(x, k, y) = s(x)^{-1} \cdot k \cdot s(y)$, which is clearly smooth. Therefore $\Phi = (\Phi_M, \mathbf{1}_{\mathrm{Obj}(\mathbf{K})})$ is an isomorphism from $\mathbf{K}$ to $\mathrm{Triv}(X, G)$. By the way, we also proved the proposition A′. □

8.6. A simple nonfibrating groupoid. It is of course not difficult to find a non-fibrating groupoid, but the following example is simple and instructive. Let $\mathcal{R}$ be an equivalence relation defined on a diffeological space X. We can regard the graph of $\mathcal{R}$ as the groupoid $\Gamma(\mathcal{R})$ given by

$$\mathrm{Obj}(\Gamma(\mathcal{R})) = X \quad \text{and} \quad \mathrm{Mor}(\Gamma(\mathcal{R})) = \{(x, x') \in X \times X \mid x \mathcal{R} x'\}.$$

The multiplication of (x, x') by (x'', x''') is defined only if $x' = x''$ by

$$(x, x') \cdot (x', x''') = (x, x''').$$

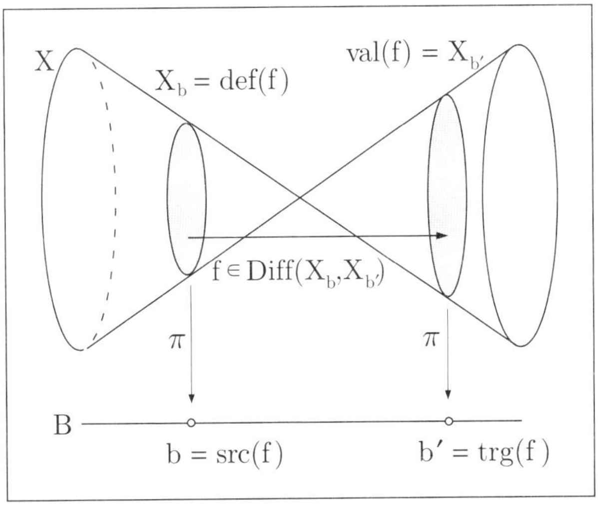

FIGURE 8.1. The groupoid associated with a surjection.

Equipped with their natural respective diffeologies these spaces define a diffeological groupoid. The space $\mathrm{Mor}(\Gamma(\mathcal{R}))$ is just the graph $\mathrm{Gr}(\mathcal{R})$ of the relation $\mathcal{R}$. The components of the groupoid are just the classes $X/\mathcal{R}$ of the equivalence relation $\mathcal{R}$, and it is fibrating if and only if there is only one class. In this case it coincides with the trivial groupoid with base X and trivial group $\{\mathbf{1}\}$.

8.7. Structure groupoid of a smooth surjection. Let X and B be two diffeological spaces. Let $\pi : X \to B$ be a smooth surjection. Let us define

$$\mathrm{Obj}(\mathbf{K}) = B \quad \text{and for all } b, b' \in B, \quad \mathrm{Mor}_{\mathbf{K}}(b, b') = \mathrm{Diff}(X_b, X_{b'}),$$

where the $X_b = \pi^{-1}(b)$, $b \in B$, are equipped with the subset diffeology. Let us define on

$$\mathrm{Mor}(\mathbf{K}) = \bigcup_{b,b' \in B} \mathrm{Mor}_{\mathbf{K}}(b, b')$$

the natural multiplication $f \cdot g = g \circ f$, for $f \in \mathrm{Mor}_{\mathbf{K}}(b, b')$ and $g \in \mathrm{Mor}_{\mathbf{K}}(b', b'')$, $\mathbf{K}$ is clearly a groupoid. The source and target maps are given by

$$\mathrm{src}(f) = \pi(\mathrm{def}(f)) \quad \text{and} \quad \mathrm{trg}(f) = \pi(\mathrm{val}(f)).$$

The groupoid $\mathbf{K}$ is then equipped with a functional diffeology of $\mathbf{K}$ defined as follows. Let X_{src} be the total space of the pullback of π by src (art. 8.1), that is,

$$X_{\mathrm{src}} = \{(f, x) \in \mathrm{Mor}(\mathbf{K}) \times X \mid x \in \mathrm{def}(f)\}.$$

We define the evaluation map ev as usual (see (art. 1.57))

$$\mathrm{ev} : X_{\mathrm{src}} \to X \quad \text{with} \quad \mathrm{ev}(f, x) = f(x).$$

There exists a coarsest diffeology on $\mathrm{Mor}(\mathbf{K})$, which gives to $\mathbf{K}$ the structure of a diffeological groupoid and such that the evaluation map ev is smooth. It will be called the *functional diffeology*. Equipped with this functional diffeology, the groupoid $\mathbf{K}$ captures the *smooth structure* of the projection π. It is why we define $\mathbf{K}$ as the *structure groupoid* of the surjection π.

NOTE 1. If B is reduced to a point, $\mathrm{Obj}(\mathbf{K}) = \{\star\}$, this diffeology coincides with the usual functional diffeology of $\mathrm{Diff}(X) = \mathrm{Mor}(\mathbf{K})$ (art. 1.61).

NOTE 2. This construction also applies when we have just a partition $\mathcal{P}$ on a diffeological space X. We can equip the quotient $Q = X/\mathcal{P}$ with the quotient diffeology, and we get the structure groupoid of the partition as the structure groupoid of the projection $\pi : X \to Q$.

PROOF. Let $P : U \to \mathrm{Mor}(\mathbf{K})$ be some parametrization, and let us define

$$\begin{cases} X_{\mathrm{src}\circ P} = \{(r, x) \in U \times X \mid x \in \mathrm{def}(P(r))\}, \\ X_{\mathrm{trg}\circ P} = \{(r, x) \in U \times X \mid x \in \mathrm{val}(P(r))\}. \end{cases}$$

The space $X_{\mathrm{src}\circ P}$ is the total space of the pullback of $\pi : X \to B$ by the map $\mathrm{src} \circ P$, that is, $(\mathrm{src} \circ P)^*(X)$. It is equivalent to the total space $P^*(\mathrm{src}^*(X))$, thanks to the identification $(r, x) \mapsto (r, (P(r), x))$. But, under this identification, the second projection mutes into $P \times \mathbf{1}_X$, that is, $(r, x) \mapsto (P(r), x)$. The situation is summarized in the following diagram, where the maps are restricted to the indicated domains, it applies *mutatis mutandis* to $X_{\mathrm{trg}\circ P}$.

$$\begin{array}{ccccc} X_{\mathrm{src}\circ P} & \xrightarrow{P \times \mathbf{1}_X} & X_{\mathrm{src}} & \xrightarrow{\mathrm{pr}_2} & X \\ \downarrow{\scriptstyle \mathrm{pr}_1} & & \downarrow{\scriptstyle \mathrm{pr}_1} & & \downarrow{\scriptstyle \pi} \\ U & \xrightarrow[P]{} & \mathrm{Mor}(\mathbf{K}) & \xrightarrow[\mathrm{src}]{} & B \end{array}$$

Now, let us introduce the following maps

$$\begin{cases} P_{\mathrm{src}} : X_{\mathrm{src}\circ P} \to X \quad \text{with} \quad P_{\mathrm{src}}(r, x) = P(r)(x), \\ P_{\mathrm{trg}} : X_{\mathrm{src}\circ P} \to X \quad \text{with} \quad P_{\mathrm{trg}}(r, x) = P(r)^{-1}(x). \end{cases}$$

Then, let us define $\mathcal{D}(U, \mathrm{Mor}(\mathbf{K}))$ as the set of all the parametrizations $P : U \to \mathrm{Mor}(\mathbf{K})$ such that:

$$(\clubsuit) \quad \chi \circ P \in \mathcal{C}^\infty(U, B \times B) \quad \text{and} \quad \begin{cases} (\diamondsuit) & P_{\mathrm{src}} \in \mathcal{C}^\infty(X_{\mathrm{src}\circ P}, X), \\ (\heartsuit) & P_{\mathrm{trg}} \in \mathcal{C}^\infty(X_{\mathrm{trg}\circ P}, X)), \end{cases}$$

where $X_{\mathrm{src}\circ P}$ and $X_{\mathrm{trg}\circ P}$ are equipped with the subset diffeology of the product $U \times X$. Let us show now that the union $\mathbf{D}$ of all the families $\mathcal{D}(U, \mathrm{Mor}(\mathbf{K}))$, where U runs over the set of real domains, is a diffeology. Note first that the condition $(\clubsuit)$ defines already a diffeology: the pullback of the diffeology of B by χ (art. 1.26). So, we need only check that the conditions $(\diamondsuit)$ and $(\heartsuit)$ define a diffeology. And since the condition $(\heartsuit)$ is just the condition $(\diamondsuit)$ inverting the arrows, it will be sufficient to prove that the condition $(\diamondsuit)$ defines a diffeology.

Covering axiom. Let $\mathbf{f} = [0 \mapsto f]$, where $f \in \mathrm{Diff}(X_b, X_{b'})$.We have

$$X_{\mathrm{src}\circ\mathbf{f}} = X_b \quad \text{and} \quad \mathbf{f}_{\mathrm{src}} = f.$$

But $\mathbf{f}_{\mathrm{src}} : X_{\mathrm{src}\circ\mathbf{f}} \to X$ is smooth, thus $\mathbf{f}$ belongs to $\mathbf{D}$.

Compatibility axiom. Let $P \in \mathbf{D}$ and F be a smooth parametrization in $\mathrm{def}(P)$. The map $(P \circ F)_{\mathrm{src}}$ decomposes as follows

$$\begin{array}{ccccc} X_{\mathrm{src}\circ P\circ F} & \to & X_{\mathrm{src}\circ P} & \to & X \\ (r, x) & \mapsto & (F(r), x) & \mapsto & P(F(r))(x). \end{array}$$

As a composite of smooth maps, $(\mathrm{P}\circ\mathrm{F})_{\mathrm{src}}$ is a smooth map. Thus, $\mathrm{P}\circ\mathrm{F}\in\mathbf{D}$.

Locality axiom. Let $\mathrm{P}:\mathrm{U}\to\mathrm{Mor}(\mathbf{K})$ be a parametrization such that, for every $\mathrm{r}\in\mathrm{U}$ there exists an open neighborhood $\mathrm{A}\subset\mathrm{U}$ of r such that $(\mathrm{P}\upharpoonright\mathrm{A})\in\mathbf{D}$. So, $(\mathrm{P}\upharpoonright\mathrm{A})_{\mathrm{src}}$, defined on $\mathrm{X}_{\mathrm{src}\circ(\mathrm{P}\upharpoonright\mathrm{A})}=\mathrm{X}_{\mathrm{src}\circ\mathrm{P}}\upharpoonright\mathrm{A}=\mathrm{pr}_1^{-1}(\mathrm{A})$ is smooth for the subset diffeology. Since $\mathrm{X}_{\mathrm{src}\circ\mathrm{P}}\subset\mathrm{U}\times\mathrm{X}$ is equipped with the subset diffeology and since $\mathrm{pr}_1:\mathrm{U}\times\mathrm{X}\to\mathrm{U}$ is smooth, its restriction $\mathrm{pr}_1\upharpoonright\mathrm{X}_{\mathrm{src}\circ\mathrm{P}}$ is smooth (art. 1.33), therefore D-continuous (art. 2.8) (art. 2.9). Thus, $(\mathrm{pr}_1\upharpoonright\mathrm{X}_{\mathrm{src}\circ\mathrm{P}})^{-1}(\mathrm{A})$ is D-open. Now, since $(\mathrm{P}\upharpoonright\mathrm{A})_{\mathrm{src}}$ is smooth for the subset diffeology and defined on a D-open, $(\mathrm{P}\upharpoonright\mathrm{A})_{\mathrm{src}}$ is a local smooth map (art. 2.10). Therefore, $\mathrm{P}_{\mathrm{src}}$ is everywhere locally smooth, so $\mathrm{P}_{\mathrm{src}}$ is smooth (art. 2.3), and P belongs to $\mathbf{D}$. As we claimed above, what we said for the src part applies *mutatis mutandis* to the trg part. Hence, $\mathbf{D}$ is a diffeology of $\mathrm{Mor}(\mathbf{K})$. We know already that the characteristic function χ is smooth, let us show now that the evaluation map is smooth. Let Φ be a plot of $\mathrm{X}_{\mathrm{src}}$, that is, $\Phi(\mathrm{r})=(\mathrm{P}(\mathrm{r}),\mathrm{Q}(\mathrm{r}))$ where P is a plot of $\mathrm{Mor}(\mathbf{K})$ and Q is a plot of X, such that $\mathrm{Q}(\mathrm{r})\in\mathrm{def}(\mathrm{P}(\mathrm{r}))$. The map $\mathrm{ev}\circ\Phi$ decomposes into the product $\mathrm{P}_{\mathrm{src}}\circ\bar{\mathrm{Q}}$, where $\bar{\mathrm{Q}}:\mathrm{r}\mapsto(\mathrm{r},\mathrm{Q}(\mathrm{r}))$,

$$\mathrm{r}\mapsto(\mathrm{r},\mathrm{Q}(\mathrm{r}))\mapsto\mathrm{P}(\mathrm{r})(\mathrm{Q}(\mathrm{r}))=\mathrm{P}_{\mathrm{src}}(\mathrm{r},\mathrm{Q}(\mathrm{r})).$$

But $\bar{\mathrm{Q}}$ is as smooth as Q and, by the very definition of $\mathbf{D}$, $\mathrm{P}_{\mathrm{src}}$ is smooth, thus ev is smooth. Moreover, note that for any diffeology on $\mathrm{Mor}(\mathbf{K})$ such that the evaluation is smooth, for any plot P of $\mathrm{Mor}(\mathbf{K})$, the map $\mathrm{P}_{\mathrm{src}}$ is smooth. Let us prove now that $\mathbf{D}$ is a groupoid diffeology. Let us begin by showing that the multiplication is smooth. Let $\Phi:\mathrm{r}\mapsto(\mathrm{P}(\mathrm{r}),\mathrm{P}'(\mathrm{r}))$ be a plot of the domain of the multiplication, that is, a plot of the square $\mathrm{Mor}(\mathbf{K})^2$ such that $\mathrm{trg}\circ\mathrm{P}=\mathrm{src}\circ\mathrm{P}'$. Let us denote $\mathrm{P}''(\mathrm{r})=\mathrm{P}(\mathrm{r})\cdot\mathrm{P}'(\mathrm{r})=\mathrm{P}'(\mathrm{r})\circ\mathrm{P}(\mathrm{r})$ for all $\mathrm{r}\in\mathrm{def}(\Phi)$.

1. Since $\chi\circ\mathrm{P}''(\mathrm{r})=(\mathrm{src}(\mathrm{P}(\mathrm{r})),\mathrm{trg}(\mathrm{P}'(\mathrm{r})))$ and since $\mathrm{src}\circ\mathrm{P}$ and $\mathrm{trg}\circ\mathrm{P}'$ are smooth, χ is smooth.

2. First of all, let us notice that $\mathrm{X}_{\mathrm{src}\circ\mathrm{P}''}=\mathrm{X}_{\mathrm{src}\circ\mathrm{P}}$ and $\mathrm{X}_{\mathrm{trg}\circ\mathrm{P}''}=\mathrm{X}_{\mathrm{trg}\circ\mathrm{P}'}$. Now, the map $\mathrm{P}''_{\mathrm{src}}:(\mathrm{r},\mathrm{x})\mapsto\mathrm{P}''(\mathrm{r})(\mathrm{x})$ is the composite of two smooth maps $(\mathrm{r},\mathrm{x})\mapsto(\mathrm{r},\mathrm{P}(\mathrm{r})(\mathrm{x}))=(\mathrm{r},\mathrm{P}_{\mathrm{src}}(\mathrm{r},\mathrm{x}))\mapsto\mathrm{P}'(\mathrm{r})(\mathrm{P}(\mathrm{r})(\mathrm{x}))=\mathrm{P}'_{\mathrm{src}}(\mathrm{r},\mathrm{P}_{\mathrm{src}}(\mathrm{r},\mathrm{x}))$. Thus, the multiplication is smooth.

3. Let P be a plot of $\mathrm{Mor}(\mathbf{K})$, lets us show that $\mathrm{Q}=[\mathrm{r}\mapsto\mathrm{P}(\mathrm{r})^{-1}]$ is also a plot of $\mathrm{Mor}(\mathbf{K})$. First of all, since $\chi(\mathrm{Q}(\mathrm{r}))=\chi(\mathrm{P}(\mathrm{r})^{-1})$ and

$$\chi(\mathrm{P}(\mathrm{r})^{-1})=(\mathrm{src}(\mathrm{P}(\mathrm{r})^{-1}),\mathrm{trg}(\mathrm{P}(\mathrm{r})^{-1}))=(\mathrm{trg}(\mathrm{P}(\mathrm{r})),\mathrm{src}(\mathrm{P}(\mathrm{r}))),$$

$\chi\circ\mathrm{Q}$ is smooth. Now, $\mathrm{X}_{\mathrm{src}\circ\mathrm{Q}}=\mathrm{X}_{\mathrm{trg}\circ\mathrm{P}}$ and $\mathrm{X}_{\mathrm{trg}\circ\mathrm{Q}}=\mathrm{X}_{\mathrm{src}\circ\mathrm{P}}$, with $\mathrm{Q}_{\mathrm{src}}=\mathrm{P}_{\mathrm{trg}}$ and $\mathrm{Q}_{\mathrm{trg}}=\mathrm{P}_{\mathrm{src}}$. Thus, $\mathrm{Q}_{\mathrm{src}}$ and $\mathrm{Q}_{\mathrm{trg}}$ are smooth.

4. Finally, let $\mathrm{i}_{\mathrm{Obj}(\mathbf{K})}=\mathrm{i}_{\mathrm{B}}$ be the inclusion of $\mathrm{B}=\mathrm{Obj}(\mathbf{K})$ into $\mathrm{Mor}(\mathbf{K})$, $\mathrm{i}_{\mathrm{B}}(\mathrm{b})=\mathbf{1}_{\mathrm{X}_{\mathrm{b}}}$. Let $\mathrm{F}:\mathrm{U}\to\mathrm{B}$ be a plot and let $\mathrm{P}=\mathrm{i}_{\mathrm{B}}\circ\mathrm{F}$, that is, $\mathrm{P}(\mathrm{r})=\mathbf{1}_{\mathrm{X}_{\mathrm{F}(\mathrm{r})}}$. Let us show that P is a plot of $\mathrm{Mor}(\mathbf{K})$. First of all, we have $\chi\circ\mathrm{P}(\mathrm{r})=(\mathrm{F}(\mathrm{r}),\mathrm{F}(\mathrm{r}))$, which is smooth. Next, for all $\mathrm{r}\in\mathrm{def}(\mathrm{F})$, we have

$$\mathrm{X}_{\mathrm{src}\circ\mathrm{P}}=\mathrm{X}_{\mathrm{trg}\circ\mathrm{P}}=\mathrm{F}^*(\mathrm{X})=\{(\mathrm{r},\mathrm{x})\in\mathrm{U}\times\mathrm{X}\mid\pi(\mathrm{x})=\mathrm{F}(\mathrm{r})\}.$$

Then, $\mathrm{P}_{\mathrm{src}}(\mathrm{r},\mathrm{x})=\mathrm{P}(\mathrm{r})(\mathrm{x})=\mathbf{1}_{\mathrm{X}_{\mathrm{F}(\mathrm{r})}}(\mathrm{x})=\mathrm{x}$, and as well $\mathrm{P}_{\mathrm{trg}}(\mathrm{r},\mathrm{x})=\mathrm{x}$. Thus, $\mathrm{P}_{\mathrm{src}}$ and $\mathrm{P}_{\mathrm{trg}}$ are smooth because they are the restrictions, to a subspace of $\mathrm{U}\times\mathrm{X}$, of the second projection, which is smooth. Hence, $\mathbf{D}$ defines a groupoid diffeology for which the evaluation map is smooth. We also saw that, for any groupoid diffeology on $\mathrm{Mor}(\mathbf{K})$ such that the evaluation is smooth, $\mathrm{P}_{\mathrm{src}}$ and $\mathrm{P}_{\mathrm{trg}}$ are smooth, for all

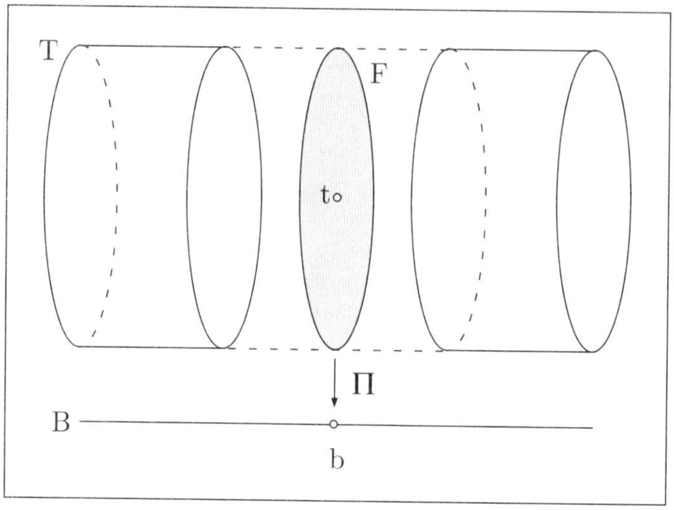

FIGURE 8.2. A fibration.

plots P of Mor(**K**). Therefore, **D** is the coarsest groupoid diffeology such that the evaluation map is smooth. □

8.8. Diffeological fibrations. Let $\pi : T \to B$ be a smooth projection, where T and B are two arbitrary diffeological spaces.

Definition. We say that π is a *diffeological fibration*, or simply a *fibration*, if the structure groupoid **K** (art. 8.7) is fibrating, that is, if and only if the characteristic map $\chi : \mathrm{Mor}(\mathbf{K}) \to B \times B$ is a subduction (art. 8.4).

The space T is called the *total space* of the fibration, the space B is called the *base space* and π the *projection*. We also say that T is *fibered* over B by π, or T is a *fiber bundle* over M. The preimages $T_b = \pi^{-1}(b)$, necessarily all diffeomorphic since χ is surjective, are called the *fibers* of the fibration, or of the fiber bundle. The fiber T_b is called the *fiber over* b. The diffeological class of the fibers is called the *type* of the fiber.

8.9. Fibrations and local triviality along the plots. A smooth map $\pi : T \to B$, where T and B are two diffeological spaces, is a fibration (art. 8.8) if and only if there exists a diffeological space F such that the pullback of π by any plot P of B is locally trivial, with fiber F (art. 8.1). The space F represents the type of the fibers of π, it is why it is also called *the fiber* of the fibration. We say that π is *locally trivial along the plots* (of B).

NOTE 1. A diffeological fibration $\pi : T \to B$ is, in particular, a local subduction (art. 2.16). Precisely, for every plot $P : U \to B$, for all $r \in U$ and for all $t \in T_b = \pi^{-1}(b)$, with $b = P(r)$, there exists a plot Q of T defined on some open neighborhood V of r lifting $P \restriction V$, that is, $P \restriction V = \pi \circ Q$, and such that $Q(r) = t$.

NOTE 2. There is a hierarchy in the various notions of fiber bundles: trivial bundles are locally trivial (with respect to the D-topology), locally trivial bundles are locally trivial along the plots. The converse is not true. To be locally trivial along the plots does not mean that the fibration itself is locally trivial, as many examples will point it out. Look for example at the irrational torus T_α, quotient of T^2 by the Kronecker flow (art. 8.38).

NOTE 3. If the base of a diffeological fiber bundle is a manifold, then the fiber bundle is locally trivial. This comes immediately from the definition, consider the pullback by local charts. If moreover the fiber is a manifold, then the diffeological fiber bundle is a fiber bundle in the category of manifolds. This shows in particular that the classical notion of fiber bundle can also be defined directly in diffeological terms as a property of its associated groupoid, but of course this definition leads to leave an instant the category of manifolds.

PROOF. Let us begin by assuming that π is a fibration, according to the definition (art. 8.8). Let $P : U \to B$ be a plot, and let $F = X_b = \pi^{-1}(b)$, for some base point $b \in B$. The map $\Phi : r \mapsto (b, P(r))$ is obviously a plot of $B \times B$. Let χ be the characteristic map of the groupoid $\mathbf{K}$ associated with π. Since χ is a subduction, by hypothesis, for every $r_0 \in U$ there exist an open neighborhood V of r_0 and a plot $\phi : V \to \mathrm{Mor}(\mathbf{K})$ such that $\chi \circ \phi = \Phi \restriction V$. So, for all $r \in V$, $\mathrm{src}(\phi) = b$ and $\mathrm{trg}(\phi) = P(r)$, that is, $\phi \in \mathrm{Diff}(F, X_{P(r)})$, see diagram ($\clubsuit$). Let us then define $\psi : (r, \xi) \mapsto (r, \phi(r)(\xi))$, where $(r, \xi) \in V \times F$. A priori this map takes its values in $V \times X$, but since $\phi(r) \in \mathrm{Diff}(F, X_{P(r)})$, $\mathrm{val}(\psi) \subset P^*(X)$. The map ψ is obviously bijective, and smooth. Indeed, ψ decomposes into $(r, \xi) \mapsto (r, \phi(r), \xi) \mapsto (r, \phi(r)(\xi))$, the first map $(r, \xi) \mapsto (r, \phi(r), \xi)$ is clearly smooth, and the second map $(r, (\phi(r), \xi)) \mapsto (r, \phi(r)(\xi))$ is just $1_V \times \mathrm{ev}$ which is smooth by definition of the functional diffeology. Its inverse is given by $\psi^{-1} : (r, x) \mapsto (r, \phi(r)^{-1}(x))$ which is again smooth, thanks to the smoothness of the inverse map in $\mathrm{Mor}(\mathbf{K})$ and to the smoothness of the evaluation map ev (art. 8.7).

$$
\begin{array}{ccccc}
V \times F & \xrightarrow{\ \psi\ } & P^*(X) & \xrightarrow{\ \mathrm{pr}_2\ } & X \\
 & \searrow{\scriptstyle \mathrm{pr}_1} & \downarrow{\scriptstyle \mathrm{pr}_1} & & \downarrow{\scriptstyle \pi} \\
 & & V & \xrightarrow{\ P\ } & B
\end{array}
\qquad (\clubsuit)
$$

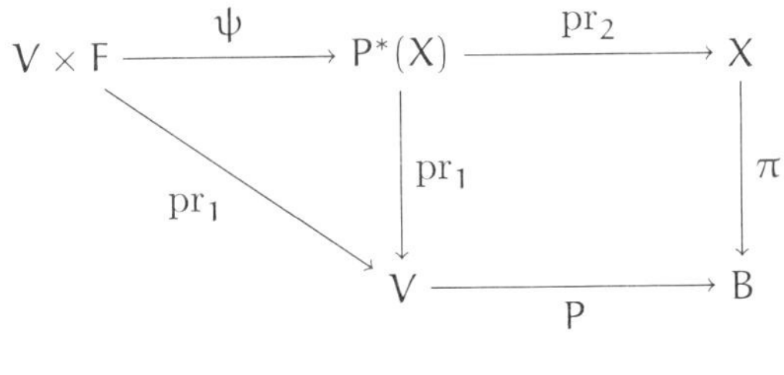

Thus, ψ is a diffeomorphism from $V \times F$ to $P^*(X) \restriction V$. Therefore $P^*(X)$ is locally trivial, with fiber F.

Conversely, let us assume that π is locally trivial along the plots, with fiber F. Let $\Phi : r \mapsto (\phi(r), \phi'(r))$ be a plot of $B \times B$, defined on U. Let $r_0 \in U$, since the pullbacks of π by ϕ and ϕ' are locally trivial, let us choose two open neighborhoods V and V' of r_0 over which $\phi^*(P)$ and $\phi'^*(P)$ are trivial. Let us consider then $W = V \cap V'$. Let Ψ and Ψ' be the respective trivializations, so Ψ and Ψ' write $\Psi(r, \xi) = (r, \psi(r)(\xi))$ and $\Psi'(r, \xi) = (r, \psi'(r)(\xi))$. Since the restriction of a diffeomorphism to a subspace is a diffeomorphism onto its image (art. 1.37), $\psi(r)$ and $\psi'(r)$ take their values respectively in $\mathrm{Diff}(W, X_{\phi(r)})$, and in $\mathrm{Diff}(W, X_{\phi'(r)})$. Then, let us define $Q : r \mapsto \psi'(r) \circ \psi(r)^{-1}$, Q is defined on W with values in $\mathrm{Mor}(\mathbf{K})$ and satisfies $\chi \circ Q(r) = (\phi(r), \phi'(r))$. Now, the spaces $X_{\mathrm{srco}Q}$ and $X_{\mathrm{trgo}Q}$ of (art. 8.7) are given by

$$
\begin{cases}
X_{\mathrm{srco}Q} & = \ \{(r, x) \in W \times X \mid x \in \phi(r)\}, \\
X_{\mathrm{trgo}Q} & = \ \{(r, x) \in W \times X \mid x \in \phi'(r)\}.
\end{cases}
$$

The maps $Q_{\rm src}$ and $Q_{\rm trg}$ of (art. 8.7) decompose into

$$\begin{cases} Q_{\rm src} : (r,x) \mapsto (r, \psi(r)^{-1}(x)) \mapsto (r, \psi'(r)(\psi(r)^{-1}(x))) \mapsto \psi'(r)(\psi(r)^{-1}(x)), \\ Q_{\rm trg} : (r,x) \mapsto (r, \psi'(r)^{-1}(x)) \mapsto (r, \psi(r)(\psi'(r)^{-1}(x))) \mapsto \psi(r)(\psi'(r)^{-1}(x)), \end{cases}$$

that is, $Q_{\rm src} = {\rm pr}_2 \circ \Psi' \circ \Psi^{-1}$ and $Q_{\rm trg} = {\rm pr}_2 \circ \Psi \circ \Psi'^{-1}$. Therefore, $Q_{\rm src}$ and $Q_{\rm trg}$ are smooth, and since $\chi \circ Q$ is also smooth, Q is a plot of ${\rm Mor}(\mathbf{K})$ for the functional diffeology. So, Q is a smooth local lift of the plot Φ of $B \times B$, therefore $\chi : {\rm Mor}(\mathbf{K}) \to B \times B$ is a subduction and π is a fibration, according to (art. 8.8). □

8.10. Diffeological fiber bundles category. Diffeological fibrations form a subcategory of the category of smooth surjections (art. 8.1). But note that for any two diffeological fiber bundles $\pi : X \to B$ and $\pi' : X' \to B'$, a morphism from p to p' is just a smooth map $\phi : X \to X'$ mapping fiber to fiber, since π is a subduction the projection $\phi = {\rm pr}(\Phi) : B \to B'$ is necessarily smooth.

1. *Pullbacks.* The pullback of any diffeological fibration by any smooth map of the base is a diffeological fibration, with same fiber.

2. *Fibered products.* Let $\pi : T \to B$ and $\pi' : T' \to B$ be two diffeological fibrations with same base, F. The *fibered product* is the pullback of the product $\pi \times \pi'$ by the diagonal map $b \mapsto (b, b)$, it is denoted by $\pi \times_B \pi' : T \times T' \to B$.

3. *Products.* Let $\pi : T \to B$ and $\pi' : T' \to B'$ be two diffeological fibrations with fibers F and F'. Then, the product $\pi \times \pi' : T \times T' \to B \times B'$ is a diffeological fibration with fiber $F \times F'$.

4. *Restriction.* Let $\pi : T \to B$ be a diffeological fibration and $A \subset B$ a subset, equipped with the subset diffeology. The restriction of π over A (art. 8.1) is a diffeological fibration.

5. *Subbundles.* Let $\pi : T \to B$ be a diffeological fibration. We call *subbundle* the reduction of π to some subset $\Sigma \subset T$ (art. 8.1) when it happens that this restriction is still a diffeological fibration.

PROOF. We only check the first assertion, the other ones are obvious or direct consequences of the first. Let $f : B' \to B$ be a smooth map and $\pi : X \to B$ be a fiber bundle. Let $P : U \to B'$ be a plot, the pullbacks $P^*(f^*(X))$ and $(f \circ P)^*(X)$ are equivalent. Indeed, an element of the first is a triple $(r, b', x) \in U \times B' \times X$ such that $P(r) = b'$ and $f(b') = \pi(x)$, an element of the second is a pair $(r, x) \in U \times X$ such that $f \circ P(r) = \pi(x)$, the equivalence of the two pullbacks is given by $(r, x) \mapsto (r, b' = P(r), x)$ and conversely $(r, b', x) \mapsto (r, x)$. Now, π is a fiber bundle and $f \circ P$ is a plot of B, the pullback ${\rm pr}_1 : (f \circ P)^*(X) \to B$ is locally trivial (art. 8.9), let $(r, \xi) \mapsto (r, \varphi(r)(\xi))$ be a local trivialization. By composition with the above equivalence we get a local trivialization of the pullback ${\rm pr}_1 : P^*(f^*(X)) \to U$, that is, $(r, \xi) \mapsto (r, P(r), \varphi(r)(\xi))$. □

Exercise

✎ EXERCISE 129 (Groupoid associated with $x \mapsto x^3$). Describe the groupoid associated with the real map $x \mapsto x^3$ and its standard diffeology (art. 8.7). Is this groupoid fibrating?

Principal and Associated Fiber Bundles

The most important subcategory of the category of diffeological fiber bundles is the category of principal fiber bundles. *They are bundles fibered by free actions of diffeological groups* (art. 8.11). *They play a central role because in diffeology every diffeological fiber bundle is "associated" with a principal fiber bundle* (art. 8.16), *even the principal fiber bundles themselves. Another important aspect of fiber bundles and principal fiber bundles is the notion of structure. The structure of the fiber bundle is understood as an equivalent structure for all of its fibers, for example a vector space structure, an affine structure, a group structure, etc. The groupoid approach and the fact that every fiber bundle is defined through a general structure groupoid* (art. 8.8), *which characterizes its* smooth structure, *makes this concept clear: a* structure *is defined by a wide fibrating subgroupoid of the structure groupoid of the fiber bundle, the structure group of the structure groupoid being the structure group of the fiber bundle structure* (art. 8.17).

8.11. Principal diffeological fiber bundles. Let X be a diffeological space and $g \mapsto g_X$ be a smooth action of a diffeological group G on X, that is, a smooth homomorphism from G to Diff(X) (art. 7.4). Let F be the *action map*,

$$F : X \times G \to X \times X \quad \text{with} \quad F(x, g) = (x, g_X(x)).$$

PROPOSITION. If F is an induction, then the projection π from X to its quotient X/G is a diffeological fibration, with the group G as fiber. We shall say that the action of G on X is *principal.*

a) If F is inductive, then it is in particular injective, which implies that the action of G on X is free, which is indeed a necessary condition.

b) If a projection $\pi : X \to Q$ is equivalent to class : $X \to G/H$, that is, if there exists a diffeomorphism $\varphi : G/H \to Q$ such that $\pi = \varphi \circ \text{class}$, we shall say that π is a *principal fibration*, or a *principal fiber bundle*, with structure group G.

NOTE 1. The main assertion is based on the following constructions. If F is an induction, then the subgroupoid $\mathbf{K}_G$ of equivariant arrows of the groupoid $\mathbf{K}$ associated with the projection $\pi : X \to Q$ (art. 8.7),

$$\begin{cases} \text{Obj}(\mathbf{K}_G) & = & Q, \\ \text{Mor}(\mathbf{K}_G) & = & \{f \in \text{Mor}(\mathbf{K}) \mid f \circ g_X = g_X \circ f, \text{ for all } g \in G\}, \end{cases}$$

is fibrating, which implies immediately that $\mathbf{K}$ is fibrating, and then that π is a diffeological fibration. The groupoid $\mathbf{K}_G$ is the *principal fiber bundle structure groupoid*, for the action of G on X (art. 8.17). In order to prove that the groupoid $\mathbf{K}_G$ is fibrating, we introduce a third groupoid denoted by $\pi \times_G \pi$ and defined by

$$\text{Obj}(\pi \times_G \pi) = Q \quad \text{and} \quad \text{Mor}(\pi \times_G \pi) = X \times_G X,$$

where $X \times_G X$ is the quotient of the product $X \times X$ by the diagonal action of G, that is, $g_{X \times X}(x, y) = (g_X(x), g_X(y))$. Let $[x, y] \in X \times_G X$ be the class of the pair (x, y), and let

$$\xi : [x, y] \mapsto (\pi(x), \pi(y))$$

be the characteristic map. The groupoid multiplication is then defined by

$$[x, y] \cdot [y', z'] = [x, g_X^{-1}(z')] \quad \text{where} \quad g_X(y) = y', \ g \in G.$$

The identities and inverses are given by

$$\mathbf{1}_q = [x, x] \quad \text{where} \quad \pi(x) = q, \quad \text{and} \quad [x, y]^{-1} = [y, x].$$

There exists a natural smooth faithful functor Φ from $\pi \times_G \pi$ to $\mathbf{K}_G$ defined by

$$\Phi_O = \mathbf{1}_Q \quad \text{and} \quad \Phi_M : [x, y] \mapsto R_y \circ R_x^{-1},$$

where R_x is the orbit map $g \mapsto g_X(x)$, with $x \in X$ and $g \in G$. Next, we check that $\pi \times_G \pi$ is fibrating. Then, since Φ is smooth, $\mathbf{K}_G$, and therefore $\mathbf{K}$ is also fibrating. Actually, the functor Φ is an equivalence of diffeological groupoid, its inverse Φ^{-1} from $\mathbf{K}_G$ to $\pi \times_G \pi$ is smooth too.

NOTE 2. In the above construction, if $\pi : X \to Q$ and $\pi' : X' \to Q$ are two G-principal fiber bundles and $\psi : X \to X'$ is a G-equivariant diffeomorphism, $\psi(g_X(x)) = g_{X'}(\psi(x))$, then the map $\Psi : [x, y] \mapsto [\psi(x), \psi(y)]'$, from $X \times_G X$ to $X' \times_G X'$, is well defined and it is an isomorphism of diffeological groupoids.

PROOF. a) Let us prove first that $\pi \times_G \pi$ is a well defined groupoid. Let $[x, y]$ and $[s, t]$ be two elements of $X \times_G X$ such that $s = g_X(y)$, $g \in G$. Let x', y', s', t' such that $[x', y'] = [x, y]$ and $[s', t'] = [s, t]$. So, there exists $k, h \in G$ such that $x' = k_X(x)$, $y' = k_X(y)$, $s' = h_X(s)$, $t' = h_X(t)$. By definition we should have $[x, y] \cdot [s, t] = [x, g_X^{-1}(t)]$, and we have $s' = h_X(s) = h_X \circ g_X(y) = h_X \circ g_X \circ k_X^{-1}(y')$, then $[x', y'] \cdot [s', t'] = [x', k_X \circ g_X^{-1} \circ h_X^{-1}(t')] = [k_X(x), k_X(g_X^{-1}(t))] = [x, g_X^{-1}(t)] = [x, y] \cdot [s, t]$. Therefore, the multiplication is well defined. The inverse and the identities are then obvious and obviously smooth.

b) Let p be the natural projection from $X \times X$ to its quotient $X \times_G X$. Thanks to the following diagram and to the fact that π, and thus $\pi \times \pi$, is a subduction, the characteristic map ξ is a subduction.

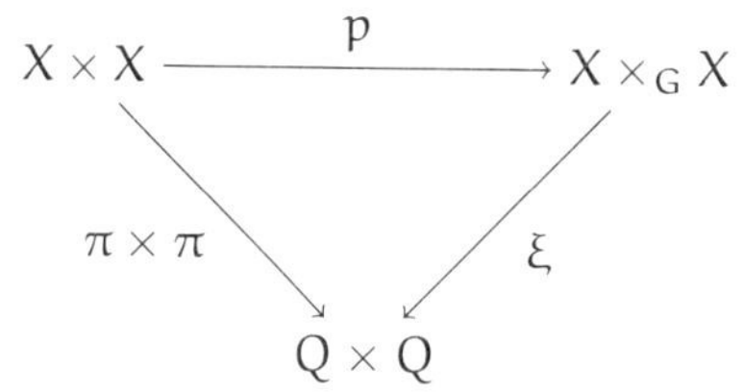

c) Let us prove now that the multiplication is smooth. Let Φ be a plot of the domain of the groupoid multiplication, that is, $\Phi(r) = (\phi(r), \phi'(r))$, where ϕ and ϕ' are plots of $X \times_G X$ and $\mathrm{trg}(\phi(r)) = \mathrm{src}(\phi'(r))$ for all $r \in U = \mathrm{def}(\Phi)$. Since p is a subduction, ϕ and ϕ' can be lifted locally in $X \times X$. Let $\Psi : r \mapsto (\psi_1(r), \psi_2(r))$ and $\Psi' : r \mapsto (\psi_2'(r), \psi_3'(r))$ be two smooth local lifts of ϕ and ϕ', respectively. The condition $\mathrm{trg} \circ \phi = \mathrm{src} \circ \phi'$ writes $\pi \circ \psi_2 = \pi \circ \psi_2'$, for all r in the common domain of the lifts. Since the action of G on X is free, there exists a map γ such that $\psi_2'(r) = \gamma(r)_X(\psi_2(r))$ wherever the two functions are defined. The map $r \mapsto (\psi_2(r), \psi_2'(r))$ is smooth with values in the image of F, since F is an induction and $(\psi_2(r), \gamma(r)) = F^{-1}(\psi_2(r), \psi_2'(r))$, γ is smooth. And since G is a diffeological group, the parametrization $r \mapsto \gamma(r)^{-1}$ is smooth too. Now, let $\psi_3(r) = \gamma(r)^{-1}(\psi_3'(r))$, the map $\Psi'' : r \mapsto (\psi_2(r), \psi_3(r))$ is another smooth local lift of ϕ' and then $\phi(r) \cdot \phi'(r) = p(\psi_1(r), \psi_3(r))$. Therefore, the map $r \mapsto \phi(r) \cdot \phi'(r)$ is locally smooth, and thus smooth. Hence, the groupoid $\pi \times_G \pi$ is a diffeological groupoid, and since ξ is a subduction, $\pi \times_G \pi$ is a fibrating groupoid.

d) We shall exhibit a smooth morphism from $\pi \times_G \pi$ to $\mathbf{K}_G$, which will be used to prove that $\mathbf{K}$ is fibrating. Let φ be the map, from $X \times X$ to $\mathrm{Mor}(\mathbf{K}_G)$, defined by

$$\varphi(x, y) = R_y \circ R_x^{-1} : G_X(x) \to G_X(y), \text{ for all } x, y \in X.$$

This is the only equivariant map f from the orbit $G_X(x)$ to $G_X(y)$ mapping x to y. Indeed, if another equivariant map $f' : G_X(x) \to G_X(y)$ maps x to y, then for all $g \in G$, $f(g_X(x)) = g_X(f(x)) = g_X(f'(x)) = f'(g_X(x))$, and since every $x' \in G_X(x)$ writes in a unique way $g_X(x)$ for some $g \in G$, $f(x') = f'(x')$ for all $x' \in G_X(x)$, and $f = f'$. Thus, $\mathrm{val}(\Phi_M) = \{f \in \mathrm{Mor}(\mathbf{K}) \mid f \circ g_X = g_X \circ f, \text{ for all } g \in G\} = \mathrm{Mor}(\mathbf{K}_G)$. Now, $\varphi(x, y) = \varphi(x', y')$ if and only if $(x', y') = (g_X(x), g_X(y))$ for some $g \in G$. Indeed, if $R_y \circ R_x^{-1} = R_{y'} \circ R_{x'}^{-1}$, then $x' = g_X(x)$ for some $g \in G$, and $R_{y'} \circ R_{x'}^{-1}(x') = R_y \circ R_x^{-1}(g_X(x))$ gives $y' = g_X(y)$. Thus, there exists a map

$$\bar\varphi : X \times_G X = \mathrm{Mor}(\pi \times_G \pi) \to \mathrm{Mor}(\mathbf{K}_G)$$

such that $\bar\varphi \circ p = \varphi$. Let us check that φ is smooth, since p is a subduction it will imply that $\bar\varphi$ also is smooth.

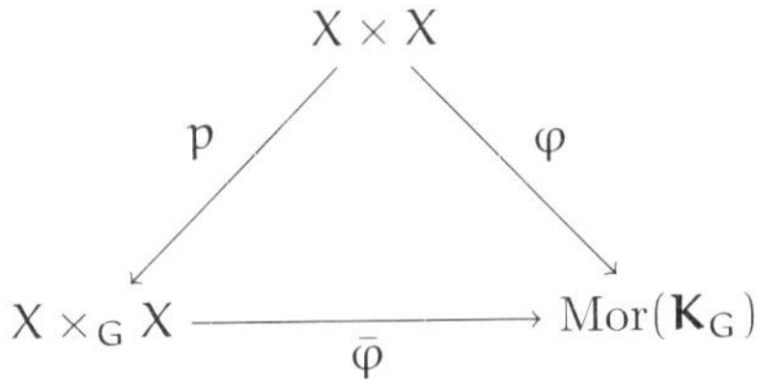

(♣)

Let $H : r \mapsto (\eta(r), \eta'(r))$ be a plot of $X \times X$, let $\mathbf{H} = \varphi \circ H$. The spaces $X_{\mathrm{src}\circ\mathbf{H}}$, $X_{\mathrm{trg}\circ\mathbf{H}}$ and the maps $\mathbf{H}_{\mathrm{src}}$ and $\mathbf{H}_{\mathrm{trg}}$, see (art. 8.7, proof), are given by

$$\begin{aligned} X_{\mathrm{src}\circ\mathbf{H}} &= \{(r, x) \in \mathrm{def}(H) \times X \mid \pi \circ \eta(r) = \pi(x)\}, & \mathbf{H}_{\mathrm{src}}(r, x) &= \mathbf{H}(r)(x), \\ X_{\mathrm{trg}\circ\mathbf{H}} &= \{(r, x) \in \mathrm{def}(H) \times X \mid \pi \circ \eta'(r) = \pi(x)\}, & \mathbf{H}_{\mathrm{trg}}(r, x) &= \mathbf{H}(r)^{-1}(x). \end{aligned}$$

Next, $\pi \circ \eta(r) = \pi(x)$ means that there exists

$$\gamma : X_{\mathrm{src}\circ\mathbf{H}} \to G \quad \text{such that} \quad \gamma(r, x)_X(\eta(r)) = x.$$

As well, there exists γ' for η'. Let us check that γ is smooth. Let $\psi : t \mapsto (R(t), P(t))$ be a plot of $X_{\mathrm{src}\circ\mathbf{H}}$, then $\gamma \circ \psi(t) = \gamma(R(t), P(t))$ satisfies $\gamma(R(t), P(t))_X(\eta(R(t))) = P(t)$, but the maps $\eta \circ R$ and P are smooth and with values in $\mathrm{val}(F)$. Since F is an induction, the map $t \mapsto \gamma(R(t), P(t))$ is smooth, and thus γ is smooth. Therefore, the map $\mathbf{H}_{\mathrm{src}}$, which writes $\mathbf{H}_{\mathrm{src}}(r, x) = \mathbf{H}(r)(\gamma(r, x)_X(\eta(r)))$, is smooth because it is the composite of smooth maps. Actually, thanks to the equivariance of the map $\mathbf{H}(r) \in \mathrm{Mor}(\pi \times_G \pi)$ we have $\mathbf{H}_{\mathrm{src}}(r, x) = \gamma(r, x)_X(\mathbf{H}(r)(\eta(r)))$, but $\mathbf{H}(r)(\eta(r)) = \eta'(r)$, so $\mathbf{H}_{\mathrm{src}}(r, x) = \gamma(r, x)_X(\eta'(r))$. The same holds *mutatis mutandis* for η', and we can check that the map $\mathbf{H}_{\mathrm{trg}}$ is smooth, using moreover the smoothness of the inversion in G. Now, on the one hand we have $\bar\varphi([x, y] \cdot [y', z']) = \bar\varphi([x, y] \cdot [y, z])$, where $y' = g_X(y)$ and $z = g_X^{-1}(z')$, and $\bar\varphi([x, y] \cdot [y, z]) = \bar\varphi([x, z]) = R_z \circ R_x^{-1}$, and on the other hand $\bar\varphi([x, y]) \cdot \bar\varphi([y, z]) = (R_z \circ R_y^{-1}) \circ (R_y \circ R_x^{-1}) = R_z \circ R_x^{-1}$. Then, $\Phi = (\Phi_O = \mathbf{1}_Q, \Phi_M = \bar\varphi)$ is a (smooth faithful) functor from $\pi \times_G \pi$ to $\mathbf{K}_G$. Moreover, by the next commutative diagram, where p and $\bar\varphi$ are smooth and $\pi \times \pi$ and ξ are subductions, we deduce that χ is smooth and is a subduction. Thus, $\mathbf{K}_G$ is a fibrating groupoid, and so is $\mathbf{K}$. Let us check now that Φ is an equivalence of diffeological groupoid. Let $r \mapsto f_r$ be a plot of $\mathrm{Mor}(\mathbf{K}_G)$. Thus, $r \mapsto q_r = \mathrm{src}(f_r)$ is a plot of Q. Let $r \mapsto x_r$ be a smooth local lift of $r \mapsto q_r$, so $r \mapsto x'_r = f_r(x_r)$ is a plot of X, and $f_r = R_{x'_r} \circ R_{x_r}^{-1}$, since f_r is only determined by its value at one point. Then, the map φ of the diagram (♣) is a subduction, and $\bar{\varphi}$ is a diffeomorphism. So, $\mathbf{K}_G$ and $\pi \times_G \pi$ are equivalent diffeological groupoids.

$$
\begin{array}{ccccc}
X \times X & \xrightarrow{\quad p \quad} & X \times_G X & \xrightarrow{\quad \bar{\varphi} \quad} & \mathrm{Mor}(\mathbf{K}_G) \\
& \searrow & \downarrow{\scriptstyle \xi} & \swarrow & \\
{\scriptstyle \pi \times \pi} & & & {\scriptstyle \chi} & \\
& & Q \times Q & &
\end{array}
$$

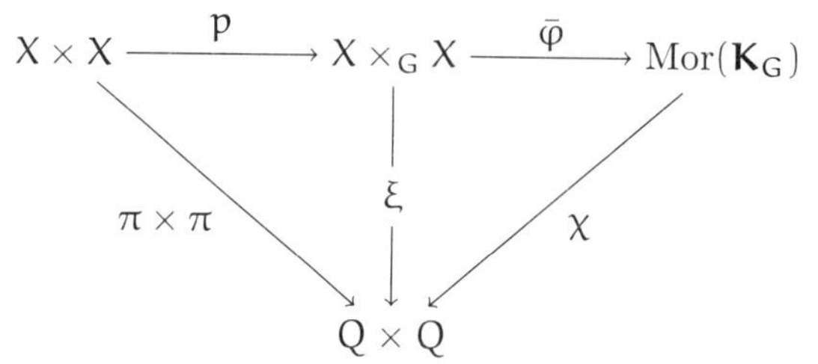

e) Concerning the projection $\pi : X \mapsto Q$, if the preimages of π are the orbits of G, and if π is a subduction, then Q is equivalent to X/G, that is, the map $\varphi : X/G \to Q$ defined by $\varphi(\mathrm{class}(x)) = \pi(x)$ is a natural diffeomorphism. Then, the couple of morphisms $(\mathbf{1}_X, \varphi)$ is an equivalence between the projection class $: X \to X/G$ and $\pi : X \to Q$, π is thus a fibration. $\qquad\square$

8.12. Category of principal fiber bundles. From the definition of morphisms between diffeological groupoids (art. 8.2) we get a natural category of principal fiber bundles. Let $\pi : X \to Q$ and $\pi' : X' \to Q'$ be two principal fiber bundles with structure groups G and G'. A *morphism of principal fiber bundles* from π to π' is a smooth map $\phi : X \to X'$, and a smooth homomorphism $h : G \to G'$ such that

$$\phi \circ g_X = h(g)_{X'} \circ \phi, \text{ for all } g \in G.$$

This defines a map $\varphi : Q \to Q'$ by $\varphi(q) = \pi'(\phi(x))$, with $\pi(x) = q$, and since π is a subduction, φ is smooth. When ϕ is a diffeomorphism and h an isomorphism, they define an isomorphism of principal fiber bundle, φ is then a diffeomorphism.

NOTE 1. When $Q = Q'$ and $\varphi = \mathbf{1}_Q$, the definition above gives the subcategory of Q-principal fiber bundles.

NOTE 2. Fixing the structure group G introduces the subcategory of G-*principal fiber bundles*. In this category, the trivial principal bundle is the trivial projection with fiber G, $\mathrm{pr}_1 : Q \times G \mapsto Q$. A principal fiber bundle is trivial if and only if it has a global smooth section.

NOTE 3. Pullbacks of diffeological principal fiber bundles are naturally diffeological principal fiber bundles. Let $\pi : X \to Q$ be a principal fiber bundle with structure group G, and let $f : Q' \to Q$ be a smooth map. The pullback of π by f is the projection $\mathrm{pr}_1 : f^*(X) \to Q'$ where $f^*(X)$ is the set of pairs $(q', x) \in Q' \times X$ such that $f(q') = \pi(x)$, and this projection is a fiber bundle (art. 8.10). Then, the pullback $\mathrm{pr}_1 : f^*(X) \to Q'$ is a principal fiber bundle for the natural action of G on $f^*(X)$ defined by $g_{f^*(X)}(q', x) = (q', g_X(x))$, where $g \in G$ and g_X denotes the action of G on X. The second projection $\mathrm{pr}_2 : f^*(X) \to X$ is clearly a morphism of a principal fiber bundle.

PROOF. Let us prove the second note. A fiber bundle $\pi : X \to Q$ with fiber G is trivial if and only if there exists a diffeomorphism $\psi : Q \times G \to X$ over the identity of Q, that is, $\pi \circ \psi = \mathrm{pr}_1$ (art. 8.1). It is trivial as a G-principal bundle if and only if this diffeomorphism is equivariant. In this case we get a smooth section $\sigma : q \mapsto \psi(q, \mathbf{1}_G)$. Conversely, any smooth section $\sigma : Q \to X$ defines an equivariant diffeomorphism $\psi : Q \times G \to X$ by $\psi(q, g) = g_X(\sigma(q))$. For the third note. The proof of the assertion is a direct consequence of (art. 8.10, point 1) and the characterization (art. 8.13) of diffeological principal bundles by equivariant trivializations along plots. □

8.13. Equivariant plot trivializations. Let $\pi : X \to Q$ be a G-principal fiber bundle. There exists a family of *equivariant trivializations* along the plots. Let $q : U \to Q$ be a plot lifted in X by a plot P, $\pi \circ P = q$. We can restrict to some subdomain of U if necessary, it is always possible since π is a subduction. Thus, the map $\psi : U \times G \to q^*(X)$, defined by

$$\psi(r, g) = (r, g_X(P(r))),$$

is an equivariant diffeomorphism from $U \times G$ to $q^*(X)$, over the identity of U. Conversely, if there exits such a family of equivariant local trivializations along the plots, then $\pi : X \to Q$ is a principal fibration with structure group G, that is, the action map $F : (x, g) \mapsto (x, g_X(x))$ is an induction (art. 8.11).

NOTE 1. For any two plots $P : U \to X$ and $P' : U \to X$ such that $\pi \circ P = \pi \circ P'$, there exists a plot $\gamma : U \to G$ such that $P'(r) = \gamma(r)_X(P(r))$.

NOTE 2. Let $\pi' : X' \to Q'$ be a G'-principal fiber bundle. Let ϕ be a smooth morphism of principal bundle from π to π' such that $\pi' \circ \phi = \varphi \circ \pi$, and $h : G \to G'$ be a smooth morphism such that $\phi(g_X(x)) = h(g)_{X'}(\phi(x))$. If φ is a diffeomorphism and h an isomorphism of diffeological groups, then ϕ is an isomorphism of diffeological principal fiber bundle.

PROOF. We assume that there exists such a family of equivariant trivializations, which implies in particular that the action of G on X is free. Let $r \mapsto (P(r), P'(r))$ be a smooth parametrization defined on some real domain U, with values in the image of the action map F. Thus, there exists a unique parametrization $r \mapsto g(r)$ of G such that $P'(r) = g(r)_X(P(r))$, for all $r \in U$. Now, $\pi \circ P = \pi \circ P'$ is a plot of Q, let us denote it by q. Let $\phi : q^*(X) \to U \times G$ be an equivariant trivialization. The plots P and P' define two sections of $q^*(X)$, $r \mapsto (r, P(r))$ and $r \mapsto (r, P'(r))$. Thus, there exist two unique smooth parametrizations γ and γ' of G such that $\phi(r, P(r)) = (r, \gamma(r))$ and $\phi(r, P'(r)) = (r, \gamma'(r))$, then $(r, \gamma'(r)) = \phi(r, P'(r)) = \phi(r, g(r)_X(P(r))) = g(r)_{U\times G}(\phi(r, P(r))) = g(r)_{U\times G}(r, \gamma(r)) = (r, g(r)\gamma(r))$. Hence, $\gamma'(r) = g(r)\gamma(r)$ and finally $g(r) = \gamma'(r)\gamma(r)^{-1}$. Since γ and γ' are smooth and the inversion in G is smooth, $r \mapsto g(r)$ is a smooth parametrization, and F is an induction.

Let us check Note 1. Since $\pi : X \to Q$ is a principal fiber bundle, for all $r \in U$, there exists a unique $\gamma(r) \in G$ such that $P'(r) = \gamma(r)_X(P(r))$. Let $q = \pi \circ P = \pi \circ P'$, and let us restrict $q^*(X)$ over an open subset V of U where $\mathrm{pr}_1 : q^*(X) \to U$ is trivial. Let $\psi : (q \upharpoonright V)^*(X) \to V \times G$ be an equivariant trivialization. Let $r \mapsto g_r$ and $r \mapsto g'_r$ be the two plots of G such that $\psi(r, P(r)) = (r, g_r)$ and $\psi(r, P'(r)) = (r, g'_r)$. Thus, thanks to the equivariance property, $\psi(r, P(r)) = \psi(r, \gamma(r)_X(P(r)))$, with $\gamma(r) = g_r'^{-1} g_r$. Therefore, $\gamma \upharpoonright V$ is a plot of G, and so is γ.

For Note 2, we check immediately that, under these assumptions φ a diffeomorphism and h an isomorphism, ϕ is bijective, that is, a smooth bijection. Now, let us check that ϕ^{-1} is smooth. Let $P : U \to X'$ be a plot, thus $\varphi^{-1} \circ \pi' \circ P$ is a plot of Q. Hence, there exists a local smooth lift F such that $\pi \circ F = \varphi^{-1} \circ \pi' \circ P$ on the domain of F. Thus, $\pi' \circ \phi \circ F = \varphi \circ \pi \circ F = \pi' \circ P$, and thanks to Note 1, there exists a plot γ' of G' such that $\phi \circ F(r) = \gamma'(r)_{X'}(P(r))$. Let $\gamma(r) = h^{-1}(\gamma'(r))$, γ is a plot of G and $\phi \circ F(r) = h(\gamma(r))_{X'}(P(r))$, that is, $\phi(\gamma(r)_X(F(r))) = P(r)$, then $\phi^{-1} \circ P(r) = h(\gamma(r))\chi(F(r))$. Hence, $\phi^{-1} \circ P$ is locally smooth, thus smooth. Therefore, ϕ is an isomorphism of diffeological principal fiber bundle. □

8.14. Principal bundle attached to a fibrating groupoid. Let $\mathbf{K}$ be a fibrating groupoid (art. 8.4). Let $Q = \mathrm{Obj}(\mathbf{K})$ and pick a point $o \in Q$, then define

$$X = \mathrm{src}^{-1}(o), \quad \pi = \mathrm{trg} \upharpoonright X, \quad G = \mathbf{K}_o = \chi^{-1}(o, o),$$

and denote by $g \mapsto g_X$ the natural left action of G on X,

$$\text{for all } (g, x) \in G \times X, \quad g_X(x) = g \cdot x.$$

Then, the projection $\pi : X \to Q$ is a principal fibration with structure group G. We shall call it the *principal fiber bundle attached to* $\mathbf{K}$ *at the point* o.

NOTE 1. The principal structure bundles π and π' of a fibrating groupoid, attached at different basepoints o and o′, are naturally isomorphic. Let f be any element of $\mathrm{Mor}_{\mathbf{K}}(o', o)$, the map $x \mapsto f \cdot x$ from X to X′ is such an isomorphism, with the structure groups G and G′ naturally conjugated by $g \mapsto f \cdot g \cdot f^{-1}$.

NOTE 2. Every diffeological fibration $p : T \to Q$ has a natural family of *principal structure bundles*, the principal fiber bundle attached to the structure groupoid of p (art. 8.7) at the various points of Q. As well as the structure groupoid, they capture equivalently the smooth structure of the fibration p.

PROOF. First of all, let us notice that trg is a subduction and then $Q = X/G$. Indeed, let $P : U \to Q$ be a plot, thus $r \mapsto (o, P(r))$ is a plot of $Q \times Q$. Since χ is a subduction, this plot can be locally lifted into Mor(K) by a plot $r \mapsto \bar{P}(r)$, but since $\mathrm{src}(\bar{P}(r)) = o$, $\bar{P}$ is a plot of X, lifting locally P. Now, the action map $F : X \times G \to X \times X$, defined by $F(x, g) = (x, g_X(x))$ just writes $F(x, g) = (x, g \cdot x)$, it is obviously smooth and its inverse is $F^{-1}(x, x') = (x, x' \cdot x^{-1})$, which is smooth thanks to the very property of diffeological groupoid. Hence, F is an induction and π is a principal fibration (art. 8.11). □

8.15. Fibrations of groups by subgroups. Let G be a diffeological group, and let $H \subset G$ be any subgroup. The projection $\pi : G \to G/H$, is a principal diffeological fibration with structure group H, where H acts on G by left or right multiplication and the coset G/H is equipped with the quotient diffeology.

NOTE. In the example where $G = T^2 = \{(e^{ix}, e^{iy}) \mid (x, y) \in \mathbf{R}^2\}$ and $H = \{(e^{2\pi i t}, e^{2\pi i \alpha t}) \mid t \in \mathbf{R}\}$, with $\alpha \in \mathbf{R} - \mathbf{Q}$, the quotient G/H is the *irrational torus* T_α *of slope* α; see Exercise 31, p. 31. Since T_α is not a manifold, the projection $\pi : G \to G/H$ is not a fibration in the category {Manifolds}, but it is a fibration in the category {Diffeology}.

PROOF. Let $\pi : G \to X$ be the projection from G onto its quotient $X = G/H$. We choose H acting on G by left multiplication, *i.e.*, $g \sim hg$, $g \in G$ and $h \in H$. Let $P : U \to X$ be a plot and $r_0 \in U$. By definition of the quotient diffeology, there exist an open neighborhood V of r_0 and a plot $Q : V \to G$ such that $\pi \circ Q = P \upharpoonright V$. Then the map $\Phi : (r, h) \mapsto (r, h \cdot Q(r))$, defined on $V \times H$, takes its values into $(P \upharpoonright V)^*(G) \subset P^*(G)$, and it is smooth because the multiplication is smooth. The inverse is given by $\Phi^{-1}(r, g) = (r, g \cdot Q(r)^{-1})$, and it is smooth because the inversion is smooth. Thus, $\mathrm{pr}_1 : (P \upharpoonright V)^*(G) \to V$ is trivial, hence $\mathrm{pr}_1 : P^*(G) \to U$ is locally trivial, and the projection π is a fibration (art. 8.9). Now, let $L(g')(r, g) = (r, g'g)$

for every $r \in U$, and $g, g' \in G$. Then, for all $h \in H$, we have $\Phi \circ L(h) = L(h) \circ \Phi$. The projection $\pi : G \to G/H$ is thus a principal fibration with structure group H. □

8.16. Associated fiber bundles. Let $\pi : T \to B$ be a principal fiber bundle (art. 8.11) with structure group G, and let E be a diffeological space together with a smooth action of G, that is, a smooth homomorphism $g \mapsto g_E$ from G to $\mathrm{Diff}(E)$. Let $X = T \times_G E$ be the quotient of $T \times E$ by the diagonal action of G, $g_{T \times E} : (t, e) \mapsto (g_T(t), g_E(e))$. Let $p : X \to B$ be the projection $\mathrm{class}(t, e) \mapsto \pi(t)$. Then the projection p is a diffeological fiber bundle, with fiber E. We call it the fiber bundle *associated with* π *by the action of* G *on* E. We remark that if the action of G on E is trivial, then the associated fiber bundle is also trivial. The converse is given by the following proposition.

(♣) Every diffeological fiber bundle $p : X \to B$ is associated with its principal structure bundles (art. 8.14, Note 2).

Precisely, let $\mathbf{K}$ be the structure groupoid of the projection p. Let $\pi : T \to B$ be the principal fiber bundle attached to $\mathbf{K}$ at some point $o \in B$, and let $E = p^{-1}(o)$ be the fiber. Let $G = \pi^{-1}(o) = \mathrm{Diff}(E)$ be the structure group, acting on T by right action, $(g, f) \mapsto f \circ g^{-1}$, $(g, f) \in G \times T$, and acting on E naturally by $(g, e) \mapsto g(e)$, $(g, e) \in G \times E$. Therefore, $p : X \to B$ is equivalent to the associated fiber bundle $T \times_G E \to B$.

$$\begin{array}{ccc} T \times E & \xrightarrow{\ \mathrm{class}\ } & X = T \times_G E \\ {\scriptstyle \mathrm{pr}_1}\downarrow & & \downarrow{\scriptstyle p} \\ T & \xrightarrow[\ \pi\ ]{} & B \end{array}$$

NOTE 1. If a principal fiber bundle $\pi : T \to B$ is trivial, then the associated fiber bundle $p : T \times_G E \to B$ is also trivial. Conversely, if a diffeological fiber bundle $p : X \to B$ is trivial, then its principal structure bundles, the ones which characterize its smooth structure, are trivial.

NOTE 2. If $f : B' \to B$ is a smooth map, then $f^*(T \times_G E)$ is equivalent to $f^*(T) \times_G E$, where the action of G on $f^*(T)$ is on the second factor.

NOTE 3. Every smooth section σ of the associated bundle $p : T \times_G E$ corresponds to a smooth equivariant map $s : T \to E$, that is, $s(g_X(t)) = g_E(s(t))$, with $\sigma(b) = \mathrm{class}(t, s(t))$, for any t such that $\pi(t) = b$. This is summarized by $\mathrm{Sec}^\infty(p : T \times_G E \to B) \simeq \mathrm{Eq}^\infty(T, E)$, with a clear meaning of the notations.

PROOF. We prove the first note independently of the main proposition. Let $\pi = \mathrm{pr}_1 : T = Q \times G \to Q$. The diagonal action of G on $T \times E$ writes $g_{T \times E}(q, k, e) = (q, gk, g_E(e))$, and the map $\psi : \mathrm{class}(q, k, e) \mapsto (q, k_E^{-1}(e))$ from $T \times_Q E$ to $Q \times E$ is a realization of the quotient space $X = T \times_G E$.

For the second note, the pullback $f^*(T \times_G E)$ and $f^*(T) \times_G E$ are equivalent to the quotient of $\{(b', t, e) \in B' \times T \times E \mid f(b') = \pi(t)\}$ by the diagonal action of G on the last two variables.

Let us check now that the associated bundle $p : T \times_G E \to B$ is a diffeological fiber bundle. Let P be a plot of B. Thanks to Note 2, the pullback $P^*(T \times_G E)$ is equivalent to $P^*(T) \times_G E$, but $P^*(T)$ is locally trivial (art. 8.13), thus $P^*(T) \times_G E$ is also locally trivial, thanks to Note 1. Therefore $P^*(T \times_G E)$ is locally trivial, and $p : T \times_G E \to B$ is a diffeological fiber bundle (art. 8.9).

Conversely, let $p : X \to B$ be a diffeological fiber bundle. Let $\pi : T \to B$ be a principal bundle attached to its structure groupoid $\mathbf{K}$, at some point $o \in B$ (art. 8.14), (art. 8.7), and let $E = X_o$. Thus, $T = \{f \in \mathrm{Diff}(E, T_b) \mid b \in B\}$ and $G = T_o = \mathrm{Diff}(E)$ is the structure group. Now, let $\phi : T \times E \to X$ be defined by $\phi(f, e) = f(e)$. Since p is a diffeological fibration, p is surjective and all the fibers are equivalent. Now, $f' = f \circ g^{-1}$ and $e' = g(e)$ if and only if $\phi(f, e) = \phi(f', e')$. Then, ϕ is the composite $\varphi \circ \mathrm{class}$, where $\mathrm{class} : T \times E \to T \times_G E$ is the projection, and $\varphi : T \times_G E \to X$ is a bijection. But ϕ is a subduction. Indeed, let P be a plot of X and $\beta = p \circ P$, since β is a plot of B and $\pi : T \to B$ is a subduction, β lifts locally in a plot τ of T, in particular $\tau(r) \in \mathrm{Diff}(E, X_{\beta(r)})$. Then, since $P(r) \in X_{\beta(r)}$, and since τ and P are smooth, $r \mapsto e(r) = \tau(r)^{-1}(P(r))$ is smooth. Hence, $r \mapsto (\tau(r), e(r))$ is a plot, lifting locally P along ϕ. Therefore, ϕ is a subduction. Next, since ϕ and class are subductions, φ is a diffeomorphism and it projects onto the identity of B. Therefore, X is equivalent to the associated fiber bundle $T \times_G E \to B$.

For the third note, if s is an equivariant map, it is clear that the map $\bar\sigma : t \mapsto \mathrm{class}(t, s(t))$ from T to $T \times_G E$, is invariant by the action of G, that is, $\bar\sigma(g_T(t)) = \bar\sigma(t)$. And since $\pi : T \to B$ is a subduction, there exists a smooth map, which is clearly a section, $\sigma : B \to T \times_G E$ such that $\sigma \circ \pi = \bar\sigma$. Conversely, let σ be a smooth section, since the action of G on T is free, there exists a unique $s(t) \in E$ such that $\sigma(\pi(t)) = \mathrm{class}(t, s(t))$, and the map s is clearly equivariant. Now, let us show that s is smooth. Let $r \mapsto t_r$ be a plot of T, since class is a subduction, there exists a local lift $r \mapsto (t'_r, e'_r)$ such that $\bar\sigma(t_r) = \mathrm{class}(t'_r, e'_r)$, but $\pi(t'_r) = \pi(t_r)$, then $t'_r = (g_r)_X(t_r)$, where $r \mapsto g_r$ is a plot of G, because π is locally trivial along the plots. Now, let $e_r = (g_r^{-1})_E(e'_r)$, then $r \mapsto (t_r, e_r)$ is another smooth local lift of $\bar\sigma$, and by unicity $e_r = s(t_r)$, hence s is smooth. □

8.17. Structures on fiber bundles. Let $p : X \to B$ be a diffeological fiber bundle. Let $\mathbf{K}$ be the associated structure groupoid (art. 8.7). We call the *structure* of the fiber bundle any fibrating subgroupoid Γ of $\mathbf{K}$.

1. Every principal fiber bundle $\pi : T \to B$, attached to Γ at some point $o \in B$ (art. 8.14), is a *principal fiber bundle structure* for the Γ-structure, and the structure group G of π is the *structure group* of the *structure groupoid* Γ.

2. Together with a structure Γ, a fiber bundle $p : X \to B$ is called a *structured fiber bundle*. The structured fiber bundles form a category, let $p' : X' \to B'$ be a fiber bundle structured by Γ'. A *structured morphism* $\Phi : X \to X'$ is a fiber bundle morphism (art. 8.10) which defines a functor from Γ to Γ' by

$$f \mapsto f' = (\Phi \restriction X_{b_2}) \circ f \circ (\Phi \restriction X_{b_1})^{-1}, \text{ for all } f \in \mathrm{Mor}_\Gamma(b_1, b_2). \qquad (\diamondsuit)$$

For every morphism $f \in \mathrm{Mor}_\Gamma(b_1, b_2)$, f' belongs to $\mathrm{Mor}_{\Gamma'}(\phi(b_1), \phi(b_2))$, where $\phi : B \to B'$ is the factorization of Φ, $p' \circ \Phi = \phi \circ p$.

EXAMPLE. The main example of structured bundle, except the structure of principal fiber bundle that we have already seen, is the structure of vector fiber bundle for which the structure consists in a groupoid of linear maps.

NOTE 1. The notion of structure introduced in this paragraph is the diffeological version of F-structure, for topological fiber bundles, one can find in [**FuRo81**]. The structure groupoid Γ replaces the family of “marked” homeomorphisms.

NOTE 2. Every diffeological fiber bundle has a favorite structure, the *smooth structure*, defined by the groupoid $\mathbf{K}$ itself. The structured smooth morphisms ($\diamondsuit$) coincide with the fiber bundles morphisms defined in (art. 8.10).

NOTE 3. Principal fiber bundles deserve a special attention. We define the *principal fiber bundle structure* of any principal fiber bundle $p : X \to B$, with structure group G, as the groupoid $\mathbf{K}_G$ defined in (art. 8.11). So, every principal structure fiber bundle $\pi : T \to B$, attached at any point $o \in B$ (art. 8.14), is equivalent to $p : X \to B$, and the principal structure morphisms ($\diamondsuit$) coincide with the morphisms of principal fiber bundles.

NOTE 4. Let $f : B' \to B$ be a smooth map. There exists, on the pullback $p' : f^*(X) \to B'$ (art. 8.10), a natural structure Γ', induced by Γ, defined by

$$\mathrm{Obj}(\Gamma') = B' \quad \text{and} \quad \mathrm{Mor}_{\Gamma'}(b'_1, b'_2) = \mathrm{Mor}_\Gamma(f(b'_1), f(b'_2)).$$

NOTE 5. Let $p : X \to B$ be a fiber bundle, let b be a point of B, let $F = p^{-1}(b)$, and let Γ be a subgroupoid of its smooth structure groupoid $\mathbf{K}$. The groupoid Γ defines a structure on p if and only if, for any plot $P : U \to B$, there exists a system of local trivializations $\{\Psi_i\}_{i \in \mathcal{J}}$ of the pullback $\mathrm{pr}_1 : P^*(X) \to U$ such that, for all $(r, u) \in U_i \times F - \mathrm{def}(\Psi_i)$,

$$\Psi_i(r, \xi) = (r, \psi_i(r)(\xi)), \text{ with } \begin{cases} \psi_i \in \mathcal{C}^\infty(U_i, \mathrm{Mor}(\Gamma)), \\ \psi_i(r) \in \mathrm{Mor}_\Gamma(F, X_{P(r)}). \end{cases}$$

This is the specialization, in the presence of a Γ-structure, of the criterion of local triviality along the plots for diffeological fiber bundles (art. 8.9).

NOTE 6. As we have seen, every diffeological fiber bundle is associated with any one of its smooth structure principal fiber bundles (art. 8.16, ($\clubsuit$)), by the same process by which every structured diffeological fiber bundle is associated with any one of its structure principal bundles.

PROOF. The assertion of the second note is clear, the restriction of every global diffeomorphism to a subspace is a diffeomorphism from the subspace to its image. Now, let $p : X \to B$ be a principal fiber bundle with group G. Let $\chi : X \times_G X \to B \times B$ be the characteristic map of the groupoid $p \times_G p$, equivalent to $\mathbf{K}_G$ (art. 8.11, Note 1). The total space of the principal structure bundle, attached at $o \in B$, is $T = \chi^{-1}(\{o\} \times B)$, that is, the subspace $T = \{\mathrm{class}(x, x') \mid x, x' \in X \text{ and } p(x) = o\}$, thus $T = X_o \times_G X$. But, by definition of a principal fiber bundle, the action map $F : X \times G \to X \times X$ is an induction (art. 8.11) and $X_o \times X$ is equivalent to $G \times X$ where the diagonal action of G transmutes into $g_{G\times X}(g', x') = (gg', g_X(x'))$, hence the quotient T is equivalent to X, thanks to the subduction $(g', x') \mapsto g'^{-1}_X(x')$. That proves the first part. Next, let us consider a structure morphism Φ, from $p : X \to B$ to $p' : X' \to B'$. Since, by definition, Φ maps fibers into fibers and since the action map $F' : X' \times G' \to X' \times X'$ is an induction (art. 8.11), there exists a smooth map $h : G \times X \to G'$ such that $\Phi(g_X(x)) = h(g, x)_{X'}(\Phi(x))$. So, for every $f \in \mathrm{Mor}(\mathbf{K}_G)$ with $f : X_{b_1} \to X_{b_2}$, $\Phi(g_X(f(x))) = h(g, f(x))_{X'}(\Phi(f(x)))$. But $g_X \circ f = f \circ g_X$, then $\Phi(f(g_X(x))) = h(g, f(x))_{X'}(\Phi(f(x)))$. Now, by hypothesis, there exists $f' \in \mathrm{Mor}(\mathbf{K}_{G'})$ such that $\Phi_{b_2} \circ f = f' \circ \Phi_{b_1}$, where $\Phi_{b_i} = \Phi \restriction X_{b_i}$, thus $f'(\Phi(g_X(x))) = h(g, f(x))_{X'}(f'(\phi(x))) = f'(h(g, f(x))_{X'}(\phi(x)))$. Then, since f' is a diffeomorphism, $\Phi(g_X(x)) = h(g, f(x))_{X'}(\phi(x))$, but $\Phi(g_X(x)) = h(g, x)_{X'}(\phi(x))$, therefore $h(g, f(x))_{X'}(\phi(x)) = h(g, x)_{X'}(\phi(x))$. But the action of G' is free, then

$h(g, f(x)) = h(g, x)$. Now, for every pair of points x, y in X, there exists a morphism f such that $f(x) = y$, thus $h(g, x)$ does not depend on x, and $\Phi(g_X(x)) = h(g)_{X'}(\Phi(x))$. It is then clear that h is a homomorphism. ☐

8.18. Space of structures of a diffeological fiber bundle. We know that every diffeological fiber bundle $p : X \to B$, with some fiber F, is associated with a diffeological principal fiber bundle $\pi : T \to B$, with some group G as structure group (art. 8.17, Note 6), and with structure groupoid Γ equivalent to $\pi \times_G \pi$ (art. 8.11, Note 1). For the smooth structure, the group G is the full group $\mathrm{Diff}(F)$. So, equipping the fiber bundle p with a special structure consists in reducing the principal bundle π to a principal subbundle for some subgroup $H \subset G$.

Definition. A *reduction* of a principal fiber bundle $\pi : T \to B$, with structure group G, to a subgroup $H \subset G$ is any H-principal subbundle $\pi_S : S \to B$, where $S \subset T$.

Let us consider the quotient $\pi/H : T/H \to B$, where T/H is the quotient of T by the action of H, induced from the action of G.

$$\begin{array}{ccc} T & \xrightarrow{\quad q \quad} & T/H \\ & {\scriptstyle \pi} \searrow \quad \swarrow {\scriptstyle \pi/H} & \\ & B & \end{array}$$

1. Every reduction S of π to H corresponds to a smooth section $\sigma : B \to T/H$ and conversely, $S = q^{-1}(\mathrm{val}(\sigma))$.

2. The projection $\pi/H : T/H \to B$ is a diffeological fiber bundle with fiber G/H. It is equivalent to the associated bundle $\pi_H : T \times_G (G/H) \to B$, where G acts on G/H by $g_{G/H}(Hg') = Hg'g^{-1}$. Hence, the space of reductions of π to H is in bijection with the space $\mathrm{Eq}^\infty(T, G/H)$ (art. 8.16, Note 3).

PROOF. First of all, let us check that T/H is a diffeological bundle. Let $P : U \to B$ be a plot, the pullback $P^*(T/H)$ is equivalent to $P^*(T)/H$. Now let $\Phi : (r, g) \mapsto (r, \phi_r(g))$ be a local trivialization of $\mathrm{pr}_1 : P^*(T) \to U$, thus the map $\Psi : (r, Hg) \mapsto (r, H_T(\phi_r(g)))$ is a local trivialization of $P^*(T)/H \sim P^*(T/H)$. Therefore, π/H satisfies the local triviality condition along the plots (art. 8.9).

Now, let q be the projection from T to T/H, and let σ be a smooth section of π/H. On the one hand, it is clear that, since $q(h_T(t)) = q(t)$, the preimage $S = q^{-1}(\mathrm{val}(\sigma))$ is invariant by H. Next, let t and t' be two points of S such that $\pi(t) = \pi(t')$, since $\pi/H \circ q = \pi$, $\pi/H(q(t)) = \pi/H(q(t'))$, but $q(t), q(t') \in \sigma(B)$ and σ is a section, then $q(t) = q(t')$. Thus, there exists $h \in H$ such that $t' = h_T(t)$, and S is a H-principal subbundle, that is, a reduction of π to H.

Conversely, let $S \subset T$ be a H-subbundle. Because S is H-invariant, there exists a map $\sigma : B \to T/H$ such that $\sigma \circ (\pi \restriction S) = q \restriction S$. This map is smooth because π is a subduction. But since, for all $b \in B$, $\pi^{-1}(b) \cap S$ is only one H-orbit, σ is a section of π/H.

For the second note, let us remark that T is also equivalent to $T \times_G G$, where G acts on the product by $g'_{T \times G}(t, g) = (g'_T(t), gg'^{-1})$, thanks to the factorization of the map $\phi : (t, g) \mapsto g_T(t)$.

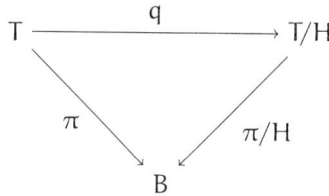

$$\begin{array}{ccc} \mathsf{T} \times \mathsf{G} & \xrightarrow{\quad \phi \quad} & \mathsf{T} \times_{\mathsf{G}} \mathsf{G} \sim \mathsf{T} \\ \big\downarrow & & \big\downarrow \\ \mathsf{T} \times \mathsf{G}/\mathsf{H} & \xrightarrow[\quad \varphi \quad]{} & (\mathsf{T} \times_{\mathsf{G}} \mathsf{G})/\mathsf{H} \sim \mathsf{T} \times_{\mathsf{G}} (\mathsf{G}/\mathsf{H}) \sim \mathsf{T}/\mathsf{H} \end{array}$$

Then, taking on both sides of the quotient by the left action of H, on the second factor, we get a factorization φ of ϕ which realizes the quotient $\mathsf{T} \times_{\mathsf{G}} (\mathsf{G}/\mathsf{H}) \sim (\mathsf{T} \times_{\mathsf{G}} \mathsf{G})/\mathsf{H}$ as the quotient T/H. □

Exercises

✎ EXERCISE 130 (Polarized smooth functions). Let $\mathsf{P}^1(\mathbf{R})$ be the space of all lines in $\mathbf{R}^2$ passing through the origin. Let

$$\mathsf{T} = \{(\mathbf{D}, \mathsf{f}) \in \mathsf{P}^1(\mathbf{R}) \times \mathcal{C}^\infty(\mathbf{R}^2, \mathbf{C}) \mid \mathsf{f}(\mathsf{x}+\mathsf{v}) = \mathsf{f}(\mathsf{x}),\ \forall \mathsf{x} \in \mathbf{R}^2,\ \forall \mathsf{v} \in \mathbf{D}\}.$$

In other words, if $(\mathbf{D}, \mathsf{f}) \in \mathsf{T}$, then f is constant on all affine lines parallel to $\mathbf{D}$. First, describe a good diffeology for $\mathsf{P}^1(\mathbf{R})$. Then, show that $\pi : \mathsf{T} \to \mathsf{P}^1(\mathbf{R})$, with $\pi(\mathbf{D}, \mathsf{f}) = \mathbf{D}$, is a nontrivial fiber bundle with fiber $\mathsf{E} = \mathcal{C}^\infty(\mathbf{R}, \mathbf{C})$.

✎ EXERCISE 131 (Playing with SO(3)). Let SO(3) be the group of direct rotations of the space $\mathbf{R}^3$, that is, the group of real 3×3 matrices A such that $\bar{\mathsf{A}}\mathsf{A} = \mathbf{1}_{\mathbf{R}^3}$ and $\det(\mathsf{A}) = +1$, where the bar denotes the transposition. Let $\mathbf{e}_1 \in \mathbf{R}^3$ be the vector of coordinates $(1,0,0)$, and let $\mathrm{SO}(2, \mathbf{e}_1)$ be the stabilizer of $\mathbf{e}_1$, that is, the subgroup of elements $\mathsf{k} \in \mathrm{SO}(3)$ such that $\mathsf{k}\mathbf{e}_1 = \mathbf{e}_1$.

1) Show that $\mathsf{p} : \mathrm{SO}(3) \to \mathsf{S}^2$ defined by $\mathsf{A} \mapsto \mathsf{A}\mathbf{e}_1$ is a principal fibration with structure group $\mathrm{SO}(2, \mathbf{e}_1)$, for the action $(\mathsf{A}, \mathsf{k}) \mapsto \mathsf{A}\mathsf{k}^{-1}$, where $(\mathsf{A}, \mathsf{k}) \in \mathrm{SO}(3) \times \mathrm{SO}(2, \mathbf{e}_1)$.

2) Show that the tangent space TS^2 of the 2-sphere, that is, $\mathrm{TS}^2 = \{(\mathsf{u}, \mathsf{v}) \in \mathsf{S}^2 \times \mathbf{R}^3 \mid \bar{\mathsf{u}}\mathsf{v} = 0\}$, is a fiber bundle with fiber $\mathbf{R}^2$, for the first projection $\mathrm{pr}_1 : (\mathsf{u}, \mathsf{v}) \mapsto \mathsf{u}$. Use the first question, and consider the associated fiber bundle $\mathrm{SO}(3) \times_{\mathrm{SO}(2,\mathbf{e}_1)} \mathbf{e}_1^\perp$, where $\mathrm{SO}(2, \mathbf{e}_1)$ acts naturally on the subspace $\mathbf{e}_1^\perp$ of $\mathbf{e}_1$-orthogonal vectors.

✎ EXERCISE 132 (Homogeneity of manifolds). Consider $\mathbf{R}^n \times \mathbf{R}$, let C be a cylinder $\mathsf{B} \times [-\delta, 1+\delta] \subset \mathbf{R}^n \times \mathbf{R}$, where B is a ball, centered at $0_n \in \mathbf{R}^n$, of radius $\mathsf{r} > 0$, and $\delta > 0$. Build a diffeomorphism of $\mathbf{R}^n \times \mathbf{R}$ which coincides with the identity out of C and maps $(0_n, 0)$ to $(0_n, 1)$. Next, let M be a connected Hausdorff manifold, deduce first that for any two points x and x' belonging to the values of a chart $\mathsf{F} : \mathcal{B} \to \mathsf{M}$, where $\mathcal{B}$ is some open ball, there exists a compactly supported diffeomorphism mapping x to x', and extend this result for any two points of M. Let $\mathrm{Diff}_{\mathsf{K}}(\mathsf{M})$ be the group of compactly supported diffeomorphisms, equipped with the functional diffeology (art. 1.61). Let $\hat{\mathsf{x}}_0 : \mathrm{Diff}_{\mathsf{K}}(\mathsf{M}) \to \mathsf{M}$ be the map $\hat{\mathsf{x}}_0(\varphi) = \varphi(\mathsf{x}_0)$, $\mathsf{x}_0 \in \mathsf{M}$. Prove that $\hat{\mathsf{x}}_0$ is a principal diffeological fibration with structure group $\mathrm{Diff}_{\mathsf{K}}(\mathsf{M}, \mathsf{x}_0)$, the subgroup of diffeomorphisms preserving x_0, and thus that M is a homogeneous space of $\mathrm{Diff}_{\mathsf{K}}(\mathsf{M}, \mathsf{x}_0)$. Remark that M is by consequence a homogeneous space of its whole group of diffeomorphisms, as well. Can you imagine the tangent space of M as an associated fiber bundle?

Homotopy of Diffeological Fiber Bundles

The main result concerning homotopy and diffeological fiber bundles is the exact homotopy sequence which links the homotopy of the base and the homotopy of the fiber with the homotopy of the total space. For the notations used thereafter, let us recall that $\text{Paths}(X) = \mathcal{C}^\infty(\mathbf{R}, X)$ *denotes the set of paths in* X, *with* $\hat{0}(\gamma) = \gamma(0)$, $\hat{1}(\gamma) = \gamma(1)$, *and* $\text{ends}(\gamma) = (\gamma(0), \gamma(1))$, *for all paths* γ. *Let us also recall that* X *is said to be* connected *if* ends *is surjective on* $X \times X$, *and that* ends *is then a* subduction (art. 5.6). *Let us finally recall that* $\text{Loops}(X)$ *denotes the subspace of* $\ell \in \text{Paths}(X)$ *such that* $\hat{0}(\ell) = \hat{1}(\ell)$, *and that for every point* x *in* X, $\text{Loops}(X, x)$ *denotes the subspace of* $\ell \in \text{Paths}(X)$ *such that* $\hat{0}(\ell) = \hat{1}(\ell) = x$, *and more generally that* $\text{Paths}(X, x, \star)$ *denotes the subspace of* $\gamma \in \text{Paths}(X)$ *such that* $\hat{0}(\gamma) = x$. *We end with the following inclusions:* $\text{Loops}(X, x) \subset \text{Paths}(X, x, \star) \subset \text{Paths}(X)$ *and* $\text{Loops}(X, x) \subset \text{Loops}(X) \subset \text{Paths}(X)$.

8.19. Triviality over global plots. Every diffeological fibration with base space $\mathbf{R}^n$ is trivial, for all $n \in \mathbf{N}$. Equivalently, let $\pi : T \to X$ be some principal fiber bundle. Every global plot $f : \mathbf{R}^n \to X$ admits a global smooth lift $F : \mathbf{R}^n \to T$, that is, $F \in \mathcal{C}^\infty(\mathbf{R}^n, T)$ and $\pi \circ F = f$.

NOTE 1. This property is closely related to what is called, in some other contexts and for some other kind of objects, the *homotopy lifting property*. This is a crucial property in the establishment of the exact homotopy sequence of the diffeological fiber bundles (art. 8.21).

NOTE 2. Actually the proof shows that any diffeological fiber bundle, over an open ball of a real vector space, is trivial.

PROOF. We shall prove this proposition in a few steps.

LEMMA 1. Let $\pi : T \to X \times]a, b[$ be a principal fibration with structure group G. Let $a < b' < a' < b$ such that π is trivial over $X \times]a, a'[$ and trivial over $X \times]b', b[$. Then, π is trivial over $X \times]a, b[$.

Proof of Lemma 1. Let us denote by $[g \mapsto g_T] \in \text{Hom}^\infty(G, \text{Diff}(T))$ the action of G on T. Thanks to (art. 8.11) there exist two sections $s_1 : X \times]a, a'[\to T$ and $s_2 : X \times]b', b[\to T$, and the isomorphisms (equivariant diffeomorphisms) associated write $\Phi_i(y, g) = g_T(s_i(y))$, where $y = (x, t)$ is in the domain of definition of the section s_i. Now, thanks to the equivariance of the trivializations Φ_1 and Φ_2, $\Phi_1^{-1} \circ \Phi_2 : X \times]b', a'[\times G \to X \times]b', a'[\times G$ writes $\Phi_1^{-1} \circ \Phi_2(y, g) = (y, g\gamma(y))$, and

$$\gamma : X \times]b', a'[\to G, \text{ defined by } \gamma(y) = \text{pr}_2 \circ \Phi_1^{-1} \circ \Phi_2(y, 1_G)$$

is clearly smooth. Then, for all $y = (x, t) \in X \times]b', a'[$, $\Phi_2(y, g) = \Phi_1(y, g\gamma(y))$, that is, $g_T(s_2(y)) = g_T(\gamma(y)_T(s_1(y)))$, thus $s_2(y) = \gamma(y)_T(s_1(y))$. Now, let $c \in]b', a'[$, and let μ be a real smooth function, described by Figure 8.3, satisfying

$$\mu :]a - \varepsilon, a' + \varepsilon[\to]b' - \varepsilon, a' + \varepsilon[\quad \text{with} \quad \begin{cases} \mu(a) = b' \text{ and } \mu(a') = a', \\ \mu \restriction]c, a'[= \mathbf{1}_{]c, a'[}, \end{cases}$$

where ε is some positive number. Now, since for all $(x, t) \in X \times]a, a'[$, $(x, \mu(t)) \in X \times]b', a'[= \text{def}(\gamma)$, the map $(x, t) \mapsto \gamma(x, \mu(t))$ is defined on $X \times]a, a'[$, and since s_1 is also defined on $X \times]a, a'[$, we have a new section given by

$$s'_1 : X \times]a, a'[\to T \quad \text{with} \quad s'_1(x, t) = \gamma(x, \mu(t))_T(s_1(x, t)).$$

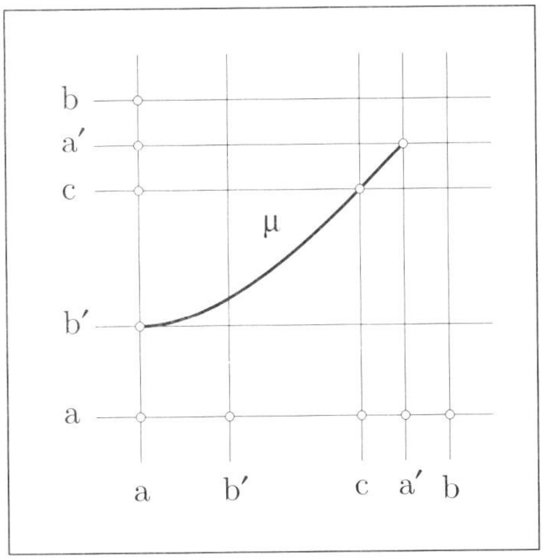

FIGURE 8.3. The extension function.

Then, let us define

$$s_2' : X \times]c, b[\to T \quad \text{with} \quad s_2' = s_2 \restriction X \times]c, b[.$$

The common domain of s_1' and s_2' is $X \times]c, a'[$. Now, for every $(x,t) \in X \times]c, a'[$, $s_1'(x,t) = \gamma(x, \mu(t))_T(s_1(x,t)) = \gamma(x,t)_T(s_1(x,t)) = s_2(x,t) = s_2'(x,t)$, and the sections s_1' and s_2' coincide on the intersection (which is D-open) of their domains. Hence, they are the restriction, to their domains, of a smooth section $s : X \times]a, b[\to T$. Therefore, the principal fibration π is trivial.

LEMMA 2. Every principal fibration $\pi : T \to \mathbf{R}^n$, for any structure group G, is trivial over every paracompact open subset of $\mathbf{R}^n$. In particular, it is trivial over every ball $\mathcal{B}$ of any radius, centered at the origin $0 \in \mathbf{R}^n$.

Proof of the Lemma 2. Since the identity of $\mathbf{R}^n$ is a plot, the fibration $\pi : T \to \mathbf{R}^n$ is locally trivial (art. 8.9). So, there exists a family $\{s_i : U_i \to T\}_{i \in \mathcal{I}}$ of smooth sections of π whose domains are a cover of $\mathbf{R}^n$, that is,

$$\operatorname{def}(s_i) = U_i, \quad \bigcup_{i \in \mathcal{I}} U_i = \mathbf{R}^n, \quad s_i \in \mathcal{C}^\infty(U_i, T), \quad \text{and} \quad \pi \circ s_i = \mathbf{1}_{U_i}.$$

Let K_L be the cube with side $2L$ centered at the origin $0 \in \mathbf{R}^n$, $K_L = [-L, L]^n$. Let us consider the subcover $\{U_a\}_{a \in \mathcal{A}}$, $\mathcal{A} \subset \mathcal{I}$, of K_L defined by $U_a \cap K_L \neq \varnothing$. Since K_L is compact, there exists a finer cover of K_L made of small cubes of side ℓ arranged along a lattice directed by the canonical basis, such that each cube meets only its closest neighbours, and such that their union is a cube K_L' containing K_L. See Figure 8.4. Such a cover can be built using the Lebesgue number of the cover [**EOM02**]. Indeed, for every compact metric space (here K_L) and an open cover (of K_L) is given, then there exists a number ℓ such that every subset of K_L of diameter ℓ is contained in some member of the cover. Let us denote this cover of K_L by $\{C_{i_1,\ldots,i_n}\}_{i_k=1}^N$, then

$$K_L' = \bigcup_{i_1,\ldots,i_n=1}^{N} C_{i_1,\ldots,i_n}, \text{ and } \forall i_1, \ldots, i_n, \ \exists a \in \mathcal{A}, \ C_{i_1 \cdots i_n} \in U_a.$$

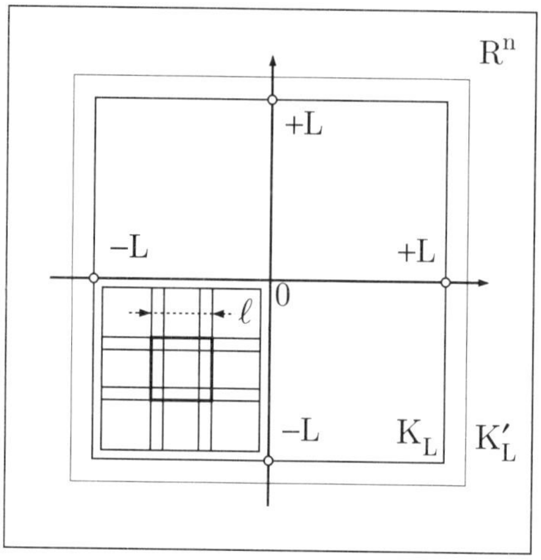

FIGURE 8.4. Cubes filling cubes.

The restriction of the fiber bundle $\pi : T \to \mathbf{R}^n$ over each of these cubes is trivial. Thus, we can apply recursively the previous lemma to the first two cubes of a row $C_{1,1,\ldots,1}$ and $C_{2,1,\ldots,1}$, then to the $C_{1,1,\ldots,1} \cup C_{2,1,\ldots,1}$ and $C_{3,1,\ldots,1}$ and so on, until $C_{N,1,\ldots,1}$, we construct a section of π over the union $\bigcup_{i=1}^{N} C_{i,1,\ldots,1}$. Repeating this process on the first index $i_1 = 1, \ldots, N$, we construct a smooth section over every domain of this type $C_{\star,i_2,\ldots,i_n} = \bigcup_{j=1}^{N} C_{j,i_2,\ldots,i_n}$. Then, by recurrence, we construct a smooth section of the fibration $\pi : T \to \mathbf{R}^n$ over the cube K'_L. Thus, the fibration is trivial over K'_L and therefore on every relatively compact open subset. We also say that π is trivial over the closed cube K_L, and this is what we mean when we say in general that the fibration is trivial over a closed subset $K \subset \mathbf{R}^n$. In particular, π is trivial over every ball of any radius, centered at the origin.

LEMMA 3. Let $\pi : T \to \mathbf{R}^n$ be a principal fibration with structure group G. There exists a family of smooth sections $s_k : \mathcal{B}_k \to T$, with $k \in \mathbf{N}$, such that $s_k \prec s_{k+1}$, that is, $s_{k+1} \upharpoonright \mathcal{B}_k = s_k$, for all k.

Proof of Lemma 3. Thanks to the previous lemma, the fibration $\pi : T \to \mathbf{R}^n$ is trivial over every open ball $\mathcal{B}_k$. Let us choose once and for all a family of smooth sections $s'_k : \mathcal{B}_k \to T$, for every positive integer k, and let us introduce two series of maps, $\varepsilon_k : [0, k+1] \to [0, k]$ and $\lambda_k : \mathcal{B}_{k+1} \to \mathcal{B}_k$, for k integer and $k \geq 2$. The function ε_k, described by Figure 8.5, is the restriction of an increasing smooth map satisfying the following conditions,

$$\varepsilon_k \upharpoonright [0, k-1] = \mathbf{1}_{[0,k-1]} \quad \text{and} \quad \varepsilon_k(k+1) = k.$$

The maps λ_k are then defined by

$$\lambda_k(x) = \varepsilon_k(\|x\|) \times \frac{x}{\|x\|}, \quad \text{for all } x \in \mathcal{B}_{k+1}.$$

Since $\varepsilon_k \upharpoonright [0, k-1] = \mathbf{1}_{[0,k-1]}$, $\lambda_k \upharpoonright \mathcal{B}_{k-1} = \mathbf{1}_{\mathcal{B}_{k-1}}$, and the map λ_k is smooth. Now, s'_2 is given, let us define s_1 and s''_2 by

$$s_1 = s''_2 \upharpoonright \mathcal{B}_1 \quad \text{with} \quad s''_2 = s'_2.$$

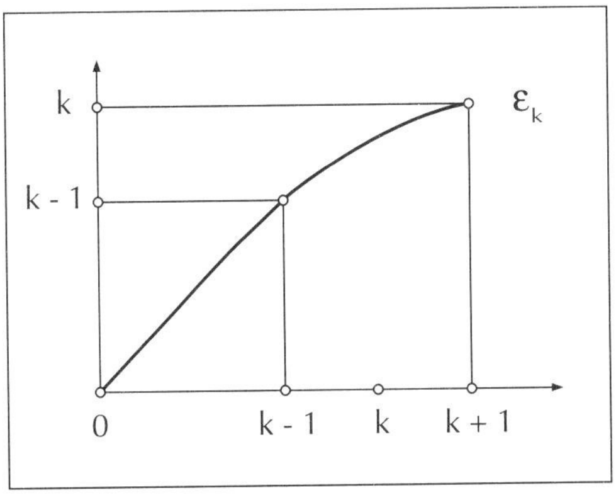

FIGURE 8.5. The smashing balls function.

Let us then consider the series of sections $s_1, s_2'', s_3', s_4', \ldots$, and let us focus on s_2'' and s_3'. There exists a smooth map

$$\gamma_{2,3} : \mathcal{B}_2 \to G \quad \text{such that} \quad s_2''(x) = \gamma_{2,3}(x)_T(s_3'(x)), \text{ for all } x \in \mathcal{B}_2.$$

Let

$$s_2 = s_3'' \upharpoonright \mathcal{B}_2 \quad \text{with} \quad s_3'' = \gamma_{2,3} \circ \lambda_2(x)_T(s_3'(x)).$$

The function λ_2 maps $\mathcal{B}_3$ onto $\mathcal{B}_2$ and is the identity on $\mathcal{B}_1$. So, since $\gamma_{2,3}$ is defined on $\mathcal{B}_2$, $\gamma_{2,3} \circ \lambda_2 : \mathcal{B}_3 \to \mathcal{B}_2 \to G$ and coincides with $\gamma_{2,3}$ on $\mathcal{B}_1$. Thus, for all $x \in \mathcal{B}_1$, $s_2(x) = s_3''(x) = \gamma_{2,3}(x)_T(s_3'(x)) = s_2''(x) = s_1(x)$. Therefore, $s_1 = s_2 \upharpoonright \mathcal{B}_1$, and we have the series of sections, $s_1, s_2, s_3'', s_4', s_5', \ldots$, for which $s_1 \prec s_2 \prec s_3''$. The recurrence is now clear, for $k \geq 3$ let $s_1 \prec \cdots \prec s_{k-1} \prec s_k''$ being a series of sections over the balls $\mathcal{B}_1 \subset \cdots \subset \mathcal{B}_{k-1} \subset \mathcal{B}_k$. So, there exist two sections s_k and s_{k+1}'' over the balls $\mathcal{B}_k$ and $\mathcal{B}_{k+1}$ such that $s_{k-1} \prec s_k \prec s_{k+1}''$. Indeed, let

$$\gamma_{k,k+1} : \mathcal{B}_k \to G \quad \text{such that} \quad s_k''(x) = \gamma_{k,k+1}(x)_T(s_{k+1}'(x)), \text{ for all } x \in \mathcal{B}_k.$$

The map $\gamma_{k,k+1}$ is smooth, and we define

$$s_k = s_{k+1}'' \upharpoonright \mathcal{B}_k \quad \text{with} \quad s_{k+1}'' = \gamma_{k,k+1} \circ \lambda_k(x)_T(s_{k+1}'(x)).$$

The function λ_k maps $\mathcal{B}_{k+1}$ onto $\mathcal{B}_k$ and is the identity on $\mathcal{B}_{k-1}$. Since $\gamma_{k,k+1}$ is defined on $\mathcal{B}_k$, $\gamma_{k,k+1} \circ \lambda_k : \mathcal{B}_{k+1} \to \mathcal{B}_k \to G$ and coincides with $\gamma_{k,k+1}$ on $\mathcal{B}_{k-1}$. Thus, for all $x \in \mathcal{B}_{k-1}$, $s_{k+1}''(x) = \gamma_{k,k+1}(x)_T(s_{k+1}'(x)) = s_k''(x) = s_{k-1}(x)$, that is to say, $s_{k-1} \prec s_k$, and by construction, $s_k \prec s_{k+1}''$. Lemma 3 is thus proved.

Now, the rest of the proposition is clear, once we have an exhaustion of $\mathbf{R}^n$ by the open balls of radius the integers, and a series of compatible sections of the G-principal fibration $\pi : T \to \mathbf{R}^n$ over these balls, we have a global section, global by compatibility and smooth by the locality axiom of diffeology. This completes the proof of the theorem.

Note that the triviality of fiber bundles in diffeology is reduced to the triviality of principal fiber bundles (art. 8.16). Thus every diffeological fiber bundle over $\mathbf{R}^n$ is trivial.

Finally, considering a smooth map $f : \mathbf{R}^n \to X$, a smooth lifting F of f in T is equivalent to a smooth section of $\mathrm{pr}_1 : f^*(T) \to \mathbf{R}^n$. We just built one for every principal bundle over $\mathbf{R}^n$. ☐

8.20. Associated loops fiber bundles. Let $\pi : X \to B$ be a diffeological fiber bundle. Let $x \in X$, $b = \pi(x)$, $F = \pi^{-1}(b)$ and

$$\pi_* : \mathrm{Loops}(X, x) \to \mathrm{Loops}(B, b) \quad \text{with} \quad \pi_*(\ell) = \pi \circ \ell.$$

Let $\mathrm{Loops}^\bullet(B, b) = \mathrm{comp}(\mathbf{b}) \subset \mathrm{Loops}(B, \mathbf{b})$ be the connected component of the constant loop $\mathbf{b} = [t \mapsto b]$, and let us introduce

$$\mathrm{Loops}^*(X, x) = (\pi_*)^{-1}(\mathrm{Loops}^\bullet(B, b)) \quad \text{and} \quad \pi_*^\bullet = \pi_* \restriction \mathrm{Loops}^*(X, x).$$

The map π_* satisfies the following properties.

($\Diamond$) The map π_* is onto, that is, every loop in B, based at b and null-homotopic, can be lifted along π_* by a loop in X, based at x.

($\heartsuit$) The projection $\pi_*^\bullet : \mathrm{Loops}^*(X, x) \to \mathrm{Loops}^\bullet(B, b)$ is a diffeological fibration with fiber $\mathrm{Loops}(F, x)$.

We call this diffeological fiber bundle, the *Loops bundle* associated with π at the point x. This situation is summarized by the short fiber bundle sequence,

$$x \longrightarrow \mathrm{Loops}(F, x) \xrightarrow{j_*} \mathrm{Loops}^*(X, x) \xrightarrow{\pi_*^\bullet} \mathrm{Loops}^\bullet(B, b) \longrightarrow \mathbf{b}.$$

PROOF. Let us prove ($\Diamond$). Let φ be a homotopy from $\ell \in \mathrm{Loops}^\bullet(B, b)$ to the constant loop $\mathbf{b}$. Thus, $\varphi \in \mathrm{Paths}(\mathrm{Loops}(B, b))$ with $\varphi(0) = \ell$ and $\varphi(1) = \mathbf{b}$. We identify φ with $\bar\varphi \in \mathcal{C}^\infty(\mathbf{R}^2, B)$ such that $\bar\varphi(t, s) = \varphi(t)(s)$ (art. 1.64), we have

$$\bar\varphi \restriction_{s=0} = \mathbf{b}, \quad \bar\varphi \restriction_{s=1} = \mathbf{b}, \quad \bar\varphi \restriction_{t=0} = \ell, \quad \bar\varphi \restriction_{t=1} = \mathbf{b}.$$

The pullback π_φ of π by φ is a diffeological fiber bundle over $\mathbf{R}^2$, thus it is trivial (art. 8.19).

$$\begin{array}{ccccc}
\mathbf{R}^2 \times F & \xrightarrow{\Psi} & P^*(X) & \xrightarrow{\mathrm{pr}_2} & X \\
& \searrow^{\mathrm{pr}_1} & \downarrow \mathrm{pr}_1 & & \downarrow \pi \\
& & \mathbf{R}^2 & \xrightarrow[\bar\varphi]{} & B
\end{array} \qquad (\spadesuit)$$

Let Ψ be a trivialization and let ψ be defined, for all $(t, s; \xi) \in \mathbf{R}^2 \times F$, by

$$\Psi(t, s; \xi) = (t, s; \psi(t)(s)(\xi)) \quad \text{thus} \quad \psi(t)(s) \in \mathrm{Diff}(F, X_{\varphi(t)(s)}).$$

By definition of the involved diffeologies (art. 1.57) (art. 8.14), ψ is a smooth lift of the plot $(t, s) \mapsto (b, \varphi(t)(s))$ of $B \times B$, in the total space

$$T = \{f \in \mathrm{Diff}(F, X_b) \mid b \in B\}$$

of the principal fiber bundle associated with the smooth structure groupoid of π, attached at the point b. Now, since $\varphi(t)(s) = b$ when $t = 1$ or $s = 0$ or $s = 1$, $\psi(t)(s) \in \mathrm{Diff}(F)$ for $t = 1$ or $s = 0$ or $s = 1$. Thus, $\psi(0)(0)$ and $\psi(1)(0)$ are two diffeomorphisms of F, and since they can be connected by the family of smooth paths $\bar\psi \restriction_{s=0}$, $\bar\psi \restriction_{s=1}$, and $\bar\psi \restriction_{t=1}$ (with the same convention for the over bar as previously for φ), they belong to the same component. Then, there exists a smooth path σ in $\mathrm{Diff}(F)$ such that $\sigma(0) = \psi(0)(0)$ and $\sigma(1) = \psi(1)(0)$ (art. 5.7). Let us consider the restriction Ψ_0 of Ψ to $\{0\} \times \mathbf{R} \times F$, then for all $(s, \xi) \in \mathbf{R} \times F$,

$$\Psi_0(s, \xi) = (s, \psi_0(s)(\xi)) \text{ with } \psi_0(s) = \psi(0)(s), \text{ thus } \psi_0(s) \in \mathrm{Diff}(F, X_{\ell(s)}).$$

Actually, the path ψ_0 is a smooth lift in T of the loop ℓ, Ψ_0 is a trivialization of the pullback of $\pi : X \to B$ by ℓ. Let us introduce then Ψ', for all $(s, \xi) \in \mathbf{R} \times F$,

$$\Psi'(s, \xi) = (s, \psi'(s)(\xi)) \quad \text{with} \quad \psi'(s) = \psi_0(s) \circ \sigma(s)^{-1}.$$

The path ψ' is again a smooth path in T, and Ψ' is a trivialization of $\mathrm{pr}_1 : \ell^*(X) \to \mathbf{R}$, but ψ' satisfies now $\psi'(0) = \psi'(1) = \mathbf{1}_F$. Finally, mapping the x-constant section $s \mapsto (s, x)$ to X, by the trivialization Ψ', we get a path in X, that is, $\ell' : s \mapsto \psi'(x)$, which is a loop because $\psi'(0) = \psi'(1) = \mathbf{1}_F$, and which lifts ℓ because $\psi' : F \to X_{\ell(s)}$.

Now, let us prove (♡). Let Φ be a global n-plot in $\mathrm{Loops}^\bullet(B, b)$, and let $\bar\Phi(r, t) = \Phi(r)(t)$, for all $(r, t) \in \mathbf{R}^n \times \mathbf{R}$, then $\bar\Phi \in \mathcal{C}^\infty(\mathbf{R}^n \times \mathbf{R}, B)$ and $\bar\Phi$ satisfies $\bar\Phi(r, 0) = \bar\Phi(r, 1) = b$ for all $r \in \mathbf{R}^n$. Since the pullback of $\pi : X \to B$ by $\bar\Phi$ is trivial (art. 8.19), let ψ be a global smooth lift of $\bar\Phi$ in the principal bundle T associated with π (see above), and let Ψ be the associated trivialization,

$$\Psi(r, t; \xi) = (r, t; \psi(r, t)(\xi)) \quad \text{with} \quad \psi(r, t) \in \mathrm{Diff}(F, X_{\Phi(r)(t)}),$$

for all $(r, t; \xi) \in \mathbf{R}^n \times \mathbf{R} \times F$. Let $\psi_0 = \psi \restriction \mathbf{R}^n \times \{0\}$ and $\psi_1 = \psi \restriction \mathbf{R}^n \times \{1\}$, ψ_0 and ψ_1 belong to $\mathcal{C}^\infty(\mathbf{R}^n, \mathrm{Diff}(F))$ because $\Phi(r)(0) = \Phi(r)(1) = F$. Since $\mathbf{R}^n$ is connected, $\mathrm{val}(\psi_0)$, and $\mathrm{val}(\psi_1)$, are contained in some connected components of $\mathrm{Diff}(F)$. But $\psi_0(r)$ and $\psi_1(r)$ are contained in a same connected component of $\mathrm{Diff}(F)$, for all $r \in \mathbf{R}^n$, because $\Phi(r) \in \mathrm{Loops}^\bullet(B, b)$, *i.e.*, is null-homotopic, see above the proof of (♢). Thus, $\mathrm{val}(\psi_0)$ and $\mathrm{val}(\psi_1)$ are contained in a same connected component of $\mathrm{Diff}(F)$. But, because $\mathbf{R}^n$ is contractible and because ψ_0 and ψ_1 take their values in the same connected component, there exists a homotopy in $\mathcal{C}^\infty(\mathbf{R}^n, \mathrm{Diff}(F))$ connecting ψ_0 with ψ_1 (art. 5.11, Note 3), let us denote it by σ, that is, $\sigma \in \mathrm{Paths}(\mathcal{C}^\infty(\mathbf{R}^n, \mathrm{Diff}(F)))$ with $\sigma(0) = \psi_0$ and $\sigma(1) = \psi_1$. We define now $\psi'(r)(t) = \psi(r)(t) \circ \sigma(t)(r)^{-1}$ which satisfies $\psi' \in \mathcal{C}^\infty(\mathbf{R}^n \times \mathbf{R}, T)$ with $\psi'(r)(0) = \psi'(r)(1) = \mathbf{1}_F$ for all $r \in \mathbf{R}^n$. The map Ξ from $\mathbf{R}^n \times \mathrm{Loops}(F, x)$ to $\mathbf{R}^n \times \mathrm{Paths}(X)$, defined by

$$\Xi(r, \ell) = (r, [t \mapsto \psi'(r)(t)(\ell(t))]),$$

takes its values in $\Phi^*(\mathrm{Loops}^*(X, x))$ and is a trivialization of the pullback by Φ of the smooth surjection $\pi_*^\bullet$. Therefore, according to (art. 8.9), $\pi_*^\bullet$ is a diffeological fibration with fiber $\mathrm{Loops}(F, x)$. □

8.21. Exact homotopy sequence of a diffeological fibration. Let $p : X \to B$ be a diffeological fiber bundle. Let $x \in X$, $b = p(x)$ and $F = p^{-1}(b)$. We consider the relative homotopy of the pair (X, F), at the point x, as it is described in (art. 5.18) and (art. 5.19) and we use the same notations. Let $i_{k\#}$ and $p_{k\#}$ be the maps in homotopy, induced by the inclusion $i : F \to X$ (art. 5.17) and by the projection p,

$$i_{k\#} : \pi_k(F, x) \to \pi_k(X, x) \quad \text{and} \quad p_{k\#} : \pi_k(X, x) \to \pi_k(B, b).$$

Now, since the projection p maps F to the point b, the map $p_k : \mathrm{Loops}_k(X, x) \to \mathrm{Loops}_k(B, b)$ (art. 5.17) maps $\mathrm{Loops}_k(F, x)$ to $\mathbf{b}_k$, then every element of

$$\mathrm{Paths}_k(X, F, x)$$

is mapped into an element of $\mathrm{Loops}_k(B, b)$. We still denote by $p_{k\#}$ its induced map in homotopy for all $k \geq 1$,

$$p_{k\#} : \pi_k(X, F, x) \to \pi_k(B, b).$$

Then, $p_{k\#}$ satisfies the following property.

(♣) The map $p_{k\#}$ is an isomorphism of group for $k > 1$ and an isomorphism of pointed spaces for $k = 1$.

Let us define then, for all integers k, the *connector* Δ_k by

$$\Delta_{k+1} : \pi_{k+1}(B, b) \to \pi_k(F, x) \quad \text{with} \quad \Delta_k = \hat{0}_\# \circ p^{-1}_{(k+1)\#},$$

where $\hat{0}_\# : \pi_{k+1}(X, F, x) \to \pi_k(F, x)$ is the connector defined in (art. 5.18). By connecting then each degree $k+1$ of the exact sequence of the pair (X, F), at the point x, to the degree k, with the connector Δ_{k+1}, we get the *exact homotopy sequence of the fiber bundle* $p : X \to B$. Omitting the indices of the maps since they are clearly defined by their source and target, the sequence writes

$$\left.\begin{array}{l} \cdots \xrightarrow{p_\#} \pi_{k+1}(B, b) \xrightarrow{\Delta} \pi_k(F, x) \xrightarrow{i_\#} \pi_k(X, x) \xrightarrow{p_\#} \pi_k(B, b) \xrightarrow{\Delta} \cdots \\ \cdots \xrightarrow{p_\#} \pi_1(B, b) \xrightarrow{\Delta} \pi_0(F, x) \xrightarrow{i_\#} \pi_0(X, x) \xrightarrow{p_\#} \pi_0(B, b). \end{array}\right\} \quad (\spadesuit)$$

NOTE. The following propositions are immediate consequences of the exact homotopy sequence.

a) If the homotopy of the fiber F vanishes, in particular if F is contractible, then $\pi_k(X, x) = \pi_k(B, b)$ for all k.

b) If the fiber F is a deformation retract of X, then $\pi_k(B, b) = \{0\}$ for all k.

c) If $p : X \to B$ is a principal fibration with group G, then we can replace the fiber F pointed at x by the group G pointed at $\mathbf{1}_G$ and the inclusion $i : F \to X$ by the orbit map $R(x) : G \to X$. The exact sequence writes then

$$\cdots \xrightarrow{\Delta_{k+1}} \pi_k(G, \mathbf{1}_G) \xrightarrow{R(x)_{k\#}} \pi_k(X, x) \xrightarrow{p_{k\#}} \pi_k(B, b) \xrightarrow{\Delta_k} \cdots.$$

This replacement is not really necessary but makes the sequence looks prettier. This can be used to compute the homotopy of some quotients which are not manifolds and cannot be computed otherwise, such as the irrational torus.

PROOF. Let us prove first (♣) for $k = 1$, that is, for $p_1 : \text{Paths}(X, F, x) \to \text{Loops}(B, b)$. Let $\ell \in \text{Loops}(B, b)$, since the pullback $\text{pr}_1 : \ell^*(X) \to \mathbf{R}$ is trivial, we can lift ℓ by a path $\bar{\ell}$ by mapping any section of the pullback to X. We can even choose $\bar{\ell}$ such that $\bar{\ell}(1) = x$ and, since $\ell(0) = \ell(1) = b$, $\bar{\ell}(0) \in p^{-1}(b) = F$. This shows that p_1 is surjective, and therefore $p_{1\#}$. Now let us prove the injectivity. Since $p_{1\#}$ is a group morphism, we shall prove that its kernel is reduced to $\text{class}(\mathbf{x}) \in \pi_1(X, F, x)$, that is, every path $\gamma' \in \text{Paths}(X, F, x)$ which projects to a null-homotopic loop $\ell \in \text{Loops}(B, b)$ is relatively homotopic to $\mathbf{x}$ and, which is sufficient, relatively homotopic to a path in F. So, let Φ be a homotopy connecting γ' to $\mathbf{b}$, $\Phi(0) = \gamma'$ and $\Phi(1) = \mathbf{b}$. As usual we identify Φ with a map, denoted here by the same letter, $\Phi \in \mathcal{C}^\infty(\mathbf{R}^2, B)$, that is, $\Phi(s, t) = \Phi(s)(t)$. Let Ψ be a trivialization of the pullback $\text{pr}_1 : \Phi^*(X) \to \mathbf{R}^2$ (art. 8.19), and let $\Psi(s, t; \xi) = (s, t; \psi(s)(t)(\xi))$, where $(s, t) \mapsto \psi(s)(t)$ is a plot of the principal smooth structure bundle $\pi : T \to B$ associated with $p : X \to B$, attached at the point b (art. 8.18). In particular $\psi(s)(t) \in \text{Diff}(F, X_{\Phi(s)(t)})$. We can even choose $\psi(s)(1) = \mathbf{1}_F$ by just replacing $\psi(s)(t)$ by $\psi(s)(t) \circ \psi(s)(1)^{-1}$. Now, let

$$\gamma(t) = \psi(0)(t)^{-1}(\gamma'(t)).$$

Since $\psi(0)(t) \in \mathrm{Diff}(F, X_{\Phi(0)(t)})$ and $\Phi(0)(t) = \gamma'(t)$, $\psi(0)(t)^{-1}(\gamma'(t)) \in F$ and therefore $\gamma \in \mathrm{Paths}(F)$. Next, let us define

$$\Phi' = [s \mapsto [t \mapsto \psi(s)(t)(\gamma(t))]],$$

which satisfies

$$\Phi'(0) = \gamma', \ \Phi'(1) \in \mathrm{Paths}(F) \text{ and } \Phi'(s)(1) = x \text{ for all } s \in \mathbf{R}.$$

So, Φ' is a fixed-ends homotopy connecting γ' to a path $\gamma'' \in \mathrm{Paths}(F, \star, x)$. But, every path in $\mathrm{Paths}(F, \star, x)$ is relatively homotopic to the constant loop $\mathbf{x}$, consider for example the homotopy $s \mapsto \gamma''_s = [t \mapsto (\gamma''((1-t)s+t))]$. Thus, $p_{1\#}$ is injective and therefore bijective. Now, by recurrence over k, on the one hand we have

$$\pi_{k+1}(X, F, x) = \pi_1(\mathrm{Loops}_k(X, x), \mathrm{Loops}_k(F, x), \mathbf{x}_k),$$

$$\pi_{k+1}(X, x) = \pi_1(\mathrm{Loops}_k(X, x), \mathbf{b}_k) \text{ and } \pi_{k+1}(B, b) = \pi_1(\mathrm{Loops}_k(B, b), \mathbf{b}_k).$$

On the other hand, by definition,

$$\begin{cases} \mathrm{Loops}_{k+1}(X, x) = \mathrm{Loops}(\mathrm{Loops}_k(X, x), \mathbf{x}_k), \\ \mathrm{Loops}_{k+1}(B, b) = \mathrm{Loops}(\mathrm{Loops}_k(B, b), \mathbf{b}_k), \end{cases}$$

and the projection $p_k : \mathrm{Loops}_{k+1}(X, x) \to \mathrm{Loops}_{k+1}(B, b)$ is the associated loops fiber bundle (art. 8.20) of the fiber bundle $p_k : \mathrm{Loops}_k(X, x) \to \mathrm{Loops}_k(B, b)$ with fiber $\mathrm{Loops}(\mathrm{Loops}_k(F, x), \mathbf{x}_k) = \mathrm{Loops}_{k+1}(F, x)$. By denoting $X_k = \mathrm{Loops}_k(X, x)$, $B_k = \mathrm{Loops}_k(B, b)$, $F_k = \mathrm{Loops}_k(F, x)$, we are in the same situation as previously, replacing X by X_k, B by B_k, F by F_k, x by x_k and b by b_k, that is,

$$p_{(k+1)\#} = (p_k)_{1\#} : \pi_1(X_k, F_k, \mathbf{x}_x) \to \pi_1(B_k, \mathbf{b}_k).$$

Then, because $p_{1\#}$ is an isomorphism, $p_{k\#}$ is an isomorphism for all k.

Now, to complete the proof about the exactness of the homotopy sequence of diffeological fiber bundles, it is sufficient to check only the case not covered by the exact sequence of a pair, that is,

$$\pi_0(F, x) \xrightarrow{\quad i_\# \quad} \pi_0(X, x) \xrightarrow{\quad p_\# \quad} \pi_0(B, b) \longrightarrow 0.$$

But $\mathrm{val}(p_\#) = \pi_0(B, b)$ is a simple consequence of the surjectivity of p. Then,

$$\mathrm{val}(i_\#) = \ker(p_\#) \text{ means } p^{-1}(\mathrm{comp}_B(b)) = \bigcup_{y \in F} \mathrm{comp}_X(y).$$

This means firstly that connected points are mapped to connected points by p, which is clear, and secondly that, for every point b' connected to b, if a point x' in X is mapped to b' by p, then it is connected to F, which is a direct consequence of the global triviality of the pullback of $p : X \to B$ by any path in B. □

Coverings of Diffeological Spaces

The definition of diffeological discrete spaces (art. 1.20) *leads naturally to the definition of diffeological coverings. In this section we shall see, first of all, that every diffeological space has a simply connected universal covering, unique up to equivalence, and then that the various coverings are classified, up to equivalence, by the conjugacy classes of the subgroups of the fundamental group. The universal covering is built actually thanks to the Poincaré groupoid, introduced previously* (art. 5.15). *Moreover, we shall establish the important monodromy theorem concerning the lifting of maps from simply connected spaces to coverings. It has already been applied for lifting actions of groups on coverings* (art. 7.11). *Also note that, applied to the category of manifolds, these constructions and theorems give again the classical results, and that the covering of a manifold as a diffeological space is a covering as a manifold, which is satisfactory.*

8.22. What is a covering? Let X be a connected diffeological space, a *covering* of X is any fiber bundle $p : Y \to X$ with a discrete fiber F. The space Y is called the *covering space.*

1. If Y is connected, we say that p is a *connected covering.*
2. If the fiber F is made of N points, we say that p is an N*-folds covering.*
3. If p is a principal bundle, we say that p is a *Galoisian covering.*

Coverings over a diffeological space X form naturally a full subcategory of the category of fiber bundles (art. 8.10) with base X.

NOTE. As a first example of covering we have all the manifolds coverings, but also the quotients of any diffeological group G by any discrete subgroup Γ, since a quotient of a group by a subgroup is always a diffeological fiber bundle (art. 8.15). If Γ is normal, these coverings are Galoisian. For example, $\mathbf{R} \to \mathbf{R}/\mathbf{Q}$, $\mathbf{R} \to \mathbf{R}/(\mathbf{Z} + \alpha\mathbf{Z})$, etc. Every strict subgroup $\Gamma \subset \mathbf{R}$ defines a Galoisian covering from $\mathbf{R}$ to $\mathbf{R}/\Gamma$.

8.23. Lifting global plots on coverings. Let $p : Y \to X$ be a covering with fiber $F = p^{-1}(x)$. Let $\varphi : \mathbf{R}^n \to X$ be a plot centered at x, that is, $\varphi(0) = x$. There exists a trivialization $\Psi : (r, y) \mapsto (r, \psi(r)(y))$, $(r, y) \in \mathbf{R}^n \times F$, of the pullback $\mathrm{pr}_1 : \varphi^*(Y) \to \mathbf{R}$ such that $\Psi \restriction \{0\} \times F = \mathbf{1}_F$. Moreover, for all $y \in F$ there exists a unique lift $\bar\varphi : \mathbf{R}^n \to Y$, $\bar\varphi : r \mapsto \psi(r)(y)$, such that $\bar\varphi(0) = y$.

NOTE. We get the same conclusion *mutatis mutandis*, by choosing any basepoint $r_0 \in \mathbf{R}^n$ instead of the origin $0 \in \mathbf{R}^n$.

PROOF. Thanks to (art. 8.19) the pullback $\mathrm{pr}_1 : \varphi^*(Y) \to \mathbf{R}$ is trivial. Let $\Psi : \mathbf{R}^n \times F \to \varphi^*(Y)$ be a trivialization. Then, for all $(r, \xi) \in \mathbf{R}^n \times F$,

$$\Psi(r, \xi) = (r, \psi(r)(\xi)) \text{ with } \psi \in \mathcal{C}^\infty(\mathbf{R}^n, T),\ \psi(r) \in \mathrm{Diff}(F, Y_{\varphi(r)}),$$

where $\pi : T \to X$ is the principal structure bundle associated with p, attached at the point x (art. 8.16). Now since $F = Y_{\varphi(0)}$, replacing $\psi(r)$ by $\psi(r) \circ \psi(0)^{-1}$ we get $\psi(0) = \mathbf{1}_F$ and an isomorphism Ψ satisfying the condition $\Psi \restriction \{0\} \times F = \mathbf{1}_F$. Now, the map $\bar\varphi : r \mapsto \psi(r)(y)$ is a lift of φ satisfying $\bar\varphi(0) = y$. Next, let σ be another lift of φ such that $\sigma(0) = y$, then $\Psi^{-1}(r, \sigma(r)) = (r, f(r))$ where $f \in \mathcal{C}^\infty(\mathbf{R}^n, F)$ and $f(0) = y$. But F is discrete, thus f is constant, $f(r) = y$ for all r, and $\sigma = \bar\varphi$. □

8.24. Fundamental group acting on coverings. Let X be a connected diffeological space, $\pi_0(X) = \{X\}$. Let $p : Y \to X$ be a covering with fiber $F = p^{-1}(x)$. Let $\gamma \in \mathrm{Paths}(X, \star, x)$, and let y be some point in F, there exists a unique path $\mathrm{lift}_y(\gamma) \in \mathrm{Paths}(Y, F, y)$ lifting γ, that is,

$$p \circ \mathrm{lift}_y(\gamma) = \gamma \quad \text{and} \quad \mathrm{lift}_y(\gamma)(1) = y.$$

The map $\ell \mapsto [y \mapsto \mathrm{lift}_y(\ell)(0)]$, where $\ell \in \mathrm{Loops}(X, x)$ and $y \in F$ is smooth and depends only on the homotopy class τ of ℓ. This defines an action of $\pi_1(X, x)$ on F denoted by

$$\tau_F(y) = \mathrm{lift}_y(\ell)(0) \quad \text{with} \quad \tau = \mathrm{class}(\ell). \tag{$\clubsuit$}$$

Since F is discrete there is a natural identification between F and $\pi_0(\mathsf{F})$ or between $\pi_0(\mathsf{F}, y)$ and (F, y), so the exact homotopy sequence of the fiber bundle p (art. 8.21) splits into

$$\left.\begin{array}{l} 0 \longrightarrow \pi_k(Y, y) \xrightarrow{p_{k\#}} \pi_k(X, x) \longrightarrow 0 \quad \text{for } k \geq 2, \\ 0 \longrightarrow \pi_1(Y, y) \xrightarrow{p_{1\#}} \pi_1(X, x) \xrightarrow{R(y)} (\mathsf{F}, y) \xrightarrow{i_\#} \pi_0(Y, y) \longrightarrow \{X\}, \end{array}\right\}$$

where $R(y)$ denotes the orbit map for the $\pi_1(X, x)$ action above (♣).

1. The map $p_{k\#}$ is an isomorphism for $k \geq 2$ and a monomorphism for $k = 1$.
2. The image of $\pi_1(Y, y)$ by $p_{1\#}$ is the stabilizer of y for the action of $\pi_1(X, x)$. The space Y is connected if and only if $\pi_1(X, x)$ is transitive on F.
3. If Y is connected and $p_{1\#}$ surjective, thus bijective, then p is a diffeomorphism. If X is simply connected, then all of its coverings are trivial.
4. The space Y is connected and simply connected if and only if the action of $\pi_1(X, x)$ is free and transitive on F.

PROOF. The path $\text{lift}_y(\gamma)$ is explicitly given by $\text{lift}_y(\gamma) : t \mapsto \psi(t)(y)$ where $(t, y) \mapsto (t, \psi(t)(y))$, $(t, y) \in \mathbf{R} \times \mathsf{F}$, is a trivialization of $\text{pr}_1 : \ell^*(Y) \to \mathbf{R}$ such that $\psi(1) = 1_\mathsf{F}$, and this lift is unique; see (art. 8.23). Actually, ψ is a smooth lift of ℓ in the principal structure bundle of $\text{pr}_1 : Y \to X$. Then, since the fiber F is discrete, any plot in F is locally constant and the composite with $y \mapsto \tau_\mathsf{F}(y) = \text{lift}_y(\ell)(0) = \psi(0)(y)$ is locally constant, thus smooth. Therefore, τ_F is smooth. Let $\bar{\ell}$ be the reverse of ℓ, we get a trivialization of $\text{pr}_1 : \bar{\ell}^*(Y) \to \mathbf{R}$, by $(t, y) \mapsto (t, \bar{\psi}(t)(y))$ with $\bar{\psi}(t) = \psi(1-t) \circ \psi(0)^{-1}$, and $\bar{\psi}(1) = 1_\mathsf{F}$, and $\text{lift}_y(\bar{\ell})(t) = \bar{\psi}(t)(y)$. So, let $\bar{\tau}_\mathsf{F}(y) = \text{lift}_y(\bar{\ell})(0) = \psi(1) \circ \psi(0)^{-1}(y)$, we get then $\bar{\tau}_\mathsf{F} \circ \tau_\mathsf{F}(y) = \psi(1) \circ \psi(0) \circ \psi(0)^{-1}(y) = y$, and $\tau_\mathsf{F} \circ \bar{\tau}_\mathsf{F}(y) = \psi(0) \circ \psi(1) \circ \psi(0)^{-1}(y) = y$. Thus, $\bar{\tau}_\mathsf{F} = \tau_\mathsf{F}^{-1}$, and τ_F is a diffeomorphism of F. Now let us consider two elements $\tau = \text{class}(\ell)$ and $\tau' = \text{class}(\ell')$ of $\pi_1(X, x)$, let ψ and ψ' be the lifts of ℓ and ℓ' in the principal structure bundle of $\pi : Y \to X$. On one hand we have $\tau_\mathsf{F}(\tau'_\mathsf{F}(y)) = \psi(0)(\psi'(0)(y))$. On the other hand, assuming as usual, that the loops ℓ and ℓ' are stationary, the lifts ψ and ψ' can also be chosen stationary, thus the concatenation $(\psi \circ \psi'(0)) \vee \psi'$ is well defined. This is a lift of $\ell \vee \ell'$, in the principal structure bundle of $\pi : Y \to X$, satisfying $((\psi \circ \psi'(0)) \vee \psi')(1) = 1_\mathsf{F}$. Next, $\tau \cdot \tau' = \text{class}(\ell \vee \ell')$, thus $(\tau \cdot \tau')_\mathsf{F}(y) = [(\psi \circ \psi'(0)) \vee \psi'](0)(y) = \psi(0)(\psi'(0)(y))$. Hence, $\tau \mapsto \tau_\mathsf{F}$ is a homomorphism from $\pi_1(X, x)$ to $\text{Diff}(\mathsf{F})$, that is, an action of $\pi_1(X, x)$ on F.

Now, let us check that the connecting homomorphism $\Delta_0 : \pi_1(X, x) \to \pi_0(\mathsf{F}, y)$ of the exact homotopy sequence (art. 8.21) coincides with the orbit map $R(y)$ of the action of $\pi_1(X, x)$ on F. Remember that $\Delta_0 = \hat{0}_\# \circ p_{1\#}^{-1}$, where $p_{1\#} : \pi_1(Y, \mathsf{F}, y) \to \pi_1(X, x)$. Then, the map Δ_0 is exactly the map associating the homotopy class of every loop $\ell \in \text{Loops}(X, x)$ with the connected component of the origin of the lift $\bar{\ell}$ ending at y, that is, $\text{lift}_y(\ell)(0)$. Thus, the connecting homomorphism from $\pi_1(X, x)$ to $\pi_0(\mathsf{F}, y)$, identified to the pointed space (F, y), is just the orbit map $R(y)$ of the action of $\pi_1(X, x)$ on the fiber F. Then,

1. Reading the homotopy exact sequence of $\pi : Y \to X$ gives immediately that $p_{k\#}$ is an isomorphism for $k \geq 2$, and a monomorphism for $k = 1$.

2. We read from the exact homotopy sequence that the kernel of $R(y)$, which is the stabilizer of y in $\pi_1(X, x)$, is the image of $\pi_1(Y, y)$ by $p_{1\#}$. Then, Y is connected, that

is, $\pi_0(Y, y) = \{Y\}$, or $i_\#(F, y) = \{Y\}$, if and only if $\pi_1(X, x)(F) = R(y)(\pi_1(X, x)) = F$, that is, if and only if $\pi_1(Y, y)$ acts transitively on F.

3. If $p_{1\#}$ is surjective, then $\ker(R(y)) = \pi_1(X, x)$ and then $\mathrm{val}(R(y)) = \ker(i_\#) = \{y\}$, but if Y is connected, then $\ker(i_\#) = (F, y)$, thus $(F, y) = \{y\}$, that is, $F = \{y\}$. The fiber F is reduced to a point, and p is an injective subduction, hence a diffeomorphism (art. 1.49). Now, if X is simply connected, then $(F, y) = \pi_0(Y, y)$, that is, F intersects each component of Y in one and only one point. Thus, the restriction of π to each component is an injective subduction, that is, a diffeomorphism. Hence, p is equivalent to the direct product $\mathrm{pr}_1 : X \times F \to X$.

4. If $\pi_1(X, x)$ acts freely on F, then $\ker(R(y)) = \mathbf{1}_{\pi_1(X,x)}$, and $\pi_1(X, x) = \mathrm{val}(R(y)) = F$. Thus, $\pi_1(Y, y) = 0$, $\pi_0(Y, y) = \{Y\}$, and conversely. □

8.25. Monodromy theorem. Let X and X′ be two connected diffeological spaces, let $p : Y \to X$ be a covering, and let $f : X' \to X$ be a smooth map.

1. Let φ_1 and φ_2 be two smooth lifts of f in Y. If they coincide somewhere they coincide everywhere.
2. If X′ is simply connected, then for all smooth maps $f : X' \to X$, for every $x' \in X'$ and every $y \in Y_x = p^{-1}(x)$ such that $x = f(x')$, there exists one and only one smooth lift φ such that $\varphi(x') = y$.

PROOF. Let $x' \in X'$ such that $\varphi_1(x') = \varphi_2(x')$, and let $x'' \in X'$. Since X′ is connected, there exists a path γ in X′ connecting x′ to x″. Thus, the paths $\varphi_1 \circ \gamma$ and $\varphi_2 \circ \gamma$ are two lifts of $f \circ \gamma$ coinciding in 0, thanks to (art. 8.23), they coincide everywhere. Therefore $\varphi_1(x'') = \varphi_1(\gamma(1)) = \varphi_2(\gamma(1)) = \varphi_2(x'')$, that is, $\varphi_1 = \varphi_2$. Now let us assume that X′ is simply connected, the pullback $\mathrm{pr}_1 : f^*(Y) \to X'$ is a covering, it is trivial because $\pi_1(X') = 0$ (art. 8.24, point 3). Then, the composite of any section of pr_1 with $\mathrm{pr}_2 : \mathrm{pr}_1^*(Y) \to Y$ gives a lift φ of f in Y. Choosing a fold of the covering, we can map x′ wherever we want over $f(x')$. Thanks to the first item this choice adjusts the lift. □

8.26. The universal covering. Let X be a connected diffeological space. Let $\mathbf{X}$ be the Poincaré groupoid defined in (art. 5.15). Remember that $\mathrm{Obj}(\mathbf{X}) = X$ and $\mathrm{Mor}(\mathbf{X}) = \Pi(X)$ is the space of fixed-ends homotopy classes of paths in X. Let x be any point of X, and let us denote

$$\tilde{X} = \mathrm{Mor}_{\mathbf{X}}(X, x, \star) \quad \text{and} \quad \pi : \tilde{X} \to X, \quad \text{with} \quad \pi = \hat{1} \restriction \mathrm{Mor}_{\mathbf{X}}(X, x, \star).$$

Then, $\tilde{X}$ satisfies the following properties.

1. The space $\tilde{X}$ is connected and simply connected. The projection $\pi : \tilde{X} \to X$ is a Galoisian covering, with structure group $\pi_1(X, x)$.
2. Every other connected and simply connected covering is isomorphic to the covering $\pi : \tilde{X} \to X$.
3. Every connected covering $p : Y \to X$ is a quotient of $\pi : \tilde{X} \to X$ by the action of a subgroup $H \subset \pi_1(X, x)$.

Thanks to these universal properties, the covering $\pi : \tilde{X} \to X$ is called *universal covering* of X.

NOTE 1. To paraphrase the above proposition: *every connected diffeological space has a connected and simply connected covering, unique up to isomorphism. Any other covering is isomorphic to one of its quotients.*

NOTE 2. The set of equivalence classes of coverings of X is in a one-to-one correspondence with the conjugacy classes of the subgroups of $\pi_1(X, x)$.

NOTE 3. The universal covering of a product is the product of the universal coverings of the factors.

PROOF. First of all, the structure group of the Poincaré groupoid at a point $x \in X$ is, by definition, the fundamental group $\pi_1(X, x)$. We have seen that the concatenation of paths is smooth, as well as the inversion. The map $x \mapsto \mathrm{class}(\mathbf{x})$, where $\mathbf{x}$ is the constant loop $t \mapsto x$ is an induction, indeed a plot in the units of $\Pi(X)$ lifts locally as a plot of $\mathrm{Paths}(X)$, then taking the value for $t = 0$, which is a smooth map, we get a plot of X. So, $\Pi(X)$ is a diffeological groupoid according to (art. 8.3). Moreover, since the characteristic map $\mathrm{ends} : \Pi(X) \to X$ is a subduction (art. 5.15), the Poincaré groupoid is a fibrating groupoid (art. 8.4). Then, the structure bundle attached at a point x (art. 8.14) is a principal fibration with structure group $\pi_1(X, x)$. Next, since the fundamental group $\pi_1(X, x)$ acts freely and transitively on the fiber $F = \pi^{-1}(x)$, and thanks to (art. 8.24, point 4), we conclude that $\tilde{X}$ is connected and simply connected. Now let $p : Y \to X$ be another covering, and let $F = p^{-1}(x)$. The pullback $\mathrm{pr}_1 : \pi^*(Y) \to X$ is trivial because $\tilde{X}$ is simply connected (art. 8.24, point 3). Let $\Psi : \tilde{X} \times F \to \pi^*(Y)$ be a trivialization such that, for all $(\xi, y) \in \tilde{X} \times F$,

$$\Psi(\xi, y) = (\xi, \psi(\xi)(y)) \text{ with } \begin{cases} \psi(\xi) \in \mathrm{Diff}(F, Y_{\pi(\xi)}), \\ \psi \upharpoonright \pi^{-1}(x) = \mathbf{1}_F. \end{cases}$$

The map ψ is a lift of π in the principal structure bundle associated with the fibration p. The commutative diagram ($\clubsuit$) describes the situation. Let ξ be the homotopy class of a path γ in X, with origin x. The lift $\tilde{\gamma}$ of γ in $\tilde{X}$ is given by $\tilde{\gamma} : s \mapsto \mathrm{class}(t \mapsto \gamma(st))$, and $\gamma_y : t \mapsto \psi(\tilde{\gamma}(t))(y)$ is the lift of the path γ in Y, with origin $y \in F$. We deduce then the fundamental property of ψ, for all $\tau \in \pi_1(X, x)$ and all $\xi \in \tilde{X}$, for all $y \in F$, $\psi(\tau \cdot \xi)(y) = \psi(\xi)(\tau_F^{-1}(y))$ and then $\psi(\tau)(y) = \tau_F^{-1}(y)$, where the action of $\pi_1(X, x)$ on F has been described in (art. 8.24). Next, let us choose a basepoint $y \in p^{-1}(x)$, the map $\xi \mapsto \psi(\xi)(y)$ is a smooth lift of $\pi : \tilde{X} \to X$. This lift is the composite of the section $\xi \mapsto (\xi, \psi(\xi)(y))$ of $\pi^*(Y)$ with pr_2. Note first that it is surjective in particular because the action of $\pi_1(X, x)$ is transitive on the fiber F, thanks to its connectedness (art. 8.24, point 2). Moreover this lift is a subduction because p and π are two subductions.

$$\begin{array}{ccccc}
\tilde{X} \times F & \xrightarrow{\ \Psi\ } & \pi^*(Y) & \xrightarrow{\ \mathrm{pr}_2\ } & Y \\
& \searrow_{\mathrm{pr}_1} & \downarrow_{\mathrm{pr}_1} & & \downarrow_p \\
& & \tilde{X} & \xrightarrow[\ \pi\ ]{} & X
\end{array} \qquad (\clubsuit)$$

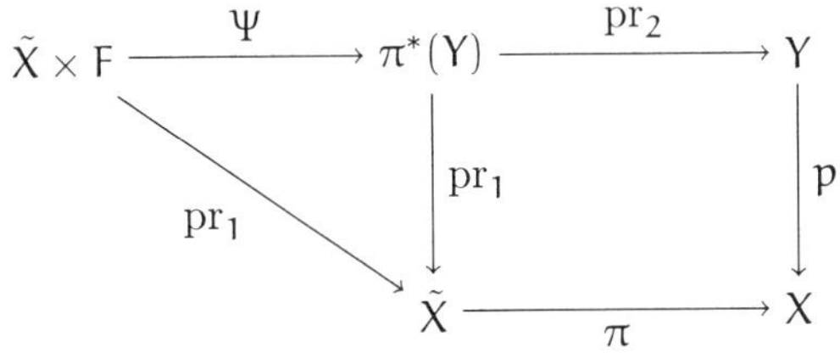

Now, if two points ξ and ξ' have the same image $\psi(\xi)(y) = \psi(\xi')(y)$, then $\pi(\xi) = \pi(x')$, and there exists $\tau \in \pi_1(X, x)$ such that $\xi' = \tau \cdot \xi$, thus $\psi(\xi')(y) = \psi(\tau \cdot \xi)(y) = \psi(\xi)(\tau_F^{-1}(y))$, and therefore $\tau_F(y) = y$. Conversely, if $\tau \in \pi_1(X, x)$ stabilizes y, then $\psi(\tau \cdot \xi)(y) = \psi(\xi)(y)$. Let $H = \{\tau \in \pi_1(X, x) \mid \tau_F(y) = y\}$ be the stabilizer of y, the space Y identifies now naturally with the quotient $\tilde{X}/H$. This proves that every covering is the quotient of the universal covering $\pi : \tilde{X} \to X$. This also proves, in particular, that there exists only one connected and simply connected covering, up to isomorphism.

For Note 2, it is purely algebraic and easy to prove that two conjugate subgroups H and H' will give two equivalent coverings, and conversely two equivalent connected coverings will give two conjugate stabilizers in $\pi_1(X, x)$.

For Note 3, the product of the universal coverings is still a covering and still simply connected, thus it is the universal covering of the product. □

8.27. Coverings and differential forms. Let X be a connected diffeological space, and let $\pi : \tilde{X} \to X$ be a universal covering.

1. A differential k-form α on $\tilde{X}$ is the pullback of a k-form β of X if and only if α is invariant by $\pi_1(X)$. In this case α is said to be *basic*, and β is the pushforward of α, also denoted by $\pi_*(\alpha)$.
2. Let $p : Y \to X$ be a covering, a differential k-form $\bar{\alpha}$ on Y is the pullback of a k-form β of X if and only if $\bar{\alpha}$ is the pushforward on Y of a $\pi_1(X)$-invariant k-form α of $\tilde{X}$.

NOTE. If the covering $p : Y \to X$ is Galoisian with structure group H, the differential form α is basic if and only if α is invariant by H.

PROOF. Let $x \in X$, and let us identify $\tilde{X}$ with the principal bundle, associated with the Poincaré groupoid, attached at the point x (art. 8.26). Since $\pi \circ \tau = \pi$ for all $\tau \in \pi_1(X, x)$, it is clear that if $\alpha = \pi^*(\beta)$, then α is invariant under the action of $\pi_1(X, x)$. Now, let us assume that α is invariant under $\pi_1(X, x)$. Let us apply the criterion of (art. 6.38). Let $P : U \to \tilde{X}$ and $P' : U \to \tilde{X}$ be two plots of $\tilde{X}$ such that $\pi \circ P = \pi \circ P'$. Because π is a principal fibration, for all $r_0 \in U$, there exist an open ball $\mathcal{B}$ centered at r_0 and a plot $r \mapsto \tau_r$ in the structure group $\pi_1(X, x)$ such that, for all $r \in \mathcal{B}$, $P'(r) = \tau_r \cdot P(r)$. But $\pi_1(X, x) = \pi^{-1}(x)$ is discrete and $\mathcal{B}$ is connected, then τ_r is constant and $P'(r) = \tau \cdot P(r)$ on $\mathcal{B}$, thus $\alpha(P') = \alpha(\tau \cdot P) = \tau^*(\alpha(P)) = \alpha(P)$. Therefore, $\alpha(P) = \alpha(P')$ on $\mathcal{B}$ and hence on U. Thanks to the criterion cited above, there exists a k-form β on X such that $\alpha = \pi^*(\beta)$. The second assertion is a consequence of the first item, and because every connected covering is a quotient of the universal covering by a subgroup of $\pi_1(X)$ (art. 8.26). □

Exercise

✎ EXERCISE 133 (Covering tori). Let $T_\Gamma = \mathbf{R}^n/\Gamma$ be a torus, that is, $\Gamma \subset \mathbf{R}^n$ is any discrete subgroup generating $\mathbf{R}^n$, dense or not. Give a canonical representative for every class of homotopic loops of the torus T_Γ. Show, in particular, that, for the circle S^1, every loop based at a point x is homotopic to the loop $t \mapsto \mathcal{R}(2\pi kt)(x)$, where $\mathcal{R}(\theta)$ is the rotation of angle θ, and $k \in \mathbf{Z}$. Use the diffeomorphism $S^1 \simeq \mathbf{R}/2\pi\mathbf{Z}$ built in Exercise 27, p. 27.

Integration Bundles of Closed 1-Forms

How are differential forms connected to fiber bundles? The first case comes with closed 1*-forms. To be a closed differential form is not far from being exact. Where can we read the lack of exactness of a closed form? How can we represent* *it? We have seen that a closed* 1*-form* α*, on a connected diffeological space* X*, is exact if and only if its integral vanishes on every loop* (art. 6.89). *We can do more, we can construct, explicitly, the minimal space* X_α *over* X*, on which the pullback of* α *is exact. This space is a covering over* X*, and a Galoisian covering. The group of the covering is the* group of periods of α, $P_\alpha \subset (\mathbf{R}, +)$, *made with the integrals of* α *on the loops in* X*. The idea is that, thanks to the proposition in* (art. 6.91), *the pullback of* α *on the space of pointed paths in* X, $\mathrm{Paths}(X, x, \star)$, *where* $x \in X$ *is any base point, is exact* (art. 6.91). *So, we shall find the smallest quotient of* $\mathrm{Paths}(X, x, \star)$, *covering* X*, on which the pullback of* α *can be pushed forward, and remains exact. Conversely, cohomology classes of closed* 1*-forms can be regarded as* characteristic classes *of such coverings.*

8.28. Periods of closed 1-forms. Let X be a diffeological space and α be a closed 1-form, $\alpha \in \Omega^1(X)$ and $d\alpha = 0$. Let $\mathcal{K}$ be the Chain-Homotopy operator (art. 6.83) and consider $\mathcal{K}\alpha$, the smooth real function, defined on Paths(X) by

$$\mathcal{K}\alpha(\gamma) = \int_\gamma \alpha = \int_0^1 \alpha(\gamma)_t(1)\, dt, \quad \text{for all} \quad \gamma \in \mathrm{Paths}(X).$$

1. The restriction of $\mathcal{K}\alpha$ to the subspace Loops(X) is closed,

$$d\,[\mathcal{K}\alpha \upharpoonright \mathrm{Loops}(X)] = 0.$$

Thus, $\mathcal{K}\alpha \upharpoonright \mathrm{Loops}(X)$ is constant on the connected components of Loops(X).

2. The group of periods P_α of α (art. 6.74, (♣)) coincides with the group of periods of $\mathcal{K}\alpha \upharpoonright \mathrm{Loops}(X)$ and it is generated by the following set of periods

$$\mathrm{Periods}(\alpha) = \left\{ \mathcal{K}\alpha(\ell) = \int_\ell \alpha \;\middle|\; \ell \in \mathrm{Loops}(X) \right\}. \qquad (\clubsuit)$$

3. If X is connected, $\pi_0(X) = \{X\}$ (art. 5.6), then $P_\alpha = \mathrm{Periods}(\alpha)$, and the periods can be computed on Loops(X, x), the subspace of loops based at some point x of X. The result does not depend on x and the group of periods $P_\alpha \subset \mathbf{R}$ is a homomorphic image of the *Abelianized fundamental group* $\pi_1^{\mathrm{Ab}}(X)$.

4. If X is not connected, a 1-form α on X is any family of 1-forms $\alpha_i = \alpha \upharpoonright X_i$ defined on the component $X_i \in \pi_0(X)$. The form α is closed if and only if each α_i is closed. In this case, the group of periods of α is generated by the union of the groups of periods on all components

$$\mathrm{Periods}(\alpha) = \bigcup_{X_i \in \pi_0(X)} \mathrm{Periods}(\alpha \upharpoonright X_i).$$

For each component X_i the periods $P_{\alpha_i} = \mathrm{Periods}(\alpha \upharpoonright X_i)$ are a subgroup of $\mathbf{R}$, a homomorphic image of the Abelianized fundamental group of the component X_i.

PROOF. The space X is assumed to be connected. Thanks to the identity $d \circ \mathcal{K} + \mathcal{K} \circ d = \hat{1}^* - \hat{0}^*$, applied to α, and restricted to the subspace Loops(X), for which $\hat{1}^* = \hat{0}^*$, and given that α is closed, we get $d[\mathcal{K}\alpha \upharpoonright \mathrm{Loops}(X)] = 0$. Now, let us prove that the set Periods(α) is a subgroup of $(\mathbf{R}, +)$. Let ℓ and ℓ' be two loops, based in x and x′, that is, $\ell(0) = \ell(1) = x$, $\ell'(0) = \ell'(1) = x'$. Let $a = \int_\ell \alpha$ and $a' = \int_{\ell'} \alpha$, we want to find a loop ℓ'' such that $\int_{\ell''} \alpha = a + a'$. Since X is connected, there exists a path γ connecting x to x′, that is, $\gamma(0) = x$ and $\gamma(1) = x'$. Now let us consider the smashed concatenation (art. 5.5, Note) $\ell'' = \ell \star (\gamma \star (\ell \star \bar{\gamma}))$, where

$\bar\gamma(t) = \gamma(1-t)$, ℓ'' is a loop based at x and $\int_{\ell''} \alpha = \int_\ell \alpha + \int_\gamma \alpha + \int_{\ell'} \alpha + \int_{\bar\gamma} \alpha$, but since $\int_{\bar\gamma} \alpha = -\int_\gamma \alpha$, $\int_{\ell''} = a + a'$. Now, if $a = \int_\ell \alpha$, then $\int_{\bar\ell} \alpha = -a$. Hence, Periods($\alpha$) is a subgroup of $\mathbf{R}$. Next, let ℓ and ℓ' be two homotopic (free or not) loops, that is, such that there exists a path $[s \mapsto \ell_s] \in \mathcal{C}^\infty(\mathbf{R}, \text{Loops}(X))$ with $\ell_0 = \ell$ and $\ell_1 = \ell'$. Thanks to the homotopic invariance of the De Rham cohomology (art. 6.88), we get $\ell_1^*\alpha = \ell_0^*\alpha + d\beta$, then $\int_{\ell'} \alpha = \int_\ell \alpha + \int_\ell d\beta$, but thanks to Stokes' theorem (art. 6.69) $\int_\ell d\beta = \int_{\partial\ell} \beta = 0$ since $\partial\ell = 0$. Hence, $\int_{\ell'} \alpha = \int_\ell \alpha$. Therefore, the group of periods $P_\alpha = \text{Periods}(\alpha)$ is the homomorphic image of $\pi_1(X, x)$ by the integration morphism. ☐

8.29. Integrating closed 1-forms. Let X be a connected diffeological space, $\pi_0(X) = \{X\}$ (art. 5.6). Let α be a closed 1-form, $\alpha \in \Omega^1(X)$ and $d\alpha = 0$. Let $\mathcal{K}\alpha \in \Omega^0(\text{Paths}(X)) = \mathcal{C}^\infty(\text{Paths}(X), \mathbf{R})$, where $\mathcal{K}$ is the Chain-Homotopy operator (art. 6.83). Restricted to the subspace $\text{Paths}(X, x_0, \star)$ of the paths in X with origin $x_0 \in X$, it satisfies $d\,[\mathcal{K}\alpha \restriction \text{Paths}(X, x_0, \star)] = \hat{1}^*(\alpha)$. Let X_α be the quotient of $\text{Paths}(X, x_0, \star)$ by the equivalence relation

$$\gamma \sim \gamma' \quad \text{if} \quad \hat{1}(\gamma) = \hat{1}(\gamma') \quad \text{and} \quad \mathcal{K}\alpha(\gamma) = \mathcal{K}\alpha(\gamma'). \qquad (\diamondsuit)$$

Let class : $\text{Paths}(X, x_0, \star) \to X_\alpha$ be the projection, and $\text{pr}_\alpha : X_\alpha \to X$ be the factor defined by $\text{pr}_\alpha \circ \text{class} = \hat{1}$. Let $F_\alpha : X_\alpha \to \mathbf{R}$ be defined by $F_\alpha \circ \text{class} = \mathcal{K}\alpha$, then

$$F_\alpha(\hat{x}) = \int_\gamma \alpha = \int_0^1 \alpha(\gamma)_t(1)dt \quad \text{with} \quad \hat{x} = \text{class}(\gamma),$$

and $F_\alpha \in \mathcal{C}^\infty(X_\alpha, \mathbf{R})$. The commutative diagram (♠) expresses the relationship between all these maps.

1. The projection $\text{pr}_\alpha : X_\alpha \to X$ is a Galoisian covering (art. 8.22) with structure group P_α and the function F_α is a primitive of the pullback $\text{pr}_\alpha^*(\alpha)$, that is, $dF_\alpha = \text{pr}_\alpha^*(\alpha)$. The bundle pr_α will be called the *integration covering* of α.

Let P_α be the group of periods of α (art. 8.28), and let us define the *torus of periods* of α as the quotient group (art. 7.1)

$$T_\alpha = \mathbf{R}/P_\alpha \quad \text{with projection} \quad \text{class}_\alpha : \mathbf{R} \to T_\alpha.$$

Then, the map $F_\alpha : X_\alpha \to \mathbf{R}$ projects onto a smooth map $f_\alpha : X \to T_\alpha$, that is, $\text{class}_\alpha \circ F_\alpha = f_\alpha \circ \text{pr}_\alpha$, which is summarized by the following diagram.

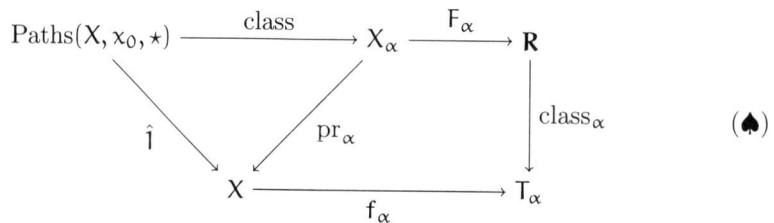

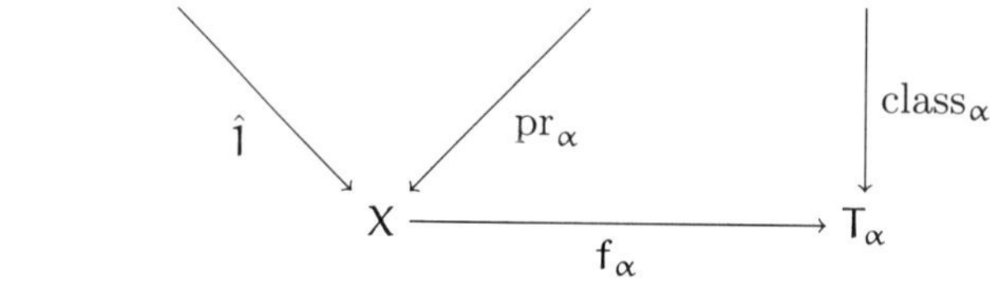

(♠)

2. If the group of periods P_α is a discrete diffeological subgroup of $\mathbf{R}$, which is equivalent to $P_\alpha \neq \mathbf{R}$, then there exists a unique 1-form

$$\theta_\alpha \in \Omega^1(T_\alpha) \quad \text{such that} \quad dt = \text{class}_\alpha^*(\theta_\alpha). \qquad (\clubsuit)$$

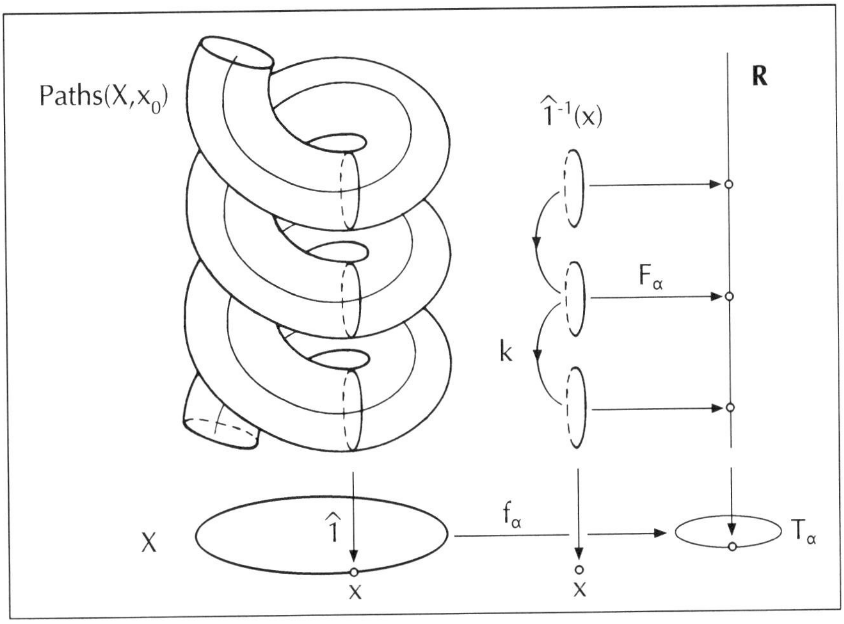

FIGURE 8.6. Integration of a closed 1-form.

The form θ_α is the *canonical* 1-*form* on T_α. Moreover, the form α coincides with the pullback of θ_α by f_α, that is,

$$\alpha = f_\alpha^*(\theta_\alpha) \quad \text{with} \quad f_\alpha(x) = \text{class}_\alpha\left(\int_{x_0}^{x} \alpha\right), \tag{$\heartsuit$}$$

where the integral is computed on any path γ connecting x_0 to x. We shall say, in this case, that f_α *integrates* α, or that f_α is an *integration function* of α.

This whole construction, integration bundle and integration function, is illustrated by Figure 8.6.

NOTE 1. The proposition in (art. 6.89) follows directly from this construction. If $P_\alpha = \{0\}$, then $T_\alpha = \mathbf{R}$ and $\alpha = f_\alpha^*(dt) = df_\alpha$. In this case, the integration function is a primitive of α.

NOTE 2. Another important case occurs in differential geometry, when the group of periods P_α is generated by one number $a \in \mathbf{R}$, the form is said to be *integral*. In this case, $T_\alpha \sim S^1_a = \mathbf{R}/a\mathbf{Z}$ is a manifold, the circle of perimeter a, and f_α is a smooth map from X to S^1_a.

NOTE 3. The projection $\text{class}_\alpha : X_\alpha \to X$ is the smallest diffeological covering over X, for which the pullback of α is exact on the total space. It is equivalent to the pullback $f_\alpha^*[\text{class}_\alpha : \mathbf{R} \to T_\alpha] = \{(x, t) \in X \times \mathbf{R} \mid f_\alpha(x) = \text{class}_\alpha(t)\}$.

NOTE 4. When the group of periods P_α is discrete, the solution of the equation $\alpha = f_\alpha^*(\theta_\alpha)$ ($\clubsuit$) is unique up to a constant. Every solution f'_α writes $f'_\alpha = f_\alpha + \tau$, with $\tau \in T_\alpha$, where the group operation on T_α has been denoted additively and f_α has been defined above ($\heartsuit$) by the choice of the base point x_0.

NOTE 5. It may happen that the group of periods P_α is not discrete, that is, $P_\alpha = \mathbf{R}$, then the torus of periods is reduced to $\{0\}$. The map $\text{pr}_\alpha : X_\alpha \to X$ defined above (♠) continues to be a covering of X, with a fiber equivalent to the set of real numbers equipped with the discrete diffeology. But there is no longer an integration function, for the form α, because the projection from $\mathbf{R}$ to $\{0\}$ is no more a covering; see also Exercise 135, p. 272.

PROOF. Let us check that $\mathrm{pr}_\alpha : X_\alpha \to X$ is a principal covering with P_α as structure group. Since α is closed, the integral $\mathcal{K}\alpha(\gamma) = \int_\gamma \alpha$ depends only on the fixed-ends homotopy class (art. 6.88). So, the map $\mathcal{K}\alpha$ factorizes through the space of fixed-end homotopy classes of paths with origin x_0, that is, the universal covering $\pi : \tilde{X} \to X$ (art. 8.26). Let $\tilde{F}_\alpha$ be defined by $\tilde{F}_\alpha(\mathrm{class}(\gamma)) = \mathcal{K}\alpha(\gamma)$. Since the projection class : $\mathrm{Paths}(X, x_0, \star) \to \tilde{X}$ is a subduction, $\tilde{F}_\alpha \in \mathcal{C}^\infty(\tilde{X}, \mathbf{R})$. Now let $\eta_\alpha : \pi_1(X, x_0) \to \mathbf{R}$ be the homomorphism defined by the integration, $\eta_\alpha(\mathrm{class}(\ell)) = \int_\ell \alpha$, so $X_\alpha = \tilde{X}/\ker(\eta_\alpha)$. Indeed, let $k = \mathrm{class}(\ell) \in \pi_1(X, x_0)$, and $\tilde{x} = \mathrm{class}(\gamma) \in \tilde{X}$, so $\tilde{F}_\alpha(k(\tilde{x})) = \int_{\ell\star\gamma} \alpha = \int_\ell \alpha + \int_\gamma \alpha$, where $\star$ is the smashed concatenation of paths. So, if $k \in \ker(\eta_\alpha)$, that is, if $\int_\ell \alpha = 0$, then $\tilde{F}_\alpha(k(\tilde{x})) = F_\alpha(\tilde{x})$. Now, if $\tilde{F}_\alpha(\tilde{x}) = \tilde{F}_\alpha(\tilde{x}')$, with $\tilde{x}' = \mathrm{class}(\gamma')$ and $\hat{1}(\gamma') = \hat{1}(\gamma)$, then $\int_\gamma \alpha = \int_{\gamma'} \alpha$, but $\int_{\gamma'} \alpha - \int_\gamma \alpha = \int_\ell \alpha$ with $\ell = \gamma' \star \bar{\gamma} \in \mathrm{Loops}(X, x_0)$. So, $\tilde{x}' = k(\tilde{x})$ with $k = \mathrm{class}(\ell)$ and $\eta_\alpha(k) = 0$. Now, since η_α is a homomorphism, the projection $\mathrm{pr}_\alpha : X_\alpha = \tilde{X}/\ker(\eta_\alpha) \to X$ is a covering (art. 8.26), and since η_α is a homomorphism to an Abelian group, $\ker(\eta_\alpha)$ is an invariant subgroup of $\pi_1(X, x_0)$ and the covering $\mathrm{pr}_\alpha : X_\alpha \to X$ is a principal covering, with fiber $\pi_1(X)/\ker(\eta_\alpha) = \mathrm{val}(\eta_\alpha) = P_\alpha$. Now, $d[\mathcal{K}\alpha] = \hat{1}^*(\alpha)$ writes $d[F_\alpha \circ \mathrm{class}] = (\mathrm{pr}_\alpha \circ \mathrm{class})^*(\alpha)$, that is, $\mathrm{class}^*[dF_\alpha)] = \mathrm{class}^*[\mathrm{pr}_\alpha^*(\alpha)]$, and since class is a subduction, $\mathrm{pr}_\alpha^*(\alpha) = dF_\alpha$ (art. 6.39). Finally, if P_α is a discrete diffeological subgroup of $\mathbf{R}$, then $\mathrm{class}_\alpha : \mathbf{R} \to T_\alpha$ is a covering, and since the 1-form $dt \in \Omega^1(\mathbf{R})$ is invariant by P_α there exists a unique 1-form $\theta_\alpha \in \Omega^1(T_\alpha)$ such that $\mathrm{class}_\alpha^*(\theta_\alpha) = dt$ (art. 8.27). Now, $(\mathrm{class}_\alpha \circ F_\alpha)^*(\theta_\alpha) = (f_\alpha \circ \mathrm{pr}_\alpha)^*(\theta_\alpha)$, thus $F_\alpha^*(dt) = F_\alpha^*(\mathrm{class}_\alpha^*(\theta_\alpha)) = \mathrm{pr}_\alpha^*(f_\alpha^*(\theta_\alpha))$, but $F_\alpha^*(dt) = dF_\alpha = \mathrm{pr}_\alpha^*(\alpha)$, hence $\mathrm{pr}_\alpha^*(\alpha) = \mathrm{pr}_\alpha^*(f_\alpha^*(\theta_\alpha))$. Then, since pr_α is a subduction, $f_\alpha^*(\theta_\alpha) = \alpha$ (art. 6.39). □

8.30. The cokernel of the first De Rham homomorphism. Let X be a connected diffeological space and $\tilde{X}$ be its universal covering (art. 8.26). The injectivity of the first De Rham homomorphism η (art. 6.89, Note 2) can be reformulated as the short sequence

$$0 \longrightarrow H^1_{\mathrm{dR}}(X) \xrightarrow{\ \eta\ } \mathrm{Hom}(\pi_1(X), \mathbf{R}).$$

We would close now this sequence, on the right, by interpreting geometrically the cokernel

$$\mathrm{coker}(\eta) \simeq \mathrm{Hom}(\pi_1(X), \mathbf{R})/H^1_{\mathrm{dR}}(X).$$

Let ρ be any homomorphism $\rho \in \mathrm{Hom}(\pi_1(X), \mathbf{R})$. Let us consider the diagonal action of $\pi_1(X)$ on the direct product $\tilde{X} \times \mathbf{R}$, defined by

$$k(\tilde{x}, t) = (k(\tilde{x}), t + \rho(k)),$$

for all $k \in \pi_1(X)$ and $(\tilde{x}, t) \in \tilde{X} \times \mathbf{R}$, where we have denoted by $(k, \tilde{x}) \mapsto k(\tilde{x})$ the natural action of $\pi_1(X)$ on $\tilde{X}$. The diagonal action of $\pi_1(X)$ on $\tilde{X} \times \mathbf{R}$ is free and we shall denote the quotient by

$$X_\rho = (\tilde{X} \times \mathbf{R})/\pi_1(X).$$

Let $\mathbf{p}$ be the projection from $\tilde{X} \times \mathbf{R}$ to $\tilde{X}_\rho$. The diffeological space X_ρ, quotient of the product $\tilde{X} \times \mathbf{R}$ by ρ, is the total space of the associated fibration (art. 8.16)

$$p : X_\rho \to X \quad \text{with} \quad p : \mathbf{p}(\tilde{x}, t) \mapsto \pi(\tilde{x}).$$

This construction is summarized by the next commutative diagram. The projection $p : X_\rho \to X$ is a principal bundle with structure group $(\mathbf{R}, +)$ (art. 8.11), whose action is given by

$$s(\mathbf{p}(\tilde{x}, t)) = \mathbf{p}(x, t + s),$$

for all $s \in \mathbf{R}$ and all $\mathbf{p}(\tilde{x}, t) \in X_\rho$.

$$
\begin{array}{ccc}
\tilde{X} \times \mathbf{R} & \xrightarrow{\ \mathbf{p}\ } & X_\rho \\
\downarrow{\scriptstyle \mathrm{pr}_1} & & \downarrow{\scriptstyle p} \\
\tilde{X} & \xrightarrow[\ \pi\ ]{} & X
\end{array}
$$

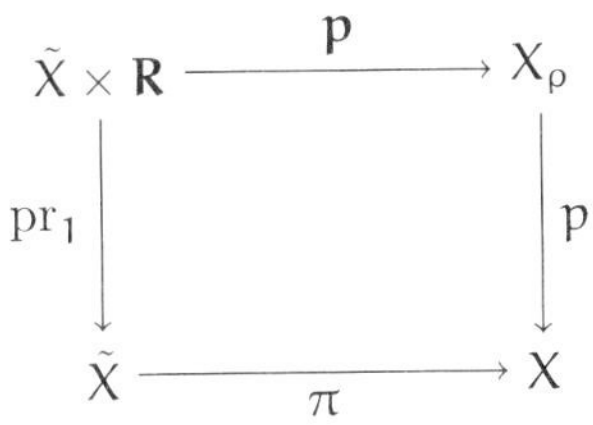

Moreover, this fiber bundle $p : X_\rho \to X$ is trivial if and only if the homomorphism ρ is the image of a closed 1-form by the De Rham homomorphism, that is, if and only if there exists a closed 1-form $\alpha \in Z_{dR}^1(X)$ such that

$$\rho(k) = \int_\ell \alpha,$$

for all $k \in \pi_1(X)$ with $k = \mathrm{class}(\ell)$ and $\ell \in \mathrm{Loops}(X)$. Therefore, $\mathrm{coker}(\eta)$ can be interpreted as the group of equivalence classes of principal flat $(\mathbf{R}, +)$-bundles over X, that is, the ones whose pullback on the universal covering $\tilde{X}$ is trivial.

PROOF. Let us begin by assuming that ρ is associated with the closed 1-form α on X. The pullback $\pi^*(\alpha) \in \Omega^1(\tilde{X})$, of α by the projection π, is closed. But since $\tilde{X}$ is simply connected, $\pi^*(\alpha)$ is exact (art. 6.76), let $f \in C^\infty(\tilde{X}, \mathbf{R})$ be a primitive of $\pi^*(\alpha)$. Since $k^*(df) = df$, for all $k \in \pi_1(X)$, f satisfies

$$f(k(\tilde{x})) = f(\tilde{x}) + \rho(k).$$

Then, the smooth map Σ, defined by

$$\Sigma : \tilde{x} \mapsto (\tilde{x}, f(\tilde{x})) \in C^\infty(\tilde{X}, \tilde{X} \times \mathbf{R}), \quad \text{satisfies} \quad \Sigma(k(\tilde{x})) = k(\Sigma(\tilde{x})).$$

Therefore, Σ projects on a map $\sigma : X \to \tilde{X}_\rho$ such that $\sigma \circ \pi = \mathbf{p} \circ \Sigma$. Since π and $\mathbf{p}$ in the diagram ($\clubsuit$) are subductions, and since Σ is smooth, the map σ is smooth. Now, the map σ is a smooth section of a principal fiber bundle, $p \circ \sigma = 1_X$, and the bundle $p : X_\rho \to X$ is trivial (art. 8.11).

$$
\begin{array}{ccc}
\tilde{X} \times \mathbf{R} & \xrightarrow{\ \mathbf{p}\ } & X_\rho \\
\uparrow{\scriptstyle \Sigma} & & \uparrow{\scriptstyle \sigma} \\
\tilde{X} & \xrightarrow[\ \pi\ ]{} & X
\end{array}
\qquad (\clubsuit)
$$

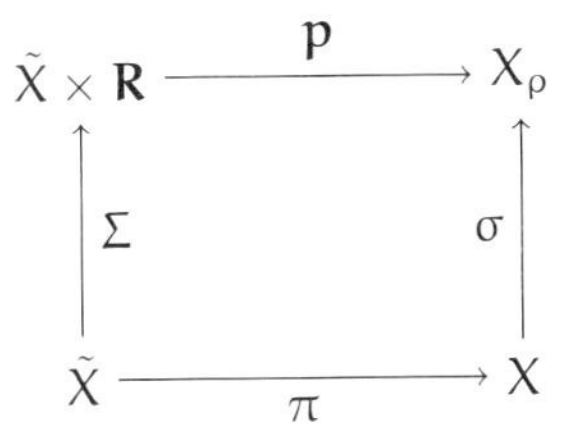

Conversely, let us assume that the bundle $p : X_\rho \to X$ is trivial. Then, there exists a smooth section $\sigma : X \to \tilde{X}_\rho$. But the map $\sigma \circ \pi$ is smooth and $\tilde{X}$ is simply connected, thus $\mathbf{p}$ is the universal covering of X_ρ. Now, thanks to the monodromy theorem (art. 8.25), there exists a smooth map $\Sigma : \tilde{X} \to \tilde{X} \times \mathbf{R}$ lifting $\sigma \circ \pi$, that is, such that $\sigma \circ \pi = \mathbf{p} \circ \Sigma$. This lift is unique with the condition $\sigma(x_0) = [\tilde{x_0}, 0]$, where x_0 is some point in X and $\tilde{x_0}$ is any point of $\tilde{X}$ above x_0. The map Σ is a smooth section of $\mathrm{pr}_1 : \tilde{X} \times \mathbf{R} \to \tilde{X}$. Thus, $\mathbf{p} \circ \Sigma = \sigma \circ \pi$ implies $p \circ \mathbf{p} \circ \Sigma = p \circ \sigma \circ \pi$, but $p \circ \sigma = 1_X$ and $p \circ \mathbf{p} = \pi \circ \mathrm{pr}_1$, hence $\pi \circ \mathrm{pr}_1 \circ \Sigma = \pi$. Therefore, for any $\tilde{x} \in \tilde{X}$ there exists some $\kappa(\tilde{x}) \in \pi_1(X)$ such that $\mathrm{pr}_1 \circ \Sigma(\tilde{x}) = \kappa(\tilde{x})(\tilde{x})$. But since the maps pr_1 and Σ are smooth, the map κ is smooth, and since $\pi_1(X)$ is discrete and $\tilde{X}$ connected, κ is constant. By the choice made above, $\sigma(\tilde{x_0}) = [\tilde{x_0}, 0]$, we get that $\kappa(\tilde{x})$ is the identity for all $\tilde{x}$. Now, since the map Σ is a section, there exists a smooth map $f \in \mathcal{C}^\infty(\tilde{X}, \mathbf{R})$ such that $\Sigma(\tilde{x}) = (\tilde{x}, f(\tilde{x}))$. Let us consider $k \in \pi_1(X)$, and let us apply $\mathbf{p} \circ \Sigma = \sigma \circ p$ to $k(\tilde{x})$, we get $\mathbf{p} \circ \Sigma(k(\tilde{x})) = \sigma \circ p(k(\tilde{x})) = \sigma \circ p(x) = \mathbf{p} \circ \Sigma(\tilde{x})$. Then, $\Sigma(k(\tilde{x}))$ and $\Sigma(\tilde{x})$ are on the same orbit of $\pi_1(X)$. Thus, there exists $h \in \pi_1(X)$ such that $\Sigma(k(\tilde{x})) = h(\Sigma(\tilde{x}))$, that is, $(k(\tilde{x}), f(k(\tilde{x})) = (h(\tilde{x}), f(\tilde{x}) + \rho(h))$. But since the action of $\pi_1(X)$ on $\tilde{X}$ is free, $h = k$ and $f(k(\tilde{x})) = f(\tilde{x}) + \rho(k)$. Hence, the differential $\mathbf{d}f$ is invariant by the action of $\pi_1(X)$, $d[f \circ k] = d[f + \rho(k)] = df$. Therefore, there exists a closed 1-form α on X such that $\pi^*(\alpha) = \mathbf{d}f$ (art. 6.38). Finally, the fiber bundle $p : \tilde{X}_\rho \to X$ is associated with the 1-form α. ☐

8.31. The first De Rham homomorphism for manifolds. It is known classically that every bundle with a contractible fiber over a manifold M admits a smooth section [**Die70c**]. Then, any $(\mathbf{R}, +)$ principal bundle over a manifold admits a section and is thus trivial. Therefore, the cokernel of the first De Rham homomorphism is trivial, and the De Rham homomorphism η is an isomorphism. This is a part of the De Rham theorem for manifolds, which says the general De Rham homomorphism is an isomorphism in each degree. As we can see in Exercise 134, p. 272, this is not always the case in diffeology. But the construction above (art. 8.30) shows that the direct proof of the De Rham isomorphism in degree 1, at least, is reduced to the existence of a section on fiber bundles with contractible fiber, which is by itself an interesting information.

Exercises

✎ EXERCISE 134 (De Rham homomorphism and irrational tori). Let $T_\alpha = \mathbf{R}/(\mathbf{Z} + \alpha\mathbf{Z})$ be the irrational torus with slope $\alpha \in \mathbf{R} - \mathbf{Q}$. We know that $H^1_{\mathrm{dR}}(T_\alpha) = \mathbf{R}$, every 1-forms on T_α is closed and proportional to $\theta = \pi_*(dt)$, where $\pi : \mathbf{R} \to T_\alpha$ is the projection; see Exercise 119, p. 202. We shall admit that $\pi_1(T_\alpha) = \mathbf{Z} \times \mathbf{Z}$ is included in the universal covering $\tilde{T}_\alpha = \mathbf{R}$ as the numbers $\mathbf{Z} + \alpha\mathbf{Z}$. Then, any morphism $\rho \in \mathrm{Hom}(\pi_1(T_\alpha), \mathbf{R})$ writes $\rho : (n, m) \mapsto rn + sm$ with $(r, s) \in \mathbf{R}^2$. Now, for any closed 1-form $c \times \theta$ of T_α, the homomorphism $\rho_c : \mathrm{class}(\ell) \mapsto \int_\ell c \times \theta$ is given by $\rho_c(n, m) = c(n + \alpha m)$. Describe the space $\mathbf{R} \times_\rho \mathbf{R} = [\mathbf{R} \times \mathbf{R}]/\rho_c$ and its fibration on T_α.

✎ EXERCISE 135 (The fiber of the integration bundle). Prove directly that the fiber of the projection $\mathrm{class}_\alpha : X_\alpha \to X$ (art. 8.29) is discrete. Use the formula of the variation of the integral of a differential form (art. 6.70).

Connections on Diffeological Fiber Bundles

In this section we suggest what should be a connection in diffeology. This is more a program than a strict immutable definition. The general idea behind the concept of connection on a principal fiber bundle is to give a way for lifting paths from the base space to the total space, in an equivariant way, and such that, once a source point of the lift is given, the lift is unique. The main and simplest example of a connection is the lifting of paths in coverings. Once a connection is chosen in a principal bundle, it gives a way for lifting paths in every associated fiber bundle. Naturally, this lifting process must satisfy some locality and compatibility properties. We shall suggest a series of conditions, fulfilled by the standard connections, which seem appropriate to be required in a generalization to diffeology. We treat in particular the case of principal bundles with generalized tori $\mathbf{R}/\Gamma$ *as structure groups, where* Γ *is any subgroup of* $\mathbf{R}$ *and for which the connection is defined by a* 1*-form. These principal bundles play a major role in the geometric integration of closed* 2*-forms* (art. 8.42). *Also, an important property is that the pullbacks by homotopic smooth maps of a principal bundle, equipped with a connection, are equivalent* (art. 8.34). *This property, always satisfied in the geometry of manifolds, is related to the existence of connections on ordinary principal fiber bundles.*

This section does not intend to exhaust all the consequences of such a definition and leaves the field open for future investigations. This approach of connections in diffeology follows the classical theory exposed in [**KoNo63**] *and what I began in* [**Igl87**]. *It is possible to give a more synthetic approach, by reduction of fibrating groupoids, which I leave for later.*

8.32. Connections on principal fiber bundles. Let Y be a diffeological space. We denote by $\mathrm{Paths}_{\mathrm{loc}}(Y)$ the space of *local paths* in Y, that is, the set of 1-plots of Y defined on open intervals,

$$\mathrm{Paths}_{\mathrm{loc}}(Y) = \{\tilde{c} \in \mathcal{C}^\infty(]a, b[, Y) \mid a, b \in \mathbf{R}\},$$

equipped with the functional diffeology induced by the functional diffeology of the 1-plots of Y (art. 1.63). Let us denote then by $\mathrm{tbPaths}_{\mathrm{loc}}(Y)$ the *tautological bundle*, equipped naturally with the subset diffeology of the product,

$$\mathrm{tbPaths}_{\mathrm{loc}}(Y) = \{(\tilde{c}, t) \in \mathrm{Paths}_{\mathrm{loc}}(Y) \times \mathbf{R} \mid t \in \mathrm{def}(\tilde{c})\}.$$

Next, let $\pi : Y \to X$ be a principal diffeological fibration with structure group G, and let $(g, y) \mapsto g_Y(y)$ denote the action of G on Y.

Definition. We shall call a *connection* on the G-principal fiber bundle $\pi : Y \to X$ every smooth map

$$\Theta : \mathrm{tbPaths}_{\mathrm{loc}}(Y) \to \mathrm{Paths}_{\mathrm{loc}}(Y) \qquad (\spadesuit)$$

satisfying the following series of conditions.

1. *Domain.* $\mathrm{def}(\Theta(\tilde{c}, t)) = \mathrm{def}(\tilde{c})$.
2. *Lifting.* $\pi \circ \Theta(\tilde{c}, t) = \pi \circ \tilde{c}$.
3. *Basepoint.* $\Theta(\tilde{c}, t)(t) = \tilde{c}(t)$.
4. *Reduction.* $\Theta(\gamma \cdot \tilde{c}, t) = \gamma(t)_Y \circ \Theta(\tilde{c}, t)$, where $\gamma : \mathrm{def}(\tilde{c}) \to G$ is any smooth path and $\gamma \cdot \tilde{c} = [s \mapsto \gamma(s)_Y(\tilde{c}(s))]$.
5. *Locality.* $\Theta(\tilde{c} \circ f, s) = \Theta(\tilde{c}, f(s)) \circ f$, where f is any smooth local path defined on an open domain with values in $\mathrm{def}(\tilde{c})$.

6. *Projector.* $\Theta(\Theta(\tilde{c},t),t) = \Theta(\tilde{c},t)$.

The local path $\Theta(\tilde{c},t)$ is the *horizontal projection* of $\tilde{c}$ pointed at t; it is a *horizontal path* for the connection Θ. The set of horizontal local paths will be denoted by

$$\mathrm{HorPaths_{loc}}(Y,\Theta) = \{\Theta(\tilde{c},t) \mid \tilde{c} \in \mathrm{Paths_{loc}}(Y) \text{ and } t \in \mathrm{def}(\tilde{c})\}.$$

We may also denote by $\Theta_t(\tilde{c})$ or $\Theta(\tilde{c})(t)$ the horizontal path $\Theta(\tilde{c},t)$. Then, these diffeological connections satisfy the following series of properties.

a) Let $\bar{c}$ and $\bar{c}'$ be two horizontal paths in Y defined on the same domain, such that $\pi \circ \bar{c} = \pi \circ \bar{c}'$. If they have one value in common, then they coincide.

b) Let $\tilde{c}$ be a local path in Y, and let s,t be two points in $\mathrm{def}(\tilde{c})$. If $\tilde{c}(t) = g_Y(\Theta(\tilde{c},s)(t))$, for some g in G, then $\Theta(\tilde{c},t) = g_Y \circ \Theta(\tilde{c},s)$.

c) Let $\tilde{c}$ and $\tilde{c}'$ be two local paths in Y, and let us assume that $\tilde{c}$ and $\tilde{c}'$ are compatible, that is, $\tilde{c} \restriction \mathrm{def}(\tilde{c}) \cap \mathrm{def}(\tilde{c}') = \tilde{c}' \restriction \mathrm{def}(\tilde{c}) \cap \mathrm{def}(\tilde{c}')$. Let $\tilde{c} \cup \tilde{c}'$ be the smallest common extension. Let $t \in \mathrm{def}(\tilde{c})$ and $s \in \mathrm{def}(\tilde{c}) \cap \mathrm{def}(\tilde{c}')$, there exists a unique $g \in G$ such that $\Theta(\tilde{c},t)(s) = g_Y(\tilde{c}'(s))$, then $\Theta(\tilde{c},t)$ and $g_Y \circ \Theta(\tilde{c}',s)$ are compatible and

$$\Theta(\tilde{c} \cup \tilde{c}',t) = \Theta(\tilde{c},t) \cup g_Y \circ \Theta(\tilde{c}',s).$$

d) The constant paths are fixed by the connection. For all $y \in Y$ and all t,

$$\Theta([s \mapsto y],t) = [s \mapsto y].$$

e) For any path $c \in \mathrm{Paths_{loc}}(X)$, any $t \in \mathrm{def}(c)$, any $y \in \pi^{-1}(c(t))$, there exists a unique path $\bar{c}$ in Y, such that

$$\bar{c} \in \mathrm{HorPaths_{loc}}(Y,\Theta) \quad \text{with} \quad \pi \circ \bar{c} = c \quad \text{and} \quad \bar{c}(t) = y.$$

The path $\bar{c}$ is called the *horizontal lift* of c, pointed at y at time t. Therefore, we get a map

$$\mathrm{hor}_\Theta : (c,t,y) \mapsto \bar{c} \in \mathrm{HorPaths_{loc}}(Y,\Theta) \subset \mathrm{Paths}(Y), \qquad (\clubsuit)$$

defined on

$$\mathrm{ev}^*(Y) = \{(c,t,y) \in \mathrm{Paths_{loc}}(X) \times \mathbf{R} \times Y \mid t \in \mathrm{def}(c) \text{ and } c(t) = \pi(y)\},$$

where $\mathrm{ev} : \mathrm{tbPaths_{loc}}(X) \to X$ is the evaluation $\mathrm{ev}(c,t) = c(t)$. The *horizontal lifting* is smooth and G-equivariant,

$$\mathrm{hor}_\Theta(c,t)(g_Y(y)) = g_Y \circ \mathrm{hor}_\Theta(c,t)(y) \quad \text{for all} \quad g \in G.$$

Thanks to the axiom of locality, the horizontal lift of a composite is given by

$$\mathrm{hor}_\Theta(c \circ f, s, y) = \mathrm{hor}_\Theta(c, f(s), y) \circ f.$$

We also remark that, thanks to (d), the horizontal lift of a constant path is a constant path, $\mathrm{hor}_\Theta([s \mapsto x],t)(y) = [s \mapsto y]$.

f) Let c and c' be two paths in X such that the concatenation $c \vee c'$ (art. 5.2) is smooth. Let $x = c(0)$, $y \in \pi^{-1}(x)$, and $y' = \mathrm{hor}_\Theta(c,0,y)(1)$, then the horizontal lifts $\mathrm{hor}_\Theta(c,0,y)$ and $\mathrm{hor}_\Theta(c',0,y')$ can be concatenated and the result is the horizontal lift of the concatenation $c \vee c'$ starting at y, that is,

$$\mathrm{hor}_\Theta(c \vee c',0,y) = \mathrm{hor}_\Theta(c,0,y) \vee \mathrm{hor}_\Theta(c',0,\mathrm{hor}_\Theta(c,0,y)(1)).$$

In particular, this applies for all pairs of stationary paths c and c' (art. 5.4) such that $c(1) = c'(0)$.

NOTE 1. Let E be a diffeological space with a smooth action of G denoted by $(g,v) \mapsto g_E(v)$. Let c be a local path in X, $t \in \mathrm{def}(c)$, $(y,v) \in Y \times E$ such that

$y \in \pi^{-1}(c(t))$. Let $pr : Y \times_G E \to X$ be the associated fiber bundle (art. 8.16), and let class $: Y \times E \to Y \times_G E$ be the projection. Let $\bar{c}$ be the horizontal lift of c in Y such that $\bar{c}(t) = y$, so the local path $t \mapsto class(\bar{c}(t), v)$ in $Y \times_G E$ covers c, it is the horizontal lifting of c associated with the chosen initial conditions.

NOTE 2. We denote by $HorPaths(Y, \Theta)$ the set of global horizontal paths, that is, the set of all $\Theta(\tilde{c}, t)$ such that $\tilde{c} \in Paths(Y)$ and $t \in \mathbf{R}$. Note that, thanks to the point (b), it is sufficient to consider $t = 0$,

$$HorPaths(Y, \Theta) = \{\Theta(\tilde{c}, 0) \mid \tilde{c} \in Paths(Y)\}.$$

Now, if X is connected, then the projection $\chi_{Y,\Theta} = \pi \times \pi \circ ends_Y$, from $HorPaths(Y, \Theta)$ to $X \times X$, which associates with a horizontal path $\bar{c}$ the projections $\pi \circ \bar{c}(0)$ and $\pi \circ \bar{c}(1)$ of its ends, is a subduction. Actually, the projection $\pi_* : Paths(Y) \to Paths(X)$ is a principal fibration with structure group $Paths(G)$. The restriction $\pi_* \upharpoonright HorPaths(Y, \Theta)$ is a reduction of π_* to a $G \subset Paths(G)$ principal subbundle (art. 8.11), where G injects itself in $Paths(G)$ as the subgroup of constant paths. And the map, which associates the points of Y and X with the associated constant paths, is a strict morphism of G-principal fiber bundles from π to the restriction $\pi_* \upharpoonright HorPaths(Y, \Theta)$.

PROOF. a) Let $\mathcal{I} = def(\bar{c}) = def(\bar{c}')$, the paths $\bar{c}$ and $\bar{c}'$ are horizontal, so $\Theta(\bar{c}, t) = \bar{c}$ and $\Theta(\bar{c}', t) = \bar{c}'$. Next, since $\pi \circ \bar{c} = \pi \circ \bar{c}'$ there exists a local smooth path $\gamma : \mathcal{I} \to G$ such that $\bar{c}'(t) = \gamma(t)_Y(\bar{c}(t))$ for all $t \in \mathcal{I}$ (art. 8.19), thus $\bar{c}' = \gamma(t)_Y \circ \Theta(\bar{c}, t) = \gamma(t)_Y \circ \bar{c}$ (reduction and projector axioms). Hence, we choose $t = s$ for which $\bar{c}(s) = \bar{c}'(s)$, so $\gamma(s) = \mathbf{1}_G$ and $\bar{c} = \bar{c}'$.

b) By the reduction axiom, $g_Y(\Theta(\tilde{c}, s)(t)) = \Theta(g_Y \circ \tilde{c}, s)(t)$, and by assumption $\tilde{c}(t) = g_Y(\Theta(\tilde{c}, s)(t))$, then since $\tilde{c}(t) = \Theta(\tilde{c}, t)(t)$, $\Theta(\tilde{c}, t)(t) = \Theta(g_Y \circ \tilde{c}, s)(t)$. Now, the paths $\Theta(\tilde{c}, t)$ and $\Theta(g_Y \circ \tilde{c}, s)$ are horizontal, they project on the same path $\pi \circ \tilde{c}$ and they coincide at one point, thus they coincide (a). Therefore, $\Theta(\tilde{c}, t) = \Theta(g_Y \circ \tilde{c}, s) = g_Y \circ \Theta(\tilde{c}, s)$.

c) For all $t \in def(\tilde{c}) \cap def(\tilde{c}')$, $\tilde{c}(t) = \tilde{c}'(t)$ and $\pi(\tilde{c}'(t)) = \pi(\tilde{c}(t)) = \pi(\Theta(\tilde{c}, s)(t))$ (lifting axiom), then there exists a unique $g \in G$ such that $\Theta(\tilde{c}, s)(t) = g_Y(\tilde{c}'(t)) = g_Y(\tilde{c}(t))$. Now, on the one hand $\Theta(\tilde{c}, s) \upharpoonright def(\tilde{c}) \cap def(\tilde{c}') = \Theta(\tilde{c} \upharpoonright def(\tilde{c}) \cap def(\tilde{c}'), s)$ (axiom of locality), and on the other hand, $g_Y \circ \Theta(\tilde{c}', t) \upharpoonright def(\tilde{c}) \cap def(\tilde{c}') = \Theta(g_Y \circ \tilde{c}' \upharpoonright def(\tilde{c}) \cap def(\tilde{c}'), t)$ (axiom of locality and reduction). Thus, these two restrictions are horizontal paths. But we have $\Theta(\tilde{c}, s)(t) = g_Y(\tilde{c}'(t)) = \Theta(g_Y \circ \tilde{c}', t)(t) = g_Y \circ \Theta(\tilde{c}', t)(t)$ (basepoint axiom), hence $\Theta(\tilde{c}, s) \upharpoonright def(\tilde{c}) \cap def(\tilde{c}')$ and $g_Y \circ \Theta(\tilde{c}', t) \upharpoonright def(\tilde{c}) \cap def(\tilde{c}')$ are two horizontal paths having a common value at t, thanks to point a) they coincide, and the local paths $\Theta(\tilde{c}, s)$ and $g_Y \circ \Theta(\tilde{c}', t)$ are compatible. Now, thanks again to point a), $\Theta(\tilde{c}, s) = g_Y \circ \Theta(\tilde{c}, t)$ and $g_Y \circ \Theta(\tilde{c} \cup \tilde{c}', t) = \Theta(\tilde{c} \cup \tilde{c}', s)$. Thus, $\Theta(\tilde{c}, s) \cup g_Y \circ \Theta(\tilde{c}', t) = g_Y \circ \Theta(\tilde{c}, t) \cup g_Y \circ \Theta(\tilde{c}', t) = g_Y \circ \Theta(\tilde{c} \cup \tilde{c}', t) = \Theta(\tilde{c} \cup \tilde{c}', s)$.

d) Let $c :]a, b[\to X$, there exists a lift $\tilde{c} :]a, b[\to Y$ covering c (art. 8.19), and we can choose $\tilde{c}(t) = y$. Now, $\bar{c} = \Theta(\tilde{c}, t)$ satisfies the condition. Thanks to point a), this lift is unique.

e) If $\tilde{c}$ covers c with $c(t) = y$, then $\tilde{c} \circ f$ covers $c \circ f$ with $t = f(s)$ and $\tilde{c} \circ f(s) = y$, thus $hor_\Theta(c \circ f, s, y) = \Theta(\tilde{c} \circ f, s) = \Theta(\tilde{c}, f(s)) \circ f = hor_\Theta(c, f(s), y) \circ f$, thanks to the axiom of locality. Now, let $\mathbf{0} : s \mapsto 0$, and $\mathbf{y} : s \mapsto y$, by the locality axiom we have $\Theta(\mathbf{y} \circ \mathbf{0}, s)(t) = \Theta(\mathbf{y}, \mathbf{0}(s))(\mathbf{0}(t))$ for all t, that is, $\Theta(\mathbf{y}, s)(t) = \Theta(\mathbf{y}, 0)(0) = \mathbf{y}(0) = y$, i.e., $\Theta(\mathbf{y}, s) = \mathbf{y}$.

f) The paths $\mathrm{hor}_\Theta(c,0,y)$ and $\mathrm{hor}_\Theta(c',0,\mathrm{hor}_\Theta(c,0,y)(1))$ are smooth and can be concatenated since $\mathrm{hor}_\Theta(c,0,y)(1) = \mathrm{hor}_\Theta(c',0,\mathrm{hor}_\Theta(c,0,y)(1))(0)$. Then, thanks to point e) and to the unicity of horizontal lift given a starting point, $\mathrm{hor}_\Theta(c \vee c',0,y)(s) = \mathrm{hor}_\Theta(c,0,y)(2s)$ for $s < 1/2$, as well $\mathrm{hor}_\Theta(c \vee c',0,y)(s) = \mathrm{hor}_\Theta(c',0,\mathrm{hor}_\Theta(c,0,y)(1))(2s-1)$ for $s > 1/2$. Now,

$$\mathrm{hor}_\Theta(c,0,y) \vee \mathrm{hor}_\Theta(c',0,\mathrm{hor}_\Theta(c,0,y)(1))(1/2) = y' = \mathrm{hor}_\Theta(c \vee c',0,y)(1/2),$$

thus concatenation and horizontal lifts commute.

For Note 2, $\chi_{Y,\Theta}$ is surjective since for every pair of points $(x,x') \in X \times X$ there exists a path c connecting x to x' and any horizontal lift $\bar{c}$ satisfies $\pi \circ \bar{c}(0) = x$ and $\pi \circ \bar{c}(1) = x'$. Then, since $\mathrm{ends}_Y : \mathrm{Paths}(Y) \to Y \times Y$ is a subduction and $\pi \times \pi : Y \times Y \to X \times X$ is also a subduction, the composite $\pi \times \pi \circ \mathrm{ends}_Y$ is a subduction. Now, the connection $\Theta : \mathrm{Paths}(Y) \to \mathrm{HorPaths}(Y,\Theta)$ is smooth (see point d)) thus every plot of $X \times X$ lifts locally to $\mathrm{Paths}(Y)$, and composed with Θ lifts locally in $\mathrm{HorPaths}(Y,\Theta)$, thus $\chi_{Y,\Theta}$ is a subduction. □

8.33. Pullback of a connection. Let $\pi : Y \to X$ be a principal fiber bundle with structure group G, and let Θ be a connection. Let X' be a diffeological space and let $f : X' \to X$ be a smooth map. Let $\mathrm{pr}_1 : Y' \to X'$ be the pullback of π, where $Y' = f^*(Y)$ (art. 8.12). This pullback is a principal bundle with structure group G acting naturally by $g_{Y'}(x',y) = (x', g_Y(y))$, where g_Y denotes the action of G on Y and $(x',y) \in Y'$. The map $\Theta' : \mathrm{tbPaths}_{\mathrm{loc}}(Y') \to \mathrm{Paths}_{\mathrm{loc}}(Y')$, defined by

$$\Theta'(\tilde{c}',t) = [s \mapsto (\xi'(s), \Theta(\tilde{c},t)(s))] \quad \text{with} \quad \tilde{c}' : s \mapsto (\xi'(s), \tilde{c}(s)), \qquad (\clubsuit)$$

is a connection on the pullback $\mathrm{pr}_1 : Y' \to X'$, it will be called the pullback of the connection Θ by f and denoted also by $f^*(\Theta)$.

PROOF. The verification of the conditions 1–6 in (art. 8.32) is not difficult. □

8.34. Connections and equivalence of pullbacks. Let X and X' be two connected diffeological spaces. Let $\pi : Y \to X$ be a principal fiber bundle, and let $t \mapsto f_t$ be a path in $\mathcal{C}^\infty(X',X)$. If π admits a connection, then the pullbacks $\pi'_0 : f_0^*(Y) \to X'$ and $\pi'_1 : f_0^*(Y) \to X'$ are isomorphic (art. 8.12, Note 2). In particular, if X is contractible, then π is trivial.

NOTE 1. Any fiber bundle $p : T \to X$ such that X is contractible (art. 5.11), and such that its associated principal fiber bundle $\pi : Y \to X$ (art. 8.16) admits a connection, is trivial. In particular, in the subcategory of finite-dimensional manifolds every principal fiber bundle admits a connection [**KoNo63**]. Thus, in the classical category of manifolds, any fiber bundle with contractible base space is trivial.

NOTE 2. Let $\pi : Y \to X$ be a principal fiber bundle equipped with a connection Θ. Let E be a contractible space, any smooth map $f : E \to X$ admits a smooth lift ψ along π, that is, $\pi \circ \psi = f$.

PROOF. Let us recall that the pullback of π by f_t has total spaces

$$f_t^*(Y) = \{(x',y) \in X \times Y \mid f_t(x') = \pi(y)\}.$$

The map $x' \mapsto c_{x'} = [t \mapsto f_t(x')]$ from X' to $\mathrm{Paths}(X)$ is smooth. For all $(x',y) \in f_0^*(Y)$, let $\bar{c}_{x'}(y) = \mathrm{hor}_\Theta(c_{x'},0,y)$ be the horizontal lift of $c_{x'}$ such that $\bar{c}_{x'}(y)(0) = y$ (art. 8.32, e). By construction, the endpoint $(x', \bar{c}_{x'}(y)(1))$ belongs to $f_1^*(Y)$.

Now, the map

$$\psi : f_0^*(Y) \to f_1^*(Y), \text{ defined by } \psi(x', y) = (x', \bar{c}_{x'}(y)(1)),$$

satisfies $\mathrm{pr}_1 \circ \psi = \mathrm{pr}_1$, is equivariant with respect to the action of the structure group, and is smooth because hor_Θ is smooth. Next, let us consider $x' \mapsto c'_{x'} = [t \mapsto f_{1-t}(x')]$, and let $\bar{c}'_{x'}(y) = \mathrm{hor}_\Theta(c'_{x'}, 0, y')$ be the horizontal lift of $c'_{x'}$ such that $\bar{c}'_{x'}(y')(0) = y'$, for all $(x', y') \in f_1^*(Y)$. It is then clear that the map

$$\psi' : f_1^*(Y) \to f_0^*(Y), \text{ defined by } \psi'(x', y') = (x', \bar{c}'_{x'}(y')(1)),$$

is inverse of ψ. Therefore, ψ is an isomorphism from $f_0^*(Y)$ to $f_1^*(Y)$.

For Note 1, X is contractible if the identity $\mathbf{1}_X$ is homotopic to the constant map $x \mapsto x_0$, for a given x_0. Thus, the pullback of $p : Y \to X$ by the identity, which is equivalent to p itself, and the pullback by the constant map, which is the trivial bundle $\mathrm{pr}_1 : X \times p^{-1}(x_0) \to X$, are isomorphic.

For Note 2, the pullback of π by f is a principal fiber bundle, and the pullback of the connection Θ is a connection on $\mathrm{pr}_1 : f^*(Y) \to E$ (art. 8.33). Thus, since E is contractible, $\mathrm{pr}_1 : f^*(Y) \to E$ is trivial, and then admits a smooth section $\sigma : E \to f^*(Y)$ (art. 8.12). Now, the section σ writes necessarily $\sigma(e) = (e, \psi(e))$, where ψ is a smooth map from E to Y, ψ is a smooth lift of f. ☐

8.35. The holonomy of a connection. Let $\pi : Y \to X$ be a principal fibration with structure group G, equipped with a connection Θ. Let us choose two points, $x \in X$ and $y \in Y_x = \pi^{-1}(x)$, and let Ψ be the map associating with every path c in X, pointed at x, the end of the horizontal lift of c, pointed at y,

$$\Psi : \mathrm{Paths}(X, x) \to Y, \text{ such that } \Psi : c \mapsto \bar{c}(1), \text{ with } \bar{c} = \mathrm{hor}_\Theta(c, 0, y).$$

The map Ψ is smooth and covers a smooth map $\varphi : X \to Y/G_y$ according to the following commutative diagram,

$$\begin{array}{ccc} \mathrm{Paths}(X,x) & \xrightarrow{\ \Psi\ } & Y \\ {\scriptstyle \hat{1}}\downarrow & & \downarrow{\scriptstyle \mathrm{pr}} \\ X & \xrightarrow[\ \varphi\ ]{} & Y/G_y \end{array} \qquad (\clubsuit)$$

where G_y is roughly speaking the image by Ψ of the fiber of the subduction $\hat{1}$ over x, that is, $\hat{1}^{-1}(x) = \mathrm{Loops}(X, x)$. Precisely,

$$G_y = R(y)^{-1}\Big(\Psi(\mathrm{Loops}(X, x))\Big),$$

where $R(y) : g \mapsto g_Y(y)$ is the orbit map, identifying G with Y_x. We have denoted by g_Y the action of $g \in G$ on Y. Formally,

$$G_y = \{g \in G \mid \exists \ell \in \mathrm{Loops}(X, x), \mathrm{hor}_\Theta(\ell, 0, y)(1) = g_Y(y)\}.$$

The set G_y is a subgroup of G, depending on the choice of the point y by conjugation, it is called the *holonomy group* of Θ at the point y. Its conjugacy class, regarded as a group of automorphisms of the fiber Y_x, is called the holonomy group at the point x.

The image $Y_\Theta(y) = \Psi(\mathrm{Paths}(X, x)) \subset Y$, which is a connected subspace of Y, since Ψ is smooth and $\mathrm{Paths}(X, x)$ is connected, is precisely made of the points of Y which can be connected to y by a horizontal path, that is,

$$Y_\Theta(y) = \{\bar{c}(1) \in Y \mid \bar{c} \in \mathrm{HorPaths}(Y, \Theta) \text{ and } \bar{c}(0) = y\}. \qquad (\spadesuit)$$

The restriction $\pi_\Theta = \pi \restriction Y_\Theta(y)$ is a principal fiber bundle over the connected component $\mathrm{comp}(x) \subset X$ with structure group G_y. This is the *holonomy bundle* of the connection Θ with reference point y. Said a little bit differently, if X is connected, then the holonomy principal bundle $\pi_\Theta : Y_\Theta(y) \to X$, with structure group G_y, is a reduction of the G-principal fiber bundle $\pi : Y \to X$.

NOTE 1. If the holonomy G_y is discrete, then the connection Θ is said to be *flat*. Assuming X is connected, if Θ is flat, then $\pi_\Theta : Y_\Theta(y) \to X$ is a connected covering onto X. The G-principal fiber bundle $\pi : Y \to X$ is equivalent to the associated bundle $\mathrm{pr} : Y_\Theta \times_{G_y} G \to X$, where the holonomy G_y acts by $g : (y', g') \mapsto (g_Y(y'), g' \cdot g^{-1})$, with $g \in G_y$ and $(y', g') \in Y_\Theta \times G$. Then, Θ is equivalent to the connection associated with the natural lifting of paths in a covering; see below (art. 8.36). That said, every principal bundle equipped with a flat connection is equivalent to such an associated bundle, which can be built using the universal covering $\tilde{X}$ of X. Precisely, $Y \simeq \tilde{X} \times_{\pi_1(X,x)} G$ where $\pi_1(X, x)$ acts by multiplication on G through a homomorphism called the *monodromy representation*. In particular, every principal bundle equipped with a flat connection on a simply connected space is necessarily trivial, but also note that there exist nonflat connections on trivial principal bundles.

NOTE 2. The subset $G_y^\circ \subset G_y$, defined by the horizontal lifts of loops in X homotopic to $\mathbf{x} = [t \mapsto x]$, is an invariant subgroup and is called the *reduced holonomy group*. This is actually the holonomy of the pullback of the connection Θ (art. 8.33) on the universal covering $\mathrm{pr} : \tilde{X} \to X$. The quotient G_y/G_y° is called the *monodromy* of the connection Θ, it is a homomorphic image of $\pi_1(X, x)$.

PROOF. First of all, let us note that any loop in X based at x lifts in a unique path $\bar{c}$ of Y such that $y = \bar{c}(0)$ and $\bar{c}(1) \in Y_x$. Indeed, we use first any lift, thanks to (art. 8.19), then we apply Θ. Then, it is also clear that every horizontal path $\bar{c}$, such that $\bar{c}(0) = y$ and $\bar{c}(1) \in Y_x$, projects to a loop ℓ in X based at x. Thus, there is a one-to-one correspondence between the loops in X based at x and the horizontal paths in Y, starting at y and ending in Y_x. Now, let us prove that $G_y \subset G$ is a subgroup, let $\bar{c}$ and $\bar{c}'$ be the horizontal lifts, starting at y, of two loops ℓ and ℓ' based at x. Let $g, g' \in G_y$ such that $\bar{c}(1) = g_Y(y)$ and $\bar{c}'(1) = g_Y(y)$. Let us assume that $\bar{c}$ and $g_Y \circ \bar{c}'$ can be concatenated into a smooth path. Thus, the path $\bar{c} \vee (g_Y \circ \bar{c}')$ is horizontal and satisfies $[\bar{c} \vee (g_Y \circ \bar{c}')](1) = (g_Y \circ \bar{c}')(1) = g_Y(g'_Y(y)) = (gg')_Y(y)$ (art. 8.32, f). Therefore, gg' belongs to G_y. If c and c' cannot be concatenated, because $c \vee c'$ is not smooth, then we first smash them, $\bar{c}^\star = \bar{c} \circ \lambda$ and $\bar{c}'^\star = \bar{c}' \circ \lambda$ (art. 5.5). Thanks to the locality axiom, these paths are still horizontal, indeed $\Theta(\bar{c}^\star, 0) = \Theta(\bar{c} \circ \lambda, 0) = \Theta(\bar{c}, \lambda(0)) \circ \lambda = \Theta(\bar{c}, 0) \circ \lambda = \bar{c} \circ \lambda = \bar{c}^\star$, they have the same ends $\bar{c}^\star(0) = \bar{c}(0)$ and $\bar{c}^\star(1) = \bar{c}(1)$, and since the same holds for $\bar{c}'$, $\bar{c}^\star$ and $g_Y \circ \bar{c}'^\star$ can be concatenated. For the inverse, if $g \in G_y$, then let $\bar{c}$ be such that $g_Y(y) = \bar{c}(1)$, let $\mathrm{rev}(\bar{c}) : t \mapsto \bar{c}(1 - t)$. Thanks to the axioms of reduction and locality, the path $\bar{c}' = g_Y^{-1} \circ \mathrm{rev}(\bar{c})$ is still horizontal, it satisfies $\bar{c}'(0) = y$ and $\bar{c}'(1) = g_Y^{-1}(y)$. Thus, if $g \in G_y$, then $g^{-1} \in G_y$. Therefore G_y is a subgroup of

G. Next, let c and c' be two paths in X starting at x and ending at x', that is, $\hat{0}(c) = \hat{0}(c') = x$ and $\hat{1}(c) = \hat{1}(c') = x'$. We assume c and c' stationary, if it is not the case we smash them as above. Let $\bar{c}$ and $\bar{c}'$ be their horizontal lifts starting at y, thus there exists $g \in G$ such that $g_Y(\bar{c}'(1)) = \bar{c}(1)$, since $\pi(\bar{c}(1)) = \pi(\bar{c}'(1))$. The path $g_Y \circ \bar{c}'$ is horizontal, stationary, and the concatenation $\bar{c} \vee \mathrm{rev}(g_Y \circ \bar{c}')$ is smooth, horizontal, starts at y and ends at $\mathrm{rev}(g_Y \circ \bar{c}')(1) = g_Y \circ \bar{c}'(0) = g_Y(y)$. Thus, $\bar{c} \vee \mathrm{rev}(g_Y \circ \bar{c}')$ is the horizontal lift of a loop based at x and $g \in G_y$. Therefore, pr denoting the projection from Y to Y/G_y (see (♣)), the map

$$\varphi : x' \mapsto \mathrm{pr}(\bar{c}(1)), \text{ with } c \in \mathrm{Paths}(X, x, x') \text{ and } \bar{c} = \mathrm{hor}(c, 0, y),$$

is well defined, and smooth since Ψ is smooth and $\hat{1}$ is a subduction (art. 5.6). Consider the next diagram, the map φ is not just a smooth map but also a section of the quotient fiber bundle $\pi_y = \pi/G_y$. Then, according to (art. 8.18) the principal fiber bundle $\pi : Y \to X$ is reduced to the subbundle $\pi_\Theta : Y_\Theta(y) \to X$ with structure group G_y. And by construction, the connection Θ follows and induces a connection on π_Θ. The dependency of this construction with respect to the basepoints x and y is clear, an easy computation shows that if we change the basepoint y into $g_Y(y)$, then $G_{g_Y(y)} = g \cdot G_y \cdot g^{-1}$.

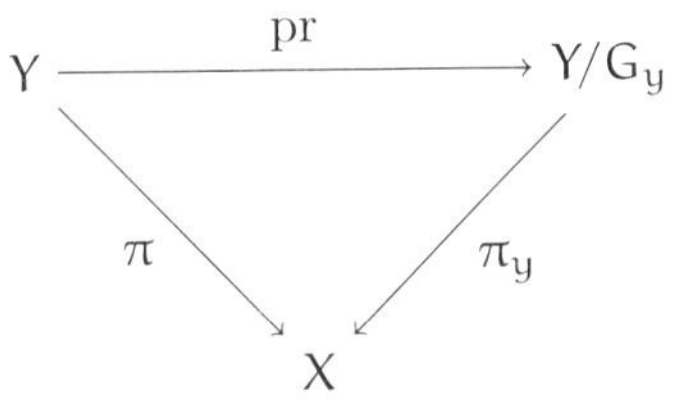

For Note 1. If the holonomy G_y is discrete, then $\pi_\Theta : Y_\Theta(y) \to X$ is a fibration with discrete fiber, that is, a covering (art. 8.22). Moreover, since π_Θ is a principal bundle, it is a Galoisian covering, and we have seen that the total space $Y_\Theta(y)$ is connected. Then, the pullback $\mathrm{pr}_1 : \pi_\Theta^*(Y) \to X$ is a G principal bundle with

$$\pi_\Theta^*(Y) = \{(y', y'') \in Y \times Y \mid y' \in Y_\Theta(y) \text{ and } \pi(y') = \pi(y'')\},$$

where G acts trivially on the first factor and naturally on the second.

Now, since $\sigma : y' \mapsto (y', y')$ is a smooth section, this pullback is trivial, and the map $\Phi : Y_\Theta(y) \times G \to \pi_\Theta^*(Y)$ defined by $\Phi(y', g') = (y', g'_Y(y'))$ is an isomorphism of G-principal bundle (art. 8.12). Then, since $\pi : Y \to X$ is a subduction, $\mathrm{pr}_2 : \pi_\Theta^*(Y) \to Y$ is a subduction and the composite $\mathrm{pr}_2 \circ \Phi$ is also a subduction. Thus, Y is equivalent to the quotient of the direct product $Y_\Theta(y) \times G$ by the lift $g : (y', g') \mapsto (g_Y(y), g' \cdot g^{-1})$ of the action of $g \in G_y$ on $Y_\Theta(y) \times G$, associated with the equivalence $g'_Y(y') = g''_Y(y'')$. It is then clear that, conversely, every G-principal fiber bundle associated this way with a Galoisian covering — and thus with the universal covering — through a morphism from the covering group to G, inherits a flat connection thanks to the natural lift of paths in coverings; see (art. 8.32, Note 1) and (art. 8.36).

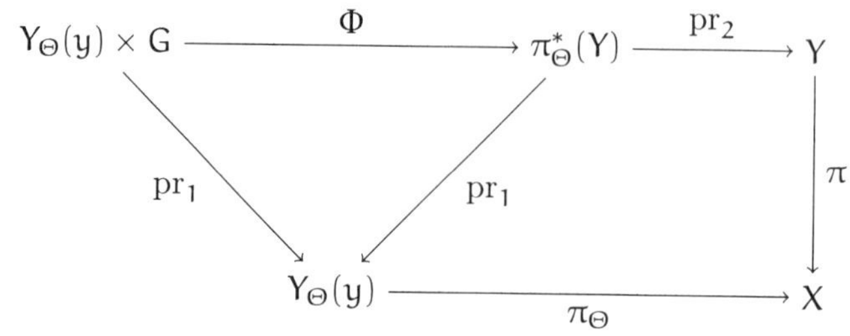

For Note 2. The map $\ell \mapsto \mathrm{hor}_\Theta(\ell, 0, y)(1)$ is smooth (art. 8.32, point e)), thus the map $\varphi : \ell \mapsto R(y)^{-1}(\mathrm{hor}_\Theta(\ell, 0, y)(1))$, from $\mathrm{Loops}(X, x)$ to G_y is smooth too. We have seen in the previous paragraph that $\varphi(\ell \vee \ell') = \varphi(\ell) \cdot \varphi(\ell')$, so the quotient map from $\pi_1(X, x) = \mathrm{Loops}(X, x)/\mathrm{comp}(\mathbf{x})$ to G_y/G_y° is a homomorphism. Note that φ maps connected components in connected components, in particular G_y° is connected since it is the image of the connected component $\mathrm{comp}(\mathbf{x})$. Now, for any two loops ℓ and ℓ' homotopic to $\mathbf{x}$, the loop $\ell \vee \ell' \vee \mathrm{rev}(\ell)$ is homotopic to $\mathbf{x}$, thus G_y° is invariant, $g \cdot G_y^\circ \cdot g^{-1} \subset G_y^\circ$. Now, let us consider the pullback $\mathrm{pr}_1 : Y' = \mathrm{pr}^*(Y) \to X$, and let us recall that Y' is the subspace of pairs $y' = (\tilde{x}, y) \in \tilde{X} \times Y$ such that $\mathrm{pr}(\tilde{x}) = \pi(y)$. Let us choose $\tilde{x} \in \mathrm{pr}^{-1}(x)$ and $y' = (\tilde{x}, y)$ as a basepoint in Y'. The holonomy $G_{y'}$ of $\Theta' = \mathrm{pr}^*(\Theta)$ is the image of $\varphi' : \tilde{\ell} \mapsto R(y')^{-1}(\mathrm{hor}_{\Theta'}(\tilde{\ell}, 0, y')(1))$, where $\tilde{\ell} \in \mathrm{Loops}(\tilde{X}, \tilde{x})$. But $\mathrm{hor}_{\Theta'}(\tilde{\ell}, 0, y')(t) = (\tilde{\ell}(t), \mathrm{hor}_\Theta(\ell, y, 0)(t))$ with $\ell = \mathrm{pr} \circ \tilde{\ell}$ and $\mathrm{hor}_\Theta(\ell, y, 0)(1) = \varphi(\ell)_Y(y)$. Thus, $\mathrm{hor}_{\Theta'}(\tilde{\ell}, 0, y')(1) = (\tilde{x}, \varphi(\ell)_Y(y)) = \varphi(\ell)_{Y'}(\tilde{x}, y) = \varphi(\ell)_{Y'}(y')$ and $\varphi'(\tilde{\ell}) = \varphi(\ell)$. Now, since the loops in $\tilde{X}$ based at $\tilde{x}$ are in one-to-one correspondence with the loops in X based at x and homotopic to the constant path $\mathbf{x} : t \mapsto x$, we get $G_{y'} = G_y^\circ$. □

8.36. The natural connection on a covering. Let X be a connected diffeological space, and let $\pi : \tilde{X} \to X$ be its universal covering (art. 8.26). Let $c :]a, b[\to X$ be a local smooth path. For all $t \in]a, b[$ and $\tilde{x} \in \pi^{-1}(c(t))$, there exists a unique smooth lift $\bar{c} :]a, b[\to \tilde{X}$ such that $\bar{c}(t) = \tilde{x}$. This defines clearly the only possible connection on the covering. Indeed, the pullback of the covering by c is the product of $]a, b[$ by $\pi_1(X)$, and $\pi_1(X)$ is discrete. By quotient, or associated bundle, this gives again the natural *lifting of paths*, on every covering of X.

8.37. Connections and connection 1-forms on torus bundles. A *diffeological* 1*-torus* is any quotient $T = \mathbf{R}/\Gamma$ where $\Gamma \subset \mathbf{R}$ is a strict subgroup, $\Gamma \neq \mathbf{R}$. In this case Γ is discrete and the projection class : $\mathbf{R} \to T$ is the universal covering (art. 8.26). These tori are Abelian diffeological groups of dimension 1; see (art. 1.78) and Exercise 49, p. 50. The differential 1-form dt defined on $\mathbf{R}$ is invariant by translation, *a fortiori* by the group Γ. Thus, according to (art. 8.27), there exists a unique canonical 1-form $\theta \in \Omega^1(T)$, which is actually a volume form (art. 6.44), defined by

$$dt = \mathrm{class}^*(\theta) \quad \text{with} \quad \mathrm{class} : \mathbf{R} \to T.$$

Now, let $\pi : Y \to X$ be a diffeological T-principal fiber bundle. Let τ_Y denote the action of $\tau \in T$ on Y, and let $R(y) : T \to Y$ be the orbit map $R(y)(\tau) = \tau_Y(y)$.

Definition. We call *connection form* on the T-principal bundle $\pi : Y \to X$ any differential 1-form λ, defined on the total space Y, such that

a) the form λ is *invariant* — for all $\tau \in T$, $\tau_Y^*(\lambda) = \lambda$,

b) the form λ is *calibrated* — for all $y \in Y$, $R(y)^*(\lambda) = \theta$.

1. First of all let us note a useful formula related to such connection forms. For any integers n and m, any n-plot $\tau : U \to T$ and any m-plot $P : V \to Y$, let $\tau \cdot P$ be the plot of Y defined by

$$\tau \cdot P : U \times V \to Y \quad \text{with} \quad \tau \cdot P : (r, s) \mapsto \tau(r)_Y(P(s)).$$

Let $(r, s) \in U \times V$, $\delta r \in \mathbf{R}^n$, $\delta s \in \mathbf{R}^m$, the value of λ on $\tau \cdot P$ is given by

$$\lambda(\tau \cdot P)_{\binom{r}{s}} \begin{pmatrix} \delta r \\ \delta s \end{pmatrix} = \tau^*(\theta)_r(\delta r) + \lambda(P)_s(\delta s). \tag{$\clubsuit$}$$

2. There exists a unique 2-form $\omega \in \Omega^2(X)$, called the *curvature* of the connection form λ, such that

$$\pi^*(\omega) = d\lambda \quad \text{and} \quad d\omega = 0. \tag{$\spadesuit$}$$

3. There exists a connection Θ associated with λ such that the horizontal paths are the ones on which the connection form vanishes,

$$\text{HorPaths}_{\text{loc}}(Y, \Theta) = \{\bar{c} \in \text{Paths}_{\text{loc}}(Y) \mid \lambda(\bar{c}) = 0\}. \tag{$\diamondsuit$}$$

Note that $\lambda(\bar{c}) = 0$ means that for all $t \in \text{def}(\bar{c})$, $\lambda(\bar{c})_t(1) = 0$. The connection Θ is explicitly given by

$$\Theta(\tilde{c}, t) : t' \mapsto \text{class}\left(-\int_t^{t'} \lambda(\tilde{c})_s(1)\, ds\right)_Y \left(\tilde{c}(t')\right). \tag{$\heartsuit$}$$

4. If the curvature ω defined above ($\spadesuit$) is zero, then the connection Θ is flat, that is, the holonomy (art. 8.35) is discrete.

NOTE. If λ is a connection form, then for any 1-form α on X, $\lambda' = \lambda + \pi^*(\alpha)$ is again a connection form. Conversely, if λ and λ' are two connection forms, then their difference $\lambda' - \lambda$ is the pullback of a 1-form α on X. The space of connection forms of a T-principal bundle, if it is not empty, is an affine space of $\Omega^1(X)$.

PROOF. 1. Let us develop the 1-form $\lambda(\tau \cdot P)$ defined on $U \times V$,

$$\begin{aligned} \lambda(\tau \cdot P)_{\binom{r}{s}} \begin{pmatrix} \delta r \\ \delta s \end{pmatrix} &= \lambda[(r, s) \mapsto (\tau(r), P(s)) \mapsto \tau(r)_Y(P(s))]_{\binom{r}{s}} \begin{pmatrix} \delta r \\ \delta s \end{pmatrix} \\ &= [\lambda_{U,s}(r) \quad \lambda_{V,r}(s)] \begin{pmatrix} \delta r \\ \delta s \end{pmatrix}, \end{aligned}$$

because every 1-form on $U \times V$ at a point $(r, s) \in U \times V$ writes $[\lambda_{U,s}(r)\ \lambda_{V,r}(s)]$, where $\lambda_{U,s}$ is a 1-form of U depending on s, and $\lambda_{V,r}$ is a 1-form of V depending on r. Let $y_s = P(s)$, we have

$$\begin{aligned} \lambda(\tau \cdot P)_{\binom{r}{s}} \begin{pmatrix} \delta r \\ \delta s \end{pmatrix} &= \lambda_{U,s}(r)(\delta r) + \lambda_{V,r}(s)(\delta s) \\ &= \lambda[r \mapsto \tau(r)_Y(P(s))]_r(\delta r) + \lambda[s \mapsto \tau(r)_Y(P(s))]_s(\delta s) \\ &= \lambda(R(y_s) \circ \tau)_r(\delta r) + \tau(r)_Y^*(\lambda)(P)_s(\delta s) \\ &= \tau^*[R(y_s)^*(\lambda)]_r(\delta r) + \lambda(P)_s(\delta s) \\ &= \tau^*(\theta)_r(\delta r) + \lambda(P)_s(\delta s). \end{aligned}$$

2. Let $P : U \to Y$ and $Q : U \to Y$ be two plots such that $\pi \circ P = \pi \circ Q$. Since π is a principal fiber bundle with structural group T, for every point $r_0 \in U$ there exists at least locally, on some open neighborhood of r_0, a smooth parametrization

$\boldsymbol{\tau}$ of T such that $Q(r) = \boldsymbol{\tau}(r)_Y(P(r))$ (art. 8.11). Applying proposition 1) to Q, we get $\lambda(Q) = \boldsymbol{\tau}^*(\theta) + \lambda(P)$, and since θ is a 1-form on T which is 1-dimensional (see Exercise 49, p. 50), $d\theta = 0$ and $d\lambda(Q) = d\lambda(P)$. Then, thanks to (art. 6.38), there exists a 2-form ω on X such that $\pi^*(\omega) = d\lambda$, then $\pi^*(d\omega) = d[d\lambda] = 0$, and since π is a subduction, the 2-form ω is closed (art. 6.39).

3. Let us consider the connection as a lifting of paths. Let c be a local path in X, let $x = c(t)$ for some $t \in \operatorname{def}(c)$ and $y \in Y_x$. Then, let $\tilde{c}$ be a global lift of c, which exists always thanks to (art. 8.9). We can even choose $\tilde{c}(t) = y$. Now, let $f \in \mathcal{C}^\infty(\operatorname{def}(c), \mathbf{R})$, and let us consider the new path $s \mapsto \operatorname{class}(f(s))_Y(\tilde{c}(s))$. Using the point 1) and that $\operatorname{class}^*(\theta) = dt$ means $\theta[t \mapsto \operatorname{class}(t)] = dt$, what is equivalent to $\theta[r \mapsto \operatorname{class}(F(r))] = F^*(dt) = dF$, for all smooth real parametrizations F, we get

$$\begin{aligned} \lambda\Big[s \mapsto \operatorname{class}(f(s))_Y\Big(\tilde{c}(s)\Big)\Big]_s(1) &= [s \mapsto \operatorname{class}(f(s))]^*(\theta)_s(1) + \lambda(\tilde{c})_s(1) \\ &= \frac{df(s)}{ds} + \lambda(\tilde{c})_s(1). \end{aligned}$$

Thus, the local path

$$\bar{c}(t') = \operatorname{class}(f(t'))_Y\Big(\tilde{c}(t')\Big) \quad \text{with} \quad f(t') = -\int_t^{t'} \lambda(\tilde{c})_s(1)\, ds,$$

satisfies $\bar{c}(t) = \tilde{c}(t) = y$ and $\lambda(\bar{c})_s = 0$ for all s. Now, we define $\Theta(\tilde{c}, t)$ by ($\heartsuit$) and ($\diamondsuit$) follows. To check that Θ satisfies the six axioms of connections (art. 8.32), it is enough to check the unicity of horizontal lift with a given starting point. Indeed, the other axioms are satisfied by construction or thanks to the locality of the integration of real function. So, let $\bar{c}'$ be an another horizontal lift of c starting at y at time t, we have $\bar{c}'(s) = \operatorname{class}(g(s))_Y(\bar{c}(s))$, where g is a smooth real function defined on $\operatorname{def}(c)$. Applying the same computation to $\bar{c}'$, since $\lambda(\bar{c}')_s = \lambda(\bar{c})_s = 0$ for all s, we get that $dg(s)/ds = 0$, thus g is constant and $g(s) = g(t)$, but since $\bar{c}'(t) = \bar{c}(t) = y$, we get finally that $g(t) = 0$ and hence $g(s) = 0$ for all s. Therefore, the horizontal lift of c starting at y, as defined above, is unique, and Θ defined by ($\heartsuit$) is a connection on $\pi : Y \to X$.

4. Let $x \in X$ and $y \in Y_x = \pi^{-1}(x)$, the holonomy at the point y is the subgroup T_y of the elements $\tau \in T$ for which there exists a loop ℓ based at x such that the horizontal lift $\bar{\ell}$ starting at y ends at $\tau_Y(y)$, that is, $\bar{\ell}(0) = y$ and $\bar{\ell}(1) = \tau_Y(y)$. Then, the map $\ell \mapsto \bar{\ell}$, from $\operatorname{Loops}(X, x)$ to $\operatorname{Paths}(Y, y, Y_x)$ is smooth, thus the map $\ell \mapsto \bar{\ell}(1)$ also is smooth, and also the map $\ell \mapsto R(y)^{-1}(\bar{\ell}(1))$, that is, the map $\ell \mapsto \tau$ such that $\tau_Y(y) = \bar{\ell}(1)$. Thus, for any 1-plot $s \mapsto \ell_s$, $s \mapsto \bar{\ell}_s$ is a 1-plot of $\operatorname{Paths}(Y, y, Y_x)$ and $\boldsymbol{\tau} : s \mapsto \tau_s$, such that $\bar{\ell}_s(1) = \boldsymbol{\tau}(s)_Y(y)$, is a 1-plot of T_y. Now, since $\bar{\ell}_s$ is horizontal for all s, $\lambda(\bar{\ell}_s) = 0$ ($\diamondsuit$), thus

$$\int_{\bar{\ell}_s} \lambda = \int_0^1 \lambda(\bar{\ell}_s)_t(1)\, dt = 0.$$

Then, the variation of this identity, using the formula (art. 6.70, Note 2), gives for $\delta = \partial/\partial s$

$$\delta \int_{\bar{\ell}_s} \lambda = 0, \text{ that is, } \int_0^1 d\lambda(\delta\bar{\ell}) + \Big[\lambda(\delta\bar{\ell})\Big]_0^1 = 0.$$

But

$$\begin{aligned}
\Big[\lambda(\delta\bar{\ell})\Big]_0^1 &= \lambda(\delta\bar{\ell})(1) - \lambda(\delta\bar{\ell})(0) \\
&= \lambda\left(\begin{pmatrix}s\\t\end{pmatrix} \mapsto \bar{\ell}_s(t)\right)_{\binom{s}{1}}\begin{pmatrix}1\\0\end{pmatrix} - \lambda\left(\begin{pmatrix}s\\t\end{pmatrix} \mapsto \bar{\ell}_s(t)\right)_{\binom{s}{0}}\begin{pmatrix}1\\0\end{pmatrix} \\
&= \lambda\left(s \mapsto \bar{\ell}_s(1)\right)_s(1) - \lambda\left(s \mapsto \bar{\ell}_s(0)\right)_s(1) \\
&= \lambda\left(s \mapsto \tau(s)_Y(y)\right)_s(1) - \lambda\left(s \mapsto y\right)_s(1) \\
&= \lambda\left(s \mapsto R(y)(\tau(s))\right)_s(1) - 0 \\
&= R(y)^*(\lambda)(\tau)_s(1) \\
&= \theta(\tau)_s(1).
\end{aligned}$$

Then, since $d\lambda = \pi^*(\omega)$ and $\pi \circ \bar{\ell}_s = \ell_s$, the identity above writes

$$\int_0^1 \omega(\delta\ell) + \theta(\tau)_s(1) = 0.$$

Now, it is clear that if $\omega = 0$, then $\theta(\tau)_s = 0$, which means that τ is locally constant, that is, the holonomy T_y is discrete. □

8.38. The Kronecker flows as diffeological fibrations. The Kronecker flow has been the first example of application of diffeological methods on geometrical constructions regarded by classical differential geometry as *singular* [**DoIg85**]. We consider the 2-dimensional torus T^2, equipped with the quotient diffeology, $T = \mathbf{R}/\mathbf{Z}$ to be a manifold as well as the square $T^2 = T \times T$, equipped with the product diffeology. It is also a group and a diffeological group (actually a Lie group) (art. 7.1). Then, we pick a 1-parameter subgroup $A = \{[t, \alpha t] \in T^2 \mid t \in \mathbf{R}\}$, for some real number α, where the brackets denote the class modulo $\mathbf{Z}^2$. The subgroup A is just the projection of the line $\Delta \subset \mathbf{R}^2$ with equation $y = \alpha x$. The nature of the quotient $T_\alpha = T^2/A$ depends on the rationality of α: if $\alpha \in \mathbf{Q}$, then T_α is a Lie group diffeomorphic to T, else if $\alpha \in \mathbf{R} - \mathbf{Q}$, then T_α is still a 1-dimensional diffeological group but not anymore a manifold, we say that T_α is an *irrational torus*. We have already seen many aspects of this irrational torus in Exercise 4, p. 8; Exercise 19, p. 20; Exercise 31, p. 31; Exercise 74, p. 84. As a direct application of the theory of diffeological fiber bundles developed in this part, we get the following properties.

1. Considering the homotopy of T_α, since the fiber $\mathbf{R}$ of the fibration $\pi : T^2 \to T_\alpha$ is contractible, the homotopy sequence (art. 8.21) tells us that its homotopy coincides with the homotopy of T^2, that is, $\pi_0(T_\alpha) = \{T_\alpha\}$, $\pi_1(T_\alpha) \simeq \mathbf{Z}^2$, and $\pi_k(T_\alpha) = \{0\}$ for all $k > 1$.

2. The 1-form $\Lambda = dy/\alpha$ defined on $\mathbf{R}^2$ is invariant by $\mathbf{Z}^2$ and defines a 1-form λ on T^2 (art. 8.27). The 1-form λ satisfies the conditions to be a connection form for the $\mathbf{R}$-principal fibration $\pi : T^2 \to T_\alpha$; it is invariant by the Kronecker flow and calibrated (art. 8.37) with $\Gamma = \{0\}$. Then, since $\dim(T_\alpha) = 1$, the curvature of λ vanishes and the connection defined by λ is flat (art. 8.37, point 4). Every path in T^2 writes locally $t \mapsto [x(t), y(t)]$, the path is horizontal if and only if $d(y(t))/dt = 0$, that is, if and only if $y(t)$ is locally constant. Therefore, the holonomy bundle total space, with reference to the point $[0,0]$ (art. 8.35, (♠)), is the torus $T \times \{0\} = \{[x, 0] \mid x \in \mathbf{R}\} \subset T^2$. Now, the holonomy is the subgroup $H \subset \mathbf{R}$ made of t such that $t_{T^2}[x, 0] = [x + t, 0 + \alpha t] = [x', 0]$, that is, if and only if

$t = m/\alpha$, with $m \in \mathbf{Z}$, *i.e.*, $H \simeq \mathbf{Z}$. Applying (art. 8.35, Note 1), we must find the torus T^2 again by quotient of the space of pairs $([x, 0], t) \simeq ([x], t)$, where $[x] \in T$ and $t \in \mathbf{R}$, by the action of $\mathbf{Z}$ given by

$$m([x], t) = \left(\left[x + \frac{m}{\alpha}\right], t - \frac{m}{\alpha}\right),$$

where the brackets denote the class modulo $\mathbf{Z}$. The quotient of the product of the holonomy fiber bundle by the structure group $\mathbf{R}$, under the action of the holonomy, is then given by the projection $([x], t) \mapsto ([x+t], [\alpha t])$, and we find the torus T^2 again as claimed.

NOTE. To end this discussion, note that, as a group T_α is far to be *singular*, to the contrary there is nothing more homogeneous/regular than a group, and nothing smoother than a diffeological group, whether it is a Lie group or not. Diffeology does justice to this intuition.

8.39. R-bundles over irrational tori and small divisors. In the classical category of differential manifolds, every fiber bundle with contractible fiber has a global smooth section; see for example [**Die70c**]. Therefore, every principal bundle over a manifold, with structure group the additive group of real numbers $\mathbf{R}$, is trivial. But this is not true anymore in diffeology, as we have seen with the example of the irrational torus, the fibration $\pi : T^2 \to T_\alpha$, is a nontrivial $\mathbf{R}$-principal fibration. This remark suggests, for every diffeological space X, a new class of invariant: the set $\mathrm{Fb}(X, \mathbf{R})$ of (classes of) $\mathbf{R}$-principal bundles with base X. This question has been raised first in [**Igl85**] and [**Igl86**], let us summarize its main aspects.

First of all, the set $\mathrm{Fb}(X, \mathbf{R})$ has a structure of an Abelian group. Let us describe the group operation. Let $\pi : Y \to X$ and $\pi' : Y' \to X$ be two $\mathbf{R}$-principal bundles. Let $\pi \otimes \pi' : Y \otimes Y' \to X$ be the fiber product defined by $Y \otimes Y' = \pi^*(Y') = \{(y, y') \in Y \times Y' \mid \pi(y) = \pi'(y')\}$ and $\pi \otimes \pi'(y, y') = \pi(y) = \pi'(y')$. But $\pi \otimes \pi'$ is an $\mathbf{R}^2$-principal bundle for the product of the action on each factor, $(t, t')_{Y \otimes Y'}(y, y') = (t_Y(y), t'_{Y'}(y'))$. Let us denote by $\overline{\mathbf{R}}$ the antidiagonal, that is, the subgroup of pairs $(t, -t) \in \mathbf{R}^2$. Now, the quotient $\pi'' : Y'' \to X$, where $Y'' = (Y \otimes Y')/\overline{\mathbf{R}}$ and $\pi''(y'') = \pi(y) = \pi'(y')$, with $y'' = [y, y']$, is again an $\mathbf{R}$-principal bundle on X, the brackets denoting the orbits of the action of $\overline{\mathbf{R}}$. The map $(\pi, \pi') \mapsto \pi''$ passes to the equivalence classes of the $\mathbf{R}$-principal bundles over X.

a) Let $p = \mathrm{class}(\pi)$, $p' = \mathrm{class}(\pi')$, and $p'' = \mathrm{class}(\pi'')$. Then the operation $(p, p') \mapsto p'' = p + p'$ is an Abelian group operation on $\mathrm{Fb}(X, \mathbf{R})$.

b) The identity element $0 \in \mathrm{Fb}(X, \mathbf{R})$ is represented by the trivial bundle $\mathrm{pr}_1 : X \times \mathbf{R} \to X$.

c) The inverse $-p$ of $p = \mathrm{class}(\pi)$, $\pi : Y \to X$, is the class of the principal bundle $\bar{\pi} : \overline{Y} \to X$, with same total space $\overline{Y} = Y$ and same projection, but with the inverse $\mathbf{R}$-action, $t_{\overline{Y}}(y) = (-t)_Y(y)$.

Let us exemplify this construction by making explicit the group $\mathrm{Fb}(T, \mathbf{R})$ for $T = \mathbf{R}/\Gamma$, where Γ is a strict subgroup of $\mathbf{R}$. Let $\pi : Y \to T$ be a $\mathbf{R}$-principal bundle, its pullback by the projection $\mathrm{pr} : \mathbf{R} \to T$ is trivial, there exists actually a smooth lift $\varphi : \mathbf{R} \to Y$ of pr, that is, $\pi \circ \varphi = \mathrm{pr}$. The map $\Phi : (x, t) \mapsto (x, t_Y(\varphi(x)))$ is an isomorphism from $\mathbf{R} \times \mathbf{R}$ to $\mathrm{pr}^*(Y)$. Then, since π and pr are two subductions, the second projection $\mathrm{pr}_2 : \mathrm{pr}^*(Y) \to Y$ is a subduction and Y is equivalent to the quotient of $\mathbf{R} \times \mathbf{R}$ by the relation $(x, t) \sim (x', t')$ if and only if $\mathrm{pr}_2 \circ \Phi(x, t) = \mathrm{pr}_2 \circ \Phi(x', t')$, that is, $t_Y(\varphi(x)) = t'_Y(\varphi(x'))$. This implies in particular that $x' = x + k$, for some $k \in \Gamma$, and there exists $\tau(k)(x) \in \mathbf{R}$ such that $\varphi(x) = \tau(k)(x)_Y(\varphi(x+k))$. We then get a map $\tau : \Gamma \mapsto \mathcal{C}^\infty(\mathbf{R})$ such that

$$\tau(k + k')(x) = \tau(k)(x + k') + \tau(k')(x) \qquad (\diamondsuit)$$

and (x', t') is equivalent to (x, t) if and only if $x' = x + k$ and $t' = t + \tau(k)(x)$. Hence, the classes of the equivalence relation $\sim$ are the orbits of the group Γ acting on the product $\mathbf{R} \times \mathbf{R}$ by

$$k : (x, t) \mapsto (x + k, t + \tau(k)(x)),$$

and Y is equivalent to $\mathbf{R} \times_\Gamma \mathbf{R}$, the quotient of $\mathbf{R} \times \mathbf{R}$ by this action of Γ.

Conversely, for all τ satisfying $(\diamondsuit)$, let us define Y as the quotient $\mathbf{R} \times_\Gamma \mathbf{R}$ and $\pi : Y \to T$ by $\pi([x, t]) = [x]$, where the brackets denote the class under the action of Γ. We get a natural free action of $\mathbf{R}$ on Y by $t'_Y([x, t]) = [x, t + t']$. This projection π, together with this action of $\mathbf{R}$, is a principal fibration.

Next, τ defines a trivial principal bundle if and only if there is a global smooth section $\sigma : T \to Y$. Then, $\mathrm{pr} \circ \sigma : \mathbf{R} \to Y$ has a smooth lift $x \mapsto (x, \sigma(x))$ in $\mathbf{R} \times \mathbf{R}$ such that $\sigma([x]) = [x, \sigma(x)]$. This can be regarded as a consequence of the monodromy theorem (art. 8.25), since the projection from $\mathbf{R} \times \mathbf{R}$ to its quotient Y is a covering, actually the universal covering. Now, for all $k \in \Gamma$, on the one hand $\sigma([x]) = \sigma([x + k])$ gives $[x, \sigma(x)] = [x + k, \sigma(x + k)]$, and on the other hand $[x, \sigma(x)] = [x + k, \sigma(x) + \tau(k)(x)]$. Thus, $\sigma(x+k) = \sigma(x) + \tau(k)(x)$, that is, $\tau(k)(x) = \sigma(x + k) - \sigma(x)$. Therefore, two maps τ and τ' satisfying $(\diamondsuit)$ define two equivalent $\mathbf{R}$-principal bundles if and only if there exists a map $\sigma \in \mathcal{C}^\infty(\mathbf{R})$ such that

$$\tau'(k)(x) = \tau(k)(x) + \sigma(x + k) - \sigma(x). \qquad (\heartsuit)$$

Thus, the group $\mathrm{Fb}(X, \mathbf{R})$ of classes of $\mathbf{R}$-principal bundles over T is equivalent to the set of maps τ defined by $(\diamondsuit)$ modulo the equivalence defined by $(\heartsuit)$. This can be interpreted in terms of group cohomology.

The map τ is actually a 1-cocycle of Γ with coefficients in $\mathcal{C}^\infty(\mathbf{R})$, where Γ acts on $\mathcal{C}^\infty(\mathbf{R})$ by translations $(k, f) \mapsto T_k^*(f) = f \circ T_k$, with $(k, f) \in \Gamma \times \mathcal{C}^\infty(\mathbf{R})$ and $T_k(x) = x + k$. Indeed, the identity $(\diamondsuit)$ writes equivalently

$$\tau(k + k') = T_{k'}^*(\tau(k)) + \tau(k').$$

Next, the cocycle τ defines a trivial bundle if $\tau(k)(x) = \sigma(x + k) - \sigma(x)$, that is, if τ is a coboundary, $\tau(k) = T_k^*(\sigma) - \sigma$. Thus the group $\mathrm{Fb}(X, \mathbf{R})$ is equivalent to the first cohomology group

$$\mathrm{Fb}(X, \mathbf{R}) \simeq H^1(\Gamma, \mathcal{C}^\infty(\mathbf{R})).$$

Now, let us come to the particular example of $\Gamma = \mathbf{Z} + \alpha\mathbf{Z}$, where $\alpha \in \mathbf{R} - \mathbf{Q}$, $\mathbf{R}/\Gamma = T_\alpha$. The space T_α is equivalent to the quotient $T/\mathbf{Z}$, where $T = \mathbf{R}/\mathbf{Z}$, with $n(x) = x + n$, $x \in \mathbf{R}$ and $n \in \mathbf{Z}$, and then $\mathbf{Z}$ acts on T by $n\,[x] = [x + n\alpha]$, where $[x] \in T$ denotes the class of x modulo $\mathbf{Z}$. Let us denote by $\mathrm{pr} : T \to T_\alpha$ the projection, and let $\pi : Y \to T_\alpha$ be an $\mathbf{R}$-principal fiber bundle. The pullback $\mathrm{pr}_1 : \mathrm{pr}^*(Y) \to T$ is an $\mathbf{R}$-principal fiber bundle over a manifold $T \simeq S^1$, thus it is trivial $\mathrm{pr}^*(Y) \simeq T \times \mathbf{R}$. And the previous reconstruction of Y, for Γ acting on $\mathbf{R} \times \mathbf{R}$, can be mimicked in this situation with $\mathbf{Z}$ acting on $T \times \mathbf{R}$ through a cocycle $\tau : \mathbf{Z} \to \mathcal{C}^\infty(T, \mathbf{R})$ satisfying

$$\tau(n + n')([x]) = \tau(n)([x + n'\alpha]) + \tau(n')([x]).$$

Let $f = \tau(1)$, then the cocycle τ satisfies

$$\tau(n)([x]) = \sum_{k=0}^{n-1} f([x + k\alpha]) \quad \text{and} \quad \tau(-n)([x]) = -\sum_{k=0}^{n-1} f([x + (k-n)\alpha]),$$

for all integers $n > 0$ and $\tau(0) = 0$. Then, f defines a coboundary if and only if there exists a function $g \in \mathcal{C}^\infty(T, \mathbf{R})$ such that

$$f([x]) = g([x + \alpha]) - g([x]) \quad \text{or} \quad F(x) = G(x + \alpha) - G(x),$$

where F and G are two 1-periodic smooth real functions on $\mathbf{R}$, representing respectively f and g. This equation in G is known as the *cohomological equation* and has been studied by many authors, beginning by [**Arn65**], [**Arn80**], [**Mos66**], [**Her79**]. The space $\mathrm{Fb}(T_\alpha, \mathbf{R})$ we are looking for, $\mathcal{C}^\infty(T, \mathbf{R})$ modulo coboundaries, has a factor $\mathbf{R}$ given by the invariant $\nu = \int_0^1 F(x)dx$. The second factor depends fundamentally on the arithmetic nature of α. In particular, if α is a diophantine, that is, if there exist an integer $k > 0$ and a real number $c > 0$ such that

$$|n\alpha - m| \geq \frac{c}{n^k} \quad \text{for all} \quad n \in \mathbf{N} - \{0\}, \; m \in \mathbf{N},$$

then the equation $F(x) = G(x + \alpha) - G(x)$, where F is normalized by $\int_0^1 F(x)dx = 0$, has a (unique) solution G. Thus, for these diophantine numbers, the space $\mathrm{Fb}(T_\alpha, \mathbf{R})$ is equivalent to $\mathbf{R}$, every element is then characterized by its invariant ν. Geometrically, every $\mathbf{R}$-principal bundle over the irrational torus T_α, with α satisfying the diophantine condition above, is either trivial or isomorphic to the 2-torus T^2 equipped with the α-Kronecker flow, but for different $\mathbf{R}$-action speeds. But this is not the case if α is not diophantine, since there exist functions F for which the cohomological equation above has no solution. So, the diffeology of the irrational torus captures not only the equivalence by $\mathrm{GL}(2, \mathbf{Z})$ or the possible quadratic character of α, as we have seen before, but also its fine arithmetic.

NOTE 1. The space $H^1(\Gamma, \mathcal{C}^\infty(\mathbf{R}))$ is actually a real vector space. In particular, for every cocycle τ and real number s, $s \cdot \tau$ is again a cocycle. Geometrically speaking, if $s \neq 0$, then $s \cdot \tau$ represents the same total space with a $\mathbf{R}$-action dilated by s.

NOTE 2. Every $\mathbf{R}$-principal bundle $\pi : Y \to T$ over a torus $T = \mathbf{R}/\Gamma$, defined by a cocycle τ, can be naturally equipped with a connection associated with the connection of the covering $\mathrm{pr} : \mathbf{R} \to T$ (art. 8.36). However, not all these bundles support a connection form λ (art. 8.37), only those whose cocycle τ defining π is equivalent to a homomorphism from Γ to $\mathbf{R}$; see Exercise 139, p. 287. In other words, if the cocycle τ is not cohomologous to a homomorphism, then there is no connection which can be defined by a connection form. In particular, for $\Gamma = \mathbf{Z} + \alpha\mathbf{Z}$, the only $\mathbf{R}$-principal bundles equipped with connections defined by a connection form are the Kronecker flows.

PROOF. The above construction of $\pi : Y = \mathbf{R} \times_\Gamma \mathbf{R} \to T$, together with its action of $\mathbf{R}$, is not exactly an associated bundle to the covering $\mathrm{pr} : \mathbf{R} \to T$ as described in (art. 8.16). Let us prove directly that it is indeed an $\mathbf{R}$-principal bundle. Since pr is a generating function of the quotient diffeology, it is sufficient to check that the pullback of π by pr is a (trivial) $\mathbf{R}$-principal bundle, or, which is equivalent, that the map $\Phi : \mathbf{R} \times \mathbf{R} \to \mathrm{pr}^*(Y)$ defined by $\Phi(x, t) = (x, [x, t])$ is a (clearly equivariant) diffeomorphism. The map Φ is obviously smooth and injective. It is also surjective. Let $(x, [x', t']) \in \mathrm{pr}^*(Y)$, then there exists $k \in \Gamma$ such that $x' = x + k$, thus $[x', t'] =$

$[x, t = t' - \tau(k)(x)]$ and $(x, [x', t']) = \Phi(x, t)$. Let us check that Φ^{-1} is smooth. A plot of $\mathrm{pr}^*(Y)$ writes locally $r \mapsto (x(r), [x'(r), t'(r)])$, with $[x(r)] = [x'(r)]$. Thus, there exists a plot $r \mapsto k(r)$ of Γ such that $x'(r) = x(r) + k(r)$, but since Γ is discrete, $k(r)$ is locally constant, that is, $k(r) =_{\mathrm{loc}} k$. Hence, $r \mapsto t(r) = t'(r) - \tau(k)(x(r))$ is a plot of $\mathbf{R}$ and $\Phi(x(r), t(r)) = (x(r), [x'(r), t'(r)])$. Therefore, Φ^{-1} is smooth and Φ is a diffeomorphism.

Now, let us check that the projection $(x, t) \mapsto [x, t]$ is a covering as claimed. Let $P : r \mapsto y_r$ be a plot of Y, since $(x, t) \mapsto [x, t]$ is a subduction the plot P has a local lift $r \mapsto (x_r, t_r)$. Then, $\Psi : (r, k) \mapsto (r, (x_r + k, t_r + \tau(k)(x_r)))$ is a smooth bijection from $\mathrm{def}(P) \times \Gamma$ to $P^*(Y) = \{(r, (x, t)) \in \mathrm{def}(P) \times Y \mid P(r) = [x, t]\}$. Its inverse is given by $\Psi^{-1}(r, (x, t)) = (r, x - x_r)$ which is clearly smooth. Thus, $(x, t) \mapsto [x, t]$ is a covering with group Γ, and since $\mathbf{R} \times \mathbf{R}$ is simply connected, it is the universal covering. ☐

Exercises

✎ EXERCISE 136 (Spheric periods on toric bundles). Let $\pi : Y \to X$ be a principal fiber bundle with structure group a torus $T = \mathbf{R}/\Gamma$ (art. 8.37). Let $x \in X$ and σ be any homotopy of the constant loop $\mathbf{x} : t \mapsto x$ to itself, that is, $\sigma \in \mathrm{Loops}(\mathrm{Loops}(X, x), \mathbf{x})$. Show that $\int_\sigma \omega \in \Gamma$.

✎ EXERCISE 137 (Fiber bundles over tori). Let $T_\Gamma = \mathbf{R}^n/\Gamma$ be a torus as defined in Exercise 101, p. 160. Show that every fiber bundle over T_Γ is "associated" with a covering, as in (art. 8.39).

✎ EXERCISE 138 (Flat connections on toric bundles). Show the converse of (art. 8.37, point 4), that is, if a connection on a principal bundle, with structure group a 1-dimensional torus, is flat, then the curvature of the connection vanishes.

✎ EXERCISE 139 (Connection forms over tori). Let τ be a cocycle as defined in (art. 8.39, (♢)). Show that there exists a connection 1-form on the $\mathbf{R}$-principal fiber bundle $\pi : Y \to T$, with $T = \mathbf{R}/\Gamma$ and $Y = \mathbf{R} \times_\Gamma \mathbf{R}$, if and only if τ is equivalent to a homomorphism from Γ to $\mathbf{R}$. Make explicit the subspace of $\mathrm{Fb}(T_\alpha, \mathbf{R})$ made of classes of such bundles.

Integration Bundles of Closed 2-Forms

Integral closed 2-forms play an important role in theory of fiber bundles, because they are used to classify complex line bundles [**Mil74**]. *They are also crucial in symplectic mechanics because of geometric quantization; see for example* [**Sou70**], [**Bry93**], [**BaWe97**]. *We shall see in this section a universal construction, associating every closed 2-form with some* integration fiber bundle, *with structure group the* torus of periods *of the 2-form, whether the 2-form is integral or not. The 2-form will be the curvature of a connection 1-form on this bundle. This construction generalizes for diffeological spaces a previous construction, on manifolds, described in* [**Igl95**]. *We only address in this section the case of simply connected diffeological spaces, and show that—in this case at least—every closed 2-form is, in diffeology, the curvature of a connection. The general case and the relation of this construction with symplectic diffeology (see (Chap. 9)) is a work in progress.*

8.40. Torus bundles over diffeological spaces. Let X be a connected diffeological space, $\pi_0(X) = \{X\}$. Let $T = \mathbf{R}/\Gamma$, where Γ is a strict subgroup of $\mathbf{R}$, be a 1-dimensional diffeological torus. We denote by θ the canonical 1-form defined on T by pushing forward the canonical 1-form dt of $\mathbf{R}$. Let $\pi : Y \to X$ be a T-principal bundle, and let λ be a connection form on Y with curvature $\omega \in \Omega^2(X)$ (art. 8.37). The action of T on Y is denoted as usual by $\tau_Y(y)$ for $\tau \in T$ and $y \in Y$. Let $x \in X$ be a basepoint in X, and let Paths(X, x) be the space of paths in X based at x. Let us consider the T-principal bundle, pullback of π by the target map $\hat{1} : \text{Paths}(X, x) \to X$, with total space

$$\hat{1}^*(Y) = \{(\gamma, y) \in \text{Paths}(X) \times Y \mid \gamma(0) = x \text{ and } \pi(y) = \gamma(1)\}.$$

Since λ is a connection form on Y, $\text{pr}_1 : \hat{1}^*(Y) \to \text{Paths}(X, x)$ can be equipped by pullback with a connection, and since Paths(X, x) is contractible (Exercise 83, p. 111), $\text{pr}_1 : \hat{1}^*(Y) \to \text{Paths}(X, x)$ is a trivial T-principal bundle (art. 8.34).

1. There exists an equivariant diffeomorphism

$$\Psi : \text{Paths}(X, x) \times T \to \hat{1}^*(Y) \text{ such that } (\text{pr}_2 \circ \Psi)^*(\lambda) = \mathcal{K}\omega \oplus \theta,$$

where $\mathcal{K}$ is the Chain-Homotopy Operator (art. 6.83). Note that $\mathcal{K}\omega \oplus \theta$ is a connection form on the trivial T-principal fiber bundle $\text{pr}_1 : \text{Paths}(X, x) \times T \to \text{Paths}(X, x)$ with curvature $\hat{1}^*(\omega)$.

2. The equivariant diffeomorphism Ψ writes $\Psi(\gamma, \tau) = (\gamma, \tau_Y(\psi(\gamma)))$, where $\psi \in \mathcal{C}^\infty(\text{Paths}(X, x), Y)$ is a lift of $\hat{1}$ along π, that is, $\pi \circ \phi = \hat{1}$. The projection $\text{pr}_2 \circ \Psi$ is a subduction. This means that the total space Y of the T-principal bundle π can be regarded as a quotient of the product $\text{Paths}(X, x) \times T$, onto which the connection form $\mathcal{K}\omega + \theta$ descends.

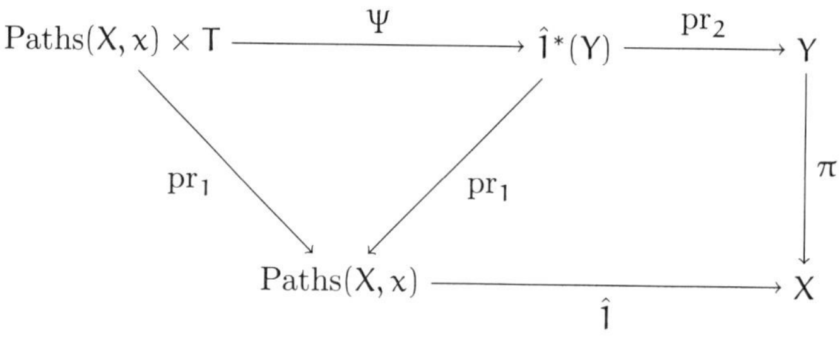

3. There exists a smooth map Φ defined on the tautological pullback

$$\hat{1}^*(\text{Paths}(X, x)) = \{(\gamma, \gamma') \in \text{Paths}(X, x)^2 \mid \gamma(1) = \gamma'(1)\}$$

with values in T, $\Phi \in \mathcal{C}^\infty(\hat{1}^*(\text{Paths}(X, x)), T)$, satisfying the cocycle condition

$$\Phi(\gamma, \gamma') + \Phi(\gamma', \gamma'') = \Phi(\gamma, \gamma''), \qquad (\diamondsuit)$$

for every couple of pairs of paths (γ, γ') and (γ', γ'') in $\hat{1}^*(\text{Paths}(X, x))$, and such that

$$\psi(\gamma) = \psi(\gamma') \quad \text{if and only if} \quad \Phi(\gamma, \gamma') = 0. \qquad (\heartsuit)$$

Note that ψ takes its values in Y but Φ takes its values in T. Then, the equivalence relation ϕ identifying Y as a quotient of $\text{Paths}(X, x) \times T$ writes

$$(\gamma, \tau) \ \phi \ (\gamma', \tau') \text{ if } \hat{1}(\gamma) = \hat{1}(\gamma') \text{ and } \tau' - \tau = \Phi(\gamma, \gamma'), \qquad (\clubsuit)$$

with (γ, τ) and (γ', τ') in $\text{Paths}(X, x) \times T$. The function Φ above $(\diamondsuit)$ will be called a *paths* 1-*cocycle*. It characterizes the class of the T-principal bundle $\pi : Y \to X$, together with its connection λ. Modulo coboundaries $\Delta F(\gamma, \gamma') = F(\gamma') - F(\gamma)$, the cocycle Φ characterizes the class of the T-principal bundle $\pi : Y \to X$ only.

$$\begin{array}{ccc}
\mathrm{Paths}(X, x) \times T & \xrightarrow{\ \mathrm{pr}_2 \circ \Psi\ } & Y \simeq \mathrm{Paths}(X, x)/\phi \\
\Big\downarrow \mathrm{pr}_1 & & \Big\downarrow \pi \\
\mathrm{Paths}(X, x) & \xrightarrow[\ \hat{1}\ ]{} & X
\end{array}$$

NOTE. The map Φ above is the restriction to $\hat{1}^*(\mathrm{Paths}(X, x))$ of a smooth function from $\mathrm{ends}^*(\mathrm{Paths}(X)) = \{(\gamma, \gamma') \in \mathrm{Paths}(X)^2 \mid \mathrm{ends}(\gamma) = \mathrm{ends}(\gamma')\}$ to T, satisfying the same cocycle property ($\diamondsuit$). The cohomology group defined by these paths cocycles, modulo coboundaries, may be denoted by $H^1(\mathrm{Paths}(X), T)$. Thus, every T-principal bundle, which can be equipped with a connection 1-form, defines a unique class in $H^1(\mathrm{Paths}(X), T)$ which can be regarded as its *characteristic class.*

PROOF. As it is claimed, the pullback of λ by pr_2 on $\hat{1}^*(Y)$ is a connection form on a T-principal bundle over a contractible base, thus the fiber bundle is trivial (art. 8.34). Let $\Psi : \mathrm{Paths}(X, x) \to \hat{1}^*(Y)$ be a trivialization of T-principal bundle, that is, an equivariant diffeomorphism. Thus, $\Psi(\gamma, \tau) = (\gamma, \tau_Y(\psi(\gamma)))$, where $\psi : \mathrm{Paths}(X, x) \to Y$ is a lift of $\hat{1}$ along π, $\pi \circ \psi = \hat{1}$. Then, the pullback of λ by $\mathrm{pr}_2 \circ \Psi$ is a connection form of the trivial bundle $\mathrm{pr}_1 : \mathrm{Paths}(X, x) \times T$ with curvature $\hat{1}^*(\omega) = d[\mathcal{K}\omega]$. But the 1-form $\mathcal{K}\omega \oplus \theta$ also is a connection form on $\mathrm{Paths}(X, x) \times T$ with curvature $\hat{1}^*(\omega)$, thus the difference of these two connections is the pullback of a closed 1-form of $\mathrm{Paths}(X, x)$ (art. 8.37, Note), but since $\mathrm{Paths}(X, x)$ is contractible, this closed 1-form is exact. Therefore, there exists a smooth function $f : \mathrm{Paths}(X, x) \to \mathbf{R}$ such that

$$(\mathrm{pr}_2 \circ \Psi)^*(\lambda) = (\mathcal{K}\omega + df) \oplus \theta \quad \text{and} \quad \psi^*(\lambda) = \mathcal{K}\omega + df.$$

Now, let us denote by $[t] \in T = \mathbf{R}/\Gamma$ the class of $t \in \mathbf{R}$, and let us define

$$\psi'(\gamma) = [-f(\gamma)]_Y(\psi(\gamma)).$$

The map $\psi' : \mathrm{Paths}(X, x) \to Y$ is smooth and satisfies

$$\psi'^*(\lambda) = \mathcal{K}\omega.$$

Indeed, let $P : r \mapsto \gamma_r$ be a plot of $\mathrm{Paths}(X, x)$ and $\boldsymbol{\tau} : r \mapsto [-f(\gamma_r)]$. Thus, according to (art. 8.37, ($\clubsuit$)), and since $\boldsymbol{\tau}^*(\theta) = -df$, we get

$$\begin{aligned}
\psi'^*(\lambda)(P) &= \lambda(\psi' \circ P) \\
&= \lambda\left(r \mapsto [-f(\gamma_r)]_Y(\psi(\gamma_r))\right) \\
&= \lambda\left(r \mapsto \boldsymbol{\tau}(r)_Y(\psi \circ P(r))\right) \\
&= \boldsymbol{\tau}^*(\theta) + \lambda(\psi \circ P) \\
&= \boldsymbol{\tau}^*(\theta) + \psi^*(\lambda)(P) \\
&= -df + (\mathcal{K}\omega + df) \\
&= \mathcal{K}\omega.
\end{aligned}$$

Hence, $\Psi'(\gamma, \tau) = (\gamma, \tau_Y(\psi'(\gamma)))$ is a trivialization of the principal bundle $\mathrm{pr}_1 : \hat{1}^*(Y) \to \mathrm{Paths}(X, x)$ such that $(\mathrm{pr}_2 \circ \Psi')^*(\lambda) = \mathcal{K}\omega \oplus \theta$. Therefore, we can rename ψ' by ψ, Ψ' by Ψ, and we get what is claimed.

Next, let $r \mapsto y_r$ be a plot of Y and $x_r = \pi(y_r)$. Since π is a subduction, there exists locally $r \mapsto \gamma_r$, a plot such that $\gamma_r(1) = x_r$ (art. 5.6, (♡)). Since $\pi(y_r) = \pi \circ \psi(\gamma_r)$ and since π is a diffeological bundle, that is, locally trivial along the plots, there exists a plot $r \mapsto \tau_r$ of T such that, locally, $y_r = \tau_r(\psi(\gamma_r)) = \mathrm{pr}_2 \circ \Psi(\gamma_r, \tau_r)$. That proves also that $\mathrm{pr}_2 \circ \Psi$ is surjective, thus $\mathrm{pr}_2 \circ \Psi$ is a subduction.

Now, let (γ, τ) and (γ', τ') in $\mathrm{Paths}(X, x) \times T$ projecting on the same point by $\mathrm{pr}_2 \circ \Psi$, that is, $\tau_Y(\psi(\gamma)) = \tau'_Y(\psi(\gamma'))$. This implies, first of all, that $\pi(\psi(\gamma)) = \pi(\psi(\gamma'))$, there exists then a unique $\Phi(\gamma, \gamma') \in T$ such that $\psi(\gamma) = \Phi(\gamma, \gamma')_Y(\psi(\gamma'))$, and we get $\tau' = \tau + \Phi(\gamma, \gamma')$. Therefore, Y is equivalent to the quotient defined by the relation (♣). Let us check that Φ is smooth. Let $r \mapsto (\gamma_r, \gamma'_r)$ be a plot of $\hat{1}^*(\mathrm{Paths}(X, x))$. Then, $r \mapsto x_r = \pi(\psi(\gamma_r)) = \pi(\psi(\gamma'_r))$ is a plot of X, but the pullback of Y by this plot is locally trivial, thus there exists locally everywhere a plot $r \mapsto \tau_r$ of T such that $\psi(\gamma_r) = \tau_{rY}(\psi(\gamma'_r))$. Hence, locally, $\Phi(\gamma_r, \gamma'_r) = \tau_r$, and thus Φ is smooth.

Let us assume now that Ψ and Ψ' are two isomorphisms from $\mathrm{Paths}(X, x) \times T$ to $\hat{1}^*(Y)$. They define an automorphism of $\mathrm{pr}_1 : \mathrm{Paths}(X, x) \times T \to \mathrm{Paths}(X, x)$, that is, a map $F : (\gamma, \tau) \mapsto (\gamma, \tau + F(\gamma))$ such that $\Psi = \Psi' \circ F$. Then, from $\Psi(\gamma, \tau) = \Psi'(\gamma, \tau + F(\gamma))$, and the definition of Φ and Φ', we get $\Phi' = \Phi + F(\gamma') - F(\gamma)$. Therefore, the class of Ψ (modulo the coboundaries $\Delta F(\gamma, \gamma') = F(\gamma') - F(\gamma)$) characterizes the class of the T-principal bundle π. Now, if we also consider the connection form λ, the map F must preserve $\mathcal{K}\omega \oplus \theta$, and that implies $F^*(\theta) = \theta$, that is, $F(\gamma)$ is constant. In this case $\Phi' = \Phi$, and the cocycle Φ itself characterizes the class of the T-principal bundle π together with the connection form λ. □

8.41. Integrating concatenations over homotopies. Let X be a connected diffeological space, and let ω be a closed 2-form on X. All paths in the following are assumed to be stationary (art. 5.4). Let γ_0 and γ'_0 be two paths such that $\gamma_0(1) = \gamma'_0(0)$. Then let

$$\sigma : s \mapsto \gamma_s, \quad \sigma' : s \mapsto \gamma'_s, \quad \text{and} \quad \sigma * \sigma' : s \mapsto \gamma_s \vee \gamma'_s,$$

where σ is a homotopy from γ_0 to γ_1 and σ' a homotopy from γ'_0 to γ'_1, such that $\gamma_s(1) = \gamma'_s(0)$ for all s. The homotopy $\sigma * \sigma'$ is the resultant homotopy from $\gamma_0 \vee \gamma'_0$ to $\gamma_1 \vee \gamma'_1$. Now, let $\mathcal{K}$ be the Chain-Homotopy Operator (art. 6.83), then

$$\int_{\sigma * \sigma'} \mathcal{K}\omega = \int_{\sigma} \mathcal{K}\omega + \int_{\sigma'} \mathcal{K}\omega. \qquad (\clubsuit)$$

PROOF. By definition (art. 6.65), $\int_{\sigma*\sigma'} \mathcal{K}\omega = \int_0^1 \mathcal{K}\omega(\sigma * \sigma')_t(1)\, dt$. Let us show the additivity

$$\mathcal{K}\omega(\sigma * \sigma')_t(1) = \mathcal{K}\omega(\sigma)_t(1) + \mathcal{K}\omega(\sigma')_t(1), \qquad (\star)$$

which will give the identity (♣). From the definition (art. 6.83, (♢)), we have

$$\begin{aligned}
\mathcal{K}\omega(\sigma * \sigma')_t(1) &= \int_0^1 \omega\left(\begin{pmatrix} s \\ t \end{pmatrix} \mapsto (\sigma * \sigma')(t)(s))\right)_{\binom{s}{t}} \begin{pmatrix} 1 \\ 0 \end{pmatrix} \begin{pmatrix} 0 \\ 1 \end{pmatrix} ds \\
&= \int_0^1 \omega\left(\begin{pmatrix} s \\ t \end{pmatrix} \mapsto [\gamma_t \vee \gamma'_t](s))\right)_{\binom{s}{t}} \begin{pmatrix} 1 \\ 0 \end{pmatrix} \begin{pmatrix} 0 \\ 1 \end{pmatrix} ds \\
&= \int_0^{1/2} \omega\left(\begin{pmatrix} s \\ t \end{pmatrix} \mapsto \gamma_t(2s)\right)_{\binom{s}{t}} \begin{pmatrix} 1 \\ 0 \end{pmatrix} \begin{pmatrix} 0 \\ 1 \end{pmatrix} ds \\
&+ \int_{1/2}^1 \omega\left(\begin{pmatrix} s \\ t \end{pmatrix} \mapsto \gamma'_t(2s-1)\right)_{\binom{s}{t}} \begin{pmatrix} 1 \\ 0 \end{pmatrix} \begin{pmatrix} 0 \\ 1 \end{pmatrix} ds,
\end{aligned}$$

and after a change of parameters $s' = 2s$ and $s'' = 2s - 1$, we get

$$\begin{aligned}
\mathcal{K}\omega(\sigma * \sigma')_t(1) &= \int_0^1 \omega\left(\begin{pmatrix} s' \\ t \end{pmatrix} \mapsto \gamma_t(s'))\right)_{\binom{s'}{t}} \begin{pmatrix} 1 \\ 0 \end{pmatrix} \begin{pmatrix} 0 \\ 1 \end{pmatrix} ds' \\
&+ \int_0^1 \omega\left(\begin{pmatrix} s'' \\ t \end{pmatrix} \mapsto \gamma'_t(s'')\right)_{\binom{s''}{t}} \begin{pmatrix} 1 \\ 0 \end{pmatrix} \begin{pmatrix} 0 \\ 1 \end{pmatrix} ds'' \\
&= \mathcal{K}\omega(\sigma)_t(1) + \mathcal{K}\omega(\sigma')_t(1).
\end{aligned}$$

We proved (⋆) which, by integration, gives (♣). □

8.42. Integration bundles of closed 2-forms. We have seen that every T-principal fiber bundle—T being an irrational torus—equipped with a connection form λ, with curvature ω, defines a unique paths cocycle Φ which permits us to reconstruct the bundle by quotient (art. 8.40). Conversely, we shall see in this paragraph how to associate, with any nonzero closed 2-form ω on a diffeological space X, a paths cocycle Φ and construct a T-principal bundle equipped with a connection λ whose curvature is ω. We shall however restrict ourselves to simply connected spaces, the general case is a work in progress. We assume now X connected, simply connected, here are the data:

$$\pi_0(X) = \{X\},\ \pi_1(X) = \{0\},\ \omega \in \Omega^2(X),\ \omega \neq 0 \text{ and } d\omega = 0.$$

Thanks to the fundamental property of the Chain-Homotopy Operator $\mathcal{K}$ (art. 6.83), that is, $d \circ \mathcal{K} + \mathcal{K} \circ d = \hat{1}^* - \hat{0}^*$, the restriction of the 1-form $\mathcal{K}\omega \in \Omega^1(\text{Paths}(X))$ to the subspace Loops(X) is closed, $d[\mathcal{K}\omega \restriction \text{Loops}(X)] = 0$. Then, we define $P_\omega \subset \mathbf{R}$ as the *group of periods* of $\mathcal{K}\omega \restriction \text{Loops}(X)$, that is,

$$P_\omega = \left\{ \int_\sigma \mathcal{K}\omega \mid \sigma \in \text{Loops}(\text{Loops}(X)) \right\}.$$

We shall assume now that the group P_ω is (diffeologically) discrete, or which is equivalent $P_\omega \neq \mathbf{R}$. We define the *torus of periods* T_ω as the quotient

$$T_\omega = \mathbf{R}/P_\omega. \qquad (♣)$$

Thus, T_ω is a 1-dimensional diffeological torus, see Exercise 49, p. 50. We denote by class : $t \mapsto [t]$ the canonical projection from $\mathbf{R}$ to T_ω, and by θ the canonical 1-form on T_ω, pushforward of dt by the projection class (art. 6.38),

$$\text{class} : \mathbf{R} \to T_\omega \quad \text{and} \quad \text{class}^*(\theta) = dt.$$

Now, consider the tautological pullback of $\mathrm{ends} : \mathrm{Paths}(X) \to X \times X$,

$$\mathrm{ends}^*(\mathrm{Paths}(X)) = \{(\gamma, \gamma') \in \mathrm{Paths}(X)^2 \mid \mathrm{ends}(\gamma) = \mathrm{ends}(\gamma')\}.$$

Let pr_1 and pr_2 be the projections from $\mathrm{ends}^*(\mathrm{Paths}(X))$ onto each factor.

$$\begin{array}{ccc} \mathrm{ends}^*(\mathrm{Paths}(X)) & \xrightarrow{\mathrm{pr}_2} & \mathrm{Paths}(X) \\ {\scriptstyle \mathrm{pr}_1}\downarrow & & \downarrow{\scriptstyle \mathrm{ends}} \\ \mathrm{Paths}(X) & \xrightarrow[\mathrm{ends}]{} & X \times X \end{array}$$

1. The space $\mathrm{ends}^*(\mathrm{Paths}(X))$ is homotopic to $\mathrm{Loops}(X)$, and $\mathrm{Loops}(X)$ is connected, thus $\mathrm{ends}^*(\mathrm{Paths}(X))$ is connected.

2. The following 1-form $\mathcal{K}\omega \ominus \mathcal{K}\omega$, defined on $\mathrm{ends}^*(\mathrm{Paths}(X))$, is closed,

$$\mathcal{K}\omega \ominus \mathcal{K}\omega = \mathrm{pr}_1^*(\mathcal{K}\omega) - \mathrm{pr}_2^*(\mathcal{K}\omega) \quad \text{and} \quad d[\mathcal{K}\omega \ominus \mathcal{K}\omega] = 0.$$

3. There exists, up to a constant, a unique smooth integration function Φ

$$\Phi : \mathrm{ends}^*(\mathrm{Paths}(X)) \to T_\omega \quad \text{with} \quad \begin{cases} \mathcal{K}\omega \ominus \mathcal{K}\omega = \Phi^*(\theta), \\ \Phi(\gamma, \gamma') + \Phi(\gamma', \gamma'') = \Phi(\gamma, \gamma''). \end{cases}$$

4. Let ϕ be the equivalence relation

$$\gamma \,\phi\, \gamma' \quad \text{if} \quad \mathrm{ends}(\gamma) = \mathrm{ends}(\gamma') \quad \text{and} \quad \Phi(\gamma, \gamma') = 0, \qquad (\heartsuit)$$

defined on $\mathrm{Paths}(X)$ and associated with the paths cocycle Φ. Let Y be the quotient of the space $\mathrm{Paths}(X, x)$ by the relation ϕ,

$$Y = \mathrm{Paths}(X, x)/\phi.$$

Then, Y is a T_ω-principal bundle over X for the projection $\pi : Y \to X$ defined by $\pi(\mathrm{class}_\omega(\gamma)) = \gamma(1)$, where $\gamma \in \mathrm{Paths}(X, x)$ and $\mathrm{class}_\omega : \mathrm{Paths}(X, x) \to Y$ is the projection associated with ϕ. Moreover, the 1-form $\mathcal{K}\omega$ passes to the quotient Y into a connection form λ with curvature ω, that is,

$$\mathrm{class}_\omega^*(\lambda) = \mathcal{K}\omega \quad \text{and} \quad d\lambda = \pi^*(\omega).$$

5. The T_ω-principal fiber bundle $\pi : Y \to X$, together with the connection form λ, is unique up to isomorphism. Such a pair (π, λ) will be called an *integration structure* of the closed 2-form ω.

$$\begin{array}{ccc} \mathrm{Paths}(X, x) & \xrightarrow{\mathrm{class}_\omega} & Y \\ & {\scriptstyle \hat{1}}\searrow \quad \swarrow{\scriptstyle \pi} & \\ & X & \end{array}$$

NOTE 1. The relation ϕ defines a groupoid $\mathbf{K}$, with structure group T_ω, by

$$\mathrm{Obj}(\mathbf{K}) = X \quad \text{and} \quad \mathrm{Mor}(\mathbf{K}) = \mathcal{Y} \quad \text{with} \quad \mathcal{Y} = \mathrm{Paths}(X)/\phi.$$

For all x and x' in X,

$$\mathrm{Mor}_{\mathbf{K}}(x, x') = \{\mathrm{class}_\omega(\gamma) \mid \mathrm{ends}(\gamma) = (x, x')\},$$

and the groupoid composition is the factorization of the concatenation of paths,

$$\mathrm{class}_\omega(\gamma)\cdot\mathrm{class}_\omega(\gamma') = \mathrm{class}_\omega(\gamma\vee\gamma').$$

Moreover, there exists a differential 1-form

$$\lambda \in \Omega^1(\mathcal{Y}) \quad \text{such that} \quad \mathrm{class}^*(\lambda) = \mathcal{K}\omega.$$

This 1-form is invariant by precomposition and postcomposition in $\mathcal{Y}$, that is,

$$\begin{cases} L(\tau)^*(\lambda \upharpoonright \mathrm{Mor}_{\mathbf{K}}(x,\star)) &= \lambda \upharpoonright \mathrm{Mor}_{\mathbf{K}}(x',\star), \\ R(\tau)^*(\lambda \upharpoonright \mathrm{Mor}_{\mathbf{K}}(\star,x')) &= \lambda \upharpoonright \mathrm{Mor}_{\mathbf{K}}(\star,x), \end{cases}$$

where $\tau \in \mathrm{Mor}_{\mathbf{K}}(x',x)$ acts by $L(\tau) : y \mapsto \tau\cdot y$ on the subspace $\mathrm{Mor}_{\mathbf{K}}(x,\star)$ of morphisms with source x (precomposition), and by $R(\tau) : y' \mapsto y'\cdot\tau$ on the subspace $\mathrm{Mor}_{\mathbf{K}}(\star,x')$ of morphisms with target x' (postcomposition). The groupoid $\mathbf{K}$ is fibrating (art. 8.4), and this is the characteristic groupoid of the integration bundle $\pi : Y \to X$ (art. 8.14), Y is the subspace $\mathcal{Y}_x = \mathrm{Mor}_{\mathbf{K}}(x,\star)$. The fact that λ is a connection form on Y is a direct consequence of the invariance of λ by precomposition and postcomposition in $\mathcal{Y}$. The group $\mathrm{Diff}(X,\omega)$ of diffeomorphisms of X, preserving ω, acts naturally on the groupoid $\mathbf{K}$, by precomposition. An element $\varphi \in \mathrm{Diff}(X,\omega)$ maps $x \in \mathrm{Obj}(\mathbf{K})$ naturally to $\varphi(x)$ and $\mathrm{class}_\omega(\gamma)$ to $\mathrm{class}_\omega(\varphi\circ\gamma)$. This action preserves the 1-form λ. Hence, the group of automorphisms of the structure (X,ω) has a natural representation in the group of automorphisms of the structure $(\mathbf{K},\lambda)$.

NOTE 2. If $\omega = d\alpha$, then $P_\omega = \{0\}$ and $T_\omega = \mathbf{R}$. The integration structure is made up of the trivial bundle $\mathrm{pr}_1 : Y = X\times\mathbf{R} \to X$ and of the connection form $\lambda = \alpha\oplus dt$. If ω is not exact, then the principal bundle $\pi : Y \to X$ is not trivial.

PROOF. Let us check first that $\mathrm{ends}^*(\mathrm{Paths}(X))$ is homotopy equivalent to $\mathrm{Loops}(X)$. Let us recall that $\mathrm{Paths}(X)$ and $\mathrm{Loops}(X)$ are homotopy equivalent, respectively, to the stationary paths $\mathrm{stPaths}(X)$ and $\mathrm{stLoops}(X)$; see (art. 5.4) and Exercise 84, p. 111. Thus, we shall work with stationary paths. We consider the subspace $\mathrm{stLoops}_{1/2}(X)$ of stationary loops that are stationary at $t = 1/2$, that is, constant on an open interval $]1/2-\varepsilon, 1/2+\varepsilon[$. The proof of Exercise 84, p. 111, can be adapted to show that $\mathrm{stLoops}(X)$ and $\mathrm{stLoops}_{1/2}(X)$ are homotopy equivalent. Now, let $f : \mathrm{ends}^*(\mathrm{stPaths}(X)) \to \mathrm{stLoops}_{1/2}(X)$ defined by

$$f(\gamma,\gamma') = \gamma\vee\bar\gamma', \tag{$\diamondsuit$}$$

where $\bar\gamma'$ is the reverse path $\bar\gamma'(t) = \gamma'(1-t)$. Next, let $g : \mathrm{stLoops}_{1/2}(X) \to \mathrm{ends}^*(\mathrm{stPaths}(X))$ defined by $g(\ell) = (\gamma,\gamma')$ with $\gamma(t) = \ell(t/2)$ and $\gamma'(t) = \ell(1-t/2)$ if $t \in]0,1[$ and $\gamma'(t) = 0$ otherwise. These two maps, f and g, are homotopic inverse to each other, thus $\mathrm{ends}^*(\mathrm{Paths}(X))$ is homotopy equivalent to $\mathrm{Loops}(X)$. Now, since $\pi_1(X,x) = \pi_0(\mathrm{Loops}(X,x)) = \{0\}$ and $\pi_0(\mathrm{Loops}(X))$ is a quotient of $\pi_0(\mathrm{Loops}(X,x))$ (Exercise 87, p. 123), then $\mathrm{Loops}(X)$ is connected and so is $\mathrm{ends}^*(\mathrm{Paths}(X))$. Moreover, the map f defined by ($\diamondsuit$) satisfies the identity

$$\mathrm{pr}_1^*(\mathcal{K}\omega) - \mathrm{pr}_2^*(\mathcal{K}\omega) = f^*(\mathcal{K}\omega). \tag{$\star$}$$

Indeed, let (P,P') be a plot of $\mathrm{ends}^*(\mathrm{stPaths}(X))$, then $f^*(\mathcal{K}\omega)(P,P') = \mathcal{K}\omega(f\circ(P,P')) = \mathcal{K}\omega(P*\bar P')$, where the operation $*$ has been defined in (art. 8.41) and $\bar P'(r)(t) = P'(r)(1-t)$. Thanks to (art. 8.41, ($\clubsuit$)), $\mathcal{K}\omega(P*\bar P') = \mathcal{K}\omega(P) + \mathcal{K}\omega(\bar P') = \mathcal{K}\omega(P) - \mathcal{K}\omega(P')$, that is, $f^*(\mathcal{K}\omega) = \mathrm{pr}_1^*(\mathcal{K}\omega) - \mathrm{pr}_2^*(\mathcal{K}\omega)$. Therefore,

$\mathcal{K}\omega \ominus \mathcal{K}\omega$ is closed and has the same periods as $\mathcal{K}\omega \upharpoonright \mathrm{Loops}(\mathrm{X})$, as a consequence of the homotopic invariance of the De Rham cohomology (art. 6.88). Now, we assumed that the periods P_ω of the 1-form $\mathcal{K}\omega \upharpoonright \mathrm{ends}^*(\mathrm{Paths}(\mathrm{X}))$ are discrete. Then, since $\mathrm{ends}^*(\mathrm{Paths}(\mathrm{X}))$ is connected, there exists an integration function $\Phi : \mathrm{ends}^*(\mathrm{Paths}(\mathrm{X})) \to \mathrm{T}_\omega$ (art. 8.29), unique up to a constant, such that

$$[\mathrm{pr}_1^*(\mathcal{K}\omega) - \mathrm{pr}_2^*(\mathcal{K}\omega)] = \Phi^*(\theta).$$

This function is explicitly given by

$$\Phi(\gamma, \gamma') = \mathrm{class}\left(\int_{\gamma'}^{\gamma} \mathcal{K}\omega\right) \in \mathrm{T}_\omega, \qquad (\spadesuit)$$

where the integral is computed along a path σ in $\mathrm{Paths}(\mathrm{X}, x, x')$, connecting γ to γ', with $(x, x') = \mathrm{ends}(\gamma) = \mathrm{ends}(\gamma')$. Indeed, we know that

$$\Phi(\gamma, \gamma') = \mathrm{class}\left(\int_{(\mathbf{x},\mathbf{x})}^{(\gamma,\gamma')} \mathrm{pr}_1^*(\mathcal{K}\omega) - \mathrm{pr}_2^*(\mathcal{K}\omega)\right)$$

is a solution of $\mathrm{pr}_1^*(\mathcal{K}\omega) - \mathrm{pr}_2^*(\mathcal{K}\omega) = \Phi^*(\theta)$ (art. 8.29), where $\mathbf{x} : t \mapsto x$ and $x \in \mathrm{X}$ is a fixed basepoint, and the integral is computed along any path in $\mathrm{ends}^*(\mathrm{stPaths}(\mathrm{X}))$ connecting $(\mathbf{x}, \mathbf{x})$ to (γ, γ'). But, thanks to $(\diamondsuit)$ and $(\star)$,

$$\mathrm{class}\left(\int_{(\mathbf{x},\mathbf{x})}^{(\gamma,\gamma')} \mathrm{pr}_1^*(\mathcal{K}\omega) - \mathrm{pr}_2^*(\mathcal{K}\omega)\right) = \mathrm{class}\left(\int_{\mathbf{x}}^{\gamma\vee\bar{\gamma}'} \mathcal{K}\omega\right),$$

where the integral is computed along a path in $\mathrm{stLoops}(\mathrm{X}, x)$, connecting $\mathbf{x}$ to $\gamma \vee \bar{\gamma}'$. Now, let us consider the map $\nu : \gamma' \mapsto \gamma \vee \bar{\gamma}'$ from $\mathrm{Paths}(\mathrm{X}, x, x')$ to $\mathrm{Loops}(\mathrm{X}, x)$. Thanks to the invariance of $\mathcal{K}\omega$ by concatenation, $\nu^*(\mathcal{K}\omega) = -\mathcal{K}\omega$, the "$-$" sign is due to reversing path. Thus,

$$\mathrm{class}\left(\int_{\mathbf{x}}^{\gamma\vee\bar{\gamma}'} \mathcal{K}\omega\right) = \mathrm{class}\left(\int_{\gamma'}^{\gamma} \mathcal{K}\omega\right),$$

and this confirms the expression of Φ given above. Then, Φ is clearly additive and $\Phi(\gamma, \gamma') + \Phi(\gamma', \gamma'') = \Phi(\gamma, \gamma'')$ for any triple of paths γ, γ' and γ'' such that $\mathrm{ends}(\gamma) = \mathrm{ends}(\gamma') = \mathrm{ends}(\gamma'')$.

Next, let us check that the 1-form $\mathcal{K}\omega \in \Omega^1(\mathrm{Paths}(\mathrm{X}))$ passes to the quotient $\mathcal{Y} = \mathrm{Paths}(\mathrm{X})/\phi$, where ϕ is the equivalence relation defined by $(\heartsuit)$. According to (art. 6.38, Note 2), we need only check that $\mathrm{pr}_1^*(\mathcal{K}\omega) - \mathrm{pr}_2^*(\mathcal{K}\omega)$ vanishes on the total space of the tautological pullback of the projection $\mathrm{class}_\omega : \mathrm{Paths}(\mathrm{X}) \to \mathcal{Y}$, that is,

$$\mathrm{class}_\omega^*(\mathrm{Paths}(\mathrm{X})) = \{(\gamma, \gamma') \in \mathrm{ends}^*(\mathrm{Paths}(\mathrm{X})) \mid \Phi(\gamma, \gamma') = 0\}.$$

But $\mathrm{class}_\omega^*(\mathrm{Paths}(\mathrm{X})) = \Phi^{-1}(0)) \subset \mathrm{ends}^*(\mathrm{Paths}(\mathrm{X}))$ and $\mathrm{pr}_1^*(\mathcal{K}\omega) - \mathrm{pr}_2^*(\mathcal{K}\omega) = \Phi^*(\theta)$ implies immediately that $\mathrm{pr}_1^*(\mathcal{K}\omega) - \mathrm{pr}_2^*(\mathcal{K}\omega) \upharpoonright \mathrm{class}_\omega^*(\mathrm{Paths}(\mathrm{X})) = 0$. Therefore, there exists $\lambda \in \Omega^1(\mathcal{Y})$ such that $\mathrm{class}_\omega^*(\lambda) = \mathcal{K}\omega$.

Now, $\mathrm{Y} = \mathrm{Paths}(\mathrm{X}, x)/\phi$ is the subset $\mathcal{Y}_x \subset \mathcal{Y}$ made of classes of paths based at $x \in \mathrm{X}$, thus $\lambda = \lambda \upharpoonright \mathcal{Y}_x$. To prove that λ is a connection form on a T_ω-principal bundle (art. 8.37), we shall prove first that $\mathbf{K}$, defined in the Note 1, is a groupoid with structure group T_ω, objects X, and morphisms $\mathcal{Y}$.

Concatenation. Let γ_0 and γ_0' be two stationary paths such that $\gamma_0(1) = \gamma_0'(0)$. Then, let γ_1 and γ_1' be two other paths such that $\mathrm{class}_\omega(\gamma_0) = \mathrm{class}_\omega(\gamma_1)$ and

$\text{class}_\omega(\gamma'_0) = \text{class}_\omega(\gamma'_1)$. Because X is simply connected, there exists a fixed-ends homotopy σ connecting γ_0 to γ_1, and another one σ' connecting γ'_0 to γ'_1, that is, $\int_\sigma \mathcal{K}\omega \in P_\omega$ and $\int_{\sigma'} \mathcal{K}\omega \in P_\omega$. Thanks to (art. 8.41), $\int_{\sigma*\sigma'} \mathcal{K}\omega = \int_\sigma \mathcal{K}\omega + \int_{\sigma'} \mathcal{K}\omega$, then $\int_{\sigma*\sigma'} \mathcal{K}\omega \in P_\omega$, where $\sigma * \sigma'$ is a homotopy from $\gamma_0 \vee \gamma'_0$ to $\gamma_1 \vee \gamma'_1$. Thus $\text{class}_\omega(\gamma_0 \vee \gamma'_0) = \text{class}_\omega(\gamma_1 \vee \gamma'_1)$, and hence the composition $\text{class}(\gamma) \cdot \text{class}(\gamma')$ is well defined on $\mathcal{Y}$.

Associativity, identities, and inverses. The associativity of the concatenation, the fact that $\mathbf{1}_x = \text{class}_\omega(\mathbf{x})$ and $\text{class}_\omega(\gamma)^{-1} = \text{class}_\omega(\bar{\gamma})$, where $\bar{\gamma}$ is the reverse of γ, are all based on the homotopies described in (art. 5.15, Proof), connecting $(\gamma_1 \vee \gamma_2) \vee \gamma_3$ to $\gamma_1 \vee (\gamma_2 \vee \gamma_3)$, γ to $\mathbf{x} \vee \gamma$ and $\gamma \vee \bar{\gamma}$ to $\mathbf{x}$. The integral of $\mathcal{K}\omega$ on these homotopies vanishes. Indeed, the integrand of

$$\mathcal{K}\omega(\sigma)_t(1) = \int_0^1 \omega\left(\begin{pmatrix} s \\ t \end{pmatrix} \mapsto \sigma(t)(s)\right)_{\binom{s}{t}} (1)\ ds$$

itself vanishes, because the homotopy σ factorizes through a path, that is, $\sigma(t)(s) = \gamma'(\varphi(t,s))$ for some path γ' and some real function φ, but the pullback of a 2-form on $\mathbf{R}$ vanishes. Thus, $\mathbf{K}$ is a groupoid. Then, since the concatenation is smooth, as well as the inversion (art. 5.4, 2), and since $x \mapsto \text{class}_\omega(\mathbf{x})$ is clearly an induction, $\mathbf{K}$ is a diffeological groupoid. Moreover, since ends : Paths(X) $\to$ X $\times$ X is a subduction and class_ω : Paths(X) $\to \mathcal{Y}$ is smooth, the factorization $\pi : \mathcal{Y}_\omega \to X \times X$ is a subduction (art. 1.51). Therefore, $\mathbf{K}$ is a fibrating groupoid (art. 8.4), and then $Y = \mathcal{Y}_x$ is the principal fiber bundle attached at the point x (art. 8.14).

Let us examine now the structure group of $\mathbf{K}$ at a point $x \in X$. By construction, $\mathbf{K}_x = \text{Mor}_{\mathbf{K}}(x, x)$ is the quotient Loops$(X, x)/\phi$, but the restricted cocycle $\Phi \restriction$ Loops(X, x) is trivial. Indeed, for any pair of loops ℓ and ℓ' of X based at x,

$$\Phi(\ell, \ell') = F(\ell) - F(\ell') \quad \text{with} \quad F(\ell) = \text{class}\left(\int_{\mathbf{x}}^{\ell} \mathcal{K}\omega\right) \in T_\omega.$$

Therefore, $\Phi(\ell, \ell') = 0$ if and only if $F(\ell) = F(\ell')$. Hence, set theoretically, F : Loops$(X, x) \to T_\omega$ identifies val(F) with $\mathbf{K}_x$. Let us show first that val(F) can be *a priori* either $\{0\}$ or T_ω, but since $\omega \neq 0$, then val(F) $= T_\omega$. The map F is the projection, on Loops(X, x), of the smooth map

$$F : \text{Paths}(\text{Loops}(X, x), \mathbf{x}, \star) \to \mathbf{R} \quad \text{with} \quad F(\sigma) = \int_\sigma \mathcal{K}\omega.$$

Let σ and σ' be two elements of Paths(Loops$(X, x), \mathbf{x}, \star)$, connecting $\mathbf{x}$ to ℓ and $\mathbf{x}$ to ℓ'. Let $\hat{\sigma}$ defined by $\hat{\sigma}_t(s) = \sigma_t(1 - s)$, hence $\hat{\sigma}_t(s)$ is a path in Loops(X, x) connecting $\mathbf{x}$ to the reverse $\bar{\ell}$. Thanks to (art. 8.41), and to the fact that $\mathcal{K}\omega(\hat{\sigma}) = -\mathcal{K}\omega(\sigma)$, we get

$$F(\sigma * \sigma') = F(\sigma) + F(\sigma') \quad \text{and} \quad F(\hat{\sigma}) = -F(\sigma). \tag{$*$}$$

Thus, val(F) is a subgroup of $\mathbf{R}$, but val(F) is connected because Loops(X, x) is connected (by hypothesis) and F is smooth. Hence, either val(F) $= \{0\}$ or val(F) $= \mathbf{R}$, that gives val(F) $= \{0\}$ or val(F) $= T_\omega$. Let us examine the two different cases.

The zero case, val(F) $= \{0\}$. Then, $\mathbf{K}$ is trivial and $\omega = 0$. Indeed, in this condition the groupoid $\mathbf{K}_x$ is reduced to the identity $\{1_x\}$, and the principal fiber bundle attached to the point x (art. 8.14) is just the identity map, $Y = X$ and $\pi = \mathbf{1}_X$: $X \to X$. Thus, ω is exact $\omega = d\lambda$. But returning to $\mathcal{K}\omega$, we have $\mathcal{K}\omega = \mathcal{K}(d\lambda) =$

$\hat{1}^*(\lambda) - \hat{0}^*(\lambda) - d(\mathcal{K}\lambda)$, then $\mathcal{K}\omega \restriction \mathrm{Loops}(X,x) = d(\mathcal{K}\lambda)$ and the condition $F = 0$ writes, for all $\ell \in \mathrm{Loops}(X,x)$,

$$\int_x^\ell d\mathcal{K}\lambda = \mathcal{K}\lambda(\ell) = \int_\ell \lambda = 0.$$

Therefore, λ is a 1-form vanishing on every loop, thus λ is closed and $\omega = 0$; see Exercise 118, p. 202, and this is not permitted by the hypothesis.

The full case, $\mathrm{val}(F) = T_\omega$. Then, $\omega \neq 0$ and $F : \mathrm{Loops}(X,x) \to T_\omega$ is a subduction projecting to a smooth isomorphism $j : \mathbf{K}_x \to T_\omega$. Indeed, F is already a smooth surjection, let us check that F is a subduction, what will imply that F itself is a subduction, and therefore that its projection $j : \mathbf{K}_x \to T_\omega$ is a smooth isomorphism. Let us choose any $\sigma \in \mathrm{Paths}(\mathrm{Loops}(X,x), \mathbf{x}, \star)$ such that $\mathrm{F}(\sigma) \neq 0$, and let $\sigma_s(t) = \sigma(st)$. Then, $\varphi : s \mapsto \mathrm{F}(\sigma_s)$ is a smooth parametrization such that $\varphi(0) = \mathrm{F}(\mathbf{x}) = 0$ and $\varphi(1) = \mathrm{F}(\sigma) \neq 0$. Thus, there exists $s_0 \in]0,1[$ such that $\varphi'(0) \neq 0$. Let $\mathfrak{S}(s) = \hat{\sigma}_{s_0} * \sigma_{s+s_0}$, where $\hat{\sigma}_{s_0}(t) : t' \mapsto \sigma_{s_0}(t)(1-t')$, and $\psi(s) = \mathrm{F}(\mathfrak{S}(s))$, we have $\mathrm{F}(\hat{\sigma}_{s_0} * \sigma_{s+s_0}) = \mathrm{F}(\sigma_{s+s_0}) - \mathrm{F}(\sigma_{s_0})$ (see above), and then $\psi(s) = \varphi(s+s_0) - \varphi(s_0)$. Thus, $\psi(0) = 0$ and $\psi'(0) \neq 0$. Therefore, there exists $\varepsilon > 0$ such that $\psi \restriction]-\varepsilon,+\varepsilon[$ is a local diffeomorphism by the implicit function theorem. Hence, $\mathfrak{S} \circ \psi^{-1} :]-\varepsilon,+\varepsilon[\to \mathrm{Paths}(\mathrm{Loops}(X,x), \mathbf{x}, \star)$ is a smooth local section of $\mathrm{F} : \mathrm{Paths}(\mathrm{Loops}(X,x), \mathbf{x}, \star) \to \mathbf{R}$. By additivity with respect to the $*$ operation, there exists a local smooth section of F everywhere and therefore F is a subduction. Then, by projection, F is itself a subduction, and the projection $j : \mathbf{K}_x \to T_\omega$, which is injective by construction, is a diffeomorphism and therefore a smooth isomorphism from $\mathbf{K}_x$ to T_ω. Therefore, $\mathbf{K}$ is a fibrating groupoid with structure group T_ω.

Next, the invariance of $\boldsymbol{\lambda}$, with respect to precomposition and postcomposition, is the translation on $\mathcal{Y}_\omega = \mathrm{Paths}(X)/\phi$ of the invariance of $\mathcal{K}\omega$ by preconcatenation and postconcatenation, proved in (art. 6.85). Considering the 1-form $\lambda = \boldsymbol{\lambda} \restriction Y$, the invariance of $\boldsymbol{\lambda}$ by precomposition and postcomposition writes in particular for $x' = x$, $\tau_Y^*(\lambda) = \lambda$, $R(y)^*(\lambda) = \theta$ and $\pi^*(\omega) = d\lambda$, for all τ in T_ω, where τ_Y denotes the action of τ on Y, that is, $\tau_Y(y) = j^{-1}(\tau) \cdot y$. Therefore, the 1-form λ is precisely a connection form on Y with curvature ω.

Let us consider now the special case $\omega = d\alpha$. First of all, $P_\omega = \{0\}$ and $T_\omega = \mathbf{R}$. Now let γ and γ' be two paths such that $\mathrm{ends}(\gamma) = \mathrm{ends}(\gamma')$, and σ a homotopy connecting γ' to γ in $\mathrm{Paths}(X,x,x')$. Then,

$$\Phi(\gamma,\gamma') = \int_\sigma \mathcal{K}\omega = \int_\sigma \mathcal{K}[d\alpha] = \int_\sigma (\hat{1}^* - \hat{0}^*)(\alpha) - \int_\sigma d[\mathcal{K}\alpha] = \mathcal{K}\alpha(\gamma') - \mathcal{K}\alpha(\gamma).$$

Thus, the cocycle Φ is trivial, and hence the principal bundle $\pi : Y \to X$ also is trivial (art. 8.40). Therefore, $\mathrm{pr}_1 : X \times \mathbf{R} \to X$, equipped with the connection $\lambda = \alpha \oplus dt$, is a representative of the integration structure.

Let us prove that this construction is essentially unique. Let $\pi' : Y' \to X$ be some T_ω-principal fiber bundle, equipped with a connection λ' of curvature $\omega \neq 0$. In (art. 8.40), we have seen that there exists a smooth map $\psi : \mathrm{Paths}(X,x) \to Y'$ such that $\psi^*(\lambda') = \mathcal{K}\omega$, and that Y' can be then reconstructed by quotient. Let us begin by proving that there exists a smooth injection $J : Y = \mathrm{Paths}(X,x)/\phi \to Y'$ such that $J \circ \mathrm{class}_\omega = \psi$, that is, $\psi(\gamma) = \psi(\gamma')$ if and only if $\mathrm{class}_\omega(\gamma) = \mathrm{class}_\omega(\gamma')$. First of all, $\psi(\gamma) = \psi(\gamma')$ implies $\gamma(1) = \gamma'(1)$, let $y' = \psi(\gamma) = \psi(\gamma')$ and $x' = \pi(y') = \gamma(1) = \gamma'(1)$. Then, since by hypothesis X is simply connected, there exists a smooth path σ in Paths(X, x, x') connecting γ to γ'. Hence,

$$\int_\sigma \mathcal{K}\omega = \int_\sigma \psi^*(\lambda) = \int_{\psi\circ\sigma} \lambda.$$

Since $\psi(\sigma(0)) = \psi(\gamma) = \psi(\gamma') = \psi(\sigma(1))$, the path $\psi \circ \sigma : t \mapsto y'_t$ is a loop in $Y_{x'} = \pi'^{-1}(x')$, based at y'. But, since π is a principal fibration, there exists a smooth loop τ in T_ω, based at the origin, such that $y'_t = \tau(t)_Y(y') = R(y')(\tau(t))$, where $R(y')$ denotes the orbit map of the point y'. But λ is a connection form, so $R(y')^*(\lambda) = \theta$, where θ is the canonical 1-form on T_ω, thus

$$\int_{\psi\circ\sigma} \lambda = \int_{R(y')\circ\tau} \lambda = \int_\tau R(y')^*(\lambda) = \int_\tau \theta = \int_0^p \frac{d}{ds}(t_s)\,ds = p,$$

where $s \mapsto t_s$ is a lift in $\mathbf{R}$ of τ, with $t_0 = 0$ and $t_1 = p$. But, since τ is a loop, $p \in P_\omega$, thus $\Phi(\gamma, \gamma') = 0$, that is, by definition, $\text{class}_\omega(\gamma) = \text{class}_\omega(\gamma')$. Conversely, if $\text{class}_\omega(\gamma) = \text{class}_\omega(\gamma')$, then the same computation shows that $\psi(\gamma) = \psi(\gamma')$. Therefore, J is well defined and is a smooth injection.

Now, let us consider the restriction of ψ to Loops(X, x) with values in $Y'_x = \pi'^{-1}(x)$. Let us identify the fiber Y'_x with T_ω, $\psi(x)$ then becoming the identity. By a computation analogous to the previous one and thanks to (art. 8.41), we get $\psi(\ell \vee \ell') = \psi(\ell) + \psi(\ell')$ and $\psi(\bar{\ell}) = -\psi(\ell)$, where $\bar{\ell}$ is the reverse of ℓ. Thus, $\psi(\text{Loops}(X, x))$ is a connected subgroup of T_ω, that is, either T_ω or $\{0\}$. If $\psi(\text{Loops}(X, x)) = T_\omega$, then the injection J is surjective, thus bijective, and moreover an equivariant diffeomorphism. The integration structures (π, λ) and (π', λ') are equivalent. Now, if $\psi(\text{Loops}(X, x)) = \{0\}$, we note first that $\psi(\text{Loops}(X, x))$ gives exactly the holonomy of the connection λ' defined in (art. 8.35), and we know that the fibration is reducible to its holonomy. Thus, the fibration is reduced to a covering, but since X is simply connected the covering is trivial. Hence, $\omega = d\alpha$ $P_\omega = \{0\}$, $T_\omega = \mathbf{R}$ and $\pi' : Y' \mapsto X$, equipped with the connection λ', is equivalent to $\text{pr}_1 : X \times \mathbf{R} \to X$ equipped with the connection $\lambda = \alpha \oplus dt$. But this is also a representative of the standard integration structure for an exact 2-form. Therefore, there exists essentially only one integration structure for the closed 2-form ω. ☐

Exercises

✎ Exercise 140 (Loops in the torus). Consider the torus $T^2 = \mathbf{R}^2/\mathbf{Z}^2$, and let class : $\mathbf{R}^2 \to T^2$ be the projection. Let $\omega = \text{class}_*(\mathbf{e}^1 \wedge \mathbf{e}^2)$ be the canonical volume on T^2. Let $\ell : t \mapsto \text{class}(t, 0) \in \text{Loops}(T^2)$, use the 1-form $\mathcal{K}\omega$ to show that the connected component $\text{comp}(\ell) \subset \text{Loops}(T^2)$ is not simply connected.

✎ Exercise 141 (Periods of a surface). Consider the torus $T^2 = \mathbf{R}^2/\mathbf{Z}^2$. Justify the fact that every connected component of Loops(T^2) is the component of a loop $\ell_{n,m} : t \mapsto \text{class}(nt, mt)$, for some $(n, m) \in \mathbf{Z}^2$. Show that every loop $\sigma : s \mapsto \sigma_s$ in $\text{comp}(\ell_{n,m})$, based at $\ell_{n,m}$, is fixed-ends homotopic to a loop $s \mapsto \sigma'_s = [t \mapsto \text{class}(nt + ks, mt + k's)]$, for some $(k, k') \in \mathbf{Z}^2$. Let ω be the 2-form defined on T^2 in Exercise 140, p. 297, show that the periods of $\mathcal{K}\omega$ on the component of $\ell_{n,m}$ are the group

$$\text{Periods}(\mathcal{K}\omega \upharpoonright \text{comp}(\ell_{n,m})) = \{nk' - mk \mid k, k' \in \mathbf{Z}\}.$$

Comment on the result.

CHAPTER 9

Symplectic Diffeology

The generalization of classical symplectic geometry to diffeology needs first an appropriate extension of the classical notion of moment map.[1] We know already that diffeology is suitable for describing, in a unique and satisfactory way, manifolds or infinite dimensional spaces, as well as singular quotients. But, if diffeology excels with covariant objects, like differential forms, it is more subtle when it is a question of contravariant objects like vector fields, Lie algebra,[2] kernels, etc. Thus, in order to build a good diffeological theory of the moment map and to avoid useless debates, we need to get rid from everything related to contravariant geometrical objects.

Actually, the notion of moment map is not really an object of the symplectic world, but relates more generally to the category of spaces equipped with closed 2-forms. The nondegeneracy condition is secondary and can be first skipped from the data. This has been underlined explicitly by Souriau in his symplectic formulation of Noether's theorem, which involves presymplectic manifolds. On symplectic manifolds Noether's theorem is silent.[3] The moment map is just an object of the world of differential closed forms, and there is no reason *a priori* that it could not be extended to diffeology which offers a pretty well developed framework for Cartan-De Rham calculus.

In order to generalize the moment map in diffeology, we need to understand its meaning, and this meaning lies in the following simplest possible case. Let M be a manifold equipped with a closed 2-form ω. Let G be a Lie group acting smoothly on M and preserving ω, that is, $g_M^*(\omega) = \omega$ for all elements g of G, where g_M denotes the action of g on M. Let us assume that ω is exact, $\omega = d\lambda$, and moreover that λ also is invariant by the action of G. Then, for every point m of M, the pullback of λ by the orbit map $\hat{m} : g \mapsto g_M(m)$ is a left-invariant 1-form of G, that is, an element of the dual of the Lie algebra $\mathcal{G}^*$. The map $\mu : m \mapsto \hat{m}^*(\lambda)$ is exactly the moment map of the action of G on the pair (M, ω)—at least one of the moment maps, since they are defined up to constants. As we can see, this construction does not involve really the Lie algebra of G but the space $\mathcal{G}^*$ of left-invariant 1-forms on G. Since this space is well defined in diffeology, we just have to replace "manifold" by "diffeological space", and "Lie group" by "diffeological group", and everything works the same way. Thus, let us change the manifold M for a diffeological space[4]

[1]The notion of moment map in the framework of classical symplectic geometry was originally introduced in the early 1970s by Souriau; see [**Sou70**].

[2]Several authors, beginning with Souriau, proposed some generalizations of Lie algebra in diffeology. But it does not seem to exist a unique good choice. Such generalizations rely actually on the kind of problem treated.

[3]Noether's theorem states that the moment map is constant on the characteristics of the 2-form; if the form is nondegenerate, then the characteristics are trivial.

[4]The space X will be assumed to be connected, as many results need this hypothesis.

X, and let G be some diffeological group. Let us continue to denote the space of left-invariant 1-forms on G by $\mathcal{G}^*$, even if the star does not refer *a priori* to some duality, and let us simply call it the *space of momenta* of the group G. Note that the group G continues to act on $\mathcal{G}^*$ by pullback of its adjoint action $\mathrm{Ad} : (g, k) \mapsto gkg^{-1}$, so we do not lose the notions of coadjoint action and coadjoint orbits.

Next, if we got the good space of momenta, which is the space where the moment maps are assumed to take their values, the problem remains that not every G-invariant closed 2-form is exact. And moreover, even if such form is exact, there is no reason for some of its primitives to be G-invariant. We shall pass over this difficulty by introducing an intermediary, on which we can realize the simple case described above. This intermediary is the space Paths(X), of all the smooth paths in X, where the group G acts naturally by composition. And since Paths(X) carries a natural functional diffeology, it is legitimate to consider its differential forms, and this is what we do. By integrating ω along the paths, we get a differential 1-form defined on Paths(X), and invariant by the action of G. The exact tool used here is the Chain-Homotopy Operator $\mathcal{K}$. The 1-form $\Lambda = \mathcal{K}\omega$, defined on Paths(X), is a G-invariant primitive of the 2-form $\Omega = (\hat{1}^* - \hat{0}^*)(\omega)$, where $\hat{1}$ and $\hat{0}$ map every path in X to its ends. Thus, thanks to the construction described above, we get a moment map Ψ for the 2-form $\Omega = d\Lambda$ and the action of G on Paths(X). But this *paths moment map* Ψ is not the one we are waiting for. We need to push it down on X, or rather on $X \times X$. Now, if we get this way a *2-points moment map* ψ well defined on $X \times X$, it no longer takes its value in $\mathcal{G}^*$, as Ψ does, but in the quotient $\mathcal{G}^*/\Gamma$, where Γ is the image by Ψ of all the loops in X. Fortunately, $\Gamma = \Psi(\mathrm{Loops}(X))$ is a subgroup of $(\mathcal{G}^*, +)$ and depends on the loops only through their free homotopy classes. In other words, Γ is a homomorphic image of the Abelianized fundamental group $\pi_1^{\mathrm{Ab}}(X)$ of X. Well, it is not a big deal to have the moment map taking its values in some quotient of the space of momenta, we can live with that, especially if the group Γ is invariant under the coadjoint action of G, which is actually the case.[5] But we are not completely done: the usual moment map is not a 2-points function, but a 1-point function. Hence, we have to extract our usual moment maps from this 2-points function ψ. This is fairly easy, thanks to its very definition, the moment map Ψ satisfies an additive property for the concatenation of paths, and the moment map ψ inherits this property as a cocycle condition: for any three points x, x' and x'' of X we have $\psi(x, x') + \psi(x', x'') = \psi(x, x'')$. Therefore, for X connected, there exists always a map μ such that $\psi(x, x') = \mu(x') - \mu(x)$, and any two such maps differ just by a constant. We get finally our wanted *moment maps* μ, defined in the diffeological framework. The only difference, with the simplest case described above, is that a moment map takes its values in some quotient of the space of momenta, instead the space of momenta itself. But this is in fact already the case in the classical theory. It does not appear explicitly because people focus more on Hamiltonian actions than just on symplectic actions. Actually, the group Γ represents the very obstruction, for the action of G on (X, ω), to be *Hamiltonian*. We shall call Γ the *holonomy* of the action of G.

Now, let us come back to some properties of the various moment maps introduced above. The paths moment map Ψ and its projection ψ are equivariant with respect to the action of G on X and the coadjoint action of G on $\mathcal{G}^*$, or the projection

[5] More precisely, the elements of Γ are not just elements of $\mathcal{G}^*$ but are moreover closed, and therefore invariant, each of them, by the coadjoint action of G.

of the coadjoint action on $\mathcal{G}^*/\Gamma$. But this is no longer the case for the moment maps μ. The variance of the maps μ reveals a family of cocycles θ from G to $\mathcal{G}^*/\Gamma$ differing just by coboundaries, and generalizing the *Souriau cocycles* [**Sou70**]. Their common class σ belongs to the cohomology group $H^1(G, \mathcal{G}^*/\Gamma)$, and will be called the *Souriau class* of the action of G of (X, ω). The Souriau class σ is precisely the obstruction for the 2-points moment map ψ to be exact, that is, for some moment map μ to be equivariant. Actually, σ is just the pullback, at the group level, of the class of ψ regarded as a cocycle. In parallel with the classical situation, every Souriau cocycle θ defines a new action of G on $\mathcal{G}^*/\Gamma$, which we still call the affine coadjoint action (associated with θ). And the image of a moment map μ is a collection of coadjoint orbits for this action. We call these orbits the (Γ, θ)-coadjoint orbits of G. Two different cocycles give two families of orbits translated by the same constant.

Let us remark that the holonomy group Γ and the Souriau class σ appear clearly on a different level of meaning: the first one is responsible for the non-Hamiltonian character of the action of G and the second characterizes the lack of equivariance of the moment maps.

Well, until now we did not use all the facilities offered by the diffeological framework. Since we do not restrict ourselves to the category of Lie groups, nothing prevents us from considering the group of all the *automorphisms* of the pair (X, ω), that is, the group $\mathrm{Diff}(X, \omega)$ of all the diffeomorphisms of X preserving ω. This group is a natural diffeological group, acting smoothly on X. Thus, everything built above applies to $\mathrm{Diff}(X, \omega)$, and every other action preserving ω, of any diffeological group, passes through $\mathrm{Diff}(X, \omega)$, and through the associated object of the theory developed here. Therefore, considering the whole group of automorphisms of the closed 2-form ω of X, we get a natural notion of universal moment maps Ψ_ω, ψ_ω and μ_ω, universal holonomy Γ_ω, universal Souriau cocycles θ_ω, and universal Souriau class σ_ω. By the way, this universal construction suggests a simple and new characterization, for any diffeological space X equipped with a closed 2-form ω, of the group of *Hamiltonian diffeomorphisms* $\mathrm{Ham}(X, \omega)$, as the largest connected subgroup of $\mathrm{Diff}(X, \omega)$ whose holonomy vanishes.

It is interesting to notice that, contrary to the original constructions [**Sou70**] and most of their generalizations, the theory described above is essentially global, more or less algebraic, does not refer to any differential or partial differential, equation, and does not involve any notion of vector field or functional analysis techniques.

Considering the classical case of a closed 2-form ω defined on a manifold M, we show in particular that ω is nondegenerate if and only if the group $\mathrm{Diff}(M, \omega)$ is transitive on M and if a universal moment map μ_ω is injective. In other words, symplectic manifolds are identified, by the universal moment maps, with some coadjoint orbits—in our general sense—of their group of symplectomorphisms. This idea that *every symplectic manifold is a coadjoint orbit* is not new, it is actually very natural and suggested by a well known classification theorem for symplectic homogeneous Lie group actions [**Kir76**], [**Kos70**], [**Sou70**]. This has been stated already in a different context, for example in [**Omo86**]. But the real question is, How can we make this statement rigorous at a good price, without involving the heavy functional analysis apparatus? This is what brings diffeology.

The examples and exercises at the end of this chapter show several situations involving diffeological groups which are not Lie groups, or involving diffeological spaces which are not manifolds. We can see, for example, the general theory applying meaningfully to the singular *symplectic irrational tori* for which topology is irrelevant. These general constructions of moment maps are also applied to a few examples in infinite dimension, and also when finite and infinite dimensions are mixed. Finally, two exercises on orbifolds exhibit a strong difference between classical symplectic geometry and what we expect from its diffeological counterpart. These examples show without any doubt the ability of this theory to treat correctly, in a unique framework, avoiding heuristic arguments, the large variety of situations we can find in the mathematical literature today. Infinite dimensional heuristic examples can be found for instance in [**Dnl99**]. The solutions of some of the exercises need tedious computations, which just shows diffeology at work in this particular field.

In conclusion, besides the point that the construction developed in this chapter is a first step in the elaboration of the *symplectic diffeology program*, I would emphasize the fact that, since {Manifolds} is a full subcategory of {Diffeology}, all the constructions developed here apply to manifolds and give a faithful description of the classical theory of moment maps. As we have seen, there is no mention, and no use, of Lie algebra or vector fields in this exposé. This reveals the fact that these objects also are useless in the traditional approach, and can be avoided. And, I would add, they should be avoided. Not just because they can then be extended to larger categories, but because the use of contravariant objects hides the deep fact that the theory of moment maps is a pure covariant theory. Let us take an example. We know that since coadjoint orbits of Lie groups are symplectic they are even dimensional. This is often regarded as a miracle, since it is not necessarily the case for adjoint orbits. But if we think that the Lie algebra has little to do with the space of momenta of a Lie group, there is no more miracle, just different behaviors for different objects, which is unsurprising. Moreover I would add, but this can appear as more or less subjective, that avoiding all this *va-et-vient* between Lie algebra and dual of Lie algebra, the diffeological approach of the moment maps is much simpler, and even deeper, than the classical approach. Compare for example the Souriau cocycle constructions in the original "Structure des systèmes dynamiques" [**Sou70**] and in diffeology. The only crucial property used here is connectedness, that is, the existence of enough smooth paths connecting points in spaces. Finally, it goes without saying that the diffeological approach respects scrupulously the principle of minimality required by mathematics.

The Paths Moment Map

We shall introduce the various flavors of moment map in diffeology step by step. The first step consists, in this section, in defining the paths moment map.

9.1. Definition of the paths moment map. Let X be a diffeological space, and let ω be a closed 2-form defined on X. Let G be a diffeological group, and let $\rho : \mathrm{G} \to \mathrm{Diff}(\mathrm{X})$ be a smooth action. Let us denote by the same letter the natural action of G on $\mathrm{Paths}(\mathrm{X})$, induced by the action ρ of G on X, that is, for all $g \in \mathrm{G}$, for all $p \in \mathrm{Paths}(\mathrm{X})$,

$$\rho(g)(p) = \rho(g) \circ p = [t \mapsto \rho(g)(p(t))].$$

Let us assume now that the action ρ of G on X preserves ω, that is, for all $g \in G$,

$$\rho(g)^*(\omega) = \omega \quad \text{or} \quad \rho \in \mathrm{Hom}^\infty(G, \mathrm{Diff}(X, \omega)).$$

Let $\mathcal{K}$ be the Chain-Homotopy Operator (art. 6.83), so $\mathcal{K}\omega$ is a 1-form on $\mathrm{Paths}(X)$, and the action of G on $\mathrm{Paths}(X)$ preserves $\mathcal{K}\omega$. This is a consequence of the variance of the Chain-Homotopy Operator (art. 6.84). Thus, for all $g \in G$,

$$\rho(g)^*(\mathcal{K}\omega) = \mathcal{K}\omega.$$

Now, let p be a path in X, and let $\hat{p} : G \to \mathrm{Paths}(X)$ be the orbit map, $\hat{p}(g) = \rho(g) \circ p$. Then, the pullback $\hat{p}^*(\mathcal{K}\omega)$ is a left-invariant 1-form of G, that is, an element of $\mathcal{G}^*$. The map

$$\Psi : \mathrm{Paths}(X) \to \mathcal{G}^*, \text{ defined by } \Psi(p) = \hat{p}^*(\mathcal{K}\omega),$$

is smooth with respect to the functional diffeology, $\Psi \in \mathcal{C}^\infty(\mathrm{Paths}(X), \mathcal{G}^*)$. The map Ψ will be called the *paths moment map*.

9.2. Evaluation of the paths moment map. Let X be a diffeological space and ω be a closed 2-form defined on X. Let G be a diffeological group, and let ρ be a smooth action of G on X, preserving ω. Let p be a path in X. Thanks to the explicit expression of the Chain-Homotopy Operator (art. 6.83), we get the evaluation of the momentum $\Psi(p)$ on any n-plot P of G,

$$\Psi(p)(P)_r(\delta r) = \int_0^1 \omega \left[\begin{pmatrix} s \\ u \end{pmatrix} \mapsto (\rho \circ P)(u)(p(s+t)) \right]_{\left(\substack{s=0 \\ u=r} \right)} \begin{pmatrix} 1 \\ 0 \end{pmatrix} \begin{pmatrix} 0 \\ \delta r \end{pmatrix} dt, \qquad (\heartsuit)$$

for all r in $\mathrm{def}(P)$ and all δr in $\mathbf{R}^n$. Now, as a differential 1-form, $\Psi(p)$ is characterized by its values on the 1-plots (art. 6.37). Then, let $f : t \mapsto f_t$ be a 1-plot of G centered at the identity $\mathbf{1}_G$, that is, $f \in \mathrm{Paths}(G)$ and $f(0) = \mathbf{1}_G$. For every $t \in \mathbf{R}$, let F_t be the path in $\mathrm{Diff}(X, \omega)$—centered at the identity $\mathbf{1}_X$—defined by

$$F_t : s \mapsto \rho(f_t^{-1} \circ f_{t+s}).$$

We have then

$$\Psi(p)(f)_t(1) = -\int_p i_{F_t}(\omega) = -\int_0^1 i_{F_t}(\omega)(p)_s(1)ds, \qquad (\clubsuit)$$

where $i_{F_t}(\omega)$ is the contraction of ω by F_t (art. 6.56). But, as an invariant 1-form on G, the moment $\Psi(p)$ is characterized by its *value at the identity*, for $t = 0$,

$$\Psi(p)(f)_0(1) = -\int_p i_F(\omega) = -\int_0^1 i_F(\omega)(p)_t(1) \, dt \quad \text{with} \quad F = \rho \circ f. \qquad (\diamondsuit)$$

NOTE. Let $f \in \mathrm{Hom}^\infty(\mathbf{R}, G)$, then $\Psi(p)(f)$ is an invariant 1-form on $\mathbf{R}$ whose coefficient is just $\int_p i_F(\omega)$, that is,

$$\Psi(p)(f) = h_f(p) \times dt \quad \text{where} \quad h_f(p) = -\int_p i_F(\omega).$$

The smooth map $h_f : \mathrm{Paths}(X) \to \mathbf{R}$ is the *Hamiltonian* of f, or the Hamiltonian of the 1-parameter group $f(\mathbf{R})$. Also note that the map $h : \mathrm{Hom}^\infty(\mathbf{R}, G) \to \mathcal{C}^\infty(\mathrm{Paths}(X), \mathbf{R})$, defined above, is smooth.

PROOF. Let us prove (♡). Let us recall that for every $p \in \mathrm{Paths}(X)$ and every $g \in G$, $\hat{p}(g) = \rho(g) \circ p = [t \mapsto \rho(g)(p(t))]$. Thus, by definition,

$$\begin{aligned}
\Psi(p)(P)_r(\delta r) &= \hat{p}^*(\mathcal{K}\omega)(P)_r(\delta r) \\
&= \mathcal{K}\omega(\hat{p} \circ P)_r(\delta r) \\
&= \int_0^1 \omega\left[\begin{pmatrix} s \\ r \end{pmatrix} \mapsto \hat{p} \circ P(r)(s+t)\right]_{\binom{0}{r}} \begin{pmatrix} 1 \\ 0 \end{pmatrix} \begin{pmatrix} 0 \\ \delta r \end{pmatrix} dt \\
&= \int_0^1 \omega\left[\begin{pmatrix} s \\ r \end{pmatrix} \mapsto (\rho \circ P)(r)(p(s+t))\right]_{\binom{0}{r}} \begin{pmatrix} 1 \\ 0 \end{pmatrix} \begin{pmatrix} 0 \\ \delta r \end{pmatrix} dt.
\end{aligned}$$

Let us prove (♣). Let us apply the general formula (♡) for $P = f$. Introducing $u' = u - t$ and $s'' = s + s'$, using the compatibility property of $\omega(P \circ Q) = Q^*(\omega(P))$ and the $\rho(f_t)$ invariance of ω, we get

$$\begin{aligned}
\Psi(p)(f)_t(1) &= \int_0^1 \omega\left[\begin{pmatrix} s \\ u \end{pmatrix} \mapsto \rho(f_u)(p(s+s'))\right]_{\binom{s=0}{u=t}} \begin{pmatrix} 1 \\ 0 \end{pmatrix} \begin{pmatrix} 0 \\ 1 \end{pmatrix} ds' \\
&= \int_0^1 \omega\left[\begin{pmatrix} s'' \\ u' \end{pmatrix} \mapsto \rho(f_{t+u'})(p(s''))\right]_{\binom{s''=s'}{u'=0}} \begin{pmatrix} 1 \\ 0 \end{pmatrix} \begin{pmatrix} 0 \\ 1 \end{pmatrix} ds' \\
&= \int_0^1 \omega\left[\begin{pmatrix} s'' \\ u' \end{pmatrix} \mapsto \rho(f_t \circ f_t^{-1} \circ f_{t+u'})(p(s''))\right]_{\binom{s''=s'}{u'=0}} \begin{pmatrix} 1 \\ 0 \end{pmatrix} \begin{pmatrix} 0 \\ 1 \end{pmatrix} ds' \\
&= \int_0^1 \omega\left[\begin{pmatrix} s'' \\ u' \end{pmatrix} \mapsto \rho(f_t)\Big(F_t(u')(p(s''))\Big)\right]_{\binom{s''=s'}{u'=0}} \begin{pmatrix} 1 \\ 0 \end{pmatrix} \begin{pmatrix} 0 \\ 1 \end{pmatrix} ds' \\
&= \int_0^1 \omega\left[\begin{pmatrix} s'' \\ u' \end{pmatrix} \mapsto F_t(u')(p(s''))\right]_{\binom{s''=s'}{u'=0}} \begin{pmatrix} 1 \\ 0 \end{pmatrix} \begin{pmatrix} 0 \\ 1 \end{pmatrix} ds' \\
&= \int_0^1 \omega\left[\begin{pmatrix} u' \\ s'' \end{pmatrix} \mapsto F_t(u')(p(s''))\right]_{\binom{u'=0}{s''=s'}} \begin{pmatrix} 0 \\ 1 \end{pmatrix} \begin{pmatrix} 1 \\ 0 \end{pmatrix} ds' \\
&= -\int_0^1 \omega\left[\begin{pmatrix} u' \\ s'' \end{pmatrix} \mapsto F_t(u')(p(s''))\right]_{\binom{u'=0}{s''=s'}} \begin{pmatrix} 1 \\ 0 \end{pmatrix} \begin{pmatrix} 0 \\ 1 \end{pmatrix} ds' \\
&= -\int_0^1 i_{F_t}(\omega)(p)_{s'}(1)ds' \\
&= -\int_p i_{F_t}(\omega).
\end{aligned}$$

Let us prove the Note. Let $f \in \mathrm{Hom}^\infty(\mathbf{R}, G)$. By definition of differential forms and pullbacks, $\Psi(p)(f) = f^*(\Psi(p))$, but since f is a homomorphism from $\mathbf{R}$ to $\mathrm{Diff}(X, \omega)$ and $\Psi(p)$ is a left-invariant 1-form on $\mathrm{Diff}(X, \omega)$, $f^*(\Psi(p))$ is an invariant 1-form of $\mathbf{R}$, then $\Psi(p)(f) = f^*(\Psi(p)) = a \times dt$, for some real a. Thus, $\Psi(p)(f)_r = \Psi(p)(f)_0(1) \times dt = h_f(p) \times dt$, with $h_f(p) = \Psi(p)(f)_0(1) = -\int_p i_F(\omega)$, where dt is the canonical 1-form on $\mathbf{R}$. ☐

9.3. Variance of the paths moment map. Let X be a diffeological space, and let ω be a closed 2-form defined on X. Let G be a diffeological group and ρ be a smooth action of G on X, preserving ω. The paths moment map Ψ, defined in (art. 9.1), is equivariant under the action of G, that is, for all $g \in G$,

$$\Psi \circ \rho(g) = \mathrm{Ad}(g)_* \circ \Psi.$$

PROOF. Let us denote here the orbit map $\hat{p}$ of $p \in \mathrm{Paths}(X)$ by $R(p)$, that is, $R(p)(g) = \rho(g) \circ p$ and $\Psi(p) = R(p)^*(\mathcal{K}\omega)$. Then, $\Psi(\rho(g)(p)) = \Psi(\rho(g) \circ p) = R(\rho(g) \circ p)^*(\mathcal{K}\omega)$. But $R(\rho(g) \circ p)(g') = \rho(g')(\rho(g) \circ p) = \rho(g') \circ \rho(g) \circ p = \rho(g'g) \circ p = R(p)(g'g) = R(p) \circ R(g)(g')$, thus $R(\rho(g) \circ p) = R(p) \circ R(g)$, and $\Psi(\rho(g)(p)) = (R(p) \circ R(g))^*(\mathcal{K}\omega) = R(g)^*(R(p)^*(\mathcal{K}\omega)) = R(g)^*(\Psi(p))$. But since $\Psi(p)$ is left-invariant, $R(g)^*(\Psi(p)) = \mathrm{Ad}(g)_*(\Psi(p))$. □

9.4. Additivity of the paths moment map. Let X be a diffeological space and ω be a closed 2-form defined on X. Let G be a diffeological group and ρ be a smooth action of G on X, preserving ω. The paths moment map Ψ, defined in (art. 9.1), satisfies the following additive property,

$$\Psi(p \vee p') = \Psi(p) + \Psi(p') \text{ and } \Psi(\bar{p}) = -\Psi(p), \text{ with } \bar{p}(t) = p(1-t),$$

for any two juxtaposable paths p and p' in X.

PROOF. This is a direct application of the expression given in (art. 9.2, (♢)), and of the additivity of the integral of differential forms on paths. □

9.5. Differential of the paths moment map. Let X be a diffeological space, and let ω be a closed 2-form defined on X. Let G be a diffeological group, and let ρ be a smooth action of G on X, preserving ω. Let p be a path in X. Then, the exterior derivative of the paths momentum $\Psi(p)$ is given by

$$d(\Psi(p)) = \hat{x}_1^*(\omega) - \hat{x}_0^*(\omega),$$

where $x_0 = p(0)$ and $x_1 = p(1)$, and the $\hat{x}_i$ denote the orbit maps.

PROOF. This is a direct application of the main property of the Chain-Homotopy Operator, $d \circ \mathcal{K} + \mathcal{K} \circ d = \hat{1}^* - \hat{0}^*$. Since $d\omega = 0$, $d(\mathcal{K}\omega) = \hat{1}^*(\omega) - \hat{0}^*(\omega)$, composed with $\hat{p}^*$ we get $\hat{p}^* \circ d(\mathcal{K}\omega) = \hat{p}^* \circ \hat{1}^*(\omega) - \hat{p}^* \circ \hat{0}^*(\omega)$, that is, $d(\hat{p}^*(\mathcal{K}\omega)) = (\hat{1} \circ \hat{p})^*(\omega) - (\hat{0} \circ \hat{p})^*(\omega)$. Thus, $d(\Psi(p)) = \hat{x}_1^*(\omega) - \hat{x}_0^*(\omega)$. □

9.6. Homotopic invariance of the paths moment map. Let X be a diffeological space and ω be a closed 2-form defined on X. Let G be a diffeological group, and let ρ be a smooth action of G on X, preserving ω. Let p_0 and p_1 be any two paths in X. If p_0 and p_1 are fixed-ends homotopic, then $\Psi(p_0) = \Psi(p_1)$. In other words, Ψ passes on the Poincaré groupoid of X (art. 5.15).

PROOF. Let $s \mapsto p_s$ be a fixed-ends homotopy connecting p_0 to p_1, for example let $p_s(0) = x_0$ and $p_s(1) = x_1$, for all s. Let f be a 1-plot of G centered at the identity 1_G, that is, $f(0) = 1_G$, and let $F = \rho \circ f$. We use the fact that the moment of paths is characterized by its value at the identity, $\Psi(p_s)(f)_0(1) = -\int_{p_s} i_F(\omega)$; see (art. 9.2, (♢)). Let us differentiate this equality with respect to s,

$$\frac{\partial}{\partial s}\Big(\Psi(p_s)(f)_0(1)\Big) = -\delta \int_{p_s} i_F(\omega), \quad \text{with} \quad \delta = \frac{\partial}{\partial s}.$$

The variation of the integral of differential forms on cubes (art. 6.70) gives

$$\begin{aligned}\delta\int_{p_s} i_F(\omega) &= \int_0^1 d\,[i_F(\omega)](\delta p_s) + \int_0^1 d[i_F(\omega)(\delta p_s)] \\ &= \int_0^1 d\,[i_F(\omega)](\delta p_s) + \Big[i_F(\omega)(\delta p_s)\Big]_0^1.\end{aligned}$$

The second summand of the right term vanishes because δp_s vanishes at the ends: $p_s(0) = \text{cst}$ and $p_s(1) = \text{cst}$. Now, thanks to the Cartan formula (art. 6.72), $d[i_F(\omega)] = \pounds_F(\omega) - i_F(d\omega)$. But ω is invariant under the action of G, thus $\pounds_F(\omega) = 0$, and $d\omega = 0$, so $d[i_F(\omega)] = 0$. Thus, $\delta\int_{p_s} i_F(\omega) = 0$ and therefore $\Psi(p_0) = \Psi(p_s) = \Psi(p_1)$, for all s. □

9.7. The holonomy group. Let X be a connected diffeological space, and let ω be a closed 2-form defined on X. Let G be a diffeological group, and let ρ be a smooth action of G on X, preserving ω. Let Ψ be the paths moment map (art. 9.1). We define the *holonomy* Γ of the action ρ by

$$\Gamma = \{\Psi(\ell) \mid \ell \in \text{Loops}(X)\}.$$

1. The holonomy Γ is an additive subgroup of the subspace of closed momenta, $\Gamma \subset Z$ (art. 7.17), that is,

   $$d\gamma = 0 \quad \text{and} \quad \gamma - \gamma' \in \Gamma,$$

   for any two elements γ and γ' in Γ.
2. The paths moment map Ψ, restricted to Loops(X), factorizes through a homomorphism from $\pi_1(X)$ to $\mathcal{G}^*$. Thus, Γ is a homomorphic image of the Abelianized fundamental group $\pi_1^{\text{Ab}}(X)$.
3. In particular, every element γ of Γ is invariant by the coadjoint action of G on $\mathcal{G}^*$. For all g in G,

   $$\text{Ad}_*(g)(\gamma) = \gamma.$$

The holonomy Γ is the obstruction for the action ρ to be *Hamiltonian*. Precisely, the action of G on X will be said to be Hamiltonian if $\Gamma = \{0\}$. Note that if $\pi_1^{\text{Ab}}(X) = \{0\}$, or if the group G has no Ad_*-invariant 1-form except 0, the action ρ is necessarily Hamiltonian.

Proof. We get immediately that $\gamma \in \Gamma$ is closed, by application of (art. 9.5). Indeed, for all paths $p \in \text{Paths}(X)$, $d(\Psi(p)) = \hat{x}_1^*(\omega) - \hat{x}_0^*(\omega)$, where $x_0 = p(0)$ and $x_1 = p(1)$. Thus, for a loop ℓ, since $\ell(0) = \ell(1)$, $d(\Psi(\ell)) = 0$. Now, let us choose a basepoint $x_0 \in X$. For every loop $\ell \in \text{Loops}(X, x_0)$, the momentum $\Psi(\ell)$ depends on ℓ only through its homotopy class (art. 9.6), so Γ is the image of $\pi_1(X, x_0)$. And, thanks to the additive property of Ψ (art. 9.4), the map $\text{class}(\ell) \mapsto \Psi(\ell)$ is a homomorphism. Now, since X is connected, for every other point x_1 of X, there exists a path c connecting x_0 to x_1, $\bar{c} = t \mapsto c(1-t)$ connects x_1 to x_0. Thanks again to the additive property of Ψ, $\Psi(\bar{c} \vee \ell \vee c) = \Psi(\bar{c}) + \Psi(\ell) + \Psi(c) = -\Psi(c) + \Psi(\ell) + \Psi(c) = \Psi(\ell)$. Then, since the map $\text{class}(\ell) \mapsto \text{class}(\bar{c} \vee \ell \vee c)$ is a conjugation from $\pi_1(X, x_0)$ to $\pi_1(X, x_1)$, Γ is the same homomorphic image of $\pi_1(X, x)$, for every point $x \in X$. Hence, we proved points 1 and 2. Point 3 is a direct consequence of (art. 7.17). □

Exercise

✎ EXERCISE 142 (Compact supported real functions I). Let us denote by X the space of compact supported real functions defined on $\mathbf{R}$, equipped with the diffeology defined in Exercise 25, p. 23. Precisely, $P : U \to X$ is a plot if and only if a) $(r,t) \mapsto P(r)(t)$ is a real smooth map defined on $U \times \mathbf{R}$, and b) for every $r_0 \in U$ there exist an open neighborhood $V \subset U$ of r_0 and a compact $K \subset \mathbf{R}$ such that for every $r \in V$, $P(r)$ and $P(r_0)$ coincide out of K. We want to consider the following bilinear form $\bar{\omega}$ as a differential 2-form on X,

$$\bar{\omega}(f,g) = \int_{-\infty}^{+\infty} \dot{f}(t) g(t)\, dt,$$

where the dot denotes the derivative with respect to the parameter t of f. For all $n \in \mathbf{N}$, for every n-plot $P : U \to X$, for all $r \in U$ and $\delta r, \delta' r \in \mathbf{R}^n$, we define

$$\omega(P)_r(\delta r, \delta' r) = \int_{-\infty}^{+\infty} \frac{\partial}{\partial r}\left(\frac{\partial P(r)(t)}{\partial t}\right)(\delta r)\, \frac{\partial P(r)(t)}{\partial r}(\delta' r)\, dt\,.$$

1) Show that ω is a 2-form on X.

2) Check that ω realizes $\bar{\omega}$, that is, for any $f, g \in X$,

$$\bar{\omega}(f,g) = \omega\left(\begin{pmatrix} s \\ s' \end{pmatrix} \mapsto sf + s'g\right)_{\binom{s}{s'}} \begin{pmatrix} 1 \\ 0 \end{pmatrix} \begin{pmatrix} 0 \\ 1 \end{pmatrix}.$$

Let us consider now the group $G = (X, +)$ acting on itself by translation, that is, for all $u \in X$,

$$T_u(f) = f + u \text{ for all } f \in X.$$

3) Show that ω is invariant by G.

4) Say why the holonomy group Γ, associated with the action of G on (X, ω), must vanish.

5) Show that, for any path p connecting $f = p(0)$ to $g = p(1)$, the paths moment map is given, for any plot F of G, that is, a plot of X, by

$$\Psi(p)(F)_r(\delta r) = \int_{-\infty}^{+\infty} \left(\dot{g}(t) - \dot{f}(t)\right) \frac{\partial F(r)(t)}{\partial r} \delta r\, dt\,.$$

The 2-points Moment Map

The definition of the paths moment map leads immediately to the 2-points moment map. *The 2-points moment map satisfies a cocycle condition inherited from the additive property of the paths moment map. This is the second step in the general construction.*

9.8. Definition of the 2-points moment map. Let X be a connected diffeological space, and let ω be a closed 2-form defined on X. Let G be a diffeological group and ρ be a smooth action of G on X, preserving ω. Let Ψ be the paths moment map (art. 9.1), and let Γ be the holonomy of the action ρ (art. 9.7). Then, there exists a smooth map $\psi : X \times X \to \mathcal{G}^*/\Gamma$ such that the following diagram commutes,

$$\begin{array}{ccc} \mathrm{Paths}(X) & \xrightarrow{\Psi} & \mathcal{G}^* \\ \text{ends} \downarrow & & \downarrow \text{pr} \\ X \times X & \xrightarrow[\psi]{} & \mathcal{G}^*/\Gamma \end{array}$$

where pr is the canonical projection from $\mathcal{G}^*$ onto its quotient, and ends maps p to $(p(0), p(1))$. The map $\psi \in \mathcal{C}^\infty(X \times X, \mathcal{G}^*/\Gamma)$ will be called the *2-points moment map.*

1. The 2-points moment map ψ satisfies the Chasles cocycle relation, for any three points x, x', x'' of X,
$$\psi(x, x'') = \psi(x, x') + \psi(x', x''). \qquad (\heartsuit)$$
2. The 2-points moment map ψ is equivariant under the action of G. Precisely, for any $g \in G$, and any pair of points x and x' of X,
$$\psi(\rho(g)(x), \rho(g)(x')) = \mathrm{Ad}_*^\Gamma(g)(\psi(x, x')).$$

PROOF. By construction ψ is defined by $\psi(x, x') = \mathrm{class}_\Gamma(\Psi(p))$, where $p \in \mathrm{Paths}(X)$, $x = p(0)$, $x' = p(1)$, and $\mathrm{class}_\Gamma(\alpha)$ denotes the class of $\alpha \in \mathcal{G}^*$ in $\mathcal{G}^*/\Gamma$. The map ψ is smooth simply by the general properties of subductions in diffeology. Next, the first point is a consequence of the additivity of the paths moment map (art. 9.4). The second point is a consequence of the equivariance of the paths moment map of the Ad_* invariance of Γ (art. 9.3), and of the definition of the Ad_*^Γ action (art. 7.16). □

The Moment Maps

From the construction of the paths moment map given in (art. 9.1) *and the 2-points moment map given in* (art. 9.8) *we get the notion of* 1*-point moment maps, or simply moment maps. This is the third step of the general construction, and the generalization of the classical notion of moment maps.*

9.9. Definition of the moment maps. Let X be a connected diffeological space, and let ω be a closed 2-form defined on X. Let G be a diffeological group and ρ be a smooth action of G on X, preserving ω. Let ψ be the 2-points moment map defined in (art. 9.8). There exists always a smooth map $\mu : X \to \mathcal{G}^*/\Gamma$, called a *primitive* of ψ, such that, for any two points x and x' of X,
$$\psi(x, x') = \mu(x') - \mu(x).$$
For every point $x_0 \in X$, for every constant $c \in \mathcal{G}^*/\Gamma$, the map μ defined by
$$\mu(x) = \psi(x_0, x) + c$$
is a primitive of ψ. Every primitive μ of ψ is of this kind, and any two primitives μ and μ' of ψ differ only by a constant. The 2-points moment map ψ will be said to be *exact* if there exists a primitive μ, *equivariant* by the action of G, that is, if there exists a primitive μ such that, for all $g \in G$,
$$\mu \circ \rho(g) = \mathrm{Ad}_*^\Gamma(g) \circ \mu.$$

The primitives μ of ψ, equivariant or not, will be called the *moment maps*.[6]

NOTE. By the identity (♡) of (art. 9.8), ψ is a 1-cocycle of the G-equivariant cohomology of X with coefficients in $\mathcal{G}^*/\Gamma$, twisted by the coadjoint action. Two cocycles ψ and ψ' are cohomologous if and only if there exists a smooth equivariant map $\mu : X \to \mathcal{G}^*/\Gamma$ such that $\psi'(x, x') = \psi(x, x') + \Delta\mu(x, x')$, where $\Delta\mu(x, x') = \mu(x') - \mu(x)$, and $\Delta\mu$ is a coboundary. So, the 2-points moment map ψ defines a class belonging to $H^1_G(X, \mathcal{G}^*/\Gamma)$ which depends only on the form ω and on the action ρ of G on X. If the moment map ψ is exact, that is, if $\mathrm{class}(\psi) = 0$, we shall say that the action ρ of G on X is *exact*, with respect to ω. In this case, there exists a point x_0 of X and a constant c such that $\mu : x \mapsto \psi(x_0, x) + c$ is an equivariant primitive for ψ.

PROOF. Let us choose a basepoint $x_0 \in X$. Since X is connected, for any $x \in X$ there exists always a path $p \in X$ such that $p(0) = x_0$ and $p(1) = x$. Thus, defining $\mu(x) = \psi(x_0, x) = \mathrm{class}(\Psi(p))$, and thanks to the cocycle properties of ψ, we have $\psi(x, x') = \psi(x, x_0) + \psi(x_0, x') = \psi(x_0, x') - \psi(x_0, x) = \mu(x') - \mu(x)$. Now, since ψ is smooth, μ is smooth. Therefore, the equation $\psi(x', x) = \mu(x') - \mu(x)$ always has a solution in μ. Now, let μ and μ' be two primitives of ψ. For each pair x, x' of points of X we have $\mu'(x') - \mu'(x) = \mu(x') - \mu(x)$, that is, $\mu'(x') - \mu(x') = \mu'(x) - \mu(x)$. So, the map $x \mapsto \mu'(x) - \mu(x)$ is constant. There exists $c \in \mathcal{G}^*/\Gamma$ such that $\mu'(x) - \mu(x) = c$, that is, $\mu'(x) = \mu(x) + c$. Since the map $x \mapsto \psi(x_0, x)$ is a special solution of the equation in μ: $\psi(x', x) = \mu(x') - \mu(x)$, any solution writes $\mu(x) = \psi(x_0, x) + c$ for some point $x_0 \in X$ and some constant $c \in \mathcal{G}^*/\Gamma$. □

9.10. The Souriau cocycle. Let X be a connected diffeological space, and let ω be a closed 2-form defined on X. Let G be a diffeological group and ρ be a smooth action of G on X, preserving ω. Let ψ be the 2-points moment map defined in (art. 9.8), and let μ be a primitive of ψ as defined in (art. 9.9). Then, there exists a map $\theta \in \mathcal{C}^\infty(G, \mathcal{G}^*/\Gamma)$ such that

$$\mu(\rho(g)(x)) = \mathrm{Ad}_*^\Gamma(g)(\mu(x)) + \theta(g).$$

This map θ is a $(\mathcal{G}^*/\Gamma)$-cocycle, as defined in (art. 7.16). For all $g, g' \in G$,

$$\theta(gg') = \mathrm{Ad}_*^\Gamma(g)(\theta(g')) + \theta(g).$$

We shall call the cocycle θ the *Souriau cocycle* of the moment μ.

1. Two Souriau cocycles θ and θ', associated with two moment maps μ and μ' are *cohomologous*, they differ by a *coboundary*
$$\Delta c : g \mapsto \mathrm{Ad}_*^\Gamma(g)(c) - c, \quad \text{where} \quad c \in \mathcal{G}^*/\Gamma.$$
2. For the affine coadjoint action of G on $\mathcal{G}^*/\Gamma$ defined by θ (art. 7.16), the moment map μ is equivariant. For all $g \in G$,
$$\mu \circ \rho(g) = \mathrm{Ad}_*^{\Gamma,\theta}(g) \circ \mu.$$
3. For every cocycle θ, associated with some moment map μ, there always exist a point $x_0 \in X$ and a constant $c \in \mathcal{G}^*/\Gamma$ such that,
$$\theta(g) = \psi(x_0, \rho(g)(x_0)) + \Delta c(g) \quad \text{for all } g \in G.$$

[6]These maps should have been called "1-point moment maps", but to conform to the usual denomination, we chose to call them simply "moment maps".

4. The cohomology class σ of θ belongs to a cohomology group denoted by $H^1(G, \mathcal{G}^*/\Gamma)$. It depends only on the cohomology class of the 2-points moment map ψ. This class σ will be called the *Souriau cohomology class.*

NOTE 1. Let x_0 be some point of X. The 2-points moment map ψ (which can also be regarded as a 1-cocycle) defines a 1-cocycle f from G to $\mathcal{G}^*/\Gamma$ by $f(g, g') = \psi(\rho(g)(x_0), \rho(g')(x_0))$. The cocycle f' associated with another point x_0' will differ only by a coboundary. So, the Souriau cocycle σ represents just the class of the pullback $f = \hat{x}_0^*(\psi)$ by the orbit map $\hat{x}_0$, where $\hat{x}_0^* : H^1_\rho(X, \mathcal{G}^*/\Gamma) \to H^1(G, \mathcal{G}^*/\Gamma)$. And, by the way, it depends only on the restriction of ω to any one orbit of G on X. Hence, a good choice of the point x_0 can simplify the computation of σ.

NOTE 2. The nature of the action ρ has strong consequences on the Souriau class. For example, thanks to the third item, if the group G has a fixed point x_0, that is, $\rho(g)(x_0) = x_0$ for all g in G, then the Souriau class is zero and the cocycle ψ is exact, *i.e.*, there exists an equivariant primitive μ of ψ.

PROOF. Thanks to (art. 9.9), every moment map μ writes $\mu(x) = \psi(x_0, x) + c$, where x_0 is some fixed point of X and $c \in \mathcal{G}^*/\Gamma$. Thus, $\mu(\rho(g)(x)) - \mathrm{Ad}_*^\Gamma(g)(\mu(x)) = \psi(x_0, \rho(g)(x)) + c - \mathrm{Ad}_*^\Gamma(g)(\psi(x_0, x) + c) = \psi(x_0, \rho(g)(x)) + c - \mathrm{Ad}_*^\Gamma(g)(\psi(x_0, x)) - \mathrm{Ad}_*^\Gamma(g)(c) = \psi(x_0, \rho(g)(x)) - \psi(\rho(g)(x_0), \rho(g)(x)) - \Delta c(g) = \psi(x_0, \rho(g)(x)) + \psi(\rho(g)(x), \rho(g)(x_0)) - \Delta c(g) = \psi(x_0, \rho(g)(x_0)) - \Delta c(g)$. Therefore, $\mu(\rho(g)(x)) - \mathrm{Ad}_*^\Gamma(g)(\mu(x))$ is constant with respect to x. That proves points 1 and 4. Now, the variance of θ, with respect to the multiplication of G, is a classical result of cohomology (see for example [**Kir76**]). It is then obvious that, since two moment maps μ and μ' differ only by a constant, the associated cocycles θ and θ' differ by a coboundary. The remaining items are just the results of elementary algebraic computations. □

Exercise

✎ EXERCISE 143 (Compact supported real functions, II). Let us consider the data and notations of Exercise 142, p. 307.

1) Show that the moment maps of the action of G on (X, ω) are given, for any plot F of G (that is, X) by

$$\mu(f)(F)_r(\delta r) = \int_{-\infty}^{+\infty} \dot{f}(t) \frac{\partial F(r)(t)}{\partial r} \delta r \, dt + \text{cst}.$$

2) Compute the associated Souriau cocycle θ. Is it trivial?

The Moment Maps for Exact 2-Forms

The special case where a closed 2-form is the exterior derivative of an invariant 1-form deserves special care, since it justifies a posteriori *the constructions above, by analogy with the classical moment maps* [**Sou70**].

9.11. The exact case. Let X be a connected diffeological space, and let ω be a closed 2-form defined on X. Let G be a diffeological group, and let ρ be a smooth action of G on X, preserving ω. Let us assume that $\omega = d\alpha$ and that α also is invariant under the action of G, that is, $\rho(g)^*(\alpha) = \alpha$ for all g in G. Let Ψ be the paths moment map defined in (art. 9.1), and ψ be the 2-points moment map defined in (art. 9.8). Then, for every $p \in \mathrm{Paths}(X)$

$$\Psi(p) = \psi(x, x') = \hat{x}_1^*(\alpha) - \hat{x}_0^*(\alpha),$$

where $x_1 = p(1)$ and $x_0 = p(0)$. Moreover, the 2-points moment map ψ is exact, and every equivariant moment map is cohomologous to

$$\mu : x \mapsto \hat{x}^*(\alpha).$$

The action of G is Hamiltonian, $\Gamma = \{0\}$ and exact $\sigma = 0$; see (art. 9.7) and (art. 9.10). This shows, in particular, the coherence of the general constructions developed until now.

PROOF. By definition of the paths moment map $\Psi(p) = \hat{p}^*(\mathcal{K}\omega)$, that is, $\Psi(p) = \hat{p}^*(\mathcal{K}(d\alpha))$. But $\mathcal{K}(d\alpha) + d(\mathcal{K}\alpha) = \hat{1}^*(\alpha) - \hat{0}^*(\alpha)$, hence $\mathcal{K}(d\alpha) = \hat{p}^*[\hat{1}^*(\alpha) - \hat{0}^*(\alpha) - d(\mathcal{K}\alpha)]$, and $\Psi(p) = (\hat{1} \circ \hat{p})^*(\alpha) - (\hat{0} \circ \hat{p})^*(\alpha) - d[\hat{p}^*(\mathcal{K}(\alpha))]$. But $\hat{1} \circ \hat{p} = \hat{x}_1$ and $\hat{0} \circ \hat{p} = \hat{x}_0$, then $\Psi(p) = \hat{x}_1^*(\alpha) - \hat{x}_0^*(\alpha) - d[\hat{p}^*(\mathcal{K}\alpha)]$. Now, $\mathcal{K}\alpha$ is the real function

$$\mathcal{K}\alpha : p \mapsto \int_p \alpha,$$

since $\hat{p}^*(\mathcal{K}\alpha) = \mathcal{K}\alpha \circ \hat{p}$, for all $g \in G$,

$$\mathcal{K}\alpha(\hat{p}(g)) = \int_{\rho(g)\circ p} \alpha = \int_p \rho(g)^*(\alpha) = \int_p \alpha.$$

Thus, the function $\hat{p}^*(\mathcal{K}\alpha) : G \to \mathbf{R}$ is constant and equal to $\int_p \alpha$. Then, $d[\hat{p}^*(\mathcal{K}\alpha)] = 0$, and $\Psi(p) = \hat{x}_1^*(\alpha) - \hat{x}_0^*(\alpha)$. Hence, $\Psi(p) = \psi(x_0, x_1)$ and $\Gamma = \{0\}$.

Next, the function $\mu : x \mapsto \hat{x}^*(\alpha)$ is clearly a primitive of ψ, that is, $\psi(x_0, x_1) = \mu(x_1) - \mu(x_0)$. But $R(\rho(g)(x)) = \hat{x} \circ R(g)$, where $R(\rho(g)(x))$ is the orbit map of $\rho(g)(x)$, $g \in G$. Thus, $\mu(\rho(g)(x)) = (\hat{x} \circ R(g))^*(\alpha) = R(g)^*(\hat{x}^*(\alpha)) = R(g)^*(\mu(x)) = \mathrm{Ad}_*(g)(\mu(x))$. Hence, μ is an equivariant primitive of ψ and $\sigma = 0$. □

Functoriality of the Moment Maps

In this section we focus on the behavior of the moment maps, and the various associated objects, under natural transformations.

9.12. Images of the moment maps by morphisms. Let X be a connected diffeological space, and let ω be a closed 2-form defined on X. Let G be a diffeological group, and let ρ be a smooth action of G on X, preserving ω. Let G' be another diffeological group, and let $h : G' \to G$ be a smooth homomorphism. Let $\rho' = \rho \circ h$ be the induced action of G' on X. Let us recall that the pullback $h^* : \mathcal{G}^* \to \mathcal{G}'^*$ is a linear smooth map.

1. Let $\Psi : \mathrm{Paths}(X) \to \mathcal{G}$, and $\Psi' : \mathrm{Paths}(X) \to \mathcal{G}'$ be the paths moment maps with respect to the actions of G and G' on X. Then, $\Psi' = h^* \circ \Psi$.
2. Let Γ and Γ' be the holonomy groups with respect to the actions of G and G' on X. Then, $\Gamma' = h^*(\Gamma)$.
3. The linear map h^* projects on a smooth homomorphism $h_\Gamma^* : \mathcal{G}/\Gamma \to \mathcal{G}'^*/\Gamma'$, such that the following diagram commutes.

$$\begin{array}{ccc} \mathcal{G}^* & \xrightarrow{\ h^*\ } & \mathcal{G}'^* \\ {\scriptstyle \mathrm{pr}}\downarrow & & \downarrow{\scriptstyle \mathrm{pr}'} \\ \mathcal{G}^*/\Gamma & \xrightarrow[\ h^*_\Gamma\ ]{} & \mathcal{G}'^*/\Gamma' \end{array}$$

4. Let ψ and ψ' be the 2-points moment maps with respect to the actions of G and G'. Then, $\psi' = h^*_\Gamma \circ \psi$.
5. Let μ be a moment map relative to the action ρ of G. Then, $\mu' = h^*_\Gamma \circ \mu$ is a moment map relative to the action ρ' of G'.
6. Let μ be a moment map relative to the action ρ of G, and let $\mu' = \mu \circ h^*_\Gamma$ be the associated moment map relative to the action ρ' of G'. Then, the associated Souriau cocycles satisfy $\theta' = h^*_\Gamma \circ \theta \circ h$, which is summarized by the following commutative diagram.

$$\begin{array}{ccc} G & \xleftarrow{\ h\ } & G' \\ {\scriptstyle \theta}\downarrow & & \downarrow{\scriptstyle \theta'} \\ \mathcal{G}^*/\Gamma & \xrightarrow[\ h^*_\Gamma\ ]{} & \mathcal{G}'^*/\Gamma' \end{array}$$

Said differently, if θ is the Souriau cocycle associated with a moment μ of the action ρ of G, and μ' is a moment of the action ρ' of G', then θ' and $h^*_\Gamma \circ \theta \circ h$ are cohomologous.

NOTE. Thanks to the identification between the space of momenta of a diffeological group and any of its extensions by a discrete group (art. 7.13), the moment maps of the action of a group or the moment map of the restriction of this action to its identity component coincide. Said differently, the moment maps do not say anything about actions of discrete groups.

PROOF. To avoid confusion, let us denote by $R(p)$ and $R'(p)$ the orbit maps of G and G' of $p \in \mathrm{Paths}(X)$, that is, $R(p)(g) = \rho(g) \circ p$ and $R'(p)(g) = \rho'(g) \circ p$. We have then, $R'(p)(g) = \rho'(g) \circ p = \rho(h(g)) \circ p = (R(p) \circ h)(g)$. Thus, $R'(p) = R(p) \circ h$.

1. By definition of the paths moment map, $\Psi'(p) = R'(p)^*(\mathcal{K}\omega) = (R(p) \circ h)^*(\mathcal{K}\omega) = h^*(R(p)^*(\mathcal{K}\omega)) = h^*(\Psi(p))$, that is, $\Psi' = h^* \circ \Psi$.

2. Since $\Gamma' = \Psi'(\mathrm{Loops}(X))$, and thanks to item 1, $\Gamma' = h^*(\Gamma)$.

3. The map h^*_Γ is defined by $\mathrm{class}_\Gamma(\alpha) \mapsto \mathrm{class}_{\Gamma'}(h^*(\alpha))$, for all $\alpha \in \mathcal{G}^*$. If $\beta = \alpha + \gamma$, with $\gamma \in \Gamma$, then $h^*(\beta) = h^*(\alpha) + \gamma'$, with $\gamma' = h^*(\gamma) \in \Gamma'$ (item 2). Thus, $\mathrm{class}_{\Gamma'}(h^*(\beta)) = \mathrm{class}_{\Gamma'}(h^*(\alpha))$ and h^*_Γ is well defined. Thanks to the linearity of h^*, h^*_Γ is clearly a homomorphism. For $\mathcal{G}^*/\Gamma$ and $\mathcal{G}'^*/\Gamma'$ equipped with the quotient diffeologies, h^*_Γ is naturally smooth.

4. With the notations above, ψ and ψ' are defined by $\mathrm{pr} \circ \Psi = \psi \circ \mathrm{ends}$ and $\mathrm{pr}' \circ \Psi' = \psi' \circ \mathrm{ends}$, where $\mathrm{ends}(p) = (p(0), p(1))$, with $p \in \mathrm{Paths}(X)$. Thus, by item 1 and 3, $\mathrm{pr}' \circ h^* \circ \Psi = h^*_\Gamma \circ \psi \circ \mathrm{pr}$, that is, $\mathrm{pr}' \circ \Psi' = (h^*_\Gamma \circ \psi) \circ \mathrm{pr}$. Hence, $h^*_\Gamma \circ \psi = \psi'$.

5. Let $\mu' = h_\Gamma^* \circ \mu$ and $x, y \in X$. Then, $\mu'(y) - \mu'(x) = h_\Gamma^* \circ \mu(y) - h_\Gamma^* \circ \mu(y) = h_\Gamma^*(\mu(y) - \mu(x)) = h_\Gamma^* \circ \psi(y,x) = \psi'(y,x)$. Thus, μ' is a moment map for the action ρ' of G.

6. According to (art. 9.10), there exists a point $x_0 \in X$ such that, for all $g' \in G'$, $\theta'(g') = \psi'(x_0, \rho'(g')(x_0))$. Thus, thanks to the previous items, $\theta'(g') = (h_\Gamma^* \circ \psi)(x_0, \rho(h(g'))(x_0)) = h_\Gamma^*(\psi(x_0, \rho(h(g'))(x_0))) = h_\Gamma^*(\theta(h(g'))) = (h_\Gamma^* \circ \theta \circ h)(g')$. Hence, $\theta' = h_\Gamma^* \circ \theta \circ h$. ☐

9.13. Pushing forward moment maps. Let X and X' be two connected diffeological spaces. Let ω and ω' be two closed 2-forms, defined respectively on X and X'. Let G be a diffeological group, let ρ be a smooth action of G on X preserving ω, and let ρ' be a smooth action, of G on X', preserving ω'. Let $f : X \to X'$ be a smooth map such that $\omega = f^*(\omega')$ and $f \circ \rho(g) = \rho'(g) \circ f$, for all $g \in G$.

1. Let $f_* : \mathrm{Paths}(X) \to \mathrm{Paths}(X')$ defined by $f_*(p) = f \circ p$. Then, the paths moment maps Ψ and Ψ' relative to the actions ρ and ρ' are related by
$$\Psi = \Psi' \circ f_*,$$
and the associated holonomy groups Γ and Γ' satisfy
$$\Gamma = \{\Psi'(f \circ \ell) \mid \ell \in \mathrm{Loops}(X)\} \subset \Gamma'.$$
2. Let $\phi : \mathcal{G}^*/\Gamma \to \mathcal{G}^*/\Gamma'$ be the projection induced by the inclusion $\Gamma \subset \Gamma'$. Let ψ and ψ' be the 2-points moment maps relative to the actions ρ and ρ'. Then, for any two points of X, x_1 and x_2,
$$\psi'(f(x_1), f(x_2)) = \phi(\psi(x_1, x_2)).$$
3. For every moment map μ relative to the action ρ, there exists a moment map μ' relative to the action ρ', such that
$$\mu' \circ f = \phi \circ \mu.$$
4. Let θ and θ' be two Souriau cocycles relative to the actions ρ and ρ'. Then, the map $\phi \circ \theta$ is a Souriau cocycle, cohomologous to θ' and the two Souriau classes σ and σ' satisfy $\sigma' = \phi_*(\sigma)$, where ϕ_* denotes the action of ϕ on cohomology, $\phi_*(\mathrm{class}(\theta)) = \mathrm{class}(\phi \circ \theta)$.

PROOF. 1. By definition $\Psi(p) = \hat{p}^*(\mathcal{K}\omega)$, that is, $\Psi(p) = \hat{p}^*(\mathcal{K}(f^*(\omega')))$. Thanks to the variance of the Chain-Homotopy Operator $\mathcal{K} \circ f^* = (f_*)^* \circ \mathcal{K}'$ (art. 6.84), $\Psi(p) = \hat{p}^* \circ (f_*)^*(\mathcal{K}'\omega') = (f_* \circ \hat{p})^*(\mathcal{K}'\omega')$. But for all $g \in G$, $f_* \circ \hat{p}(g) = f \circ \rho(g) \circ p = \rho'(g) \circ f \circ p = \hat{p}'(g)$, where $p' = f \circ p$. Then, $\Psi(p) = \hat{p}'^*(\mathcal{K}'\omega') = \Psi'(p') = \Psi'(f_*(p))$. Therefore, $\Psi = \Psi' \circ f_*$. Now, by definition of the holonomy groups, $\Gamma = \Psi(\mathrm{Loops}(X)) = \Psi'(f_*(\mathrm{Loops}(X)))$, and since $f_*(\mathrm{Loops}(X)) \subset \mathrm{Loops}(X')$, $\Gamma \subset \Gamma'$.

2. Since $\Gamma \subset \Gamma'$, the map $\phi : \mathrm{class}_\Gamma(\alpha) \mapsto \mathrm{class}_{\Gamma'}(\alpha)$, from $\mathcal{G}^*/\Gamma$ to $\mathcal{G}^*/\Gamma'$, is well defined. Now, let $x_1' = f(x_1)$ and $x_2' = f(x_2)$. There exists then $p \in \mathrm{Paths}(X)$ connecting x_1 to x_2, and the path $f_*(p)$ connects x_1' to x_2'. By definition of ψ', $\psi'(x_1', x_2') = \mathrm{class}_{\Gamma'}(\Psi'(p')) = \mathrm{class}_{\Gamma'}(\Psi' \circ f_*(p))$, and thanks to the first item, $\mathrm{class}_{\Gamma'}(\Psi'(p')) = \mathrm{class}_{\Gamma'}(\Psi(p)) = \phi(\mathrm{class}_\Gamma(\Psi(p)))$. But $\mathrm{class}_\Gamma(\Psi(p)) = \psi(x_1, x_2)$. Hence, $\psi'(x_1', x_2') = \phi(\psi(x_1, x_2))$, that is, $\psi'(f(x_1), f(x_2)) = \psi(x_1, x_2)$.

3. According to (art. 9.9), for every moment map μ there exist a point $x_0 \in X$ and a constant $c \in \mathcal{G}^*/\Gamma$ such that $\mu(x) = \psi(x_0, x) + c$. Let us define μ' by $\mu'(x') = \psi'(x_0', x') + c'$, where $x_0' = f(x_0)$ and $c' = \phi(c)$. Then, thanks to item 2,

$\psi'(f(x_0), f(x)) = \phi(\psi(x_0, x))$, and $\mu'(f(x)) = \phi(\psi(x_0, x)) + \phi(c) = \phi(\psi(x_0, x) + c) = \phi(\mu(x))$. Hence, μ' satisfies $\mu' \circ f = \phi \circ \mu$.

4. Let θ be a Souriau cocycle for the action ρ. According to (art. 9.10), θ is cohomologous to $\vartheta : g \mapsto \psi(x_0, \rho(g)(x))$, where x_0 is some point of X. Then, let $x'_0 = f(x_0)$, and $\vartheta' : g \mapsto \psi'(x'_0, \rho'(g)(x'_0))$. Thus, $\vartheta'(g) = \psi'(f(x_0), \rho'(g)(f(x_0))) = \psi'(f(x_0), f(\rho(g)(x_0))) = \phi(\psi(x_0, \rho(g)(x_0))) = \phi \circ \vartheta(g)$. Now, since all Souriau cocycles, with respect to a given action of G, are cohomologous, the cocycle θ' is cohomologous to ϑ', and then cohomologous to $\phi \circ \vartheta$, that is, cohomologous to $\phi \circ \theta$. Hence, $\sigma' = \text{class}(\theta') = \text{class}(\phi \circ \theta) = \phi_*(\text{class}(\theta)) = \phi_*(\sigma)$. □

The Universal Moment Maps

In this section we build the universal moment maps, and related objects, associated with the whole group of automorphisms $\text{Diff}(X, \omega)$, *where* ω *is a closed 2-form on a diffeological space* X. *We show then how these objects, associated with a smooth automorphic action of some arbitrary diffeological group* G, *relate to the universal construction.*

9.14. Universal moment maps. Let X be a connected diffeological space, and let ω be a closed 2-form defined on X. We consider the group $\text{Diff}(X, \omega)$, of all the automorphisms of (X, ω), equipped with the functional diffeology of group of diffeomorphisms. We shall also denote this group by G_ω. Every construction introduced in the previous sections—the space of momenta, the paths moment map, the holonomy group, the 2-points moment map, the moment maps, the Souriau cocycles, and the Souriau class—apply for G_ω. We shall distinguish these objects by the index ω. So, we shall denote by $\mathcal{G}^*_\omega$ the space of momenta of G_ω, by $\Psi_\omega : \text{Paths}(X) \to \mathcal{G}^*_\omega$ the paths moment map, by $\Gamma_\omega = \Psi_\omega(\text{Loops}(X))$ the holonomy group, by ψ_ω the 2-points moment map, by μ_ω the moment maps, by θ_ω the Souriau cocycles, and by σ_ω the Souriau class. Since G_ω and its action on X are uniquely defined by ω, these objects depend only on the 2-form ω.

Now, let G be a diffeological group, and let ρ be a smooth action of G on X, preserving ω, that is, a smooth homomorphism ρ from G to G_ω. The values of the various objects Ψ, Γ, ψ, μ, θ, with respect to the action ρ of G on X, depend only on the pullback ρ^* and on Ψ_ω, Γ_ω, ψ_ω, μ_ω, and θ_ω, as it is described in (art. 9.12),

$$\left\{ \begin{array}{lcl} \Psi & = & \rho^* \circ \Psi_\omega \\ \Gamma & = & \rho^*(\Gamma_\omega) \\ \psi & = & \rho^*_{\Gamma_\omega} \circ \psi_\omega \end{array} \right. \quad \text{and} \quad \left\{ \begin{array}{lcl} \mu & \simeq & \rho^*_{\Gamma_\omega} \circ \mu_\omega \\ \theta & \simeq & \rho^*_{\Gamma_\omega} \circ \theta_\omega \circ \rho. \end{array} \right.$$

In this sense the objects G_ω, Γ_ω, Ψ_ω, Γ_ω, ψ_ω, μ_ω, θ_ω, and σ_ω are *universal*. That is why we shall call Ψ_ω the *universal paths moment map*, Γ_ω the *universal holonomy*, ψ_ω the *universal 2-points moment map*, μ_ω the *universal moment maps*, θ_ω the *universal Souriau cocycles*, and σ_ω the *universal Souriau class* of ω.

NOTE. The universal holonomy leads naturally to the notion of *Hamiltonian spaces*, the ones for which, for one reason or another, $\Gamma_\omega = \{0\}$.

9.15. The group of Hamiltonian diffeomorphisms. Let X be a connected diffeological space, equipped with a closed 2-form ω. There exists a largest connected subgroup $\text{Ham}(X, \omega) \subset \text{Diff}(X, \omega)$ whose action is Hamiltonian, that is, whose holonomy is trivial. The elements of $\mathrm{Ham}(X, \omega)$ are called *Hamiltonian diffeomorphisms*. An action ρ of a diffeological group G on X is Hamiltonian if and only if, restricted to the identity component of G, ρ takes its values in $\mathrm{Ham}(X, \omega)$.

The group $\mathrm{Ham}(X, \omega)$ is precisely built as follows. Let us denote by G_ω the group $\mathrm{Diff}(X, \omega)$ and by G_ω° its identity component. Let $\pi : \tilde{G}_\omega^\circ \to G_\omega^\circ$ be the universal covering. Since the universal holonomy Γ_ω is made of closed momenta, every $\gamma \in \Gamma_\omega$ defines a unique homomorphism $k(\gamma)$ from $\tilde{G}_\omega^\circ$ to $\mathbf{R}$ such that $\pi^*(\gamma) = d[k(\gamma)]$ (art. 7.17). Let

$$\widehat{H}_\omega = \bigcap_{\gamma \in \Gamma_\omega} \ker(k(\gamma)),$$

and let $\widehat{H}_\omega^\circ$ be its identity component. Then,

$$\mathrm{Ham}(X, \omega) = \pi(\widehat{H}_\omega^\circ).$$

NOTE 1. The map $f : \tilde{G}_\omega^\circ \to \mathrm{Hom}(\pi_1(X), \mathbf{R})$, defined by $f(\tilde{g}) = [\tau \mapsto k(\gamma)(\tilde{g})]$, with $\tau = \mathrm{class}(\ell)$ and $\gamma = \Psi(\ell)$, is a homomorphism, and $\widehat{H}_\omega = \ker(f)$. In classical symplectic geometry, the image $F = \mathrm{val}(f)$ is called, by some authors, the *group of flux* of ω.

NOTE 2. Since the Hamiltonian nature of a group of automorphisms depends only on its identity component (see (art. 7.13) and (art. 7.14)) every extension $H \subset \mathrm{Diff}(X, \omega)$ of $\mathrm{Ham}(X, \omega)$, such that $H/\mathrm{Ham}(X, \omega)$ is discrete,[7] is Hamiltonian. In particular $\pi(\widehat{H}_\omega)$ is Hamiltonian, or if $\Gamma_\omega = \{0\}$, then $\mathrm{Diff}(X, \omega)$ is Hamiltonian, and $\mathrm{Ham}(X, \omega)$ is the identity component of $\mathrm{Diff}(X, \omega)$.

NOTE 3. Let us choose a point x_0 in X, and let μ be the moment map with respect to the group $\mathrm{Ham}(X, \omega)$, defined by $\mu(x_0) = 0$. Let f be a 1-parameter subgroup of $\mathrm{Ham}(X, \omega)$. Applying (art. 9.2, Note), we get the expression of $\mu(x)$, for all $x \in X$, evaluated on f

$$\mu(x)(f) = h_f(x) \times dt \quad \text{with} \quad h_f(x) = -\int_{x_0}^{x} i_f(\omega).$$

The smooth function $h_f : X \to \mathbf{R}$ is the *Hamiltonian* (vanishing at x_0) of the 1-parameter subgroup f.

PROOF. Let us remark first of all that for every $\gamma \in \Gamma_\omega$, $\pi^*(\gamma) \upharpoonright \widehat{H}_\omega = 0$. Indeed, $\pi^*(\gamma) \upharpoonright \widehat{H}_\omega = d[k(\gamma)] \upharpoonright \widehat{H}_\omega = d[k(\gamma) \upharpoonright \widehat{H}_\omega]$. But, by the very definition of $\widehat{H}_\omega$, $k(\gamma) \upharpoonright \widehat{H}_\omega = 0$, thus $\pi^*(\gamma) \upharpoonright \widehat{H}_\omega = 0$.

a) Let us prove that the holonomy of $\mathrm{Ham}(X, \omega)$ is trivial. Let $H_\omega = \pi(\widehat{H}_\omega)$, and let us denote by j_{H_ω} the inclusion $H_\omega \subset G_\omega$, by $j_{\widehat{H}_\omega}$ the inclusion $\widehat{H}_\omega \subset \tilde{G}_\omega^\circ$, and by $\pi_{H_\omega} : \widehat{H}_\omega \to H_\omega$ the projection, so that $j_{H_\omega} \circ \pi_{H_\omega} = \pi \circ j_{\widehat{H}_\omega}$. Let Γ_{H_ω} be the holonomy of H_ω, then, according to (art. 9.12), $\Gamma_{H_\omega} = j_{H_\omega}^*(\Gamma_\omega)$. Thus, for every $\bar{\gamma} \in \Gamma_{H_\omega}$ there exists $\gamma \in \Gamma_\omega$ such that $\bar{\gamma} = \gamma \upharpoonright H_\omega = j_{H_\omega}^*(\gamma)$. Hence, for all $\bar{\gamma} \in \Gamma_{H_\omega}$, $\pi_{H_\omega}^*(\bar{\gamma}) = \pi_{H_\omega}^*(j_{H_\omega}^*(\gamma)) = (j_{H_\omega} \circ \pi_{H_\omega})^*(\gamma) = (\pi \circ j_{\widehat{H}_\omega})^*(\gamma) = j_{\widehat{H}_\omega}^*(\pi^*(\gamma)) = \pi^*(\gamma) \upharpoonright \widehat{H}_\omega$. But $\pi^*(\gamma) \upharpoonright \widehat{H}_\omega = 0$, thus $\pi_{H_\omega}^*(\bar{\gamma}) = 0$, and since π_{H_ω} is a subduction, $\bar{\gamma} = 0$. Therefore, the holonomy of H_ω trivial, $\Gamma_{H_\omega} = \{0\}$.

[7] where H and $\mathrm{Ham}(X, \omega)$ are equipped with the subset diffeology of the functional diffeology of $\mathrm{Diff}(X, \omega)$.

b) Let us prove now that every connected subgroup $H \subset G_\omega$ whose action is Hamiltonian is a subgroup of $\mathrm{Ham}(X, \omega)$. Let $\widehat{H} = \pi^{-1}(H)$ and $\widehat{H}^\circ$ be its identity component. Let j_H be the inclusion $H \subset G_\omega$, and $j_{\widehat{H}^\circ}$ be the inclusion $\widehat{H}^\circ \subset \tilde{G}^\circ_\omega$. Let $\pi_H = \pi \upharpoonright \widehat{H}^\circ$, so that $j_H \circ \pi_H = \pi \circ j_{\widehat{H}^\circ}$. Let Γ_H be the holonomy of H. Since $\Gamma_H = j_H^*(\Gamma_\omega)$ and $\Gamma_H = \{0\}$, for all $\gamma \in \Gamma_\omega$, $j_H^*(\gamma) = 0$. Thus, for all $\gamma \in \Gamma_\omega$, $\pi_H^*(j_H^*(\gamma)) = 0$. But $\pi_H^*(j_H^*(\gamma)) = (j_H \circ \pi_H)^*(\gamma) = (\pi \circ j_{\widehat{H}^\circ})^*(\gamma) = j^*_{\widehat{H}^\circ}(\pi^*(\gamma)) = \pi^*(\gamma) \upharpoonright \widehat{H}^\circ$, thus, for all $\gamma \in \Gamma_\omega$, $\pi^*(\gamma) \upharpoonright \widehat{H}^\circ = 0$. Now, $\pi^*(\gamma) = d[k(\gamma)]$, hence $d[k(\gamma) \upharpoonright \widehat{H}^\circ] = 0$. Then, since H° is connected, $k(\gamma)$ is constant on $\widehat{H}^\circ$, and since $k(\gamma)$ is a homomorphism to $\mathbf{R}$, this constant is necessarily 0. Thus, $\widehat{H}^\circ \subset \ker(k(\gamma))$, for all $\gamma \in \Gamma_\omega$, that is, $\widehat{H}^\circ \subset \widehat{H}_\omega$. But since H° is connected, $\widehat{H}^\circ \subset \widehat{H}^\circ_\omega \subset H_\omega$, and thus $H = \pi(\widehat{H}^\circ) \subset \mathrm{Ham}(X, \omega) = \pi(\widehat{H}^\circ_\omega)$. □

9.16. Time-dependent Hamiltonian. Let X be a connected diffeological space, and let ω be a closed 2-form defined on X. A diffeomorphism f of X belongs to $\mathrm{Ham}(X, \omega)$ if and only if the two following conditions are fulfilled.

1. There exists a smooth path $t \mapsto f_t$ in $\mathrm{Diff}(X, \omega)$ connecting the identity $\mathbf{1}_M = f_0$ to $f = f_1$.
2. There exists a smooth path $t \mapsto \Phi_t$ in $\mathcal{C}^\infty(X, \mathbf{R})$ such that

$$i_{F_t}(\omega) = -d\Phi_t \quad \text{with} \quad F_t : s \mapsto f_t^{-1} \circ f_{t+s},$$

for all t. According to the tradition of classical symplectic geometry, the path $t \mapsto \Phi_t$ may be called a *time-dependent Hamiltonian* of the 1-parameter family of Hamiltonian diffeomorphisms $t \mapsto f_t$.

PROOF. Let us assume first that f satisfies the condition above, that there exists a smooth path $t \mapsto f_t$ in $\mathrm{Diff}(X, \omega)$ such that $f_0 = \mathbf{1}_M$, $f_1 = f$, and there exists a smooth path $t \mapsto \Phi_t$ in $\mathcal{C}^\infty(X, \mathbf{R})$ such that $i_{F_t}(\omega) = -d\Phi_t$, for all t, where $F_t : s \mapsto f_t^{-1} \circ f_{t+s}$. Let us recall that $\mathrm{Ham}(X, \omega) = \pi(\widehat{H}^\circ_\omega)$, with $\widehat{H}^\circ_\omega$ the identity component of $\widehat{H}_\omega = \bigcap_{\gamma \in \Gamma_\omega} \ker(k(\gamma))$, and let $\tilde{f} \in G^\circ_\omega$ be the homotopy class of the path $t \mapsto f_t$, notations of (art. 9.15). Then, let $\gamma \in \Gamma_\omega$, that is, $\gamma = \Psi_\omega(\ell)$, where ℓ is some loop in M. By definition,

$$k(\gamma)(\tilde{f}) = \int_{[t \mapsto f_t]} \gamma = \int_{[t \mapsto f_t]} \Psi_\omega(\ell) = \int_0^1 \Psi_\omega(\ell)([t \mapsto f_t])_t(1)\, dt.$$

Now, thanks to (art. 9.2, (♣)),

$$\Psi_\omega(\ell)([t \mapsto f_t])_t(1) = -\int_\ell i_{F_t}(\omega) = \int_\ell d\Phi_t = \int_{\partial \ell} \Phi_t = 0.$$

Thus, $k(\gamma)(\tilde{f}) = 0$ for all $\gamma \in \Gamma_\omega$, and $\tilde{f}$ belongs to $\widehat{H}_\omega$, more precisely to the identity component of $\widehat{H}_\omega$. Therefore, $f \in \mathrm{Ham}(X, \omega)$.

Conversely, let $f \in \mathrm{Ham}(M, \omega)$. Since $\mathrm{Ham}(M, \omega)$ is connected, there exists a path $t \mapsto f_t$ in $\mathrm{Ham}(M, \omega)$ connecting $\mathbf{1}_M$ to f. And, since the projection $\pi \upharpoonright \widehat{H}^\circ_\omega : \widehat{H}^\circ_\omega \to \mathrm{Ham}(M, \omega)$ is a covering, there exists a (unique) lift $t \mapsto \tilde{f}_t$ of $t \mapsto f$ in $\widehat{H}^\circ_\omega$, along $\pi \upharpoonright \widehat{H}^\circ_\omega$, such that $\tilde{f}_0 = \mathbf{1}_{\widehat{H}_\omega}$. This lift is actually given by $\tilde{f}_t = \mathrm{class}(p_t)$, with $p_t : s \mapsto f_{st}$. Thus, for all t, $\tilde{f}_t \in \widehat{H}^\circ_\omega \subset \widehat{H}_\omega = \bigcap_{\gamma \in \Gamma_\omega} \ker(k(\gamma))$, that is, for all $\gamma \in \Gamma_\omega$, $k(\gamma)(\tilde{f}_t) = 0$. In other words, for all $\ell \in \mathrm{Loops}(M)$,

$k(\Psi_\omega(\ell))(\tilde{f}_t) = 0$. But

$$\begin{aligned} k(\Psi_\omega(\ell))(\tilde{f}_t) &= \int_{p_t} \Psi_\omega(\ell) \\ &= \int_0^1 \Psi_\omega(\ell)(s \mapsto f_{st})_s(1)ds \\ &= \int_0^1 \Psi_\omega(\ell)(s \mapsto st \mapsto f_{st})_s(1)ds \\ &= \int_0^1 [\Psi_\omega(\ell)(u \mapsto f_u)]_{u=st} \left(\frac{d(st)}{ds}\right) ds \\ &= \int_0^t \Psi_\omega(\ell)(u \mapsto f_u)_u(1)\, du. \end{aligned}$$

Thus, $k(\Psi_\omega(\ell))(\tilde{f}_t) = 0$, which implies $\frac{1}{t}\int_0^t \Psi_\omega(\ell)(u \mapsto f_u)_u(1)\, du = 0$, and $\Psi_\omega(\ell)(t \mapsto f_t)_t(1) = \lim_{t\to 0} \frac{1}{t}\int_0^t \Psi_\omega(\ell)(u \mapsto f_u)_u(1)\, du = 0$. Next, since $\Psi_\omega(\ell)([t \mapsto f_t])_t(1) = -\int_\ell i_{F_t}(\omega)$ (art. 9.2, (♣)), for all t and all $\ell \in$ Loops(X), $\int_\ell i_{F_t}(\omega) = 0$. Then, since F_t is a path in $\text{Diff}(X, \omega)$ centered at the identity, the Lie derivative of ω by F_t vanishes, and by application of the Cartan formula (art. 6.72), we get $\pounds_{F_t}\omega = 0$, which implies $d[i_{F_t}(\omega)] + i_{F_t}(d\omega) = d[i_{F_t}(\omega)] = 0$. Thus, the 1-form $i_{F_t}(\omega)$ is closed and its integral on any loop ℓ in X vanishes, hence $i_{F_t}(\omega)$ is exact (art. 6.89). Therefore, for all real number t, there exists a real function $\Phi_t \in \mathcal{C}^\infty(X, \mathbf{R})$ such that $i_{F_t}(\omega) = -d\Phi_t$. The fact that $t \mapsto \Phi_t$ is a smooth map from $\mathbf{R}$ to $\mathcal{C}^\infty(X, \mathbf{R})$, for the functional diffeology, is a consequence of the explicit construction of the function Φ_t by integration along the paths. □

9.17. The characteristics of moment maps. Let X be a connected diffeological space equipped with a closed 2-form ω. Let G be a diffeological group, and let ρ be a smooth action of G on X, preserving ω. Let ψ be the 2-points moment map (art. 9.8). Thanks to the additive property of ψ, the relation R, defined on X by

$$x \, R \, x' \quad \text{if} \quad \psi(x, x') = 0_{\mathcal{G}^*/\Gamma},$$

is an equivalence relation. The classes of this equivalence relation are the preimages of the values of a moment map μ, solution of $\psi(x, x') = \mu(x') - \mu(x)$ (art. 9.9). We shall define the *characteristics of the moment map* μ (or ψ) as the connected components of the equivalence classes of R, that is, the connected components of the preimages of μ.

NOTE. This definition applies obviously to the universal moment map μ_ω. Since, in this case, the characteristics depend only on ω, it is tempting to call them the *characteristics of the 2-form* ω, especially when we have in mind the particular case of homogeneous manifolds, treated in (art. 9.26). We get then a general picture: by equivariance, the image of the universal moment map μ_ω is a union of coadjoint orbits of $\text{Diff}(X, \omega)$, images of its orbits in X. The moment map factorizes then through the space of characteristics of ω, denoted here by $\text{Chars}(X, \omega)$, by a map $\bar{\mu}_\omega : \text{Chars}(X, \omega) \to \mathcal{G}^*_\omega/\Gamma_\omega$. The preimages by $\bar{\mu}_\omega$ are the connected components of the preimages by μ_ω, that is, $\bar{\mu}_\omega^{-1}(m) = \pi_0(\mu_\omega^{-1}(m))$. The projection char_ω, associating with each $x \in X$ the characteristic passing through x, is a kind of *symplectic reduction*. But we do not know if, in general, the 2-form ω passes to the quotient, especially when the group of automorphisms has more than one orbit. This is still an open question, but the framework is here.

$$\begin{array}{ccc} X & \xrightarrow{\ \mu_\omega\ } & \mathcal{G}^*_\omega/\Gamma_\omega \\ {\scriptstyle \mathrm{char}_\omega}\downarrow & \swarrow {\scriptstyle \bar{\mu}_\omega} & \\ \mathrm{Chars}(X,\omega) & & \end{array}$$

This construction is related to the well-known Marsden-Weinstein symplectic reduction [**MaWe74**], when it is applied to some subspaces $W \subset X$ for the restriction $\omega \upharpoonright W$. There is, however, a small difference: we reduce first by the characteristics of ω and then by the moment map, which can be interpreted as a *regularization* of the reduction by the characteristics.[8]

The Homogeneous Case

Because of its simple character, the case of a homogeneous action of a diffeological group G *on a space* X, *preserving a closed 2-form* ω, *deserves a special attention. We shall see in* (art. 9.23) *how this applies to classical symplectic geometry.*

9.18. The homogeneous case. Let X be a connected diffeological space equipped with a closed 2-form ω. Let ρ be a smooth action of a diffeological group G on X, preserving ω. Let us assume that X is homogeneous for this action (art. 7.8). Let Γ be the holonomy of the action ρ, let μ be a moment map, and let θ be the cocycle associated to μ. Let x_0 be any point of X, and let $\mu_0 = \mu(x_0)$. Let $\mathrm{St}_{\mathrm{Ad}_*^{\Gamma,\theta}}(\mu_0)$ be the stabilizer of μ_0 for the affine coadjoint action of G on $\mathcal{G}^*/\Gamma$. Thanks to the equivariance of the moment map μ, with respect to the θ-affine coadjoint action of G on $\mathcal{G}^*/\Gamma$, $\mu \circ \rho(g) = \mathrm{Ad}_*^{\Gamma,\theta}(g) \circ \mu$, the image $\mathcal{O} = \mu(X)$ is a (Γ, θ)-orbit of G, and $\mathrm{St}_\rho(x_0) \subset \mathrm{St}_{\mathrm{Ad}_*^{\Gamma,\theta}}(\mu_0)$. Let us equip $\mathcal{O}$ with the pushforward of the diffeology of G by the orbit map $\hat{\mu}_0 : g \mapsto \mathrm{Ad}_*^{\Gamma,\theta}(g)(\mu_0)$. Then, the orbit map $\hat{x}_0 : G \to X$ is a principal fibration with structure group $\mathrm{St}_\rho(x_0)$, the orbit map $\hat{\mu}_0 : G \to \mathcal{O}$ is a principal fibration with structure group $\mathrm{St}_{\mathrm{Ad}_*^{\Gamma,\theta}}(\mu_0)$, and the moment map $\mu : X \to \mathcal{O}$ is a fibration, with fiber the homogeneous space $\mathrm{St}_{\mathrm{Ad}_*^{\Gamma,\theta}}(\mu_0)/\mathrm{St}_\rho(x_0)$.

NOTE 1. The moment maps μ are defined up to a constant. But the characteristics of μ, that is, the connected components of the subspaces defined by $\mu(x) = \mathrm{cst}$, are not (art. 9.17).

NOTE 2. Let us consider the simple case $d\alpha$, where $\alpha \in \mathcal{G}^*$. Its moment map $\mu : G \to \mathcal{G}^*$ is the coadjoint orbit map $\mu : g \mapsto \mathrm{Ad}_*(g)(\alpha)$. A natural question is then, is there a closed 2-form ω, defined on the coadjoint orbit $\mathcal{O}_\alpha = \mu(G)$, such that $\mu^*(\omega) = d\alpha$? I do not yet have a definitive answer to this question; the best I can say is contained in the following proposition. Let us consider $\mathcal{O}_\alpha$, equipped with the quotient diffeology of G.

PROPOSITION. There exists a closed 2-form ω on $\mathcal{O}_\alpha$ such that $\mu^*(\omega) = d\alpha$, if and only if $\alpha \upharpoonright \mathrm{St}_{\mathrm{Ad}_*}(\alpha)$ is closed.

---

[8]It is not impossible that in particular, for the two-bodies problem [**Sou83**], it be precisely the universal moment map which regularizes the space of motions.

There are particular cases, or special group diffeologies, for which the restriction of α on its stabilizer $\mathrm{St}_{\mathrm{Ad}_*}(\alpha)$ is closed. For example, if G is a Lie group, then we can use the Cartan formula to check it. But even simpler, if the stabilizer $\mathrm{St}_{\mathrm{Ad}_*}(\alpha)$ is discrete or 1-dimensional, then every 1-form is closed. Paul Donato has given in [**Don94**] an interesting example of a discrete stabilizer. Let us give here the diffeological version of his construction. We consider $G = \mathrm{Diff}(S^1)$ and its universal covering $\tilde{G}$, described in Exercise 28, p. 27, as the subgroup of diffeomorphisms φ of $\mathbf{R}$ such that $\varphi(x + 2\pi) = \varphi(x) + 2\pi$. The kernel of the projection $\pi : \tilde{G} \to G$ is the subgroup $2\pi\mathbf{Z}$ of the translations $T_{2\pi k} : x \mapsto x + 2\pi k$, commuting with all $\varphi \in \tilde{G}$. Let $P : U \to \tilde{G}$ be an n-plot, and let $r \in U$, $\delta r \in \mathbf{R}^n$, and α be defined by

$$\alpha(P)_r(\delta r) = \sum_{n=0}^{\infty} \frac{1}{2^n} \frac{\partial}{\partial s} \left\{ P(r)^{-1} \circ P(s)(n) \right\}_{s=r} (\delta r).$$

One can check that α is a left invariant 1-form, and by the way an element of $\mathcal{G}^*$. The moment map μ is then given by

$$\mu(\varphi)(P)_r(\delta r) = \sum_{n=0}^{\infty} \frac{1}{2^n} \frac{\partial}{\partial s} \left\{ \varphi^{-1} \circ P(r)^{-1} \circ P(s) \circ \varphi(n) \right\}_{s=r} (\delta r).$$

Because a momentum is characterized by its values on arcs centered at the identity, and because every arc γ centered at the identity in $\tilde{G}$ is tangential to some ray $h \in \mathrm{Hom}^\infty(\mathbf{R}, \tilde{G})$, the computation of the stabilizer of α, for the coadjoint action, is reduced to Donato's computation in his paper, and it coincides with the orbits of $2\pi\mathbf{Z}$. Thus, $d\alpha$ passes to the coadjoint orbit $\mathcal{O}_\alpha$, which is actually diffeomorphic to $\mathrm{Diff}(S^1)$ itself, and symplectic according to the meaning we define below (art. 9.19).

PROOF. The triple fibration is an application of (art. 8.15).

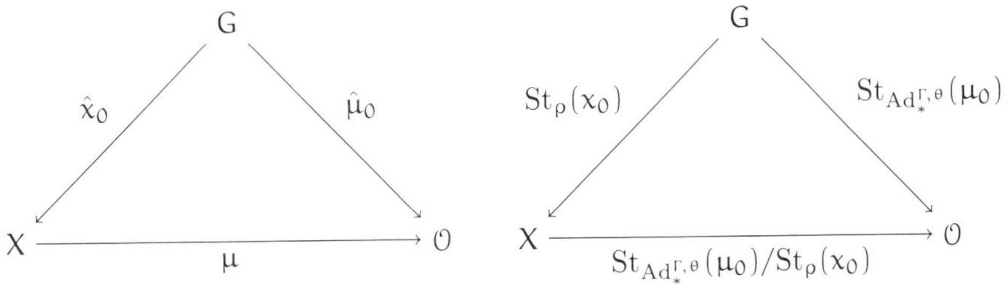

Let us focus on Note 2. There exists a (closed) 2-form ω on $\mathcal{O}_\alpha$ such that $\mu^*(\omega) = d\alpha$ if and only if, for two plots P and P' of G such that $\mu \circ P = \mu \circ P'$, then $d\alpha(P) = d\alpha(P')$ (art. 6.38). But $\mu \circ P = \mu \circ P'$ means that there exists a plot κ of $\mathrm{St}_{\mathcal{G}^*}(\alpha)$ such that $P'(r) = P(r) \cdot \kappa(r)$. Then, thanks to Exercise 127, p. 227, $\alpha[r \mapsto P(r) \cdot \kappa(r)]_r = [L(P(r))^*(\alpha)](\kappa)_r + [R(\kappa(r))^*(\alpha)](P)_r = \alpha(\kappa)_r + [\mathrm{Ad}_*(\kappa(r))(\alpha)](P)_r$, but since κ is a plot of the stabilizer of α for the coadjoint action, $\mathrm{Ad}_*(\kappa(r))(\alpha) = \alpha$ and $\alpha(P') = \alpha(P) + \alpha(\kappa)$. Thus, $d\alpha$ passes to the quotient $\mathcal{O}_\alpha \simeq G/\mathrm{St}_{\mathcal{G}^*}(\alpha)$ if and only if $d\alpha(\kappa) = 0$ for all plots κ of $\mathrm{St}_{\mathcal{G}^*}(\alpha)$, *i.e.*, if and only if $d[\alpha \restriction \mathrm{St}_{\mathcal{G}^*}(\alpha)] = 0$. □

9.19. Symplectic homogeneous diffeological spaces. Let X be a connected diffeological space and ω be a closed 2-form defined on X.

Definition. We say that (X, ω) is a *homogeneous symplectic space* if it is homogeneous under the action of $\mathrm{Diff}(X, \omega)$ and if the characteristics of a universal moment map μ_ω are reduced to points (art. 9.17).

For example, the 2-form $d\alpha$ on $\mathrm{Diff}(S^1)$ described in the previous article (art. 9.18) is symplectic in this sense. Let us recall that being homogeneous under the action of $\mathrm{Diff}(X, \omega)$ means that the orbit map $R(x) : \mathrm{Diff}(X, \omega) \to X$, defined by $R(x)(\varphi) = \varphi(x)$, is a subduction, where $x \in X$ and $\mathrm{Diff}(X, \omega)$ is equipped with the functional diffeology (art. 1.61).

NOTE 1. The characteristics of a universal moment map μ_ω are reduced to points if and only if it is a covering onto its image, equipped with the quotient diffeology. If it is the case for one universal moment map, then it is the case for every one.

NOTE 2. The homogeneous situation where a moment map μ_ω is not a covering onto its image can be regarded as the *presymplectic homogeneous case*, as suggested by (art. 9.26).

NOTE 3. Let G be a diffeological group, and let ρ be a smooth action of G on X, preserving ω. If the action ρ of G on X is homogeneous, then X is a homogeneous space of $\mathrm{Diff}(X, \omega)$. And, if a moment map $\mu : X \to \mathcal{G}^*/\Gamma$ is a covering onto its image, then any universal moment map $\mu_\omega : X \to \mathcal{G}^*_\omega/\Gamma_\omega$ is a covering onto its image. Thus, to check that a homogeneous pair (X, ω) is symplectic it is sufficient to find a homogeneous smooth action, of some diffeological group G, for which a moment map is a covering onto its image.

PROOF. Note 1 is obvious, by definition of the characteristics (art. 9.17) and by homogeneity (art. 9.18). Let us prove Note 3. To be homogeneous under the action of G means that, for some point (and thus for every point) $x \in X$, the orbit map $\hat{x} : G \to X$, defined by $\hat{x}(g) = \rho(g)(x)$, is a subduction. So, $\hat{x}$ is surjective and, for any plot $P : U \to X$, for any $r_0 \in U$, there exist an open neighborhood V of r_0 and a plot $Q : V \to G$ such that $P \upharpoonright V = \hat{x} \circ Q$, that is, $P(r) = \rho(Q(r))(x)$ for all $r \in V$. Since ρ is smooth, $\bar{Q} = \rho \circ Q$ is a plot of $\mathrm{Diff}(X, \omega)$, and $P \upharpoonright V = \hat{x} \circ \bar{Q}$. Since, $\hat{x} : \mathrm{Diff}(X, \omega) \to X$ is surjective, it is a subduction and X is a homogeneous space of $\mathrm{Diff}(X, \omega)$. Now, let us remark that, since the moment maps differ only by a constant, if a moment map μ is a covering onto its image $\mathcal{O}$ equipped with the quotient diffeology of G, then every other moment map $\mu' = \mu + \mathrm{cst}$ is a covering onto its image $\mathcal{O}' = \mathcal{O} + \mathrm{cst}$. Then, let x_0 be a point of X, and let $\mu(x) = \psi(x_0, x)$, where ψ is the 2-points moment map. Let $\mu_\omega = \psi_\omega(x_0, x)$. According to (art. 9.14), $\mu = \rho^*_{\Gamma_\omega} \circ \mu_\omega$. Let $\mathcal{O} = \mu(X) = \rho^*_{\Gamma_\omega}(\mu_\omega(X)) = \rho^*_{\Gamma_\omega}(\mathcal{O}_\omega)$, with $\mathcal{O}_\omega = \mu_\omega(X)$. Let $m_\omega \in \mathcal{O}_\omega$ and $m = \rho^*_{\Gamma_\omega}(m_\omega)$. If $\mu_\omega(x) = m_\omega$, then $\rho^*_{\Gamma_\omega}(\mu_\omega(x)) = \rho^*_{\Gamma_\omega}(m_\omega)$, that is, $\mu(x) = m$. Thus, $\mu_\omega^{-1}(m_\omega) \subset \mu^{-1}(m)$. Therefore, if $\mu^{-1}(m)$ is discrete, then $\mu_\omega^{-1}(m_\omega)$ is discrete, *a fortiori*, and if μ is injective, then μ_ω is injective. □

Exercise

✎ EXERCISE 144 (Compact supported real functions III). Let us consider the space (X, ω) as defined in Exercise 142, p. 307. Show that (X, ω) is a homogeneous symplectic space.

About Symplectic Manifolds

The case of classical symplectic manifolds (M, ω) *deserves special care. We shall see in this section that, in this case, any universal moment map* μ_ω *is injective and therefore identifies* M *with a coadjoint orbit of* $\mathrm{Diff}(M, \omega)$, *in the general meaning that we gave in* (art. 7.16).

9.20. Value of the moment maps for manifolds. Let M be a connected manifold equipped with a closed 2-form ω. In this context, the paths moment map Ψ_ω takes a special expression. Let p be a path in M and $F : U \to \mathrm{Diff}(M, \omega)$ be an n-plot, then

$$\Psi_\omega(p)(F)_r(\delta r) = \int_0^1 \omega_{p(t)}(\dot{p}(t), \delta p(t))\, dt, \qquad (\diamondsuit)$$

for all $r \in U$ and $\delta r \in \mathbf{R}^n$, where δp is the lift in the tangent space TM of the path p, defined by

$$\delta p(t) = [D(F(r))(p(t))]^{-1} \frac{\partial F(r)(p(t))}{\partial r}(\delta r). \qquad (\heartsuit)$$

PROOF. By definition, $\Psi(p)(F) = \hat{p}^*(\mathcal{K}\omega)(F) = \mathcal{K}\omega(\hat{p} \circ F)$. The expression of the operator $\mathcal{K}$ (art. 6.83), applied to the plot $\hat{p} \circ F : r \mapsto F(r) \circ p$ of Paths(X), gives

$$(\mathcal{K}\omega)(\hat{p} \circ F)_r(\delta r) = \int_0^1 \omega \left[\begin{pmatrix} s \\ u \end{pmatrix} \mapsto (\hat{p} \circ F)(u)(s+t) \right]_{\binom{s=0}{u=r}} \begin{pmatrix} 1 \\ 0 \end{pmatrix} \begin{pmatrix} 0 \\ \delta r \end{pmatrix} dt.$$

But $(\hat{p} \circ F)(u)(s+t) = F(u)(p(s+t))$, let us denote temporarily by Φ_t the plot $(s, u) \mapsto F(u)(p(s+t))$, so $F(u)(p(s+t))$ writes $\Phi_t(s, u)$. Now, let us denote by $\mathcal{I}$ the integrand of the right term of this expression. We have

$$\begin{aligned}
\mathcal{I} &= \omega \left[\begin{pmatrix} s \\ u \end{pmatrix} \mapsto \Phi_t(s, u) \right]_{\binom{s=0}{u=r}} \begin{pmatrix} 1 \\ 0 \end{pmatrix} \begin{pmatrix} 0 \\ \delta r \end{pmatrix} \\
&= \Phi_t^*(\omega)_{\binom{0}{r}} \begin{pmatrix} 1 \\ 0 \end{pmatrix} \begin{pmatrix} 0 \\ \delta r \end{pmatrix} \\
&= \omega_{\Phi_t\binom{0}{r}} \left(D(\Phi_t)_{\binom{0}{r}} \begin{pmatrix} 1 \\ 0 \end{pmatrix}, D(\Phi_t)_{\binom{0}{r}} \begin{pmatrix} 0 \\ \delta v \end{pmatrix} \right) \\
&= \omega_{F(r)(p(t))} \left(\frac{\partial}{\partial s} \Big\{ F(r)(p(s+t)) \Big\}_{s=0}, \frac{\partial}{\partial r} \Big\{ F(r)(p(t)) \Big\}(\delta r) \right).
\end{aligned}$$

But,

$$\begin{aligned}
\frac{\partial}{\partial s} \Big\{ F(r)(p(s+t)) \Big\}_{s=0} &= D(F(r))(p(t)) \left(\frac{\partial p(s+t)}{\partial s} \Big|_{s=0} \right) \\
&= D(F(r))(p(t))(\dot{p}(t)).
\end{aligned}$$

Then, using this last expression and the fact that F is a plot of $\mathrm{Diff}(M, \omega)$, that is, for all r in U, $F(r)^*\omega = \omega$, we have

$$\begin{aligned}
\mathcal{I} &= \omega_{F(r)(p(t))} \left(D(F(r))(p(t))(\dot{p}(t)), \frac{\partial F(r)(p(t))}{\partial r}(\delta r) \right) \\
&= \omega_{p(t)} \left(\dot{p}(t), [D(F(r))(p(t))]^{-1} \frac{\partial F(r)(p(t))}{\partial r}(\delta r) \right) \\
&= \omega_{p(t)}(\dot{p}(t), \delta p(t)).
\end{aligned}$$

Therefore, $\Psi_\omega(p)(F)_r(\delta r) = \mathcal{K}\omega(\hat{p} \circ F)_r(\delta r) = \int_0^1 \omega_{p(t)}(\dot{p}(t), \delta p(t))\, dt.$ □

9.21. The paths moment map for symplectic manifolds. Let M be a Hausdorff manifold, and let ω be a nondegenerate closed 2-form defined on M. Let m_0 and m_1 be two points of M connected by a path p. Let $f \in \mathcal{C}^\infty(M, \mathbf{R})$ with compact support. Let F be the exponential of the symplectic gradient[9] $\mathrm{grad}_\omega(f)$, F is a 1-plot of Diff(M, ω), and precisely a 1-parameter homomorphism. Then, the universal paths moment map Ψ_ω, computed at the path p, evaluated on the 1-plot F, is the constant 1-form of $\mathbf{R}$

$$\Psi_\omega(p)(F) = [f(m_1) - f(m_0)] \times dt \quad \text{with} \quad F : t \mapsto e^{t\, \mathrm{grad}_\omega(f)},$$

where dt is the standard 1-form of $\mathbf{R}$. Note that we are in the special case where F is actually a 1-parameter homomorphism of Ham$(M, \omega) \subset$ Diff(M, ω), and the function f is one *Hamiltonian* of F.

PROOF. Let us remark that, in our case, the lift δp defined by (art. 9.20, (♡)) writes simply

$$\delta p(t) = [D(e^{r\xi})(p(t))]^{-1} \frac{\partial e^{r\xi}(p(t))}{\partial r}(\delta r) = \xi(p(t)) \times \delta r,$$

with $\xi = \mathrm{grad}_\omega(f)$, and where r and δr are real numbers. Thus, the expression (art. 9.20, (◇)) becomes

$$\begin{aligned}
\Psi_\omega(p)(F)_r(\delta r) &= \int_0^1 \omega_{p(t)}(\dot{p}(t), \xi(p(t)))\, dt \times \delta r \\
&= \int_0^1 \omega_{p(t)}(\dot{p}(t), \mathrm{grad}_\omega(f)(p(t)))\, dt \times \delta r \\
&= \int_0^1 df\left(\frac{dp(t)}{dt}\right) dt \times \delta r \\
&= [f(p(1)) - f(p(0))] \times \delta r,
\end{aligned}$$

that is, $\Psi_\omega(p)(F) = [f(m_1) - f(m_0)] \times dt.$ □

9.22. Hamiltonian diffeomorphisms of symplectic manifolds. Let (M, ω) be a connected Hausdorff symplectic manifold. According to Banayaga [**Ban78**], a diffeomorphism f is said to be *Hamiltonian* if it can be connected to the identity 1_M by a smooth path $t \mapsto f_t$ in Diff(M, ω) such that

$$\omega(\dot{f}_t, \cdot) = d\phi_t \quad \text{with} \quad \dot{f}_t(x) = \frac{d}{ds}\left\{f_s \circ f_t^{-1}(x)\right\}_{s=t},$$

where $(t, x) \mapsto \phi_t(x)$ is a smooth real function. If f is Hamiltonian according to this definition, then it belongs to Ham(M, ω), as defined in (art. 9.15). Conversely, any element f of Ham(M, ω) satisfies the above Banyaga's condition. Thus, the definition of Hamiltonian diffeomorphisms given in (art. 9.15) is a faithful generalization of the usual definition for symplectic manifolds.

PROOF. This proposition is a direct consequence of the general statement given in (art. 9.16) and the following comparison between the above 1-parameter family

[9] Let us recall that the symplectic gradient is defined by $\omega(\mathrm{grad}_\omega(f), \cdot) = -df$.

of vector fields $\dot{f}_t$ and the family F_t of the (art. 9.16). Since $f_{t'} \circ f_t^{-1} = f_t \circ (f_t^{-1} \circ f_{t'}) \circ f_t^{-1}$, the vector fields $\dot{f}_t$ and F_t are conjugated by f_t, precisely

$$\dot{f}_t = (f_t)_*(F_t) \quad \text{or} \quad \dot{f}_t(x) = D(f_t)(f_t^{-1}(x))(F_t(f_t^{-1}(x))).$$

This implies in particular that if the vector field $\dot{f}_t$ satisfies Banyaga's condition for the function ϕ_t, then the vector field F_t satisfies Banyaga's condition for the function $\Phi_t = -\phi_t \circ f_t$, and conversely, that is,

$$\omega(\dot{f}_t, \cdot) = d\phi_t \quad \Leftrightarrow \quad \omega(F_t, \cdot) = -d\Phi_t \quad \text{with} \quad \Phi_t = -\phi_t \circ f_t.$$

Let us check it. Let $x \in M$, $x' = f_t(x)$, $\delta x \in T_x M$, and $\delta x' = D(f_t)(x)(\delta x)$, then

$$\begin{aligned}
\omega_{x'}(\dot{f}_t(x'), \delta x') &= [d\phi_t]_{x'}(\delta x') \\
\omega_{f_t(x)}(\dot{f}_t(f_t(x)), D(f_t)(x)(\delta x)) &= [d\phi_t]_{f_t(x)}(D(f_t)(x)(\delta x)) \\
\omega_{f_t(x)}(D(f_t)(x)(F_t(x)), D(f_t)(x)(\delta x)) &= [f_t^*(d\phi_t)]_x(\delta x) \\
[f_t^*(\omega)]_x(F_t(x), \delta x) &= d[f_t^*(\phi_t)]_x(\delta x) \\
\omega_x(F_t(x), \delta x) &= d[\phi_t \circ f_t]_x(\delta x).
\end{aligned}$$

Thus, $\Phi_t = -\phi_t \circ f_t$. □

9.23. Symplectic manifolds are coadjoint orbits. Let M be a manifold, and let ω be a closed 2-form defined on M. We assume M connected and Hausdorff. Then, the form ω is nondegenerate, thus symplectic, if and only if the two following conditions are fulfilled.

A. The manifold M is a homogeneous space of $\mathrm{Diff}(M, \omega)$.

B. The universal moment map $\mu_\omega : M \to \mathcal{G}_\omega^* / \Gamma_\omega$ is injective.

In that case, the image $\mathcal{O}_\omega = \mu_\omega(M) \in \mathcal{G}_\omega^*/\Gamma_\omega$ of the universal moment map[10] (art. 9.14) is a $(\Gamma_\omega, \theta_\omega)$-coadjoint orbit of $\mathrm{Diff}(M, \omega)$ (art. 7.16), and μ_ω identifies M with $\mathcal{O}_\omega$, where $\mathcal{O}_\omega$ is equipped with the quotient diffeology of $\mathrm{Diff}(M, \omega)$. In other words, *every symplectic manifold is a coadjoint orbit.*

NOTE 1. Let $\mathrm{Ham}(M, \omega)$ be the group of Hamiltonian diffeomorphisms, and let $\mathcal{H}_\omega^*$ be the space of its momenta. Let $\mu_\omega^\star : M \to \mathcal{H}_\omega^*$ be a moment map associated with the action of $\mathrm{Ham}(M, \omega)$, and let $\theta_\omega^\star$ be the associated Souriau cocycle. Then, $\mu_\omega^\star$ is also injective, and identifies M to a $\theta_\omega^\star$-coadjoint orbit $\mathcal{O}_\omega^\star \subset \mathcal{H}_\omega^*$ of $\mathrm{Ham}(M, \omega)$.

NOTE 2. Let us consider the example $M = \mathbf{R}^2$ and $\omega = (x^2 + y^2)\, dx \wedge dy$. Since $\mathbf{R}^2$ is contractible, the holonomy Γ_ω is trivial (art. 9.7). Next, ω is nondegenerate on $\mathbf{R}^2 - \{0\}$, but degenerates at the point $(0, 0)$. Thus, $(0, 0)$ is an orbit of the group $\mathrm{Diff}(\mathbf{R}^2, \omega)$, and actually $\mathbf{R}^2 - \{0\}$ is the other orbit. Hence, the universal moment map μ_ω such that $\mu_\omega(0, 0) = 0_{\mathcal{G}_\omega^*}$ is equivariant (art. 9.10, Note 2). Moreover, μ_ω is injective. The closed 2-form ω not being symplectic, with an injective universal moment map, shows that the hypothesis of transitivity of $\mathrm{Diff}(M, \omega)$ on M is not superfluous is this proposition.

PROOF. Let us assume first that ω is nondegenerate, that is, symplectic. Then, the group $\mathrm{Diff}(M, \omega)$ is transitive on M [**Boo69**]. Moreover, for every $m \in M$, the orbit map $\hat{m} : \varphi \mapsto \varphi(m)$ is a subduction [**Don84**]. Thus, the image of the moment moment map μ_ω is one orbit $\mathcal{O}_\omega$ of the affine coadjoint action of G_ω on

[10]The universal moment maps are defined up to a constant, but if one is injective, then they are all injectives.

$\mathcal{G}^*_\omega/\Gamma_\omega$, associated with the cocycle θ_ω. Hence, the orbit $\mathcal{O}_\omega$ being equipped with the quotient diffeology of G_ω, the moment map μ_ω is a subduction.

Now, let m_0 and m_1 be two points of M such that $\mu_\omega(m_0) = \mu_\omega(m_1)$, that is, $\psi_\omega(m_0, m_1) = \mu_\omega(m_1) - \mu_\omega(m_0) = 0$. Let $p \in \text{Paths}(M)$ such that $p(0) = m_0$ and $p(1) = m_1$. Thus, $\psi_\omega(m_0, m_1) = 0$ is equivalent to $\Psi_\omega(p) = \Psi_\omega(\ell)$, where ℓ is some loop in M, we can choose $\ell(0) = \ell(1) = m_0$. Now, let us assume that $m_0 \neq m_1$. Since M is Hausdorff there exists a smooth real function $f \in \mathcal{C}^\infty(M, \mathbf{R})$, with compact support, such that $f(m_0) = 0$ and $f(m_1) = 1$. Let us denote by ξ the symplectic gradient field associated with f and by F the exponential of ξ. Thanks to (art. 9.21), on the one hand we have $\Psi(p)(F) = [f(m_1) - f(m_0)]\,dt = dt$, and on the other hand $\Psi_\omega(\ell)(F) = [f(m_0) - f(m_0)]\,dt = 0$. But $dt \neq 0$, thus $\psi_\omega(m_0, m_1) \neq 0$, and the moment map μ_ω is injective. Therefore, μ_ω is an injective subduction on $\mathcal{O}_\omega$, that is, a diffeomorphism.

Conversely, let us assume that M is a homogeneous space of $\text{Diff}(M, \omega)$ and that μ_ω is injective. Let us notice first that since $\text{Diff}(M, \omega)$ is transitive, the rank of ω is constant. In other words, $\dim(\ker(\omega)) = \text{cst}$. Now, let us assume ω is degenerate, that is, $\dim(\ker(\omega)) \geq 1$. Since $m \mapsto \ker(\omega_m)$ is a smooth foliation, for every point m of M there exists a smooth path p in M such that $p(0) = m$ and for t belonging to a small interval around $0 \in \mathbf{R}$, $\dot p(t) \neq 0$ and $\dot p(t) \in \ker(\omega_{p(t)})$ for all t in this interval. Then, we can reparametrize the path p and assume now that p is defined on the whole $\mathbf{R}$ and satisfies $p(0) = m$, $p(1) = m'$ with $m \neq m'$, and $\dot p(t) \in \ker(\omega_{p(t)})$ for all t. Next, since $\dot p(t) \in \ker(\omega_{p(t)})$ for all t, using the expression (art. 9.20, ($\diamondsuit$)), we get $\Psi_\omega(p) = 0_{\mathcal{G}^*_\omega}$, that is, $\mu_\omega(m) = \mu_\omega(m')$. But since $m \neq m'$ and we assumed μ_ω injective, this is a contradiction. Thus, the kernel of ω is reduced to $\{0\}$, and ω is nondegenerate, that is, symplectic.

Let us prove Note 1. According to a Boothby's theorem, the group $\text{Ham}(M, \omega)$ acts transitively on M [**Boo69**]. With respect to this group, and by construction, the holonomy is trivial: the associated paths moment map $\Psi^\star_\omega$ and the moment maps $\mu^\star_\omega$ take their values in the space $\mathcal{H}^*_\omega$. Let $j : \text{Ham}(M, \omega) \to \text{Diff}(M, \omega)$ be the inclusion, thus the universal holonomy Γ_ω is in the kernel of j^*, and we get a natural mapping $j^*_{\Gamma_\omega} : \mathcal{G}^*_\omega/\Gamma_\omega \to \mathcal{H}^*_\omega$. Now, the paths moment maps satisfy $\Psi^\star_\omega = j^*_{\Gamma_\omega} \circ \Psi_\omega$, and $\mu^\star_\omega = j^*_{\Gamma_\omega} \circ \mu_\omega$ (art. 9.14). Then, since (art. 9.21) involves only plots of $\text{Ham}(X, \omega)$, the proof above applies *mutatis mutandis* to the Hamiltonian case, and we deduce that the moment maps $\mu^\star_\omega$ are injective. By transitivity, they identify M with a $\theta^\star_\omega$-coadjoint orbit of $\text{Ham}(M, \omega)$.

Let us finish by proving the second note, that is, the universal moment map μ_ω of $\omega = (x^2 + y^2)\,dx \wedge dy$ is injective. First of all $\mu_\omega(0, 0) = 0_{\mathcal{G}^*}$. Now, if $z = (x, y)$ and $z' = (x', y')$ are two different points of $\mathbf{R}^2$ and different from $(0, 0)$, then there is a smooth function with compact support contained in a small ball, not containing $(0, 0)$ nor z, such that $f(z') = 1$. Then, the 1-parameter group generated by $\text{grad}_\omega(f)$ belongs to $\text{Diff}(\mathbf{R}^2, \omega)$, and a similar argument as the one of the proof above shows that $\mu_\omega(z) \neq \mu_\omega(z')$. We still need to prove that if $z \neq (0, 0)$, then $\mu_\omega(z) \neq 0_{\mathcal{G}^*}$. Let us consider $p(t) = tz$ and $F(r)$ be the positive rotation of angle $2\pi r$, where $r \in \mathbf{R}$. The application of (art. 9.20, ($\diamondsuit$)), computed at the point $r = 0$ and applied to the vector $\delta r = 1$, gives $(2\pi/3)(x^2 + y^2)^2$ which is not zero. Therefore, the moment map μ_ω is injective. □

9.24. The classical homogeneous case. Let (M, ω) be a symplectic manifold. Let G be a Lie group together with a homogeneous Hamiltonian action on (M, ω), that is, the holonomy Γ of G is trivial. For the sake of simplicity we assume M connected and G a Lie subgroup of $\mathrm{Diff}(M, \omega)$. By functoriality of the moment maps (art. 9.12), we know that if a moment map μ of G is injective, then every universal moment map μ_ω is injective (art. 9.23). But we are now in the opposite case, since ω is symplectic every universal moment map μ_ω is injective, but what about μ? This is actually the original case treated by Souriau in [**Sou70**]. He showed that the moment map μ is a covering onto its image, which is some coadjoint orbit $\mathcal{O} \subset \mathcal{G}^*$ (affine or not) of G. We give here the proof of this theorem, according to the present framework. This case is illustrated by Exercise 146, p. 328.

PROOF. Let p be a path in M such that $\mu \circ p = \mathrm{cst}$, that is, $\Psi(p) = 0_{\mathcal{G}^*}$, where Ψ is the paths moment map of G. Then, for any 1-parameter subgroup $F \in \mathrm{Hom}^\infty(\mathbf{R}, G)$, $\Psi(p)(F)_r(\delta r) = 0$, for all r and all δr belonging to $\mathbf{R}$. Adapting to our case the expression of Ψ given in (art. 9.20), we get

$$\int_0^1 \omega_{p(t)}(\dot{p}(t), Z(p(t)))\, dt = 0 \quad \text{where} \quad Z(m) = \left.\frac{\partial F(t)(m)}{\partial t}\right|_{t=0}.$$

Now, considering the 1-parameter family of paths $p_s : t \mapsto p(st)$, the derivative of the above expression gives $\omega_{p(0)}(\dot{p}(0), Z(p(0))) = 0$. But since G is transitive, by running over all the 1-parameter subgroups F of G we describe the whole tangent space $T_m M$, where $m = p(0)$. And since ω is nondegenerate, $\dot{p}(0) = 0$. The path p is thus constant, $p(t) = m$ for all t in $\mathbf{R}$. Therefore the preimages of the values of the moment map μ are discrete. But, μ is a fibration (see (art. 9.18)), thus μ is a covering (art. 8.24) onto its image which is, by transitivity, a coadjoint orbit. □

9.25. The Souriau-Nœther theorem. Let M be a manifold, and let ω be a closed 2-form on M. We say that two points m and m' are on the same *characteristic*[11] of ω if there exists a path p connecting m to m' such that $\dot{p}(t) \in \ker(\omega_{p(t)})$ for all t. Then, the universal moment maps μ_ω are constant on the characteristics of ω.

NOTE 1. In particular, for any smooth action of a diffeological group G, preserving ω, the moment maps μ are constant on the characteristics of ω.

NOTE 2. This is an analogue to the first Nœther theorem, relating symmetries to conserved quantities for Lagrangian systems. For a comprehensive exposé on the subject, see the book of Y. Kosmann-Schwarzbach [**YKS10**].

PROOF. We have $\mu_\omega(m') - \mu_\omega(m) = \psi_\omega(m, m') = \mathrm{class}(\Psi_\omega(p)) \in \mathcal{G}^*_\omega/\Gamma_\omega$, where p is any (smooth) path connecting m to m'. Then, thanks to the hypothesis, we can use a path p contained in the characteristic of ω containing m and m', that is, $\dot{p}(t) \in \ker(\omega_{p(t)})$ for all t. Now, thanks to the explicit formula of (art. 9.20), for every n-plot F of $\mathrm{Diff}(M, \omega)$, $n \in \mathbf{N}$, for every $r \in \mathrm{def}(F)$, for every $\delta r \in \mathbf{R}^n$,

$$\Psi_\omega(p)(F)_r(\delta r) = \int_0^1 \omega_{p(t)}(\dot{p}(t), \delta p(t))\, dt = \int_0^1 0 \times dt = 0.$$

Thus, $\Psi_\omega(p) = 0$ and therefore $\mu_\omega(m') = \mu_\omega(m)$. The note is a consequence of the functoriality of the moment map (art. 9.14). Indeed, let ρ be the morphism from G to $\mathrm{Diff}(M, \omega)$, then the paths moment map Ψ, relative to G, satisfies $\Psi = \rho^* \circ \Psi_\omega$.

[11] This is the classical definition of the characteristics in the case of a closed 2-form ω defined on a manifold M. They are the integral submanifolds of the distribution $m \mapsto \ker(\omega_m)$.

Thus, $\Psi_\omega(p) = 0$ implies $\Psi(p) = 0$ which implies $\mu(m) = \mu(m')$, for any moment map μ relative to G. ☐

9.26. Presymplectic homogeneous manifolds. Let M be a connected Hausdorff manifold, and let ω be a closed 2-form on M. Let $G \subset \mathrm{Diff}(M,\omega)$ be a connected subgroup. If M is a homogeneous space of G, then the characteristics of ω are the connected components of the preimages of the moment maps μ.

NOTE. In particular, if M is a homogeneous space of $\mathrm{Diff}(M,\omega)$, and thus of its identity component (art. 7.9), then the characteristics of ω are the connected components of the preimages of the values of a universal moment map μ_ω. This justifies *a posteriori* the definition of the characteristics of moment maps, for the general case of homogeneous diffeological spaces, in (art. 9.17).

PROOF. The Souriau-Nœther theorem states that if m and m′ are on the same characteristic, then $\mu(m) = \mu(m')$ (art. 9.25). We shall prove the converse in a few steps.

a) Let us consider first the case when the holonomy of Γ is trivial, $\Gamma = \{0\}$. Let us assume m and m′ connected by a path p such that $\mu(p(t)) = \mu(m)$ for all t. Then, let $s \mapsto p_s$ be defined by $p_s(t) = p(st)$, for all s and t. We have $\mu(p_s(1)) = \mu(p_s(0))$, that is, $\Psi(p_s) = 0^*_{\mathfrak{g}}$, for all s. Thus, for all n-plots F of G, for all $r \in \mathrm{def}(F)$ and all $\delta r \in \mathbf{R}^n$, $\Psi(p_s)(F)_r(\delta r) = 0$, and hence

$$0 = \frac{\partial \Psi(p_s)(F)_r(\delta r)}{\partial s} = \frac{\partial}{\partial s}\int_0^1 \omega_{p_s(t)}(\dot p_s(t), \delta p_s(t))\, dt = \omega_{p(s)}(\dot p(s), \delta p(s)),$$

where $\delta p(t)$ is given by (art. 9.20, (♡)). Next, let $v \in T_{p(t)}(M)$, then there exists a path c of M such that $c(0) = p(t)$ and $dc(s)/ds|_{s=0} = v$. Since M is assumed homogeneous under the action of G, there exists a 1-plot $s \mapsto F(s)$ centered at the identity, that is, $F(0) = \mathbf{1}_M$, such that $F(s)(p(t)) = c(s)$. Then, for $s = 0$ and $\delta s = 1$, we get from (art. 9.20, (♡)),

$$\delta p(t) = \mathbf{1}_{T_{p(t)}M} \left.\frac{dF(s)(p(t))}{ds}\right|_{s=0} = \left.\frac{dc(s)}{ds}\right|_{s=0} = v.$$

Hence, for every $v \in T_{p(t)}M$, $\omega(\dot p(t), v) = 0$, *i.e.*, $\dot p(t) \in \ker(\omega_{p(t)})$ for all t. Therefore, the connected components of the preimages of the values of the moment map μ are the characteristics of ω.

b) Let us consider the general case. Let $\widetilde{M}$ be the universal covering of M, $\pi : \widetilde{M} \to M$ the projection, and let $\tilde\omega = \pi^*(\omega)$. Let $\widehat{G}$ be the group defined by

$$\widehat{G} = \{\hat g \in \mathrm{Diff}(\widetilde{M}, \tilde\omega) \mid \exists g \in G \text{ and } \pi \circ \hat g = g \circ \pi\}.$$

Let $\rho : \widehat{G} \to G$ be the morphism $\hat g \mapsto g$. The group $\widehat{G}$ is an extension of G by the homotopy group $\pi_1(M)$

$$\mathbf{1} \longrightarrow \pi_1(M) \longrightarrow \widehat{G} \xrightarrow{\ \rho\ } G \longrightarrow \mathbf{1}.$$

1. The morphism ρ is surjective. Let $g \in G$, let $t \mapsto g_t$ be a smooth path in G connecting $\mathbf{1}_G$ to g. Let $\tilde m \in \widetilde{M}$ and $m = \pi(\tilde m)$, the path $t \mapsto g_t(m)$ has a unique lift $t \mapsto \tilde m_t$ in $\widetilde{M}$ starting at $\tilde m$ (art. 8.25). We can check that, $\tilde g : \tilde m \mapsto \tilde m_1$ is a diffeomorphism of $\widetilde{M}$, satisfying by construction $\pi \circ \tilde g = g \circ \pi$. Next, $\tilde g^*(\tilde\omega) =$

$\tilde{g}^*(\pi^*(\omega)) = \tilde{g}^* \circ \pi^*(\omega) = (\pi \circ \tilde{g})^*(\omega) = (g \circ \pi)^*(\omega) = \pi^*(g^*(\omega)) = \pi^*(\omega) = \tilde{\omega}$. Thus, $\tilde{g} \in \widehat{G}$.

2. The group $\widehat{G}$ is transitive on $\widetilde{M}$. Let $\tilde{m}$ and $\tilde{m}'$ be two points of $\widetilde{M}$, let $m = \pi(\tilde{m})$ and $m' = \pi(\tilde{m}')$. Since G is transitive on M there exists $g \in G$ such that $g(m) = m'$. The lift $\tilde{g}$ defined in part 1 maps $\tilde{m}$ to $\tilde{m}_1$, and we have $\pi(\tilde{m}_1) = \pi(\tilde{m}') = m'$. So there exists an element $k \in \pi_1(M)$ such that $k_{\widetilde{M}}(\tilde{m}_1) = \tilde{m}'$ (art. 8.26). Let $\hat{g} = k_{\widetilde{M}} \circ \tilde{g}$, since $\pi \circ \hat{g} = g \circ \pi$ and $\hat{g}^*(\tilde{\omega}) = (k_{\widetilde{M}} \circ \tilde{g})^*(\tilde{\omega}) = \tilde{g}^*(k^*_{\widetilde{M}}(\tilde{\omega})) = \tilde{g}^*(\tilde{\omega}) = \tilde{\omega}$, $\hat{g}$ belongs to $\widehat{G}$, and maps $\tilde{m}$ to $\tilde{m}'$.

3. The kernel of ρ is reduced to $\pi_1(M)$. Let $\tilde{g} \in \widehat{G}$ such that $\rho(\tilde{g}) = \mathbf{1}_M$, that is, $\pi \circ \tilde{g} = \pi$. Thus, for every $\tilde{m} \in M$ there exists $\kappa(\tilde{m}) \in \pi_1(M)$ such that $\tilde{g}(\tilde{m}) = \kappa(\tilde{m})_M(\tilde{m})$. But the map $\kappa : \widetilde{M} \to \pi_1(M)$ is smooth, and $\pi_1(M)$ discrete, so κ is constant. Therefore, there exists $k \in \pi_1(M)$ such that $\tilde{g}(\tilde{m}) = k_M(\tilde{m})$, for all $\tilde{m} \in \widetilde{M}$.

4. Since $\widetilde{M}$ is simply connected, $\widehat{G}$ has no holonomy. Let $\hat{\mu}$ be a moment map of the action of $\widehat{G}$ on $\widetilde{M}$. Since $\widehat{G}$ is a discrete extension of G, their space of momenta coincide (art. 7.13), thus $\hat{\mu}$ takes its values in $\mathcal{G}^*$, and since the action of G on M is the image by the morphism ρ of the action of $\widehat{G}$ on $\widetilde{M}$, for every $\tilde{m} \in \widetilde{M}$, $\mu(\pi(\tilde{m})) = \mathrm{class}(\hat{\mu}(\tilde{m})) \in \mathcal{G}^*/\Gamma$ (art. 9.13). Next, let $c = \mu(m) = \mathrm{class}(\hat{\mu}(\tilde{m})) \in \mathcal{G}^*/\Gamma$, $m = \pi(\tilde{m})$, and $C = \mu^{-1}(c)$. The preimage $\pi^{-1}(C) = \pi^{-1}(\mu^{-1}(c))$ is equal to $(\mu \circ \pi)^{-1}(c) = (\mathrm{class} \circ \hat{\mu})^{-1}(c) = \hat{\mu}^{-1}(\mathrm{class}^{-1}(c)) = \hat{\mu}^{-1}(\hat{\mu}(\tilde{m}) + \Gamma)$. Thus,

$$\mu^{-1}(c) = \pi\Bigg(\bigcup_{\gamma \in \Gamma} \hat{\mu}^{-1}(\alpha + \gamma)\Bigg) \quad \text{with} \quad \alpha = \hat{\mu}(\tilde{m}) \quad \text{and} \quad c = \mathrm{class}(\alpha).$$

Since $\widehat{G}$ is transitive on $\widetilde{M}$ and since there is no holonomy, we can apply the result of part a): for every $\gamma \in \Gamma$, $\hat{\mu}^{-1}(\alpha + \gamma)$ is a union of characteristics of $\tilde{\omega}$. Thus, the union over all the $\gamma \in \Gamma$ is still a union of characteristics of ω. Hence, $\mu^{-1}(c)$ is the π-projection of a union of characteristics of $\tilde{\omega}$. But, since $\pi : \widetilde{M} \to M$ is a covering and since $\tilde{\omega} = \pi^*(\omega)$, the π-projection of a characteristic of $\tilde{\omega}$ is a characteristic of ω. Thus $\mu^{-1}(c)$ is a union of characteristics of ω, and the connected components of the preimages of the values of μ are the characteristics of ω. □

Exercises

✎ EXERCISE 145 (The classical moment map). Let M be a connected Hausdorff manifold equipped with a closed 2-form ω. Let G be a Lie group (that is, a diffeological group which is a manifold). Let $\rho : g \mapsto g_M$ be a Hamiltonian action of G on M. Let us recall that a 1-parameter subgroup $F \in \mathrm{Hom}^\infty(\mathbf{R}, G)$ is uniquely defined by

$$Z = \frac{dF(t)}{dt}\bigg|_{t=0} \quad \text{and} \quad Z \in \mathcal{G} = T_{1_G}(G).$$

We denote $F(t) = e^{tZ}$. The *fundamental vector field* Z_M is defined on M by

$$Z_M(m) = \frac{\partial e^{tZ}_M(m)}{\partial t}\bigg|_{t=0} \in T_m(M).$$

Let μ be a moment map of the action of G on M, we shall denote

$$\mu_Z(m) = \mu(m)(t \mapsto e^{tZ})_0(1).$$

1) Show that the moment map μ is defined, up to a constant, by the differential equation

$$i_{Z_M}(\omega) = -d\mu_Z, \quad \text{that is,} \quad \omega_m(Z_M(m), \delta m) = -d[\mu_Z]_m(\delta m),$$

for all $m \in M$ and all $\delta m \in T_m(M)$.

2. Show then, using a basis of $\mathcal{G}$, that $Z \mapsto \mu_Z$ is linear.

✎ EXERCISE 146 (The cylinder and SL(2, R)). We consider the space $\mathbf{R}^2$, equipped with the standard symplectic form $\omega = dx \wedge dy$, with $X = (x, y) \in \mathbf{R}^2$. Check that the special linear group $\mathrm{SL}(2, \mathbf{R})$ preserves ω, and that its action on $\mathbf{R}^2$ has two orbits, the origin $0 \in \mathbf{R}^2$ and the *cylinder* $M = \mathbf{R}^2 - \{0\}$.

1) Justify, without computation, that the action of $\mathrm{SL}(2, \mathbf{R})$ on $\mathbf{R}^2$ is Hamiltonian and exact.

2) For every $X \in \mathbf{R}^2$, let $\gamma_X = [t \mapsto tX] \in \mathrm{Paths}(\mathbf{R}^2)$ connecting 0 to X. Use the general expression of the paths moment map given in (art. 9.20), for $p = \gamma_X$ and $F_\sigma = [s \mapsto e^{s\sigma}]$, with $\sigma \in \mathfrak{sl}(2, \mathbf{R})$ — the Lie algebra of $\mathrm{SL}(2, \mathbf{R})$, that is, the space of 2×2 traceless matrices — to show that

$$\mu(X)(F_\sigma) = \frac{1}{2}\omega(X, \sigma X) \times dt.$$

3) Deduce that the moment map $\mu_M = \mu \upharpoonright M$ of $\mathrm{SL}(2, \mathbf{R})$ on M is a nontrivial double sheet covering onto its image.

Examples of Moment Maps in Diffeology

The few following examples want to illustrate how the theory of moment maps in diffeology can be applied to the field of infinite dimensional situations, but also to the less familiar case of singular spaces. It is, at the same time, the opportunity to familiarize ourselves with the computational techniques in diffeology.

9.27. On the intersection 2-form of a surface, I. Let Σ be a closed surface oriented by a 2-form Surf, chosen once and for all. Let us consider $\Omega^1(\Sigma)$, the infinite dimensional vector space of 1-forms of Σ, equipped with functional diffeology. Let us consider the antisymmetric bilinear map defined on $\Omega^1(\Sigma)$ by

$$(\alpha, \beta) \mapsto \int_\Sigma \alpha \wedge \beta,$$

for all α, β in $\Omega^1(\Sigma)$. Since the wedge-product $\alpha \wedge \beta$ is a 2-form of Σ, there exists a real smooth function $\varphi \in \mathcal{C}^\infty(\Sigma, \mathbf{R})$ such that $\alpha \wedge \beta = \varphi \times \mathrm{Surf}$. Thus, by definition, $\int_\Sigma \alpha \wedge \beta = \int_\Sigma \varphi \times \mathrm{Surf}$.

1. A well defined differential 2-form ω of $\Omega^1(X)$ is naturally associated with the above bilinear form. For every n-plot $P : U \to X$, for all $r \in U$, δr and $\delta' r$ in $\mathbf{R}^n$,

$$\omega(P)_r(\delta r, \delta' r) = \int_\Sigma \frac{\partial P(r)}{\partial r}(\delta r) \wedge \frac{\partial P(r)}{\partial r}(\delta' r).$$

2. The 2-form ω is the differential of the 1-form λ defined on $\Omega^1(\Sigma)$ by

$$\lambda(P)_r(\delta r) = \frac{1}{2}\int_\Sigma P(r) \wedge \frac{\partial P(r)}{\partial r}(\delta r) \quad \text{and} \quad \omega = d\lambda.$$

3. Consider now the additive group $(\mathcal{C}^\infty(\Sigma, \mathbf{R}), +)$ of smooth real functions on Σ, and let us define the following action of $\mathcal{C}^\infty(\Sigma, \mathbf{R})$ on $\Omega^1(\Sigma)$,

$$\text{for all } f \in \mathcal{C}^\infty(\Sigma, \mathbf{R}), \quad f \mapsto \bar{f} = [\alpha \mapsto \alpha + df].$$

Then, the additive group $\mathcal{C}^\infty(\Sigma, \mathbf{R})$ acts by automorphisms on the pair $(\Omega^1(\Sigma), \omega)$,

$$\text{for all f in } \mathcal{C}^\infty(\Sigma, \mathbf{R}), \quad f^*(\omega) = \omega.$$

Note that the kernel of the action $f \mapsto \bar{f}$ is the subgroup of constant maps, and the image of $\mathcal{C}^\infty(\Sigma, \mathbf{R})$ is the group $B^1_{\mathrm{dR}}(\Sigma)$ of exact 1-forms of Σ.

4. Let $p \in \mathrm{Paths}(\Omega^1(\Sigma))$ be a path connecting α_0 to α_1. The paths moment map $\Psi(p)$ is then given by

$$\Psi(p) = \left(\hat{\alpha}_1^*(\lambda) + d\left[f \mapsto \frac{1}{2}\int_\Sigma f \times d\alpha_1\right]\right) - \left(\hat{\alpha}_0^*(\lambda) + d\left[f \mapsto \frac{1}{2}\int_\Sigma f \times d\alpha_0\right]\right).$$

On this expression, we get immediately the 2-points moment map $\psi(\alpha_0, \alpha_1) = \Psi(p)$, for any path p connecting α_0 to α_1. Note that, since $\Omega^1(\Sigma)$ is contractible, the holonomy of the action of $\mathcal{C}^\infty(\Sigma, \mathbf{R})$ is trivial, $\Gamma = \{0\}$, and the action of $\mathcal{C}^\infty(\Sigma, \mathbf{R})$ is Hamiltonian.

5. The moment maps of this action of $\mathcal{C}^\infty(\Sigma, \mathbf{R})$ on $\Omega^1(\Sigma)$ are, up to a constant, equal to

$$\mu : \alpha \mapsto d\left[f \mapsto \int_\Sigma f \times d\alpha\right].$$

Moreover, the moment map μ is equivariant, that is, invariant, since the group $\mathcal{C}^\infty(\Sigma, \mathbf{R})$ is Abelian,

$$\text{for all } f \in \mathcal{C}^\infty(\Sigma, \mathbf{R}), \quad \mu \circ \bar{f} = \mu.$$

In summary, the action of $\mathcal{C}^\infty(\Sigma, \mathbf{R})$ on $\Omega^1(\Sigma)$ is exact and Hamiltonian.

NOTE. The moment map $\mu(\alpha)$ is fully characterized by $d\alpha$. This is why we find in the mathematical literature on the subject that the moment map for this action is the exterior derivative (or curvature, depending on the authors) $\alpha \mapsto d\alpha$. But as we see again on this example, the diffeological framework gives to this assertion a precise meaning. Let us also remark that the moment map μ is linear, for all real numbers t and s, and for all α and β in $\Omega^1(\Sigma)$, $\mu(t\,\alpha + s\,\beta) = t\,\mu(\alpha) + s\,\mu(\beta)$. The kernel of μ is the subspace of closed 1-forms,

$$\ker(\mu) = Z^1_{\mathrm{dR}}(\Sigma) = \left\{\alpha \in \Omega^1(\Sigma) \mid d\alpha = 0\right\}.$$

The orbit of the zero form $0 \in \Omega^1(\Sigma)$ by $\mathcal{C}^\infty(\Sigma, \mathbf{R})$ is just the subspace $B^1_{\mathrm{dR}}(\Sigma) \subset Z^1_{\mathrm{dR}}(\Sigma)$, see (art. 9.29, Note 3) for a discussion about that.

PROOF. 1. Let us check that ω defines a differential 1-form on $\Omega^1(\Sigma)$. Note that, for any $r \in U = \mathrm{def}(P)$, $P(r)$ is a section of the ordinary cotangent bundle $T^*\Sigma$, $P(r) = [x \mapsto P(r)(x)] \in \mathcal{C}^\infty(\Sigma, T^*\Sigma)$, where $P(r)(x) \in T^*_x(\Sigma)$. Thus,

$$\frac{\partial P(r)}{\partial r}(\delta r) = [x \mapsto \frac{\partial P(r)(x)}{\partial r}(\delta r)] \quad \text{and} \quad \frac{\partial P(r)(x)}{\partial r}(\delta r) \in T^*_x(\Sigma),$$

where $\partial P(r)(x)/\partial r$ denotes the tangent linear map $D(r \mapsto P(r)(x))(r)$. The formula giving ω is then well defined. Now, $\omega(P)_r$ is clearly antisymmetric and depends smoothly on r. Hence, $\omega(P)$ is a smooth 2-form of U. Let us check that $P \mapsto \omega(P)$ defines a 2-form on $\Omega^1(\Sigma)$, that is, satisfies the compatibility condition $\omega(P \circ F) =$

$F^*(\omega(P))$, for all $F \in \mathcal{C}^\infty(V, U)$, where V is a real domain. Let $s \in V$, δs and $\delta' s$ two tangent vectors to V at s, let $r = F(s)$, and compute

$$\begin{aligned}
\omega(P \circ F)_s(\delta s, \delta' s) &= \int_\Sigma \frac{\partial P \circ F(s)}{\partial s}(\delta s) \wedge \frac{\partial P \circ F(s)}{\partial s}(\delta' s) \\
&= \int_\Sigma \frac{\partial P(r)}{\partial r}\frac{\partial F(s)}{\partial s}(\delta s) \wedge \frac{\partial P(r)}{\partial r}\frac{\partial F(s)}{\partial s}(\delta' s) \\
&= \omega(P)_{F(s)}(DF_s(\delta s), DF_s(\delta' s)) \\
&= F^*(\omega(P))_s(\delta s, \delta' s).
\end{aligned}$$

Thus, $\omega(P \circ F) = F^*(\omega(P))$, and ω is a well defined 2-form on $\Omega^1(\Sigma)$.

2. First of all, the proof that the map $P \mapsto \lambda(P)$ is a well defined differential 1-form of $\Omega^1(\Sigma)$ is analogous to the proof of the first item. Now, let us recall that $\omega = d\lambda$ means $d(\lambda(P)) = \omega(P)$, for all plots P of $\Omega^1(\Sigma)$. Let us apply the usual formula of differentiation of 1-forms on real domains,

$$d\epsilon_r(\delta r, \delta' r) = \delta(\epsilon_r(\delta' r)) - \delta'(\epsilon_r(\delta r)),$$

where δ and δ' are two independent variations. For the sake of simplicity let us denote

$$\alpha = P(r), \quad \delta\alpha = \frac{\partial P(r)}{\partial r}(\delta r), \quad \delta'\alpha = \frac{\partial P(r)}{\partial r}(\delta' r).$$

Then,

$$\begin{aligned}
d(\lambda(P))_r(\delta r, \delta' r) &= \frac{1}{2}\left[\delta \int_\Sigma \alpha \wedge \delta'\alpha - \delta' \int_\Sigma \alpha \wedge \delta\alpha\right] \\
&= \frac{1}{2}\left[\int_\Sigma \delta\alpha \wedge \delta'\alpha + \alpha \wedge \delta\delta'\alpha - \int_\Sigma \delta'\alpha \wedge \delta\alpha + \alpha \wedge \delta'\delta\alpha\right],
\end{aligned}$$

but $\delta\delta'\alpha = \delta'\delta\alpha$, thus

$$\begin{aligned}
d(\lambda(P))_r(\delta r, \delta' r) &= \frac{1}{2}\left[\int_\Sigma \delta\alpha \wedge \delta'\alpha - \int_\Sigma \delta'\alpha \wedge \delta\alpha\right] \\
&= \frac{1}{2}\left[\int_\Sigma \delta\alpha \wedge \delta'\alpha + \int_\Sigma \delta\alpha \wedge \delta'\alpha\right] \\
&= \int_\Sigma \delta\alpha \wedge \delta'\alpha \\
&= \omega_r(\delta r, \delta' r).
\end{aligned}$$

3. Let us compute the pullback of λ by the action of $f \in \mathcal{C}^\infty(\Sigma, \mathbf{R})$. Let $P : U \to \Omega^1(\Sigma)$ be an n-plot, and let $r \in U$ and $\delta r \in \mathbf{R}^n$,

$$\begin{aligned}
\bar{f}^*(\lambda)(P)_r(\delta r) &= \lambda(\bar{f} \circ P)_r(\delta r) \\
&= \lambda(r \mapsto P(r) + df)_r(\delta r) \\
&= \frac{1}{2}\int_\Sigma (P(r) + df) \wedge \frac{\partial P(r)}{\partial r}(\delta r) \\
&= \frac{1}{2}\int_\Sigma P(r) \wedge \frac{\partial P(r)}{\partial r}(\delta r) + \frac{1}{2}\int_\Sigma df \wedge \frac{\partial P(r)}{\partial r}(\delta r) \\
&= \lambda(P)_r(\delta r) + \frac{\partial}{\partial r}\left\{\frac{1}{2}\int_\Sigma df \wedge P(r)\right\}(\delta r) \\
&= \lambda(P)_r(\delta r) - \frac{\partial}{\partial r}\left\{\frac{1}{2}\int_\Sigma f \times d(P(r))\right\}(\delta r).
\end{aligned}$$

Next, we define, for every $f \in \mathcal{C}^\infty(\Sigma, \mathbf{R})$, $\varphi(f) : \Omega^1(\Sigma) \to \mathbf{R}$ by

$$\varphi(f) : \alpha \mapsto \frac{1}{2}\int_\Sigma f \times d\alpha.$$

Then,

$$d(\varphi(f))(P)_r(\delta r) = \frac{\partial}{\partial r}\left\{\frac{1}{2}\int_\Sigma f \times d(P(r))\right\}(\delta r).$$

Thus,

$$\bar{f}^*(\lambda)(P)_r(\delta r) = \lambda(P)_r(\delta r) - (d\varphi(f))(P)_r(\delta r),$$

that is,

$$\bar{f}^*(\lambda) = \lambda - d(\varphi(f)).$$

Hence, $d[\bar{f}^*(\lambda)] = d\lambda$, and $\omega = d\lambda$ is invariant by the action of $\mathcal{C}^\infty(\Sigma, \mathbf{R})$.

4. Let p be a path in $\Omega^1(\Sigma)$ connecting α_0 to α_1. By definition $\Psi(p) = \hat{p}^*(\mathcal{K}\omega)$. Applying the property of the Chain-Homotopy Operator $d \circ \mathcal{K} + \mathcal{K} \circ d = \hat{1}^* - \hat{0}^*$ to $\omega = d\lambda$, we get

$$\begin{aligned}
\Psi(p) &= \hat{p}^*(\mathcal{K}d\lambda) \\
&= \hat{p}^*(\hat{1}^*(\lambda) - \hat{0}^*(\lambda) - d(\mathcal{K}\lambda)) \\
&= (\hat{1} \circ \hat{p})^*(\lambda) - (\hat{0} \circ \hat{p})^*(\lambda) - d[(\mathcal{K}\lambda) \circ \hat{p}] \\
&= \hat{\alpha}_1^*(\lambda) - \hat{\alpha}_0^*(\lambda) - d[f \mapsto \mathcal{K}\lambda(\hat{p}(f))].
\end{aligned}$$

But $\mathcal{K}\lambda(\hat{p}(f)) = \mathcal{K}\lambda(\bar{f} \circ p) = \int_{\bar{f}\circ p} \lambda = \int_p \bar{f}^*(\lambda)$, and since $\bar{f}^*(\lambda) = \lambda - d(\varphi(f))$, $\mathcal{K}\lambda(\hat{p}(f)) = \int_p \lambda - \int_p d(\varphi(f)) = \int_p \lambda - \varphi(f)(\alpha_1) + \varphi(f)(\alpha_0)$. Therefore,

$$\begin{aligned}
\Psi(p) &= \hat{\alpha}_1^*(\lambda) - \hat{\alpha}_0^*(\lambda) - d[f \mapsto -\varphi(f)(\alpha_1) + \varphi(f)(\alpha_0)] \\
&= \hat{\alpha}_1^*(\lambda) - \hat{\alpha}_0^*(\lambda) + d\left[f \mapsto \frac{1}{2}\int_\Sigma f \times d\alpha_1 - \frac{1}{2}\int_\Sigma f \times d\alpha_0\right].
\end{aligned}$$

We get then the paths moment map Ψ,

$$\Psi(p) = \left(\hat{\alpha}_1^*(\lambda) + d\left[f \mapsto \frac{1}{2}\int_\Sigma f \times d\alpha_1\right]\right) - \left(\hat{\alpha}_0^*(\lambda) + d\left[f \mapsto \frac{1}{2}\int_\Sigma f \times d\alpha_0\right]\right).$$

Concerning the 2-points moment map ψ, we clearly have $\psi(\alpha_0, \alpha_1) = \Psi(p)$, for any path connecting α_0 to α_1.

5. The 1-point moment maps are given by $\mu(\alpha) = \psi(\alpha_0, \alpha)$ for any origin α_0. Let us choose $\alpha_0 = 0$. So,

$$\mu(\alpha) = \hat{\alpha}^*(\lambda) + d\left[f \mapsto \frac{1}{2}\int_\Sigma f \times d\alpha\right] - \hat{0}^*(\lambda).$$

But $\hat{0}^*(\alpha)$ is not necessarily zero. Let us compute generally $\hat{\alpha}^*(\lambda)$. Let $P : U \to \Omega^1(\Sigma)$ be an n-plot, $\hat{\alpha}^*(\lambda)(P) = \lambda(\hat{\alpha} \circ P) = \lambda(r \mapsto \hat{\alpha}(P(r)) = \lambda(r \mapsto \alpha + d(P(r)))$, and

$$\begin{aligned}\lambda(r \mapsto \alpha + d(P(r))) &= \frac{1}{2}\int_\Sigma (\alpha + P(r)) \wedge \frac{\partial}{\partial r}(\alpha + d(P(r))) \\ &= \frac{1}{2}\int_\Sigma (\alpha + P(r)) \wedge \frac{\partial d(P(r))}{\partial r} \\ &= \frac{1}{2}\int_\Sigma \alpha \wedge \frac{\partial d(P(r))}{\partial r} + \frac{1}{2}\int_\Sigma P(r) \wedge \frac{\partial d(P(r))}{\partial r}.\end{aligned}$$

Then,

$$(\hat{\alpha}^*(\lambda) - \hat{0}^*(\lambda))(P) = \frac{1}{2}\int_\Sigma \alpha \wedge \frac{\partial d(P(r))}{\partial r}.$$

Therefore,

$$\begin{aligned}\mu(\alpha)(P)_r &= (\hat{\alpha}^*(\lambda) - \hat{0}^*(\lambda))(P)_r + d\left[f \mapsto \frac{1}{2}\int_\Sigma f \times d\alpha\right](P)_r \\ &= \frac{1}{2}\int_\Sigma \alpha \wedge \frac{\partial d(P(r))}{\partial r} + \frac{\partial}{\partial r}\left\{\frac{1}{2}\int_\Sigma P(r) \times d\alpha\right\} \\ &= \frac{1}{2}\frac{\partial}{\partial r}\left\{\int_\Sigma \alpha \wedge d(P(r)) + P(r) \times d\alpha\right\} \\ &= \frac{\partial}{\partial r}\left\{\int_\Sigma P(r) \times d\alpha\right\},\end{aligned}$$

which gives finally

$$\mu(\alpha) = d\left[f \mapsto \int_\Sigma f \times d\alpha\right].$$

Now, let us express the variance of μ. Let $f \in \mathcal{C}^\infty(\Sigma, \mathbf{R})$ and $F(\alpha)$ be the real function $F(\alpha) : f \mapsto \int_\Sigma f \times d\alpha$, such that $\mu(\alpha) = dF(\alpha)$. We have $\mu(\bar{f}(\alpha)) = \mu(\alpha + df) = dF(\alpha + df)$ but, for every $h \in \mathcal{C}^\infty(\Sigma, \mathbf{R})$, $F(\alpha + df)(h) = \int_\Sigma h \times d(\alpha + df) = \int_\Sigma h \times d\alpha = F(\alpha)(h)$. Then, for all $f \in \mathcal{C}^\infty(\Sigma, \mathbf{R})$, $\mu \circ \hat{f} = \mu$. The moment map μ is invariant by the group $\mathcal{C}^\infty(\Sigma, \mathbf{R})$. The Souriau class is zero, the action of $\mathcal{C}^\infty(\Sigma, \mathbf{R})$ is then exact and Hamiltonian.

Let us end with the computation of the kernel of the moment map μ. Clearly, $\mu(\alpha) = 0$ if and only if $dF(\alpha) = 0$. But since $\mathcal{C}^\infty(\Sigma, \mathbf{R})$ is connected (actually contractible as a diffeological vector space), $dF(\alpha) = 0$ if and only if $F(\alpha) = \text{cst} = F(\alpha)(0) = 0$. But $F(\alpha) = 0$ if and only if, for all $f \in \mathcal{C}^\infty(\Sigma, \mathbf{R})$, $\int_\Sigma f \times d\alpha = 0$, that is, if and only if $d\alpha = 0$. □

9.28. On the intersection 2-form of a surface, II. We continue with the example of (art. 9.27) using the same notations. Let us introduce the group G of positive diffeomorphisms of (Σ, Surf), that is,

$$G = \left\{ g \in \text{Diff}(\Sigma) \;\middle|\; \frac{g^*(\text{Surf})}{\text{Surf}} > 0 \right\}.$$

The group G acts by pushforward on $\Omega^1(\Sigma)$. For all $g \in G$, for all $\alpha \in \Omega^1(\Sigma)$, $g_*(\alpha) \in \Omega^1(\Sigma)$, and for all pairs g, g' of elements of G, $(g \circ g')_* = g_* \circ g'_*$, and this action is smooth.

1. The pushforward action of G on $\Omega^1(\Sigma)$ preserves the 1-form λ, and thus the 2-form ω. For all $g \in G$, $(g_*)^*(\lambda) = \lambda$, and $(g_*)^*(\omega) = \omega$. Thus, the action of G is exact, $\sigma = 0$, and Hamiltonian, $\Gamma = \{0\}$.

2. The moment maps are, up to a constant, equal to the moment μ,

$$\mu(\alpha)(P)_r(\delta r) = \frac{1}{2}\int_\Sigma \alpha \wedge P(r)^*\left(\frac{\partial P(r)_*(\alpha)}{\partial r}(\delta r)\right),$$

for all $\alpha \in \Omega^1(\Sigma)$, for all n-plots P, where $r \in \mathrm{def}(P)$ and $\delta r \in \mathbf{R}^n$. In particular, applied to any 1-plot F centered at the identity $\mathbf{1}_G$, that is, $F(0) = \mathbf{1}_G$, we get the special expression

$$\mu(\alpha)(F)_0(1) = -\frac{1}{2}\int_\Sigma \alpha \wedge \pounds_F(\alpha) = -\int_\Sigma i_F(\alpha) \times d\alpha,$$

where $\pounds_F(\alpha)$ and $i_F(\alpha)$ are the Lie derivative and the contraction of α by F. Note that it is not surprising that the Lie derivative of α is closely associated with the moment map of the action of the group of diffeomorphisms.

PROOF. 1. Let us compute the pullback of λ by the action of $g \in G$, that is, $(g_*)^*(\lambda)$. Let $P : U \to \Omega^1(\Sigma)$ be an n-plot, let $r \in U$, and $\delta r \in \mathbf{R}^n$, then

$$\begin{aligned}
(g_*)^*(\lambda)(P)_r(\delta r) &= \lambda(g_* \circ P)_r(\delta r) \\
&= \frac{1}{2}\int_\Sigma g_*(P(r)) \wedge \frac{\partial g_*(P(r))}{\partial r}(\delta r) \\
&= \frac{1}{2}\int_\Sigma g_*(P(r)) \wedge g_*\left(\frac{\partial P(r)}{\partial r}(\delta r)\right) \\
&= \frac{1}{2}\int_\Sigma g_*\left(P(r) \wedge \frac{\partial P(r)}{\partial r}(\delta r)\right) \\
&= \frac{1}{2}\int_{g^*(\Sigma)} P(r) \wedge \frac{\partial P(r)}{\partial r}(\delta r) \\
&= \frac{1}{2}\int_\Sigma P(r) \wedge \frac{\partial P(r)}{\partial r}(\delta r) \\
&= \lambda(P)_r(\delta r).
\end{aligned}$$

Thus, λ is invariant by G, and so is $\omega = d\lambda$.

2. Since the 1-form λ is invariant by the action of G, we can use directly the results of the exact case detailed in (art. 9.11). The moment maps are, up to a constant, equal to $\mu : \alpha \mapsto \hat{\alpha}^*(\lambda)$. Then, let $P : U \to G$ be an n-plot, let $r \in U$, $\delta r \in \mathbf{R}^n$ and

let us compute,

$$\begin{aligned}\mu(\alpha)(P)_r(\delta r) &= \alpha^*(\lambda)(P)_r(\delta r)\\ &= \lambda(\hat{\alpha}\circ P)_r(\delta r)\\ &= \lambda(r\mapsto P(r)_*(\alpha))_r(\delta r)\\ &= \frac{1}{2}\int_\Sigma P(r)_*(\alpha)\wedge \frac{\partial P(r)_*(\alpha)}{\partial r}(\delta r)\\ &= \frac{1}{2}\int_\Sigma \alpha\wedge P(r)^*\Big(\frac{\partial P(r)_*(\alpha)}{\partial r}(\delta r)\Big).\end{aligned}$$

Now, let $P = F$ be a 1-plot centered at the identity, $F(0) = 1_G$. Let us change the variable r for the variable t. The previous expression, computed at $t = 0$ and applied to the vector $\delta t = 1$ gives immediately

$$\mu(\alpha)(F)_0(1) = \frac{1}{2}\int_\Sigma \alpha\wedge \left.\frac{\partial F(t)_*(\alpha)}{\partial t}\right|_{t=0}.$$

But, by definition of the Lie derivative,

$$\left\{\frac{\partial F(t)_*(\alpha)}{\partial t}\right\}_{t=0} = \left\{\frac{\partial (F(t)^{-1})^*(\alpha)}{\partial t}\right\}_{t=0} = -\pounds_F(\alpha).$$

Thus, we get the first expression of the moment map μ applied to F, that is,

$$\mu(\alpha)(F)_0(1) = -\frac{1}{2}\int_\Sigma \alpha\wedge \pounds_F(\alpha).$$

Now, on a surface $\alpha\wedge d\alpha = 0$, and $i_F(\alpha\wedge d\alpha) = i_F(\alpha)\times d\alpha - \alpha\wedge i_F(d\alpha)$, thus $i_F(\alpha)\times d\alpha = \alpha\wedge i_F(d\alpha)$. Then, using the Cartan-Lie formula $\pounds_F(\alpha) = i_F(d\alpha) + d(i_F(\alpha))$,

$$\begin{aligned}\int_\Sigma \alpha\wedge\pounds_F(\alpha) &= \int_\Sigma \alpha\wedge[i_F(d\alpha) + d(i_F(\alpha))]\\ &= \int_\Sigma i_F(\alpha)d\alpha + \int_\Sigma \alpha\wedge d(i_F(\alpha))\\ &= \int_\Sigma i_F(\alpha)d\alpha + \int_\Sigma i_F(\alpha)d\alpha - \int_\Sigma d[\alpha\wedge i_F(\alpha)]\\ &= 2\int_\Sigma i_F(\alpha)d\alpha.\end{aligned}$$

And finally, we get the second expression for the moment map, that is,

$$\mu(\alpha)(F)_0(1) = -\int_\Sigma i_F(\alpha)\times d\alpha,$$

for any 1-plot of the group of positive diffeomorphisms of the surface Σ, centered at the identity. □

9.29. On the intersection 2-form of a surface, III. We continue again with the example of (art. 9.27), using the same notations. Let us consider the space $\Omega^1(\Sigma)$ as an additive group acting onto itself by translations. Let us denote by t_β the translation $t_\beta : \alpha\mapsto\alpha+\beta$, where α and β belong to $\Omega^1(\Sigma)$.

1. The 2-form ω is invariant by translation, that is, $t_\alpha^*(\omega) = \omega$ for all $\alpha\in\Omega^1(\Sigma)$. This action of $\Omega^1(\Sigma)$ onto itself is Hamiltonian but not exact.

2. The moment maps of the additive action of $\Omega^1(\Sigma)$ onto itself are equal, up to a constant, to

$$\mu : \alpha \mapsto d\left[\beta \mapsto \int_\Sigma \alpha \wedge \beta\right].$$

In other words, $\mu(\alpha) = d[\omega(\alpha)]$, where ω is regarded as the smooth linear function $\omega(\alpha) : \beta \mapsto \omega(\alpha, \beta)$, defined on $\Omega^1(\Sigma)$. Moreover, the moment map μ is linear and injective.

3. The moment map μ is its own Souriau cocycle, $\theta = \mu$. The moment map μ identifies $\Omega^1(\Sigma)$ with the θ-affine coadjoint orbit of $0 \in \Omega^1(\Sigma)^*$. Be aware that $\Omega^1(\Sigma)^*$ denotes the space of invariant 1-forms of the Abelian group $\Omega^1(\Sigma)$, and not its algebraic dual.

NOTE 1. This example is analogous to finite dimension symplectic vector spaces. The 2-form ω can be regarded as a real 2-cocycle of the additive group $\Omega^1(\Sigma)$. This cocycle builds up a central extension by $\mathbf{R}$,

$$(\alpha, t) \cdot (\alpha', t') = \left(\alpha + \alpha', t + t' + \int_\Sigma \alpha \wedge \alpha'\right)$$

for all (α, t) and (α', t') in $\Omega^1(\Sigma) \times \mathbf{R}$. This central extension acts on $\Omega^1(\Sigma)$, preserving ω. This action is Hamiltonian, and now exact. The lack of equivariance, characterized by the Souriau class, has been absorbed in the extension. This group could be named, by analogy, the *Heisenberg group* of the oriented surface (Σ, Surf).

NOTE 2. According to (art. 9.19), the space $\Omega^1(\Sigma)$ equipped with the 2-form ω is a homogeneous symplectic space. Thus, we have a first simple example of an infinite dimensional *symplectic diffeological space*, avoiding any consideration on the *kernel* of ω.

NOTE 3. The preimage of zero by the moment map of the Abelian group $\mathcal{C}^\infty(\Sigma, \mathbf{R})$ on $\Omega^1(\Sigma)$ in (art. 9.27) is the subgroup of closed forms $Z^1_{\mathrm{dR}}(\Sigma) \subset \Omega^1(\Sigma)$. This group acts homogeneously on itself and its moment map for the restriction of the 2-form ω is the projection of the moment map of the action of $\mathcal{C}^\infty(\Sigma, \mathbf{R})$, that is, $\mu' : \alpha \mapsto d\left[\beta \mapsto \int_\Sigma \alpha \wedge \beta\right]$, but now α and β are closed. The characteristics of this moment map are the orbits of the subgroup $\mu'^{-1}(0)$, that is, the subgroup of $\alpha \in Z^1_{\mathrm{dR}}(\Sigma)$ such that $\int_\Sigma \alpha \wedge \beta = 0$ for all $\beta \in Z^1_{\mathrm{dR}}(\Sigma)$. This is the subgroup of exact 1-forms $B^1_{\mathrm{dR}}(\Sigma)$. Then, the image of μ' identifies naturally with the quotient space $H^1_{\mathrm{dR}}(\Sigma) = Z^1_{\mathrm{dR}}(\Sigma)/B^1_{\mathrm{dR}}(\Sigma)$. Moreover, the closed 2-form passes to this quotient, it is the well known *intersection form*, denoted here by ω_{int}, $\omega \upharpoonright Z^1_{\mathrm{dR}}(\Sigma) = \mu'^*(\omega_{\mathrm{int}})$. This is an example of symplectic reduction in diffeology (see also (art. 9.17, Note)), the full general case will be addressed in a future work.

PROOF. Let us compute the pullback of λ by a translation. Let $P : U \to X$ be an n-plot, let $r \in U$, and let $\delta r \in \mathbf{R}^n$,

$$\begin{aligned} t_\alpha^*(\lambda)(P)_r(\delta r) &= \lambda(t_\alpha \circ P)_r(\delta r) \\ &= \lambda[r \mapsto P(r) + \alpha]_r(\delta r) \\ &= \frac{1}{2}\int_\Sigma (P(r)+\alpha) \wedge \frac{\partial(P(r)+\alpha)}{\partial r}(\delta r) \\ &= \frac{1}{2}\int_\Sigma P(r) \wedge \frac{\partial P(r)}{\partial r}(\delta r) + \frac{1}{2}\int_\Sigma \alpha \wedge \frac{\partial P(r)}{\partial r}(\delta r) \\ &= \lambda(P)_r(\delta r) + d\left[\beta \mapsto \frac{1}{2}\int_\Sigma \alpha \wedge \beta\right](P)_r(\delta r). \end{aligned}$$

Let us define next, for all $\alpha \in \Omega^1(\Sigma)$, the smooth real function $F(\alpha)$ by

$$F(\alpha) : \beta \mapsto \frac{1}{2}\int_\Sigma \alpha \wedge \beta.$$

Then,

$$t_\alpha^*(\lambda) = \lambda + d(F(\alpha)) \quad \text{and} \quad t_\alpha^*(\omega) = \omega.$$

Next, $\Omega^1(\Sigma)$, as an additive group, acts on itself by automorphisms. Let us compute the moment maps. Let p be a path in $\Omega^1(\Sigma)$, connecting α_0 to α_1, then

$$\begin{aligned} \Psi(p) &= \hat{\alpha}_1^*(\lambda) - \hat{\alpha}_0^*(\lambda) - d\left[\beta \mapsto \int_p d(F(\beta))\right] \\ &= \hat{\alpha}_1^*(\lambda) - \hat{\alpha}_0^*(\lambda) - d[\beta \mapsto F(\beta)(\alpha_1) - F(\beta)(\alpha_0)] \\ &= \{\alpha_1^*(\lambda) - d[\beta \mapsto F(\beta)(\alpha_1)]\} - \{\alpha_0^*(\lambda) - d[\beta \mapsto F(\beta)(\alpha_0)]\} \\ &= \{\hat{\alpha}_1^*(\lambda) + d(F(\alpha_1))\} - \{\hat{\alpha}_0^*(\lambda) + d(F(\alpha_0))\}. \end{aligned}$$

Thus, the 2-points moment map is clearly given by $\psi(\alpha_0, \alpha_1) = \Psi(p)$. Now, the moment maps are, up to a constant, equal to

$$\mu(\alpha) = \psi(0, \alpha) = \hat{\alpha}_1^*(\lambda) + d(F(\alpha)) - \hat{0}^*(\lambda).$$

But for any plot $P : U \to \Omega^1(\Sigma)$,

$$\begin{aligned} \hat{\alpha}^*(\lambda)(P) - \hat{0}^*(\lambda)(P) &= \lambda(\hat{\alpha} \circ P) - \lambda(\hat{0} \circ P) \\ &= \lambda(r \mapsto P(r) + \alpha) - \lambda(r \mapsto P(r)) \\ &= d\left[\beta \mapsto \frac{1}{2}\int_\Sigma \alpha \wedge \beta\right](P) \\ &= d(F(\alpha))(P). \end{aligned}$$

Hence, $\hat{\alpha}^*(\lambda)(P) - \hat{0}^*(\lambda) = d(F(\alpha))$ and μ is finally given by

$$\mu(\alpha) = 2d(F(\alpha)) = d\left[\beta \mapsto \int_\Sigma \alpha \wedge \beta\right].$$

The moment map μ is not equivariant, the Souriau cocycle θ is given by

$$\mu(t_\alpha^*(\beta)) = \mu(\alpha + \beta) = \mu(\beta) + \theta(\alpha) \quad \text{with} \quad \theta(\alpha) = \mu(\alpha).$$

Considering the Note, the moment map μ is clearly smooth and linear. Let $\alpha \in \ker(\mu)$, $\mu(\alpha) = 0$ if and only if $d(F(\alpha)) = 0$, that is, if and only if $F(\alpha) = \text{cst} = F(\alpha)(0) = 0$. But $F(\alpha)(\beta) = 0$, for all $\beta \in \Omega^1(\Sigma)$, implies $\alpha = 0$. Therefore, the moment map μ is injective. ☐

9.30. On symplectic irrational tori. Consider the smooth space $\mathbf{R}^n$, for some integer n. For all $u \in \mathbf{R}^n$, let t_u be the translation by u, that is, $t_u : x \mapsto x + u$. Let ω be a 2-form of $\mathbf{R}^n$ invariant by translation, that is, for all $u \in \mathbf{R}^n$, $t_u^*(\omega) = \omega$. Thus, ω is a constant bilinear 2-form, thus closed, $d\omega = 0$. Let us consider the moment maps associated with the translations $(\mathbf{R}^n, +)$. Since $\mathbf{R}^n$ is simply connected, the holonomy is trivial, $\Gamma = \{0\}$. Let p be a path in $\mathbf{R}^n$ connecting $x = p(0)$ to $y = p(1)$, the paths moment map $\Psi(p)$, and the 2-points moment map $\psi(p)$ are given by

$$\Psi(p) = \psi(x, y) = \omega(y - x),$$

where $\omega(u)$ is regarded as the linear 1-form $\omega(u) : v \mapsto \omega(u, v)$. The moment maps are, up to constant, equal to the linear map

$$\mu : x \mapsto \omega(x),$$

and the Souriau cocycle θ associated with μ is equal to μ. For all $u \in \mathbf{R}^n$,

$$\theta(u) = \mu(u) = \omega(u).$$

Consider now a discrete diffeological subgroup $K \subset \mathbf{R}^n$. Let us denote by Q the quotient $Q = \mathbf{R}^n/K$ and by $\pi : \mathbf{R}^n \to Q$ the projection. Let us continue to denote by t_u the translation on Q, by $u \in \mathbf{R}^n$, that is, $t_u(q) = \pi(x + u)$, for any x such that $q = \pi(x)$. Now, since ω is invariant by translation, ω is invariant by K, and since K is discrete, ω projects on Q as a $\mathbf{R}^n$-invariant closed 2-form denoted by ω_Q, that is,

$$\omega_Q = \pi_*(\omega) \quad \text{or} \quad \omega = \pi^*(\omega_Q).$$

Note that the translation by any vector u of $\mathbf{R}^n$ on Q is still an automorphism of ω_Q, that is, $t_u^*(\omega_Q) = \omega_Q$.

1. The holonomy Γ_Q of the action of $(\mathbf{R}^n, +)$ on (Q, ω_Q) is the image of the subgroup K by μ,

$$\Gamma_Q = \mu(K), \quad \Gamma_Q \subset \mathbf{R}^{n*}.$$

Thus, if $\omega \neq 0$ and if K is not reduced to $\{0\}$, then the action of $(\mathbf{R}^n, +)$ on (Q, ω_Q) is not Hamiltonian and not exact.

2. The moment map $\mu : \mathbf{R}^n \to \mathbf{R}^{n*}$ projects on a moment μ_Q such that the following diagram commutes.

$$\begin{array}{ccc} \mathbf{R}^n & \xrightarrow{\ \mu\ } & \mathbf{R}^{n*} \\ \pi \downarrow & & \downarrow \mathrm{pr} \\ Q = \mathbf{R}^n/K & \xrightarrow[\ \mu_Q\ ]{} & \mathbf{R}^{n*}/\mu(K) \end{array}$$

For all $q \in Q$, $\mu_Q(q) = \mathrm{pr}(\omega(x))$ for any x such that $q = \pi(x)$. The Souriau cocycle θ_Q associated with μ_Q, for all $u \in \mathbf{R}^n$, is given by

$$\theta_Q(u) = \mu_Q(\pi(u)).$$

Hence, if we consider the space Q as an additive group acting on itself by translations, then the moment map μ_Q coincides with its Souriau cocycle θ_Q.

3. The map μ is a fibration onto its image whose fiber is the kernel of μ, that is, $\mathrm{val}(\mu) \simeq \mathbf{R}^n/E$, $E = \ker(\mu)$. And, the map μ_Q is a fibration onto its image

$\mu(\mathbf{R}^n)/\mu(K)$ whose fiber is $\ker(\mu_Q) = E/(K \cap E)$. If $\omega : \mathbf{R}^n \to \mathbf{R}^{n*}$ is injective (which implies that n is even), then the moment map μ_Q is a diffeomorphism which identifies Q with its image $\mathbf{R}^{n*}/\mu(K)$.

NOTE 1. Regarded as a group, $Q = \mathbf{R}^n/K$ acts onto itself by projection of the translations of $\mathbf{R}^n$. Since the pullback by $\pi : \mathbf{R}^n \to Q$ is an isomorphism from $\mathcal{Q}^*$ to $\mathbf{R}^{n*}$ ($\mathbf{R}^n$ is the universal covering of Q), the moment maps computed above give the moment maps associated with this action.

NOTE 2. This construction applies to the torus $T^2 = \mathbf{R}^2/\mathbf{Z}^2$. The action of $(\mathbf{R}^2, +)$, is obviously not Hamiltonian, but the moment map μ_{T^2} is well defined. And, μ_{T^2} identifies T^2 with the quotient of $\mathbf{R}^{2*}$ — the (Γ_Q, θ_Q)-coadjoint orbit of the point 0 — by the holonomy $\Gamma_Q = \omega(\mathbf{Z}^2) \subset \mathbf{R}^{2*}$, according to the meaning of a coadjoint orbit we gave in (art. 7.16). The torus T^2, equipped with the standard symplectic form ω, is a coadjoint orbit of $\mathbf{R}^2$, or even a coadjoint orbit of itself. This is a special case of the (art. 9.23) discussion.

NOTE 3. All this construction above also applies to situations regarded as more singular that the simple quotient of $\mathbf{R}^n$ by a lattice. It applies, for example, to the product of any irrational tori. An (n-dimensional) irrational torus T_K is the quotient of $\mathbf{R}^n$ by any generating discrete strict subgroup K of $\mathbf{R}^n$. See for example [**IgLa90**] for an analysis of 1-dimensional irrational tori. For example, we can consider the product of two 1-dimensional irrational tori $Q = T_H \times T_K$, quotient of $\mathbf{R}^2 = \mathbf{R} \times \mathbf{R}$ by the discrete subgroup $\alpha_H(\mathbf{Z}^p) \times \alpha_K(\mathbf{Z}^q)$, where $\alpha_H : \mathbf{R}^p \to \mathbf{R}$ and $\alpha_K : \mathbf{R}^q \to \mathbf{R}$ are two linear 1-forms with coefficients independent over $\mathbf{Q}$. In this case, the moment map μ_Q will also identify $T_H \times T_K$ with the quotient of $\mathbf{R}^{2*}$ — (Γ_Q, θ_Q)-coadjoint orbit of 0 — by $\Gamma_Q = \omega(\alpha_H(\mathbf{Z}^p) \times \alpha_K(\mathbf{Z}^q))$. This is the simplest example of *totally irrational symplectic space*, and *totally irrational coadjoint orbit.* Note that these cases escape completely the usual analysis, of course, but also the analysis in terms of Sikorski or Frölicher spaces; see Exercise 80, p. 99.

PROOF. First of all, the fact that there exists a closed 2-form ω_Q on $\mathbf{R}/K$ such that $\pi^*(\omega_Q) = \omega$ is an application of the criterion of pushing forward forms, in the special case of a covering (art. 8.27). Now, the computation of the moment map of a linear antisymmetric form ω on $\mathbf{R}^n$ is well known, and independently of the method gives the same result $\mu(x) = \omega(x)$. The additive constant is set here by the condition $\mu(0) = 0$. But the value of the paths moment map $\Psi(p)$ can be computed as well by the general method, applying the particular expression

$$\mathcal{K}\omega_p(\delta p) = \int_0^1 \omega_{p(t)}(\dot{p}(t), \delta p(t))\, dt, \text{ with } \dot{p}(t) = \frac{dp(t)}{dt},$$

of the Chain-Homotopy operator for manifold, where p is a path and δp is a *variation* of p. Then, since the result depends only on the ends of the path, choose, for any two points x and y in $\mathbf{R}^n$, the connecting path $p : t \mapsto x + t(y - x)$. Let us recall that $\Psi(p) = \hat{p}^*(\mathcal{K}\omega)$, let u and δu in $\mathbf{R}^n$, note that $\hat{p}_*(t_u) = t_u \circ p = [t \mapsto p(t) + u]$.

Then,

$$\begin{aligned}\Psi(p)_u(\delta u) &= \hat{p}^*(\mathcal{K}\omega)_u(\delta u)\\ &= (\mathcal{K}\omega)_{t_u\circ p}(\delta(t_u\circ p)), \quad \text{with} \quad \delta p = 0\\ &= \int_0^1 \omega(\dot{p}(t),\delta u)\,dt\\ &= \omega(y-x,\delta u).\end{aligned}$$

Thus, $\Psi(p) = \psi(x,y) = \omega(y-x) = \omega(y)-\omega(x)$, and $\mu : x \mapsto \omega(x)$, $x \in \mathbf{R}^n$. Consider now ω_Q, since $\mathbf{R}^n$ is the universal covering of Q, every loop $\ell \in \mathrm{Loops}(Q,0)$ can be lifted into a path p in $\mathbf{R}^n$ starting at 0 and ending in K. In other words,

$$\Gamma = \{\Psi(\ell) \mid \ell \in \mathrm{Loops}(Q)\} = \{\Psi(t \mapsto tk) \mid k \in K\} = \omega(K).$$

The other propositions are then a direct application of the functoriality of the moment map described in (art. 9.13), and standard analysis on quotients and fibrations. □

9.31. The corner orbifold. Let us consider the quotient $\mathcal{Q}$ of $\mathbf{R}^2$ by the action of the finite subgroup $K \simeq \{\pm 1\}^2$, embedded in $\mathrm{GL}(2,\mathbf{R})$ by

$$K = \left\{ \begin{pmatrix} \varepsilon & 0 \\ 0 & \varepsilon' \end{pmatrix} \;\middle|\; \varepsilon, \varepsilon' \in \{\pm 1\} \right\}.$$

The space $\mathcal{Q} = \mathbf{R}^2/K$ (Figure 9.1) is an orbifold, according to [**IKZ05**]. It is diffeomorphic to the quarter space $[0,\infty[\, \times\, [0,\infty[\, \subset \mathbf{R}^2$, equipped with the pushforward of the standard diffeology of $\mathbf{R}^2$ by the map $\pi : \mathbf{R}^2 \to [0,\infty[\, \times\, [0,\infty[$, defined by,

$$\pi(x,y) = (x^2,y^2) \quad \text{and} \quad \mathcal{Q} \simeq \pi_*(\mathbf{R}^2).$$

The letter $\mathcal{Q}$ will denote indifferently the quotient $\mathbf{R}^2/K$ or the quarter space $\pi_*(\mathbf{R}^2)$, and the meaning of the letter π follows. Be aware that the corner orbifold is not a manifold with boundary, $\mathcal{Q}$ is not diffeomorphic to the corner equipped with the induced diffeology of $\mathbf{R}^2$. That said, we remark that the decomposition of $\mathcal{Q}$ in terms of point's structure is given by

$$\mathrm{Str}(0,0) = \{\pm 1\}^2, \quad \mathrm{Str}(x,0) = \mathrm{Str}(0,y) = \{\pm 1\}, \quad \text{and} \quad \mathrm{Str}(x,y) = \{1\},$$

where x and y are positive real numbers. Then, since the structure group of a point is preserved by diffeomorphisms [**IKZ05**], there are at least three orbits of $\mathrm{Diff}(\mathcal{Q})$, the point $0_{\mathcal{Q}} = (0,0)$, the regular stratum $\dot{\mathcal{Q}} =]0,\infty[^2$ and the union of the two axes, ox and oy. In particular, any diffeomorphism of $\mathcal{Q}$ preserves the origin $0_{\mathcal{Q}}$. Actually, these are exactly the orbits of $\mathrm{Diff}(\mathcal{Q})$. Let us remark that, since $\dim(\mathcal{Q}) = 2$ (art. 1.78), every 2-form is closed.

1. Every 2-form of $\mathcal{Q}$ is proportional to the 2-form ω defined on $\mathcal{Q}$ by

$$\pi^*(\omega) : \begin{pmatrix} x \\ y \end{pmatrix} \mapsto 4xy \times dx \wedge dy,$$

that is, for any other 2-form ω' there exists a smooth function $\phi \in \mathcal{C}^\infty(\mathcal{Q},\mathbf{R})$ such that $\omega' = \phi \times \omega$.

2. The space $(\mathcal{Q},\omega)$ is Hamiltonian, $\Gamma_\omega = \{0\}$, and the action of G_ω is exact, that is, $\sigma_\omega = 0$. In particular, the universal moment map μ_ω defined by $\mu_\omega(0_{\mathcal{Q}}) = 0$, is equivariant.

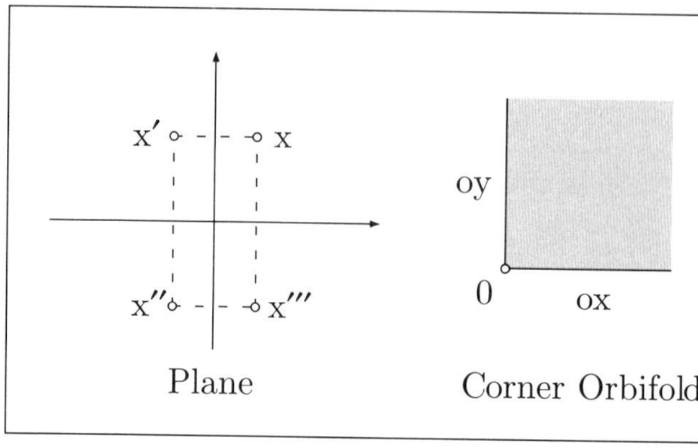

FIGURE 9.1. The corner orbifold $\mathcal{Q}$.

3. The universal equivariant moment map μ_ω vanishes on the singular strata $\{0\}$, $\mathrm{o}x$ and $\mathrm{o}y$, and is injective on the regular stratum $\dot{\mathcal{Q}}$. Therefore, the image $\mu_\omega(\mathcal{Q})$ is diffeomorphic to an open disc with a point attached somehow on the boundary.

PROOF. 1. Let ω' be a 2-form on $\mathcal{Q}$, and let $\tilde{\omega}'$ be its pullback by π, $\tilde{\omega}' = \pi^*(\omega')$. Thus, there exists a smooth real function F such that $\tilde{\omega}' = F \times dx \wedge dy$. But since $\pi \circ k = \pi$, for all $k \in K$, we get $\varepsilon\varepsilon' F(\varepsilon x, \varepsilon' y) = F(x,y)$, for all $(x,y) \in \mathbf{R}^2$ and all ε, ε' in $\{\pm 1\}$. Thus, $F(-x,y) = -F(x,y)$ and $F(x,-y) = -F(x,y)$. In particular, $F(0,y) = 0$ and $F(x,0) = 0$. Therefore, since F is smooth, there exists $f \in \mathcal{C}^\infty(\mathbf{R}^2, \mathbf{R})$ such that $F(x,y) = 4xyf(x,y)$, with $f(\varepsilon x, \varepsilon' y) = f(x,y)$. Therefore, $\tilde{\omega}' = f \times \tilde{\omega}$, with $\tilde{\omega} = 4xy \times dx \wedge dy$, that is, $\tilde{\omega} = d(x^2) \wedge d(y^2)$, but $x \mapsto x^2$ and $y \mapsto y^2$ are invariant by K so, they are the pullback by π of some smooth real functions on $\mathcal{Q}$. Thus, $d(x^2)$ and $d(y^2)$ are the pullback of 1-forms on $\mathcal{Q}$, let us say $d(x^2) = \pi^*(ds)$ and $d(y^2) = \pi^*(dt)$, so $\tilde{\omega} = \pi^*(\omega)$, where $\omega = ds \wedge dt$ is a well defined 2-form on $\mathcal{Q}$. Now, since $f(\epsilon x, \epsilon' y) = f(x,y)$ means just that f is the pullback of a smooth real function ϕ on $\mathcal{Q}$, it follows that any 2-form ω' on $\mathcal{Q}$ is proportional to ω, that is, $\omega' = \phi \times \omega$, with $\phi \in \mathcal{C}^\infty(\mathcal{Q}, \mathbf{R})$.

2. The orbifold is contractible. The deformation retraction $(s,x,y) \mapsto (sx, sy)$ of $\mathbf{R}^2$ to $\{(0,0)\}$ projects on a smooth deformation retraction of $\mathcal{Q}$. Thus, the holonomy is trivial, $\Gamma = \{0\}$. Now, since the origin $0_\mathcal{Q}$ is the only point with structure $\{\pm 1\}$, every diffeomorphism of $\mathcal{Q}$ preserves the origin $0_\mathcal{Q}$. Then, the 2-point moment map is exact, see (art. 9.10, Note 2), the Souriau class is zero, $\sigma_\omega = 0$. Let q be any point of $\mathcal{Q}$ and let $\mu_\omega(q) = \psi(0_\mathcal{Q}, q)$. This is an equivariant moment map and $\mu_\omega(0_\mathcal{Q}) = \psi(0_\mathcal{Q}, 0_\mathcal{Q}) = 0$.

3. Let $q \in \mathcal{Q}$, thus $\mu_\omega(q) = \Psi(p)$ for any path p connecting $0_\mathcal{Q}$ to q. Let q belong to a semi-axis $\mathrm{o}x$ or $\mathrm{o}y$, and let us choose $p = t \mapsto \lambda(t)q$, where λ is a smashing function equal to 0 on $]-\infty, 0]$ and equal to 1 on $[1, +\infty[$. Thus, for all $t \in \mathbf{R}$, $p(t)$ belongs to the same semi-axis as q. Thanks to (art. 9.2, ($\heartsuit$)), for any 1-plot ϕ of $\mathrm{Diff}(\mathcal{Q}, \omega_\omega)$ centered at the identity,

$$\Psi(p)(\phi)_0(1) = \int_0^1 \omega\left[\begin{pmatrix} s \\ r \end{pmatrix} \mapsto \phi(r)(\lambda(s+t)q)\right]_{\binom{0}{0}} \begin{pmatrix} 1 \\ 0 \end{pmatrix} \begin{pmatrix} 0 \\ 1 \end{pmatrix} dt.$$

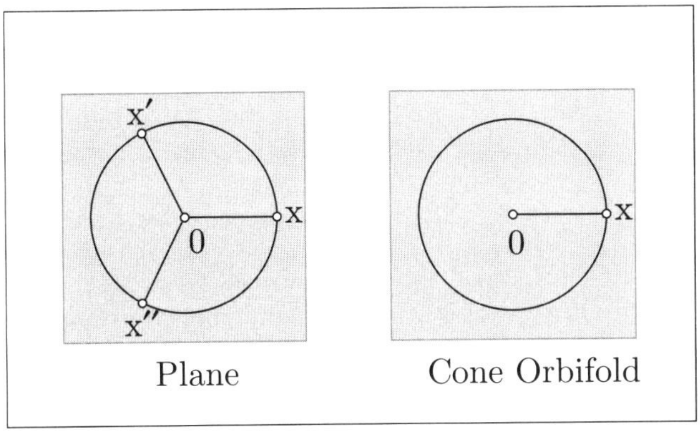

FIGURE 9.2. The cone orbifold $\mathcal{Q}_3$.

But $(s, r) \mapsto \phi(r)(\lambda(s+t)q)$ is a plot of the semi-axis, and thanks to item 1, the form ω vanishes on the semi-axis. Thus, the integrand vanishes and $\Psi(p)(\phi)_0(1) = 0$. Then, since 1-forms are characterized by 1-plots and since momenta are characterized by centered plots, $\mu_\omega(q) = 0$ for all $q \in \mathcal{Q}$ belonging to any semi-axis.

On the other hand, let q and q' be two points of the regular stratum $\dot{\mathcal{Q}}$. Since $\pi \upharpoonright \{(x,y) \mid x > 0 \text{ and } y > 0\}$ is a diffeomorphism, and since $\tilde{\omega} \upharpoonright \{(x,y) \mid x > 0$ and $y > 0\}$ is a symplectic Hausdorff manifold, there exists always a symplectomorphism ϕ with compact support $\mathcal{S} \subset \{(x,y) \mid x > 0$ and $y > 0\}$ which exchanges q and q'. Then, the image of this diffeomorphism on $\dot{\mathcal{Q}}$ can be extended by the identity on the whole $\mathcal{Q}$. Therefore, the automorphisms of ω are transitive on the regular stratum. The fact that the universal moment map is injective on the regular stratum comes from what we know already on symplectic manifolds (art. 9.23). □

9.32. The cone orbifold. Let $\mathcal{Q}_m$ be the quotient of the smooth complex plane $\mathbf{C}$ by the multiplicative action of the cyclic subgroup

$$Z_m \simeq \{\zeta \in \mathbf{C} \mid \zeta^m = 1\} \quad \text{with} \quad m > 1.$$

The space $\mathcal{Q}_m$ (Figure 9.2) is an orbifold, according to [**IKZ05**]. We identify $\mathcal{Q}_m$ to the complex plane $\mathbf{C}$, equipped with the pushforward of the standard diffeology by the map $\pi_m : z \mapsto z^m$. Precisely, the plots of $\mathcal{Q}_m$ are the parametrizations P of $\mathbf{C}$ which write locally $P(r) = \phi(r)^m$, where ϕ is a smooth parametrization in $\mathbf{C}$. Let us remark first that the decomposition of $\mathcal{Q}_m$, in terms of structure group, is given by

$$\operatorname{Str}(0) = Z_m, \quad \text{and} \quad \operatorname{Str}(z) = \{1\} \quad \text{if} \quad z \neq 0.$$

And secondly that there are two orbits of $\operatorname{Diff}(\mathcal{Q}_m)$, the point 0 and the regular stratum $\dot{\mathcal{Q}}_m = \mathbf{C} - \{0\}$. In particular any diffeomorphism of $\mathcal{Q}_m$ preserves the origin 0. It is not difficult to check that $\{\pi_m\}$ is a minimal generating family, thus $\dim(\mathcal{Q}_m) = 2$ (art. 1.78), and every 2-form on $\mathcal{Q}_m$ is closed.

1. Every 2-form of $\mathcal{Q}_m$ is proportional to the 2-form ω defined by

$$\pi_m^*(\omega) : z \mapsto dx \wedge dy \quad \text{with} \quad z = x + iy.$$

For every other 2-form ω' there exists a smooth function $f \in \mathcal{C}^\infty(\mathcal{Q}_m, \mathbf{R})$ such that $\omega' = f \times \omega$.

2. The space $(\mathcal{Q}, \omega)$ is Hamiltonian, $\Gamma_\omega = \{0\}$, and the action of G_ω is exact, that is, $\sigma_\omega = 0$. In particular, the universal moment map μ_ω, defined by $\mu_\omega(0) = 0$, is equivariant.

3. The universal moment map μ_ω is injective. Its image is the reunion of two coadjoint orbits, the point $0 \in \mathcal{G}^*_\omega$, value of the origin of $\mathcal{Q}_m$, and the image of the regular stratum $\dot{\mathcal{Q}}_m$.

PROOF. Let us first prove that the usual surface form Surf $= dx \wedge dy$ is the pullback of a 2-form ω defined on $\mathcal{Q}_m$. We shall apply the standard criterion and prove that for any two plots ϕ_1 and ϕ_2 of $\mathbf{C}$ such that $\pi_m \circ \phi_1 = \pi_m \circ \phi_2$, we have $\mathrm{Surf}(\phi_1) = \mathrm{Surf}(\phi_2)$, that is, $\phi_1(r)^m = \phi_2(r)^m$ implies $\mathrm{Surf}(\phi_1) = \mathrm{Surf}(\phi_2)$. First of all let us recall that, since we are dealing with 2-forms, is is sufficient to consider 2-plots. So, let the ϕ_i be defined on some real domain $U \subset \mathbf{R}^2$. Let $r_0 \in U$, we split the problem into two cases.

1. $\phi_1(r_0) \neq 0$ — Thus $\phi_2(r_0) \neq 0$, there exists an open disc B centered at r_0 on which the ϕ_i do not vanish. Thus, the map $r \mapsto \zeta(r) = \phi_2(r)/\phi_1(r)$ defined on B is smooth with values in Z_m. But, since Z_m is discrete there exists $\zeta \in Z_m$ such that $\phi_2(r) = \zeta \times \phi_1(r)$ on B. Now, Surf is invariant by $U(1) \supset Z_m$. Therefore $\mathrm{Surf}(\phi_1) = \mathrm{Surf}(\phi_2)$ on B.

2. $\phi_1(r_0) = 0$ — Thus, $\phi_2(r_0) = 0$. Now, we have $\mathrm{Surf}(\phi_i) = \det(D(\phi_i)) \times \mathrm{Surf}$, where $D(\phi_i)$ denotes the tangent map of ϕ_i. We split this case in two subcases.

2.a. $D(\phi_1)_{r_0}$ is not degenerate — Thus, thanks to the implicit function theorem, there exists a small open disc B around r_0 where ϕ_1 is a local diffeomorphism onto its image. Since $\phi_1(r)^m = \phi_2(r)^m$, the common zero r_0 of both ϕ_1 and ϕ_2 is isolated. Thus, the map $r \mapsto \zeta(r) = \phi_2(r)/\phi_1(r)$ defined on $B - \{r_0\}$ is smooth, and for the same reason as in the first case, ζ is constant. Then, $\phi_2(r) = \zeta \times \phi_1(r)$ on $B - \{r_0\}$. But since $\phi_i(r_0) = 0$, this equality extends on B. Therefore $\mathrm{Surf}(\phi_1) = \mathrm{Surf}(\phi_2)$ on B.

2.b. $D(\phi_1)_{r_0}$ is degenerate — Let u be in the kernel of $D(\phi_1)_{r_0}$. We have $\phi_1(r_0+su)^m = \phi_2(r_0+su)^m$ for s small enough. Then, differentiating this equality m times with respect to s, we get at $s = 0$, $D(\phi_1)_{r_0}(u)^m = D(\phi_2)_{r_0}(u)^m = 0$. Therefore, $D(\phi_2)_{r_0}$ is also degenerate at r_0 and thus $\mathrm{Surf}(\phi_1)_{r_0} = \mathrm{Surf}(\phi_2)_{r_0} = 0$.

Thus, we proved that for all $r \in U$, $\mathrm{Surf}(\phi_1)_r = \mathrm{Surf}(\phi_2)_r$. Therefore, there exists a 2-form ω on $\mathcal{Q}_m$ such that $\pi_m^*(\omega) = \mathrm{Surf}$, and this form ω is completely defined by its pullback. Now, since the pullback by π_m of any other 2-form ω' on $\mathcal{Q}_m$ is proportional to Surf, the form ω' is proportional to ω. Now, for the same reasons as in (art. 9.31), the universal holonomy Γ_ω is trivial, the Souriau class σ_ω is zero, and the universal moment map μ_ω defined by $\mu_\omega(0) = 0_{\mathcal{G}^*}$ is equivariant. Moreover, the regular stratum $\dot{\mathcal{Q}}$ is just a symplectic manifold for the restriction of ω. Every symplectomorphism with compact support which does not contain 0 can be extended to an automorphism of $(\mathcal{Q}, \omega)$. Thus, since the compactly supported symplectomorphisms of a connected symplectic manifold are transitive [**Boo69**], the regular stratum $\dot{\mathcal{Q}}$ is an orbit of $\mathrm{Diff}(\mathcal{Q}, \omega)$ and, the universal moment map is injective on this stratum (art. 9.23). Therefore, the moment map μ_ω maps injectively $\mathcal{Q}$ onto the two orbits, $\{0_{\mathcal{G}^*}\}$ and $\mu_\omega(\dot{\mathcal{Q}})$. ☐

9.33. The infinite projective space. Let $\mathcal{H}$ be the Hilbert space of the square summable complex series

$$\mathcal{H} = \left\{ Z = (Z_i)_{i=1}^{\infty} \;\middle|\; \sum_{i=1}^{n} Z_i \cdot Z_i < \infty \right\},$$

where the dot denotes the Hermitian product. The space $\mathcal{H}$ is equipped with the *fine structure* of complex diffeological vector space (art. 3.15). Let $P : U \to \mathcal{H}$ be a plot, then for every $r_0 \in U$ there exist an integer n, an open neighborhood $V \subset U$ of r_0 and a finite family $\mathcal{F} = \{(\lambda_a, Z_a)\}_{a \in A}$, where the $Z_a \in \mathcal{H}$, and the $\lambda_a \in \mathcal{C}^\infty(V, \mathbf{C}^n)$, such that $P \restriction V : r \mapsto \sum_{a \in A} \lambda_a(r) \times Z_a$. Such a family $\{(\lambda_a, Z_a)\}_{a \in A}$ is called a *local family* of P at the point r_0. We introduced the symbol dZ in Exercise 99, p. 159, which associates every local family $\mathcal{F} = \{(\lambda_a, Z_a)\}_{a \in A}$, defined on the domain V, with the complex valued 1-form of V

$$dZ(\mathcal{F}) : r \mapsto \sum_{a \in A} d\lambda_a(r) Z_a.$$

For every $\lambda_a = x_a + i y_a$, where x_a and y_a are real smooth parametrizations, $d\lambda_a = dx_a + i dy_a$. There exists on $\mathcal{H}$ a 1-form α defined by

$$\alpha = \frac{1}{2i}[Z \cdot dZ - dZ \cdot Z].$$

1. As an additive group $(\mathcal{H}, +)$ acts on itself, preserving $d\alpha$. Let $Z \in \mathcal{H}$, and let t_Z be the translation by Z, then $t_Z^*(d\alpha) = d\alpha$. This action is Hamiltonian but not exact. Let μ be the moment map of the translations $(\mathcal{H}, +)$, defined by $\mu(0_{\mathcal{H}}) = 0$,

$$\mu(Z) = 2d[w(Z)] \quad \text{with} \quad w(\zeta) : Z \mapsto \frac{1}{2i}[\zeta \cdot Z - Z \cdot \zeta] \in \mathcal{C}^\infty(\mathcal{H}, \mathbf{R}).$$

The moment map μ is injective and $(\mathcal{H}, d\alpha)$ is a homogeneous symplectic space.

2. Let $\mathbf{U}(\mathcal{H})$ be the group of unitary transformations of $\mathcal{H}$, equipped with the functional diffeology. The group $\mathbf{U}(\mathcal{H})$ acts on $\mathcal{H}$ preserving α. The action of $\mathbf{U}(\mathcal{H})$ on $(\mathcal{H}, d\alpha)$ is exact and Hamiltonian. Let $P : U \to \mathbf{U}(\mathcal{H})$ be an n-plot. The value of the moment map μ, for the action of $\mathbf{U}(\mathcal{H})$ on $(\mathcal{H}, d\alpha)$, evaluated on P, is given by

$$\mu(Z)(P)_r(\delta r) = \frac{1}{2i}\left[P(r)(Z) \cdot \frac{\partial P(r)(Z)}{\partial r}(\delta r) - \frac{\partial P(r)(Z)}{\partial r}(\delta r) \cdot P(r)(Z)\right],$$

where, $r \in U$, $\delta r \in \mathbf{R}^n$ and if

$$P(r)(Z) =_{\mathrm{loc}} \sum_{\alpha \in A} \lambda_\alpha(r) Z_\alpha, \text{ then } \frac{\partial P(r)(Z)}{\partial r}(\delta r) =_{\mathrm{loc}} \sum_{\alpha \in A} \frac{\partial \lambda_\alpha(r)}{\partial r}(\delta r) Z_\alpha.$$

3. The unit sphere $\mathcal{S} \subset \mathcal{H}$ is a homogeneous space of $\mathbf{U}(\mathcal{H})$; see Exercise 126, p. 221. The fibers of the equivariant moment map μ of the action of $\mathbf{U}(\mathcal{H})$ on $(\mathcal{S}, d\alpha \restriction \mathcal{S})$ are the fibers of the infinite Hopf fibration $\pi : \mathcal{S} \to \mathcal{P} = \mathcal{S}/S^1$, where $S^1 \in \mathbf{C}$ acts multiplicatively on $\mathcal{S}$. There exists a symplectic form ω on $\mathcal{P}$, such that $\pi^*(\omega) = d\alpha \restriction \mathcal{S}$; see Exercise 100, p. 160. The equivariant moment map of the induced action of $\mathbf{U}(\mathcal{H})$ on $\mathcal{P}$ is injective. Therefore, the infinite projective space $\mathcal{P}$, equipped with the Fubini-Study form, is a homogeneous symplectic space and can be regarded as a coadjoint orbit of $\mathbf{U}(\mathcal{H})$.

PROOF. Let us prove what is claimed here and has not been already proved in a previous paragraph or exercise.

1. Since $\mathcal{H}$ is contractible, there is no holonomy. Now, let $\zeta \in \mathcal{H}$, and let t_ζ be the translation $t_\zeta(Z) = Z + \zeta$. A direct computation shows that $t_\zeta^*(\alpha) = \alpha + d[w(\zeta)]$. Thus, $d\alpha$ is invariant by translation $t_\zeta^*(d\alpha) = d\alpha$. Now, let p be a path connecting $0_{\mathcal{H}}$ to Z, we have $\mu(Z) = \Psi(p) = \hat{p}^*\mathcal{K}(d\alpha) = \hat{Z}^*(\alpha) - \hat{0}_{\mathcal{H}}^*(\alpha) - d[\mathcal{K}\alpha \circ \hat{p}]$. But on the one hand we have $\hat{Z} = t_Z$, thus $\hat{Z}^*(\alpha) - \hat{0}_{\mathcal{H}}^*(\alpha) = t_Z^*(\alpha) - \mathbf{1}_{\mathcal{H}}^*(\alpha) = \alpha + d[w(Z)] - \alpha = d[w(Z)]$, and on the other hand, $\hat{p}(\zeta) = t_\zeta \circ p$. Thus, $\mathcal{K}\alpha \circ \hat{p} = \int_{t_\zeta \circ p} \alpha = \int_p t_\zeta^*(\alpha) = \int_p \alpha + \int_p d[w(\zeta)] = \int_p \alpha + w(\zeta)(Z)$, since $w(\zeta)(0_{\mathcal{H}}) = 0$. Hence, $\mu(Z) = d[w(Z)] - d[\zeta \mapsto w(\zeta)(Z)]$. But $w(\zeta)(Z) = -w(Z)(\zeta)$, then $\mu(Z) = d[w(Z)] - d[\zeta \mapsto -w(Z)(\zeta)] = 2d[w(Z)]$. Next, let Z be in the kernel of μ, thus $w(Z) = \mathrm{cst} = w(0_{\mathcal{H}}) = 0$. But $w(Z)(Z') = 0$ for all $Z' \in \mathcal{H}$ if and only if $Z = 0_{\mathcal{H}}$, we have just to decompose Z into real and imaginary parts and use the fact that the Hermitian norm on $\mathcal{H}$ is nondegenerate. Therefore, μ is injective.

2. Since the 1-form α is invariant by $\mathbf{U}(\mathcal{H})$, this statement is a direct application of (art. 9.11). ☐

9.34. The Virasoro coadjoint orbits. Let $\mathrm{Imm}(S^1, \mathbf{R}^2)$ be the space of all the immersions of the circle $S^1 = \mathbf{R}/2\pi\mathbf{Z}$ into $\mathbf{R}^2$, equipped with the functional diffeology.

1. For all integers n and all n-plots $P : U \to \mathrm{Imm}(S^1, \mathbf{R}^2)$, let $\alpha(P)$ be the 1-form defined on U by

$$\alpha(P)_r(\delta r) = \int_0^{2\pi} \frac{1}{\|P(r)'(t)\|^2} \left\langle P(r)''(t) \;\middle|\; \frac{\partial P(r)'(t)}{\partial r}(\delta r) \right\rangle dt,$$

where $r \in U$ and $\delta r \in \mathbf{R}^n$. The prime denotes the derivative with respect to the parameter t, and the brackets $\langle \cdot \mid \cdot \rangle$ denote the ordinary scalar product. Then, α is a 1-form on $\mathrm{Imm}(S^1, \mathbf{R}^2)$.

2. We consider now the group $\mathrm{Diff}_+(S^1)$, of positive diffeomorphisms of the circle, and its action on $\mathrm{Imm}(S^1, \mathbf{R}^2)$ by reparametrization. For all $\varphi \in \mathrm{Diff}_+(S^1)$, for all $x \in \mathrm{Imm}(S^1, \mathbf{R}^1)$, we denote by $\bar{\varphi}(x)$ the pushforward of x by φ,

$$\bar{\varphi}(x) = \varphi_*(x) = x \circ \varphi^{-1}.$$

Let $F : \mathrm{Diff}_+(S^1) \to \mathcal{C}^\infty(\mathrm{Imm}(S^1, \mathbf{R}^2), \mathbf{R})$ be the map defined, for all $\varphi \in \mathrm{Diff}_+(S^1)$, by

$$F(\varphi) : x \mapsto \int_0^{2\pi} \log \|x'(t)\| \, d\log(\varphi'(t)).$$

Then, the map F is smooth and for every $\varphi \in \mathrm{Diff}(S^1)$,

$$\bar{\varphi}^*(\alpha) = \alpha - d[F(\varphi)].$$

It follows that the exact 2-form $\omega = d\alpha$, defined on $\mathrm{Imm}(S^1, \mathbf{R}^2)$, is invariant by the action of $\mathrm{Diff}(S^1)$. Moreover, the action of $\mathrm{Diff}(S^1)$ is Hamiltonian.

3. Let $x_0 : \mathrm{class}(t) \mapsto (\cos(t), \sin(t))$ be the *standard immersion* from S^1 to $\mathbf{R}^2$. The moment maps for ω restricted to the connected component of $x_0 \in \mathrm{Imm}(S^1, \mathbf{R}^2)$, relative to $\mathrm{Diff}_+(S^1)$, are given by

$$\mu(x)(r \mapsto \varphi)_r(\delta r) = \int_0^{2\pi} \left\{ \frac{\|x''(u)\|^2}{\|x'(u)\|^2} - \frac{d^2}{du^2} \log \|x'(u)\|^2 \right\} \delta u \, du \; + \; \mathrm{cst},$$

where $r \mapsto \varphi$ is any plot of $\mathrm{Diff}_+(S^1)$ defined on some n-domain U, r is a point of U, $\delta r \in \mathbf{R}^n$, $u = \varphi^{-1}(t)$, and $\delta u = D(r \mapsto u)(r)(\delta r)$.

4. With the same conventions as in item 3, the Souriau cocycles of the $\mathrm{Diff}_+(S^1)$ action on $\mathrm{Imm}(S^1, \mathbf{R}^2)$ are cohomologous to

$$\theta(g)(r \mapsto \varphi)_r(\delta r) = \int_0^{2\pi} \frac{3\gamma''(u)^2 - 2\gamma'''(u)\gamma'(u)}{\gamma'(u)^2}\, \delta u \, du\,,$$

where $g \in \mathrm{Diff}_+(S^1)$ and $\gamma = g^{-1}$. We recognize in the integrand of the right hand side the Schwartzian derivative of γ.

5. Let β be the function defined, for all g and h in $\mathrm{Diff}_+(S^1)$, by

$$\beta(g, h) = \int_0^{2\pi} \log(g \circ h)'(t)\; d \log h'(t).$$

Then, for all g and h in $\mathrm{Diff}_+(S^1)$,

$$F(g \circ g') = F(g) \circ \bar{g}' + F(g') - \beta(g, g').$$

This function β is known as the *Bott cocycle* [**Bot78**]. The central extension of $\mathrm{Diff}_+(S^1)$ by β is the Virasoro group. Its action on $\mathrm{Imm}(S^1, \mathbf{R}^2)$, through $\mathrm{Diff}_+(S^1)$, is still Hamiltonian, and now exact. This is a well known construction which will not be more developed here.

NOTE. This example, built on purpose [**Igl95**], gathers the main ingredients found in the literature on the construction of Virasoro's group. It illustrates the whole theory, connecting objects which appeared originally in disorder.

PROOF. The proof is actually a long and tedious series of computations. To make it as clear as possible, we shall split the computations into a few steps.

The 1-form α. We prove first that α is a well defined 1-form on $\mathrm{Imm}(S^1, \mathbf{R}^2)$. Let $F : U \to U$ be a smooth m-parametrization. Let $s \in V$, $\delta s \in \mathbf{R}^m$. Denoting by r the point $F(s)$, we have

$$\begin{aligned}
\alpha(P \circ F)_s(\delta s) &= \int_0^{2\pi} \frac{1}{\|(P \circ F)(s)'(t)\|^2} \left\langle (P \circ F)(s)''(t) \,\middle|\, \frac{\partial (P \circ F)(s)'(t)}{\partial s}(\delta s) \right\rangle dt \\
&= \int_0^{2\pi} \frac{1}{\|P(F(s))'(t)\|^2} \left\langle P(F(s))''(t) \,\middle|\, \frac{\partial P(F(s))'(t)}{\partial s}(\delta s) \right\rangle dt \\
&= \int_0^{2\pi} \frac{1}{\|P(r)'(t)\|^2} \left\langle P(r)''(t) \,\middle|\, \frac{\partial P(r)'(t)}{\partial r}\left(\frac{\partial F(s)}{\partial s}(\delta s)\right) \right\rangle dt \\
&= \alpha(P)_{r=F(s)}\left(\frac{\partial F(s)}{\partial s}(\delta s)\right) \\
&= F^*(\alpha(P))_s(\delta s).
\end{aligned}$$

Thus, $\alpha(P \circ F) = F^*(\alpha(P))$, α satisfies the compatibility condition and is then a differential 1-form on $\mathrm{Imm}(S^1, \mathbf{R}^2)$.

Let us consider now the action of $\mathrm{Diff}_+(S^1)$ on $\mathrm{Imm}(S^1, \mathbf{R}^2)$. This action is obviously smooth from the very definition of the functional diffeology of $\mathrm{Diff}_+(S^1)$. Let us denote φ^{-1} by ϕ such that

$$\bar{\varphi}^*(\alpha)(P) = \alpha(\bar{\varphi} \circ P) = \alpha[r \mapsto P(r) \circ \varphi^{-1}] = \alpha[r \mapsto P(r) \circ \phi].$$

Note that $\mathrm{Diff}_+(S^1)$ acts on *speed* and *acceleration* of any immersion x, by

$$\begin{aligned}(x\circ\phi)'(t) &= x'(\phi(t))\cdot\phi'(t),\\ (x\circ\phi)''(t) &= x''(\phi(t))\cdot\phi'(t)^2 + x'(\phi(t))\cdot\phi''(t).\end{aligned} \qquad (\heartsuit)$$

Let us denote by Q the plot $\bar{\varphi}\circ P$, that is, $Q = [r \mapsto P(r)\circ\phi]$, such that

$$\alpha(\bar{\varphi}\circ P)_r(\delta r) = \int_0^{2\pi} \frac{1}{\|Q(r)'(t)\|^2}\left\langle Q(r)''(t) \,\middle|\, \frac{\partial Q(r)'(t)}{\partial r}(\delta r)\right\rangle dt$$

for all $r \in U$ and all $\delta r \in \mathbf{R}^n$. Now, from $(\heartsuit)$ we have

$$\begin{aligned}Q(r)'(t) &= (P(r)\circ\phi)'(t) = P(r)'(\phi(t))\cdot\phi'(t),\\ Q(r)''(t) &= (P(r)\circ\phi)''(t) = P(r)''(\phi(t))\cdot\phi'(t)^2 + P(r)'(\phi(t))\cdot\phi''(t).\end{aligned}$$

Let us write $\alpha(\bar{\varphi}\circ P)_r(\delta r)$ according to this decomposition,

$$\alpha(\bar{\varphi}\circ P)_r(\delta r) = \int_0^{2\pi} A\,dt + \int_0^{2\pi} B\,dt.$$

One has first,

$$A = \frac{1}{\|P(r)'(\phi(t))\cdot\phi'(t)\|^2}\left\langle P(r)''(\phi(t))\cdot\phi'(t)^2 \,\middle|\, \frac{\partial P(r)'(\phi(t))\cdot\phi'(t)}{\partial r}(\delta r)\right\rangle,$$

that is,

$$A = \frac{1}{\|P(r)'(\phi(t))\|^2}\left\langle P(r)''(\phi(t)) \,\middle|\, \frac{\partial P(r)'(\phi(t))}{\partial r}(\delta r)\right\rangle \phi'(t).$$

Since φ, and thus ϕ, is a positive diffeomorphism, after the change of variable $t \mapsto \phi(t)$ under the integral, we get already

$$\int_0^{2\pi} A\,dt = \alpha(P)_r(\delta r).$$

Next,

$$B = \frac{1}{\|P(r)'(\phi(t))\cdot\phi'(t)\|^2}\left\langle P(r)'(\phi(t))\cdot\phi''(t) \,\middle|\, \frac{\partial P(r)'(\phi(t))\cdot\phi'(t)}{\partial r}(\delta r)\right\rangle,$$

then,

$$\int_0^{2\pi} B\,dt = \int_0^{2\pi} \frac{1}{\|P(r)'(\phi(t))\|^2}\left\langle P(r)'(\phi(t)) \,\middle|\, \frac{\partial P(r)'(\phi(t))}{\partial r}(\delta r)\right\rangle \frac{\phi''(t)}{\phi'(t)}\,dt.$$

Let us then denote for short,

$$x = P(r), \quad x' = P(r)', \quad \text{and} \quad \delta x' = \left[t \mapsto \frac{\partial P'(r)(t)}{\partial r}(\delta r)\right].$$

Using that, for any variation δ, we have the identities

$$\delta\|v\| = \frac{1}{\|v\|}\langle v \mid \delta v\rangle \quad \text{and} \quad \delta \log\|v\| = \frac{1}{\|v\|}\delta\|v\| = \frac{1}{\|v\|^2}\langle v \mid \delta v\rangle,$$

we get, with a change of variable $s = \varphi^{-1}(t)$ in the middle,

$$\begin{aligned}
\int_0^{2\pi} B\,dt &= \int_0^{2\pi} \frac{1}{\|x'(\phi(t))\|^2}\langle x'(\phi(t)) \mid \delta x'(\phi(t))\rangle \frac{\phi''(t)}{\phi'(t)}\,dt \\
&= \int_0^{2\pi} \delta \log \|x'(\phi(t))\|\, d\log(\phi'(t)) \\
&= \delta \int_0^{2\pi} \log \|x'(\phi(t))\|\, d\log(\phi'(t)) \\
&= \delta \int_0^{2\pi} \log \|x'(\varphi^{-1}(t))\|\, d\log((\varphi^{-1})'(t)) \\
&= \delta \int_0^{2\pi} \log \|x'(s)\|\, d\log[(\varphi^{-1})'(\varphi(s))] \\
&= -\,\delta \int_0^{2\pi} \log \|x'(s)\|\, d\log(\varphi'(s)) \\
&= -\,\frac{\partial}{\partial r}\left\{ \int_0^{2\pi} \log \|P(r)'(s)\|\, d\log(\varphi'(s)) \right\}(\delta r) \\
&= \int_0^{2\pi} \delta \log \|x'(\phi(t))\|\, d\log(\phi'(t)) \\
&= -\,\frac{\partial}{\partial r}\left\{ F(\varphi)(P(r)) \right\}(\delta r) \\
&= -\, d[F(\varphi)](P)_r(\delta r).
\end{aligned}$$

Coming back to $\alpha(\bar{\varphi} \circ P)_r(\delta r)$, we get finally

$$\alpha(\bar{\varphi} \circ P)_r(\delta r) = \alpha(P)_r(\delta r) - d[F(\varphi)](P)_r(\delta r),$$

that is,

$$\bar{\varphi}^*(\alpha) = \alpha - d[F(\varphi)].$$

Hence, the exterior derivative $\omega = d\alpha$ is invariant by the action of $\mathrm{Diff}_+(S^1)$, and since the difference $\bar{\varphi}^*(\alpha) - \alpha$ is exact, this action is Hamiltonian.

The 2-points moment map. Now, let us compute the 2-points moment map ψ of the action of $\mathrm{Diff}_+(S^1)$ on $(\mathrm{Imm}(S^1, \mathbf{R}^2), \omega)$. Let p be a path connecting two immersions x_0 and x_1. Thus, $\Psi(p) = \hat{p}^*(\mathcal{K}\omega) = \hat{p}^*(\mathcal{K}d\alpha) = \hat{p}^*(\hat{1}^*(\alpha) - \hat{0}^*(\alpha) - d(\mathcal{K}\alpha)) = \hat{x}_1^*(\alpha) - \hat{x}_0^*(\alpha) - d(\mathcal{K}\alpha \circ \hat{p})$. But, for all $\varphi \in \mathrm{Diff}_+(S^1)$,

$$\begin{aligned}
\mathcal{K}\alpha \circ \hat{p}(\varphi) &= \int_{\bar{\varphi}(p)} \alpha &&= \int_p \bar{\varphi}^*(\alpha) \\
&= \int_p \alpha - \int_p dF(\varphi) &&= \int_p \alpha - F(\varphi)(x_1) + F(\varphi)(x_0).
\end{aligned}$$

We get from there

$$\begin{aligned}
\Psi(p) &= \psi(x_0, x_1) \\
&= \{\hat{x}_1^*(\alpha) + d[\varphi \mapsto F(\varphi)(x_1)]\} - \{\hat{x}_0^*(\alpha) + d[\varphi \mapsto F(\varphi)(x_0)]\}.
\end{aligned}$$

But note that $\hat{x}^*(\alpha) + d[\varphi \mapsto F(\varphi)(x)]$ is not a momentum of $\mathrm{Diff}_+(S^1)$.

The 1-point moment maps. Let us compute the moment map $\psi(x_0, x)$. Let

$$m = \{\hat{x}^*(\alpha) + d[\varphi \mapsto F(\varphi)(x)]\}(r \mapsto \varphi)_r(\delta r).$$

Let us denote for short

$$\begin{aligned} A &= \hat{x}^*(\alpha)(r \mapsto \varphi)_r(\delta r), \\ B &= d[\varphi \mapsto F(\varphi)(x)](r \mapsto \varphi)_r(\delta r) = \frac{\partial F(\varphi)(x)}{\partial r}\delta r. \end{aligned}$$

We shall use the notation m_0, A_0 and B_0 for the immersion x_0, thus

$$\psi(x_0, x)(r \mapsto \varphi)_r(\delta r) = m - m_0 = A + B - A_0 - B_0.$$

But $\hat{x}^*(\alpha)(r \mapsto \varphi) = \alpha(\hat{x} \circ [r \mapsto \varphi]) = \alpha(r \mapsto x \circ \varphi^{-1})$, then let $\phi = \varphi^{-1}$, we get

$$A = \int_0^{2\pi} \frac{1}{\|(x \circ \phi)'(t)\|^2} \left\langle (x \circ \phi)''(t) \;\middle|\; \frac{\partial (x \circ \phi)'(t)}{\partial r}(\delta r) \right\rangle.$$

Let us now introduce

$$u = \phi(t), \quad u' = \phi(t), \quad \text{and} \quad u'' = \phi''(t).$$

Then, the decomposition given by $(\heartsuit)$ writes

$$(x \circ \phi)'(t) = x'(u) \cdot u' \quad \text{and} \quad (x \circ \phi)''(t) = x''(u) \cdot u'^2 + x'(u) \cdot u''.$$

Next, let us use the prefix δ for the various variations associated with δr, that is, $\delta\star = D(r \mapsto \star)(r)(\delta r)$. Then,

$$\frac{\partial (x \circ \phi)'(t)}{\partial r}(\delta r) = \delta[x'(u) \cdot u'] = x''(u) \cdot \delta u \cdot u' + x'(u) \cdot \delta u'.$$

Thus,

$$\begin{aligned} A &= \int_0^{2\pi} \frac{1}{\|x'(u)\|^2 u'^2} \langle x''(u)u'^2 + x'(u)u'' \mid x''(u)u'\delta u + x'(u)\delta u' \rangle \, dt \\ &= \int_0^{2\pi} \frac{\|x''(u)\|^2}{\|x'(u)\|^2} \delta u \; u' \, dt + \int_0^{2\pi} \frac{\langle x'(u), x''(u) \rangle}{\|x'(u)\|^2} \left[\delta u' + \frac{u''}{u'} \delta u \right] dt \\ &+ \int_0^{2\pi} \frac{u''}{u'} \delta u' \, dt\,. \end{aligned}$$

Now,

$$B = \frac{\partial F(\varphi)(x)}{\partial r}\delta r = -\frac{\partial \bar{F}(\phi)(x)}{\partial r}\delta r = -\delta[\bar{F}(\phi)(x)],$$

with

$$\bar{F}(\phi)(x) = \int_0^{2\pi} \log\|x'(\phi(t))\| \; d\log\phi'(t) = \int_0^{2\pi} \log\|x'(u)\| \; d\log(u').$$

Then, after the variation with respect to δr and an integration by parts, we get

$$\begin{aligned} B &= -\int_0^{2\pi} \frac{\langle x'(u), x''(u) \rangle}{\|x'(u)\|^2} \delta u \frac{u''}{u'} \, dt - \int_0^{2\pi} \log\|x'(u)\| \; \delta d\log(u') \\ &= -\int_0^{2\pi} \frac{\langle x'(u), x''(u) \rangle}{\|x'(u)\|^2} \delta u \frac{u''}{u'} \, dt + \int_0^{2\pi} \frac{\langle x'(u), x''(u) \rangle}{\|x'(u)\|^2} u' \; \delta\log(u') \, dt \\ &= -\int_0^{2\pi} \frac{\langle x'(u), x''(u) \rangle}{\|x'(u)\|^2} \delta u \frac{u''}{u'} \, dt + \int_0^{2\pi} \frac{\langle x'(u), x''(u) \rangle}{\|x'(u)\|^2} \delta u' \, dt\,. \end{aligned}$$

Therefore, grouping the terms and integrating again by parts, we get

$$\begin{aligned} A+B &= \int_0^{2\pi} \frac{\|x''(u)\|^2}{\|x'(u)\|^2}\delta u \, du + 2\int_0^{2\pi} \frac{\langle x'(u), x''(u)\rangle}{\|x'(u)\|^2}\delta u' \, dt + \int_0^{2\pi} \frac{u''}{u'}\delta u' \, dt \\ &= \int_0^{2\pi} \frac{\|x''(u)\|^2}{\|x'(u)\|^2}\delta u \, du - 2\int_0^{2\pi} \frac{d^2}{du^2}\log\|x'(u)\|\delta u \, du + \int_0^{2\pi} \frac{u''}{u'} \delta u' \, dt \\ &= \int_0^{2\pi} \left\{ \frac{\|x''(u)\|^2}{\|x'(u)\|^2} - \frac{d^2}{du^2}\log\|x'(u)\|^2 \right\} \delta u \, du + \int_0^{2\pi} \frac{u''}{u'} \delta u' \, dt. \end{aligned}$$

Now, since $\|x_0'(t)\| = 1$ we get the value of the 2-points moment map,

$$\begin{aligned} \psi(x_0, x)(r \mapsto \varphi)_r(\delta r) &= \int_0^{2\pi} \left\{ \frac{\|x''(u)\|^2}{\|x'(u)\|^2} - \frac{d^2}{du^2}\log\|x'(u)\|^2 \right\} \delta u \, du \\ &\quad - \int_0^{2\pi} \delta u \, du. \end{aligned}$$

The second term of the right hand side of this equality is a constant momentum of $\mathrm{Diff}_+(S^1)$, so it can be avoided. Thus, every moment map, up to a constant, is equal to this moment μ.

Souriau cocycles. The Souriau cocycle associated with the immersion x_0 is defined by $\theta(g) = \psi(x_0, \bar{g}(x_0))$; see (art. 9.10). We replace then, in the expression of ψ above, x by $\bar{g}(x_0) = x_0 \circ g^{-1}$, that is, $x = x_0 \circ \gamma$, and $\theta(g)(r \mapsto \varphi)_r(\delta r) = \psi(x_0, x_0 \circ \gamma)$. Let us note next that

$$(x_0 \circ \gamma)'(u) = x_0'(\gamma(u))\gamma'(u) \text{ and } (x_0 \circ \gamma)''(u) = x_0''(\gamma(u))\gamma'(u)^2 + x_0'(u)\gamma''(u),$$

and let us recall that $\|x_0'\| = \|x_0''\| = 1$, and that $\langle x_0' \mid x_0'' \rangle = 0$. We get,

$$\|x'(u)\|^2 = \gamma'(u)^2 \quad \text{and} \quad \|x''(u)\|^2 = \gamma'(u)^4 + \gamma''(u)^2,$$

which gives

$$\begin{aligned} \frac{\|x''(u)\|^2}{\|x'(u)\|^2} &= \gamma'(u)^2 + \frac{\gamma''(u)^2}{\gamma'(u)^2}, \\ \frac{d^2}{du^2}\log\|x'(u)\|^2 &= 2\frac{\gamma'''(u)\gamma'(u) - \gamma''(u)^2}{\gamma''(u)^2}. \end{aligned}$$

Thus,

$$\begin{aligned} \theta(g)(r \mapsto \varphi)_r(\delta r) &= \int_0^{2\pi} \frac{3\gamma''(u)^2 - 2\gamma'''(u)\gamma'(u)}{\gamma'(u)^2} \delta u \, du \\ &\quad + \int_0^{2\pi} \gamma'(u)^2 \, \delta u \, du - \int_0^{2\pi} \delta u \, du. \end{aligned}$$

But, after a change of variable $u \mapsto v = \gamma(u)$, we get

$$\int_0^{2\pi} \gamma'(u)^2 \, \delta u \, du = \int_0^{2\pi} (\delta u \gamma'(u)) \, \gamma'(u) \, du = \int_0^{2\pi} \delta v \, dv.$$

Then, the two last terms cancel each other, and we get the value claimed in the proposition for the Souriau cocycle θ.

Bott's cocycle. The real function $F(g \circ h) - F(g) \circ \bar{h} - F(h)$ is constant since X is connected, and its differential is equal to $(\bar{g} \circ \bar{h})^*(\alpha) - \bar{h}^*(\bar{g}^*(\alpha))$, that is, 0. Now, to make $\beta(g, g') = F(g) \circ \bar{g}' + F(g') - \beta(g, g') - F(g \circ g')$ explicit, it is sufficient to compute the right hand member on the standard immersion x_0, for which the speed norm is equal to 1, and thus $\log\|x'(t)\| = 0$ for all t. We get then

$$\begin{aligned}
\beta(g,h) &= F(g)(x_0 \circ h^{-1}) - F(h)(x_0) - F(g \circ h)(x_0) \\
&= +\int_0^{2\pi} \log\|(x_0 \circ h^{-1})'(t)\| \, d\log g'(t) \\
&= +\int_0^{2\pi} \log(h^{-1})'(t) \, d\log g'(t) \\
&= -\int_0^{2\pi} \log h'(h^{-1}(t)) \, d\log g'(t) \\
&= -\int_0^{2\pi} \log h'(s) \, d\log g'(h(s)) \\
&= +\int_0^{2\pi} \log(g \circ h)'(t) \, d\log h'(t).
\end{aligned}$$

And this is the standard expression of Bott's cocycle. □

Exercise

✎ Exercise 147 (The moment of imprimitivity). Let X be a diffeological space. Let $\Omega^1(X)$ be the vector space of 1-forms of X, equipped with the functional diffeology (art. 6.45). Let Taut be the tautological 1-form defined on $X \times \Omega^1(X)$ and let Liouv be the Liouville 1-form defined on the cotangent bundle T^*X; see (art. 6.48) and (art. 6.49). Let us consider then the additive diffeological group of smooth functions $\mathcal{C}^\infty(X, \mathbf{R})$, acting smoothly on $X \times \Omega^1(X)$ by *right action*,

$$\bar{f} : (x, \alpha) \mapsto (x, \alpha - df)$$

for all $f \in \mathcal{C}^\infty(X, \mathbf{R})$ and $(x, \alpha) \in X \times \Omega^1(X)$. This action has a natural projection on the cotangent T^*X, and this action will be denoted the same way,

$$\bar{f} : (x, a) \mapsto (x, a - df(x))$$

for all $f \in \mathcal{C}^\infty(X, \mathbf{R})$ and $(x, a) \in T^*X$.

1) Show that, for all $f \in \mathcal{C}^\infty(X, \mathbf{R})$, the variance of the tautological form and the Liouville form are given by

$$\bar{f}^*(\mathrm{Taut}) = \mathrm{Taut} - \mathrm{pr}_1^*(df) \quad \text{and} \quad \bar{f}^*(\mathrm{Liouv}) = \mathrm{Liouv} - \pi^*(df).$$

Observe that the exterior derivatives dTaut and $\omega = \mathrm{dLiouv}$ are invariant by the action of $\mathcal{C}^\infty(X, \mathbf{R})$.

2) Let p be a path in T^*X, connecting $(x_0, a_0) = p(0)$ to $(x_1, a_1) = p(1)$. Show that the paths moment map Ψ and the 2-points moment map ψ, with respect to the 2-form $\omega = \mathrm{dLiouv}$, are given by

$$\Psi(p) = \psi((x_0, a_0), (x_1, a_1)) = d[f \mapsto f(x_1)] - d[f \mapsto f(x_0)].$$

3) Check that, for all $x \in X$, the real function $[f \mapsto f(x)]$ is smooth. We call it the *Dirac function* of the point x, and we denote it by δ_x.

$$\delta_x = [f \mapsto f(x)] \in \mathcal{C}^\infty(\mathcal{C}^\infty(X, \mathbf{R}), \mathbf{R}).$$

Show that the differential $d\delta_x = d[f \mapsto f(x)]$ is an invariant 1-form[12] of the additive group $\mathcal{C}^\infty(X, \mathbf{R})$. Show that every moment map of the action of $\mathcal{C}^\infty(X, \mathbf{R})$ on T^*X is, up to a constant, equal to the invariant moment map

$$\mu : (x, a) \mapsto d\delta_x = d[f \mapsto f(x)].$$

Note that the moment μ is constant on the fibers $T_x^*X = \pi^{-1}(x)$, and if the real smooth functions separate[13] the points of X, then the image of the moment map μ is the space X, identified with the space of Dirac's functions.

4) Show that the action of $\mathcal{C}^\infty(X, \mathbf{R})$ on (T^*X, ω) is Hamiltonian and exact, that is, $\Gamma = \{0\}$ and $\sigma = 0$.

This example, in the case of differential manifolds, appears informally in Ziegler's construction of a symplectic analogue for *systems of imprimitivity* in representation theory [**Zie96**]. It is why the moment map μ may be called the *moment of imprimitivity*. The diffeological framework then gives to it a full formal status and even extends it.

Discussion on Symplectic Diffeology

Symplectic diffeology is more a program than a complete theory. The last decades of theoretical research in mechanics—in completely integrable systems, quantum mechanics, or quantum field theory etc.—have shown a special interest in structures which seem to be symplectic even if they do not live on manifolds but on spaces, generally infinite dimensional, where the formal constructions of symplectic geometry do not apply as is. Building a formal framework for these symplectic-like structures is not just a desire of formalism, but a need to embed these heuristic constructions in a well delimited and workable mathematical construction. There are different ways to approach these problems, such as functional analysis, infinite dimensional manifolds à la Banach, or maybe others. A diffeological approach is one of them, but has the virtue of involving a very light apparatus of mathematical tools. Axiomatics, reduced to only three axioms, cannot be simpler and the great stability of the category, under set theoretic operations, is a gift for this kind of problem, where infinite dimensions and what is admitted to be considered as singularities are a burden for classical geometry. On the other hand, the light structure of diffeology does not seem to be a weakness. The construction of the moment maps, associated with a closed 2-form on any diffeological space, shows the whole — and deals correctly with the — complexity of the various situations without exaggerating technicalities. This is clearly shown in the few examples given in the previous sections of this chapter. So, even if it seems clear that diffeology is adequate for such a generalization of symplectic (or presymplectic) geometry, we still need a good definition, or at least a serious discussion, about *what is a symplectic diffeological space*? Or *what does it mean to be symplectic in diffeology*?

A relatively good answer has been given at least in the case of homogeneous spaces, that is, in the case of a closed 2-form ω defined on a space X homogeneous for its group of automorphisms $\mathrm{Diff}(X, \omega)$; see (art. 9.18) and (art. 9.19). In this case the situation is clear: a universal moment map distinguishes between being

[12] This differential has nothing to do with the derivative of the Dirac distributions in the sense of De Rham's currents.

[13] which means, $f(x) = f(x')$ for all smooth real functions f if and only if $x = x'$.

presymplectic or *symplectic*. The homogeneous space (X, ω) is symplectic if a universal moment map μ_ω is a covering onto its image, or, which is equivalent, if the preimages of its values are diffeologically discrete. Otherwise it is presymplectic, and the characteristics of μ_ω can be regarded as the characteristics of the 2-form ω, by analogy with what happens in the special case of homogeneous manifolds (art. 9.26). The symplectic homogeneous case has been illustrated, in particular, by the Hilbert space $\mathcal{H}$ or the infinite dimensional projective space $\mathcal{P}$ (art. 9.33), even by the singular irrational tori (art. 9.30). The fact that these spaces are symplectic is now a well defined property, according to the definition given in (art. 9.19).

Why is the homogeneous case so important? Because for a closed 2-form ω defined on a manifold M, being symplectic implies to be homogeneous under $\mathrm{Diff}(M, \omega)$, with moreover the universal moment maps μ_ω injective (art. 9.23). The homogeneity of M under the action of $\mathrm{Diff}(M, \omega)$ is a strong consequence of two things: firstly ω is invertible, has no kernel; and secondly, every symplectic manifold is locally flat, that is, looks locally like some $(\mathbf{R}^{2n}, \omega_{st})$. There exists only one local model of symplectic manifold in each dimension, this is the famous Darboux theorem.

In diffeology, we have not a unique model to propose for all the covered cases: from singular tori to infinite projective space. But we can replace advantageously the local model of symplectic manifolds by the strong requirement of homogeneity. We remark that we could replace the global homogeneity by a local homogeneity, but this would lead to unwanted subtleties, for now. Our question is then, *Do we want to preserve the fundamental property of homogeneity for symplectic diffeological spaces?* The answer to this question is not that obvious.

If we want to preserve the property of homogeneity, the problem is just solved by (art. 9.19). But the example of the cone orbifold $\mathcal{C}_m$ (art. 9.32) is a real cultural or sociological obstacle. Many mathematicians want to consider the cone orbifold, equipped with the 2-form described in (art. 9.32), as symplectic, because it is the quotient of the symplectic space $(\mathbf{R}^2, \omega_{st})$ by a finite group preserving the standard symplectic form ω_{st}. Unfortunately the cone orbifold is not homogeneous, the origin is fixed by every diffeomorphism. However, even according to diffeology, regarding $\mathcal{C}_m$ as a symplectic diffeological space makes sense: the cone orbifold is generated by *symplectic plots* — in this case the only plot $\pi_m : \mathbf{R}^2 \to \mathcal{C}_m$. Let us remark that the corner orbifold (art. 9.31) does not satisfy this property, and cannot be symplectic even according to this definition. Yael Karshon suggested, in private discussions, to define symplectic diffeological spaces as diffeological spaces generated by symplectic plots, that is, plots P such that $\omega(P)$ is symplectic. Although this makes sense, it is a way we have not explored yet. In particular, the relationship between such a space and the moment maps needs to be clarified, as well as the relationship with the action of the group of automorphisms and the nature of its orbits. The idea behind this is to consider symplectic reductions as symplectic spaces independently of the regularity of the distribution of the moment map strata.

Another approach should consist of only requiring the universal moment maps to be injective or at least to have discrete preimages, since the preimages of a universal moment map seem to realize the characteristics of ω. Let us note that the cone orbifold satisfies this property, but not the corner orbifold, which is satisfactory. Unfortunately, the space $\mathbf{R}^2$ equipped with the 2-form $\omega = (x^2 + y^2) dx \wedge dy$

		Moment map	
		injective	non injective
Automorphisms	transitive	• Irrational tori • Infinite projective space • Symplectic manifolds	• Infinite Hilbert sphere • Homogeneous presymplectic manifolds
	intransitive	• Cone orbifold • $\mathbf{R}^2$, $\omega = (x^2+y^2)dx\wedge dy$	• Corner orbifold • Almost everything

FIGURE 9.3. Distribution of examples.

also satisfies this property, and it is clearly not symplectic. It seems that we have to either abandon this approach, or to go deeper into it.

So, what remains? If we want to include the cone orbifold in the category of symplectic diffeological spaces we have to give up homogeneity by the group of automorphisms. Or, if we do not want to give up homogeneity, we abandon the cone orbifold and certainly many other examples of diffeological spaces, generated by symplectic plots. For now, the set of examples and theorems in symplectic diffeology is not large enough to make a reasonable choice. It is why there is no general definition of symplectic diffeological spaces here. But, we can just deal with these spaces, equipped with closed 2-forms, by trying to get the maximum of information as we did in the examples above (see Figure 9.3): the nature of the characteristics of the universal moment map, the injectivity or not of the universal moment map, the transitivity of the group of automorphisms or the nature of its orbits, the computation of the universal holonomy, and the Souriau class, etc. We have all the tools needed to treat the examples given by the literature or by the physicists, and maybe to define what the word symplectic means in diffeology is not that important after all...

A few more words. Why do we need the 2-form to be nondegenerate? From a physicist's point of view, we do not need that, actually it is the opposite. A pair (X, ω), with ω a closed 2-form, represents what we could call a *dynamical structure*. The *characteristics* of ω, that is, the connected components of the preimages of the universal moment map, describe the dynamics of the system, together with the partition into orbits by the group of automorphisms. This is what physicists are interested in: they want equations describing the evolution of their systems. The fact that, in classical mechanics, the space of characteristics is symplectic is fortuitous, a consequence of the presymplectic nature of the dynamical structures involved. Then, the group of automorphisms of the space of characteristics is transitive and, by equivariance, the image of the moment map is one coadjoint orbit. In the general case of infinite dimension spaces, orbifolds, singular reductions etc., there is no reason for the automorphisms to be transitive, and the image of the universal moment map is just some union of coadjoint orbits, and then, maybe, not *symplectic*. The picture is clear, and that is what we have to deal with. Another critical question is of the *symplectic reduction* of a subspace $W \subset X$: the reduced space may be simply defined as the space of characteristics of the dynamical 2-form ω restricted to W, or its regularization, that is, the image of the universal moment map of the restriction. It is actually not different from the general case where the subspace coincides with the whole space. The task is then to describe this image and what, from the 2-form, passes to the quotient or to the image? For the structure of the space of characteristics we get two diffeologies: the first one is the natural quotient diffeology on the space of characteristics, the second one is the diffeology induced on the image of the universal moment map by the ambient space of momenta. They may coincide or not, but they both have their role to play.

Solutions to Exercises

☞ **Exercise 1, p. 6** (Equivalent axiom of covering). Consider the three axioms D1, D2, D3 (art. 1.5). From axiom D1 the constants maps cover X, thus D1' is satisfied. Hence, D1, D2, D3 imply D1', D2, D3. Conversely, consider D1', D2, and D3. Let x be a point of X. By D1' there exists a plot $P : U \to X$ such that x belongs to $P(U)$. Let r in U such that $P(r) = x$. Now let n be any integer. Let $\mathbf{r} : \mathbf{R}^n \to U$ be the constant parametrization mapping every point of $\mathbf{R}^n$ to r. The composition $P \circ \mathbf{r}$ is the constant parametrization $\mathbf{x}$ mapping $\mathbf{R}^n$ to x. Since $\mathbf{r}$ is smooth and thanks to D3, the parametrization $P \circ \mathbf{r}$ is a plot of X. Hence, D1 is satisfied and D1', D2, D3 imply D1, D2, D3. Therefore, the axioms D1, D2, D3 are equivalent to the axioms D1', D2, D3.

☞ **Exercise 2, p. 6** (Equivalent axiom of locality). Consider the three axioms D1, D2, D3 (art. 1.5). Let $P : U \to X$ be a parametrization. Assume that for any point r of U there exists an open neighborhood V_r of r such that $P_r = P \restriction V_r$ belongs to $\mathcal{D}$. The family $(P_r)_{r \in U}$ is a compatible family of elements of $\mathcal{D}$ with P as supremum. Thanks to the axiom D2, P belongs to $\mathcal{D}$, and D2' is satisfied. Hence, D1, D2, D3 imply D1, D2', D3. Conversely, consider D1, D2' and D3. Now, let $\{P_i : U_i \to X\}_{i \in \mathcal{J}}$ be a family of compatible n-parametrizations, and let $P : U \to X$ be the supremum of the family. Let r be any point of U. By definition of P, there exists $P_i : U_i \to X$ with $r \in U_i$ such that $P \restriction U_i = P_i$. Thus, the axiom D2' is satisfied. Hence, D1, D2', D3 imply D1, D2, D3. Therefore, the axioms D1, D2, D3 are equivalent to the axioms D1, D2', D3.

☞ **Exercise 3, p. 7** (Global plots and diffeology). Let $P : U \to X$ be an n-parametrization belonging to $\mathcal{D}$. For all points r in U there exists a real $\epsilon > 0$ such that the open ball $B(r, \epsilon)$, centered at r, of radius ϵ, is contained in U. Since the inclusion $B(r, \epsilon) \subset U$ is a smooth parametrization, the restriction $P \restriction B(r, \epsilon)$ belongs to $\mathcal{D}$. Now, the following parametrization

$$\varphi : B(r, \epsilon) \to B(0, 1), \quad \text{defined by} \quad \varphi : s \mapsto s' = \frac{1}{\epsilon}(s - r),$$

is a diffeomorphism. Next, let $\psi : B(0, 1) \to \mathbf{R}^n$, and then $\psi^{-1} : \mathbf{R}^n \to B(0, 1)$, given by

$$\psi(s) = \frac{s}{\sqrt{1 - \|s\|^2}} \quad \text{and} \quad \psi^{-1}(s') = \frac{s'}{\sqrt{1 + \|s'\|^2}}.$$

The parametrization ψ is a diffeomorphism. Hence, $\phi = \psi \circ \varphi$ is a diffeomorphism from $B(r, \epsilon)$ to $\mathbf{R}^n$. Then, thanks to the axiom of smooth compatibility, the global parametrization $(P \restriction B(r, \epsilon)) \circ \phi^{-1} : \mathbf{R}^n \to X$ belongs to $\mathcal{D}$. By hypothesis, it also belongs to $\mathcal{D}'$. Thus, thanks again to the axiom of smooth compatibility, the parametrization $[(P \restriction B(r, \epsilon)) \circ \phi^{-1}] \circ \phi = P \restriction B(r, \epsilon)$ belongs to $\mathcal{D}'$. Now, P being the supremum of a compatible family of elements of $\mathcal{D}'$, thanks to the axiom of locality of diffeology, P is an element of $\mathcal{D}'$. Therefore, $\mathcal{D} \subset \mathcal{D}'$, exchanging $\mathcal{D}$ and $\mathcal{D}'$ gives $\mathcal{D}' \subset \mathcal{D}$, and finally $\mathcal{D} = \mathcal{D}'$.

☞ **Exercise 4, p. 8** (Diffeomorphisms between irrational tori). Let $P : U \to T_\alpha$ be a parametrization. Let us say that P *lifts locally along* π_α, *at the point* $r \in U$, if there exist an open neighborhood V of r and a smooth parametrization $Q : V \to \mathbf{R}$ such that $\pi_\alpha \circ Q = P \upharpoonright V$. Now, the property (✱) writes P is a plot if it lifts locally along π_α, at every point of U.

1) Let us check, following (art. 1.11), that the property (✱) defines a diffeology.

D1. Since π_α is surjective, for every point $\tau \in T_\alpha$ there exists $x \in \mathbf{R}$ such that $\tau = \pi_\alpha(r)$. Then, $\mathbf{x} : r \mapsto x$ is a lift in $\mathbf{R}$ of the constant parametrization $\boldsymbol{\tau} : r \mapsto \tau$.

D2. The axiom of locality is satisfied by the very definition of $\mathcal{D}$.

D3. Let $P : U \to T_\alpha$ satisfying (✱). Let $F : U' \to U$ be a smooth parametrization. Let r' be a point of U', let $r = F(r')$, let V be an open neighborhood of r, and let Q be a smooth parametrization in $\mathbf{R}$ such that $\pi_\alpha \circ Q = P \upharpoonright V$. Let $V' = F^{-1}(V)$ and $Q' = Q \circ F$, defined on V'. Then, $\pi_\alpha \circ Q' = (P \circ F) \upharpoonright V'$.

2) Let us consider $f \in \mathcal{C}^\infty(T_\alpha, \mathbf{R})$. Since π_α and f are smooth, the map $F = f \circ \pi_\alpha$ belongs to $\mathcal{C}^\infty(\mathbf{R}, \mathbf{R})$ (art. 1.15). But since $\pi_\alpha(x + n + \alpha m) = \pi_\alpha(x)$, for every n, m in $\mathbf{Z}$, we also have $F(x + n + \alpha m) = F(x)$. Hence, F is smooth and constant on a dense subset of numbers $\mathbf{Z} + \alpha\mathbf{Z} \subset \mathbf{R}$ (for $x = 0$). Thus, F is constant and therefore f is constant. In other words, $\mathcal{C}^\infty(T_\alpha, \mathbf{R}) = \mathbf{R}$.

3) Let us consider a smooth map $f : T_\alpha \to T_\beta$. Note that, since π_α obviously satisfies (✱), π_α is a plot of T_α. Then, by definition of differentiability (art. 1.14), $f \circ \pi_\alpha$ is a plot of T_β. Hence, for every real x_0 there exist an open neighborhood V of x_0 and a smooth parametrization $F : V \to \mathbf{R}$ such that $\pi_\beta \circ F = (f \circ \pi_\alpha) \upharpoonright V$. Since V is an open subset of $\mathbf{R}$ containing x_0, we can choose V as an interval centered at x_0. For all real numbers x and all pairs (n, m) of integers such that $x + n + \alpha m$ belongs to V, the identity $\pi_\beta \circ F = (f \circ \pi_\alpha) \upharpoonright V$ writes $\pi_\beta \circ F(x + n + \alpha m) = f \circ \pi_\alpha(x + n + \alpha m) = f \circ \pi_\alpha(x) = \pi_\beta \circ F(x)$. Thus, there exist two integers n' and m' such that

$$F(x + n + \alpha m) = F(x) + n' + \beta m'. \qquad (\spadesuit)$$

Since β is irrational, for every such x, n and m, the pair (n', m') is unique. There exists an interval $\mathcal{J} \subset V$ centered at x_0 and an interval $\mathcal{O}$ centered at 0 such that for every $x \in \mathcal{J}$ and for every $n + \alpha m \in \mathcal{O}$, $x + n + \alpha m \in V$. Since F is continuous and since $\mathbf{Z} + \alpha\mathbf{Z}$ is totally discontinuous, $n' + \beta m' = F(x + n + \alpha m) - F(x)$ is constant as function of x. But F is smooth, the derivative of the identity (♠), with respect to x, at the point x_0, gives $F'(x_0 + n + \alpha m) = F'(x_0)$. Then, since α is irrational, $\mathbf{Z} + \alpha\mathbf{Z} \cap \mathcal{O}$ is dense in $\mathcal{O}$ and since F' is continuous, $F'(x) = F'(x_0)$, for all $x \in \mathcal{J}$. Hence, F restricted to $\mathcal{J}$ is affine, there exist two numbers λ and μ such that

$$F(x) = \lambda x + \mu \quad \text{for all} \quad x \in \mathcal{J}. \qquad (\clubsuit)$$

Note that, by density of $\mathbf{Z} + \alpha\mathbf{Z}$, $\pi_\alpha(\mathcal{J}) = T_\alpha$. Hence F defines completely the function f. Now, applying (♠) to the expression (♣) of F, we get for all $n + \alpha m \in \mathcal{O}$

$$\lambda \times (n + \alpha m) \in \mathbf{Z} + \beta\mathbf{Z}. \qquad (\diamondsuit)$$

Let us show that actually (♢) is satisfied for all $n + \alpha m$ in $\mathbf{Z} + \alpha\mathbf{Z}$. Let $\mathcal{O} =]-a, a[$, and let us take a not in $\mathbf{Z} + \alpha\mathbf{Z}$, even if we have to shorten $\mathcal{O}$ a little bit. Let $x \in \mathbf{Z} + \alpha\mathbf{Z}$, and $x > a$. There exists $N \in \mathbf{N}$ such that $0 < (N - 1)a < x < Na$, and then $0 < x/N < a$. Now, by density of $\mathbf{Z} + \alpha\mathbf{Z}$ in $\mathbf{R}$, for all $\eta > 0$ there exists $y \in \mathbf{Z} + \alpha\mathbf{Z}$ such that $0 < x/N - y < \eta$. Choosing $\eta < a/N$ we have $0 < x - Ny < N\eta < a$, and $0 < y < x/N < a$. Thus, since $x - Ny \in \mathbf{Z} + \alpha\mathbf{Z} \cap \mathcal{O}$, $\lambda \times (x - Ny) = \lambda x - N \times (\lambda y) \in \mathbf{Z} + \beta\mathbf{Z}$. But $y \in \mathbf{Z} + \alpha\mathbf{Z} \cap \mathcal{O}$, thus $\lambda y \in \mathbf{Z} + \beta\mathbf{Z}$, and then $N \times (\lambda y) \in \mathbf{Z} + \beta\mathbf{Z}$, therefore $\lambda x \in \mathbf{Z} + \beta\mathbf{Z}$. Now, applying successively $(\diamondsuit)$ to α and 1, we get $\lambda\alpha \in \mathbf{Z} + \beta\mathbf{Z}$ and $\lambda \in \mathbf{Z} + \beta\mathbf{Z}$. Let $\lambda\alpha = a + \beta b$ and $\lambda = c + \beta d$. If $\lambda \neq 0$, then $\alpha = (a + \beta b)/(c + \beta d)$.

4) Let us remark first that, since $\pi_\alpha(\mathcal{I}) = T_\alpha$, the map F, extended to the whole $\mathbf{R}$, still satisfies $\pi_\beta \circ F = f \circ \pi_\alpha$. Now, let us assume that f is bijective. Note that f surjective is equivalent to $\lambda \neq 0$. Let us express that f is injective: let $\tau = \pi_\alpha(x)$ and $\tau' = \pi_\alpha(x')$, if $f(\tau) = f(\tau')$, then $\tau = \tau'$, that is, $x' = x + n + \alpha m$, for some relative integers n and m. Using the lifting F, this is equivalent to if there exist two integers n' and m' such that $F(x') = F(x) + n' + \beta m'$, then there exist two integers n and m such that $x' = x + n + \alpha m$. But $F(x) = \lambda x + \mu$, with $\lambda \times (\mathbf{Z} + \alpha\mathbf{Z}) \subset \mathbf{Z} + \beta\mathbf{Z}$. Hence, the injectivity writes if $\lambda x' + \mu = \lambda x + \mu + n' + \beta m'$, then $x' = x + n + \alpha m$, which is equivalent to if $\lambda y \in \mathbf{Z} + \beta\mathbf{Z}$, then $y \in \mathbf{Z} + \alpha\mathbf{Z}$, and finally equivalent to $(1/\lambda) \times (\mathbf{Z} + \beta\mathbf{Z}) \subset \mathbf{Z} + \alpha\mathbf{Z}$. Now, let us consider the multiplication by λ, as a $\mathbf{Z}$-linear map, from the $\mathbf{Z}$-module $\mathbf{Z} + \alpha\mathbf{Z}$ to the $\mathbf{Z}$-module $\mathbf{Z} + \beta\mathbf{Z}$, defined in the respective bases $(1, \alpha)$ and $(1, \beta)$, by

$$\lambda \times 1 = c + d \times \beta \quad \text{and} \quad \lambda \times \alpha = a + b \times \beta.$$

The two modules being identified, by their bases, to $\mathbf{Z} \times \mathbf{Z}$, the multiplication by λ and the multiplication by $1/\lambda$ are represented by the matrices

$$\lambda \simeq L = \begin{pmatrix} c & a \\ d & b \end{pmatrix} \quad \text{and} \quad \frac{1}{\lambda} \simeq L^{-1}.$$

Thus, the matrix L is invertible as a matrix with coefficients in $\mathbf{Z}$, that is, $L = \mathrm{GL}(2, \mathbf{Z})$, or $ad - bc = \pm 1$.

Exercise 5, p. 9 (Smooth maps on $\mathbf{R}/\mathbf{Q}$). This exercise is similar to Exercise 4, p. 8, with solution above.

1) Let us consider a smooth map $f : E_\mathbf{Q} \to E_\mathbf{Q}$. Since $\pi : \mathbf{R} \to E_\mathbf{Q}$ is a plot, by definition of differentiability (art. 1.14), $f \circ \pi$ also is a plot of $E_\mathbf{Q}$. Hence, for every real x_0 there exist an open neighborhood V of x_0 and a smooth parametrization $f : V \to \mathbf{R}$ such that $\pi \circ F = (f \circ \pi) \restriction V$. Since V is an open subset of $\mathbf{R}$ containing x_0, we can choose V to be an interval centered at x_0. For every real number x and every rational number q such that $x + q$ belongs to V, the identity $\pi \circ F = (f \circ \pi) \restriction V$ writes $\pi \circ F(x + q) = f \circ \pi(x + q) = f \circ \pi(x) = \pi \circ F(x)$. Thus, there exists a rational number q' such that $F(x + q) = F(x) + q'$. The rational number $q' = F(x + q) - F(x)$ is smooth in x and q, thus constant in x (see also Exercise 8, p. 14). Hence, taking the derivative of this identity, at the point x_0, with respect to x, we get $F(x_0 + q) = F'(x_0)$. Let us denote $\lambda = F'(x_0)$. Now, according to the continuity of F and the density of $\mathbf{Q}$ in $\mathbf{R}$, we have $F(x) = \lambda x + \mu$, where $\mu \in \mathbf{R}$. *A priori* F is defined on a smaller neighborhood $W \subset V$ of x_0, but by density of $\mathbf{Q}$ in $\mathbf{R}$ we get, as in Exercise 4, p. 8, $\pi(W) = E_\mathbf{Q}$. Thus, F can be extended to the whole $\mathbf{R}$ by the affine map $x \mapsto \lambda x + \mu$. Now, coming back to the condition $F(x + q) = F(x) + q'$, we get $\lambda(x + q) + \mu = \lambda x + \mu + q'$, that is, $\lambda q = q'$. Thus, λ is some number mapping any rational number into another, hence it is a rational number. Let us denote it by q. Therefore, the map f being defined by $\pi \circ F = f \circ \pi$, we get $f(\tau) = q\tau + \tau'$,

where $\tau' = \pi(\mu)$. Hence, any smooth map from $E_{\mathbf{Q}}$ to $E_{\mathbf{Q}}$ is affine. Note that f is a diffeomorphism if and only if $q \neq 0$.

2) With similar arguments as in the first question, we can check that any smooth map $f : T_\alpha \to E_{\mathbf{Q}}$ is the projection of an affine map $F : x \mapsto \lambda x + \mu$. But F needs to satisfy the condition $F(x + n + \alpha m) = F(x) + q$, where n and m are integers and q is a rational number. In particular, for $n = 1$ and $m = 0$, this gives $\lambda \in \mathbf{Q}$, and for $n = 0$ and $m = 1$, $\lambda\alpha \in \mathbf{Q}$. But since $\alpha \in \mathbf{R} - \mathbf{Q}$, this is satisfied only for $\lambda = 0$. And finally F is constant.

3) The identity $\mathcal{C}^\infty(E_{\mathbf{Q}}, \mathbf{R}) = \mathbf{R}$ is analogous to $\mathcal{C}^\infty(T_\alpha, \mathbf{R}) = \mathbf{R}$ of the second question of Exercise 4, p. 8. The second part of the question is similar to the second question of this exercise, inverting $\mathbf{Q}$ and $\mathbf{Z} + \alpha\mathbf{Z}$. The map F is affine, $F : x \mapsto \lambda x + \mu$, and for every $q \in \mathbf{Q}$, there exist two numbers $n, m \in \mathbf{Z}$ such that $F(x + q) = F(x) + n + \alpha m$. So, we get $\lambda q = n + \alpha m$. In particular, for $q = 1$, we get that $\lambda = a + \alpha b$, where $a, b \in \mathbf{Z}$. Hence, for any rational number q, $(a + \alpha b)q = n + \alpha m$, that is, $aq + bq\alpha = n + \alpha m$, or $(aq - n) + (bq - m)\alpha = 0$. Since $\alpha \in \mathbf{R} - \mathbf{Q}$, we get, for all $q \in \mathbf{Q}$, $aq \in \mathbf{Z}$ and $bq \in \mathbf{Z}$. But this implies that a and b are divisible by any integer, thus $a = 0$ and $b = 0$, and therefore $\lambda = 0$.

☞ **Exercise 6, p. 9** (Smooth maps on spaces of maps). We shall denote here the derivation map d/dx^k by ϕ_k.

1) The map $\phi_k : \mathcal{C}^\infty(\mathbf{R}) \to \mathcal{C}^\infty(\mathbf{R})$ is smooth if and only if, for every plot $P : U \to \mathcal{C}^\infty(\mathbf{R})$, the parametrization $\phi_k \circ P$ is a plot of $\mathcal{C}^\infty(\mathbf{R})$. According to the definition of the functional diffeology of $\mathcal{C}^\infty(\mathbf{R})$ (art. 1.13), the parametrization $\phi_k \circ P$ is a plot of $\mathcal{C}^\infty(\mathbf{R})$ if and only if the parametrization $(r, x) \mapsto (\phi_k \circ P(r))(x)$ is a smooth parametrization of $\mathbf{R}$, that is, if the parametrization

$$\psi_k : (r, x) \mapsto \left[\frac{d^k}{dx^k}(P(r))\right](x)$$

is smooth. But ψ_k is the k-partial derivative, with respect to x, of the parametrization $\mathbf{P} : (r, x) \mapsto P(r)(x)$,

$$\psi_k(r, x) = \frac{\partial^k \mathbf{P}}{\partial x^k}(r, x).$$

Since, by the very definition of the plots of $\mathcal{C}^\infty(\mathbf{R})$ (art. 1.13), the parametrization $\mathbf{P}$ is smooth, all of its partial derivatives are smooth. They are smooth with respect to the pair of variables (r, x), by the very definition of the class $\mathcal{C}^\infty$. Therefore, ψ_k is smooth and d^k/dx^k is a smooth map from $\mathcal{C}^\infty(\mathbf{R})$ to itself.

2) The map $\hat{x}$ is smooth if and only if, for every plot $P : U \to \mathcal{C}^\infty(\mathbf{R})$, the parametrization $\hat{x} \circ P : r \mapsto P(r)(x)$ is smooth. But, by the very definition of the functional diffeology (art. 1.13), the parametrization $\mathbf{P} : (r, x) \mapsto P(r)(x)$ is smooth. Since the map $\hat{x} \circ P$ is the composition $\mathbf{P} \circ j_x$, where j_x is the (smooth) inclusion $j_x : r \mapsto (r, x)$ from U to $U \times \mathbf{R}$, the parametrization $\hat{x} \circ P$ is smooth and $\hat{x}$ belongs to $\mathcal{C}^\infty(\mathcal{C}^\infty(\mathbf{R}), \mathbf{R})$. Now, note that for every integer k, the map $f \mapsto f^{(k)}(x)$ is just

$$f \mapsto f^{(k)} \mapsto \hat{x}(f^{(k)}) = \left[\hat{x} \circ \frac{d^k}{dx^k}\right](f),$$

that is, the composition of two smooth maps, therefore it is smooth (art. 1.15).

Hence, each component of the map

$$D_x^k : f \mapsto \left(\hat{x}(f), \left(\hat{x} \circ \frac{d}{dx}\right)(f), \ldots, \left(\hat{x} \circ \frac{d^k}{dx^k}\right)(f)\right)$$

is smooth, from $\mathcal{C}^\infty(\mathbf{R})$ to $\mathbf{R}$, and $D_x^k : \mathcal{C}^\infty(\mathbf{R}) \to \mathbf{R}^{k+1}$ is smooth.

3) From differential calculus in $\mathbf{R}^n$ we know that, for all $f \in \mathcal{C}^\infty(\mathbf{R})$, the map

$$F : x \mapsto \int_0^x f(t)dt$$

is continuous and smooth, with f as derivative. Since F is continuous and $F' = f$ is smooth, the primitive F is smooth. Now, $I_{a,b}(f) = F(b) - F(a)$. To prove that $I_{a,b}$ is smooth we have just to check that the map $I : f \mapsto F$ is smooth. Let $P : U \to \mathcal{C}^\infty(R)$ be a plot, we have

$$I \circ P(r) = x \mapsto \int_0^x P(r)(t)dt = \int_0^x \mathbf{P}(r,t)dt \quad \text{with} \quad \mathbf{P}(r,x) = P(r)(x).$$

Now, $I \circ P$ is a plot of $\mathcal{C}^\infty(\mathbf{R})$ if and only if the parametrization $(r,x) \mapsto (I \circ P(r))(x)$ is smooth, that is, if and only if the parametrization

$$\mathcal{P} : (r,x) \mapsto \int_0^x \mathbf{P}(r,t)\, dt$$

is smooth. But, by the very definition of the functional diffeology of $\mathcal{C}^\infty(\mathbf{R})$, the parametrization $\mathbf{P}$ is smooth. Thus, since the partial derivatives of $\mathbf{P}$ with respect to the variables r commute with the integration, on the one hand we have

$$\frac{\partial^n}{\partial r^n}\mathcal{P}(r,x) = \int_0^x \frac{\partial^n}{\partial r^n}\mathbf{P}(r,t)\, dt.$$

On the other hand, since the partial derivatives of $\mathbf{P}$, with respect to r or x, commute, we have, for $m \geq 1$

$$\frac{\partial^n \partial^m}{\partial r^n \partial x^m}\mathcal{P}(r,x) = \frac{\partial^n}{\partial r^n}\left[\frac{\partial^m}{\partial x^m}\int_0^x \mathbf{P}(r,t)\, dt\right] = \frac{\partial^n}{\partial r^n}\frac{\partial^{m-1}}{\partial x^{m-1}}\mathbf{P}(r,x).$$

Therefore, the parametrization $\mathcal{P}$ is smooth and $I, I_{a,b} \in \mathcal{C}^\infty(\mathcal{C}^\infty(\mathbf{R}))$.

4) Checking that the condition (art. 1.13, (♢)) defines a diffeology is straightforward. The same arguments as for (art. 1.11) can be used, or the general construction of subset diffeology, described in (art. 1.33). Now, the derivative is smooth and, restricted to $\mathcal{C}_0^\infty(\mathbf{R})$, is injective. The derivative also is surjective since $F : x \mapsto \int_0^x f(t)\, dt$ satisfies $F' = f$ and $F(0) = 0$. The inverse of d/dx is just the map I defined in the previous paragraph. Since we have seen that I is smooth, the derivative is a diffeomorphism. Moreover, it is a linear diffeomorphism.

☞ **Exercise 7, p. 14** (Locally constant parametrizations). Let us first assume that $P : U \to X$ is locally constant. Let r and r' be two connected points in U and γ be a path connecting r to r', that is, $\gamma \in \mathcal{C}^\infty(\mathbf{R}, U)$ with $\gamma(0) = r$ and $\gamma(1) = r'$.

a) The parametrization $p = P \circ \gamma$ is locally constant. Indeed, let $t \in \mathbf{R}$ be any point, and let $r = \gamma(t)$. Since P is locally constant, there exists an open neighborhood V of r such that $P \restriction V$ is a constant parametrization. Since γ is smooth, thus continuous, $W = \gamma^{-1}(V)$ is a domain and $p \restriction W = P \circ \gamma \restriction \gamma^{-1}(V)$ is constant.

b) The segment $[0,1]$ can be covered with a family of open intervals such that p is constant on each of them. Since $[0,1]$ is compact, there exists a finite subcovering

$\{I_k\}_{k=1}^N$ of $[0,1]$ such that $[0,1] \subset \bigcup_{k=1}^N I_k$ and for all $k = 1 \cdots N$, $p \restriction I_k = \mathrm{cst}$. Now, there exists an interval $I \in \{I_k\}_{k=1}^N$ such that $0 \in I$. Let $\mathcal{I}_1$ be the union of all the intervals of the family $\{I_k\}_{k=1}^N$ whose intersection with I is not empty. Since $\mathcal{I}_1$ is a union of open intervals containing 0, it is itself an open interval containing 0. If $\mathcal{I}_1 = I$, then the family was reduced to $\{I\}$ and we are done: I contains 0 and 1 and $p \restriction I$ is constant. Now, let us assume that $\mathcal{I}_1 \neq I$. Note first that $p \restriction \mathcal{I}_1$ is constant with value $p(0)$. Indeed, for every interval I_k containing 0, $p \restriction I_k = [t \mapsto p(0)]$. Hence, replacing all the intervals I_k containing 0 by $\mathcal{J}_1$, we get a new finite covering of $[0,1]$ satisfying the same conditions as the previous one, but with a number of elements strictly less than N. Thus, after a finite number of steps we get a covering of $[0,1]$ made with a unique open interval on which p is constant. Therefore, $p(0) = p(1)$.

c) Conversely, let us assume that P is constant on every connected component of U. Let $r_0 \in U$, since U is open, there exists an open ball B centered at r_0 and contained in U. But, since B is path connected and contains r_0, B is contained in the connected component of r_0 in U. Thus, P is constant on B, that is, P is locally constant.

Exercise 8, p. 14 (Diffeology of $\mathbf{Q} \subset \mathbf{R}$). The fact that the plots of $\mathbf{R}$ with values in $\mathbf{Q}$ are a diffeology of $\mathbf{Q}$ is a slight adaptation of (art. 1.12), where $\mathbf{R}^2$ is replaced by $\mathbf{R}$ and the square by $\mathbf{Q}$. Now, let $P : U \to \mathbf{R}$ be a smooth map with values in $\mathbf{Q}$, let $r \in U$ and $P(r) = q$. Let us assume that P is not locally constant at the point r, that is, there exists a small ball B, centered at r, which does not contain any ball B′, centered at r, on which P would be constant. Hence, there exists $r' \in B$ with $r' \neq r$ such that $q \neq q'$, where $q' = P(r')$. Next, let $f : t \mapsto r + t(r' - r)$, which can be defined on an open neighborhood of $[0,1]$. The map f sends $[0,1]$ onto the segment $[r, r']$, $f([0,1]) \subset B$. Thus, $Q = P \circ f$ is a real continuous function mapping $[0,1]$ onto $[q, q']$, with $Q(0) = q$ and $Q(1) = q'$. By the intermediate value theorem, f takes all the real values between q and q'. But there is always an irrational number between two distinct rational numbers. This contradicts the hypothesis that f takes only rational values. Therefore, there is no such plot P, and the only plots of $\mathbf{R}$ which take their values in $\mathbf{Q}$ are locally constant. We observe that we can replace $\mathbf{Q}$ by any countable subset, the intermediate value theorem will continue to apply. Let then $A \subset U$ be a countable subset of an n-domain, the composition of any plot P of A with the n coordinate projections $\mathrm{pr}_k : U \to \mathbf{R}$ is a plot of $\mathbf{R}$ taking its values in a countable subset, so locally constant. Therefore P is locally constant, and A is discrete.

Exercise 9, p. 14 (Smooth maps from discrete spaces). The proof is contained in (art. 1.20). Let X be a discrete diffeological space, let X′ be some other diffeological space, and let $f : X \to X'$ be a map. Let P be a plot of X, that is, a locally constant map. The composition $f \circ P$ is thus locally constant, that is, a plot of the discrete diffeology. But the discrete diffeology is contained in every diffeology (art. 1.20), hence $f \circ P$ is a plot of X′ and f is smooth.

Exercise 10, p. 14 (Smooth maps to coarse spaces). The proof is contained in (art. 1.21). Let X be some diffeological space, let X′ be a coarse diffeological space, and et $f : X \to X'$ be a map. Let P be a plot of X. The composition $f \circ P$ is a parametrization of X′. Hence it is a plot of the coarse diffeology, since the coarse diffeology is the set of all the parametrizations (art. 1.21). Therefore f is smooth.

☞ **Exercise 11, p. 15** (Square root of the smooth diffeology). Let $\mathrm{sq} : x \mapsto x^2$, and let us recall that a parametrization P is a plot of the diffeology $\mathcal{D} = \mathrm{sq}^*(\mathcal{C}^\infty_\star(\mathbf{R}))$ if and only if $\mathrm{sq} \circ P$ is smooth. Since $\mathrm{sq} \in \mathcal{C}^\infty(\mathbf{R})$, for every smooth parametrization P of $\mathbf{R}$, the composite $\mathrm{sq} \circ P$ is smooth. Therefore $\mathcal{C}^\infty_\star(\mathbf{R}) \subset \mathcal{D} = \mathrm{sq}^*(\mathcal{C}^\infty_\star(\mathbf{R}))$, the diffeology $\mathcal{D}$ is coarser than the smooth diffeology. Now, the map $|\cdot|$ satisfies $\mathrm{sq} \circ |\cdot| = \mathrm{sq}$. But sq is smooth, thus $|\cdot|$ is a plot of the diffeology $\mathcal{D}$. Actually, since $\mathcal{D}$ contains the parametrization $x \mapsto |x|$, which is not smooth, the diffeology $\mathcal{D}$ is strictly coarser than $\mathcal{C}^\infty_\star(\mathbf{R})$. Finally, for every smooth parametrization Q of $\mathbf{R}$, the parametrization $Q \circ |\cdot|$ is a plot for the diffeology $\mathcal{D}$, thanks to the smooth compatibility axiom.

☞ **Exercise 12, p. 17** (Immersions of real domains). If $D(f)(r)$ is injective at the point r, then the rank of f at the point r, denoted by $\mathrm{rank}(f)_r$, and equal by definition to the rank of the tangent linear map $D(f)(r)$, is equal to n. Now, since the rank of smooth maps between real domains is semicontinuous below, $\mathrm{rank}(f)_r = n$ on some open neighborhood of r. Thus, by application of the rank theorem (see, for example, [**Die70a**, 10.3.1]) there exist an open neighborhood $\mathcal{O}$ of r, an open neighborhood $\mathcal{O}'$ of $f(r)$, a diffeomorphism φ from the open unit ball of $\mathbf{R}^n$ to $\mathcal{O}$, mapping 0 to r, and a diffeomorphism ψ from the open unit ball of $\mathbf{R}^m$ to $\mathcal{O}'$, mapping 0 to $f(r)$, such that $f \circ \varphi = \psi \circ j$, where $j : \mathbf{R}^n \to \mathbf{R}^m$ is the canonical induction from $\mathbf{R}^n$ to $\mathbf{R}^m$ $j(r_1, \ldots, r_n) = (r_1, \ldots, r_n, 0, \ldots, 0)$. Hence, $f \upharpoonright \mathcal{O}$ is conjugate to an induction by two diffeomorphisms. Thus, $f \upharpoonright \mathcal{O}$ is an induction.

☞ **Exercise 13, p. 17** (Flat points of smooth paths). Since γ is continuously differentiable and since $\gamma(t^0_n) = \gamma(t^0_{n+1}) = 0$, by application of Rolle's theorem [**Die70a**, 8.2, pb. 3], there exists a number $t^1_n \in \,]t^0_n, t^0_{n+1}[$ such that $\gamma'(t^1_n) = 0$. Thus, the sequence t^1_n converges to 0 and, by continuity, $\gamma'(0) = 0$. Now, by recursion, there exists a sequence of numbers $t^{k+1}_n \in \,]t^k_n, t^k_{n+1}[$, converging to 0, such that $\gamma^k(t^k_n) = 0$. Therefore, for any $k > 0$, $\gamma^k(0) = 0$ and γ is flat at 0.

☞ **Exercise 14, p. 17** (Induction of intervals into domains). 1) Let $\mathrm{abs} : t \mapsto |t|$, and let $F = f \circ \mathrm{abs}$. We have

$$F(t) = f(-t), \text{ if } t < 0, \quad F(0) = 0, \quad \text{and} \quad F(t) = f(+t), \text{ if } t > 0.$$

The parametrization F is smooth on $]-\varepsilon, 0[$ and on $]0, +\varepsilon[$, because, restricted to these intervals, it is the composite of two smooth parametrizations. The only question is for $t = 0$. Next, since f is flat, for all integers p we have

$$\lim_{t \to 0^\pm} \frac{f(t)}{t^p} = 0 \quad \Rightarrow \quad \lim_{t \to 0^\pm} \frac{F(t)}{t^p} = 0, \qquad (\diamondsuit)$$

in particular for $p = 1$. Thus, F is derivable at 0 and $F'(0) = 0$. But since $f'(0) = 0$, we also have

$$\begin{aligned} \lim_{t \to 0^-}(F)'(t) &= (-1)\lim_{t \to 0^-} f'(-t) = (-1) \times 0 = 0, \\ \lim_{t \to 0^+}(F)'(t) &= (+1)\lim_{t \to 0^+} f'(+t) = (+1) \times 0 = 0. \end{aligned}$$

Thus, $\lim_{t \to 0^\pm} F'(t) = F'(0) = 0$. Hence, $F \in \mathcal{C}^1(]-\varepsilon, +\varepsilon[, \mathbf{R}^n)$. Moreover, since F is $\mathcal{C}^1$, F' is derivable at 0 and its derivative is 0,

$$F''(0) = \lim_{t \to 0^\pm} \frac{F'(t)}{t} = \lim_{t \to 0^\pm} \frac{f(t)}{t^2} = 0.$$

Then, by recursion on p, and thanks to $(\diamondsuit)$, we get $F \in \mathcal{C}^p(]-\varepsilon, +\varepsilon[, \mathbf{R}^n)$, for all integers p. Thus, F is smooth. But since abs is not smooth, f is not an induction.

Now, composing with two translations at the source and at the target, the same proof applies for every point $t \in]-\varepsilon, +\varepsilon[$, and for every value $f(t)$. Therefore, an induction from $]-\varepsilon, +\varepsilon[$ to $\mathbf{R}^n$ is nowhere flat.

2) Since an induction $f :]-\varepsilon, +\varepsilon[\to \mathbf{R}^n$ is not flat at $t = 0$, there exists a smallest integer $k > 0$ such that, $f^{(j)}(0) = 0$ if $0 \leq j < k$, and $f^{(k)}(0) \neq 0$. If $k = 1$, then $p = 0$, and $\varphi = f$. Otherwise, if $k \geq 1$, then the Taylor expansion of f around 0 is reduced to

$$f(t) = t^p \times \varphi(t), \text{ with } \varphi(t) = t \times \int_0^1 \frac{(1-s)^p}{p!} f^{(p+1)}(st)ds, \text{ and } p = k - 1.$$

See, for example [**Die70a**, 8.14.3]. Since $f^{(k)}$ is smooth, the function φ is smooth and $\varphi'(0) = f^{(k)}(0)/k! \neq 0$.

☞ **Exercise 15, p. 17** (Smooth injection in the corner). Let us split the first question into two mutually exclusive cases.

1.A) If $0 \in \mathbf{R}$ is an isolated zero of γ, then there exists $\varepsilon > 0$ such that $\gamma(t) = 0$, and $t \in]-\varepsilon, +\varepsilon[$ implies $t = 0$. Since γ is continuous, γ maps $]-\varepsilon, 0[$ to the semiline $\{x\mathbf{e}_1 \mid x > 0\}$ or to the semiline $\{y\mathbf{e}_2 \mid y > 0\}$, where $\mathbf{e}_1$ and $\mathbf{e}_2$ are the vectors of the canonical basis of $\mathbf{R}^2$. But since γ is injective, these two cases are mutually exclusive. Without loss of generality, we can assume that $\gamma(]-\varepsilon, 0[) \subset \{x\mathbf{e}_1 \mid x > 0\}$ and $\gamma(]0, -\varepsilon[) \subset \{y\mathbf{e}_2 \mid y > 0\}$. Thus, for all $p > 0$, $\lim_{t \to 0^-} \gamma^{(p)}(t) = \alpha\mathbf{e}_1$ and $\lim_{t \to 0^+} \gamma^{(p)}(t) = \beta\mathbf{e}_2$. Hence, by continuity, $\alpha = \beta = 0$. Therefore, γ is flat at 0. Now, if γ is not assumed to be injective, then the parametrization $\gamma : t \to t^2\mathbf{e}_1$ is smooth, not flat, and satisfies $\gamma(0) = 0$.

1.B) If $0 \in \mathbf{R}$ is not an isolated zero of γ, then there exists a sequence $t_1^0 < \cdots < t_n^0 < \cdots$ of numbers, converging to 0, such that $\gamma(t_n^0) = 0$. The fact that γ is flat at 0 is the consequence of Exercise 13, p. 17.

2) First of all j is injective. Since the restriction of j on $]-\infty, 0[\ \cup \ ]0, +\infty[$ is smooth, the only problem is for $t = 0$. But the successive derivatives of j write

$$j^{(p)}(t) = \begin{pmatrix} q_-(t)\, e^{\frac{1}{t}} \\ 0 \end{pmatrix} \text{ if } t < 0, \quad \text{and} \quad j^{(p)}(t) = \begin{pmatrix} 0 \\ q_+(t)\, e^{-\frac{1}{t}} \end{pmatrix} \text{ if } t > 0,$$

where $q_\pm$ are two rational fractions. Then,

$$\lim_{t \to 0^\pm} q_\pm(t)\, e^{-\frac{1}{|t|}} = 0, \quad \text{thus} \quad \lim_{t \to 0^\pm} j^{(p)}(0) = \begin{pmatrix} 0 \\ 0 \end{pmatrix}.$$

Therefore, j is smooth.

3) Let abs denote the map $t \mapsto |t|$. The map $j \circ \text{abs}$ is smooth but not abs. But, for an injection, to be an induction means that for a parametrization P, $j \circ P$ is smooth if and only if P is smooth. Thus, if j is a smooth injection, it is not an induction. This example is a particular case of Exercise 14, p. 17.

☞ **Exercise 16, p. 18** (Induction into smooth maps). First of all, f is injective. Let $\varphi \in \text{val}(f)$, then $f(x, v) = \varphi$, with $x = \varphi(0)$ and $v = \varphi'(0)$. Now let $\Phi : r \mapsto \varphi_r$ be a plot of $\mathcal{C}^\infty(\mathbf{R}, \mathbf{R}^n)$ such that $\text{val}(\Phi) \subset \text{val}(f)$, thus $(r, t) \mapsto \varphi_r(t)$ is a smooth parametrization in $\mathbf{R}^n$. Now, $f^{-1} \circ \Phi(r) = (x = \varphi_r(0), v = (\varphi_r)'(0))$. Since the evaluation of a smooth function is smooth, $r \mapsto x$ is smooth. Next, since the derivative $(r, t) \mapsto (\varphi_r)'(t)$ is smooth and the evaluation is smooth, $r \mapsto v$ is smooth. Therefore, $f^{-1} : \mathrm{val}(f) \to \mathbf{R}^n \times \mathbf{R}^n$ is smooth, where val(f) is equipped with the subset diffeology of $\mathcal{C}^\infty(\mathbf{R}, \mathbf{R}^n)$, thus f is an induction.

☞ **Exercise 17, p. 19** (Vector subspaces of real vector spaces). We shall apply the criterion (art. 1.31). First of all, since the vectors $b_1, \ldots, b_k$ are free, the map $\mathcal{B}$ is injective. Then, since $\mathcal{B}$ is linear, $\mathcal{B}$ is smooth from $\mathbf{R}^k$ to $\mathbf{R}^n$. Now, let $P : U \to \mathbf{R}^n$ be a smooth parametrization of $\mathbf{R}^n$ with values in $E = \mathcal{B}(\mathbf{R}^k)$. Thanks to the theorem of the incomplete basis, we can find $n-k$ vectors $b_{k+1}, \ldots, b_n$ and a map $M \in \mathrm{GL}(n, \mathbf{R})$ such that $M(b_i) = e_i$, for $i = 1, \ldots, n$. Hence, $\mathcal{B}^{-1} \circ P = M \circ P$. Since $M \circ P$ is smooth, $\mathcal{B}^{-1} \circ P$ is smooth and $\mathcal{B}$ is an induction. Finally, every basis $\mathcal{B}$ of every k-subspace $E \subset \mathbf{R}^n$ realizes a diffeomorphism from $\mathbf{R}^k$ to E, equipped with the induced smooth diffeology.

☞ **Exercise 18, p. 19** (The sphere as diffeological subspace). The map f is clearly injective. The inverse is given by

$$f^{-1} : \{x' \in S^n \mid x' \cdot x > 0\} \to E \quad \text{with} \quad f^{-1}(x') = [1 - x\bar{x}]x',$$

where $[1 - x\bar{x}]$ is the orthogonal projector parallel to x. The notation $\bar{x}$ is for $x' \mapsto x \cdot x'$. Hence, since the map f^{-1} is the restriction of a linear map, thus smooth, to a subset, f^{-1} is smooth for the subset diffeology. And f is an induction, from the open ball $\{t \in E \mid \|t\| < 1\}$ to the semisphere $\{x' \in S^n \mid x' \cdot x > 0\}$.

☞ **Exercise 19, p. 20** (The pierced irrational torus). Let $\pi_\alpha : \mathbf{R} \to T_\alpha$ be the projection from $\mathbf{R}$ to its quotient $T_\alpha = \mathbf{R}/(\mathbf{Z} + \alpha\mathbf{Z})$; see Exercise 4, p. 8. By definition of the diffeology of T_α, this parametrization is a surjective plot of T_α. Now, by density of $\mathbf{Z} + \alpha\mathbf{Z}$ in $\mathbf{R}$, any open interval around $0 \in \mathbf{R}$ contains always a representative of every orbit of $\mathbf{Z} + \alpha\mathbf{Z}$. Hence, the plot π_α is not locally constant. Therefore, since a diffeology is discrete (art. 1.20) if and only if all its plots are locally constant, the diffeology of T_α is not discrete. Now let $\tau \in T_\alpha$ and $x \in \mathbf{R}$ such that $\pi_\alpha(x) = \tau$. Let $P : U \to T_\alpha - \tau$ be a plot, thus P is a plot of T_α such that $\tau \notin \mathrm{val}(P)$. Since P is a plot of T_α, for all $r_0 \in U$ there exist an open neighborhood V of r_0 and a smooth parametrization $Q : V \to \mathbf{R}$ such that $P \upharpoonright V = \pi_\alpha \circ Q$. But since $\tau \notin \mathrm{val}(P)$, $\mathrm{val}(Q) \cap (\mathbf{Z}+\alpha\mathbf{Z})(x) = \varnothing$, where $(\mathbf{Z}+\alpha\mathbf{Z})(x) = \{x + n + \alpha m \mid n, m \in \mathbf{Z}\}$, that is, $\mathrm{val}(Q) \subset \mathbf{R} - (\mathbf{Z} + \alpha\mathbf{Z})(x)$. Next, let $B \subset V$ be a ball centered at r_0, and let $r_1 \in B$ be any other point of B. Let $x_0 = Q(r_0)$, $x_1 = Q(r_1)$, and let us assume that $x_0 \leq x_1$; it would be equivalent to assume $x_0 \geq x_1$. Since Q is smooth and *a fortiori* continuous, the interval $[x_0, x_1]$ is contained in val(Q). But, since the orbit of x by $\mathbf{Z} + \alpha\mathbf{Z}$ is dense in $\mathbf{R}$, except if $x_1 = x_0$, the interval $[x_0, x_1]$ contains a point of the orbit $(\mathbf{Z} + \alpha\mathbf{Z})(x)$. Now, since by hypothesis Q avoids this orbit, $x_0 = x_1$. Hence, Q is locally constant, and thus P is locally constant. Therefore, the diffeology of $T_\alpha - \tau$ is discrete.

☞ **Exercise 20, p. 20** (A discrete image of $\mathbf{R}$). 1) The condition (♣) means that the parametrization is a plot of the *functional diffeology* (art. 1.13) or (art. 1.57). Then, let us consider the condition (♠).

D1. Let $\hat{\phi} : U \to \mathcal{C}^\infty(\mathbf{R}, \mathbf{R})$ be the constant parametrization $\hat{\phi}(r) = \phi$. Hence, $\hat{\phi}(r) = \hat{\phi}(r_0) = \phi$ for every r_0 and every r in U. They coincide *a fortiori* outside any interval $[a, b]$.

D2. By the very definition the condition (♠) is local.

D3. Let $P : U \to \mathcal{C}^\infty(\mathbf{R}, \mathbf{R})$ satisfying (♠). Let $F : V \to U$ be a smooth parametrization. Let $s_0 \in V$ and $r_0 = F(s_0)$. Since P satisfies (♠), there exists an open ball B centered at r_0 and for every $r \in B$ there exists an interval $[a, b]$ such that $P(r)$ and $P(r_0)$ coincide outside $[a, b]$. Since F is smooth, the pullback $F^{-1}(B)$ is a domain containing s_0. Let us consider then an open ball $B' \subset F^{-1}(B)$ centered at s_0. For every $s \in B'$, $F(s) \in B$, thus there exists an interval $[a, b]$ such that $P \circ F(s) = P(r) \in B$ and $P \circ F(s_0) = P(r_0)$ coincide outside $[a, b]$. Therefore, $P \circ F$ satisfies (♠).

Hence, the condition (♠) defines a diffeology. The two conditions (♣) and (♠) define the intersection of two diffeologies, that is, a diffeology (art. 1.22).

2) Let $f : \mathbf{R} \to \mathcal{C}^\infty(\mathbf{R}, \mathbf{R})$ be defined by $f(\alpha) = [x \mapsto \alpha x]$. Let $P : U \to \mathrm{val}(f) \subset \mathcal{C}^\infty(\mathbf{R}, \mathbf{R})$ be a plot for the diffeology defined by (♣) and (♠). Since f is injective, $f^{-1}(\phi) = \phi(1)$, there exists a unique real parametrization $r \mapsto \alpha(r)$, defined on U, such that $P(r) = [x \mapsto \alpha(r)x]$, actually $\alpha(r) = P(r)(1)$. Now let $r_0 \in U$, thanks to (♠) there exists an open ball B, centered at r_0 and for all $r \in B$, there exists an interval $[a, b]$ such that $P(r)$ and $P(r_0)$ coincide outside $[a, b]$, that is, for all $x \in \mathbf{R} - [a, b]$, $\alpha(r)x = \alpha(r_0)x$. We can choose $x \neq 0$, and thus $\alpha(r) = \alpha(r_0)$ for all $r \in B$. Therefore the plot P is locally constant, and this is the definition for $f(\mathbf{R})$ to be discrete. It follows that f is not smooth, since it is not locally constant, and therefore not a plot. But note that f is a plot for the diffeology defined only by (♣).

3) For $\mathcal{C}^\infty(\mathbf{R}^n)$, the condition (♠) must be replaced by the following:

> (♠) For any $r_0 \in U$ there exists an open ball B, centered at r_0, and for every $r \in B$ there exists a compact $K \subset \mathbf{R}^n$ such that $P(r)$ and $P(r_0)$ coincide outside K.

Now, for the same kind of reason as for the second question, the injection $j : \mathrm{GL}(n, \mathbf{R}) \to \mathcal{C}^\infty(\mathbf{R}^n, \mathbf{R}^n)$ has a discrete image, and is not smooth. Indeed, two linear maps which coincide outside a compact coincide everywhere.

☞ **Exercise 21, p. 23** (Sum of discrete or coarse spaces). Let us consider the sum $X = \coprod_{i \in \mathcal{I}} X_i$ of discrete diffeological spaces. Let $P : U \to X$ be a plot. By definition of the sum diffeology, P takes locally its values in one of the X_i (art. 1.39). Let us say that $\mathrm{val}(P \restriction V) \subset X_i$, where V is a subdomain of U. But since X_i is discrete, $P \restriction V$ is locally constant. Hence, P itself is locally constant, that is, X is discrete. Now, let X be a discrete space and $\coprod_{x \in X} \{x\}$ be the sum of its elements. Every plot P of X is locally constant, hence P is locally a plot of $\coprod_{x \in X} \{x\}$. Thus, P is a plot of $\coprod_{x \in X} \{x\}$. Conversely, every plot P of $\coprod_{x \in X} \{x\}$ is locally constant. Thus, P is a plot of X. Therefore, every discrete space is the sum of its elements. Finally, let us consider the sum of two points $X = \{0\} \coprod \{1\}$. The diffeology of $\{0\}$ and $\{1\}$ is at the same time coarse and discrete. Let us consider the parametrization $P : \mathbf{R} \to X$ which maps each rational to the point 0 and each irrational to the point 1. This is a parametrization, thus a plot of the coarse diffeology of X (art. 1.21). But, since rational (or irrational) numbers are dense in $\mathbf{R}$, this parametrization is nowhere locally equal to 0 or equal to 1. Then, this parametrization is not a plot of the sum X. Hence, the sum X is not coarse. Sum of coarse spaces may be not coarse.

☞ **Exercise 22, p. 23** (Plots of the sum diffeology). Only the following method needs to be proved. Let $P : U \to X$ be a plot of the sum diffeology. For each index $i \in \mathcal{I}$, the set $U_i = P^{-1}(X_i)$ is open. Indeed, let $r \in U_i$, by definition of the sum diffeology, there exists an open neighborhood of r, let us say an open ball B centered at r, such that $P \restriction B$ takes its values in U_i. Hence, U_i is the union of all these open

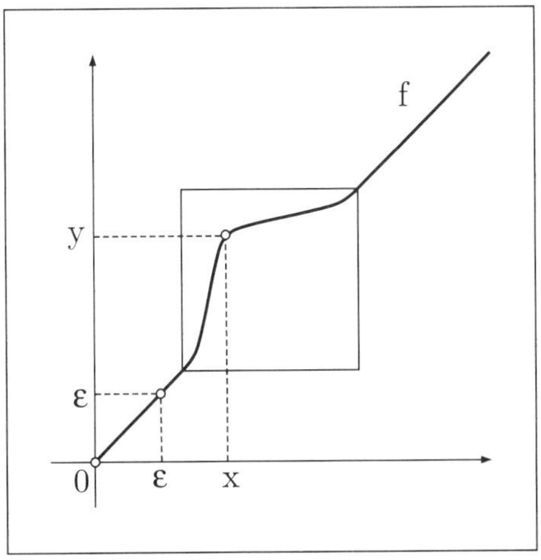

FIGURE Sol.1. The diffeomorphism f.

balls, thus U_i is a domain. Now, by construction the family $\{U_i\}_{i \in \mathcal{I}}$ is a partition of U, and for every index $i \in \mathcal{I}$, the restriction $P_i = P \restriction U_i$ is a plot of X_i.

Exercise 23, p. 23 (Diffeology of $\mathbf{R} - \{0\}$). Let $P : U \to \mathbf{R} - \{0\}$ be a plot. Let $U_- = P^{-1}(]-\infty, 0[)$ and $U_+ = P^{-1}(]0, +\infty[)$, $U_\pm$ be two domains constituting a partition of U, indeed $U = U_- \cup U_+$ and $U_- \cap U_+ = \varnothing$. Let $P_- = P \restriction U_-$ and $P_+ = P \restriction U_+$. Then, P is the supremum of P_- and P_+. Therefore, the diffeology of $\mathbf{R} - \{0\}$ is the sum of the diffeologies of $]-\infty, 0[$ and $]0, +\infty[$; see Exercise 22, p. 23.

Exercise 24, p. 23 (Klein strata of $[0, \infty[$). Let $\varphi : [0, \infty[\to [0, \infty[$ be a diffeomorphism for the subset diffeology. Let us assume that $\varphi(0) \neq 0$. Since φ is bijective, there exists a point $x_0 > 0$ such that $\varphi(x_0) = 0$. Hence, there exists a closed interval centered at x_0, let us say $I = [x_0 - \varepsilon, x_0 + \varepsilon]$, $\varepsilon > 0$, such that $f = \varphi \restriction I$ is positive, injective, continuous, and maps x_0 to 0. Let $a = f(x_0 - \varepsilon)$ and $b = f(x_0 + \varepsilon)$. We have $a > 0$ and $b > 0$, let us assume that $0 < a \leq b$ (it is not crucial). Now, since f is continuous, maps 0 to 0, and $x_0 + \varepsilon$ to b, for every point y between 0 and b there exists a point x between 0 and $x_0 + \varepsilon$ such that $f(x) = y$, and since $0 < a \leq b$, there exists $x_1 \in [0, x_0 + \varepsilon]$ such that $f(x_1) = a$. Thus, $a = f(x_0 - \varepsilon) = f(x_1)$ and $x_1 \neq x_0 - \varepsilon$. This is impossible, by hypothesis f is injective. Hence, $\varphi(0) = 0$. Therefore, the set $\{0\}$ is an orbit of $\mathrm{Diff}([0, \infty[)$, that is a Klein stratum. Next, since $\varphi(0) = 0$, $\varphi(]0, \infty[) =]0, \infty[$. Let $f = \varphi \restriction]0, \infty[$, f is bijective, smooth and its inverse is also smooth. Thus, f is a diffeomorphism of $]0, \infty[$, and moreover $\lim_{x \to 0} = 0$. Let us now try to convince ourselves that any point x of $]0, \infty[$ can be mapped to any point y of $]0, \infty[$ by a diffeomorphism φ of $[0, \infty[$. Let us assume that $0 < x \leq y$. We claim that there exist a number $\varepsilon > 0$, and a diffeomorphism f of $]0, \infty[$ such that $f \restriction]0, \varepsilon[$ is equal to the identity, and $f(x) = y$; see Figure Sol.1. The extension φ of f to $[0, \infty[$ defined by $\varphi(0) = 0$ is a diffeomorphism of $[0, \infty[$, got by gluing the identity on some interval $[0, \varepsilon[$ with f. Therefore $]0, \infty[$ is an orbit of $\mathrm{Diff}([0, \infty[)$, that is, a Klein stratum.

Exercise 25, p. 23 (Compact diffeology). First of all, let us remark that coinciding outside a compact in $\mathbf{R}$, or coinciding outside a closed interval is identical.

Now, a plot $P : U \to \mathbf{R}$ of the functional diffeology of $\mathcal{C}^\infty(\mathbf{R}, \mathbf{R})$ foliated by the relation $\sim$ is a plot of the functional diffeology of $\mathcal{C}^\infty(\mathbf{R}, \mathbf{R})$ which takes locally its values in some class of $\sim$. Precisely, if and only if P fulfills the condition (♣) of Exercise 20, p. 20, and for every $r_0 \in U$ there exists an open neighborhood V of r_0, such that for all $r \in V$, $\text{class}(P(r)) = \text{class}(P(r_0))$, that is, for all $r \in V$ there exists a closed interval $[a, b]$ such that $P(r)$ and $P(r_0)$ coincide outside $[a, b]$. This is exactly the condition (♠) of Exercise 20, p. 20.

☞ **Exercise 26, p. 25** (Square of the smooth diffeology). Except for the plots with negative values, by definition of the pushforward of a diffeology (art. 1.43), a parametrization P of $\mathcal{D} = \text{sq}_*(\mathcal{C}^\infty_\star(\mathbf{R}))$ writes locally $P(r) =_{\text{loc}} Q(r)^2$, where $\text{sq}(x) = x^2$. Hence, except for the plots with negative values (which are locally constant), every plot is locally the square of a smooth parametrization of R. Therefore, every plot of $\mathcal{D}$ is a plot of the smooth diffeology $\mathcal{C}^\infty_\star(\mathbf{R})$ of $\mathbf{R}$, thus $\mathcal{D}$ is finer than $\mathcal{C}^\infty_\star(\mathbf{R})$.

1) Let $P : U \to \mathbf{R}$ be a plot for $\mathcal{D}$. Let $r_0 \in U$ such that $P(r_0) < 0$. Since $P(r_0)$ does not belong to the set of values of sq, by application of the characterization of the plots of pushforwards of diffeologies (art. 1.43), P is locally constant around r_0. In particular, there exists an open ball B centered at r_0 such that $P \restriction B$ is constant.

2) Let $P : U \to \mathbf{R}$ be a plot for $\mathcal{D}$. Let $r_0 \in U$ such that $P(r_0) > 0$. Since $P(r_0)$ is in the set of values of sq, there exists a smooth parametrization Q of $\mathbf{R}$, defined on an open neighborhood V of r_0 such that $P(r) = Q(r)^2$. Now, since $P(r_0) > 0$, there exists an open ball B centered at r_0 such that $P \restriction B$ is strictly positive. Hence, $Q \restriction B$ does not vanish, thus Q keeps a constant sign on B and the map $\sqrt{P} : r \mapsto |Q(r)|$, defined on B, is smooth.

3) If $P(r_0) = 0$, then there exist an open neighborhood V of r_0 and a smooth parametrization $Q : V \to \mathbf{R}$ such that $P \restriction V = Q^2$. But $P(r_0) = 0$ implies $Q(r_0) = 0$. Thus, $D(P)(r_0) = 2Q(r_0) \times D(Q)(r_0) = 0$. Hence, the first derivative of P vanishes at r_0. Let u and v be two vectors of $\mathbf{R}^n$, with $n = \dim(P)$,

$$\begin{aligned} D^2(P)(r_0)(u)(v) &= D[r \mapsto 2Q(r) \times D(Q)(r)(u)](r_0)(v) \\ &= 2D(Q)(r_0)(v) \times D(Q)(r_0)(u) + 2Q(r_0) \times D^2(Q)(r_0)(u)(v) \\ &= 2D(Q)(r_0)(v) \times D(Q)(r_0)(u) \quad (\text{since } Q(r_0) = 0). \end{aligned}$$

Thus, since $H(v)(v) = 2[D(Q)(r_0)(v)]^2$, the Hessian $H = D^2(P)(r_0)$ is positive.

4) Since $f(x) = (x\sqrt{1-x})^2$, and $\sqrt{1-x}$ is smooth on $]-\infty, 1[$, $x\sqrt{1-x}$ is smooth, and f is a plot of $\mathcal{D}$.

☞ **Exercise 27, p. 27** (Subduction onto the circle). First of all, note that the map $\Pi : t \mapsto (\cos(t), \sin(t))$, from $\mathbf{R}$ to S^1, is smooth and surjective. Also note that, the function cos, restricted to the interval $]k\pi, \pi + k\pi[$, where $k \in \mathbf{Z}$, is a diffeomorphism onto $]0, 1[$. As well, the function sin, restricted to $]-\pi/2 + k\pi, \pi/2 + k\pi[$, is a diffeomorphism onto $]0, 1[$. See Figure Sol.2. For $k = 0$ the inverses are the standard functions acos and asin. Let us denote, for now, the inverses of these restrictions by

$$\text{acos}_k :]0, 1[\to]k, \pi + k[\quad \text{and} \quad \text{asin}_k :]0, 1[\to]-\pi/2 + k, \pi/2 + k[.$$

Now, let $P : U \to S^1$ be a smooth parametrization of S^1, that is, $P(r) = (x(r), y(r))$, where x and y are smooth real parametrizations and $x(r)^2 + y(r)^2 = 1$. Let $r_0 \in U$. We shall distinguish four cases.

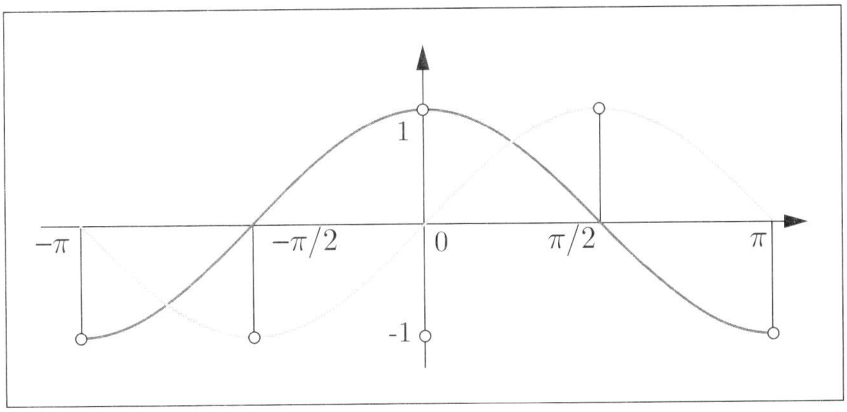

FIGURE Sol.2. The functions sine and cosine.

a) $y(r_0) \in]0,+1[$. Locally, $r \mapsto \mathrm{acos}(y(r))$ is a local lifting of P along Π.

b) $y(r_0) \in]-1,0[$. Locally, $r \mapsto \mathrm{acos}_1(y(r))$ is a local lifting of P along Π.

c) $x(r_0) = +1$. Locally, $r \mapsto \mathrm{asin}(y(r))$ is a local lifting of P along Π.

d) $x(r_0) = -1$. Locally, $r \mapsto \mathrm{asin}_1(y(r))$ is a local lifting of P along Π.

Therefore, $\Pi : \mathbf{R} \to S^1$ is a subduction. In other words, Π is strict (art. 1.54), and its factorization identifies naturally the quotient $\mathbf{R}/2\pi\mathbf{Z}$ with $S^1 \subset \mathbf{R}^2$.

Exercise 28, p. 27 (Subduction onto diffeomorphisms). Let us begin first by noting that, for all $f, g \in G$, $(g \circ f)'(x) \neq 0$, because $(g \circ f)'(x) = g'(f(x))f'(x)$, $f'(x) \neq 0$ and $g'(x) \neq 0$. Note next that, since $f'(x) \neq 0$ for all $x \in \mathbf{R}$, $f'(x)$ has a constant sign. Thus, if $f'(x) > 0$, then f is strictly increasing and $f(x + 2\pi) = f(x) + 2\pi$, else if $f'(x) < 0$, then f is strictly decreasing and $f(x + 2\pi) = f(x) - 2\pi$. Note that generally, for all $k \in \mathbf{Z}$, $f(x + 2\pi k) = f(x) + 2\pi k$ if f is increasing, and $f(x + 2\pi k) = f(x) - 2\pi k$ if f is decreasing. Now, it is immediate that, in any case, increasing or decreasing, for all $g, f \in G$, $(g \circ f)(x+2\pi) = (g \circ f)(x) \pm 2\pi$. Moreover, for all $f \in G$, f is unbounded. Therefore, $f \in \mathrm{Diff}(\mathbf{R})$, and G is a subgroup of $\mathrm{Diff}(\mathbf{R})$. Now, let us check that $(\diamondsuit)$ defines a diffeology of $\mathrm{Diff}(S^1)$. Let $r \mapsto f$ be a constant parametrization, for all plots Q of S^1, $(r, s) \mapsto f(Q(r)) = f \circ Q(r)$ is a plot of S^1 since f and Q are smooth. Axiom D1 is checked. Now, let $P : U \to \mathrm{Diff}(S^1)$ be a parametrization such that for all $r \in U$ there exists an open neighborhood W of r such that $P \restriction W$ satisfies $(\diamondsuit)$. For all plots $Q : V \to S^1$, $(r, s) \mapsto (P \restriction W)(r)(Q(s))$, defined on $W \times V$, is smooth. Thus, the parametrization $(r, s) \mapsto P(r)(Q(s))$ is locally smooth, therefore smooth. Axiom D2 is checked. Let $P : U \to \mathrm{Diff}(S^1)$ satisfying $(\diamondsuit)$, and let $F : W \to U$ be a smooth parametrization. Let $Q : V \to S^1$ be a plot, then $(t, s) \mapsto (r = F(t), s) \mapsto P(r)(Q(s)) = (P \circ F)(t)(Q(s))$ is smooth since it is the composite of smooth maps. Axiom D3 is checked. Therefore, $(\diamondsuit)$ defines a diffeology of $\mathrm{Diff}(S^1)$. Let us assume $f' > 0$. Note that $\Pi \circ f : \mathbf{R} \to S^1$ is smooth and surjective. Next, $(\Pi \circ f)(x+2\pi) = \Pi(f(x)+2\pi) = (\cos(f(x)+2\pi), \sin(f(x)+2\pi)) = (\cos(f(x)), \sin(f(x))) = \Pi \circ f(x)$. Thus $\Pi \circ f(x)$ is 2π-periodic, therefore there exists a function $\varphi : S^1 \to S^1$ defined by $\varphi(z) = \Pi \circ f(x)$ for every $x \in \mathbf{R}$ such that $z = (\cos(x), \sin(x))$.

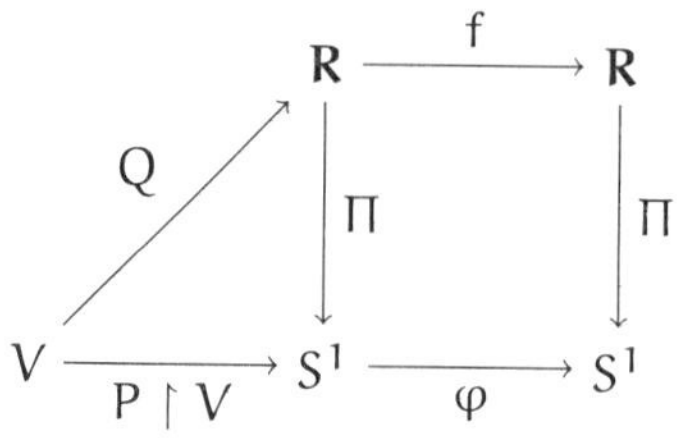

Let $P : U \to S^1$ be a plot, since Π is a subduction (Exercise 27, p. 27) for all $r \in U$ there exist an open neighborhood V of r and a smooth parametrization Q of $\mathbf{R}$ such that $\Pi \circ Q = (P \upharpoonright V)$. Thus, $\varphi \circ (P \upharpoonright V) = (\varphi \circ \Pi \circ Q) \upharpoonright V = \Pi \circ f \circ Q$, that is, a composite of smooth maps. Thus $\varphi \circ (P \upharpoonright V)$ is a plot of S^1, and therefore $\varphi \circ P$. Now, thanks to $f(x + 2\pi) = f(x) + 2\pi$, by denoting $y = f(x)$ and composing with f^{-1}, we get $f^{-1}(y + 2\pi) = f^{-1}(y) + 2\pi$. Hence, there exists a surjective smooth map $\bar{\varphi} : S^1 \to S^1$ such that $\Pi \circ f^{-1} = \bar{\varphi} \circ \Pi$. Composing the two identities we get $\varphi \circ \bar{\varphi} = \bar{\varphi} \circ \varphi = \mathbf{1}_{S^1}$, thus $\bar{\varphi} = \varphi^{-1}$. Since φ and $\bar{\varphi}$ are smooth, φ is a diffeomorphism of S^1. Next, let $\psi = \Phi(g)$, $\Pi \circ g = \psi \circ \Pi$. Thus, $\Pi \circ (g \circ f) = (\psi \circ \varphi) \circ \Pi$, therefore $\Phi(g \circ f) = \Phi(g) \circ \Phi(f)$, and $\Phi : G \to \mathrm{Diff}(S^1)$ is a homomorphism. Let $f \in \ker(\Phi)$, that is, $\Pi \circ f = \Pi$, then for all x, $f(x) = x + 2\pi k(x)$ with $k(x) \in \mathbf{Z}$, but $x \mapsto k(x) = f(x) - x$ is smooth, hence $k(x) = k$ is constant. The case $f' < 0$ is analogous. Therefore, $\ker(\Phi) = \{x \mapsto x + 2\pi k \mid k \in \mathbf{Z}\} \simeq \mathbf{Z}$. Now, let us show that Φ is surjective. Let $\varphi \in \mathrm{Diff}(S^1)$, $\varphi \circ \Pi$ is smooth and we admitted that there exists a smooth lift $f : \mathbf{R} \to \mathbf{R}$ such that $\Pi \circ f = \varphi \circ \Pi$ (it is actually a consequence of the monodromy theorem (art. 8.25)).

a) Necessarily $f(x + 2\pi) = f(x) + 2\pi k(x)$, and for the same reason as just above, $k(x) = k$ is constant. Note that k is necessarily nonzero, f cannot be periodic since a periodic function has necessarily a point where f' vanishes, and this is impossible since f projects onto a diffeomorphism of the circle. Then, note that $(1/k)f(x + 2\pi) = (1/k)f(x) + 2\pi$. Thus, there exists a smooth function $\psi : S^1 \to S^1$ such that $\psi(z) = \Pi((1/k)f(x))$, $z = \Pi(x)$. Defining $\hat{k}(z) = \Pi(kx)$, we get $\phi(z) = \hat{k} \circ \psi(x)$. But ψ and $\hat{k}$ are surjective and $\hat{k}$ is not injective if $k \neq \pm 1$, thus $k = \pm 1$ and $f(x + 2\pi) = f(x) \pm 2\pi$. Considering φ^{-1}, the same argument gives $g \in \mathcal{C}^\infty(\mathbf{R})$ such that $\varphi^{-1} \circ \Pi = \Pi \circ g$ and $g(x + 2\pi) = g(x) \pm 2\pi$.

b) Now, $\Pi \circ (f \circ g) = \Pi \circ (g \circ f) = \Pi$, and hence $f \circ g$, as well as $g \circ f$, belongs to $\ker(\Phi)$, that is, $g \circ f(x) = x + 2\pi\ell$ and $f \circ g(x) = x + 2\pi\ell'$, $\ell, \ell' \in \mathbf{Z}$. Then, since $(g \circ f)'(x) = g'(f(x))f'(x) = 1$, f and g are both strictly increasing or strictly decreasing. Next, left composition of $f \circ g(x) = x + 2\pi\ell'$ with g gives $\ell' = \ell$ if f and g are increasing, and $\ell' = -\ell$ otherwise. In both cases, let $\bar{f}(x) = g(x) - 2\pi\ell$, then $\bar{f}$ still satisfies $\varphi^{-1} \circ \Pi = \Pi \circ \bar{f}$, but now $f \circ \bar{f} = \bar{f} \circ f = \mathbf{1}_{\mathbf{R}}$. Therefore $\bar{f} = f^{-1}$, $f \in G$ and Φ is surjective.

Now, let us consider a plot $P : r \mapsto \varphi_r$ of $\mathrm{Diff}(S^1)$ defined on an open ball B. Thus, $(r, x) \mapsto \varphi_r(\Pi(x))$ is a smooth parametrization of S^1, defined on $B \times \mathbf{R}$. We admitted that there exists a smooth lift $(r, x) \mapsto f_r(x)$, from $B \times \mathbf{R}$ to $\mathbf{R}$ along Π, that is, $\Pi \circ f_r(x) = \varphi_r \circ \Pi(x)$. Applying a) to this situation, we get a smooth function $r \mapsto k_r$, defined on B to $\mathbf{Z}$ such that $2\pi k_r = f_r(x + 2\pi) - f_r(x)$. Thus, $k_r = k$ is constant. By continuing same reasoning, we get two plots $r \mapsto f_r$ and $r \mapsto g_r$ such that $f_r(x + 2\pi) = f_r(x) \pm 2\pi$, $g_r(x + 2\pi) = g_r(x) \pm 2\pi$ and $\Pi \circ (f_r \circ g_r) = \Pi \circ (g_r \circ f_r) = \Pi$, where the sign $+$ or $-$ is constant on B. Then, we get similarly $g_r \circ f_r(x) = x + 2\pi\ell_r$

and $f_r \circ g_r(x) = x + 2\pi\ell'_r$, $\ell_r, \ell'_r \in \mathbf{Z}$. For the same reason as previously, $\ell_r = \ell$ and $\ell'_r = \ell'$ are constant, and $\ell' = \pm\ell$ according to the situation. Thus, the change $\bar{f}_r = g_r - 2\pi\ell$ still defines a plot, and $\bar{f}_r = f_r^{-1}$. Therefore, thanks to b) we deduce that $f_r \in G$ for all r, and satisfies: $\Pi \circ f_r = \varphi_r \circ \Pi$ and $\Pi \circ f_r^{-1} = \varphi_r \circ \Pi$. Thus we get a plot $r \mapsto f_r$ covering $r \mapsto \varphi_r$, that is, $\Phi(f_r) = \varphi_r$, for all $r \in B$. Finally, considering a general plot $r \mapsto \varphi_r$, defined on some real domain, every point is the center of some open ball, and applying what we just checked, we can locally lift smoothly this plot in G along Φ. Therefore, Φ is a subduction. Moreover, the kernel of Φ being $\mathbf{Z}$, as we have seen above, the fiber of the projection Φ are the orbits of the action of $\mathbf{Z}$ on G, that is, $k(f) = f + 2\pi k$, $k \in \mathbf{Z}$ and $f \in G$.

For the fifth question, let us note that, since $(x, a) \mapsto x + a$ is clearly smooth, the map $a \mapsto T_a = [x \mapsto x + a]$ is smooth. Conversely if $r \mapsto T_{a(r)}$ is a plot of G, then $a(r) = T_{a(r)}(0)$ and thus $r \mapsto a(r)$ is smooth. Therefore $a \mapsto T_a$ is an induction and the image of $\mathbf{R}$ in G is diffeomorphic to $\mathbf{R}$. Now, since Φ is a subduction, the map $\phi : \mathbf{R}/2\pi\mathbf{Z} \to \mathrm{Diff}(S^1)$, defined by $\phi(\mathrm{class}(a)) = \Phi(T_a)$ is an induction. Since $\mathbf{R}/2\pi\mathbf{Z}$ is diffeomorphic to S^1, thanks to Π, the image of $\mathbf{R}$ by $a \mapsto \Phi(T_a)$, equipped with the subset diffeology, is diffeomorphic to the circle.

☞ **Exercise 29, p. 31** (Quotients of discrete or coarse spaces). Let $X_\circ$ be a discrete diffeological space: the plots of X are locally constant parametrizations (art. 1.20). Let $\sim$ be any equivalence relation on X, $Q = X/\sim$, and let $\pi : X \to Q$ be the projection. By definition of the quotient diffeology (art. 1.43), a plot $P : U \to Q$ lifts locally along a plot of X. Hence, each local lift of P is locally constant. Therefore, P is locally constant and Q is discrete.

☞ **Exercise 30, p. 31** (Examples of quotients). Let us consider the diffeology of the circle defined in (art. 1.11). The circle $S^1 \subset \mathbf{C}$ is obviously in bijection with the classes of the equivalence relation $t \sim t'$ if and only if $t' = t + k$, with $k \in \mathbf{Z}$, that is, $\mathrm{class}(t) \mapsto \exp(2\pi i t) = \cos(2\pi t) + i\sin(2\pi t)$. Then, thanks to the uniqueness of quotients (art. 1.52), to get the identification $S^1 \simeq \mathbf{R}/\mathbf{Z}$ we just have to check that the map $t \mapsto \exp(2\pi i t)$ is a subduction, but the subduction has been proved in Exercise 27, p. 27. Regarding the irrational torus T_α in Exercise 4, p. 8, or the quotient $\mathbf{R}/\mathbf{Q}$ of Exercise 5, p. 9, the diffeology defined by (✱) is, by definition, the quotient diffeology. Concerning the diffeology of $\mathrm{Diff}(S^1)$ defined in Exercise 28, p. 27, we have seen that the map $G \to \mathrm{Diff}(S^1)$ is surjective, and set theoretically $\mathrm{Diff}(S^1) \sim G/\mathbf{Z}$, where $\mathbf{Z} \sim \ker(\Phi)$. Since Φ is a subduction, by uniqueness of quotients we get that diffeologically $\mathrm{Diff}(S^1) \simeq G/\mathbf{Z}$.

☞ **Exercise 31, p. 31** (The irrational solenoid). 1) let us check that the map $q : (x, y) \mapsto (p(x), p(y))$, where $p(t) = (\cos(2\pi t), \sin(2\pi t))$, is strict. First of all, this map is clearly smooth since cos and sin are smooth. Now, according to the definition, q is strict if and only if the map

$$\mathrm{class}(x, y) \mapsto ((\cos(2\pi x), \sin(2\pi x)), (\cos(2\pi y), \sin(2\pi y)))$$

is an induction, from $\mathbf{R}^2/\mathbf{Z}^2$ to $\mathbf{R}^2 \times \mathbf{R}^2$, where $\mathrm{class} : \mathbf{R}^2 \to \mathbf{R}^2/\mathbf{Z}^2$ denotes the natural projection. We know already that $\Pi : t \mapsto (\cos(2\pi t), \sin(2\pi t))$ is strict (Exercise 27, p. 27), and q is just the product $\Pi \times \Pi$. Thus, a plot $\Phi : U \to S^1 \times S^1 \subset \mathbf{R}^2 \times \mathbf{R}^2$ is just a pair of plots P and Q from U to S^1, which can be individually smoothly lifted locally along Π, and give a local lift of q itself. Therefore, q is strict. Now, since α is irrational, $q_\alpha = q \restriction \Delta_\alpha$ is injective. Indeed, for

$t, t' \in \mathbf{R}$, $q(t, \alpha t) = q(t', \alpha t')$ means $(\cos(2\pi t'), \sin(2\pi t')) = (\cos(2\pi t), \sin(2\pi t))$ and $(\cos(2\pi\alpha t'), \sin(2\pi\alpha t')) = (\cos(2\pi\alpha t), \sin(2\pi\alpha t))$, that is, $t' = t + k$ and $\alpha t' = \alpha t + k'$ with $k, k' \in \mathbf{Z}$, that gives $\alpha k - k' = 0$, thus $k = k' = 0$ and $t' = t$.

2) Let $\Phi : U \to \mathcal{S}_\alpha \subset S^1 \times S^1 \subset \mathbf{R}^2 \times \mathbf{R}^2$ be a plot, with $\Phi(r) = (P(r), Q(r))$. Since q is strict, for all $r \in U$, there exists locally a smooth lift $r' \mapsto (x(r'), y(r'))$ in $\mathbf{R}^2$, defined on a neighborhood V of r, such that $q(x(r'), y(r')) = (P(r'), Q(r'))$ Thus, $q(x(r'), y(r')) \in \mathcal{S}_\alpha$ for all $r' \in V$. But, $r' \mapsto (x(r'), \alpha x(r')) \in \Delta_\alpha \subset \mathbf{R}^2$ is smooth, and $q(x(r'), \alpha x(r'))$ belongs to $\mathcal{S}_\alpha$ too. Therefore, there exists $r' \mapsto k(r') \in \mathbf{Z}$ such that $y(r') = \alpha x(r') + k(r')$, that is, $k(r') = y(r') - x(r')$. Thus, $r' \mapsto k(r')$ is smooth and takes its values in $\mathbf{Z}$, hence $k(r') = k$ constant. Then, $r' \mapsto (x(r'), y(r') - k)$ is a plot of $\mathcal{S}_\alpha$ with $q(x(r'), y(r') - k) = (P(r'), Q(r'))$, thus $q_\alpha : \Delta_\alpha \to \mathcal{S}_\alpha$ is an injective subduction, that is, a diffeomorphism from Δ_α to $\mathcal{S}_\alpha$, and therefore an induction.

3) We use the identification given by the factorization $h : \mathbf{R}^2/\mathbf{Z}^2 \to S^1 \times S^1$, of the strict map $q : \mathbf{R}^2 \to S^1 \times S^1$. Then, the quotient $(S^1 \times S^1)/\mathcal{S}_\alpha = h(\mathbf{R}^2/\mathbf{Z}^2)/\mathcal{S}_\alpha$, is equivalent to $\mathbf{R}^2/[\mathbf{Z}^2(\Delta_\alpha)]$ where the equivalence relation is defined by the action of the subgroup $\mathbf{Z}^2(\Delta_\alpha)$, that is, the set of $(x+n, \alpha x+m)$ with $x \in \mathbf{R}$ and $(n, m) \in \mathbf{Z}^2$. Let $\rho : \mathbf{R}^2 \to \mathbf{R}^2$ be defined by $\rho(x, y) = (0, y - \alpha x)$, it is obviously a projector, $\rho \circ \rho = \rho$, and clearly $\text{class} \circ \rho = \text{class}$, with $\text{class} : \mathbf{R}^2 \to \mathbf{R}^2/[\mathbf{Z}^2(\Delta_\alpha)]$. Now, let $X' = \text{val}(\rho)$, that is, $X' = \{0\} \times \mathbf{R}$. The restriction to X' of the equivalence relation, defined by the action of $\mathbf{Z}^2(\Delta_\alpha)$ on $\mathbf{R}^2$, is defined by the action of $\mathbf{Z}^2$, $(n, m) : (0, y) \mapsto (0, y + m - \alpha n)$. Therefore, the quotient $(S^1 \times S^1)/\mathcal{S}_\alpha$ is equivalent to $X'/(\mathbf{Z} + \alpha\mathbf{Z})$ (art. 1.53, Note), that is, equivalent to $\mathbf{R}/(\mathbf{Z} + \alpha\mathbf{Z}) = T_\alpha$.

☞ **Exercise 32, p. 31** (A minimal powerset diffeology). Let $\mathfrak{P}(X)^*$ be the set of all the nonempty subsets of X, thus $\mathfrak{P}(X) = \{\varnothing\} \cup \mathfrak{P}(X)^*$. Let $\mathcal{D}$ be the set of parametrizations of $\mathfrak{P}(X)^*$ defined as follows.

(♡) A parametrization $P : U \to \mathfrak{P}(X)^*$ belongs to $\mathcal{D}$ if, for all $r \in U$, there exist an open neighborhood $V \subset U$ of r and a plot $Q : V \to X$ such that, for all $r' \in V$, $Q(r') \in P(r')$.

Let us check that $\mathcal{D}$ is a diffeology. Let $P : r \mapsto A \in \mathfrak{P}(X)^*$ be a constant parametrization, and let $x \in A$. The constant parametrization $Q : r \mapsto x$ satisfies $Q(r) \in P(r)$. The covering axiom is thus satisfied. The locality axiom is satisfied by construction. Now, let $P : U \to \mathfrak{P}^*(X)$ belong to $\mathcal{D}$, and let $F : U' \to U$ be a smooth parametrization. Let $r' \in U'$ and $r = F(r')$, let $V \subset U$ be a neighborhood of r, and let $Q : V \to X$ be a plot such that $Q(s) \in P(s)$ for all $s \in V$, according to (♡). Let $V' = F^{-1}(V)$, since F is smooth, thus continuous, V' is an open neighborhood of r'. Now, $Q' = Q \circ F$ is a plot of X, and satisfies $Q'(s) = (Q \circ F)(s) \in (P \circ F)(s)$ for all $s \in V'$. Thus, the smooth compatibility axiom is satisfied and $\mathcal{D}$ is a diffeology of $\mathfrak{P}(X)^*$. Next, we consider $\mathfrak{P}(X)$ as the diffeological sum of the singleton $\{\varnothing\}$ and the diffeological space $\mathfrak{P}(X)^*$, equipped with $\mathcal{D}$. Then, let us consider an equivalence relation $\sim$ on X. The subset $X/\!\sim\, = \text{class}(X)$ is contained in $\mathfrak{P}(X)^*$, since no class is empty. Let $P : U \to X/\!\sim$ be a plot of $\mathfrak{P}(X)^*$. For each $r \in U$, let us choose $x_r \in X$ such that $\text{class}(x_r) = P(r)$, that is, $x_r \in P(r)$. By definition of $\mathcal{D}$, there exists — defined on a neighborhood of each point of U — a plot Q of X such that $Q(r) \in P(r)$, that is, $Q(r) \in \text{class}(x_r)$. Thus, $\text{class}(Q(r)) = \text{class}(x_r) = P(r)$, hence, Q is a local smooth lift of P along class, and that is the definition of the quotient diffeology on $X/\!\sim$.

☞ **Exercise 33, p. 31** (Universal construction). First of all, let us note that $\mathrm{ev} : \mathcal{N} \to X$ is surjective. Then, every plot $P : U \to X$ lifts naturally by $\mathbf{P} : r \mapsto (P, r)$ in $\mathcal{N}$, along ev: $\mathrm{ev} \circ \mathbf{P} = P$. Therefore $\mathrm{ev} : \mathcal{N} \to X$ is a subduction and X is the diffeological quotient of $\mathcal{N}$ by the relation $(P, r) \sim (P', r')$ if and only if $P(r) = P'(r')$. Next, let us assume that the map $\sigma : x \mapsto ([0 \mapsto x], 0) \in \mathcal{N}$ is smooth. Thus, for every plot $P : U \to X$, $\sigma \circ P : r \mapsto ([0 \mapsto P(r)], 0)$ is a plot of $\mathcal{N}$. By definition of the diffeological sum (art. 1.39), for all $r \in U$, there exists an open neighborhood V of r such that $[0 \mapsto P(r)]$ is constant on V, that is, $P(r) = x$ for some $x \in X$ and for all $r \in V$, but this is the definition of the discrete diffeology (art. 1.20). Therefore, if the set of 0-plots is a smooth section of $\mathrm{ev} : \mathcal{N} \to X$, then X is discrete, and conversely.

☞ **Exercise 34, p. 31** (Strict action of SO(3) on $\mathbf{R}^3$). If $X = 0 \in \mathbf{R}^3$, then obviously $R(0) : SO(3) \mapsto 0$ is a trivial subduction. Let us assume that $X \neq 0$, the orbit $SO(3) \cdot X$ is the sphere of vectors X' with norm $\rho = \|X\|$. By construction, SO(3) preserves the norm. Now, let $r \mapsto X_r$ be a plot of $\mathbf{R}^3$ such that $\|X_r\| = \rho$. Let $u_r = X_r/\rho$, thus $r \mapsto u_r$ is a plot of the unit sphere $S^2 \subset \mathbf{R}^3$, for the subset diffeology. Let r_0 be a point in the domain of this plot, and there exists a vector w not collinear with u_{r_0}. The parametrization $r \mapsto w_r = [\mathbf{1}_{\mathbf{R}^3} - u_r u_r^t]\,w$ is smooth, where u_r^t is the transpose of the vector u_r, and $[\mathbf{1}_{\mathbf{R}^3} - u_r u_r^t]$ is the projector orthogonal to u_r. The real function $\nu : r \mapsto \|w_r\|$ is smooth, and since w is not collinear with u_{r_0}, $\nu(r_0) \neq 0$. Thus, there exists a (possibly small) open ball B, centered at r_0, such that for all $r \in B$, $\nu(r) \neq 0$. Therefore, the parametrization $r \mapsto v_r = w_r/\nu(r)$, defined on B, is a plot of the sphere S^2 satisfying $v_r \perp u_r$. Next, let $N_r = [u_r\ v_r\ u_r \wedge v_r]$ be the matrix made by juxtaposing the three column vectors, the symbol $\wedge$ denoting the vector product. By construction, $r \mapsto N_r$ is smooth and $N_r \in SO(3)$. Now, $N_r \mathbf{e}_1 = u_r$, where $\mathbf{e}_1$ is the first vector of the canonical basis of $\mathbf{R}^3$. By the same way, we can find a unit vector v, orthogonal to $u = X/\rho$, such that $M = [u\ v\ u \wedge v] \in SO(3)$, and thus $M\mathbf{e}_1 = u$. Hence, the parametrization $r \mapsto M_r = N_r M^t$ is smooth, takes its values in SO(3) and satisfies $M_r X = X_r$. Therefore, the orbit map R(X) is strict.

NOTE. This is a particular case of a more general theorem: for a Lie group acting smoothly on a manifold, which is Hausdorff and second countable, the orbit map is always strict [**IZK10**].

☞ **Exercise 35, p. 33** (Products and discrete diffeology). Let us equip the product $X = \prod_{i \in \mathcal{J}} X_i$ with discrete diffeology. Thus, every plot $r \mapsto x$ of X is locally constant, and then any composite $r \mapsto x_i$ is locally constant too, thus smooth (first axiom of diffeology). This is an example, related to the discussion (art. 1.25), where the interesting set of diffeologies on X — the ones such that the projections π_i are smooth — is trivially bounded below. The supremum of this family, which is a maximum (the product diffeology), is therefore the distinguished diffeology. However, it is not the only reason for which that diffeology is interesting; see (art. 1.56). If we consider the sum diffeology of the family (art. 1.39), that is, $X' = \coprod_{i \in \mathcal{J}} X_i$, and if we equip X' with the coarse diffeology, then the canonical injections $j_i : X_i \to X'$, defined by $j_i(x) = (i, x)$, are smooth, simply because any map to a coarse space is smooth. In that case, the set of diffeologies such that the injections are smooth is bounded above by the coarse diffeology, the distinguished diffeology is thus the infimum of that family, that is, the sum diffeology. It is not surprising that products and coproducts are dual constructions of each other.

Exercise 36, p. 33 (Products of coarse or discrete spaces). Let us consider the product $X = \prod_{i \in \mathcal{I}} X_i$ of coarse spaces. Let P be any parametrization of X. Since for every projection π_i, the composition $\pi_i \circ P$ is a parametrization of X_i, the parametrization P is a plot of the product. Therefore, the product diffeology is coarse. Now, let $X = \prod_{i \in \mathcal{I}} X_i$ be a finite product of discrete spaces, and let $N = \#\mathcal{I}$. Let $P : U \to X$ be a plot. For every $i \in \mathcal{I}$, $\pi_i \circ P$ is locally constant. Let $r_0 \in U$, so there exist N open neighborhoods V_i of r_0 and N points x_i, such that $x_i \in X_i$ and $\pi_i \circ P \restriction V_i = [r \mapsto x_i]$. Hence, since $\#\mathcal{I}$ is finite, $V = \bigcap_{i \in \mathcal{I}} V_i$, V is still an open neighborhood of r_0, and $P \restriction V$ is constant, equal to $s = [i \mapsto (i, x_i)]$. Therefore, P is locally constant and X is discrete. Next, let us consider an arbitrary product $X = \prod_{i \in \mathcal{I}} X_i$ of discrete spaces. We cannot apply the previous method, since an arbitrary intersection of domains may be not open. Then we shall use the result of Exercise 7, p. 14. Let $P : U \to X$ be a plot. By definition of the product diffeology, for all $i \in \mathcal{I}$, $\pi_i \circ P$ is a plot of X_i, that is, locally constant. Now, let r_0 be any point of U, thanks to Exercise 7, p. 14, $\pi_i \circ P$ is constant on the path connected component V of r_0. Thus, for every $i \in \mathcal{I}$, $\pi_i \circ P \restriction V = [r \mapsto x_i]$, where $x_i = \pi_i \circ P(r_0)$. Hence, $P \restriction V = [i \mapsto (i, x_i)]$ is a constant parametrization. Therefore, P is locally constant and X is discrete.

Exercise 37, p. 33 (Infinite product of $\mathbf{R}$ over $\mathbf{R}$). The sum $X = \coprod_{t \in \mathbf{R}} \mathbf{R}$ is the set of pairs (t, s), with t and s in $\mathbf{R}$. Set theoretically, X is the product $\mathbf{R} \times \mathbf{R}$. Thus, a plot P of X is a pair (T, S) of parametrizations of $\mathbf{R}$, defined on some common domain U, such that for every $r_0 \in U$ there exist an open neighborhood V of r_0, a real t_0, with $T \restriction V = [r \mapsto t_0]$ and $S \restriction V \in \mathcal{C}^\infty(V, \mathbf{R})$. Now, let $X = \prod_{t \in \mathbf{R}} \mathbf{R}$. By definition (art. 1.55), X is the set of maps $[t \mapsto (t, s)]$ such that $s \in \mathbf{R}$. Thus, set theoretically, X is equivalent to $\mathrm{Maps}(\mathbf{R}, \mathbf{R})$, the set of maps from $\mathbf{R}$ to $\mathbf{R}$. Also, an element of X can be regarded as an indexed family $x = (x_t)_{t \in \mathbf{R}}$. A plot $P : U \to X$ is any parametrization $P : r \mapsto (x_t(r))_{t \in \mathbf{R}}$ such that for every $t \in \mathbf{R}$, the parametrization x_t is a plot of $\mathbf{R}$, that is, a smooth parametrization in $\mathbf{R}$.

Exercise 38, p. 33 (Graphs of smooth maps). Let us assume first that $f : X \to X'$ is such that $\mathrm{pr}_X : \mathrm{Gr}(f) \to X$ is a subduction. Let $P : U \to X$ be some plot, and let $r_0 \in U$. Since $\mathrm{pr}_X \restriction \mathrm{Gr}(f)$ is a subduction, there exist an open neighborhood V of r_0 and a plot $Q : V \to \mathrm{Gr}(f)$ such that $\mathrm{pr}_X \circ Q = P \restriction V$. Thus, $Q(r) = (P(r), f(P(r)))$ for every $r \in V$. But, since Q is a plot of $\mathrm{Gr}(f) \subset X \times X'$, $(f \circ P) \restriction V$ is a plot of X', by definition of the product and the subset diffeologies. So, $f \circ P$ is locally a plot of X', thus $f \circ P$ is a plot of X'. Therefore f is smooth. Conversely, let $f : X \to X'$ be a smooth map, and let $P : U \to X$ be a plot. Then, $f \circ P$ is a plot of X', and $Q : r \mapsto (P(r), f \circ P(r))$ is a plot of $X \times X'$. But $\mathrm{val}(Q) \subset \mathrm{Gr}(f)$, so Q is a plot of $\mathrm{Gr}(f)$, for the subset diffeology. Moreover $\mathrm{pr}_X \circ Q = P$, so Q is a lifting of P along pr_X. Therefore, $\mathrm{pr}_X \restriction \mathrm{Gr}(f)$ is a subduction.

Exercise 39, p. 34 (The 2-torus). In the solution of Exercise 31, p. 31, we have seen that a plot $\Phi : U \to S^1 \times S^1 \subset \mathbf{R}^2 \times \mathbf{R}^2$ is just a pair of plots (P, Q) of $S^1 \subset \mathbf{R}^2$, with $\mathrm{def}(P) = \mathrm{def}(Q) = U$, that is, by definition, a plot for the product diffeology for a finite family of spaces. Also note that the standard diffeology on $\mathbf{R}^n$ is the product diffeology of n copies of $\mathbf{R}$.

Exercise 40, p. 39 (The space of polynomials). 1) To prove that the map $j_n : (\mathbf{R}^m)^{n+1} \to \mathcal{C}^\infty(\mathbf{R}, \mathbf{R}^m)$ is an induction, we apply the criterion stated in (art. 1.31).

a) The map j_n is injective. Indeed, a polynomial is characterized by its coefficients.

b) The map j_n is smooth. Let $P : U \to (\mathbf{R}^m)^{n+1}$ be a plot, that is, $P : r \mapsto P(r) = (P_0(r), \ldots, P_n(r))$, where the P_i are smooth parametrizations of $\mathbf{R}^m$ (art. 1.55). Now, the map j_n is smooth if and only if, for every smooth parametrization $\tau : V \to \mathbf{R}$, the parametrization $(r,s) \mapsto P_0(r) + \tau(s)P_1(r) + \cdots + \tau(s)^n P^n(r)$ is a smooth parametrization of $\mathbf{R}^m$. But this is the case, since it is a sum of products of smooth parametrizations.

c) The map j_n^{-1} is smooth. Let $P : U \to \mathrm{Pol}_n(\mathbf{R}^m) \subset \mathcal{C}^\infty(R, \mathbf{R}^m)$ be a plot. Now, let $P_k(r) \in \mathbf{R}^m$ be the coefficients of $P(r)$, $r \in U$, such that $P(r)(t) = P_0(r) + tP_1(r) + \cdots + t^n P^n(r)$ for all t in $\mathbf{R}$. Or, in other words, such that $j_n^{-1} \circ P(r) = (P_0(r), \ldots, P_n(r))$. But the coefficients $P_k(r)$ are

$$P_0(r) = P(r)(0) \quad \text{and} \quad P_k(r) = \frac{1}{k!} \left.\frac{d^k P(r)(t)}{dt^k}\right|_{t=0}, \quad k = 1, \ldots, n,$$

and P being a plot of $\mathrm{Pol}_n(\mathbf{R}^m)$, the parametrization $\mathbf{P} : (r,t) \mapsto P(r)(t)$ is smooth. Hence, each coefficient P_k is a partial derivative of a smooth parametrization,

$$\left.\frac{d^k P(r)(t)}{dt^k}\right|_{t=0} = \left.\frac{\partial^k \mathbf{P}(r,t)}{\partial t^k}\right|_{t=0}.$$

Therefore, P_k is a smooth parametrization of $\mathbf{R}^m$. Thus $j_n^{-1} \circ P$ is a plot of $(\mathbf{R}^m)^{n+1}$, and j_n^{-1} is smooth. In conclusion, the space $\mathrm{Pol}_n(\mathbf{R}, \mathbf{R}^m)$, equipped with the functional diffeology, inherited from $\mathcal{C}^\infty(\mathbf{R}, \mathbf{R}^m)$, is diffeomorphic to the real vector space $(\mathbf{R}^m)^{n+1}$.

2) Let ω be a domain in $(\mathbf{R}^m)^{n+1}$, and Ω be the subset of $\mathcal{C}^\infty(\mathbf{R}, \mathbf{R}^n)$ defined by

$$\Omega = \left\{ f \in \mathcal{C}^\infty(\mathbf{R}, \mathbf{R}^m) \;\middle|\; \left(\frac{f(0)}{0!}, \frac{f'(0)}{1!}, \frac{f''(0)}{2!}, \ldots, \frac{f^{(n)}(0)}{n!} \right) \in \omega \right\}.$$

By construction, every polynomial $[t \mapsto x_0 + tx_1 + \cdots + t^n x_n]$, with coefficients $x = (x_0, x_1, \ldots, x_n)$ in ω, belongs to Ω. More precisely, $j_n(\omega) = \Omega \cap \mathrm{Pol}_n(\mathbf{R}, \mathbf{R}^n)$. Now, let $P : U \to \mathcal{C}^\infty(\mathbf{R}, \mathbf{R}^m)$ be some plot, we have

$$P^{-1}(\Omega) = \left\{ r \in U \;\middle|\; \left(P(r)(0), \frac{1}{1!} \left.\frac{\partial P(r,t)}{\partial t}\right|_{t=0}, \ldots, \frac{1}{n!} \left.\frac{\partial^n P(r,t)}{\partial t^n}\right|_{t=0} \right) \in \omega \right\}.$$

Since P is smooth, the various partial derivatives are smooth and then continuous. Hence, the following map $\phi : U \to (\mathbf{R}^m)^{n+1}$, defined by

$$\phi : r \mapsto \left(P(r,0), \frac{1}{1!} \left.\frac{\partial P(r,t)}{\partial t}\right|_{t=0}, \ldots, \frac{1}{n!} \left.\frac{\partial^n P(r,t)}{\partial t^n}\right|_{t=0} \right),$$

is continuous. Therefore, $P^{-1}(\Omega)$ is the preimage of the domain ω by the continuous map ϕ, thus a domain. The proof is complete.

Exercise 41, p. 39 (A diffeology for the space of lines). A polynomial f of degree 1, from $\mathbf{R}$ to $\mathbf{R}^n$, is a map $f : t \mapsto x + tv$, where $(x, v) \in (\mathbf{R}^n)^2$. The image of f is an (affine) line of $\mathbf{R}^n$ if and only if $v \neq 0$. The coefficients x and v are called the *origin* for x, since $x = f(0)$, and the *velocity* for v, since $v = f'(0)$. Hence, defining this subspace $\mathrm{PL}(\mathbf{R}^n)$ of polynomials as the space of *parametrized lines* of $\mathbf{R}^n$ makes sense.

1) Let $f = [t \mapsto x + tv]$ and $g = [t \mapsto x' + tv']$ be two lines having the same image in $\mathbf{R}^n$, that is, $f(\mathbf{R}) = g(\mathbf{R})$. So, $x' = g'(0) \in f(\mathbf{R})$, thus there exists a number b such that $x' = x + bv$. Now, since $f(\mathbf{R}) = g(\mathbf{R})$ the derivative of f and g are proportional. Thus, there exists a number a such that $v' = av$. But since v and v' are not zero, $a \neq 0$. Hence, $g(t) = x + bv + atv = x + (at + b)v$, and then $g(t) = f(at + b)$. Conversely if $g(t) = f(at + b)$, since $a \neq 0$, it is clear that $g(\mathbf{R}) = f(\mathbf{R})$.

2) The set $(a, b) : t \mapsto at + b$ of transformations of $\mathbf{R}$, where a and b are real numbers such that $a \neq 0$, is the affine group, denoted by $\mathrm{Aff}(\mathbf{R})$. It is isomorphic to the group of matrices

$$\begin{pmatrix} a & b \\ 0 & 1 \end{pmatrix} \quad \text{with} \quad \begin{pmatrix} a & b \\ 0 & 1 \end{pmatrix} \begin{pmatrix} t \\ 1 \end{pmatrix} = \begin{pmatrix} at + b \\ 1 \end{pmatrix}.$$

The action of the affine group on the space of lines defined in the first question, $(a, b)(f) = [t \mapsto f(at + b)]$, is the composition $(a, b)(f) = f \circ (a, b)$. It is in fact an anti-action, since $(a, b)[(a', b')(f)] = f \circ (a', b') \circ (a, b) = [(a', b') \circ (a, b)](f)$.

3) Thanks to Exercise 40, p. 39, we know that the inclusion $(x, v) \mapsto [t \mapsto x + tv]$ is an induction from $(\mathbf{R}^n)^2$ to $\mathcal{C}^\infty(\mathbf{R}, \mathbf{R}^n)$. Hence $\mathrm{PL}(\mathbf{R}^n)$, equipped with the functional diffeology, is diffeomorphic to $\mathbf{R}^n \times (\mathbf{R}^n - \{0\})$.

4) Thus, the equivalence relation defining the oriented trajectory of the parametrized line is the following, $(x, v) \sim (x + bv, av)$ where $(a, b) \in \mathrm{Aff}_+(\mathbf{R})$, that is, $a, b \in \mathbf{R}$ and $a \neq 0$. Now, the map ρ consists into two maps. The second one $v \mapsto u$ is well defined and smooth since $v \neq 0$. The first one is the orthogonal projector to u or, which is equivalent, to v. In other words $r = [\mathbf{1}_n - u\bar{u}]x$, where $\bar{u}$ is the covector $\bar{u} : w \mapsto u \cdot w$. Therefore, ρ is a smooth map from $\mathbf{R}^n \times (\mathbf{R}^n - \{0\})$ into itself. Finally, the image of ρ is clearly the subset of $\mathbf{R}^n \times \mathbf{R}^n$ made up with the pairs of vectors (r, u) such that $u \cdot r = 0$ and $\|u\| = 1$, which is equivalent to TS^{n-1}, as it has been defined. Now, let f and g be the lines defined respectively by (x, v) and (x', v'). Let us assume that $\rho(x, v) = \rho(x', v') = (r, u)$. So, $u = v/\|v\| = v'/\|v'\|$. Thus, there exists $a > 0$ such that $v' = av$. Then, $[\mathbf{1}_n - u\bar{u}]x = [\mathbf{1}_n - u\bar{u}]x' = r$ implies that the orthogonal projection to u of $x' - x$ is zero, hence $x' - x$ is proportional to u, or which is equivalent, to v. Thus, there exists a number b such that $x' = x + bv$. Therefore, if $\rho(x', v') = \rho(x, v)$, then there exists an element $(a, b) \in \mathrm{Aff}_+(\mathbf{R})$ such that $g = (a, b)(f)$, and the lines defined by (x', v') and (x, v) have the same oriented trajectory. The converse is as clear as the direct way. Moreover, if v is unitary and x is orthogonal to v, then $\rho(x, v) = (v, x)$. Hence $\rho \circ \rho = \rho$. Therefore, the map ρ satisfies the conditions of (art. 1.53). Its image, equivalent to TS^{n-1}, is diffeomorphic to the quotient space $[\mathbf{R}^n \times (\mathbf{R}^n - \{0\})]/\mathrm{Aff}_+(\mathbf{R})$, that is, diffeomorphic to $\mathrm{UL}_+(\mathbf{R}^n) = \mathrm{PL}(\mathbf{R}^n)/\mathrm{Aff}_+(\mathbf{R})$. Considering the lines in $\mathbf{R}^2$, the space TS^1 describes the oriented unparametrized lines, a point $(x, u) \in \mathrm{TS}^1$ describes the line passing through x and directed by u. Hence, the set of unparametrized and nonoriented lines is equivalent to the quotient $\mathrm{TS}^1/\{\pm 1\}$, where $\varepsilon(x, u) = (x, \varepsilon u)$, $\varepsilon \in \{\pm 1\}$. Thanks to the diffeomorphism $(x, u) \mapsto (u, r = x \cdot Ju)$ from TS^1 to $S^1 \times \mathbf{R}$, where J is the $\pi/2$ positive rotation, the action of $\{\pm 1\}$ transmutes into $\varepsilon(u, r) = (\varepsilon u, \varepsilon r)$. The quotient is a realization of the Möbius strip.

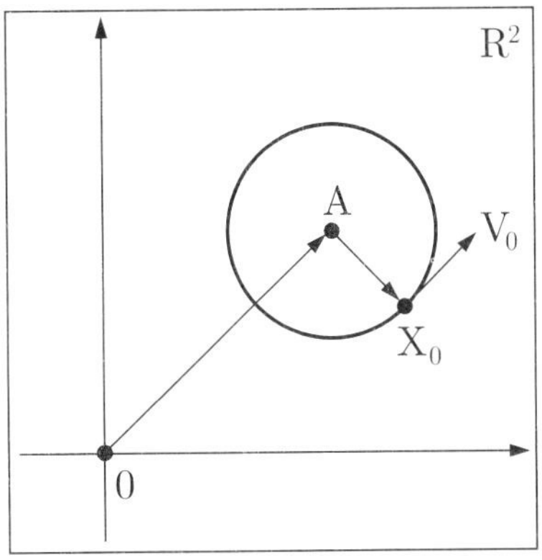

FIGURE Sol.3. Initial conditions of the ODE.

Exercise 42, p. 40 (A diffeology for the set of circles). We shall describe the set of circles in the plane $\mathbf{R}^2$ as the trajectories of the solutions of an ordinary differential equation. Let us set up the adequate differential equation. Let C be the circle centered at the point A, with radius r, where $A \in \mathbf{R}^2$ and $r \in [0, \infty[$. Let $R(\theta)$ be the rotation with angle θ, and let $X_0 \in \mathbf{R}^2$ such that $r = \|X_0\|$ (Figure Sol.3). The circle C can be described by

$$C = \{A + R(\omega t)X_0 \in \mathbf{R}^2 \mid t \in \mathbf{R}\},$$

where $\omega \in \mathbf{R}$ and $\omega \neq 0$. Thus, the circle C is the set of values of the map

$$t \mapsto X(t) = A + R(\omega t)X_0, \quad \text{with} \quad t \in \mathbf{R}. \qquad (\diamondsuit)$$

These functions are the solutions of the ordinary differential equation

$$\ddot{X}(t) + \omega^2 X(t) = \text{cst}. \qquad (\heartsuit)$$

We call the *trajectory* of the *curve* $[t \mapsto X(t)] \in \mathcal{C}^\infty(\mathbf{R}, \mathbf{R}^2)$ the set of its values, that is, $\text{traj}(X) = \{X(t) \mid t \in \mathbf{R}\}$. We must not confuse the trajectory $\text{traj}(X) \subset \mathbf{R}^2$ and the curve $X \in \mathcal{C}^\infty(\mathbf{R}, \mathbf{R}^2) \subset \mathbf{R} \times \mathbf{R}^2$. Precisely, $\text{traj}(X) = \text{pr}_2(\text{Gr}(X))$, where the graph $\text{Gr}(X)$ of X is equivalent to X. As well as the lines of Exercise 41, p. 39, are the trajectories of the solutions of the differential equation $\ddot{X}(t) = \text{cst}$, the circles are the trajectories of the solutions of the differential equation $(\heartsuit)$. Thus, an exercise about the structure of the set of circles could be the following.

Let $\text{Sol}(\heartsuit)$ *be the space of solutions of the ordinary differential equation* $(\heartsuit)$, *equipped with the functional diffeology induced by* $\mathcal{C}^\infty(\mathbf{R}, \mathbf{R}^2)$. *Show that the trajectories of the solutions are the circles, centered somewhere for some radius. Describe the spaces of circles, equipped with the quotient diffeology* $\text{Sol}(\heartsuit)/\text{traj}$, *where two solutions are identified by their trajectories.*

Exercise 43, p. 47 (Generating tori). The map $\pi : t \mapsto (\cos(t), \sin(t))$ from $\mathbf{R}$ to $S^1 \subset \mathbf{R}^2$ is a generating family for S^1. For $X = T_\alpha$ or $X = \mathbf{R}/\mathbf{Q}$, a generating family can be chosen to be the natural projections class : $\mathbf{R} \to X$.

Exercise 44, p. 47 (Global plots as generating families) Let $P : U \to X$ be an n-plot of X, with n a positive integer (for $n = 0$ there is nothing to prove). Let $r_0 \in U$. There exists $\varepsilon > 0$ such that the open ball $B(r_0, \varepsilon)$ is contained in

U. Now, it is a standard result that the ball $B(r_0, \varepsilon)$ is diffeomorphic to $\mathbf{R}^n$; let $\varphi : B(r_0, \varepsilon) \to \mathbf{R}^n$ be such a diffeomorphism. Thus, $\psi = (P \restriction B(r_0, \varepsilon)) \circ \varphi^{-1}$ is defined on $\mathbf{R}^n$ with values in X, and since ψ is the composite of a plot with a smooth map, it is a plot of X, a global plot. Then, $P \restriction B(r_0, \varepsilon) = \psi \circ \varphi$, where $\psi \in \mathcal{P}$ and φ is a smooth parametrization of $\mathbf{R}^n$. Thus, the diffeology of X is generated by its global plots.

Exercise 45, p. 47 (Generating the half-line). The pullback $j^*(\{1_{\mathbf{R}}\})$ is the set of parametrizations $F : U \to [0, \infty[$ such that $j \circ F$ is constant or there exist an element $F' \in \{1_{\mathbf{R}}\}$ and a smooth parametrization $\phi : U \to \mathrm{def}(F')$ with $j \circ F = F' \circ \phi$. Thus, $F' = 1_{\mathbf{R}}$, $\mathrm{def}(F') = \mathbf{R}$, and then $F = \phi$. Therefore, F is any smooth parametrization of $\mathbf{R}$ with values in $[0, \infty[$, and $j^*(\{1_{\mathbf{R}}\})$ is the whole diffeology of the half-line.

Exercise 46, p. 47 (Generating the sphere). Let $P : U \to S^n$ be a smooth parametrization, $r_0 \in U$, $x_0 = P(r_0)$, and $E_0 = x_0^{\perp}$. The real function $r \mapsto x_0 \cdot P(r)$ is smooth and satisfies $x_0 \cdot P(r_0) = 1$. There exists then a small open ball B_0, centered at r_0, such that for all $r \in B_0$, $x_0 \cdot P(r) > 0$. Thus, $P(B_0)$ is contained in the values of the map f_0, associated with the point r_0, of Exercise 18, p. 19. Now, let $S_0 : \mathcal{B} \to E_0$ be defined by $S_0(s_1, \ldots, s_n) = \sum_{i=1}^{n} s_i u_i$, then $F_0 = f_0 \circ S_0$. Since f_0 is an induction, $\phi = f_0^{-1} \circ (P \restriction B_0)$ is a smooth map from B_0 to E_0, and $\psi = S_0^{-1} \circ \phi$ a smooth parametrization of $\mathbf{R}^n$. But $\psi = S_0^{-1} \circ f_0^{-1} \circ (P \restriction B_0) = F_0^{-1} \circ (P \restriction B_0)$, thus $(P \restriction B_0) = F_0 \circ \psi$, where ψ is smooth. Therefore, the plots F are a generating family for the sphere S^n.

Exercise 47, p. 47 (When the intersection is empty). First of all, since $[x \mapsto x] \neq [x \mapsto 2x]$, $\mathcal{F} \cap \mathcal{F}' = \varnothing$. Thus, $\langle \mathcal{F} \cap \mathcal{F}' \rangle = \langle \varnothing \rangle = \mathcal{D}_\circ(\mathbf{R})$ (art. 1.67). Now, $\mathcal{F}$ is the family made up just with the identity of $\mathbf{R}$, so it generates the usual diffeology, $\langle \mathcal{F} \rangle = \mathcal{C}^\infty_\star(\mathbf{R})$. But, for any smooth parametrization P of $\mathbf{R}$, $2 \times P$ is smooth and $P = 2 \times Q$, with Q equal to the smooth parametrization $P/2$. Thus, $\langle \mathcal{F} \rangle = \langle \mathcal{F}' \rangle$ and $\langle \mathcal{F} \rangle \cap \langle \mathcal{F}' \rangle = \langle \mathcal{F} \rangle = \mathcal{C}^\infty_\star(\mathbf{R})$, and $\langle \mathcal{F} \cap \mathcal{F}' \rangle \neq \langle \mathcal{F} \rangle \cap \langle \mathcal{F}' \rangle$.

Exercise 48, p. 50 (Has the set $\{0, 1\}$ dimension 1?). Since $\{\pi\}$ is a generating family, the dimension of $\{0, 1\}_\pi$ is less or equal than 1, $\dim\{0, 1\}_\pi \leq 1$. Now, since the plot π is not locally constant — by density of the rational, or irrational, numbers in $\mathbf{R}$ — the space $\{0, 1\}_\pi$ is not discrete. Hence, $\dim\{0, 1\}_\pi \neq 0$ (art. 1.81), and then $\dim\{0, 1\}_\pi = 1$. This example shows how strongly the dimension of a diffeological space is related to its diffeology and not to some set theoretic considerations on the underlying set. A space consisting in a finite number of points can have an indiscrete diffeology. Remark that, in topology too, a finite set of points can be indiscrete.

Exercise 49, p. 50 (Dimension of tori). By definition, the projection $\pi : \mathbf{R} \to \mathbf{R}/\Gamma$ is a subduction (art. 1.46). But since $\mathbf{R}$ is a real domain, π is a plot of the quotient, and $\mathcal{F} = \{\pi\}$ is a generating family for $\mathbf{R}/\Gamma$, thus $\dim(\mathcal{F}) = 1$. Hence, as a direct consequence of the definition (art. 1.78) — or as a consequence of (art. 1.82), since $\dim(\mathbf{R}) = 1$ — $\dim(\mathbf{R}/\Gamma) \leq 1$. Now, if $\dim(\mathbf{R}/\Gamma) = 0$, then the diffeology of the quotient is generated by the constant parametrizations. Since the projection π is a plot, it lifts locally at the point 0 in the constant plot $0 : \mathbf{R} \to \{0\}$, but since $\mathbf{R}$ is pathwise connected, the lift is global (see Exercise 7, p. 14), and $\pi = [0] \circ 0$, where $[0] : \{0\} \to \mathbf{R}/\Gamma$ maps 0 to $[0] = \pi(0)$. Thus, since π is surjective, $\mathbf{R}/\Gamma = \{[0]\}$ and $\Gamma = \mathbf{R}$. Therefore, if $\Gamma \subset \mathbf{R}$ is a strict subgroup, that is, $\Gamma \neq \mathbf{R}$, we have necessarily

$\dim(\mathbf{R}/\Gamma) = 1$. In particular, this applies to the circle $S^1 \simeq \mathbf{R}/\mathbf{Z}$ (see Exercise 27, p. 27), or to the irrational tori $\mathbf{R}/\sum_{i=1\cdots N} \alpha_i \mathbf{Z}$, where the α_i are some numbers, independent over $\mathbf{Q}$; see Exercise 43, p. 47. It also applies to $\mathbf{R}/\mathbf{Q}$; see Exercise 5, p. 9. Thus, $\dim(S^1) = \dim(\mathbf{R}/\mathbf{Q}) = \dim(T_\alpha) = 1$.

☞ **Exercise 50, p. 50** (Dimension of $\mathbf{R}^n/O(n, \mathbf{R})$). Let $\Delta_n = \mathbf{R}^n/O(n, \mathbf{R})$, $n \in \mathbf{N}$, equipped with quotient diffeology.

1) Let us denote by $\pi_n : \mathbf{R}^n \to \Delta_n$ the projection from $\mathbf{R}^n$ onto its quotient. Since, by the very definition of $\hat{E}$ $O(n, \mathbf{R})$, $\|x'\| = \|x\|$ if and only if $x' = \hat{E}Ax$, with $A \in O(n, \mathbf{R})$, there exists a bijection $f : \Delta_n \to [0, \infty[$ such that $f \circ \pi_n = \nu_n$, where $\nu_n(x) = \|x\|^2$. Now, thanks to the uniqueness of quotients (art. 1.52), f is a diffeomorphism between Δ_n equipped with the quotient diffeology and $[0, \infty[$, equipped with the pushforward of the standard diffeology of $\mathbf{R}^n$ by the map ν_n. Now, let us denote by $\mathcal{D}_n$ the pushforward of the standard diffeology of $\mathbf{R}^n$ by ν_n. The space $([0, \infty[, \mathcal{D}_n)$ is a representation of Δ_n.

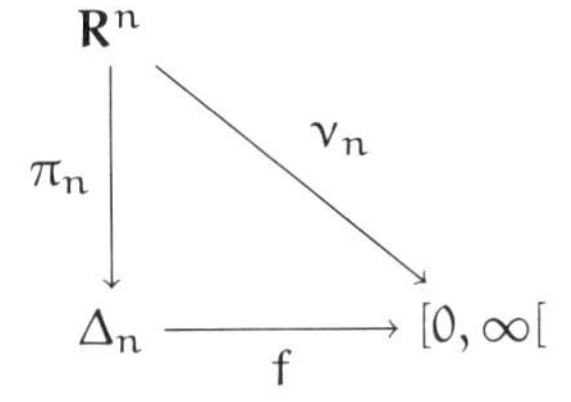

2) Let us denote by 0_k the zero of $\mathbf{R}^k$. Next, let us assume that the plot ν_n can be lifted at the point 0_n along a p-plot $P : U \to \Delta_n$, with $p < n$. Let $\phi : V \to U$ be a smooth parametrization such that $P \circ \phi = \nu_n \upharpoonright V$. We can assume without loss of generality that $P(0_p) = 0$ and $\phi(0_n) = 0_p$. If it is not the case, we compose P with a translation mapping $\phi(0_n)$ to 0_p. Now, since P is a plot of Δ_n, it can be lifted locally at the point 0_p along ν_n. Let $\psi : W \to \mathbf{R}^n$ be a smooth parametrization such that $0_p \in W$ and $\nu_n \circ \psi = P \upharpoonright W$. Let us introduce $V' = \phi^{-1}(W)$. We have then the following commutative diagram.

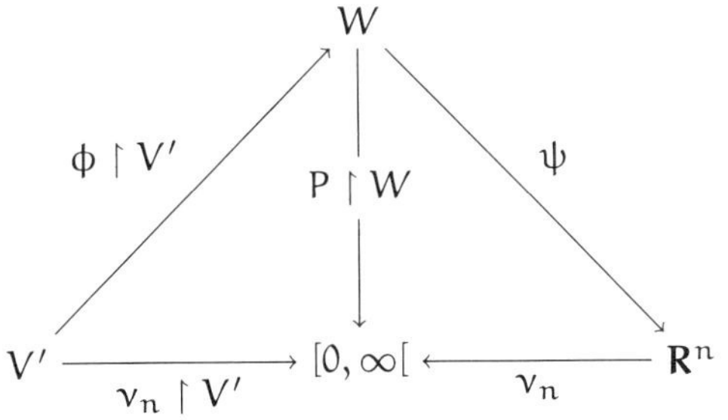

Now, denoting by $F = \psi \circ \phi \upharpoonright V'$, we get $\nu_n \upharpoonright V' = \nu_n \circ F$, with $F \in \mathcal{C}^\infty(V', \mathbf{R}^n)$, $0_n \in V'$ and $F(0_n) = 0_n$, that is,

$$\|x\|^2 = \|F(x)\|^2.$$

The derivative of this identity gives

$$x \cdot \delta x = F(x) \cdot D(F)(x)(\delta x), \text{ for all } x \in V' \text{ and for all } \delta x \in \mathbf{R}^n.$$

The second derivative, computed at the point 0_n, where F vanishes, gives then

$$1_n = M^t M \quad \text{with} \quad M = D(F)(0_n),$$

where M^t is the transposed matrix of M. But $D(F)(0_n) = D(\psi)(0_p) \circ D(\phi)(0_n)$. Let us denote $A = D(\psi)(0_p)$ and $B = D(\phi)(0_n)$, $A \in L(\mathbf{R}^p, \mathbf{R}^n)$ and $B \in L(\mathbf{R}^n, \mathbf{R}^p)$. Thus $M = AB$ and the previous identity $1_n = M^t M$ becomes $1_n = B^t A^t AB$. But the rank of B is less or equal to p which is, by hypothesis, strictly less than n, which would imply that the rank of 1_n is strictly less than n. And this is not true: the rank of 1_n is n. Therefore, the plot ν_n cannot be lifted locally at the point 0_n by a p-plot of Δ_n with $p < n$.

3) The diffeology of Δ_n, represented by $([0, \infty[, \mathcal{D}_n)$, is generated by ν_n. Hence, $\mathcal{F} = \{\nu_n\}$ is a generating family for Δ_n. Therefore, by definition of the dimension of diffeological spaces (art. 1.78), $\dim(\Delta_n) \leq n$. Let us assume that $\dim(\Delta_n) = p$ with $p < n$. Then, since ν_n is a plot of Δ_n it can be lifted locally, at the point 0_n, along an element P' of some generating family $\mathcal{F}'$ for Δ_n. The family $\mathcal{F}'$ satisfies $\dim(\mathcal{F}') = p$. But, by definition of the dimension of generating families (art. 1.77), we get $\dim(P') \leq p$, that is, $\dim(P') < n$. This is not possible, thanks to the second question. Therefore, $\dim(\Delta_n) = n$. Now, since the dimension is a diffeological invariant (art. 1.79), $\Delta_n = \mathbf{R}^n/O(n, \mathbf{R})$ is not diffeomorphic to $\Delta_m = \mathbf{R}^m/O(m, \mathbf{R})$ when $n \neq m$.

Exercise 51, p. 50 (Dimension of the half-line). First of all, let us remark that all the maps $\nu_n : \mathbf{R}^n \to \Delta_\infty$, defined by $\nu_n(x) = \|x\|^2$, are plots of Δ_∞. Indeed, these ν_n are smooth parametrizations of $\mathbf{R}$ and take their values in $[0, \infty[$. Now, let us assume that $\dim(\Delta_\infty) = N < \infty$. Hence for any integer n, the plot ν_n lifts locally at the point 0_n along some p-plot of Δ_∞, with $p \leq N$. Let us choose now $n > N$. Then, there exist a smooth parametrization $f : U \to \mathbf{R}$ such that $\mathrm{val}(f) \subset [0, \infty[$, that is, f is a p-plot of Δ_∞, and a smooth parametrization $\phi : V \to U$ such that $f \circ \phi = \nu_n \upharpoonright V$.

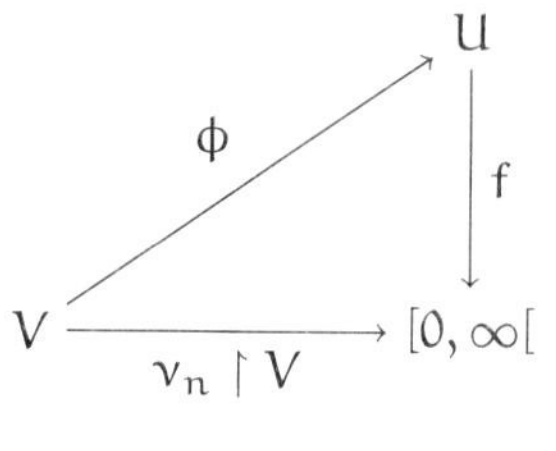

We can assume, without loss of generality, that $0_p \in U$, $\phi(0_n) = 0_p$, which implies $f(0_p) = 0$. Now, let us follow the method of Exercise 50, p. 50. The first derivative of ν_n at a point $x \in V' = \phi^{-1}(V)$ is given by

$$x = D(f)(\phi(x)) \circ D(\phi)(x).$$

Since f is smooth, positive, and $f(0) = 0$, we have in particular $D(f)(0_p) = 0$. Now, considering this property, the second derivative, computed at the point 0_n, gives, in matricial notation,

$$1_n = M^t H M, \text{ where } M = D(\phi)(0) \text{ and } H = D^2(f)(0),$$

where M^t is M transposed, and H is the Hessian of ϕ at the point 0_n, a symmetric bilinear map. The matrix M represents the tangent map of f at 0_p. Now, since we chose $n > N$ and assumed $p \leq N$, we have $p < n$. Thus the map M has a nonzero kernel and then $M^t H M$ is degenerate, which is impossible since 1_n is nondegenerate. Therefore, the dimension of Δ_∞ is unbounded, that is, infinite.

↺ **Exercise 52, p. 53** (To be a locally constant map). Let $\gamma \in \mathcal{C}^\infty(\mathbf{R}, X)$ with $\gamma(0) = x_0$ and $\gamma(1) = x_1$. For all $t \in [0, 1]$ there exists a superset V_t of $\gamma(t)$ such that $f \upharpoonright V_t$ is a local smooth map, according to the definition (art. 2.1). In particular $\mathcal{I}_t = \gamma^{-1}(V_t)$ is a 1-domain containing t and it satisfies $f(\gamma(\mathcal{I}_t)) = \mathrm{cst}$. The $\mathcal{I}_t$ are a covering of the segment $[0, 1]$ which is compact. Then, adapting the arguments of Exercise 7, p. 14, to $f \circ \gamma$, we get $f(x_0) = f(x_1)$. Therefore, f is constant on the connected components of X.

↺ **Exercise 53, p. 56** (Diffeomorphisms of the square). A diffeomorphism from the square must send corner into corner, because every smooth map into a corner must be flat (see Exercise 15, p. 17), which is not the case for the other points of the square. Thus, we can associate with every diffeomorphism φ of the square a permutation $\sigma = h(\varphi)$ of the set of four corners. The map h is clearly a homomorphism. But φ is a diffeomorphism which also permutes the edges of the square in a coherent way with σ, the image of connected edges to a corner must be connected to the image of this corner. Eventually, the image of h is the dihedral group with eight elements, generated by the rotation of angle $\pi/4$ and a reflection by an axis of symmetry.

↺ **Exercise 54, p. 56** (Smooth D-topology). Let $U \subset \mathbf{R}^n$ be a domain, that is, an ordinary open subset of $\mathbf{R}^n$. Let $A \subset U$ be open in U, that is, A open in $\mathbf{R}^n$. Let $P : V \to U$ be a plot of U, that is, any smooth parametrization. Since smooth parametrizations are continuous maps for the standard topology, the pullback $P^{-1}(A)$ is open. Then, any open set of U, for the usual topology, is D-open. Conversely, let $A \subset U$ be D-open, the identity map $\mathbf{1}_U$ being a plot of U, the subset $\mathbf{1}_U^{-1}(A) = A$ is open. Then, any D-open set of U is open for the usual topology. Thus, the standard topology and the D-topology of smooth domains coincide.

↺ **Exercise 55, p. 56** (D-topology of irrational tori). Let $\pi : \mathbf{R} \to T_\Gamma$ be the natural projection. The set T_Γ is equipped with the quotient diffeology (art. 1.50). Let $A \subset T_\Gamma$ be a nonempty D-open. Since the projection π is smooth, it is D-continuous (art. 2.9). Thus, $\pi^{-1}(A)$ is a D-open in $\mathbf{R}$, that is, $\pi^{-1}(A)$ is a domain (Exercise 54, p. 56). Let $\tau \in A$ and $x \in \mathbf{R}$ such that $\pi(x) = \tau$. So, $\pi^{-1}(A)$ contains x and its whole orbit by the action of Γ. Let us denote by $\mathcal{O}$ this orbit, thus $\pi^{-1}(A)$ is an open neighborhood of $\mathcal{O}$. But Γ being dense in $\mathbf{R}$, the orbit $\mathcal{O}$ also is dense, and $\pi^{-1}(A)$ is an open neighborhood of a dense subset of $\mathbf{R}$. Therefore, $\pi^{-1}(A) = \mathbf{R}$ and $A = T_\Gamma$. Therefore, the only nonempty D-open set of T_Γ is T_Γ itself. The D-topology of T_Γ is coarse. Now, a full functor is a functor surjective on the arrows [**McL71**]. Since the D-topology of T_Γ is coarse, any map from T_Γ to T_Γ is D-continuous. But we know by Exercise 5, p. 9, that all maps from $T_{\mathbf{Q}}$ to $T_{\mathbf{Q}}$ are not smooth, just the affine ones. Hence the D-topology functor is not surjective on the arrows, that is, not full.

↺ **Exercise 56, p. 57** ($\mathbf{Q}$ is discrete but not embedded in $\mathbf{R}$). Let us recall that $\mathbf{Q}$ is discrete in $\mathbf{R}$ (Exercise 8, p. 14), that is, the subset diffeology is discrete. The D-topology of $\mathbf{R}$ is the smooth topology Exercise 54, p. 56 and, since $\mathbf{Q} \subset \mathbf{R}$ is discrete, the D-topology of $\mathbf{Q}$ is discrete (art. 2.11). But, since any nonempty open set of the topology induced by $\mathbf{R}$ on $\mathbf{Q}$ contains always an infinite number of points (it is generated by the intersections of open intervals and $\mathbf{Q}$), the induced D-topology is not discrete. Then, $\mathbf{Q}$ is not embedded in $\mathbf{R}$.

☞ **Exercise 57, p. 57** (Embedding $\mathrm{GL}(n, \mathbf{R})$ in $\mathrm{Diff}(\mathbf{R}^n)$). Let us recall that the plots of the functional diffeology of $\mathrm{Diff}(\mathbf{R}^n)$ (art. 1.61) are the parametrizations $P : U \to \mathrm{Diff}(\mathbf{R}^n)$ such that

$$[(r, x) \mapsto P(r)(x)] \text{ and } [(r, x) \mapsto P(r)^{-1}(x)] \text{ belong to } \mathcal{C}^\infty(U \times \mathbf{R}^n, \mathbf{R}^n).$$

1) The diffeology of $\mathrm{GL}(n, \mathbf{R})$ induced by $\mathrm{Diff}(\mathbf{R}^n)$ coincides with the ordinary diffeology.

The plots of the standard diffeology of $\mathrm{GL}(n, \mathbf{R})$ are the parametrizations $P : U \to \mathrm{GL}(n, \mathbf{R})$ such that every component P_{ij} is smooth, that is,

$$P_{ij} : r \mapsto \langle \mathbf{e}_j \mid P(r)(\mathbf{e}_i) \rangle \in \mathcal{C}^\infty(U, \mathbf{R}), \text{ for all } i, j = 1, \ldots, n,$$

where we have denoted by $\mathbf{e}_1, \ldots, \mathbf{e}_n$ the vectors of the canonical basis of $\mathbf{R}^n$, and by $\langle \cdot \mid \cdot \rangle$ the ordinary scalar product of $\mathbf{R}^n$. Now, for each $(r, x) \in U \times \mathbf{R}^n$, $P(r)(x) = P(r)(\sum_{i=1}^n x^i \mathbf{e}_i) = \sum_{i=1}^n x^i P(r)(\mathbf{e}_i) = \sum_{i=1}^n x^i P_{ij}(r) \mathbf{e}_j$. If all the components P_{ij} of the parametrization P are smooth, then $(r, x) \mapsto \sum_{i=1}^n x^i P_{ij}(r) \mathbf{e}_j$ is smooth, and the map $(r, x) \mapsto P(r)(x)$ is smooth. Since the determinant of $P(r)$ never vanishes, the same holds for $(r, x) \mapsto P(r)^{-1}(x)$. Therefore, P is a plot of the functional diffeology. Conversely, if P is a plot of the functional diffeology — that is, the parametrization $(r, x) \mapsto P(r)(x)$ is smooth — then, restricting this map to $x = \mathbf{e}_i$, we get that the map $r \mapsto P(r)(\mathbf{e}_i) = \sum_{j=1}^n P_{ij}(r) \mathbf{e}_i$ is smooth. So, by contracting this parametrization to the vector $\mathbf{e}_j$, we get that all the matrix components P_{ij} are smooth. Thus, the inclusion $\mathrm{GL}(n, \mathbf{R}) \hookrightarrow \mathrm{Diff}(\mathbf{R}^n)$ is an induction.

2) The inclusion $\mathrm{GL}(n, \mathbf{R}) \hookrightarrow \mathrm{Diff}(\mathbf{R}^n)$ is an embedding.

We have to show that the topology of $\mathrm{GL}(n, \mathbf{R})$ induced by the D-topology of $\mathrm{Diff}(\mathbf{R}^n)$ coincides with the D-topology of $\mathrm{GL}(n, \mathbf{R})$, that is, the topology induced by its inclusion into $\mathbf{R}^{n \times n}$. Let $B(\mathbf{1}_n, \varepsilon)$ be the open ball in $\mathrm{GL}(n, \mathbf{R})$ centered at the identity $\mathbf{1}_n$, with radius ε. Let Ω_ε be the set of all diffeomorphisms defined by

$$\Omega_\varepsilon = \{f \in \mathrm{Diff}(\mathbf{R}^n) \mid D(f)(0) \in B(\mathbf{1}_n, \varepsilon)\},$$

where $D(f)(0)$ is the tangent linear map of f at the point 0. Now, let us prove the following.

a) The set Ω_ε is open for the D-topology of $\mathrm{Diff}(\mathbf{R}^n)$.

Let $P : U \to \mathrm{Diff}(\mathbf{R}^n)$ be a plot, that is, $[(r, x) \mapsto P(r)(x)] \in \mathcal{C}^\infty(U \times \mathbf{R}^n, \mathbf{R}^n)$. The pullback of Ω_ε by P is the set of $r \in U$ such that the tangent map $D(P(r))(0)$ is in the ball $B(\mathbf{1}_n, \varepsilon)$, formally,

$$P^{-1}(\Omega_\varepsilon) = \{r \in U \mid D(P(r))(0) \in B(\mathbf{1}_n, \varepsilon)\}.$$

Considering P as a smooth map defined on $U \times \mathbf{R}^n$, $D(P(r))(0)$ is the partial derivative of P, with respect to the second variable, computed at the point $x = 0$. The map $[r \mapsto D(P(r))(0)]$ is then continuous, by definition of smoothness. Hence, the pullback of Ω_ε by this map is open. Because the imprint of this open set on $\mathrm{GL}(n, \mathbf{R})$ is exactly the ball $B(\mathbf{1}_n, \varepsilon)$, we deduce that any open ball of $\mathrm{GL}(n, \mathbf{R})$ centered at $\mathbf{1}_n$ is the imprint of a D-open set of $\mathrm{Diff}(\mathbf{R}^n)$.

b) Every open of $\mathrm{GL}(n, \mathbf{R})$ is the imprint of a D-open set of $\mathrm{Diff}(\mathbf{R}^n)$.

By using the group operation on $\mathrm{GL}(n, \mathbf{R})$ and since any open set of $\mathrm{GL}(n, \mathbf{R})$ is a union of open balls, every open subset of $\mathrm{GL}(n, \mathbf{R})$ is the imprint of some D-open subset of $\mathrm{Diff}(\mathbf{R}^n)$. Therefore, $\mathrm{GL}(n, \mathbf{R})$ is embedded in $\mathrm{Diff}(\mathbf{R}^n)$.

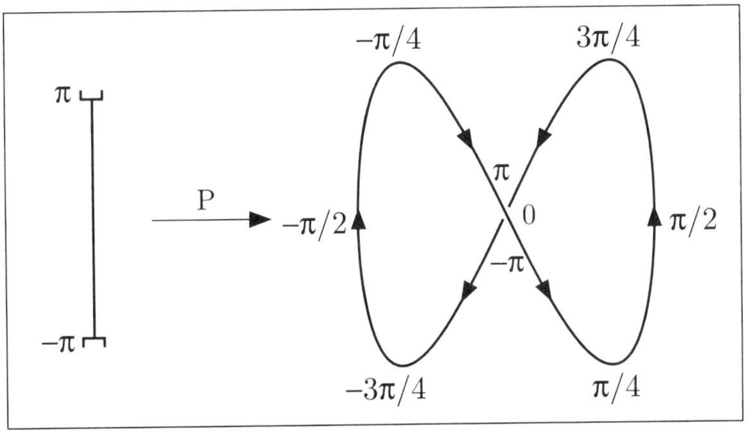

FIGURE Sol.4. The plot P.

☞ **Exercise 58, p. 57** (The irrational solenoid is not embedded). The solenoid is the subgroup $[\alpha] = \{(1, e^{i2\pi\alpha t}) \mid t \in \mathbf{R}\} \subset T^2$, with $\alpha \in \mathbf{R} - \mathbf{Q}$. It is the image of the induction $j : t \mapsto (1, e^{i2\pi\alpha t})$, from $\mathbf{R}$ to T^2, Exercise 31, p. 31. Because $[\alpha]$ is dense in T^2, the pullback of any open disc of T^2 by j is an infinite disjoint union of intervals of $\mathbf{R}$. Thus, an open interval $]a, b[\subset \mathbf{R}$ cannot be the preimage of an open subset of T^2. Therefore, the solenoid is not embedded in T^2.

☞ **Exercise 59, p. 58** (The infinite symbol). 1) If $j(t) = j(t')$, $t \neq t'$, and $t, t' \in]-\pi, \pi[$, then $t' - t = \pm\pi/2$ and $t' - t = \pm\pi/4$. Thus, $t = t'$ and thus the map j is injective.

2) The drawing of P (Figure Sol.4) clearly shows that j and P have the same image in $\mathbf{R}^2$. But the precise reason is given by (♣) in 3).

3) Comparing the figure of j and the figure of P, we see clearly that

$$j^{-1} \circ P(]-\pi/4, \pi/4[) =]-\pi, -3\pi/4[\cup \{0\} \cup]3\pi/4, \pi[.$$

The map $j^{-1} \circ P$ has a continuity gap at $t = 0$, so $j^{-1} \circ P$ is not continuous, *a fortiori* not smooth. But, we have precisely:

$$j^{-1} \circ P(t) = \begin{cases} -t - \pi & t \in]-\pi/4, 0[\\ 0 & t = 0 \\ -t + \pi & t \in]0, \pi/4[. \end{cases} \qquad (♣)$$

Hence, P is a plot of $\mathbf{R}^2$ with values in $j(]-\pi, \pi[)$, but $j^{-1} \circ P$ is not smooth. Therefore, by application of (art. 1.31), the injection j is not an induction.

4) The map j is an immersion, and its derivative never vanishes on $]-\pi, \pi[$. Thus, as an application of Exercise 12, p. 17, it is a local induction everywhere.

☞ **Exercise 60, p. 60** (Quotient by a group of diffeomorphisms). 1) Let $P : U \to \mathcal{Q}$ be a plot. By definition of the quotient diffeology of G/π, for all $r_0 \in U$ there exist an open neighborhood V of r_0 and a plot $\gamma : V \to G$ such that $P(r) = \gamma(r)(x)$ for all $r \in V$. Hence, $P \upharpoonright V = [r \mapsto (r, x) \mapsto \gamma(r)(x)]$, but $[r \mapsto (r, x)]$ is clearly smooth, and $[(r, x) \mapsto \gamma(r)(x)]$ is smooth by the very definition of the functional diffeology. Thus, $P \upharpoonright V$ is a plot of the subset diffeology. Therefore j is smooth. The quotient diffeology of $G(x)$ is finer than its subset diffeology.

2) Let $P : U \to \mathcal{Q}$ be a plot, and let $r \in U$ and $g \in G$ such that $\pi(g) = P(r)$, that is, $g(x) = P(r)$. By definition of the quotient diffeology there exist an open neighborhood V of r and a plot $\gamma : V \to G$ such that $P \restriction V = \pi \circ \gamma$. Let $g' = \gamma(r)$, then $\pi(g') = \pi(g) = P(r)$, that is, $g(x) = g'(x)$ or $g'^{-1}(g(x)) = x$. Now, let us define on V, $\gamma' = [s \mapsto \gamma(s) \circ g'^{-1} \circ g]$. On the one hand, we have $\gamma'(s)(x) = \gamma(s)(g'^{-1}(g(x)))$, but $g'^{-1}(g(x)) = x$, thus $\gamma'(s)(x) = \gamma(s)(x)$, that is, $\pi \circ \gamma' = \pi \circ \gamma = P \restriction V$. On the other hand we have $\gamma'(r) = \gamma(r) \circ g'^{-1} \circ g$, but $g' = \gamma(r)$, thus $\gamma'(r) = g$. Since, by definition of the functional diffeology of G (art. 1.61), composition and inversion are smooth, the parametrization γ' is a plot of G. Moreover, γ' satisfies the conditions $P \restriction V = \pi \circ \gamma'$ and $\gamma'(r) = g$. Therefore π is a local subduction.

3) By definition of generating families (art. 1.66), a plot P of the Tahar rug $\mathcal{T}$ writes locally $[r \mapsto (t(r), c)]$ or $[r \mapsto (c, t(r))]$, where c is some constant and t is a smooth real function. Now, let $u = (a, b) \in \mathbf{R}^2$, the composition $T_u \circ P$ writes locally either $[r \mapsto (t'(r), c')]$, with $t'(r) = t(r) + a$ and $c' = c + b$, or $[r \mapsto (c', t'(r))]$, with $t'(r) = t(r) + b$ and $c' = c + a$. Thus, T_u is smooth. Then, since $(T_u)^{-1} = T_{-u}$, T_u is a diffeomorphism of $\mathcal{T}$. Now, let $r \mapsto u(r) = (a(r), b(r))$ be a parametrization of $\mathbf{R}^2$ such that $r \mapsto T_{u(r)}$ is a plot for the functional diffeology. Composed with the 1-plots $t \mapsto (t, c)$, where c runs over $\mathbf{R}$, we must get a plot $(r, t) \mapsto (t + a(r), c + b(r))$ of $\mathcal{T}$, that is, a plot which is locally of the first or the second kind. Hence, either $(r, t) \mapsto t + a(r)$ is locally constant, or $(r, t) \mapsto c + b(r)$. But $(r, t) \mapsto t + a(r)$ is not locally constant because of its dependency on t, thus $(r, t) \mapsto c + b(r)$ is locally constant, that is, $b(r) =_{\text{loc}} b$, for some $b \in \mathbf{R}$. In the same way, composing with the 1-plots $t \mapsto (c, t)$, we get that $a(r) =_{\text{loc}} a$. Therefore, $r \mapsto T_{u(r)}$ is locally constant, that is, the group of translations, equipped with the functional diffeology, is discrete. Finally, the action of the translations on $\mathcal{T}$ is free, the orbit of $(0, 0)$ is $\mathcal{T}$, and since the diffeology of $\mathcal{T}$ is not the discrete diffeology, we get an example, for the first question, where the diffeology of $\mathcal{Q}$ is strictly finer than the one of $\mathcal{O}$.

☞ **Exercise 61, p. 60** (A not so strong subduction). As an application of (art. 1.52), the underlying set of $\mathcal{Q}$ can be represented by the half-line $[0, \infty[$ equipped with the image of the diffeology of $\mathbf{R} \coprod \mathbf{R}^2$ by the map $p : x \mapsto \|x\|^2$ see Figure Sol.5. Let 0 and $(0, 0)$ be the zeros of $\mathbf{R}$ and $\mathbf{R}^2$, and let P be the plot $p \restriction \mathbf{R}^2$. Then, $P(0, 0) = 0 \in [0, \infty[$. Let us assume now that P lifts locally at $(0, 0)$ along p, by a plot f such that $f(0, 0) = 0$. Thus, f takes its values in $\mathbf{R}$ and $p \circ f = P$, that is, $f(a, b)^2 = a^2 + b^2$, at least locally. Since f is continuous and vanishes only at $(0, 0)$, and since the complementary of $(0, 0)$ in $\mathbf{R}^2$ is connected, f keeps a constant sign, thus $f(a, b) = \pm\sqrt{a^2 + b^2}$. But none of these two cases is smooth at $(0, 0)$. Therefore, p is not a local subduction at the point 0.

☞ **Exercise 62, p. 60** (A powerset diffeology). 1) Let us check that the parametrizations defined by (♣) are a diffeology.

D1. Let $P : U \to \mathfrak{P}(X)$ be the constant parametrization $r \mapsto A \subset X$. Let $r_0 \in U$, and let $Q_0 \in \mathcal{D}$ such that $\mathrm{val}(Q_0) \subset A = P(r_0) = P(r)$ for all $r \in U$. Let $Q : U \to \mathcal{D}$ given by $Q(r) = Q_0$, for all $r \in \mathcal{D}$. This is a constant family of plots of X, hence smooth. Thus, P satisfies the condition (♣). Hence, the constant parametrizations satisfy (♣).

D2. The locality axiom is satisfied by construction: (♣) is a local property.

D3. Let us consider a parametrization $P : U \to \mathfrak{P}(X)$ satisfying (♣). Let $F : U' \to U$ be a smooth parametrization. Let $P' = P \circ F$. Let $r'_0 \in U'$ and $r_0 = F(r'_0)$. By

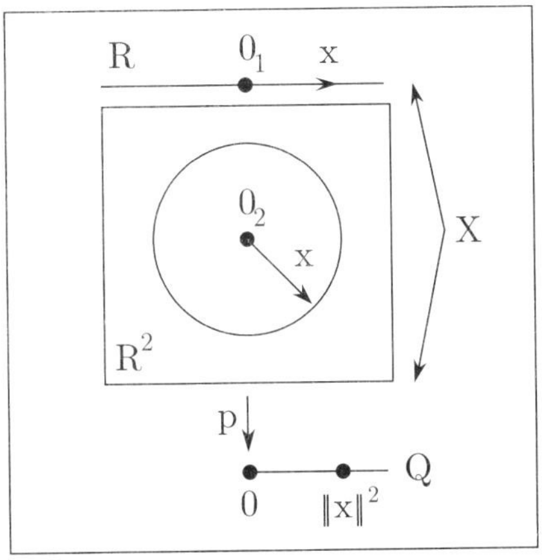

FIGURE Sol.5. The quotient $\mathcal{Q}$.

hypothesis, for every $Q_0 \in \mathcal{D}$ such that $\mathrm{val}(Q_0) \subset P(r_0) = P'(r'_0)$, there exist an open neighborhood V of r_0 and a smooth family of plots $Q : V \to \mathcal{D}$ such that $Q(r_0) = Q_0$ and $\mathrm{val}(Q(r)) \subset P(r)$, for all $r \in V$. Let us then define $V' = F^{-1}(V)$ and $Q' = Q \circ F$. Since F is smooth, V' is a domain, and since Q is a smooth family of plots, so is $Q \circ F$. Thus, for every $r'_0 \in U'$, for every $Q_0 \in \mathcal{D}$ such that $\mathrm{val}(Q_0) \subset P'(r'_0)$, we found a smooth family of plots Q' such that $Q'(r'_0) = Q(r_0) = Q_0$ and $\mathrm{val}(Q'(r')) = \mathrm{val}(Q \circ F(r')) = \mathrm{val}(Q(r)) \subset P(r) = P'(r')$, where $r = F(r')$. Therefore, $P' = P \circ F$ satisfies (♣).

2) Consider now the relation $\mathcal{R}$ from $\mathfrak{P}(X)$ to $\mathcal{D}$ defined by the inclusion

$$\mathcal{R} = \{(A, Q) \in \mathfrak{P}(X) \times \mathcal{D} \mid \mathrm{val}(Q) \subset A\}.$$

Let $P : U \to \mathfrak{P}(X)$ be a parametrization regarded as the relation

$$P = \{(r, A) \in U \times \mathfrak{P}(X) \mid A = P(r)\}.$$

The composite $\mathcal{R} \circ P$, also denoted by $P^*(\mathcal{R})$, is then given by

$$P^*(\mathcal{R}) = \{(r, Q) \in U \times \mathcal{D} \mid \mathrm{val}(Q) \subset P(r)\}.$$

The parametrization P is a plot of the powerset diffeology if and only if the first projection $\mathrm{pr}_1 : P^*(\mathcal{R}) \to U$ is everywhere a local subduction. Now, if $f : X \to Y$ is a map between two diffeological spaces, f is smooth if and only if the first projection from $f = \{(x, y) \in X \times Y \mid y = f(x)\}$ to X is a subduction, which is equivalent to being a local subduction, in this case. Note that this construction gives us an idea of the difference between a *smooth relation* and a diffeological space X to another X'. Indeed, let $\mathcal{R} \subset X \times X'$ be a relation from X to X', we declare $\mathcal{R}$ smooth if for every plot P in $\mathrm{def}(\mathcal{R}) = \mathrm{pr}_1(\mathcal{R}) \subset X$, for every $r \in U$ and every $(P(r), x') \in \mathcal{R}$, there exists a plot $Q : V \to X'$, with $V \subset U$, such that $(P(r'), Q(r')) \in \mathcal{R}$ for all $r' \in V$ and $Q(r) = x'$. In other words, $\mathrm{pr}_1 : \mathcal{R} \to \mathrm{def}(\mathcal{R})$ is a local subduction everywhere, where $\mathcal{R}$ and $\mathrm{def}(\mathcal{R})$ are equipped with the subset diffeology. With this terminology, back to the powerset diffeology, a parametrization $P : U \to \mathfrak{P}(X)$ is a plot if the composite $\mathcal{R} \circ P$ is a smooth relation from U to $\mathcal{D}$.

3) Let us check now that the map $j : x \mapsto \{x\}$ is an induction. We consider the criterion for (art. 1.31). First of all let us remark that j is injective. Next, let us

check that the map j is smooth. Let $P : U \to X$ be a plot. Thus, $j \circ P(r) = \{P(r)\}$. Let $r_0 \in U$ and $Q_0 \in \mathcal{D}$ such that $\mathrm{val}(Q_0) \subset j \circ P(r_0) = \{P(r_0)\}$. So, $\mathrm{val}(Q_0)$ is the point $P(r_0)$ of X, that is, Q_0 is the constant plot $s \mapsto P(r_0)$. Let us then define, for every $r \in U$, $Q(r)$ as the constant plot $[s \mapsto P(r)]$, with $\mathrm{def}(Q(r)) = \mathrm{def}(Q_0)$. Since, for every $r \in U$, the parametrization $[(r,s) \mapsto Q(r)(s) = P(r)]$, defined on $U \times \mathrm{def}(Q_0)$, is clearly a plot of X, then Q is a smooth family of plots of X. Therefore, $j \circ P$ is a plot of $\mathfrak{P}(X)$, and j is smooth. Then, let us check that the map $j^{-1} : j(X) \to X$ is smooth. Let $P : U \to j(X)$ be a plot of the powerset diffeology. First of all, for every $r \in U$ there exists a unique point $q(r) \in X$ such that $P(r) = \{q(r)\}$. Then, since P is a plot of $\mathfrak{P}(X)$, for all $r_0 \in U$, for every plot Q_0 of X such that $\mathrm{val}(Q_0) \subset P(r_0) = \{q(r_0)\}$, there exist an open neighborhood V of r_0, and a smooth family of plots Q of X, such that $\mathrm{val}(Q(r)) \subset P(r)$, for all $r \in V$. Then, let us choose the 0-plot $Q_0 : \mathbf{R}^0 \to X$, with $Q_0(0) = q(r_0)$, that is, $\mathrm{val}(Q_0) = P(r_0)$. Thus, Q is necessarily a smooth family of 0-plots (see (art. 1.63)). But $\mathrm{val}(Q(r)) \subset P(r) = \{q(r)\}$ means exactly that $Q(r)(0) = q(r)$. Hence, $q \restriction V = [r \mapsto Q(r)(0)]$ is a plot of X. Therefore, $q = j^{-1} \circ P$ is a plot of X, and j^{-1} is smooth. Finally, thanks to the criterion (art. 1.31), j is an induction from X into $\mathfrak{P}(X)$.

4) Let us show, now, that the Tzim-Tzum $\mathcal{T}$ is a plot of the powerset diffeology of $\mathfrak{P}(\mathbf{R}^2)$. Let $t_0 \in \mathbf{R}$ and consider $\mathcal{T}_{t_0}$. Let $Q_0 : U \to \mathcal{T}_{t_0}$ be a plot. If $t_0 < 0$, then we can choose $Q(t)(r) = Q_0(r)$ for $t \in]3t_0/2, t_0/2[$. Q is a smooth family of plots of $\mathbf{R}^2$ such that $Q(t_0) = Q_0$ and $\mathrm{val}(Q(t)) \subset \mathcal{T}(t) = \mathbf{R}^2$. For $t_0 = 0$, we choose $Q(t)(r) = (e^t + t/\|Q_0(r)\|)Q_0(r)$. Since $\mathcal{T}(0) = \mathbf{R}^2 - \{0\}$, $Q_0(r) \neq 0$ for all r, $Q(t)$ is well defined and Q is a smooth family of plots of $\mathbf{R}^2$. Next, note first that $Q(0) = Q_0$. Then, for $t \geq 0$, $\|Q(t)(r)\| = t + e^t\|Q_0(r)\| > t$, since $\|Q_0(r)\| > 0$, thus $\mathrm{val}(Q(t)) \subset \mathcal{T}(t)$. For $t < 0$ there is nothing to check since $\mathcal{T}(t) = \mathbf{R}^2$. Now, if $t_0 > 0$, we can choose $Q(t)(r) = [1 + (t - t_0)/t_0]Q_0(r)$. We have $Q(t_0) = Q_0$, then $\|Q(t)(r)\| = |1 + (t-t_0)/t_0|\|Q_0(r)\|$. But, $\|Q_0(r)\| > t_0$ implies $\|Q(t)(r)\| > |1 + (t-t_0)/t_0|t_0 = |t|$. Thus, if $t \geq 0$, then $\|Q(t)(r)\| > t$, for all r, that is, $\mathrm{val}(Q(t)) \subset \mathcal{T}(t)$. We exhausted all the cases, and therefore $\mathcal{T}$ is a plot of the powerset diffeology of $\mathfrak{P}(\mathbf{R}^2)$. As we can see, there is a blowing up for $t = 0$, the space opens up, and an empty bubble appears and grows with it. This is the reason for which we named this plot Tzim-Tzum.

Exercise 63, p. 61 (The powerset diffeology of the set of lines). First of all let us note that, given a line $\mathbf{D} \in \mathrm{Lines}(\mathbf{R}^n)$, the solution of the equation $j(u,x) = \mathbf{D}$, with $(u,x) \in \mathrm{TS}^{n-1}$, has exactly two solutions:

$$u = \pm\frac{r - r'}{\|r - r'\|} \quad \text{and} \quad x = [\mathbf{1} - u\bar{u}]r,$$

where r and r' are any two different points of $\mathbf{D}$ and $[\mathbf{1} - u\bar{u}]$ is the projector orthogonal to u. By definition, $\bar{u}(v) = u \cdot v$, then $[\mathbf{1} - u\bar{u}](v) = v - (u \cdot v)u$. Let us prove now that the map j is smooth. Let $P : s \mapsto (u(s), x(s))$ be a plot of TS^{n-1}, defined on some domain U, that is, P is a smooth parametrization of $\mathbf{R}^n \times \mathbf{R}^n$ with values in TS^{n-1}. Let $P' = j \circ P : s \mapsto x(s) + \mathbf{R}u(s)$. We want to show that P' is a plot of $\mathfrak{P}(\mathbf{R}^n)$. Let then $s_0 \in U$, $u_0 = u(s_0)$, $x_0 = x(s_0)$, $\mathbf{D}_0 = P'(s_0) = j(u_0, x_0)$ and $Q_0 : W \to \mathbf{D}_0$ be some smooth parametrization in $\mathbf{D}_0$. Since Q_0 is a plot of $\mathbf{D}_0$, for any $w \in W$, $Q_0(w) - x_0$ is proportional to u_0, that is, $Q_0(w) - x_0 = \tau u_0$. But since $u_0 \cdot x_0 = 0$, $Q_0(w) - x_0 = \tau u_0$ implies $\tau = u_0 \cdot (Q_0(w) - x_0) = u_0 \cdot Q_0(w)$. Hence, defining $\tau(w) = u_0 \cdot Q_0(w)$, we get $Q_0(w) = x_0 + \tau(w)u_0$, where $\tau \in \mathcal{C}^\infty(W, \mathbf{R})$.

Let us define now

$$Q = [s \mapsto [w \mapsto x(s) + \tau(w)u(s)]], \quad \text{where} \quad s \in U \quad \text{and} \quad w \in W.$$

Since x, u and τ are smooth, Q is a plot of the diffeology of $\mathbf{R}^n$ and satisfies $Q(s_0) = Q_0$. Hence, P' is a plot of $\mathfrak{P}(\mathbf{R}^n)$. Therefore, j is smooth. Let us now prove that j is a subduction onto its image, that is, onto the space Lines($\mathbf{R}^n$). Let $P : U \to \text{Lines}(\mathbf{R}^n)$ be a plot and $s_0 \in U$. Since $P(s_0)$ is a line of $\mathbf{R}^n$, there exists $(u_0, x_0) \in TS^{n-1}$ such that $P(s_0) = x_0 + \mathbf{R}u_0$. Let $Q_0 = [t \mapsto x_0 + tu_0]$, with $t \in \mathbf{R}$, Q_0 is a plot of $\mathbf{R}^n$ such that $\text{val}(Q_0) \subset P(s_0)$. Hence, since P is a plot for the powerset diffeology of $\mathbf{R}^n$, there exist an open neighborhood V of s_0, and a plot Q of the smooth diffeology of $\mathbf{R}^n$, such that $Q(s_0) = [t \mapsto x_0 + tu_0]$ and $\text{val}(Q(s)) \subset P(s)$. Let us choose $t = 0 \in \text{def}(Q(s_0))$, since Q is a plot of the smooth diffeology of $\mathbf{R}^n$, there exists an open neighborhood W of s_0 and there exists $\varepsilon > 0$ such that $(t, s) \mapsto Q(s)(t)$, defined on $W \times]-\varepsilon, +\varepsilon[$, is smooth. Let us then define, for all $s \in W$,

$$v(s) = \left.\frac{\partial Q(s)(t)}{\partial t}\right|_{t=0}.$$

Since Q is smooth, the parametrization v is smooth. We have $v(s_0) = u_0 \neq 0$. Thus, there exists an open neighborhood W' of s_0 on which v does not vanish. Therefore, the map

$$u : s \mapsto \frac{v(s)}{\|v(s)\|}$$

is smooth on W'. Moreover, by construction, $u(s)$ directs the line $P(s)$. Now, let

$$x(s) = Q(s)(0) - [u(s) \cdot Q(s)(0)]u(s).$$

Since $Q(s)(0) \in P(s)$ and $u(s)$ directs $P(s)$, the point $x(s)$ belongs to $P(s)$, and by construction $u(s) \cdot x(s) = 0$. So, the parametrization of TS^{n-1} defined by $\phi : s \mapsto (u(s), x(s))$ is smooth and satisfies $j \circ \phi = P$. Hence, ϕ is a local lift of P along j, defined on W'. Combined with the surjectivity and the differentiability of j, this is the criterion for j to be a subduction (art. 1.48) from TS^{n-1} onto its image Lines($\mathbf{R}^n$). Therefore, the set of lines is diffeomorphic to the quotient $TS^{n-1}/\{\pm 1\}$, where -1 acts by reversing the orientation, that is, $\pm(u, x) = (\pm u, x)$.

☞ **Exercise 64, p. 64** (The diffeomorphisms of the half-line). Let us prove first that any diffeomorphism f of $\Delta_\infty = [0, \infty[\subset \mathbf{R}$ satisfies the three points.

1) Since the dimension map is invariant under diffeomorphism (art. 2.24) and since the origin is the only point where the dimension is infinite, as shown in Exercise 51, p. 50, f fixes the origin, $f(0) = 0$.

2) Since $f(0) = 0$ and f is a bijection, we have $f(]0, \infty[) =]0, \infty[$. Now, since the restriction of a diffeomorphism to any subset is a diffeomorphism of this subset onto its image, for the subset diffeology (art. 1.33), we have $f \restriction]0, \infty[\in \text{Diff}(]0, \infty[)$. Let us recall that the induced diffeology on the open interval is the standard diffeology (art. 1.9). Moreover, since $f(0) = 0$, restricted to $]0, \infty[$, f is necessarily strictly increasing.

3) Since f is smooth, by the very definition of differentiability (art. 1.14), for any smooth parametrization P of the interval $[0, \infty[$, the composite $f \circ P$ is smooth, in particular for $P = [t \mapsto t^2]$. Hence, the map $\varphi : t \mapsto f(t^2)$ defined on $\mathbf{R}$ with values in $[0, \infty[$ is smooth. Now, by theorem 1 of [**Whi43**], since φ is smooth, f can be extended to an open neighborhood of $[0, \infty[$ by a smooth function. Hence, f

is continuously differentiable at the origin. Moreover since f is a diffeomorphism of Δ_∞, what has been said for f can be said for f^{-1}. And, since $(f^{-1})'(0) = 1/f'(0)$ and f is increasing, we have $f'(0) > 0$. Conversely, a function f satisfying the three above conditions can be extended by a smooth function to some open neighborhood of $[0, \infty[$. Hence, f is the restriction to $[0, \infty[$ of a smooth function $\tilde{f}$ such that $\tilde{f}(0) = 0$, $\tilde{f}$ is strictly increasing on $[0, \infty[$, and $\tilde{f}'(0) > 0$, then f is smooth for the subset diffeology as well as its inverse, that is, f is a diffeomorphism of Δ_∞.

Exercise 65, p. 66 (Vector space of maps into $\mathbf{K}^n$). Let us recall that a parametrization $P : U \to E = \mathcal{C}^\infty(X, \mathbf{K})$ is a plot for the functional diffeology if and only if, for every plot $Q : V \to X$, the parametrization $[(r, s) \mapsto P(r)(Q(s))]$ is a plot of $\mathbf{K}^n$ (art. 1.57). Let (P, P') be a plot of the product $E \times E$ (art. 1.55). The parametrization $[r \mapsto P(r) + P'(r)]$ satisfies, for any plot $Q : V \to X$, $[(r, s) \mapsto (P(r) + P'(r))(Q(s)) = P(r)(Q(s)) + P'(r)(Q(s))] = [(r, s) \mapsto (P(r)(Q(s)), P'(r)(Q(s))) \mapsto P(r)(Q(s)) + P'(r)(Q(s))]$. But this is the composite of two plots, thus a plot. Hence, the addition in E is smooth. Now, for any $\lambda \in \mathbf{K}$, $[(r, s) \mapsto \lambda \times P(r)(Q(s))] = [(r, s) \mapsto (\lambda, P(r)(Q(s))) \mapsto \lambda \times P(r)(Q(s))]$ also is the composite of two plots, thus a plot. Therefore, the space $\mathcal{C}^\infty(X, \mathbf{K})$, equipped with the functional diffeology, is a diffeological vector space.

Exercise 66, p. 71 (Smooth is fine diffeology). By the very definition of smooth parametrizations in $\mathbf{K}^n$, every plot $P : U \to E$ splits over the canonical basis $(\mathbf{e}_i)_{i=1}^n$, that is, $P : r \mapsto \sum_{i=1}^n P_i(r)\mathbf{e}_i$, where $P_i \in \mathcal{C}^\infty(U, \mathbf{K})$, $i = 1, \dots, n$.

Exercise 67, p. 71 (Finite dimensional fine spaces). Let $r_0 \in U$ be any point. By definition of the fine diffeology, there exist an open neighborhood V of r_0, a finite family $\{\phi_\alpha\}_{\alpha \in A}$ of smooth parametrizations of $\mathbf{K}$, defined on V, and a family $\{u_\alpha\}_{\alpha \in A}$ of vectors of E, such that $P \restriction V : r \mapsto \sum_{\alpha \in A} \phi_\alpha(r) u_\alpha$. Now, let $\mathcal{B} = \{\mathbf{e}_i\}_{i=1}^n$ be a basis of E. Each u_α writes $\sum_{i=1}^n u_\alpha^i \mathbf{e}_i$, where the u_α^i belong to $\mathbf{K}$. Hence, $P \restriction V : r \mapsto \sum_{\alpha \in A} \sum_{i=1}^n \phi_\alpha(r) u_\alpha^i \mathbf{e}_i = \sum_{i=1}^n \phi_i(r)\mathbf{e}_i$, where $\phi_i(r) = \sum_{\alpha \in A} \phi_\alpha(r) u_\alpha^i$. But the ϕ_i are still smooth parametrizations of $\mathbf{K}$, thus $P \restriction V : r \mapsto \sum_{i=1}^n \phi_i(r)\mathbf{e}_i$ with $\phi_i \in \mathcal{C}^\infty(V, \mathbf{K})$. Now, let V and V′ two such domains on which the plot P writes $P \restriction V : r \mapsto \sum_{i=1}^n \phi_i(r)\mathbf{e}_i$ with $\phi_i \in \mathcal{C}^\infty(V, \mathbf{K})$, and $P \restriction V' : r \mapsto \sum_{i=1}^n \phi_i'(r)\mathbf{e}_i$ with $\phi_i' \in \mathcal{C}^\infty(V', \mathbf{K})$. Let $r \in V \cap V'$, we have $P(r) = P \restriction V(r) = P \restriction V'(r)$, that is, $\sum_{i=1}^n \phi_i(r)\mathbf{e}_i = \sum_{i=1}^n \phi_i'(r)\mathbf{e}_i$, or $\sum_{i=1}^n (\phi_i(r) - \phi_i'(r))\mathbf{e}_i = 0$, but since $\mathcal{B} = \{\mathbf{e}_i\}_{i=1}^n$ is a basis of E, $\phi_i(r) = \phi_i'(r)$. Therefore, the ϕ_i have a unique smooth extension on U such that $P(r) = \sum_{i=1}^n \phi_i(r)\mathbf{e}_i$, for all $r \in U$, and the ϕ_i belong to $\mathcal{C}^\infty(U, \mathbf{K})$. Finally, since linear isomorphisms between fine vector spaces are smooth isomorphisms (art. 3.9), the basis $\mathcal{B}$ realizes a smooth isomorphism from $\mathbf{K}^n$ to E.

Exercise 68, p. 71 (The fine topology). The diffeology of E is generated by the linear injections $j : \mathbf{K}^n \to E$ (art. 3.8), where n runs over $\mathbf{N}$, hence Ω is D-open if and only if its preimage by each of these injections is open in $\mathbf{K}^n$. Or, equivalently, if the intersection of Ω with any vector subspace F, of finite dimension, is open for the smooth topology of F.

Exercise 69, p. 74 (Fine Hermitian vector spaces). Let E be a fine diffeological real vector space. Let $\cdot$ be any Euclidean product. Let $(P, P') : U \to E \times E$ be a plot, that is, P and P′ are plots of E. Let $r_0 \in U$. There exist two local families, $(\phi_\alpha, u_\alpha)_{\alpha \in A}$ and $(\phi'_{\alpha'}, u'_{\alpha'})_{\alpha' \in A'}$, defined on some open neighborhood V of r_0, such that $P(r) = \sum_{\alpha\in A} \phi_\alpha(r)u_\alpha$ and $P'(r) = \sum_{\alpha'\in A'} \phi'_{\alpha'}(r)u'_{\alpha'}$, for all r in V (art. 3.7). Thus, $P(r)\cdot P'(r) = \sum_{\alpha\in A}\sum_{\alpha'\in A'} \phi_\alpha(r)\phi'_{\alpha'}(r)u_\alpha \cdot u'_{\alpha'}$, for all $r \in V$. But this is a finite linear combination of smooth parametrizations of $\mathbf{R}$, thus smooth. Now, since the map $r \mapsto P(r)\cdot P'(r)$ is locally smooth, it is smooth. Therefore, the Euclidean product $\cdot$ is smooth, and $(E, \cdot)$ is an Euclidean diffeological vector space. The same arguments hold for the Hermitian case.

☞ **Exercise 70, p. 74** (Finite dimensional Hermitian spaces). Let E be a Euclidean diffeological vector space of dimension n. Let us denote by $\mathcal{D}$ its diffeology. Let $(\mathbf{e}_1, \ldots, \mathbf{e}_n)$ be an orthonormal basis of E. Let $P : U \to E$ be a plot of E. For every $r \in U$, $P(r) = \sum_{k=1}^{n}(\mathbf{e}_k \cdot P(r))\mathbf{e}_k$, since, by hypothesis, the maps $P_k : r \mapsto \mathbf{e}_k \cdot P(r)$ are smooth, the plot P is a plot of the fine diffeology. Hence, the diffeology $\mathcal{D}$ is finer that the fine diffeology, but the fine diffeology is the finest vector space diffeology. Therefore $\mathcal{D}$ is the fine diffeology. The same argument holds for the Hermitian case. The (art. 3.2) states that the coarse diffeology is always a vector space diffeology. For finite dimensional spaces, the existence of a smooth Euclidean structure reduces the set of vector space diffeologies to the unique fine diffeology. In other words, there exists only one kind of finite Euclidean, or Hermitian, diffeological vector space of dimension n, the class of $(\mathbf{R}^n, \cdot)$, or $(\mathbf{C}^n, \cdot)$.

☞ **Exercise 71, p. 74** (Topology of the norm and D-topology). Let B be the open ball, for the topology of the norm, centered at $x \in E$, and with radius ε. Let $P : U \to E$ be some plot. The preimage of B by the plot P is the preimage of $]-\infty, \varepsilon^2[$ by the map $r \mapsto \|x - P(r)\|^2$, but this map is smooth, then D-continuous (art. 2.9). Hence, the ball B is D-open. Thus, thanks to the differentiability of translations and dilations, any open set for the topology of the norm is D-open. In other words, the topology of the norm is finer than the D-topology.

☞ **Exercise 72, p. 74** (Banach's diffeology). Let us denote by $\mathcal{C}^\infty_E$ the set of class $\mathcal{C}^\infty$ parametrizations of E. Let us check first that $\mathcal{C}^\infty_E$ is a diffeology.

D1. Let $x \in E$ and $\mathbf{x} : r \mapsto x$ be the constant parametrization with value x. Then, $D(\mathbf{x})(r) = 0$ for all r. Therefore, $\mathcal{C}^\infty_E$ contains the constants.

D2. Belonging to $\mathcal{C}^\infty_E$ is by definition a local property.

D3. Let $P : U \to E$ be an element of $\mathcal{C}^\infty_E$ and $F : V \to U$ be a smooth parametrization. For the real domains equipped with the usual Euclidean norm, to be smooth and to be of class $\mathcal{C}^\infty$, in the sense of Banach spaces, coincide. Hence, $D(P \circ F)(s) = D(P)(r) \circ D(F)(s)$, where $r = F(s)$, and for any $s \in V$. Since $D(P)(r)$ and $D(F)(s)$ are together of class $\mathcal{C}^\infty$, the composite also is of class $\mathcal{C}^\infty$, and $P \circ F$ belongs to $\mathcal{C}^\infty_E$.

Therefore, $\mathcal{C}^\infty_E$ is a diffeology. Now, the fact that this diffeology is a vector space diffeology comes from the linear properties of the tangent map: $D(P + P')(r) = D(P)(r) + D(P')(r)$ and $D(\lambda \times P)(r) = \lambda \times D(P)(r)$. Next, let E and F be two Banach spaces, equipped with the Banach diffeology. Let $f : E \to F$ be a map, smooth for the Banach diffeology. Then f takes plots of E to plots of F, in particular smooth curves, since to be a smooth curve for the Banach diffeology means exactly to be Banach-smooth. Thus, thanks to Boman's theorem, f is Banach-smooth. Conversely, let f be Banach-smooth. Since every n-plot $P : U \to E$ is by definition Banach-smooth, the composite $f \circ P$ is Banach-smooth, that is, a plot of F. Hence, f is smooth for the Banach diffeology. Therefore, the functor which associates with every Banach space its diffeology is a full faithful functor.

Exercise 73, p. 75 ($\mathcal{H}_{\mathbf{C}}$ isomorphic to $\mathcal{H}_{\mathbf{R}} \times \mathcal{H}_{\mathbf{R}}$). Since for $Z = X + iY$, $Z_k^* Z_k = X_k^2 + Y_k^2$, the map ψ is an isometry. The bijectivity and the linearity of ψ are obvious. And since the diffeology is fine, to be linear implies to be smooth (art. 3.9). Therefore, ψ is a smooth linear isomorphism.

Exercise 74, p. 84 (The irrational torus is not a manifold). Let us assume that $T_\alpha = \mathbf{R}/[\mathbf{Z} + \alpha\mathbf{Z}]$ is a manifold. We know that the dimension of the torus is 1 (Exercise 49, p. 50). Thus, there should exist a family of inductions from some 1-domains to $\mathbf{R}$, satisfying the criterion (art. 4.6). Let $j : I \to \mathbf{R}$ be such an induction, where I is some interval. Since $j(I)$ cuts each orbit of $\mathbf{Z} + \alpha\mathbf{Z}$ in at most one point, and since each orbit is dense, if $j(I)$ is not empty, then $j(I)$ is just a point. But then it is not injective an cannot be an induction. Hence, T_α is not a manifold. Note that, since the D-topology contains only one nonempty D-open, T_α itself (art. 55), and since the values of any local diffeomorphism is D-open, if T_α would be a manifold it would be diffeomorphic to some 1-domain. But this cannot be for, roughly speaking, the same reasons as above.

Exercise 75, p. 84 (The sphere as paragon). 1) The map F_x is injective, its inverse is given by

$$F_x^{-1} : S_x^n \to E_x \quad \text{with} \quad u \mapsto v = \frac{1}{1 + \bar{u}x}\,[1 - x\bar{x}]\,u,$$

where $\bar{x}$ is the transposed of x, that is, the linear map from $\mathbf{R}^{n+1}$ into $\mathbf{R}$ defined by $\bar{x}y = x \cdot y$ for any $y \in \mathbf{R}^{n+1}$, and $x\bar{x}$ is the map $x\bar{x} : u \mapsto (\bar{x}u)x = (x \cdot u)x$. Since F_x is a sum and product of smooth maps (the denominator $1 + \|v\|^2$ never vanishes), it is clearly smooth. The image of F_x is the sphere S^n deprived of the point $-x$. Let us denote $S_x^n = \mathrm{val}(F_x) = S^n - \{-x\}$. The map F_x^{-1}, restricted to S_x^n, is clearly smooth, because it is the restriction of a smooth map defined on the domain $\Omega_x = \{u \in \mathbf{R}^{n+1} \mid u \cdot x \neq -1\}$ to the subspace S_x^n. Moreover S_x^n is open for the D-topology of S^n. Indeed, the pullback $P^{-1}(S_x^n)$, by any plot P of S^n, is equal to the pullback by $P^{-1}(\Omega_x)$, where P is regarded as a plot of $\mathbf{R}^{n+1}$. Since P is a plot of the smooth diffeology of $\mathbf{R}^{n+1}$, and Ω_x is a domain of $\mathbf{R}^{n+1}$, $P^{-1}(\Omega_x)$ is a domain. Therefore, thanks to (art. 2.10), F_x is a local diffeomorphism mapping E_x onto S_x^n.

2) Thus, $\bigcup_{x \in S^n} F_x(E_x) = S^n$, and for every $x \in S^n$, F_x is a local diffeomorphism with E_x. But $E_x \sim \mathbf{R}^n$, hence there exists a family of local diffeomorphisms from $\mathbf{R}^n$ to S^n whose values cover S^n. Therefore S^n is a manifold of dimension n.

3) The maps F_N and F_{-N} are local diffeomorphisms from $\mathbf{R}^n = N^\perp$ to S^n. Moreover, $\mathrm{val}(F_N) \cup \mathrm{val}(F_{-N}) = S^n$. Hence, $\{F_N, F_{-N}\}$ is a generating family of the diffeology of S^n (art. 4.2), that is, an atlas.

Exercise 76, p. 92 (The space of lines in $\mathcal{C}^\infty(\mathbf{R}, \mathbf{R})$). 1) It is immediate to check that if (f_1, g_1) and (f_2, g_2) belong to $\mathcal{E} \times (\mathcal{E} - \{0\})$ and define the same line L, then necessarily $f_2 = f_1 + \lambda g_1$ and $g_2 = \mu g_1$, where $\lambda \in \mathbf{R}$ and $\mu \in \mathbf{R} - \{0\}$. Now, let $L = \{f + sg \mid s \in \mathbf{R}\}$ and $r \in \mathbf{R}$. If $g(r) \neq 0$, then let $\beta = g/g(r)$ and $\alpha = f - f(r)\beta$. Hence, $\alpha(r) = 0$, $\beta(r) = 1$ and $L = F_r(\alpha, \beta)$. The fact that $g(r) \neq 0$ is a property of the line L, indeed for any other pair (f', g') defining the same line, $g' = \mu g$ with $\mu \neq 0$, and then $g'(r) \neq 0$. Let us denote this space, defined by $g(r) \neq 0$, by $\mathrm{Lines}_r(\mathcal{E})$. Thus, $\mathrm{val}(F_r) = \mathrm{Lines}_r(\mathcal{E})$. Note next that F_r is injective. Indeed, if (α_1, β_1) and (α_2, β_2) belong to $\mathcal{E}_r^0 \times \mathcal{E}_r^1$ and define the same line, then

$\beta_2 = \mu\beta_1$, which implies $\mu = 1$, and $\alpha_2 = \alpha_1 + \lambda\beta_1$, which implies $\lambda = 0$. Therefore F_r is a bijection from $\mathcal{E}_r^0 \times \mathcal{E}_r^1$ onto its image $\mathrm{Lines}_r(\mathcal{E})$. Moreover, since F_r is the restriction of a smooth map to $\mathcal{E}_r^0 \times \mathcal{E}_r^1 \subset \mathcal{E} \times (\mathcal{E} - \{0\})$, it is a smooth injection. Let us check now that F_r is an induction. Let $u \mapsto L_u$ be a plot of $\mathrm{Lines}_r(\mathcal{E})$, there exists locally a smooth parametrization $u \mapsto (f_u, g_u)$ in $\mathcal{E} \times (\mathcal{E} - \{0\})$, with $g_u(r) \neq 0$, such that $L_u = \{f_u + s g_u \mid s \in \mathbf{R}\}$. The map $(f_u, g_u) \mapsto (\alpha_u, \beta_u)$, defined by $\alpha_u = f_u - f_u(r)\beta_u$ and $\beta_u = g_u/g_u(r)$, being smooth, the parametrization $u \mapsto (\alpha_u, \beta_u)$ is a plot of $\mathcal{E}_r^0 \times \mathcal{E}_r^1$ such that $L_u = F_r(\alpha_u, \beta_u)$. Therefore, F_r is an induction. Finally, let us check that $\mathrm{Lines}_r(\mathcal{E})$ is D-open. Thanks to (art. 2.12), we just need to check that the subset $\mathcal{O}_r \subset \mathcal{E} \times (\mathcal{E} - \{0\})$ of (f, g) such that $g(r) \neq 0$ is D-open. Let $P : u \mapsto (f_u, g_u)$ be a plot of $\mathcal{E} \times (\mathcal{E} - \{0\})$, and let $\phi(u) = g_u(r)$. Since $(u, r) \mapsto g_u(r)$ is smooth, ϕ is smooth. Thus, $P^{-1}(\mathcal{O}_r) = \{u \in \mathrm{def}(P) \mid g_u(r) \neq 0\} = \phi^{-1}(\mathbf{R} - 0)$, is open. Therefore, $\mathcal{O}_r$ is D-open and then F_r is a local diffeomorphism from $\mathcal{E}_r^0 \times \mathcal{E}_r^1$ to $\mathrm{Lines}(\mathcal{E})$, with values $\mathrm{Lines}_r(\mathcal{E})$.

2) Since for every line L there exists (f, g) such that $L = \{f + sg \mid s \in \mathbf{R}\}$, with $g \neq 0$, there exists some $r \in \mathbf{R}$ such that $g(r) \neq 0$, thus the union of all the subsets $\mathrm{Lines}_r(\mathcal{E})$ covers $\mathrm{Lines}(\mathcal{E})$. Now, every $\mathcal{E}_r^0$ is isomorphic to $\mathcal{E}_0^0$, by $\alpha \mapsto \alpha \circ T_r = [r' \mapsto \alpha(r' + r)]$, and $\mathcal{E}_r^1$ also is isomorphic (as an affine space) to $\mathcal{E}_0^0$, by $\beta \mapsto \beta \circ T_r - 1$. Therefore, $\mathrm{Lines}(\mathcal{E})$ is a diffeological manifold modeled on $\mathcal{E}_0^0 \times \mathcal{E}_0^0$, where $\mathcal{E}_0^0 = \{f \in \mathcal{C}^\infty(\mathbf{R}, \mathbf{R}) \mid f(0) = 0\}$.

3) The space of maps from $\{1, 2\}$ to $\mathbf{R}$ is diffeomorphic to $\mathbf{R}^2$. Let $f \simeq (x_1, x_2)$ and $g \simeq (u_1, u_2)$, $g \neq 0$ means that $(u_1, u_2) \neq (0, 0)$. Now, there are four spaces $\mathcal{E}_r^i$, where $r = 1, 2$ and $i = 0, 1$, that is, $\mathcal{E}_1^0 = \{(0, x_2) \mid x_2 \in \mathbf{R}\}$, $\mathcal{E}_1^1 = \{(1, u_2) \mid x_2 \in \mathbf{R}\}$, $\mathcal{E}_2^0 = \{(x_1, 0) \mid x_1 \in \mathbf{R}\}$, $\mathcal{E}_2^1 = \{(u_1, 1) \mid u_1 \in \mathbf{R}\}$. Thus,

$$F_1\left(\begin{pmatrix} 0 \\ x_2 \end{pmatrix}, \begin{pmatrix} 1 \\ u_2 \end{pmatrix}\right) = \left\{\begin{pmatrix} s \\ x_2 + s u_2 \end{pmatrix} \mid s \in \mathbf{R}\right\},$$

$$F_2\left(\begin{pmatrix} x_1 \\ 0 \end{pmatrix}, \begin{pmatrix} u_1 \\ 1 \end{pmatrix}\right) = \left\{\begin{pmatrix} x_1 + s u_1 \\ s \end{pmatrix} \mid s \in \mathbf{R}\right\}.$$

The chart F_1 maps $\mathbf{R}^2$ to the subspace of lines not parallel to the x_2-axis and the chart F_2 maps $\mathbf{R}^2$ to the subspace of lines not parallel to the x_1-axis. We have seen that the set of unparametrized and nonoriented lines in $\mathbf{R}^2$ is diffeomorphic to the Möbius strip (Exercise 41, p. 39, question 4), thus the set $\mathcal{A} = \{F_1, F_2\}$ is an atlas of this famous manifold.

Exercise 77, p. 93 (The Hopf S^1-bundle). Let $J : \mathcal{S}_{\mathbf{C}} \to \mathcal{H}_{\mathbf{C}}$ be the natural inclusion. Since $J(z \times Z) = z \times J(Z)$, the injection J projects onto a map $j : \mathcal{S}_{\mathbf{C}}/U(1) \to \mathcal{P}_{\mathbf{C}} = \mathcal{H}_{\mathbf{C}}^\star/\mathbf{C}^\star$, according to the following commuting diagram, where $\pi_{\mathcal{S}}$ and $\pi_{\mathcal{H}}$ are the natural projections. Since J is smooth, and since $\pi_{\mathcal{S}}$ is a subduction, j is smooth. The map j is obviously injective. Then, since for every $Z \in \mathcal{H}_{\mathbf{C}}^\star$, $Z/\|Z\| \in \mathcal{S}_{\mathbf{C}}$, j is surjective. Now, since the map $\rho : Z \mapsto Z/\|Z\|$ from $\mathcal{H}_{\mathbf{C}}^\star$ to $\mathcal{S}_{\mathbf{C}}$ is smooth $(Z \neq 0)$, j^{-1} is smooth. Indeed, a plot P of $\mathcal{P}_{\mathbf{C}}$ lifts locally to $\mathcal{H}_{\mathbf{C}}^\star$. Composing the local lift with ρ we get a lift in $\mathcal{S}_{\mathbf{C}}$. Therefore, j is a diffeomorphism.

$$\begin{array}{ccc} \mathcal{S}_{\mathbf{C}} & \xrightarrow{\ \ J\ \ } & \mathcal{H}^{\star}_{\mathbf{C}} \\ \Big\downarrow{\scriptstyle \pi_{\mathcal{S}}} & & \Big\downarrow{\scriptstyle \pi_{\mathcal{H}}} \\ \mathcal{S}_{\mathbf{C}}/\mathrm{U}(1) & \xrightarrow[\ \ j\ \ ]{} & \mathcal{P}_{\mathbf{C}} = \mathcal{H}^{\star}_{\mathbf{C}}/\mathbf{C}^{\star} \end{array}$$

Now, let us transpose this construction to $\mathcal{H}_{\mathbf{R}} \times \mathcal{H}_{\mathbf{R}}$. The sphere $\mathcal{S}_{\mathbf{C}}$ is diffeomorphic to the sphere $\mathcal{S} = \{(X, Y) \in \mathcal{H}_{\mathbf{R}} \times \mathcal{H}_{\mathbf{R}} \mid \|X\|^2 + \|Y\|^2 = 1\}$. The group $\mathrm{U}(1) = \{z = \cos(\theta) + i\sin(\theta) \mid \theta \in \mathbf{R}\}$ is equivalent to the group

$$\mathrm{SO}(2, \mathbf{R}) = \left\{ \begin{pmatrix} \cos(\theta) & -\sin(\theta) \\ \sin(\theta) & \cos(\theta) \end{pmatrix} \,\middle|\, \theta \in \mathbf{R} \right\}.$$

With this identification, the action of $\mathrm{U}(1)$ on $\mathcal{S}_{\mathbf{C}}$ transmutes to the following action of $\mathrm{SO}(2, \mathbf{R})$ on $\mathcal{S}$,

$$\begin{pmatrix} \cos(\theta) & -\sin(\theta) \\ \sin(\theta) & \cos(\theta) \end{pmatrix} : \begin{pmatrix} X \\ Y \end{pmatrix} \mapsto \begin{pmatrix} \cos(\theta) & -\sin(\theta) \\ \sin(\theta) & \cos(\theta) \end{pmatrix} \begin{pmatrix} X \\ Y \end{pmatrix}.$$

Hence, the projective space $\mathcal{P}_{\mathbf{C}}$ also is diffeomorphic to $\mathcal{S}/\mathrm{SO}(2, \mathbf{R})$.

☞ **Exercise 78, p. 93** ($\mathrm{U}(1)$ as subgroup of diffeomorphisms). First of all, let us note that j is an injective homomorphism (a monomorphism), $j(zz') = j(z) \circ j(z')$, and $j(z)^{-1} = j(z^{-1})$. Moreover, j is $\mathbf{C}$-linear, thus smooth (art. 3.9). Now, let $P : U \to \mathrm{U}(1)$ be a parametrization such that $j \circ P$ is smooth for the functional diffeology of $\mathrm{GL}(\mathcal{H}_{\mathbf{C}})$. Let us apply the criterion of (art. 3.12) to the plot $j \circ P$. Let $r_0 \in U$, let $Z \in \mathcal{H}_{\mathbf{C}}$, $Z \neq 0$, and let $F = \mathbf{C}Z \subset \mathcal{H}_{\mathbf{C}}$ be the complex line generated by Z. There exist an open neighborhood V of r_0 and a finite dimensional subspace $F' \subset \mathcal{H}_{\mathbf{C}}$ such that: $(j \circ P) \restriction V \in L(F, F')$ and $r \mapsto (j \circ P(r)) \restriction F$ is a plot of $L(F, F')$. But clearly $F' = F$. Now, let $\mathcal{Z} : \mathbf{C} \to F$ be a basis, a $\mathbf{C}$-linear isomorphism. In this basis, the parametrization $r \mapsto j(P(r))$ becomes the multiplication by $P(r)$, that is, $\mathcal{Z}^{-1} \circ j(P(r)) \circ \mathcal{Z} : z \mapsto P(r)z$. Thus, $[r \mapsto [z \mapsto P(r)z]]$ being smooth means just that $P : U \to \mathrm{U}(1)$ is smooth. Therefore j is an induction. In other words, the multiplication by an element of $\mathrm{U}(1)$ is a subgroup of $\mathrm{U}(\mathcal{H})$ isomorphic, as diffeological group (art. 7.1), to $\mathrm{U}(1)$.

☞ **Exercise 79, p. 99** (Reflexive diffeologies). 1) The coarsest diffeology $\mathcal{D}$ is the intersection of the pullbacks of the smooth diffeology $\mathcal{C}^{\infty}_{\star}(\mathbf{R})$ of $\mathbf{R}$, by the elements of the family $\mathcal{F}$,

$$\mathcal{D} = \bigcap_{f \in \mathcal{F}} f^*(\mathcal{C}^{\infty}_{\star}(\mathbf{R})).$$

The plots of this diffeology are explicitly defined by,

($\diamondsuit$) $P : U \to X$ belongs to $\mathcal{D}$ if and only if, for all $f \in \mathcal{F}$, $f \circ P \in \mathcal{C}^{\infty}(U, \mathbf{R})$.

Indeed, by definition of the pullback diffeology, $f^*(\mathcal{C}^{\infty}_{\star}(\mathbf{R}))$ is the coarsest diffeology such that f is smooth (art. 1.26). Thus, every diffeology $\mathcal{D}$ such that $\mathcal{F} \subset \mathcal{D}(X, \mathbf{R})$ is contained in $f^*(\mathcal{C}^{\infty}_{\star}(\mathbf{R}))$, and therefore, is contained in their intersection over the $f \in \mathcal{F}$. Since the intersection of any family of diffeologies is a diffeology (art. 1.22), this intersection is a diffeology, and by construction, the coarsest. About the finest diffeology, we know that for the discrete diffeology on X, $\mathcal{C}^{\infty}(X_\circ, \mathbf{R}) = \mathrm{Maps}(X, \mathbf{R})$; see Exercise 9, p. 14.

2) Let $\mathcal{D}$ be the diffeology subordinated to $\mathcal{F}$, and $\mathcal{D}'$ be the diffeology subordinated to $\mathcal{D}(X, \mathbf{R})$. By construction, $\mathcal{F} \subset \mathcal{D}(X, \mathbf{R})$ and $\mathcal{D}(X, \mathbf{R}) \subset \mathcal{D}'(X, \mathbf{R})$, then $\mathcal{F} \subset \mathcal{D}'(X, \mathbf{R})$, but $\mathcal{D}$ is the coarsest diffeology such that $\mathcal{F} \subset \mathcal{D}(X, \mathbf{R})$, thus $\mathcal{D}' \subset \mathcal{D}$. Then, by definition of $\mathcal{D}(X, \mathbf{R})$, for all $\phi \in \mathcal{D}(X, \mathbf{R})$, for all plots $P \in \mathcal{D}$, $\phi \circ P$ is smooth. But this is exactly the condition $(\diamondsuit)$ for P to belong to $\mathcal{D}'$, so $\mathcal{D} \subset \mathcal{D}'$. Therefore $\mathcal{D}' = \mathcal{D}$, and the diffeology subordinated to $\mathcal{F}$ is reflexive.

3) Let X be a manifold, let $\mathcal{D}$ be its diffeology, and let $n = \dim(X)$. Let us prove first that for every $x_0 \in X$ and every local smooth function $f : \mathcal{O}' \to \mathbf{R}$, defined on an D-open neighborhood $\mathcal{O}'$ of x_0, there exists a smooth function $\bar{f} : X \to \mathbf{R}$ which coincides with f on a D-open neighborhood $\mathcal{O} \subset \mathcal{O}'$ of x_0. Indeed, let $F : V \to X$ be a chart of X, and $\xi_0 \in V$, such that $F(\xi_0) = x_0$. We can assume that $F(V) \subset \mathcal{O}'$. There exist two balls $\mathcal{B} \subset \mathcal{B}' \subset V$, centered at ξ_0, and a smooth real function $\varepsilon : V \to \mathbf{R}$ such that ε is equal to 1 on $\mathcal{B}$ and equal to 0 outside $\mathcal{B}'$. Then, the local real function $x \mapsto \varepsilon(F(x)) \times f(x)$, defined on $F(V) \subset \mathcal{O}'$, can be extended smoothly on X by 0. This extension $\bar{f}$ coincides with f on $\mathcal{O} = F(\mathcal{B})$. Now, let $P : U \to X$ be an element of $\mathcal{D}'$, $r_0 \in U$, and $x_0 = P(r_0)$. Let $F : V \to X$ be a chart, such that $F(\xi_0) = x_0$. The cochart $F^{-1} : x \mapsto (\phi_1(x), \ldots, \phi_n(x))$ is made with local smooth functions ϕ_i. Thus, the $\bar{\phi}_i \circ P$ are smooth, by definition of $\mathcal{D}'$. Then, there exists a small neighborhood $W \subset U$ of r_0 such that the $\phi_i \circ (P \restriction W)$ are smooth. By construction, $P \restriction W = F \circ Q$, where $Q : W \to \operatorname{def}(F)$ is the smooth parametrization $Q : r \mapsto (\phi_1 \circ P(r), \ldots, \phi_n \circ P(r))$. Hence, locally P belongs to $\mathcal{D}$, which implies $P \in \mathcal{D}$. Therefore, $\mathcal{D}' \subset \mathcal{D}$ and X is reflexive.

4) We know that $\mathcal{C}^\infty(T_\alpha, \mathbf{R})$ is reduced to the constants when $\alpha \notin \mathbf{Q}$; see Exercise 4, p. 8. Thus, the subordinated diffeology to $\mathcal{C}^\infty(T_\alpha, \mathbf{R})$ is the coarse diffeology, since the composite of a constant map with any parametrization is constant. But we also know that T_α is not trivial. Therefore T_α is not reflexive.

☞ **Exercise 80, p. 99** (Frölicher spaces). We assume X reflexive. If $c \in \mathcal{C} = \mathcal{C}^\infty(\mathbf{R}, X)$ and $f \in \mathcal{F} = \mathcal{C}^\infty(X, \mathbf{R})$, then, by definition of $\mathcal{C}^\infty(X, \mathbf{R})$, $f \circ c \in \mathcal{C}^\infty(\mathbf{R}, \mathbf{R})$. This gives, at the same time, $\mathcal{C} \subset \mathfrak{C}(\mathcal{F})$ and $\mathcal{F} \subset \mathfrak{F}(\mathcal{C})$. Next, let $c \in \mathfrak{C}(\mathcal{F})$, that is, for all $f \in \mathcal{F}$, $f \circ c \in \mathcal{C}^\infty(\mathbf{R}, \mathbf{R})$, but that means that $c \in \mathcal{D}(\mathbf{R}, X)$, where $\mathcal{D}$ denotes the diffeology subordinated to $\mathcal{F}$ (Exercise 79, p. 99). But since X is reflexive, $\mathcal{D}(\mathbf{R}, X) = \mathcal{C}^\infty(\mathbf{R}, X)$, and then $c \in \mathcal{C}^\infty(\mathbf{R}, X)$. Thus, if $c \in \mathfrak{C}(\mathcal{F})$ then $c \in \mathcal{C}$, that is, $\mathfrak{C}(\mathcal{F}) \subset \mathcal{C}$. Therefore, $\mathfrak{C}(\mathcal{F}) = \mathcal{C}$. Consider now $f \in \mathfrak{F}(\mathcal{C})$, that is, $f \in \operatorname{Maps}(X, \mathbf{R})$ such that for all $c \in \mathcal{C}^\infty(\mathbf{R}, X)$, $f \circ c \in \mathcal{C}^\infty(\mathbf{R}, \mathbf{R})$. Let $P \in \mathcal{C}^\infty(\mathbf{R}^n, X)$. Since P is a plot, for all $\gamma \in \mathcal{C}^\infty(\mathbf{R}, \mathbf{R}^n)$, $P \circ \gamma \in \mathcal{C}^\infty(\mathbf{R}, X)$, thus $f \circ P \circ \gamma \in \mathcal{C}^\infty(\mathbf{R}, \mathbf{R})$, let $F = f \circ P : \mathbf{R}^n \to \mathbf{R}$, then $F \circ \gamma \in \mathcal{C}^\infty(\mathbf{R}, \mathbf{R})$, for all $\gamma \in \mathcal{C}^\infty(\mathbf{R}, \mathbf{R}^n)$. By application of Boman's theorem, we get $F \in \mathcal{C}^\infty(\mathbf{R}^n, \mathbf{R})$, that is, $P \circ F \in \mathcal{C}^\infty(\mathbf{R}^n, \mathbf{R})$. Then, after localization (see Exercise 44, p. 47) it comes that for every plot P of X, $f \circ P$ is smooth, that is, $f \in \mathcal{C}^\infty(X, \mathbf{R})$. Thus $\mathfrak{F}(\mathcal{C}) \subset \mathcal{F}$, and then $\mathfrak{F}(\mathcal{C}) = \mathcal{F}$. Therefore, for every reflexive diffeological space X, $\mathcal{C} = \mathcal{C}^\infty(\mathbf{R}, X)$ and $\mathcal{F} = \mathcal{C}^\infty(X, \mathbf{R})$ satisfy the Frölicher condition.

☞ **Exercise 81, p. 111** (Connecting points). We know that X is diffeomorphic to $\mathcal{C}^\infty(\{0\}, X)$ (art. 1.64). Thus, the functor π_0 (art. 5.10) gives an isomorphism.

☞ **Exercise 82, p. 111** (Connecting segments). If x and x' are connected, then they satisfy obviously the condition of the exercise. Conversely, let $\sigma :]a', b'[\to X$ such that $\sigma(a) = x$, $\sigma(b) = x'$ and $a' < a < b < b'$. Let $f(t) = (b - a)t + a$, then $f(0) = a$ and $f(1) = b$. Thus, $\sigma \circ f(0) = x$ and $\sigma \circ f(1) = x'$. Composing then $\sigma \circ f$

with the smashing function λ (art. 5.5), we get a stationary path $\gamma = \sigma \circ f \circ \lambda$ such that $\gamma(0) = x$ and $\gamma(1) = x'$.

☞ **Exercise 83, p. 111** (Contractible space of paths). The map $\rho : s \mapsto [\gamma \mapsto \gamma_s = [t \mapsto \gamma(st)]]$, is a path in $\mathcal{C}^\infty(\text{Paths}(X, x, \star))$ such that $\rho(0)$ is the constant map with value the constant path $\mathbf{x} : t \mapsto x$, and $\rho(1)$ is the identity on $\text{Paths}(X, x, \star)$. Therefore, ρ is a deformation retraction of $\text{Paths}(X, x, \star)$ to $\mathbf{x}$. Next, the same map ρ, defined on $\text{Paths}(X, A, \star)$, gives a deformation retraction from $\text{Paths}(X, A, \star)$ to the constant paths in A, which gives a homotopy equivalence.

☞ **Exercise 84, p. 111** (Deformation onto stationary paths). We shall check that $f : \gamma \mapsto \gamma^\star$, from Paths(X) to stPaths(X), and the inclusion $j : \text{stPaths}(X) \to \text{Paths}(X)$ are homotopic inverses of each other (art. 5.10). Thus, we have to check that $f = j \circ f : \text{Paths}(X) \to \text{Paths}(X)$ is homotopic to $\mathbf{1}_{\text{Paths}(X)}$, and $f \restriction \text{stPaths}(X) = f \circ j : \text{stPaths}(X) \to \text{stPaths}(X)$ is homotopic to $\mathbf{1}_{\text{stPaths}(X)}$. Let us consider

$$f_s : \gamma \mapsto \gamma_s \quad \text{with} \quad \gamma_s : t \mapsto \gamma[\lambda(s)\lambda(t) + (1 - \lambda(s))t],$$

where λ is the smashing function (Figure 5.1). The map $s \mapsto f_s$ is clearly a smooth homotopy from $f_0 = \mathbf{1}_{\text{Paths}(X)}$ to $f_1 = f$. Now, we shall check that for all $\gamma \in \text{stPaths}(X)$, $f_s(\gamma) \in \text{stPaths}(X)$, for all s. Let $x = \gamma(0)$ and $x' = \gamma(1)$. Then, let $\varepsilon' > 0$ such that $\gamma(t) = x$ for all $t \leq \varepsilon'$, and $\gamma(t) = x'$ for all $t \geq 1 - \varepsilon'$. First of all, let $\varepsilon'' = \inf(\varepsilon, \varepsilon')$, so for all $t \leq \varepsilon''$, $\gamma(t) = x$ and $\lambda(t) = 0$, and for all $t \geq 1 - \varepsilon''$, $\gamma(t) = x'$ and $\lambda(t) = 1$.

A) If $t \leq \varepsilon''$, then $\lambda(t) = 0$, and $\lambda(s)\lambda(t) + (1 - \lambda(s))t = (1 - \lambda(s))t$ but $0 \leq 1 - \lambda(s) \leq 1$. Thus, if $t \leq 0$, then $(1 - \lambda(s))t \leq 0 < \varepsilon''$, and if $0 < t \leq \varepsilon''$, then $(1 - \lambda(s))t \leq t \leq \varepsilon''$. Hence, $\gamma_s(t) = x$.

B) If $t \geq 1 - \epsilon''$, then $\lambda(t) = 1$ and $\lambda(s)\lambda(t) + (1 - \lambda(s))t = \lambda(s) + (1 - \lambda(s))t$. If $t \geq 1$, then, since $0 \leq \lambda(s) \leq 1$, $1 \leq \lambda(s) \times 1 + (1 - \lambda(s))t \leq t$, thus $\gamma_s(t) = x'$. If $0 < 1 - t \leq \varepsilon''$, then $\lambda(s) + (1 - \lambda(s))t = t + (1 - t)\lambda(s) \geq t \geq 1 - \varepsilon''$, thus $\gamma_s(t) = x'$.

Therefore, $f_s \in \mathcal{C}^\infty(\text{stPaths}(X), \text{stPaths}(X))$ and $s \mapsto f_s$ and j are homotopic inverses of each other. Thus, Paths(X) and stPaths(X) are homotopy equivalent. Now, since $f_s(\gamma)(0) = \gamma(0)$ and $f_s(\gamma)(1) = \gamma(1)$, for all s, this equivalence also holds for the diffeology foliated by the projection ends.

☞ **Exercise 85, p. 112** (Contractible quotient). The deformation retraction $\rho : s \mapsto [z \mapsto sz]$ from $\mathbf{C}$ to $\{0\}$ is equivariant by the action of $\mathbf{Z}_m$, that is, $\rho(s) \circ \zeta_k = \zeta_k \rho(s)$, for all s. Thus, there exists a smooth map $r(s) : \mathbf{C}/\mathbf{Z}_m \mapsto \mathbf{C}/\mathbf{Z}_m$ such that $\text{class} \circ \rho(s) = r(s) \circ \text{class}$ for all s, where $\text{class} : \mathbf{C} \to \mathbf{C}/\mathbf{Z}_m$ is the projection. Moreover, considering $\bar\rho : \mathbf{R} \times \mathbf{C} \to \mathbf{R} \times \mathbf{C}$, defined by $\bar\rho(s, z) = \rho(s)(z)$, and the action of $\mathbf{Z}_m$ on $\mathbf{R} \times \mathbf{C}$ acting trivially on the first factor and accordingly on the second, we get a smooth map $\bar r : \mathbf{R} \times \mathbf{C}/\mathbf{Z}_m \to \mathbf{R} \times \mathbf{C}/\mathbf{Z}_m$, defining a deformation retraction of the quotient $\mathbf{C}/\mathbf{Z}_m$ to class(0).

☞ **Exercise 86, p. 112** (Locally contractible manifolds). Let $m \in M$, where M is a manifold. By definition of what is a manifold, there exists a chart $F : U \to M$ mapping some point r to m. Let $\Omega \subset M$ be an open neighborhood of m, its preimage by F is an open neighborhood of r. Then, there exists a small ball $B \subset F^{-1}(\Omega)$ centered at the point r. Since F is a local diffeomorphism, the image $F(B) \subset \Omega$ is a D-open neighborhood of m, and since B is contractible, $F(B)$ is contractible. Therefore, M is locally contractible.

Exercise 87, p. 123 (Homotopy of loops spaces). Fix $x \in X$, and consider the map $\mathrm{pr}_x : \mathrm{comp}_x(\ell) \mapsto \mathrm{comp}(\ell)$, where $\mathrm{comp}_x(\ell) \in \pi_0(\mathrm{Loops}(X,x)) = \pi_1(X,x)$ is the connected component of ℓ in $\mathrm{Loops}(X,x)$ and $\mathrm{comp}(\ell) \in \pi_0(\mathrm{Loops}(X))$ is the connected component of ℓ in $\mathrm{Loops}(X)$. This map is well defined. Indeed, if ℓ and ℓ' are fixed-ends homotopic, then they are *a fortiori* free-ends homotopic, or if we prefer, since the injection $\mathrm{Loops}(X,x) \subset \mathrm{Loops}(X)$ is smooth, it induces a natural map $\mathrm{pr}_x : \pi_0(\mathrm{Loops}(X,x)) \to \pi_0(\mathrm{Loops}(X))$.

First, let us check that this map is surjective. Let $\ell' \in \mathrm{Loops}(X)$ and $x' = \ell'(0)$, since X is connected there exists a path γ connecting x to x'. We can consider ℓ' and γ stationary since we know that every path is fixed-ends homotopic to a stationary path (art. 5.5). Let us consider $\ell = \gamma \vee \ell' \vee \bar{\gamma}$ with $\bar{\gamma}(t) = \gamma(1-t)$, thus $\ell \in \mathrm{Loops}(X,x)$. Now, let $\gamma_s(t) = \gamma(s + t(1-s))$, γ_s is a path connecting $\gamma_s(0) = \gamma(s)$ to $\gamma_s(1) = x'$, and $\ell_s = \gamma_s \vee \ell' \vee \bar{\gamma}_s$ is a free-ends homotopy connecting $\ell_0 = \gamma \vee \ell' \vee \bar{\gamma} = \ell$ to $\ell_1 = \mathbf{x}' \vee \ell' \vee \mathbf{x}'$, where $\mathbf{x}'$ is the constant path at x'. Now, $\mathbf{x}' \vee \ell' \vee \mathbf{x}'$ is homotopic to ℓ', then ℓ and ℓ' belong to the same connected component in $\mathrm{Loops}(X)$, thus $\mathrm{comp}(\ell) = \mathrm{comp}(\ell')$, that is, $\mathrm{pr}_x(\mathrm{comp}_x(\ell)) = \mathrm{comp}(\ell')$.

Next, let k_0 and k_1 in $\pi_1(X,x)$, and let us prove that $\mathrm{pr}_x(k_0) = \mathrm{pr}_x(k_1)$ implies $k_0 = \tau \cdot k_1 \cdot \tau^{-1}$ for some $\tau \in \pi_1(X,x)$. Let $k_0 = \mathrm{comp}_x(\ell_0)$ and $k_1 = \mathrm{comp}_x(\ell_1)$, with ℓ_0 and ℓ_1 in $\mathrm{Loops}(X,x)$. Let us assume that $\mathrm{pr}_x(k_0) = \mathrm{pr}_x(k_1)$, that is, ℓ_0 and ℓ_1 are free-ends homotopic, in other words, $\mathrm{comp}(\ell_0) = \mathrm{comp}(\ell_1)$. Let $s \mapsto \ell_s$ be a free-ends homotopy, thus $\ell' : s \mapsto \ell_s(0) = \ell_s(1)$ is a loop based at x. Indeed, $\ell'(0) = \ell_0(0) = x$ and $\ell'(1) = \ell_1(0) = x$, that is, $\ell' \in \mathrm{Loops}(X,x)$. Then, $\ell'_s : t \mapsto \ell'(st)$ is a path, connecting $\ell'_s(0) = \ell'(0) = \ell_0(0) = x$ to $\ell'_s(1) = \ell'(s) = \ell_s(0) = \ell_s(1)$. Hence, $\sigma_s = \ell'_s \vee \ell_s \vee \bar{\ell}'_s$ is a loop based at x, for all s. Thus, σ is a fixed-ends homotopy connecting $\sigma_0 = \mathbf{x} \vee \ell_0 \vee \mathbf{x}$ to $\sigma_1 = \ell' \vee \ell_1 \vee \bar{\ell}'$, *i.e.*, $\mathrm{comp}_x(\sigma_0) = \mathrm{comp}_x(\sigma_1)$. Since σ_0 is fixed-ends homotopic to ℓ_0, and since $\ell' \in \mathrm{Loops}(X,x)$, that writes again $\mathrm{comp}_x(\ell_0) = \mathrm{comp}_x(\ell' \vee \ell_1 \vee \bar{\ell}')$, that is, $k_0 = \tau \cdot k_1 \cdot \tau^{-1}$ with $\tau = \mathrm{comp}_x(\ell')$. Therefore, k_0 and k_1 are conjugate.

Conversely, let us check that $\mathrm{pr}_x(k) = \mathrm{pr}_x(\tau \cdot k \cdot \tau^{-1})$, where k and τ belong to $\pi_1(X,x)$. Let $k = \mathrm{comp}_x(\ell)$ and $\tau = \mathrm{comp}_x(\ell')$, with ℓ and ℓ' in $\mathrm{Loops}(X,x)$. Then, $\mathrm{pr}_x(k) = \mathrm{comp}(\ell)$ and $\mathrm{pr}_x(\tau \cdot k \cdot \tau^{-1}) = \mathrm{comp}(\ell' \vee \ell \vee \bar{\ell}')$. Let us define $\gamma_s : t \mapsto \ell'(s + t(1-s))$, γ_s is a path in X satisfying $\gamma_s(0) = \ell'(s)$ and $\gamma_s(1) = \ell'(1) = x$. Then, since $\gamma_s(1) = \ell(0) = x$ and $\ell(1) = \bar{\gamma}_s(0) = x$, $\sigma_s = \gamma_s \vee \ell \vee \bar{\gamma}_s$ is well defined. Now, $\sigma_s(0) = \gamma_s(0) = \ell'(s)$ and $\sigma_s(1) = \bar{\gamma}_s(1) = \gamma_s(0) = \ell'(s)$, thus $\sigma_s \in \mathrm{Loops}(X)$. Next, $\sigma_0 = \gamma_0 \vee \ell \vee \bar{\gamma}_0 = \ell' \vee \ell \vee \bar{\ell}'$ and $\sigma_1 = \gamma_1 \vee \ell \vee \bar{\gamma}_1 = \mathbf{x} \vee \ell \vee \mathbf{x}$. Hence, we got a path $s \mapsto \sigma_s$ in $\mathrm{Loops}(X)$ connecting $\ell' \vee \ell \vee \bar{\ell}'$ to $\mathbf{x} \vee \ell \vee \mathbf{x}$, that is, $\mathrm{comp}(\ell' \vee \ell \vee \bar{\ell}') = \mathrm{comp}(\mathbf{x} \vee \ell \vee \mathbf{x})$, but since $\mathrm{comp}(\mathbf{x} \vee \ell \vee \mathbf{x}) = \mathrm{comp}(\ell)$, $\mathrm{comp}(\ell' \vee \ell \vee \bar{\ell}') = \mathrm{comp}(\ell)$. Therefore, $\mathrm{pr}_x(\tau \cdot k \cdot \tau^{-1}) = \mathrm{pr}_x(k)$.

Eventually, the map $\mathrm{pr}_x : \pi_1(X,x) \to \pi_0(\mathrm{Loops}(X))$ projects onto a bijection between the set of conjugacy classes of $\pi_1(X,x)$ and $\pi_0(\mathrm{Loops}(X))$. Note that the fact that not all the paths, in the proof, are *a priori* stationary, can be addressed by using the smashing function introduced in (art. 5.5) wherever we need. Note also that, if the group $\pi_1(X,x)$ is Abelian, then $\pi_0(\mathrm{Loops}(X)) = \pi_1(X,x)$, in particular when $X = G$ is a diffeological group.

Let us consider now the inclusion $\mathrm{Loops}(X,x) \subset \mathrm{Loops}(X)$ and let us choose $\ell \in \mathrm{Loops}(X,x)$. The short exact sequence of morphisms of pointed spaces, described in (art. 5.18), applied to this situation, writes

$$\begin{cases} \hat{0}_\# : (\pi_0(\text{Paths}(\text{Loops}(X), \text{Loops}(X,x), \ell), [t \mapsto \ell])) \to (\pi_0(\text{Loops}(X,x)), \ell), \\ i_\# : (\pi_0(\text{Loops}(X,x)), \ell) \to (\pi_0(\text{Loops}(X)), \ell). \end{cases}$$

This exercise tells us that $\text{val}(\hat{0}_\#)$, which coincides with $\ker(i_\#)$ — the subset of the components of $\text{Loops}(X,x)$ which can be connected, through $\text{Loops}(X)$, to ℓ — is the subset of classes of loops of X, pointed at x, conjugated in $\pi_1(X,x) = \pi_0(\text{Loops}(X,x))$ with the class of ℓ. Note in particular that the conjugacy class of the class of the constant loop $\mathbf{x} = [t \mapsto x]$ is reduced to the class of $\mathbf{x}$, thus the intersection of the connected component of $\mathbf{x}$ in $\text{Loops}(X)$ with $\text{Loops}(X,x)$ is reduced to the connected component of $\mathbf{x}$ in $\text{Loops}(X,x)$.

Exercise 88, p. 139 (Antisymmetric 3-form). By antisymmetry, $A_{jki} = -A_{jik} = +A_{ijk}$, as well, $A_{kij} = -A_{ikj} = +A_{ijk}$. Hence, for any triple of indices, $A_{ijk} = (1/3)[A_{ijk} + A_{jki} + A_{kij}]$. Now, A is the zero tensor if and only if all its coordinates are equal to zero, that is, for every triple of indices $A_{ijk} + A_{jki} + A_{kij} = 0$.

Exercise 89, p. 140 (Expanding the exterior product). The formula for the exterior product of 1-forms is given in (art. 6.15). We get first $[\text{Ext}(b)(c)](x_2)(x_3) = b(x_2)c(x_3) - b(x_3)c(x_2)$. Then,

$$\begin{aligned} \text{Ext}(a)(\text{Ext}(b)(c))(x_1)(x_2)(x_3) &= a(x_1)[\text{Ext}(b)(c)](x_2)(x_3) \\ &- a(x_2)[\text{Ext}(b)(c)](x_1)(x_3) \\ &- a(x_3)[\text{Ext}(b)(c)](x_2)(x_1) \\ &= a(x_1)[b(x_2)c(x_3) - b(x_3)c(x_2)] \\ &- a(x_2)[b(x_1)c(x_3) - b(x_3)c(x_1)] \\ &- a(x_3)[b(x_2)c(x_1) - b(x_1)c(x_2)]. \end{aligned}$$

Developing the factors gives immediately that

$$\text{Ext}(a)(\text{Ext}(b)(c))(x_1)(x_2)(x_3) = \sum_{\sigma \in \mathfrak{S}_3} \text{sgn}(\sigma)\, a(x_{\sigma(1)})b(x_{\sigma(2)})c(x_{\sigma(3)}).$$

Exercise 90, p. 140 (Determinant and isomorphisms). First of all, if the v_i are linearly independent, then they form a basis $\mathcal{B}$, and $\text{vol} = c\, \text{vol}_\mathcal{B}$, with $c \neq 0$. Then, $\text{vol}(v_1)\cdots(v_n) = c\, \text{vol}_\mathcal{B}(v_1)\cdots(v_n) = c \neq 0$. The contraposition of this sentence is, if $\text{vol}(v_1)\cdots(v_n) = 0$, then the v_i are not linearly independent. Therefore, $\text{vol}(v_1)\cdots(v_n) = 0$ if and only if the v_i are linearly independent. Next, let us assume that M has a nonzero kernel. Let $(\mathbf{e}_1, \ldots, \mathbf{e}_k)$ be a basis of $\ker(M)$. Thanks to the incomplete basis theorem, there exists a basis $\mathcal{B} = (\mathbf{e}_1, \ldots, \mathbf{e}_n)$ of E, with $\dim(E) = n$. Then, by definition of the determinant, $\det(M) = \text{vol}_\mathcal{B}(M\mathbf{e}_1)\cdots(M\mathbf{e}_n)$ (art. 6.19, (♡)). But $M\mathbf{e}_1 = 0$, thus $\det(M) = 0$. Conversely, let $\det(M) = 0$, then for any basis $(\mathbf{e}_1, \ldots, \mathbf{e}_n)$, $\text{vol}(M_1)\cdots(M_n) = 0$, where $M_i = M\mathbf{e}_i$. Hence, the M_i are not linearly independent. There exists then a family of numbers λ_i, not all zero, such that $\sum_{i=1}^n \lambda_i M_i = 0$, that is, $M(\sum_{i=1}^n \lambda_i \mathbf{e}_i) = 0$. Hence, $v = \sum_{i=1}^n \lambda_i \mathbf{e}_i \neq 0$ and $Mv = 0$. Therefore, $\ker(M) \neq \{0\}$ and M is not a linear isomorphism.

Exercise 91, p. 140 (Determinant is smooth). Once a basis is chosen, the determinant of M is a multilinear combination of the matrix coefficients of M

(art. 6.19, (♠)), thus the determinant is smooth. The variation of the determinant is given by application of the formula (art. 6.19, (♡)).

$$\begin{aligned}\delta[\det(M)] &= \delta[\mathrm{vol}_{\mathcal{B}}(M\boldsymbol{e}_1)\cdots(M\boldsymbol{e}_n)] \\ &= \sum_{i=1}^{n} \mathrm{vol}_{\mathcal{B}}(M\boldsymbol{e}_1)\cdots(\delta M\boldsymbol{e}_i)\cdots(M\boldsymbol{e}_n) \\ &= \sum_{i=1}^{n} \mathrm{vol}_{\mathcal{B}}(M\boldsymbol{e}_1)\cdots(MM^{-1}\delta M\boldsymbol{e}_i)\cdots(M\boldsymbol{e}_n) \\ &= \det(M)\times\sum_{i=1}^{n} \mathrm{vol}_{\mathcal{B}}(\boldsymbol{e}_1)\cdots(M^{-1}\delta M\boldsymbol{e}_i)\cdots(\boldsymbol{e}_n).\end{aligned}$$

But $M^{-1}\delta M\boldsymbol{e}_i = \sum_{j=1}^{n}[M^{-1}\delta M]_i^j\boldsymbol{e}_j$. Thus,

$$\begin{aligned}\mathrm{vol}_{\mathcal{B}}(\boldsymbol{e}_1)\cdots(M^{-1}\delta M\boldsymbol{e}_i)\cdots(\boldsymbol{e}_n) &= \mathrm{vol}_{\mathcal{B}}(\boldsymbol{e}_1)\cdots\left(\sum_{j=1}^{n}[M^{-1}\delta M]_i^j\boldsymbol{e}_j\right)\cdots(\boldsymbol{e}_n) \\ &= \sum_{j=1}^{n}[M^{-1}\delta M]_i^j\,\mathrm{vol}_{\mathcal{B}}(\boldsymbol{e}_1)\cdots(\boldsymbol{e}_j)\cdots(\boldsymbol{e}_n) \\ &= [M^{-1}\delta M]_i^i.\end{aligned}$$

Therefore,

$$\delta[\det(M)] = \det(M)\times\sum_{i=1}^{n}[M^{-1}\delta M]_i^i = \det(M)\mathrm{Tr}(M^{-1}\delta M).$$

Exercise 92, p. 140 (Determinant of a product). Let $\mathcal{B} = (\boldsymbol{e}_1,\ldots,\boldsymbol{e}_n)$ be a basis of E. According to (art. 6.19), $\det(MN) = \mathrm{vol}_{\mathcal{B}}(MN\boldsymbol{e}_1)\cdots(MN\boldsymbol{e}_n) = \det(M)\,\mathrm{vol}_{\mathcal{B}}(N\boldsymbol{e}_1)\cdots(N\boldsymbol{e}_n) = \det(M)\det(N)$. Next, $\det(s\times M) = \det((s\times \mathbf{1}_n)\,M) = \det(s\times\mathbf{1}_n)\det(M)$. A direct computation shows that $\det(s\times\mathbf{1}_n) = s^n$. Therefore, $\det(s\times M) = s^n\times\det(M)$.

Exercise 93, p. 146 (Coordinates of the exterior derivative). Let us decompose ω in a basis, $\omega(x) = \sum_{i<j<\cdots<k}\omega_{ij\cdots k}(x)\,\boldsymbol{e}^i\wedge\boldsymbol{e}^j\wedge\cdots\wedge\boldsymbol{e}^k$. By definition of the exterior derivative (art. 6.24),

$$d\omega(x) = \sum_{i<j<\cdots<k}\sum_{l=1}^{n}\partial_l\omega_{ij\cdots k}(x)\,\boldsymbol{e}^l\wedge\boldsymbol{e}^i\wedge\boldsymbol{e}^j\wedge\cdots\wedge\boldsymbol{e}^k.$$

The monomial $(d\omega)_{ijk\cdots l}\,\boldsymbol{e}^i\wedge\boldsymbol{e}^j\wedge\boldsymbol{e}^k\wedge\cdots\wedge\boldsymbol{e}^l$, with $i<j<k<\cdots<l$, is then obtained by grouping the terms containing the indices $ijk\cdots l$, that is,

$$\begin{aligned}(d\omega)_{ij\cdots kl}\,\boldsymbol{e}^i\wedge\boldsymbol{e}^j\wedge\boldsymbol{e}^k\wedge\cdots\wedge\boldsymbol{e}^l &= \partial_i\omega_{jk\cdots l}\,\boldsymbol{e}^i\wedge\boldsymbol{e}^j\wedge\boldsymbol{e}^k\wedge\cdots\wedge\boldsymbol{e}^l \\ &+ \partial_j\omega_{ik\cdots l}\,\boldsymbol{e}^j\wedge\boldsymbol{e}^i\wedge\boldsymbol{e}^k\wedge\cdots\wedge\boldsymbol{e}^l \\ &+ \partial_k\omega_{ij\cdots l}\,\boldsymbol{e}^k\wedge\boldsymbol{e}^i\wedge\boldsymbol{e}^j\wedge\cdots\wedge\boldsymbol{e}^l \\ &+ \cdots \\ &+ \partial_l\omega_{ijk\cdots}\,\boldsymbol{e}^l\wedge\boldsymbol{e}^i\wedge\boldsymbol{e}^j\wedge\boldsymbol{e}^k\wedge\cdots.\end{aligned}$$

Now, let us consider a term $\partial_k\omega_{ij\cdots l}\,\boldsymbol{e}^k\wedge\boldsymbol{e}^i\wedge\boldsymbol{e}^j\wedge\cdots\wedge\boldsymbol{e}^l$, where the index k is at the rank m in $ijk\cdots l$. If $m=1$, then $k=i$ and we do not change anything. If $m=2$, then $k=j$, we transpose the first two covectors to change $\boldsymbol{e}^j\wedge\boldsymbol{e}^i\wedge\boldsymbol{e}^k\wedge\cdots\wedge\boldsymbol{e}^l$

into $\mathbf{e}^i \wedge \mathbf{e}^j \wedge \mathbf{e}^k \wedge \cdots \wedge \mathbf{e}^l$ and we do not touch the indices of $\omega_{ik\dots l}$ in the partial derivative. If $m \geq 2$, then we perform $m-1$ transpositions of the covectors to reorder $\mathbf{e}^k \wedge \mathbf{e}^i \wedge \mathbf{e}^j \wedge \cdots \wedge \mathbf{e}^l$ into $\mathbf{e}^i \wedge \mathbf{e}^j \wedge \mathbf{e}^k \wedge \cdots \wedge \mathbf{e}^l$ and, thanks to the antisymmetry of the coefficient $\omega_{ij\dots l}$, we perform $m-2$ transpositions to send the index i, which is at the rank 1, to the rank $m-1$, changing $\omega_{ij\dots l}$ into $\omega_{ji\dots l}$. Then, the cost for these transpositions is $(-1)^{m-1} \times (-1)^{m-2} = (-1)^{2m-3} = -1$. Therefore, $(d\omega)_{ijk\dots\ell} = \partial_i \omega_{jk\dots\ell} - \partial_j \omega_{ik\dots\ell} - \partial_k \omega_{ji\dots\ell} - \cdots - \partial_\ell \omega_{jk\dots i}$.

☞ **Exercise 94, p. 147** (Integral of a 3-form on a 3-cube). Let α be a 3-form on $\mathbf{R}^3$, $\alpha = f(x_1, x_2, x_3)\, dx_1 \wedge dx_2 \wedge dx_3$. Let $x \mapsto F[x_1, x_2](x)$ be a primitive of $x \mapsto f(x_1, x_2, x)$, let $x \mapsto F[x_1](x)(x_3)$ be a primitive of $x \mapsto F[x_1, x](x_3)$, and let $x \mapsto F(x)(x_2)(x_3)$ be a primitive of $x \mapsto F[x](x_2)(x_3)$. The integral of α on a 3-cube $C = [a_1, b_1] \times [a_2, b_2] \times [a_3, b_3]$ is given by

$$\begin{aligned}
\int_C \alpha &= \int_{a_1}^{b_1} dx_1 \int_{a_2}^{b_2} dx_2 \int_{a_3}^{b_3} dx_3\, f(x_1, x_2, x_3) \\
&= \int_{a_1}^{b_1} dx_1 \int_{a_2}^{b_2} dx_2 \{F[x_1, x_2](b_3) - F[x_1, x_2](a_3)\} \\
&= \int_{a_1}^{b_1} dx_1 \{F[x_1](b_2)(b_3) - F[x_1](a_2)(b_3)\} \\
&- \int_{a_1}^{b_1} dx_1 \{F[x_1](b_2)(a_3) - F[x_1](a_2)(a_3)\} \\
&= \{F(b_1)(b_2)(b_3) - F(a_1)(b_2)(b_3)\} \\
&- \{F(b_1)(a_2)(b_3) - F(a_1)(a_2)(b_3)\} \\
&- \{F(b_1)(b_2)(a_3) - F(a_1)(b_2)(a_3)\} \\
&+ \{F(b_1)(a_2)(a_3) - F(a_1)(a_2)(a_3)\}.
\end{aligned}$$

☞ **Exercise 95, p. 158** (Functional diffeology of 0-forms). Let $\phi : V \to \Omega^0(X)$ be a plot for the functional diffeology (art. 6.29), that is, for all n-plots $P : U \to X$, $n \in \mathbf{N}$, the parametrization $(s, r) \mapsto \phi(s)(P)(r)$ belongs to $\mathcal{C}^\infty(V \times U, \Lambda^0(\mathbf{R}^n))$. But $\Omega^0(X) = \mathcal{C}^\infty(X, \mathbf{R})$, thus $\phi(s)(P)(r) = \phi(s)(P(r))$, and since $\Lambda^0(\mathbf{R}^n) = \mathbf{R}$, $[(s, r) \mapsto \phi(s)(P(r))] \in \mathcal{C}^\infty(V \times U, \mathbf{R})$. But this is the very definition of the plots for the functional diffeology of $\mathcal{C}^\infty(X, \mathbf{R})$ (art. 1.57).

☞ **Exercise 96, p. 158** (Differential forms against constant plots). Let $r_0 \in U$, and let V be an open neighborhood of r_0 such that $P \upharpoonright V = [r \mapsto x_0 = P(r_0)]$. Then, $P \upharpoonright V = [r \mapsto 0 \mapsto x_0]$, where $[r \mapsto 0]$ is the only map from V to $\mathbf{R}^0 = \{0\}$. By application of the compatibility axiom we have $\alpha(P \upharpoonright V) = [r \mapsto 0]^*(\alpha(\hat{x}_0))$, where $\hat{x}_0$ is the 0-plot $0 \mapsto x_0$, but $\alpha(\hat{x}_0) = 0$, thus $\alpha(P \upharpoonright V) = 0$. Therefore, since $\alpha(P)$ vanishes locally everywhere, $\alpha(P) = 0$.

☞ **Exercise 97, p. 159** (The equi-affine plane). First of all, note that $\alpha(\gamma)$ is trilinear and smooth. Thus, $\alpha(\gamma)$ is a smooth covariant 3-tensor on $\mathrm{def}(\gamma)$. Now, let us consider a plot $P : U \to \mathbf{R}^2$ for the wire diffeology. Then, for all $r_0 \in \mathrm{def}(P)$ there exist an open neighborhood V of r_0, a smooth map $q \in \mathcal{C}^\infty(V, \mathbf{R})$, an arc γ in $\mathbf{R}^2$, defined on some interval, such that $P \upharpoonright V = \gamma \circ q$, as shown in Figure Sol.6. Now, we have to check that if we have two such decompositions, satisfying $\gamma' \circ q' = \gamma \circ q$, then $q^*(\alpha(\gamma)) = q'^*(\alpha(\gamma'))$. Let $r \in U$ and $\delta r, \delta' r, \delta'' r \in \mathbf{R}^n$, where $n = \dim(U)$,

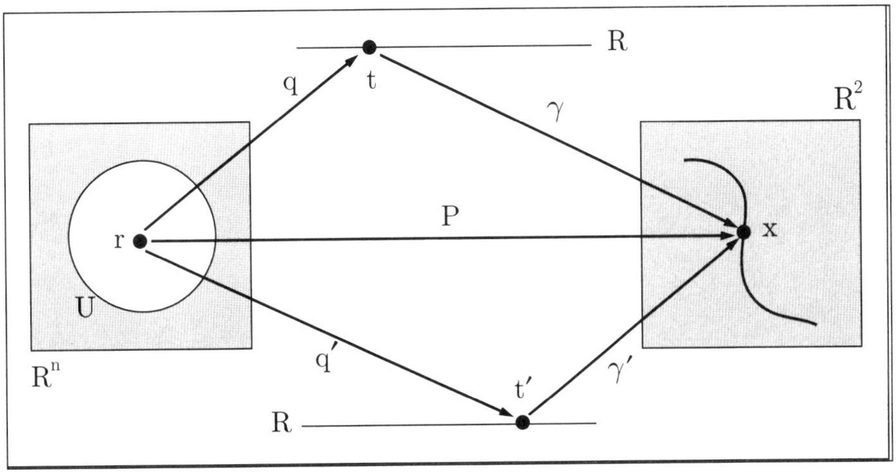

FIGURE Sol.6. Compatibility condition.

we have

$$q^*(\alpha(\gamma))_r(\delta r, \delta' r, \delta'' r) = \alpha(\gamma)_t(\delta t, \delta' t, \delta'' t),$$

with

$$t = q(r) \quad \text{and} \quad \delta t = \frac{\partial t}{\partial r}(\delta r) \quad \text{etc.},$$

and *mutatis mutandis* for γ' and q'. Next, let us derive the condition $\gamma' \circ q' = \gamma \circ q$. We get

$$D(\gamma \circ q)(r)(\delta r) = D(\gamma' \circ q')(r)(\delta r), \quad \text{that is,} \quad \dot\gamma(t)\delta t = \dot\gamma'(t')\delta t', \qquad (\diamondsuit)$$

and for the second derivative,

$$D^2(\gamma \circ q)(r)(\delta' r)(\delta'' r) = \ddot\gamma(t)(\delta' t)(\delta'' t) + \dot\gamma(t)\frac{\partial^2 t}{\partial r^2}(\delta' r)(\delta'' r).$$

We have then,

$$\ddot\gamma(t)(\delta' t)(\delta'' t) + \dot\gamma(t)\frac{\partial^2 t}{\partial r^2}(\delta' r)(\delta'' r) = \ddot\gamma'(t')(\delta' t')(\delta'' t') + \dot\gamma'(t')\frac{\partial^2 t'}{\partial r^2}(\delta' r)(\delta'' r),$$

and therefore,

$$\begin{aligned} \omega(\dot\gamma(t)(\delta t), \ddot\gamma(t)(\delta' t)(\delta'' t)) &= \omega\Big(\dot\gamma'(t')(\delta t'), \ddot\gamma'(t')(\delta' t')(\delta'' t') \\ &+ \dot\gamma'(t')\frac{\partial^2 t'}{\partial r^2}(\delta' r)(\delta'' r) - \dot\gamma(t)\frac{\partial^2 t}{\partial r^2}(\delta' r)(\delta'' r)\Big). \end{aligned}$$

But $\dot\gamma(t)$ and $\dot\gamma'(t')$ are collinear ($\diamondsuit$) and ω is antisymmetric, thus the second order derivatives in r disappear from the right hand side, and we get finally

$$\omega(\dot\gamma(t)(\delta t), \ddot\gamma(t)(\delta' t)(\delta'' t)) = \omega(\dot\gamma'(t')(\delta t'), \ddot\gamma'(t')(\delta' t')(\delta'' t')),$$

that is, $q^*(\alpha(\gamma)) = q'^*(\alpha(\gamma'))$. Thus, α is the expression, in the generating family of the wire plane, of a covariant 3-tensor.

☞ **Exercise 98, p. 159** (Liouville 1-form of the Hilbert space). 1) Let us develop the restriction of P' to V,

$$P' \upharpoonright V : r \mapsto \sum_{\alpha' \in A'} \lambda'_{\alpha'}(r)(X'_{\alpha'}, Y'_{\alpha'}).$$

Thus,

$$P \upharpoonright V = P' \upharpoonright V \Rightarrow \sum_{\alpha \in A} \lambda_\alpha X_\alpha = \sum_{\alpha' \in A'} \lambda'_{\alpha'} X'_{\alpha'} \text{ and } \sum_{\alpha \in A} \lambda_\alpha Y_\alpha = \sum_{\alpha' \in A'} \lambda'_{\alpha'} Y'_{\alpha'}.$$

Let

$$\Lambda(P \upharpoonright V) = \Big(\sum_{\alpha \in A} \lambda_\alpha X_\alpha\Big) \cdot \Big(\sum_{\beta \in A} d\lambda_\beta Y_\beta\Big) - \Big(\sum_{\alpha \in A} \lambda_\alpha Y_\alpha\Big) \cdot \Big(\sum_{\beta \in A} d\lambda_\beta X_\beta\Big).$$

Then,

$$\begin{aligned} \Lambda(P \upharpoonright V) - \Lambda(P' \upharpoonright V) &= \Big(\sum_{\alpha \in A} \lambda_\alpha X_\alpha\Big) \cdot \Big(\sum_{\alpha'' \in A''} d\lambda''_{\alpha''} Y''_{\alpha''}\Big) \\ &- \Big(\sum_{\alpha \in A} \lambda_\alpha Y_\alpha\Big) \cdot \Big(\sum_{\alpha'' \in A''} d\lambda''_{\alpha''} X''_{\alpha''}\Big), \end{aligned}$$

where A'' is the following reordering of the two sets of indices A and A', with $\lambda''_{\alpha''}$, $X''_{\alpha''}$ and $Y''_{\alpha''}$ following this reordering. Let $A = \{1, \dots, a\}$ and $A' = \{1, \dots, a'\}$. We denote $A'' = \{1, \dots, a''\}$ such that $a'' = a + a'$, with

$$\begin{array}{llll} \lambda''_{\alpha''} = \lambda_\alpha & \text{if } 1 \le \alpha'' \le a & \text{and } \lambda''_{\alpha''} = \lambda'_{\alpha'} & \text{if } a+1 \le \alpha'' \le a + a', \\ Y''_{\alpha''} = Y_\alpha & \text{if } 1 \le \alpha'' \le a & \text{and } Y''_{\alpha''} = -Y'_{\alpha'} & \text{if } a+1 \le \alpha'' \le a + a', \\ X''_{\alpha''} = X_\alpha & \text{if } 1 \le \alpha'' \le a & \text{and } X''_{\alpha''} = -X'_{\alpha'} & \text{if } a+1 \le \alpha'' \le a + a'. \end{array}$$

With this reordering we get

$$\sum_{\alpha \in A} \lambda_\alpha Y_\alpha = \sum_{\alpha' \in A'} \lambda'_{\alpha'} Y'_{\alpha'} \quad \Rightarrow \quad \sum_{\alpha'' \in A''} \lambda''_{\alpha''} Y''_{\alpha''} = 0,$$

$$\sum_{\alpha \in A} \lambda_\alpha X_\alpha = \sum_{\alpha' \in A'} \lambda'_{\alpha'} X'_{\alpha'} \quad \Rightarrow \quad \sum_{\alpha'' \in A''} \lambda''_{\alpha''} X''_{\alpha''} = 0.$$

Let us project these vectors on each factor $\mathbf{R}$ by the projection pr_k. We get

$$\sum_{\alpha'' \in A''} \lambda''_{\alpha''} Y''_{\alpha''} = 0 \quad \Rightarrow \quad \sum_{\alpha'' \in A''} \lambda''_{\alpha''} Y''_{\alpha'',k} = 0,$$

$$\sum_{\alpha'' \in A''} \lambda''_{\alpha''} X''_{\alpha''} = 0 \quad \Rightarrow \quad \sum_{\alpha'' \in A''} \lambda''_{\alpha''} X''_{\alpha'',k} = 0.$$

But $X''_{\alpha'',k}$ and $Y''_{\alpha'',k}$ are just real numbers, then both $\sum_{\alpha'' \in A''} \lambda''_{\alpha''} X''_{\alpha'',k}$ and $\sum_{\alpha'' \in A''} \lambda''_{\alpha''} Y''_{\alpha'',k}$ are smooth functions of $r \in V$. Since these functions vanish identically, their derivatives, with respect to r, vanish too. We get then

$$\sum_{\alpha'' \in A''} d\lambda''_{\alpha''} Y''_{\alpha'',k} = 0 \quad \Rightarrow \quad \sum_{\alpha'' \in A''} d\lambda''_{\alpha''} Y''_{\alpha''} = 0,$$

$$\sum_{\alpha'' \in A''} d\lambda''_{\alpha''} X''_{\alpha'',k} = 0 \quad \Rightarrow \quad \sum_{\alpha'' \in A''} d\lambda''_{\alpha''} X''_{\alpha''} = 0.$$

Finally, $\Lambda(P \upharpoonright V) = \Lambda(P' \upharpoonright V)$.

2) Let us consider a covering U_i of U such that the plot P, restricted to each U_i, is the sum of a finite linear combination of vectors, with smooth parametrizations as

coefficients. Let i and j be two indices of the covering, and let $P_i = P \restriction U_i$. By the previous statement we have

$$\Lambda(P_i) \restriction U_i \cap U_j = \Lambda(P_j) \restriction U_i \cap U_j.$$

Because a differential form is local (art. 6.36), there exists a 1-form $\Lambda(P) = \sup\{\Lambda(P_i)\}_i$, defined on U such that $\Lambda(P) \restriction U_i = \Lambda(P_i)$.

3) We still need to show that the map Λ is a 1-form on $\mathcal{H}_{\mathbf{R}} \times \mathcal{H}_{\mathbf{R}}$, that is, to check that for every plot $P : U \to \mathcal{H}_{\mathbf{R}} \times \mathcal{H}_{\mathbf{R}}$, and for any smooth parametrization $F : U' \to U$, $\Lambda(P \circ F) = F^*(\Lambda(P))$. Let $r'_0 \in U'$, $r_0 = F(r'_0)$, and

$$P \restriction V : r \mapsto \sum_{\alpha \in A} \lambda_\alpha(r)(X_\alpha, Y_\alpha),$$

as usual. Let us define now $V' = F^{-1}(V)$ and $\lambda'_\alpha = \lambda_\alpha \circ F$. We have

$$\begin{aligned}
\Lambda(P \circ F \restriction V')_{r'}(\delta r') &= \sum_{\alpha,\beta \in A} X_\alpha \cdot Y_\beta \, \lambda'_\alpha(r') d\lambda'_\beta(r')(\delta r') \\
&= \sum_{\alpha,\beta \in A} X_\alpha \cdot Y_\beta \, \lambda_\alpha(F(r')) d\lambda_\beta(F(r'))(D(F_{r'})(\delta r')) \\
&= \sum_{\alpha,\beta \in A} X_\alpha \cdot Y_\beta \, \lambda_\alpha(r) d\lambda_\beta(r)(\delta r),
\end{aligned}$$

with $r = F(r')$ and $\delta r = D(F)_{r'}(\delta r')$. But that is the definition of the pullback. Therefore, $\Lambda(P \circ F \restriction V') = F^*(\Lambda(P \restriction V))$. Then, since this is true locally, and since it is a local property, it is true globally and $\Lambda(P \circ F) = F^*(\Lambda(P))$.

☞ **Exercise 99, p. 159** (The complex picture of the Liouville form). The identity is obtained just by developing the computation as follows,

$$\begin{aligned}
(Z \cdot dZ - dZ \cdot Z)(P) &= \sum_{\alpha \in A} \lambda^*_\alpha (X_\alpha - iY_\alpha) \sum_{\beta \in A} d\lambda_\beta (X_\beta + iY_\beta) \\
&- \sum_{\alpha \in A} d\lambda^*_\alpha (X_\alpha - iY_\alpha) \sum_{\beta \in A} \lambda_\beta (X_\beta + iY_\beta) \\
&= \sum_{\alpha,\beta \in A} \lambda^*_\alpha d\lambda_\beta [X_\alpha X_\beta + Y_\alpha Y_\beta + i(X_\alpha Y_\beta - Y_\alpha X_\beta)] \\
&- \sum_{\alpha,\beta \in A} d\lambda^*_\alpha \lambda_\beta [X_\alpha X_\beta + Y_\alpha Y_\beta + i(X_\alpha Y_\beta - Y_\alpha X_\beta)] \\
&= \sum_{\alpha,\beta \in A} (X_\alpha X_\beta + Y_\alpha Y_\beta)(\lambda^*_\alpha d\lambda_\beta - d\lambda^*_\alpha \lambda_\beta) \\
&+ i \sum_{\alpha,\beta \in A} (X_\alpha Y_\beta - Y_\alpha X_\beta)(\lambda^*_\alpha d\lambda_\beta - d\lambda^*_\alpha \lambda_\beta).
\end{aligned}$$

But $\sum_{\alpha,\beta\in A}(X_\alpha X_\beta + Y_\alpha Y_\beta)(\lambda^*_\alpha d\lambda_\beta - d\lambda^*_\alpha\lambda_\beta) = 0$ for symmetry reasons. Hence, developing, for each index, $\lambda_\alpha = a_\alpha + ib_\alpha$, we get

$$\begin{aligned}(Z\cdot dZ - dZ\cdot Z)(P) \\ &= i\sum_{\alpha,\beta\in A}(X_\alpha Y_\beta - Y_\alpha X_\beta)(\lambda^*_\alpha d\lambda_\beta - d\lambda^*_\alpha\lambda_\beta)\\ &= i\sum_{\alpha,\beta\in A}(X_\alpha Y_\beta - Y_\alpha X_\beta)(a_\alpha da_\beta - a_\beta da_\alpha + b_\alpha db_\beta - b_\beta db_\alpha)\\ &\quad - \sum_{\alpha,\beta\in A}(X_\alpha Y_\beta - Y_\alpha X_\beta)(a_\alpha db_\beta + a_\beta db_\alpha - b_\alpha da_\beta - b_\beta da_\alpha).\end{aligned}$$

But the second term of the right hand side vanishes for symmetry reasons. Thus, it remains

$$\begin{aligned}(Z\cdot dZ - dZ\cdot Z)(P) &= i\sum_{\alpha,\beta\in A}(X_\alpha Y_\beta - Y_\alpha X_\beta)(a_\alpha da_\beta - a_\beta da_\alpha)\\ &\quad + i\sum_{\alpha,\beta\in A}(X_\alpha Y_\beta - Y_\alpha X_\beta)(b_\alpha db_\beta - b_\beta db_\alpha).\end{aligned}$$

Let us now come back to the map $\Phi : Z \mapsto (X, Y)$, identifying $\mathcal{H}_{\mathbf{C}}$ and $\mathcal{H}_{\mathbf{R}} \times \mathcal{H}_{\mathbf{R}}$. The plot $\Phi \circ P$ writes necessarily $\Phi\circ P(r) = \sum_{j\in\mathcal{J}}\mu_j(r)(X_j, Y_j)$. Then, developing $\sum_{\alpha\in A}\lambda_\alpha Z_\alpha$, we obtain the family $(\mu_j, (X_j, Y_j))_{j\in\mathcal{J}}$ as the union of two families,

$$(\mu_j, (X_j, Y_j))_{j\in\mathcal{J}} = (a_\alpha, (X_\alpha, Y_\alpha)_{\alpha\in A} \bigcup (b_\alpha, (-Y_\alpha, X_\alpha)_{\alpha\in A}.$$

Applying the form Λ to $\Phi\circ P$, for this family, we get

$$\begin{aligned}\Lambda(\Phi\circ P) &= \sum_{\alpha,\beta\in A}(X_\alpha Y_\beta - Y_\alpha X_\beta)(a_\alpha da_\beta - a_\beta da_\alpha)\\ &\quad + \sum_{\alpha,\beta\in A}(-Y_\alpha X_\beta + X_\alpha Y_\beta)(b_\alpha db_\beta - b_\beta db_\alpha).\end{aligned}$$

Comparing the last two expressions we get $\Lambda(\Phi\circ P) = (1/2i)(Z\cdot dZ - dZ\cdot Z)(P)$. Thanks to locality (art. 6.36), this equality is still satisfied for any plot of $\mathcal{H}_{\mathbf{R}}\times\mathcal{H}_{\mathbf{R}}$. Hence, we can conclude that

$$\Phi^*(\Lambda) = \frac{1}{2i}[Z\cdot dZ - dZ\cdot Z].$$

Exercise 100, p. 160 (The Fubini-Study 2-form). We use the notations of Exercise 99, p. 159, and $j(z)$ denotes the multiplication $Z \mapsto zZ$.

1) We have

$$j(z)^*(\varpi) = \frac{1}{2i}[(zZ)\cdot d(zZ) - d(zZ)\cdot(zZ)].$$

But since $z \in U(1)$, $z^*z = 1$, thus $(zZ)\cdot d(zZ) = Z\cdot dZ$, as well $d(zZ)\cdot(zZ) = dZ\cdot Z$. Therefore, $j(z)^*(\varpi) = \varpi$.

2) Let $P : \mathcal{O} \to \mathcal{S}_{\mathbf{C}}$ and $P' : \mathcal{O} \to \mathcal{S}_{\mathbf{C}}$ be two plots such that $\pi\circ P = \pi\circ P'$. By definition of π, there exists a unique parametrization $\zeta : \mathcal{O} \to U(1)$ such that $P'(r) = \zeta(r) \times P(r)$. We have to check that ζ is smooth. Since P and P' are plots of the fine diffeology, for every $r_0 \in \mathcal{O}$ there exist two local families (art. 3.7) $(\lambda_\alpha, Z_\alpha)_{\alpha\in A}$ and $(\lambda'_{\alpha'}, Z'_{\alpha'})_{\alpha'\in A'}$, defined on some open neighborhood V of r_0, such that $P(r) =_{\text{loc}} \sum_{\alpha\in A}\lambda_\alpha(r)Z_\alpha$ and $P'(r) =_{\text{loc}} \sum_{\alpha'\in A'}\lambda_{\alpha'}(r)Z'_{\alpha'}$. Moreover, the Z_α

and the $Z'_{\alpha'}$ can be chosen as bases of the vector subspaces, F and F', they generate (art. 3.8). Then, let us consider a basis $\mathcal{Z}'' = (Z''_{\alpha''})_{\alpha'' \in A''}$ of the sum $F'' = F + F'$. The condition $P'(r) = \zeta(r) \times P(r)$ writes in the basis $\mathcal{Z}''$,

$$\sum_{\beta'' \in A''} \lambda'_{\beta''}(r) Z''_{\beta''} = \zeta(r) \times \sum_{\beta'' \in A''} \lambda_{\beta''}(r) Z_{\beta''},$$

where the $\lambda_{\beta''}(r)$ and $\lambda'_{\beta''}(r)$ are the coordinates of $P(r)$ and $P'(r)$ in the basis $\mathcal{Z}''$. Thus, for every $\beta'' \in A''$, we have

$$\lambda'_{\beta''}(r) = \zeta(r) \times \lambda_{\beta''}(r).$$

But $P(r)$ and $P'(r)$ never vanish. There exists then some index $\beta'' \in A''$ for which $\lambda'_{\beta''}(r_0) \neq 0$. Since the function $\lambda'_{\beta''}$ is a smooth parametrization of $\mathbf{C}$, there exists an open neighborhood $W \subset V$ of r_0 such that $\lambda'_{\beta''}(r) \neq 0$ for all $r \in W$. Therefore, $\lambda_{\beta''}(r) \neq 0$ for all $r \in W$. Thus, for all $r \in W$, and for this index β'', $\zeta(r) = \lambda'_{\beta''}(r)/\lambda_{\beta''}(r)$. Hence, the function ζ is locally smooth. Therefore, ζ is smooth.

3) Let us define $dP(r)$ by its local expression $\sum_{\alpha \in A} d\lambda_\alpha(r) Z_\alpha$. Then, the form ϖ evaluated on the plot P, writes

$$\varpi(P)(r) = \frac{1}{2i}[P(r) \cdot dP(r) - dP(r) \cdot P(r)].$$

Evaluated on the plot $P' : r \mapsto \zeta(r)P(r)$, we have

$$\varpi(P')(r) = \frac{1}{2i}[(\zeta(r)P(r)) \cdot d(\zeta(r)P(r)) - d(\zeta(r)P(r)) \cdot (\zeta(r)P(r))].$$

After developing this expression, we get

$$\begin{aligned} \varpi(P')(r) &= \frac{1}{2i}[P(r) \cdot dP(r) - dP(r) \cdot P(r)] + \frac{1}{2i}[\zeta(r)^* d\zeta(r) - d(\zeta(r)^*)\zeta(r)] \\ &= \varpi(P) + \frac{1}{2i}[\zeta(r)^* d\zeta(r) - d(\zeta(r)^*)\zeta(r)]. \end{aligned}$$

But using $\zeta(r)^* = 1/\zeta(r)$, we get

$$\varpi(\zeta P)(r) = \varpi(P)(r) + \frac{d\zeta(r)}{i\zeta(r)}.$$

Then, since $d\zeta(r)/(i\zeta(r))$ is just $\zeta^*(\theta)(r)$, we get $\varpi(P') = \varpi(P) + \zeta^*(\theta)$.

4) Now, since $d\theta = 0$, we have

$$d[\varpi(P')] = d[\varpi(P) + \zeta^*(\theta)] = d[\varpi(P)] + \zeta^*(d\theta) = d[\varpi(P)].$$

Thus, by application of criterion (art. 6.38), there exists a 2-form ω on $\mathcal{P}_\mathbf{C} = \mathcal{S}_\mathbf{C}/U(1)$, such that $\pi^*(\omega) = d\varpi$. Since $d[\pi^*(\omega)] = dd\varpi = 0$, and thanks to (art. 6.39), $d\omega = 0$.

Exercise 101, p. 160 (Irrational tori are orientable). Let us consider two plots $P : U \to \mathbf{R}^n$ and $P' : U \to \mathbf{R}^n$ such that $\pi_\Gamma \circ P = \pi_\Gamma \circ P'$. Thus, the map $r \mapsto P'(r) - P(r)$ takes its values in Γ, but since Γ is discrete in $\mathbf{R}^n$, and $P' - P$ is smooth, the map $P' - P$ is locally constant. Then, there exist an open neighborhood V of r and an element $\gamma \in \Gamma$ such that $P' \restriction V = \tau_\gamma \circ P \restriction V$, where τ_γ is the translation by γ. Hence, $(P' \restriction V)^*(\mathrm{vol}_n) = (\tau_\gamma \circ P \restriction V)^*(\mathrm{vol}_n) = (P \restriction V)^*(\tau_\gamma^*(\mathrm{vol}_n)) = (P \restriction V)^*(\mathrm{vol}_n)$. Since $P^*(\mathrm{vol}_n)$ and $P'^*(\mathrm{vol}_n)$ coincide locally they coincide globally. Now, by application of (art. 6.38), there exists an n-form $\mathrm{vol}_\Gamma \in \Omega^n(T_\Gamma)$ such that $\pi_\Gamma^*(\mathrm{vol}_\Gamma) = \mathrm{vol}_n$. Since vol_n does not vanish anywhere, vol_Γ does not vanish anywhere (art. 6.39), and vol_Γ is a volume of the torus T_Γ. Hence, the torus T_Γ is orientable. Now, let vol be another volume on T_Γ. Its pullback $\pi_\Gamma^*(\mathrm{vol})$ is a volume of $\mathbf{R}^n$. Thus, there exists a smooth real function f on $\mathbf{R}^n$ such that $\pi_\Gamma^*(\mathrm{vol}) = f \times \mathrm{vol}_n$ (art. 6.44). But thanks to the invariance of $\pi_\Gamma^*(\mathrm{vol})$ and vol_n under the action of Γ, the function f also is invariant by Γ. Hence, since Γ is assumed to be dense, the function f is constant and $\mathrm{vol} = c \times \mathrm{vol}_\Gamma$, with $c \in \mathbf{R}$.

Exercise 102, p. 168 (The k-forms bundle on a real domain). For the first question, $P : V \to \Omega^k(U)$ is smooth if for all plots $Q : W \to U$, $(r, s) \mapsto [P(r)(Q)](s)$, defined on $V \times W$ is smooth. In particular, for $Q = \mathbf{1}_U$ that gives $(r, x) \mapsto [P(r)(\mathbf{1}_U)](x) = (r, x) \mapsto a_r(x)$ smooth, where $a_r(x) \in \Lambda^k(\mathbf{R}^n)$. Now, $(r, s) \mapsto [P(r)(Q)](s) = [P(r)(\mathbf{1}_U \circ Q)](s) = Q^*(P(r)(\mathbf{1}_U))(s) = Q^*(a_r)(s)$, hence, since pullback preserves smoothness, if $(r, x) \mapsto a_r(x)$ is smooth, then $(r, s) \mapsto Q^*(a_r)(s)$ also is smooth. Therefore, P is smooth if and only if $(r, x) \mapsto a_r(x)$ is smooth. For the second question, the condition $\alpha_x = \beta_x$ means that $\alpha(Q)(0) = \beta(Q)(0)$ for all plots $Q : V \to U$ centered at x, that is, $Q(0) = x$. In particular, for $Q = T_x$, where T_x is the translation $x' \mapsto x' + x$, $\alpha(T_x)(0) = \beta(T_x)(0)$, that is, $T_x^*(\alpha(\mathbf{1}_U))(0) = T_x^*(\beta(\mathbf{1}_U))(0)$, which gives $a(x) = b(x)$. Conversely, for every k-tuple of vectors of $\mathbf{R}^n$, denoted by $[v_i]$, $\alpha(Q)(0)[v_i] = Q^*(a)(0)[v_i] = a(x)[D(Q)(0)(v_i)]$. Thus, if $a(x) = b(x)$, then $\alpha(Q)(0) = \beta(Q)(0)$ and $\alpha_x = \beta_x$. Therefore, the value of a k-form α on U at the point x is defined by the value $a(x) = \alpha(\mathbf{1}_U)(x) \in \Lambda^k(\mathbf{R}^n)$. Since for every $a \in \Lambda^k(\mathbf{R}^n)$ we can choose the constant form $a(x) = a$, every $a \in \Lambda^k(\mathbf{R}^n)$ is the value of some smooth k-form α on U, and $\Lambda_x^k(U) \simeq \Lambda^k(\mathbf{R}^n)$. For the third question, note first that the map ϕ is well defined. We have seen that if $\alpha_x = \beta_x$, then $a(x) = b(x)$. Now, let $P : V \to \Lambda^k(U)$ be a plot, and let $Q : W \to U \times \Omega^k(U)$ be a smooth local lift of P along the projection $\pi : U \times \Omega^k(U) \to \Lambda^k(U)$, defined by $\pi(x, \alpha) = (x, \alpha_x)$. Let $Q(r) = (x_r, A(r))$ and $a_r = A(r)(\mathbf{1}_U)$, then $\phi \circ (P \restriction W) = \phi \circ \pi \circ Q$ gives $\phi(P(r)) = (x_r, A(r)(\mathbf{1}_U)(x_r) = a_r(x_r))$. Since $A : W \to \Omega^k(U)$ is smooth, $r \mapsto a_r = A(r)(\mathbf{1}_U)$ is smooth, and since $r \mapsto x_r$ is smooth, $r \mapsto a_r(x_r)$ is smooth. Therefore $\phi \circ (P \restriction W)$ is smooth, which implies that $\phi \circ P$ is smooth, and therefore ϕ. Conversely, let $P : r \mapsto (x_r, a_r)$ be a plot of $U \times \Lambda^k(\mathbf{R}^n)$, let $\mathbf{a}_r$ be the constant k-form on U with value a_r. Then, $\phi^{-1}(x_r, a_r) = \pi(x_r, \mathbf{a}_r)$, and since $r \mapsto x_r$, π and $r \mapsto \mathbf{a}_r$ are smooth, $\phi^{-1} \circ P$ is smooth. Therefore, ϕ is a diffeomorphism.

Exercise 103, p. 169 (The p-form bundle on a manifold). For the first question, let $x \in U$. We can choose $\varepsilon > 0$ such that the ball $B(x, \varepsilon)$ is strictly contained in U, that is, $U - B(x, \varepsilon) \neq \varnothing$. Then, there exists a smooth real function λ equal to 1 on the ball of radius $\varepsilon/2$ centered at x, and equal to zero, outside $B(x, \varepsilon)$. The smooth p-form $\bar{a} : x \mapsto \lambda(x)\, a$ satisfies $\bar{a}(x) = a$ and vanishes outside $B(x, \varepsilon)$. For the second question, Let $\mathcal{O} = F(U)$, since F is a local diffeomorphism, $\mathcal{O}$ is D-open and $F_*(\bar{a}) = (F^{-1})^*(\bar{a})$ is a p-form on $\mathcal{O}$, vanishing outside the D-open subset $F(B(x, \varepsilon))$. Since differential forms are local (art. 6.36), $F_*(\bar{a})$ can be extended by a differential p-form α on M such that $\alpha \restriction (M - \mathcal{O}) = 0$. Now, since F is a local diffeomorphism, and thanks to Exercise 102, p. 168, the value of α at the point m is characterized by the value of $\bar{a}$ at x, that is, a. More precisely $\alpha_m = F_*(\bar{a}_x)$, that is, with our identification, $\alpha_m = F_*(a)$, where F_* is defined in (art. 6.51). Therefore, $\Lambda_m^p(M) \simeq \Lambda^p(\mathbf{R}^n)$. That answers the third question. For the fourth question, we just built the map $\mathcal{F}$, from $U \times \Lambda^p(\mathbf{R}^n)$

to $\Lambda^p(M)$, defined by $\mathcal{F}(x, a) = (F(x), F_*(a))$. By construction, $\mathcal{F}$ is bijective. Let us check rapidly that $\mathcal{F}$ is a local diffeomorphism. Let $x_0 \in U$, there exists an open ball B_0 centered at x_0 and $\varepsilon > 0$ such that, for all $x \in B_0$, $B(x, \varepsilon)$ is strictly included in U. Then, by the previous construction, we get for each $(x, a) \in B_0 \times \Lambda^p(\mathbf{R}^n)$ a smooth p-form $\bar{a}$ such that $\bar{a}_x = a$. Moreover, we can choose for each $x \in B_0$ the bump-function λ depending smoothly on x, thus the map $(x, a) \mapsto (x, \bar{a})$ is smooth. Hence, $\mathcal{F} \restriction B_0$ is smooth, therefore $\mathcal{F}$. Conversely, a smooth parametrization $r \mapsto (m_r, \alpha_r)$ in $\Lambda^p(M)$ can be locally lifted by a smooth parametrization $r \mapsto (m_r, A(r))$ in $M \times \Omega^p(M)$, that is, $\alpha_r = A(r)_{m_r}$. Then, by pullback, we get a smooth parametrization $r \mapsto (F^{-1}(m_r), F^*(A(r)))$ in $U \times \Omega^p(U)$ such that $\mathcal{F}^{-1}(m_r, \alpha_r) = (x_r = F^{-1}(m_r), [F^*(A(r))]_{x_r})$. Therefore, $\mathcal{F}^{-1}$ is smooth and $\mathcal{F}$ is a local diffeomorphism. When F runs over an atlas of M the charts $\mathcal{F}$ run obviously over an atlas of $\Lambda^p(M)$. Finally, $\Lambda^p(M)$ is a manifold, and $\dim(\Lambda^p(M)) = \dim(M) + \dim(\Lambda^p(\mathbf{R}^n)) = \dim(M) + C^p_n$. Note that also shows that $\Lambda^p(M)$ is a locally trivial bundle over M (art. 8.9).

Exercise 104, p. 169 (Smooth forms on diffeological vector spaces). 1) Let $\alpha \in \Omega^1(\mathcal{O})$, and let us show that $A(x) : u \mapsto \alpha(t \mapsto x + tu)_0(1)$ is linear. Let $u, v \in E$, $A(x)(u+v) = \alpha(t \mapsto x+tu+tv)_0(1)$. Let $\phi : (t, s) \mapsto x+tu+sv$ and $\Delta : t \mapsto (t, t)$. Then, $\alpha(t \mapsto x + tu + tv)_0(1) = \alpha(\phi \circ \Delta)_0(1) = \Delta^*(\alpha(\phi))_0(1) = \alpha(\phi)_{(0,0)}(1, 1) = \alpha(\phi)_{(0,0)}(1, 0) + \alpha(\phi)_{(0,0)}(0, 1)$. Now, let $j_1 : t \mapsto (t, 0)$ and $j_2 : s \mapsto (0, s)$, then $\alpha(\phi)_{(0,0)}(1, 0) = j_1^*(\alpha(\phi))_0(1) = \alpha(\phi \circ j_1)_0(1) = \alpha(t \mapsto x + tu)_0(1) = A(x)(u)$ and $\alpha(\phi)_{(0,0)}(0, 1) = j_2^*(\alpha(\phi))_0(1) = \alpha(\phi \circ j_2)_0(1) = \alpha(s \mapsto x + sv)_0(1) = A(x)(v)$. Thus, $A(x)(u + v) = A(x)(u) + A(x)(v)$. Next, let $\lambda \in \mathbf{R}$, $A(x)(\lambda u) = \alpha(t \mapsto x + t\lambda u)_0(1) = \alpha(t \mapsto s = t\lambda \mapsto x + su)_0(1) = \alpha(s \mapsto x + su)_0(\lambda) = \lambda\alpha(s \mapsto x + su)_0(1) = \lambda A(x)(u)$. Therefore, $A(x)$ is linear, that is, $A(x) \in E^*$. Let us show that $A(x) \in E^*_\infty$, that is, for every n-plot $Q : V \to E$, $A(x) \circ Q \in \mathcal{C}^\infty(V, \mathbf{R})$. Let $T_x : u \mapsto u + x$, T_x is a diffeomorphism of E, then $A(x)(Q(s)) = \alpha(t \mapsto x + tQ(s))_0(1) = T_x^*(\alpha)(t \mapsto tQ(s))_0(1)$. But $(t, s) \mapsto tQ(s)$ is a plot of E, and $T_x^*(\alpha)$ is a differential 1-form on E, thus $T_x^*(\alpha)((t, s) \mapsto tQ(s))$ is a smooth 1-form on $\mathcal{I} \times V$, where $\mathcal{I}$ is a small interval around $0 \in \mathbf{R}$. Hence, the map $(t, s) \mapsto T_x^*(\alpha)((t, s) \mapsto tQ(s))_{(t,s)}$ is a smooth parametrization in $\Lambda^1(\mathbf{R}^{1+n})$, and $s \mapsto T_x^*(\alpha)(t \mapsto tQ(s))_0(1) = T_x^*(\alpha)((t, s) \mapsto tQ(s))_{(0,s)}(1, 0)$ is smooth. Let us show now that $A \in \mathcal{C}^\infty(\mathcal{O}, E^*_\infty)$. A is smooth if for every plot $P : U \to \mathcal{O}$ and $Q : V \to E$, the parametrization $(r, s) \mapsto A(P(r))(Q(s)) = \alpha(t \mapsto P(r) + tQ(s))_0(1)$ is smooth. But $(r, s, t) \mapsto P(r) + tQ(s)$ is a plot of E, we are in a situation analogous to the previous one and thus A is smooth.

2) Conversely, let $A \in \mathcal{C}^\infty(\mathcal{O}, E^*_\infty)$ and $\alpha(P)_r = d[A(x) \circ P]_r$, with $x = P(r)$, that is, $\alpha(P)_r = d[s \mapsto A(P(r)) \circ P(s)]_{s=r}$. Note first that, since $A(P(r)) \circ P$ is a real function defined on U, $d[s \mapsto A(P(r)) \circ P(s)]_{s=r}$ is a smooth 1-form on U. Next, let $F : V \to U$ be a smooth parametrization, let $t \in V$ and $r = f(t)$. On the one hand, $\alpha(P \circ F)_t = d[s \mapsto A(P \circ F(t)) \circ P \circ F(s)]_{s=t} = D(A(P(F(t))) \circ P \circ F)(s = t)$, and on the other hand, $F^*(\alpha(P))_t = \alpha(P)_{r=F(t)} \circ D(F)(t) = d[s \mapsto A(P(r)) \circ P(s)]_{s=r=F(t)} \circ D(F)(t) = D(A(P(F(t))) \circ P)(r = F(t)) \circ D(F)(t)$, but $D(A(P(F(t))) \circ P \circ F)(s = t) = D(A(P(F(t))) \circ P)(r = F(t)) \circ D(F)(t)$, by the chain rule. Thus, $\alpha(P \circ F)_t = F^*(\alpha(P))_t$. Therefore, α is a differential 1-form on $\mathcal{O}$.

3) Let us check now that $\sigma \circ \pi = \mathbf{1}$. Let $\sigma(A) = \alpha$ and $\pi(\alpha) = A'$, then $A'(x)(u) = \alpha(t \mapsto P(t) = x+tu)_{t=0}(1) = D(t \mapsto A(P(0)) \circ P(t))_{t=0}(1) = D[t \mapsto A(x)(x+tu) = A(x)(x) + tA(x)(u)]_{t=0}(1) = A(x)(u)$, thus $A' = A$.

4) We remark that π is linear, and let $\alpha \in \ker(\pi)$, that is, $\alpha(t \mapsto x + tu)_0(1) = 0$, for all $x \in \mathcal{O}$ and all $u \in E$. We know that the value of a 1-form is given by its values on the 1-plots (art. 6.37), that is, if $\alpha(c) = 0$, for all 1-plots c of E, then $\alpha = 0$. Now, let c be defined on an interval around $t_0 \in \mathbf{R}$, and let $x_0 = c(t_0)$. Decompose c into $t \mapsto s = t - t_0 \mapsto \bar{c}(s) = c(s + t_0) - x_0 \mapsto \bar{c}(s) + x_0$, that is, $c = T_{x_0} \circ \bar{c} \circ T_{-t_0}$. Then, $\alpha(c)_{t_0}(1) = \alpha(T_{x_0} \circ \bar{c} \circ T_{-t_0})_{t_0}(1) = T^*_{x_0}(\alpha)(\bar{c})_0(1)$. But, by hypothesis, there exists $u \in E$ such that $T^*_{x_0}(\alpha)(\bar{c})_0(1) = T^*_{x_0}(\alpha)(t \mapsto tu)_0(1)$, Then, since $\alpha \in \ker(\pi)$, $\alpha(c)_{t_0}(1) = \alpha(t \mapsto x_0 + tu)_0(1) = 0$, thus $\alpha = 0$. Therefore, π is injective and surjective, since $\pi \circ \sigma = \mathbf{1}$, that is, a linear isomorphism.

Exercise 105, p. 170 (Forms bundles of irrational tori). Let α be a p-form on T_Γ. Let $a = \pi^*(\alpha)$. Thus, for each $\gamma \in \Gamma$, $\gamma^*(a) = \gamma^*(\pi^*(\alpha)) = (\pi \circ \gamma)^*(\alpha) = \pi^*(\alpha) = a$. Hence, a is invariant under the translations $\gamma \in \Gamma$. Now, let us decompose the form a of $\mathbf{R}^n$ in the canonical basis of $\Lambda^p(\mathbf{R}^n)$ (art. 6.16),

$$a(x) = \sum_{ij\cdots k} a_{ij\cdots k}(x)\ \mathbf{e}^i \wedge \mathbf{e}^j \wedge \cdots \wedge \mathbf{e}^k,$$

where $x \in \mathbf{R}^n$ and the $a_{ij\cdots k}$ are smooth real functions. Since the monomials $\mathbf{e}^i \wedge \mathbf{e}^j \wedge \cdots \wedge \mathbf{e}^k$ are invariant by translation, we get

$$\gamma^*(a) = a \quad \Rightarrow \quad a_{ij\cdots k}(x + \gamma) = a_{ij\cdots k}(x),$$

for all families of indices $ij \cdots k$. But since Γ is a dense subgroup of $\mathbf{R}^n$, and the $a_{ij\cdots k}$ are smooth, they are constant. Therefore, a is a constant p-form of $\mathbf{R}^n$. Now, let $a \in \Lambda^p(\mathbf{R}^n)$, and let us show that there exists α, p-form of T_Γ, such that $a = \pi^*(\alpha)$. We shall apply the criterion (art. 6.38). Let P and P' be two plots of $\mathbf{R}^n$, defined on the same domain U, such that $\pi \circ P = \pi \circ P'$. Thus, for every $r \in U$, $P'(r) - P(r) \in \Gamma$. But the parametrization $P' - P$ is smooth, hence locally constant, since Γ is discrete. Thus, restricting P and P' to connected parts of U, the difference $P' - P$ is constant, that is, $P'(r) = P(r) + \gamma$. Thus, $P'^*(a) = P^*(a)$ on the whole U. The criterion is satisfied and there exists $\alpha \in \Omega^p(T_\Gamma)$ such that $a = \pi^*(\alpha)$. Conversely, the map $a \mapsto \alpha$ is injective, indeed, thanks to (art. 6.39), if $\pi^*(\alpha) = 0$, then $\alpha = 0$. Therefore, the map $a \mapsto \alpha$, defined on $\Lambda^p(\mathbf{R}^n)$ to $\Omega^p(T_\Gamma)$ is an isomorphism. Next, let us consider $T_\Gamma \times \Omega^p(T_\Gamma) \simeq T_\Gamma \times \Lambda^p(\mathbf{R}^n)$. A pair (τ, α) is equivalent to (τ', α') if and only if $\tau = \tau'$ and $\alpha_\tau = \alpha'_{\tau'}$. But $\alpha_\tau = \alpha'_{\tau'}$ implies $a = a'$, where $a = \pi^*_\Gamma(\alpha)$ and $a' = \pi^*_\Gamma(\alpha')$. Hence,

$$\Lambda^p(T_\Gamma) \simeq T_\Gamma \times \Lambda^p(\mathbf{R}^n).$$

NOTE. We can check that a smooth section $\tau \mapsto (\tau, a)$ of $\Lambda^p(T_\Gamma)$ is just a smooth map $\tau \mapsto a$, and since $\mathcal{C}^\infty(T_\Gamma, \Lambda^p(\mathbf{R}^n)) \simeq \Lambda^p(\mathbf{R}^n)$ are just the constant maps, then every smooth p-form on T_Γ is *constant*. Now, we can look at the Liouville form on $\Lambda^p(T_\Gamma)$. But, rather than looking at Liouv, let us consider its pullback $\pi^*_\Gamma(\text{Liouv})$ on $\mathbf{R}^n \times \Lambda^p(\mathbf{R}^n)$. Let $Q \times A$ be a plot of $\mathbf{R}^n \times \Lambda^p(\mathbf{R}^n)$, it is just a pair of smooth parametrizations defined on some domain U. Then, let $r \in U$ and $\delta r = (\delta_1 r) \cdots (\delta_p r)$ be a p-vector of $\mathbf{R}^n$,

$$\begin{aligned} \pi^*_\Gamma(\text{Liouv})(Q \times A)(r)(\delta r) &= A(r)(Q)(r)(\delta r) \\ &= \sum_{i<j<\cdots<k} A(r)_{ij\cdots k}(\delta_i q)(\delta_j q)\cdots(\delta_k q), \end{aligned}$$

where the $\delta_i q = D(Q)(r)(\delta_i r)$ are again vectors of $\mathbf{R}^n$.

Exercise 106, p. 170 (Vector bundles of irrational tori). Let T_Γ be the irrational torus $\mathbf{R}^n/\Gamma$ where Γ is a dense generating subgroup of $\mathbf{R}^n$ and $\Lambda^p(T_\Gamma) \simeq T_\Gamma \times \Lambda^p(\mathbf{R}^n)$; see Exercise 105, p. 170. Let Q be a global p-plot of T_Γ. There exist a small ball B, centered at $0 \in \mathbf{R}^p$, and a lifting $\tilde{Q} : B \to \mathbf{R}^n$, such that $\pi \circ \tilde{Q} = Q \upharpoonright B$. For the same reason developed in the proof of Exercise 105, p. 170, two such liftings differ from a constant element of Γ. Now, let $\alpha \in \Omega^p(T_\Gamma)$, and let $a \in \Lambda^p(\mathbf{R}^n)$ be the unique p-form such that $a = \pi^*(\alpha)$ (see Exercise 105, p. 170). We have, $\alpha(Q)(0) = \alpha(\pi \circ \tilde{Q})(0) = \pi^*(\alpha)(\tilde{Q})(0) = a(\tilde{Q})(0) = a(q_1, \ldots, q_p)$, where $q_i = D(\tilde{Q})(0)(e_i) \in \mathbf{R}^n$, $i = 1 \cdots p$. A translation of $\tilde{Q}$ by a constant element $\gamma \in \Gamma$ does not change the q_i. Hence, $\alpha(Q)(0) = \alpha(Q')(0)$ if and only if $q_i = q_i'$ for all $i = 1 \cdots p$. On the other hand, for any p vectors $(v_1, \ldots, v_p) \in (\mathbf{R}^n)^p$, there exists a p-plot $Q = (t_1, \ldots, t_p) \mapsto \pi(\sum_{i=1}^p t_i v_i)$ such that $q_i = v_i$. Hence, the map j_p from $\mathrm{Paths}_p(T_\Gamma)$ to $T_\Gamma \times L^\infty(\Omega^p(T_\Gamma), \mathbf{R}) = T_\Gamma \times L(\Lambda^p(\mathbf{R}^n), \mathbf{R})$ is given by

$$j_p : Q \mapsto (Q(0), [a \mapsto a(q_1, \ldots, q_p)]).$$

But the map $[a \mapsto a(q_1, \ldots, q_p)]$ is just an element of the dual of $\Lambda^p(\mathbf{R}^n)$. Then, since each element of the dual $\Lambda^p(\mathbf{R}^n)^*$ can be associated with a global p-plot,

$$T^p(T_\Gamma) \simeq T_\Gamma \times [\Lambda^p(\mathbf{R}^n)]^* \simeq T_\Gamma \times \mathbf{R}^{n!/p!(n-p)!}.$$

In particular,

$$T_x(T_\Gamma) \simeq \mathbf{R}^n \quad \text{and} \quad T(T_\Gamma) \simeq T_\Gamma \times \mathbf{R}^n.$$

NOTE. If we had tested the paths on differential of function, which is one of the usual ways in classical differential geometry of manifolds, we should have get $T_x(T_\Gamma) = \{0\}$, since the only real functions defined on T_Γ are constant. And this is clearly unsatisfactory. Thus, the definition suggested in (art. 6.53) is, for diffeological spaces at least, better. However, note that a section of the tangent bundle is necessarily constant $x \mapsto (x, v)$, with $v \in \mathbf{R}^n$, which is not really surprising.

Exercise 107, p. 170 (Differential 1-forms on $\mathbf{R}/\{\pm 1\}$). Let us check that the 1-form $d[t^2] = 2t \times dt$ passes to the quotient Δ. We apply the criterion (art. 6.38). Let P and P' be two plots such that $\mathrm{sq} \circ P = \mathrm{sq} \circ P'$, that is, $P(r)^2 = P'(r)^2$ for all $r \in \mathrm{def}(P) = \mathrm{def}(P')$, then $d[t^2](P)_r = D(r \mapsto P(r)^2)_r = D(r \mapsto P'(r)^2)_r = d[t^2](P')_r$. Thus, there exists a 1-form θ on Δ such that $\mathrm{sq}^*(\theta) = d[t^2]$. Now, let α be a differential 1-form on Δ, its pullback $\mathrm{sq}^*(\alpha)$ is a differential 1-form on $\mathbf{R}$, that is $\mathrm{sq}^*(\alpha) = F(t)dt$. But $\mathrm{sq}^*(\alpha)$ is invariant by $\{\pm 1\}$, hence $F(-t) = -F(t)$, then F vanishes at $t = 0$. Since F is $\mathcal{C}^\infty$, there exists a smooth function φ such that $F(t) = 2t\varphi(t)$. Thus, $\mathrm{sq}^*(\alpha) = 2t\varphi(t)dt = \varphi(t)d[t^2]$. Now, by invariance, there exists a smooth function $f \in \mathcal{C}^\infty(\Delta, \mathbf{R})$ such that $\mathrm{sq} \circ \varphi = f$. Therefore, $\mathrm{sq}^*(\alpha) = \mathrm{sq}^*(f \times \theta)$. By (art. 6.39), $\alpha = f \times \theta$. Next, every 1-plot γ of Δ such that $\gamma(0) = 0$ has a local lift $\bar{\gamma}$ in $\mathbf{R}$ such that $\bar{\gamma}(0) = 0$, and then $\alpha(\gamma)_0(1) = \mathrm{sq}^*(\alpha)(\bar{\gamma})_0(1) = \varphi(\bar{\gamma}(0)) \times 2\bar{\gamma}(0)\bar{\gamma}'(0) = 0$, where the prime $'$ denotes the derivative with respect to t. Therefore, $\alpha_0 = 0$ and by consequence $T_0(\Delta) = \{0\}$.

Exercise 108, p. 175 (Anti-Lie derivative). Let $P : U \to X$ be an n-plot, and let us shortly denote p vectors by $[v] = (v_1) \cdots (v_p)$, we have on the one hand

$$\frac{\partial}{\partial t}\left\{F(t)^*(F(t)_*(\alpha))(P)_r[v]\right\}_{t=0} = \frac{\partial}{\partial t}\left\{\alpha(P)_r[v]\right\}_{t=0} = 0,$$

and on the other hand,

$$
\begin{aligned}
\frac{\partial}{\partial t}\left\{F(t)^*(F(t)_*(\alpha))(P)_r[v]\right\}_{t=0} &= D\left(\begin{pmatrix} t \\ s \end{pmatrix} \mapsto F(t)^*(F(s)_*(\alpha))(P)_r[v]\right)_{\binom{0}{0}} \begin{pmatrix} 1 \\ 1 \end{pmatrix} \\
&= D\left(\begin{pmatrix} t \\ s \end{pmatrix} \mapsto F(t)^*(F(s)_*(\alpha))(P)_r[v]\right)_{\binom{0}{0}} \begin{pmatrix} 1 \\ 0 \end{pmatrix} \\
&+ D\left(\begin{pmatrix} t \\ s \end{pmatrix} \mapsto F(t)^*(F(s)_*(\alpha))(P)_r[v]\right)_{\binom{0}{0}} \begin{pmatrix} 0 \\ 1 \end{pmatrix} \\
&= D(t \mapsto F(t)^*(\alpha)(P)_r[v])_0(1) \\
&+ D(s \mapsto F(s)_*(\alpha)(P)_r[v])_0(1) \\
&= \frac{\partial}{\partial t}\left\{F(t)^*(\alpha)(P)_r[v]\right\}_{t=0} \\
&+ \frac{\partial}{\partial s}\left\{F(s)_*(\alpha)(P)_r[v]\right\}_{s=0}.
\end{aligned}
$$

Therefore,

$$
\frac{\partial}{\partial t}\left\{F(t)_*(\alpha)(P)_r[v]\right\}_{t=0} = -\frac{\partial}{\partial t}\left\{F(t)^*(\alpha)(P)_r[v]\right\}_{t=0},
$$

that is,

$$
\frac{\partial}{\partial t}\left\{F(t)_*(\alpha)\right\}_{t=0} = -\pounds_F(\alpha).
$$

☞ **Exercise 109, p. 175** (Multi-Lie derivative). First of all, the proof that this generalization of the Lie derivative is well defined is a slight adaptation of the first proposition of (art. 6.54). For the second question, let $P : U \to X$ be a plot. The Lie derivative decomposes in the canonical basis $(\mathbf{e}_1, \ldots, \mathbf{e}_q)$ of $\mathbf{R}^q$,

$$
\begin{aligned}
\pounds_h(\alpha)(v)(P) &= D[s \mapsto h(s)^*(\alpha(P))](0)(v) \\
&= \sum_{i=1}^{q} v^i \times D[s \mapsto h(s)^*(\alpha(P))](0)(\mathbf{e}_i) \\
&= \sum_{i=1}^{q} v^i \times D[s \mapsto h(s\mathbf{e}_i)^*(\alpha(P))](0)(1) \\
&= \sum_{i=1}^{q} v^i \times D[s \mapsto h_i(s)^*(\alpha(P))](0)(1).
\end{aligned}
$$

But

$$
D[s \mapsto h_i(s)^*(\alpha(P))](0)(1) = \frac{\partial}{\partial s}(h_i(s)^*(\alpha(P))\Big|_{s=0} = \pounds_{h_i}(\alpha)(P).
$$

Therefore,

$$
\pounds_h(\alpha)(P) = \sum_{i=1}^{q} v^i \pounds_{h_i}(\alpha)(P).
$$

☞ **Exercise 110, p. 175** (Variations of points of domains). Let F be an arc of the plot x, defined on $]-\varepsilon, +\varepsilon[$. So, $F(0) = x$, $[(s, 0) \mapsto F(s)(0)] \in \mathcal{C}^\infty(]-\varepsilon, +\varepsilon[\times \{0\}, U)$.

Thus, F (or $\bar{F}$) is just equivalent to a path $f : s \mapsto F(s)(0)$ of U, such that $f(0) = x$. Now, let $\hat{x}^i = dx^i$ be the i-th coordinate 1-form of U (art. 6.23), and let

$$v^i = \hat{x}^i(\bar{F})_{\binom{0}{0}} \begin{pmatrix} 1 \\ 0 \end{pmatrix} = \left. \frac{df^i(s)}{ds} \right|_{s=0}.$$

Now, since every smooth 1-form α of U is a combination $\alpha = \sum_{i=1}^n \alpha_i(x) dx^i$, if two paths f and f', pointed at x, are equivalent, then their derivatives v^i at 0 are equal. Conversely, let v be the vector of $\mathbf{R}^n$ with coordinates v^i. The arc $f_v : t \mapsto tv$ can be chosen as a representative of f. And, since $\mathbf{x}$ is a 0-plot, we do not have to check the equivalence on p-forms with $p > 1$. Therefore, a variation of a point x of U is just a vector v of $\mathbf{R}^n$, which we could summarize by $\delta\mathbf{x} = (x, v) \in U \times \mathbf{R}^n$.

☞ **Exercise 111, p. 176** (Liouville rays). 1) Since the p-form ω is not the zero form, there exist a plot $P : U \to X$, a point $r \in U$, and p vectors $v_1, \ldots, v_p$ of $\mathbf{R}^n$, $n = \dim(U)$, such that $\omega(P)(r)(v_1)\cdots(v_p) \neq 0$. So, $h(t)^*(\omega) = \lambda(t)\omega$ implies

$$\lambda(t) = \frac{h(t)^*(\omega)(P)(r)(v_1)\cdots(v_p)}{\omega(P)(r)(v_1)\cdots(v_p)}.$$

Since h is the smooth homomorphism from $\mathbf{R}$ to $\mathrm{Diff}(X)$, by definition of the functional diffeology, $t \mapsto h(t)^*(\omega)(P)(r)(v_1)\cdots(v_p)$ is smooth. Hence, λ is smooth.

2) Thanks to (art. 6.55), we have for all $t \in \mathbf{R}$,

$$\frac{\partial h(t)^*(\omega)}{\partial t} = h(t)^*(\pounds_h(\omega)), \quad \text{thus} \quad \lambda'(t) \times \omega = \lambda(t) \times \omega,$$

where the prime denotes the derivative. Hence, $(\lambda'(t) - \lambda(t)) \times \omega = 0$, and since $\omega \neq 0$, $\lambda'(t) = \lambda(t)$, that is, $\lambda(t) = ce^t$. But $h(0) = 1_X$, thus $\omega = h(0)^*(\omega) = \lambda(0) \times \omega = c \times \omega$. Therefore, $c = 1$ and $\lambda(t) = e^t$.

☞ **Exercise 112, p. 186** (The boundary of a 3-cube). Using the notation $\{t_1t_2t_3\}$ for $\sigma(t_1)(t_2)(t_3)$ and formulas in (art. 6.59), we have

$$\partial\sigma(t_1)(t_2) = \{1t_1t_2\} - \{0t_1t_2\} - \{t_11t_2\} + \{t_10t_2\} + \{t_1t_21\} - \{t_1t_20\},$$

and therefore

$$\begin{aligned}
\partial(\partial\sigma)(t) &= [\partial\sigma(1)(t) - \partial\sigma(0)(t)] - [\partial\sigma(t)(1) - \partial\sigma(t)(0)] \\
&= \{11t\} - \{01t\} - \{11t\} + \{10t\} + \{1t1\} - \{1t0\} \\
&- \{10t\} + \{00t\} + \{01t\} - \{00t\} - \{0t1\} + \{0t0\} \\
&- \{1t1\} + \{0t1\} + \{t11\} - \{t01\} - \{t11\} + \{t10\} \\
&+ \{1t0\} - \{0t0\} - \{t10\} + \{t00\} + \{t01\} - \{t00\} \\
&= 0.
\end{aligned}$$

☞ **Exercise 113, p. 186** (Cubic homology of a point). First of all let us note that $\mathrm{Cub}_p(\star) = \{\hat{0} = [t \mapsto 0]\}$, thus $C_p(\star) = \{n\hat{0} \mid n \in \mathbf{Z}\} \simeq \mathbf{Z}$, for all $p \in \mathbf{N}$. Since $\partial\hat{0} = 0$ (the zero chain), $\partial[C_p(\star)] = \{0\}$, that is, $Z_p(\star) = C_p(\star) \simeq \mathbf{Z}$ and $B_p(\star) = \{0\}$, for all $p \in \mathbf{N}$. Therefore, for all $p \in \mathbf{N}$, $H_p(\star) \simeq \mathbf{Z}$. Now, let us consider the reduced cubic chains. For $p = 0$, there is no degenerate 0-chain, $C_0^\bullet(X) = \{0\}$, then $\mathbf{C}_0(\star) = C_0(X)/C_0^\bullet(X) = C_0(X) \simeq \mathbf{Z}$. For $p \geq 1$, since every cubic chain is constant, every cubic chain is degenerate, that is, $C_p(X) = C_p^\bullet(X)$, and then $\mathbf{C}_p(\star) = C_p(X)/C_p^\bullet(X) = \{0\}$. Therefore, $\mathbf{H}_0(\star) = \mathbf{Z}$ and $\mathbf{H}_p(\star) = \{0\}$, for all $p \geq 1$.

☞ **Exercise 114, p. 198** (Liouville rays and closed forms). By application of the Cartan formula (art. 6.72), $\pounds_h(\omega) = i_h(d\omega) + d(i_h(\omega))$. Since $d\omega = 0$, and $\pounds_h(\omega) = \omega$, we get $\omega = d[i_h(\omega)]$. The p-form ω is exact and $i_h(\omega)$ is one of its primitives.

☞ **Exercise 115, p. 198** (Integrals on homotopic cubes). In our case, $d\alpha = 0$ and $\delta\alpha = 0$, the variation of the integral on a p-cube σ (art. 6.70) is just

$$\delta \int_\sigma \alpha = \int_{\partial I^p} \alpha(\delta\sigma),$$

with

$$\alpha(\delta\sigma)_r(v_2)\cdots(v_p) = \alpha(\sigma)_{\binom{0}{r}} \begin{pmatrix}1\\0\end{pmatrix}\begin{pmatrix}0\\v_2\end{pmatrix}\cdots\begin{pmatrix}0\\v_p\end{pmatrix},$$

$\sigma : (s, r) \mapsto \sigma_s(r)$, and $v_i \in \mathbf{R}^p$. But the variation involves the restriction of σ to the boundary ∂I^p, and by hypothesis, $\sigma(s, r) \restriction \partial I^p = \sigma(r) \restriction \partial I^p$, for all s. Thus, restricted to the boundary,

$$\begin{aligned}
\alpha(\sigma)_{\binom{0}{r}} \begin{pmatrix}1\\0\end{pmatrix}\begin{pmatrix}0\\v_2\end{pmatrix}\cdots\begin{pmatrix}0\\v_p\end{pmatrix} &= \alpha(\sigma\circ \mathrm{pr}_2)_{\binom{0}{r}} \begin{pmatrix}1\\0\end{pmatrix}\begin{pmatrix}0\\v_2\end{pmatrix}\cdots\begin{pmatrix}0\\v_p\end{pmatrix}\\
&= \mathrm{pr}_2^*[\alpha(\sigma)]_{\binom{0}{r}} \begin{pmatrix}1\\0\end{pmatrix}\begin{pmatrix}0\\v_2\end{pmatrix}\cdots\begin{pmatrix}0\\v_p\end{pmatrix}\\
&= \alpha(\sigma)_r(0)(v_2)\cdots(v_p)\\
&= 0.
\end{aligned}$$

Therefore, the variation vanishes, and the integral of a closed p-form is constant along a fixed-boundary homotopy of p-cubes.

☞ **Exercise 116, p. 198** (Closed 1-forms on connected spaces). We have seen in Exercise 115, p. 198, that the integral of a closed p-form on a p-cube does not depend on the fixed-boundary homotopy class of the p-cube. Applied to 1-forms it just says that the integral $\int_\ell \alpha$, where $\alpha \in \Omega^1(X)$, $d\alpha = 0$ and $\ell \in \mathrm{Loops}(X, x)$, does not depend on the fixed-ends homotopy class of ℓ. Now let $\ell' \in \mathrm{Loops}(X, x)$, since ℓ and ℓ' are always fixed-ends homotopic to two stationary loops (which do not change the integrals), we can assume that ℓ and ℓ' are stationary. Then,

$$\begin{aligned}
\int_{\ell\vee\ell'} \alpha &= \int_0^1 \alpha(\ell\vee\ell')_t(1)\,dt\\
&= \int_0^{1/2} \alpha(\ell\vee\ell')_t(1)\,dt + \int_{1/2}^1 \alpha(\ell\vee\ell')_t(1)\,dt\\
&= \int_0^{1/2} \alpha(t\mapsto \ell(2t))_t(1)\,dt + \int_{1/2}^1 \alpha(t\mapsto \ell'(2t-1))_t(1)\,dt\\
&= \int_0^1 \alpha(t\mapsto \ell(t))_t(1)\,dt + \int_0^1 \alpha(t\mapsto \ell'(t))_t(1)\,dt\\
&= \int_\ell \alpha + \int_{\ell'} \alpha.
\end{aligned}$$

Thus, $\mathrm{class}(\ell) \mapsto \int_\ell \alpha$ is a homomorphism from $\pi_1(X, x)$ to $\mathbf{R}$ and P_α is the image of this homomorphism. Now, let $x' \in X$, since X is connected, there exists a stationary path c connecting x to x′, and clearly $\int_{c\vee\ell\vee\bar c} \alpha = \int_c \alpha + \int_\ell \alpha - \int_c \alpha = \int_\ell \alpha$, where

$\ell \in \mathrm{Loops}(X, x')$ and $\bar{c}(t) = 1 - t$ is the reverse of c. Therefore, the group P_α does not depend on the point where it is computed.

☞ **Exercise 117, p. 198** (Closed 1-forms on simply connected spaces). Let us assume that the connected component of x is simply connected. Let $s \mapsto \ell_s$ be a smooth path in X connecting $\ell = \ell_0$ to the constant path $\ell_1 : t \mapsto x$. We have

$$\frac{\partial}{\partial s}\left\{\int_\ell \alpha\right\}_s = \int_0^1 d\alpha(\delta\ell_s) + \int_0^1 d[\alpha(\delta\ell_s)] = 0 + \left[\alpha(\delta\ell_s)\right]_{t=0}^{t=1}.$$

Now, let $\bar{\ell} : (s,t) \mapsto \ell_s(t)$ and $j_t : s \mapsto (s,t)$, we have

$$\begin{aligned} \alpha(\delta\ell_s)(0) &= \alpha(\bar{\ell})_{\binom{s}{0}}\begin{pmatrix}1\\0\end{pmatrix} = j_0^*[\alpha(\bar{\ell})](s)(1) \\ &= \alpha(\bar{\ell} \circ j_0)(s)(1) = \alpha(s \mapsto \ell_s(0) = x)(s)(1) \\ &= 0, \end{aligned}$$

and the same holds for $t = 1$, $\alpha(\delta\ell_s)(1) = 0$. Thus,

$$\frac{\partial}{\partial s}\left\{\int_\ell \alpha\right\}_s = 0 \quad \Rightarrow \quad \int_{\ell_s} \alpha = \mathrm{cst} \quad \Rightarrow \quad \int_\ell \alpha = \int_{[t \mapsto x]} \alpha = 0.$$

So, if the integral of a closed 1-form on a loop is nonzero, then the space cannot be simply connected.

☞ **Exercise 118, p. 202** (1-forms vanishing on loops). Let $\sigma : \mathbf{R}^2 \to X$ be a 2-cube, and let its boundary $\partial\sigma$ be a loop. It is the concatenation ℓ of the four paths $\gamma_1 : t \mapsto \sigma(t,0)$, $\gamma_2 : t \mapsto \sigma(1,t)$, $\gamma_3 : t \mapsto \sigma(1-t,1)$ and $\gamma_4 : t \mapsto \sigma(0,1-t)$. Then, because the integral of α vanishes on loops, and thanks to Stokes' theorem,

$$\int_\sigma d\alpha = \int_{\partial\sigma} \alpha = \int_{\gamma_1} \alpha + \int_{\gamma_2} \alpha + \int_{\gamma_3} \alpha + \int_{\gamma_4} \alpha = \int_\ell \alpha = 0.$$

Thus, since the integral of $d\alpha$ vanishes on every 2-cube, $d\alpha = 0$ (art. 6.66). Next, since α is closed and vanishes on every loop, α is exact (art. 6.89). If this proof is essentially correct, we could have been more careful, indeed the concatenation of the four paths may be not smooth. We should have smashed the paths before concatenation, that is to say, exchanged γ_i into $\gamma_i^\star = \gamma_i \circ \lambda$, where λ is the smashing function described in Figure 5.1. Since this operation leads to a change of variable under the integral, it does not change the result.

☞ **Exercise 119, p. 202** (Forms on irrational tori are closed). Let α be a p-form on T_Γ. Let $a = \pi_\Gamma^*(\alpha)$ be the pullback of α by the projection $\pi_\Gamma : \mathbf{R}^n \to T_\Gamma$. By construction, the p-form a is invariant by the action of Γ, that is, for all $\gamma \in \Gamma$, $\gamma^*(a) = a$, where $\gamma(x) = x + \gamma$, $x \in \mathbf{R}^n$. But the invariance of a under Γ and the density of Γ in $\mathbf{R}^n$ imply that every component $a_{i\cdots k}$ of a is constant. Hence $a \in \Lambda^p(\mathbf{R}^n)$ and therefore, $\Omega^p(T_\Gamma) \simeq \Lambda^p(\mathbf{R}^n)$. Now, since the components of any $a = \pi_\Gamma^*(\alpha)$ are constant, the form a is closed, $da = 0$. But, since π_Γ is a subduction (art. 6.38), $da = 0$ and $da = \pi_\Gamma^*(d\alpha)$ imply $d\alpha = 0$. Hence, all the differential forms of T_Γ are closed, $Z_{dR}^p(T_\Gamma) = \Omega^p(T_\Gamma) \simeq \Lambda^p(\mathbf{R}^n)$. Now, if $\alpha = d\beta$, $\beta \in \Omega^{p-1}(T_\Gamma, \mathbf{R})$, then $\alpha = 0$ since $d\beta = 0$ for any form on T_Γ. Then, $B_{dR}^p(T_\Gamma) = \{0\}$ and $H_{dR}^p(T_\Gamma) \simeq \Lambda^p(\mathbf{R}^n)$.

☞ **Exercise 120, p. 202** (Is the group $\mathrm{Diff}(S^1)$ simply connected?) Clearly, $\alpha(P)$ is a 1-form on U. Let $F : V \to U$ be a smooth m-parametrization. We have

$$\alpha(P \circ F)_s(\delta s) = \int_0^{2\pi} \left\langle J\{[P(F(s))](X(\theta))\}, \frac{\partial[P(F(s))](X(\theta))}{\partial s}(\delta s) \right\rangle d\theta.$$

Let $r = F(s)$, then

$$\alpha(P \circ F)_s(\delta s) = \int_0^{2\pi} \left\langle J[P(r)(X(\theta))], \frac{\partial P(r)(X(\theta))}{\partial r} \frac{\partial r}{\partial s}(\delta s) \right\rangle d\theta,$$

where $\partial r/\partial s = D(F)(s)$. Thus,

$$\alpha(P \circ F)_s(\delta s) = \alpha(P)_{F(s)}(D(F)(s))(\delta s) = F^*(\alpha(P))_s(\delta s).$$

Therefore, α is a 1-form on $\mathrm{Diff}(S^1)$. Now, let us consider the canonical 1-form Θ on S^1 defined by

$$\Theta(Q)_s(\delta s) = \left\langle J[Q(s)], \frac{\partial Q(s)}{\partial s}(\delta s) \right\rangle,$$

where Q is a plot of S^1, with the same kind of notation as above. Next, let $z \in S^1$ and $R(z) : \mathrm{Diff}(S^1) \to S^1$ be the orbit map, $R(z)(\varphi) = \varphi(z)$. The pullback of Θ by $R(z)$ is then given, on the plot P, by

$$[R(z)^*(\Theta)](P)_r(\delta r) = \left\langle J[P(r)(z)], \frac{\partial P(r)(z)}{\partial r}(\delta r) \right\rangle,$$

and then

$$\alpha(P)_r(\delta r) = \int_0^{2\pi} [R(X(\theta))^*(\Theta)](P)_r(\delta r)\, d\theta.$$

Thus, by additivity of the integral and since $d\Theta = 0$, $d\alpha = 0$. Note that this last expression of α proves directly that α is a 1-form on $\mathrm{Diff}(S^1)$. Next, to compute $\int_\sigma \alpha$, we need

$$\sigma(t)(X(\theta)) = \begin{pmatrix} \cos(2\pi t) & \sin(2\pi t) \\ -\sin(2\pi t) & \cos(2\pi t) \end{pmatrix} \begin{pmatrix} \cos(\theta) \\ \sin(\theta) \end{pmatrix} = \begin{pmatrix} \cos(2\pi t + \theta) \\ \sin(2\pi t + \theta) \end{pmatrix}.$$

Then,

$$\begin{aligned} \int_\sigma \alpha &= \int_0^1 \alpha(\sigma)_t(1)\, dt \\ &= \int_0^1 dt \int_0^{2\pi} d\theta \left\langle J \begin{pmatrix} \cos(2\pi t + \theta) \\ \sin(2\pi t + \theta) \end{pmatrix}, \frac{\partial}{\partial t} \begin{pmatrix} \cos(2\pi t + \theta) \\ \sin(2\pi t + \theta) \end{pmatrix} \right\rangle \\ &= 2\pi \int_0^1 dt \int_0^{2\pi} d\theta \\ &= 4\pi^2. \end{aligned}$$

Therefore, the identity component of $\mathrm{Diff}(S^1)$ is not simply connected; see Exercise 117, p. 198. Actually, it was not necessary to integrate the pullback $R(z)^*(\Theta)$ on the loop X, we could have considered just the pullback $R(\mathbf{e}_1)^*(\Theta)$.

☞ **Exercise 121, p. 212** (The Fubini-Study form is locally exact). We know that the infinite projective space $\mathcal{P}_{\mathbf{C}}$ is a diffeological manifold modeled on the Hilbert space $\mathcal{H}_{\mathbf{C}}$ (art. 4.11). Every point $p \in \mathcal{P}_{\mathbf{C}}$ is in the range of some chart $F_k : \mathcal{H}_{\mathbf{C}} \to \mathcal{P}_{\mathbf{C}}$, for some $k \in \mathbf{N}$, (art. 4.11, item 2). Since $F_k(\mathcal{H}_{\mathbf{C}})$ is D-open and contractible (because $\mathcal{H}_{\mathbf{C}}$ is contractible), every closed differential form on $\mathcal{P}_{\mathbf{C}}$ is locally exact, in particular the Fubini-Study form ω.

Exercise 122, p. 213 (Closed but not locally exact). From Exercise 119, p. 202, we know that every differential form on an irrational torus $T_\Gamma = \mathbf{R}^n/\Gamma$, where Γ is a dense discrete generating subgroup of $\mathbf{R}^n$, $n \geq 1$, is closed. But since the D-topology of T_Γ is the coarse topology, and since T_Γ is not simply connected, $\pi_1(T_\Gamma) = \Gamma$, T_Γ is also not locally simply connected. Hence, none of the closed forms, except the form 0, is locally exact.

Exercise 123, p. 213 (A morphism from $H^\star_{\mathrm{dR}}(X)$ to $H^\star_{\mathrm{dR}}(\mathrm{Diff}(X))$). Let x_0 and x_1 be two points of X connected by a path $t \mapsto x_t$, the map $\hat{x}_t : \varphi \mapsto \varphi(x_t)$ is a homotopy from $\hat{x}_0$ to $\hat{x}_1$. Then, thanks to (art. 6.88), they induce the same map in cohomology, $\hat{x}^*_{0\mathrm{dR}} = \hat{x}^*_{1\mathrm{dR}}$. Next, let us consider $X = S^1 \subset \mathbf{R}^2$. Note first that S^1 is connected and $H^p_{\mathrm{dR}}(S^1) = \{0\}$ for all $p \geq 1$, then the morphism from $H^p_{\mathrm{dR}}(S^1)$ to $H^p_{\mathrm{dR}}(\mathrm{Diff}(S^1))$ is unique and obviously injective for $p = 0$ and $p \geq 1$. Let $\hat{\mathbf{e}}_1 : \mathrm{Diff}(S^1) \to S^1$ be the orbit map of the point $\mathbf{e}_1 = (1, 0)$. Let α be a closed 1-form on S^1 such that $\hat{\mathbf{e}}^*_{1\mathrm{dR}}(\mathrm{class}(\alpha)) = 0$, that is, $\hat{\mathbf{e}}^*_1(\alpha) = dF$, where $F \in \mathcal{C}^\infty(\mathrm{Diff}(S^1), \mathbf{R})$. Now, since α is closed, the integral $\int_\ell \alpha$, where $\ell \in \mathrm{Loops}(S^1, \mathbf{e}_1)$, depends only on the homotopy class of ℓ (see Exercise 116, p. 198), and thanks to Exercise 133, p. 266, we know that there exists $k \in \mathbf{Z}$ such that $\ell \sim [t \mapsto \mathcal{R}(2\pi kt)(\mathbf{e}_1)]$, where $\mathcal{R}(\theta)$ is the rotation of angle θ. Let $\sigma_k(t) = \mathcal{R}(2\pi kt)$, then σ_k belongs to $\mathrm{Loops}(\mathrm{Diff}(S^1), \mathbf{1}_{S^1})$, and $\ell \sim [t \mapsto \sigma_k(t)(\mathbf{e}_1)]$. Thus, on the one hand, $\int_{\sigma_k} dF = 0$, and on the other hand, $\int_{\sigma_k} dF = \int_{\sigma_k} \hat{\mathbf{e}}^*_1(\alpha) = \int_{\hat{\mathbf{e}}_1 \circ \sigma_k} \alpha = \int_{[t \mapsto \sigma_k(t)(\mathbf{e}_1)]} \alpha = \int_\ell \alpha$. Hence, $\int_\ell(\alpha) = 0$ for all loops in S^1, and thanks to (art. 6.89), α is exact. Therefore, $\hat{\mathbf{e}}^*_{1\mathrm{dR}}$ is injective. Compared with Exercise 120, p. 202, this also shows that the identity component of $\mathrm{Diff}(S^1)$ is not simply connected.

This example is a particular case in which a smooth map $f : X \to X'$ induces a surjection from $\pi_1(X, x)$ onto $\pi_1(X', x')$, where $x' = f(x)$. We assume X and X′ connected. In this situation, for all $\alpha \in \Omega^1(X)$ such that $d\alpha = 0$, if $f^*(\alpha)$ is exact, that is, if $\int_{\ell'} f^*(\alpha) = 0$ for all $\ell' \in \mathrm{Loops}(X', x')$, then $\int_\ell \alpha = 0$ for all $\ell \in \mathrm{Loops}(X, x)$, and thus α is exact. Therefore, the associated homomorphism $f^*_{\mathrm{dR}} : H^1_{\mathrm{dR}}(X') \to H^1_{\mathrm{dR}}(X)$ is injective.

Exercise 124, p. 221 (Subgroups of $\mathbf{R}$). For the first question, there are two possibilities:

a) There exists $\varepsilon > 0$ such that $]-\varepsilon, +\varepsilon[\subset K$.

b) For all $\varepsilon > 0$ there exists $t \in \mathbf{R}$ such that $0 < t < \varepsilon$ and $t \notin K$.

If we are in the first situation, then $K = \mathbf{R}$. Indeed, for every $t \in \mathbf{R}$, there exists $N \in \mathbf{N}$ such that $N\varepsilon \leq t < (N+1)\varepsilon$. So, $x = t/(N+1) \in K$. And, since K is a group for the addition, $t = (N+1)x$ belongs to K. Therefore, in this case K is not a strict subgroup of $\mathbf{R}$. Thus, we are in the second case. Let $P : U \to K$ be a plot, that is $P \in \mathcal{C}^\infty(U, \mathbf{R})$ and $P(U) \subset K$. Let us assume that $0 \in U$ and that $P(0) = 0$. If it is not the case, we can compose P at the source and the target such that it will be the case. Since U is open, there exists $R > 0$ such that for every $\rho < R$ the open balls $B(\rho)$ of radius ρ, centered at $0 \in U$, are contained in U. Then, for every $0 < \rho < R$ let us choose $r \in B(\rho)$ and let $k = P(r)$. Thus, $k \in K$, but also all the $P(sr)$, $0 \leq s \leq 1$, are elements of K. Since $p : s \mapsto P(sr)$ is smooth, therefore continuous, p takes all the values between $0 = p(0)$ and $k = p(1)$. But by hypothesis a) this is not possible except if p is constant and $k = 0$. Therefore, P is locally constant and the group K is discrete.

Next, let K be a strict subgroup of $\mathbf{R}$, K is discrete and therefore its D-topology also is discrete (art. 2.11). There are two cases: either K is generated by one number a, $K = a\mathbf{Z}$; or there exist two numbers $a \neq 0$ and $b \neq 0$, independent over $\mathbf{Q}$, such that $\{na\}_{n\in\mathbf{Z}} \subset K$, $\{mb\}_{m\in\mathbf{Z}} \subset K$, and $\{na\}_{n\in\mathbf{Z}} \cap \{mb\}_{m\in\mathbf{Z}} = \{0\}$. In the first case, the group is embedded in $\mathbf{R}$. Indeed, any subset A of a discrete space is D-open, if $K = a\mathbf{Z}$, then one can find an open interval I_x centered around each point x of A such that $I_x \cap K = \{x\}$, and the intersection of this union of intervals with A is just A. In the second case, K contains the dense subgroup $a \times [\mathbf{Z} + \alpha\mathbf{Z}]$, with $\alpha = b/a \in \mathbf{R} - \mathbf{Q}$. Then, the intersection of every open interval of $\mathbf{R}$ with K contains an infinite number of points of $\mathbf{K}$. Therefore, $\mathbf{K}$ is not embedded in $\mathbf{R}$.

☞ **Exercise 125, p. 221** (Diagonal diffeomorphisms). A plot of $\Delta(\mathrm{Diff}(X)) \subset \mathrm{Diff}(X^N)$ writes, in a unique way, $r \mapsto \Delta(\varphi_r)$, where $r \mapsto \varphi_r$ is some parametrization in $\mathrm{Diff}(X)$. Now, $r \mapsto \Delta(\varphi_r)$ is smooth only if $(r, x_1, \ldots, x_N) \mapsto (\varphi_r(x_1), \ldots, \varphi_r(x_N))$ is smooth, which means that, for every $k = 1, \ldots, N$, the map $(r, x_1, \ldots, x_N) \mapsto \varphi_r(x_k)$ is smooth. Since the projection $\mathrm{pr}_k : (x_1, \ldots, x_N) \mapsto x_k$ is a subduction, that is equivalent to $(r, x_k) \mapsto \varphi_r(x_k)$ being smooth, which means then that $r \mapsto \varphi_r$ is a plot of $\mathrm{Diff}(X)$. Therefore, Δ is an induction.

☞ **Exercise 126, p. 221** (The Hilbert sphere is homogeneous). Let us give the proof in two steps.

1. *The map π is surjective.* Let Z and Z' be two elements of $\mathcal{S}_{\mathbf{C}}$. If Z and Z' are collinear, then there exists $\tau \in S^1 \simeq \mathbf{U}(\mathbf{C})$ such that $Z' = \tau Z$, and the map $Z \mapsto \tau Z$ belongs to $\mathbf{U}(\mathcal{H})$. Otherwise, let E be the plane spanned by these two vectors, and let F be its orthogonal for the Hermitian product. According to Bourbaki [**Bou55**], E and F are supplementary $\mathcal{H} = E \oplus F$. The vectors Z and Z' are vectors of the unit sphere $S^3 \subset E \simeq \mathbf{C}^2$, now the group $\mathbf{U}(\mathbf{C}^2)$ acts transitively on S^3, there exists $A \in \mathbf{U}(\mathbf{C}^3)$ such that $Z' = AZ$. This map, extended to $\mathcal{H}$ by the identity on F, belongs to $\mathbf{U}(\mathcal{H})$ and maps Z to Z'. Therefore, the action of $\mathbf{U}(\mathcal{H})$ is transitive on $\mathcal{S}_{\mathbf{C}}$, which is equivalent to the assertion that π is surjective.

2. *The map π is a subduction.* Let $Q : U \to \mathcal{S}_{\mathbf{C}}$ be a plot. We want to lift Q locally along the projection π, that is, for any $r_0 \in U$, to find a plot $P : V \to \mathbf{U}(\mathcal{H})$, defined on some open neighborhood V of r_0, such that $P(r)(\mathbf{e}_1) = Q(r)$, for all $r \in V$. So, let $r_0 \in U$, let V be an open neighborhood of r_0, let $j : \mathbf{C}^m \to \mathcal{H}$ be an injection, and let $\phi : V \to \mathbf{C}^m$ be a smooth parametrization such that $Q \upharpoonright V = j \circ \phi$. Let us denote $E = j(\mathbf{C}^m)$. The plot Q of $\mathcal{S}_{\mathbf{C}}$ takes its values in E, and hence in the unit sphere of E: $S(E) = E \cap \mathcal{S}_{\mathbf{C}}$. The diffeology induced on $S(E)$ is the standard diffeology: $S(E) \simeq S^{2m-1}$. Thus, $Q \upharpoonright V$ is an ordinary smooth map from V into $S(E)$. But we know that the projection from $\mathbf{U}(m)$ onto $S(\mathbf{C}^m)$ is a submersion, *a fortiori* a subduction. Thus, for any $r_0 \in V$ there exist a domain $W \subset V$ and a smooth lifting $\varphi : W \to \mathbf{U}(m)$ such that $Q(r) = \varphi(r)(\mathbf{e}_1^m)$, for all $r \in W$, where $\mathbf{e}_1^m$ is the vector $(1, 0, \ldots, 0) \in \mathbf{C}^m$. Let us assume that $\mathbf{e}_1 = j(\mathbf{e}_1^m)$, if it is not the case we conjugate everything with some suitable linear map. Now, let F be the orthogonal of E. The space $\mathcal{H}$ is the direct sum of E and F, *i.e.*, $\mathcal{H} = E \oplus F$. Every vector $Z \in \mathcal{H}$ has a unique decomposition $Z = Z_E + Z_F$ such that $Z_E \in E$ and $Z_F \in F$. Let then

$$P(r)(Z) = \varphi(r)(Z_E) + Z_F,$$

for all $r \in W$ and all $Z \in \mathcal{H}$. For all $r \in W$, the map $P(r)$ is smooth because the decomposition $Z \mapsto (Z_E, Z_F)$ is linear, and then smooth for the fine diffeology. Moreover $P(r)$ clearly preserves the Hermitian product, and is obviously invertible. The map P lifts Q locally,

$$P(r)(\mathbf{e}_1) = \varphi(r)(\mathbf{e}_1^m) + 0 = Q(r),$$

for all $r \in W$. It remains then to check that P is a plot of the functional diffeology of $\mathbf{U}(\mathcal{H})$. But this is quite clear—a finite family of vectors decomposes into components belonging to E and to F, and because the family is finite, one has only a finite intersection of open sets which is open, we get the property we are looking for. The inverse of $P(r)$ does not give more problems. Therefore, we get that $\mathcal{S}_C$ is homogeneous under $U(\mathcal{H}_C)$.

Now, $\mathcal{P}_C$ is the quotient of $\mathcal{S}_C$ by $\mathbf{U}(1)$. Since the composite of subductions $\mathbf{U}(\mathcal{H}_C) \to \mathcal{S}_C \to \mathcal{P}_C$ is a subduction (art. 1.47), and since the action of $\mathbf{U}(\mathcal{H}_C)$ on $\mathcal{S}_C$ passes to $\mathcal{P}_C$, $\mathcal{P}_C$ is a homogeneous space of $\mathbf{U}(\mathcal{H}_C)$.

Exercise 127, p. 227 (Pullback of 1-forms by multiplication). Let us develop the form $m^*(\alpha)(P \times Q)$,

$$\begin{aligned} m^*(\alpha)(P \times Q)_{\binom{r}{s}} \begin{pmatrix} \delta r \\ \delta s \end{pmatrix} &= \alpha[(r,s) \mapsto (P(r), Q(s)) \mapsto P(r)\cdot Q(s)]_{\binom{r}{s}} \begin{pmatrix} \delta r \\ \delta s \end{pmatrix} \\ &= [\alpha_{U,s}(r) \quad \alpha_{V,r}(s)] \begin{pmatrix} \delta r \\ \delta s \end{pmatrix}, \end{aligned}$$

because any 1-form on $U \times V$, at a point (r,s), writes $[\alpha_{U,s}(r)\ \alpha_{V,r}(s)]$, where $\alpha_{U,s}$ is a 1-form of U depending on s, and $\alpha_{V,r}$ is a 1-form of V depending on r. Thus,

$$\begin{aligned} m^*(\alpha)(P \times Q)_{\binom{r}{s}} \begin{pmatrix} \delta r \\ \delta s \end{pmatrix} &= \alpha_{U,s}(r)(\delta r) + \alpha_{V,r}(s)(\delta s) \\ &= \alpha[r \mapsto P(r)\cdot Q(s)]_r(\delta r) \\ &+ \alpha[s \mapsto P(r)\cdot Q(s)]_s(\delta s) \\ &= (R(Q(s))^*\alpha)(P)_r(\delta r) \\ &+ (L(P(r))^*\alpha)(Q)_s(\delta s). \end{aligned}$$

Each term of the right sum above is computed by considering successively, in $m^*(\alpha)(P \times Q)_{(r,s)}(\delta r, \delta s)$, s constant and $\delta s = 0$, then r constant and $\delta r = 0$. We get finally, considering the diagonal map $\Delta : r \mapsto (r,r)$,

$$\begin{aligned} \alpha[r \mapsto P(r)\cdot Q(r)]_r(\delta r) &= \Delta^*(m^*(\alpha)(P \times Q))_r(\delta r) \\ &= (R(Q(r))^*\alpha)(P)_r(\delta r) \\ &+ (L(P(r))^*\alpha)(Q)_r(\delta r). \end{aligned}$$

Exercise 128, p. 227 (Liouville form on groups). Let $F : V \to U$ be a smooth parametrization, then $Q \circ F = (P \circ F, A \circ F)$. On the one hand,

$$\begin{aligned} \lambda(Q \circ F)_s(\delta s) &= A(F(s))(P \circ F)_s(\delta s) \\ &= F^*[A(F(s))(P)]_s(\delta s) \\ &= A(F(s))(P)_{F(s)}(D(F)(s)(\delta s)), \end{aligned}$$

and on the other hand,

$$\begin{aligned} F^*[\lambda(Q)]_s(\delta s) &= \lambda(Q)_{F(s)}(D(F)(s)(\delta s)) \\ &= A(F(s))(P)_{F(s)}(D(F)(s)(\delta s)). \end{aligned}$$

Therefore, $\lambda(Q \circ F) = F^*(\lambda(Q))$, and λ defines a differential 1-form on $G \times \mathcal{G}^*$. Now, $j_\alpha^*(\lambda)(P)_r(\delta r) = \lambda(j_\alpha \circ P)_r(\delta r) = \lambda(r \mapsto (P(r), \alpha))_r(\delta r) = \alpha(P)_r(\delta r)$. Thus, $j_\alpha^*(\lambda) = \alpha$. Next, let $g' \in G$, recalling that $\mathrm{Ad}_*(g') = \mathrm{Ad}(g'^{-1})^*$, we have

$$\begin{aligned} g_{G\times\mathcal{G}^*}^*(\lambda)(Q)_r(\delta r) &= \lambda(g_{G\times\mathcal{G}^*} \circ Q)_r(\delta r) \\ &= \lambda(r \mapsto (\mathrm{Ad}(g')(P(r)), \mathrm{Ad}_*(g')[A(r)]))_r(\delta r) \\ &= [\mathrm{Ad}_*(g')(A(r))](\mathrm{Ad}(g') \circ P)_r(\delta r) \\ &= \mathrm{Ad}(g')^*[\mathrm{Ad}(g'^{-1})^*(A(r))](P)_r(\delta r) \\ &= A(r)(P)_r(\delta r). \end{aligned}$$

Therefore, λ is invariant under this action of G on $G \times \mathcal{G}^*$.

☞ **Exercise 129, p. 242** (Groupoid associated with $x \mapsto x^3$). Let $X = \mathbf{R}$ and $Q = \mathbf{R}$, equipped with the standard diffeology, and let $\pi : X \to Q$, $\pi(x) = x^3$. Let $\mathbf{K}$ be the groupoid associated with π. Since π is injective, $X_q = \pi^{-1}(q)$ is equal to the singleton $\{x = \sqrt[3]{q}\}$. Hence, $\mathrm{Mor}(q, q')$ is itself reduced to the singleton $\{[\sqrt[3]{q} \mapsto \sqrt[3]{q'}]\}$. Then,

$$\mathrm{Mor}(\mathbf{K}) = \{[x \mapsto x'] \mid x, x' \in \mathbf{R}\},$$

and set theoretically $\mathrm{Mor}(\mathbf{K}) \simeq \mathbf{R} \times \mathbf{R}$. The question is about the diffeology, on $\mathbf{R} \times \mathbf{R}$, induced by the diffeology of $\mathrm{Mor}(\mathbf{K})$. Let $P : U \to \mathrm{Mor}(\mathbf{K})$ be a plot, and let $P(r) = \{[x_r \mapsto x'_r]\}$. Let us make explicit the spaces $X_{\mathrm{src}\circ P}$ and $X_{\mathrm{trg}\circ P}$,

$$X_{\mathrm{src}\circ P} = \{(r, x_r) \in U \times \mathbf{R} \mid r \in U\} \quad \text{and} \quad X_{\mathrm{trg}\circ P} = \{(r, x'_r) \in U \times \mathbf{R} \mid r \in U\}.$$

The maps P_{src} and P_{trg} are then given by

$$P_{\mathrm{src}}(r, x_r) = P(r)(x_r) = x'_r \quad \text{and} \quad P_{\mathrm{trg}}(r, x'_r) = P(r)^{-1}(x'_r) = x_r.$$

These maps are smooth if and only if the parametrizations $r \mapsto x'_r$ and $r \mapsto x_r$ are smooth, which implies in particular that $\chi \circ P : r \mapsto (x_r^3, x_r'^3)$ is smooth. Therefore, the diffeology induced on $\mathbf{R} \times \mathbf{R}$ by the groupoid diffeology of $\mathbf{K}$ is the standard diffeology. Note that the injection $i_Q : Q \to \mathrm{Mor}(\mathbf{K})$ is indeed smooth, even if the presence of the cubic root is disturbing. Let $r \mapsto q_r$ be a plot of $Q = \mathbf{R}$, defined on U, and let $P : r \mapsto \mathbf{1}_{X_{q_r}}$ be the composite with i_Q. We have to check that P is a plot of $\mathrm{Mor}(\mathbf{K}) \simeq \mathbf{R} \times \mathbf{R}$. In this case $X_{\mathrm{src}\circ P} = X_{\mathrm{trg}\circ P} = \{(r, \sqrt[3]{q_r}) \mid r \in U\}$ and $P_{\mathrm{src}}(r, \sqrt[3]{q_r}) = P_{\mathrm{trg}}(r, \sqrt[3]{q_r}) = \sqrt[3]{q_r}$. A parametrization $s \mapsto (x_{r_s}, \sqrt[3]{q_{r_s}})$ of $X_{\mathrm{src}\circ P}$ is a plot if and only if $s \mapsto x_{r_s}$ and $s \mapsto \sqrt[3]{q_{r_s}}$ are smooth, and thus P_{src} and P_{trg} are smooth. Eventually, the real question is, Is this groupoid fibrating? The answer is No! Because the map $\chi : (x, x') \mapsto (x^3, x'^3)$ is not a subduction, and now the reason is exactly because $q \mapsto \sqrt[3]{q}$ is not smooth.

☞ **Exercise 130, p. 253** (Polarized smooth functions). A line $\mathbf{D}$ passing through the origin has two unit direction vectors $\{\pm u\}$. So, we can equip $P^1(\mathbf{R})$ with the quotient diffeology of $S^1 \subset \mathbf{R}^2$ by $\{\pm 1\}$. This diffeology is equivalent to the powerset diffeology; see Exercise 63, p. 61. Therefore, we can regard T as

$$T = \{(\pm u, f) \in P^1(\mathbf{R}) \times \mathcal{C}^\infty(\mathbf{R}^2, \mathbf{C}) \mid f(x + su) = f(x),\ \forall x \in \mathbf{R}^2,\ \forall s \in \mathbf{R}\}.$$

Now, let us consider the pullback of the projection $\pi : T \to P^1(\mathbf{R})$ by the projection $p : S^1 \to P^1(\mathbf{R})$, that is,

$$p^*(T) = \{(u, f) \in S^1 \times \mathcal{C}^\infty(\mathbf{R}^2, \mathbf{C}) \mid f(x + su) = f(x)\ \forall x \in \mathbf{R}^2,\ \forall s \in \mathbf{R}\}.$$

Thus, T is now equivalent to the quotient of $p^*(T)$ by the equivalence relation $(u, f) \sim (\pm u, f)$. Then, let us introduce the $\pi/2$ rotation J in the plane $\mathbf{R}^2$, and let us consider the map

$$\phi : (u, F) \mapsto (u, f = [x \mapsto F(u \cdot Jx)]), \quad \text{for all} \quad (u, F) \in S^1 \times \mathcal{C}^\infty(\mathbf{R}, \mathbf{C}),$$

where the dot $\cdot$ denotes the usual scalar product. Next, since $u \cdot Ju = 0$, $f(x+su) = F(u \cdot J(x + su)) = F((u \cdot Jx) + (su \cdot Ju)) = F(u \cdot Jx) = f(x)$, and $f = [x \mapsto F(u \cdot Jx)]$ is constant on all the lines parallel to $\mathbf{R}u$. Moreover, ϕ is bijective, for all $u \in S^1$ and $f \in \mathcal{C}^\infty(\mathbf{R}, \mathbf{C})$ such that $f(x + su) = f(x)$,

$$\phi^{-1}(u, f) = (u, F = [t \mapsto f(t \times J^{-1}u)]).$$

The maps ϕ and ϕ^{-1} are clearly smooth, thus $p^*(T)$ is trivial, equivalent to $\mathrm{pr}_1 : S^1 \times \mathcal{C}^\infty(\mathbf{R}, \mathbf{C}) \to S^1$. This is sufficient to prove that $\pi : T \to P^1$ is a diffeological fiber bundle.

Then, the pullback of the action of $\varepsilon \in \{\pm 1\}$, by ϕ, writes $\varepsilon(u, F) = (\varepsilon u, F \circ \hat{\varepsilon})$, where $\hat{\varepsilon} : t \mapsto \varepsilon \times t$. Hence, T is equivalent to the quotient of $S^1 \times \mathcal{C}^\infty(\mathbf{R}, \mathbf{C})$ by this action. Finally, let us consider the induction of $\mathbf{R}$ into $\mathcal{C}^\infty(\mathbf{R}, \mathbf{C})$ by $t \mapsto [x \mapsto tx]$. The subbundle $\phi(S^1 \times \mathbf{R})/\{\pm 1\} \subset T$ is the quotient of the product $S^1 \times \mathbf{R}$ by the action $\varepsilon(u, t) = (\varepsilon u, \varepsilon t)$, $\varepsilon \in \{\pm 1\}$. But this is exactly the Möbius strip, and the Möbius strip is not trivial over $P^1(\mathbf{R})$, thus the fiber bundle T also is not trivial.

Exercise 131, p. 253 (Playing with SO(3)). Since $SO(2, \mathbf{e}_1)$ is a subgroup of SO(3), the projection $\pi : SO(3) \to SO(3)/SO(2, \mathbf{e}_1)$ is a principal fibration, (art. 8.15). Now, let us prove that the map $p : SO(3) \to S^2$, defined by $A \mapsto A\mathbf{e}_1$, is a subduction and then identifies $SO(3)/SO(2, \mathbf{e}_1)$ with S^2. First of all, the map p is surjective. Indeed, let $u \in S^2$, there exists a vector $v \in S^2$ orthogonal to u. Then, the matrix $A = [u\ v\ u \times v]$, where $\times$ denotes the vector product, belongs to SO(3), and $A\mathbf{e}_1 = u$. Now, let $r \mapsto u_r$ be a smooth parametrization of S^2, let r_0 be a point in its domain, and let $u_0 = u_{r_0}$. Let $w \in \mathbf{R}^3$ such that $[1 - u_0\bar{u}_0]w \neq 0$, where $[1 - u_0\bar{u}_0]$ is the orthogonal projector along u_0. Since the map $r \mapsto [1 - u_r\bar{u}_r]w$ is smooth, there exists a small open ball B centered at r_0 such that, for all $r \in B$, the vector $w_r = [1 - u_r\bar{u}_r]w$ is not zero. Now, let $v_r = w_r/\|w_r\|$. Since w_r is not zero, $r \mapsto v_r$ is smooth, it belongs to S^2 and it is orthogonal to u_r. Thus, $r \mapsto A_r = [u_r\ v_r\ u_r \times v_r]$ is a plot of SO(3) such that $A_r(\mathbf{e}_1) = u_r$. Therefore p is a subduction. Now we just observe that $p(A) = p(A')$ if and only if there exists $k \in SO(2, \mathbf{e}_1)$ such that $A' = Ak^{-1}$. Therefore, thanks to the uniqueness of quotients (art. 1.52), we conclude that p is a fibration, and moreover a principal fibration. Finally, the map $(A, v) \mapsto (A\mathbf{e}_1, Av)$ from $SO(3) \times \mathbf{e}_1^\perp$ to $S^2 \times \mathbf{R}^3$ takes its values in TS^2 and represents the associated fiber bundle $SO(3) \times_{SO(2,\mathbf{e}_1)} \mathbf{e}_1^\perp$.

Exercise 132, p. 253 (Homogeneity of manifolds). Let ε_n be a smooth bump-function defined on $\mathbf{R}^n$, equal to 1 on the ball B' of radius $r/2$, centered at 0_n, and equal to 0 outside the ball B of radius r. Let ε be a smooth bump-function defined on $\mathbf{R}$, equal to 1 on the interval $[-\delta/2, 1 + \delta/2]$ and equal to 0 outside the interval $]-\delta, 1 + \delta[$. Let f be the vector field defined on $\mathbf{R}^n \times \mathbf{R}$ by $f(x, t) = (0_n, \varepsilon_n(x) \times \varepsilon(t))$. The vector field f is equal to zero outside the cylinder

C and equal to $(0_n, 1)$ into the cylinder $C' = B' \times [-\delta/2, 1+\delta/2]$. Since the support of f is contained in C, f is integrable, and the time 1 of its exponential is a compactly supported diffeomorphism, whose support is contained in C. Moreover, since inside C' the vector field f is constant equal to $(0_n, 1)$, its exponential is the translation $\exp(sf)(x,t) = (x, t+s)$ as soon as (x,t) and $(x, t+s)$ belong to C'. Thus $\exp(f)(0_n, 0) = (0_n, 1)$, and $\exp(f)$ satisfies the conditions of the exercise. Now, let $F : \mathcal{B} \to M$ be some chart of M, where $\mathcal{B}$ is an open ball. Let $r, r' \in \mathcal{B}$, let $x = F(r)$ and $x' = F(r')$. There exists a small open cylinder $\mathcal{C}$ containing the segment $\{r + s(r'-r)\}_{s=0}^{1}$ and contained in $\mathcal{B}$. According to the first part of the exercise (modulo smooth equivalence) there exists a diffeomorphism of $\mathcal{B}$ with compact support contained in $\mathcal{C}$, and mapping r to r'. Thus, the image of this diffeomorphism by the chart F defines a local diffeomorphism of M, defined on the open subset $F(\mathcal{B})$, mapping x to x' and which is the identity outside a closed subset. It can thus be extended, by the identity, into a global compactly supported diffeomorphism of M, mapping x to x'. Next, let us choose x and x', any two points in M. Let us note first that there exists an atlas of M made of charts whose domains are open balls, let us call these charts *round charts*. Since M is connected, there exists a path γ connecting x to x', let $\{F_t\}_{t=0}^{1}$ be a family of round charts such that $\gamma(t) \in \mathrm{val}(F_t)$. The set $\{\gamma^{-1}(\mathrm{val}(F_t))\}_{t=0}^{1}$ is an open covering of $[0,1]$, by compacity, after re-indexation, there exists a finite family $\{F_i\}_{i=1}^{N}$ such that $\{\gamma^{-1}(\mathrm{val}(F_i))\}_{i=1}^{N}$ is an open covering of $[0,1]$. We can even assume that for every index $i = 1, \ldots, N$, $J_i = \gamma^{-1}(\mathrm{val}(F_i))$ is an open interval of $\mathbf{R}$ such that only two successive intervals intersect. Choosing a point $t_i \in J_i \cap J_{i+1}$, $i = 1, \ldots, N-1$, we get a family of points $x_i = \gamma(t_i)$ such that two consecutive points (x_i, x_{i+1}) belong to the values of one round chart F_i. Now, thanks to the previous result we get a finite family of compactly supported diffeomorphisms of M mapping every point of this family to its successor, the first point being x and the last x'. Thus, after composition we get a compactly supported diffeomorphism mapping x to x'. And we conclude that the map $\hat{x}_0 : \mathrm{Diff}_K(M) \to M$ is surjective. Now, to prove that the projection $\hat{x}_0$ is a principal fibration, with structure group the stabilizer $\mathrm{Diff}_K(M, x_0)$, we need to lift locally any plot of M. But since M is a manifold, it is sufficient to lift locally any chart $F : U \to M$. Without loss of generality, we can assume that $0 \in U$, and for simplicity that U is connected. Let us prove first that there exists a smooth map $r \mapsto \Psi_r$, where r belongs to an open ball centered at 0, contained in U, and Ψ_r is a compactly supported diffeomorphism of U, mapping 0 to r. For that, let us chose two balls B and B$'$ of radii $R < R'$, contained in U and centered at 0. Let λ be a smooth bump-function defined on U, equal to 1 in B and equal to 0 outside B$'$. Let us define $f_r(r') = \lambda(r') \times r$, for $r' \in U$ and $|r| < R$. The map $r \mapsto f_r$ is a smooth family of vector fields on U, with supports contained in B$'$. For all t, r and r' such that r' and $r' + tr$ belong to B, the exponential of f_r coincides with the translation $\exp(tf_r)(r') = r' + tr$. Thus, for $t = 1$, $r' = 0$, and $|r| < R$, we get $\exp(f_r)(0) = r$. Thanks to the differentiability of $r \mapsto f_r$ and to the differentiability of the solutions of an ordinary differential equation with respect to the parameters, the map $r \mapsto \exp(f_r)$ is smooth for the functional diffeology. Now, since the support of $\exp(f_r)$ is contained in U, $F \circ \exp(f_r) \circ F^{-1}$ is a local compactly supported diffeomorphism of M mapping $F(U)$ into itself, therefore it can be extended, by the identity, into a global compactly supported diffeomorphism Φ_r of M, mapping $x_0' = F(0)$ to $x = F(r)$, that is, $\Phi_r(x_0') = F(r)$. Next, we know that there exists a diffeomorphism

φ mapping x_0 to x_0', thus the diffeomorphism $\phi_r = \Phi_r \circ \varphi$ maps x_0 to $F(r)$, that is, $\hat{x}_0(\phi_r) = \phi_r(x_0) = F(r)$. Finally, since $r \mapsto \exp(f_r)$ is smooth and the composition with smooth maps is a smooth operation, the local lift $r \mapsto \phi_r$ of F along $\hat{x}_0$, in $\mathrm{Diff}_K(M)$, is smooth. Therefore, $\hat{x}_0$ is a diffeological principal fibration. The homogeneity of manifolds has first been proved by Donato in his dissertation [**Don84**].

Let $T(M)$ be the tangent bundle to M, that is, the space of 1-jets of paths in M. Precisely, $T(M)$ is the quotient of $\mathrm{Paths}(M)$ by the equivalence relation $\gamma \sim \gamma'$ if $\gamma(0) = \gamma'(0)$, and for all differential 1-forms α on M, $\alpha(\gamma)(0) = \alpha(\gamma')(0)$. Then, $T(M)$ is the associated fiber bundle $\mathrm{Diff}_K(M) \times_{\mathrm{Diff}_K(M,x_0)} E$, where $E = T_{x_0}(M) = \mathrm{Paths}(M, x_0, \star)/\sim$.

☞ **Exercise 133, p. 266** (Covering tori). First of all, since $(\mathbf{R}^n, +)$ is a group and $\Gamma \subset \mathbf{R}^n$ is a subgroup, the projection $p : \mathbf{R}^n \to T_\Gamma$ is a fibration (art. 8.15), a principal fibration. Then, thanks to the unicity, up to equivalence, of universal coverings (art. 8.26), since $\mathbf{R}^n$ is simply connected and Γ is discrete, $p : \mathbf{R}^n \to T_\Gamma$ is the universal covering of T_Γ, and Γ identifies with $\pi_1(T_\Gamma)$. Now every path $t \mapsto t\gamma$, where $\gamma \in \Gamma$, projects into a loop in T_Γ. This family of loops, one for each element of Γ, gives a favorite representative for each element of $\pi_1(T_\Gamma, \mathbf{1}_{T_\Gamma})$, that is, each class of homotopy of loop based at $\mathbf{1}_{T_\Gamma} = p(0)$. Next, considering the circle, we know that the map $p : t \mapsto (\cos(t), \sin(t))$ is a subduction, making the circle $S^1 \subset \mathbf{R}^2$ diffeomorphic to $\mathbf{R}/2\pi\mathbf{Z}$; see Exercise 27, p. 27. Each loop of S^1 is then homotopic to some $t \mapsto p(2\pi kt)$, for $k \in \mathbf{Z}$. But $p(2\pi tk) = (\cos(2\pi kt), \sin(2\pi kt))$ is also equal to $\mathcal{R}(2\pi kt)(\mathbf{1}_{S^1})$. By translation with an element of S^1, we get the general statement.

☞ **Exercise 134, p. 272** (De Rham homomorphism and irrational tori). There are two different cases, $s = \alpha r$ and $s \neq \alpha r$.

1. CASE $s = \alpha r$. The action of $\mathbf{Z} \times \mathbf{Z}$ on $\mathbf{R} \times \mathbf{R}$ is given by $(x, y) \mapsto (x + n + \alpha m, y + r(n + \alpha m))$, with $(n, m) \in \mathbf{Z} \times \mathbf{Z}$. The following map

$$\Phi : (x, y) \mapsto ([x], y - ax),$$

defined from $\mathbf{R} \times \mathbf{R}$ to $T_\alpha \times \mathbf{R}$, where $[x] \in T_\alpha$, is a realization of the quotient, $\Phi(x, y) = \Phi(x', y') \Leftrightarrow (x', y') = (x + n + \alpha m, y + r(n + \alpha m))$. We find again the situation with $\rho = \rho_c$, and the quotient is trivial.

2. CASE $s \neq \alpha r$. Let us consider the following linear map,

$$M : \begin{pmatrix} x \\ y \end{pmatrix} \mapsto \begin{pmatrix} u \\ v \end{pmatrix} = \frac{1}{s - \alpha r} \begin{pmatrix} sx - \alpha y \\ -rx + y \end{pmatrix}.$$

The map M is a linear isomorphism of $\mathbf{R} \times \mathbf{R}$ with determinant 1. The image by M of the action of $\mathbf{Z} + \alpha\mathbf{Z}$ is the standard action of $\mathbf{Z} \times \mathbf{Z}$, $(n, m) : (u, v) \mapsto (u+n, v+m)$. Thus, the quotient $\mathbf{R} \times_\rho \mathbf{R}$ is diffeomorphic to the 2-torus $T^2 = T \times T$. The projection from $\mathbf{R} \times_\rho \mathbf{R} \simeq T \times T$ onto the irrational torus T_α is given in terms of (u, v), above, by

$$p : T^2 \to T_\alpha \quad \text{with} \quad p([u, v]) = [u + \alpha v].$$

Then, the action of $(\mathbf{R}, +)$ on the variables (u, v) is generated by the translation along the vector

$$\xi = \frac{1}{s - \alpha r} \begin{pmatrix} -\alpha \\ 1 \end{pmatrix}.$$

Note that the various choices of homomorphisms ρ change only the speed of the action of $\mathbf{R}$ on the 2-torus.

Exercise 135, p. 272 (The fiber of the integration bundle). Let $r \mapsto \hat{x}_r$ be a plot of X_α with values in the fiber $\mathrm{pr}_\alpha^{-1}(x_0)$. By definition of the quotient diffeology, there exists, at least locally, a smooth lift $r \mapsto \ell_r$ such that $\hat{x}_r = \mathrm{class}(\ell_r)$. Let us represent the quotient by the values of the integrals $\int_\ell \alpha$, $\ell \in \mathrm{Loops}(X, x_0)$, and let us prove that the map $r \mapsto \int_{\ell_r} \alpha$ is locally constant. Let us thus compute the variation of the integral, according to (art. 6.70). We can reduce the question to a 1-parameter variation $s \mapsto \ell_s$, and because $d\alpha = 0$, the variation is reduced to

$$\delta \int_{\ell_s} \alpha = \int_{\partial I} \alpha(\delta\ell) = \Big[\alpha(\delta\ell)\Big]_0^1 = \alpha(\delta\ell)(1) - \alpha(\delta\ell)(0).$$

But $\alpha(\delta\ell)$ is given (art. 6.56) by

$$\begin{aligned}
\alpha(\delta\ell)(t) &= \alpha\left(\begin{pmatrix} s \\ t \end{pmatrix} \mapsto \ell_s(t)\right)_{\binom{s=0}{t}} \begin{pmatrix} 1 \\ 0 \end{pmatrix} \\
&= \alpha\left(\begin{pmatrix} s \\ t \end{pmatrix} \mapsto \ell_s(t)\right)_{\binom{s=0}{t}} \left[D\left(s \mapsto \begin{pmatrix} s \\ t \end{pmatrix}\right)(s=0)(1)\right] \\
&= \alpha\left(s \mapsto \begin{pmatrix} s \\ t \end{pmatrix} \mapsto \ell_s(t)\right)_{s=0} (1) \\
&= \alpha\,(s \mapsto \ell_s(t))_{s=0}\,(1).
\end{aligned}$$

Now, since $\ell_s(0) = \ell_s(1)$ for all s, we get $\alpha(\delta\ell)(1) = \alpha(\delta\ell)(0)$, and hence $\delta \int_{\ell_s} \alpha = 0$. Therefore, the map $r \mapsto \hat{x}_r$ is locally constant and the fiber in X_α over x_0, equipped with the subset diffeology, is discrete.

Exercise 136, p. 287 (Spheric periods on toric bundles). Since σ belongs to $\mathrm{Loops}(\mathrm{Loops}(X, x), x)$, we have $\sigma(0)(t) = \sigma(1)(t) = \sigma(s)(0) = \sigma(s)(1) = x$ for all s, t, that is, $\sigma \circ j_1(0) = \sigma \circ j_1(1) = \sigma \circ j_2(0) = \sigma \circ j_2(1) = x$, where j_k is defined in (art. 6.59). We know that the pullback $\mathrm{pr}_1 : \sigma^*(Y) \to \mathbf{R}^2$ is trivial (art. 8.9), so there exists a global lifting $\tilde{\sigma}$ of σ, $\pi_*(\tilde{\sigma}) = \pi \circ \tilde{\sigma} = \sigma$. Now,

$$\int_\sigma \omega = \int_{\pi_*(\tilde{\sigma})} \omega = \int_{\tilde{\sigma}} \pi^*(\omega) = \int_{\tilde{\sigma}} d\lambda = \int_{\partial\tilde{\sigma}} \lambda = \sum_{k=1}^{2} (-1)^k \left[\int_{\tilde{\sigma}\circ j_k(0)} \lambda - \int_{\tilde{\sigma}\circ j_k(1)} \lambda\right].$$

Let $\gamma_{k,a} = \tilde{\sigma} \circ j_k(a)$, with $k = 1, 2$ and $a = 0, 1$, and let $\bar{\gamma}_{k,a}(t) = \gamma_{k,a}(1-t)$. We have

$$\sum_{k=1}^{2} (-1)^k \left[\int_{\tilde{\sigma}\circ j_k(0)} \lambda - \int_{\tilde{\sigma}\circ j_k(1)} \lambda\right] = \int_{\gamma_1 = \gamma_{2,0}} \lambda + \int_{\gamma_2 = \gamma_{1,1}} \lambda + \int_{\gamma_3 = \bar{\gamma}_{2,1}} \lambda + \int_{\gamma_4 = \bar{\gamma}_{1,0}} \lambda.$$

Next, since $\sigma \circ j_k(a) = x$, the paths γ_i are paths in $Y_x = \pi^{-1}(x)$, and they describe a closed circuit: $\tilde{\sigma}(1)(0) = \gamma_1(1) = \gamma_2(0)$, $\tilde{\sigma}(1)(1) = \gamma_2(1) = \gamma_3(0)$, $\tilde{\sigma}(0)(1) = \gamma_3(1) = \gamma_4(0)$, $\tilde{\sigma}(0)(0) = \gamma_4(1) = \gamma_1(0)$. Then, choosing a point $y_0 \in Y_x$ and identifying Y_x with T, thanks to the orbit map $\tau \mapsto \tau_Y(y_0)$, we can regard the γ_i as paths in T, describing a close circuit starting and ending at $0 \in T$. But since $\mathbf{R}$ is a covering of T (actually, the universal covering), thanks to the monodromy theorem, we can lift each path γ_i by a path x_i in $\mathbf{R}$, $\gamma_i(t) = \mathrm{class}(x_i(t))$. Moreover, we can choose these liftings such that $x_1(0) = 0$, $x_1(1) = x_2(0)$, $x_2(1) = x_3(0)$, $x_3(1) = x_4(0)$, and since $\gamma_4(1) = 0$, $x_4(1) \in \Gamma$. Then,

$$\int_{\gamma_i} \lambda = \int_0^1 \lambda(\gamma_i)_t(1)\, dt = \int_0^1 \theta[t \mapsto \text{class}(x_i(t))]_t(1)\, dt.$$

Let us recall that $\text{class}^*(\theta) = dt$ means $\theta[t \mapsto \text{class}(t)] = dt$, that is, $\theta[r \mapsto \text{class}(x(r))] = x^*(dt) = dx$, for all smooth real parametrizations x, thus

$$\int_{\gamma_i} \lambda = \int_0^1 \frac{dx_i(t)}{dt}\, dt = x_i(1) - x_i(0),$$

and, finally

$$\int_\sigma \omega = \sum_{i=1}^4 \int_{\gamma_i} \lambda = \gamma_4(1) - \gamma_1(0) = \gamma_4(1) \in \Gamma.$$

☞ **Exercise 137, p. 287** (Fiber bundles over tori). Let us recall that every fiber bundle is associated with a principal fiber bundle (art. 8.16). Thus, it is sufficient, for this exercise, to assume the fiber bundle to be principal. Let $\pi : Y \to T_\Gamma$ be a principal fiber bundle with structure group G. Let $p : \mathbf{R}^n \to T_\Gamma$ be the universal covering. The pullback $\text{pr}_1 : p^*(Y) \to \mathbf{R}^n$ is a G-principal fiber bundle, with the action $g(x, y) = (x, g_Y(y))$, where g_Y denotes the action of $g \in G$ on Y. Since the base of the pullback is $\mathbf{R}^n$, the fibration is trivial (art. 8.19). Therefore, it admits a smooth section, that is, there exists $\varphi \in \mathcal{C}^\infty(\mathbf{R}^n, Y)$ such that $\pi \circ \varphi = p$. Then, for each $x \in \mathbf{R}^n$ and each $\gamma \in \Gamma$, there exists $h(\gamma)(x) \in G$ such that $\varphi(x) = h(\gamma)(x)_Y(\varphi(x+\gamma))$. Thus, $h \in \text{Maps}(\Gamma, \mathcal{C}^\infty(\mathbf{R}^n, G))$ and $h(\gamma+\gamma')(x) = h(\gamma)(x) \cdot h(\gamma')(x+\gamma)$. Note also that, since Γ is Abelian, $h(\gamma)(x) \cdot h(\gamma')(x+\gamma) = h(\gamma')(x) \cdot h(\gamma)(x+\gamma')$. Next, $\text{pr}_2 : p^*(Y) \to Y$ is a subduction, indeed if $r \mapsto y_r$ is a plot of Y, then locally $y_r = \text{pr}_2(x_r, y_r)$, where $r \mapsto x_r$ is a local lift in $\mathbf{R}^n$ of $r \mapsto \pi(y_r)$. Then, Y is equivalent to the quotient of $\mathbf{R}^n \times G$ by the action of Γ induced by h, that is, $\gamma(x, g) = (x+\gamma, g \cdot h(\gamma)(x))$. Conversely, given such a map h, we get a G-principal fiber bundle by quotient. These maps h are kinds of cocycles, the above construction gives a trivial bundle if and only if h is trivial, that is, if and only if there exists $f \in \mathcal{C}^\infty(\mathbf{R}^n, G)$ such that $h(\gamma)(x) = f(x)^{-1} \cdot f(x+\gamma)$, such h can be regarded as coboundaries.

☞ **Exercise 138, p. 287** (Flat connections on toric bundles). Let $\pi : Y \to X$ be a T-principal bundle, with T a 1-dimensional torus. Let λ be a connection 1-form and let us assume that Θ, the associated connection, is flat, which means, by definition, that the holonomy group is discrete (art. 8.35). Thus, the holonomy bundle $p : Y_\Theta(y) \to X$ is a covering, (art. 8.35, (♠)) and (art. 8.35, Note 1). Let us recall that $Y_\Theta(y) \subset Y$ is made of the ends of horizontal paths starting at y, and let us denote by $j : Y_\Theta(y) \to Y$ the inclusion. Let $\gamma \in \text{Paths}(Y_\Theta(y), y)$, the horizontal path $\Theta(\gamma, 0)$ takes necessarily its values in $Y_\Theta(y)$ and projects on the same path $\pi \circ \gamma$ as γ in X. But since $Y_\Theta(y)$ is a covering, there is one and only one lift of $\pi \circ \gamma$ in $Y_\Theta(y)$ starting at y, thus $\Theta(\gamma, 0) = \gamma$. Hence, all the 1-plots in $Y_\Theta(y)$ are horizontal and $\lambda(\gamma) = 0$ for all $\gamma \in \text{Paths}(Y_\Theta(y))$. Therefore, the connection form λ vanishes on $Y_\Theta(y)$ (art. 6.37), that is, $j^*(\lambda) = 0$. Now $\pi \circ j = p$ implies, on the one hand, $j^*(d\lambda) = d(j^*(\lambda)) = 0$, and on the other hand $j^*(d\lambda) = j^*(\pi^*(\omega)) = (\pi \circ j)^*(\omega) = p^*(\omega)$, where ω is the curvature of λ. Thus, $p^*(\omega) = 0$, and since p is a subduction, $\omega = 0$ (art. 6.39).

☞ **Exercise 139, p. 287** (Connection forms over tori). Let λ be a connection 1-form on Y. Let $\mathrm{pr} : (x, t) \mapsto [x, t]$ be the projection from $\mathbf{R} \times \mathbf{R}$ to Y, the pullback $\Lambda = \mathrm{pr}^*(\lambda)$ is a connection 1-form on $\mathbf{R} \times \mathbf{R}$ and thus writes, with our usual notation for points and vectors,

$$\Lambda_{\binom{x}{t}} \begin{pmatrix} \delta x \\ \delta t \end{pmatrix} = a(x)\delta x + \delta t,$$

where a is a smooth real function. Now, Λ is invariant by Γ. Thus, for all $k \in \Gamma$,

$$\begin{aligned} [k^*_{\mathbf{R}\times\mathbf{R}}(\Lambda)]_{\binom{x}{t}} \begin{pmatrix} \delta x \\ \delta t \end{pmatrix} &= \Lambda_{\binom{x}{t}} \begin{pmatrix} \delta x \\ \delta t \end{pmatrix}, \\ \Lambda_{\binom{x+k}{t+\tau(k)(x)}} \begin{pmatrix} \delta(x+k) \\ \delta(t+\tau(k)(x)) \end{pmatrix} &= \Lambda_{\binom{x}{t}} \begin{pmatrix} \delta x \\ \delta t \end{pmatrix}, \\ a(x+k)\delta x + \delta t + \tau(k)'(x)\delta x &= a(x)\delta x + \delta t, \\ a(x+k)\delta x + \tau(k)'(x)\delta x &= a(x)\delta x. \end{aligned}$$

Hence, $\tau(k)'(x) = a(x) - a(x+k)$. Let A be a primitive of a. We get by integration, $\tau(k)(x) - \tau(k)(0) = A(x) - A(0) - A(x+k) + A(k)$. Then, $\tau(k)(x) = B(k) + \sigma(x+k) - \sigma(x)$, for some $B : \Gamma \to \mathbf{R}$ and $\sigma \in \mathcal{C}^\infty(\mathbf{R})$. Thus, τ is equivalent to B (art. 8.39, (♡)). The condition of cocycle writes then $B(k+k') = B(k) + B(k')$, and B is a homomorphism from Γ to $\mathbf{R}$. Therefore, if there exists a connection 1-form on Y, then the cocycle τ defining π is equivalent to a homomorphism. Conversely, if τ is equivalent to a homomorphism B, we just consider $\tau = B$. Then, any 1-form $\Lambda_{(x,t)} : (\delta x, \delta t) \mapsto a(x)\delta x + \delta t$, where a is a Γ-invariant real function on $\mathbf{R}$, is invariant by the action of Γ on $\mathbf{R} \times \mathbf{R}$, and since $\mathbf{R} \times \mathbf{R}$ is a covering of the quotient $Y = \mathbf{R} \times_\Gamma \mathbf{R}$, there exists a 1-form λ on Y such that $\mathrm{pr}^*(\lambda) = \Lambda$ (art. 6.38). Next, since Λ is clearly a connection 1-form, so is λ.

In the special case $\Gamma = \mathbf{Z} + \alpha\mathbf{Z}$, a homomorphism writes $B(n + \alpha m) = an + bm$, but considering the function $\sigma(x) = ax$, B is equivalent to $\beta(n + \alpha m) = cm$ for some constant c. If $c \neq 0$, then we can choose $c = 1$, for simplicity. Thus, the quotient of $\mathbf{R} \times \mathbf{R}$ by the action $(n, (x, t)) \mapsto (x + n + \alpha m, t + m)$ of $\mathbf{Z}$ is equivalent to the 2-torus T^2, with $T = \mathbf{R}/\mathbf{Z}$, for the projection $(x, t) \mapsto ([t], [x - \alpha t])$.

☞ **Exercise 140, p. 297** (Loops in the torus). The 1-form $\mathcal{K}\omega$ restricted to $\mathrm{Loops}(T^2)$ is closed, thanks to the fundamental property of the Chain-Homotopy operator. In particular, the restriction of $\mathcal{K}\omega$ to each component of $\mathrm{Loops}(T^2)$ is closed. Let $\mathrm{comp}(\ell)$ be the connected component of ℓ in $\mathrm{Loops}(T^2)$. The path $\sigma : s \mapsto \sigma_s$, with $\sigma_s(t) = \mathrm{class}(t, s)$, is a loop in $\mathrm{Loops}(T^2)$, based at ℓ, thus a loop in $\mathrm{comp}(\ell)$. Let us compute the integral of $\mathcal{K}\omega$ on σ, that is,

$$\int_\sigma \mathcal{K}\omega = \int_0^1 \mathcal{K}\omega(\sigma)_s(1)\, ds.$$

Let $\bar{\sigma}(t,s) = \sigma_s(t)$, and let $\tilde{\sigma}$ be the lift of $\bar{\sigma}$ in $\mathbf{R}^2$, $\text{class} \circ \tilde{\sigma} = \bar{\sigma}$, such that $\tilde{\sigma}(0,0) = (0,0)$, by definition,

$$\begin{aligned}
\mathcal{K}\omega(\sigma)_s(1) &= \int_0^1 \omega\left(\begin{pmatrix} t \\ s \end{pmatrix} \mapsto \sigma_s(t)\right)_{\binom{t}{s}} \begin{pmatrix} 1 \\ 0 \end{pmatrix} \begin{pmatrix} 0 \\ 1 \end{pmatrix} dt \\
&= \int_0^1 \omega(\bar{\sigma})_{\binom{t}{s}} \begin{pmatrix} 1 \\ 0 \end{pmatrix} \begin{pmatrix} 0 \\ 1 \end{pmatrix} dt \\
&= \int_0^1 \bar{\sigma}^*(\omega)_{\binom{t}{s}} \begin{pmatrix} 1 \\ 0 \end{pmatrix} \begin{pmatrix} 0 \\ 1 \end{pmatrix} dt \\
&= \int_0^1 (\text{class} \circ \tilde{\sigma})^*(\omega)_{\binom{t}{s}} \begin{pmatrix} 1 \\ 0 \end{pmatrix} \begin{pmatrix} 0 \\ 1 \end{pmatrix} dt \\
&= \int_0^1 (\tilde{\sigma})^*(\mathbf{e}^1 \wedge \mathbf{e}^2)_{\binom{t}{s}} \begin{pmatrix} 1 \\ 0 \end{pmatrix} \begin{pmatrix} 0 \\ 1 \end{pmatrix} dt \\
&= \int_0^1 \det\left[D(\tilde{\sigma})_{\binom{t}{s}}\right] dt.
\end{aligned}$$

Hence,

$$\int_\sigma \mathcal{K}\omega = \int_0^1 ds \int_0^1 dt \det\left[D(\tilde{\sigma})_{\binom{t}{s}}\right]. \qquad (\diamondsuit)$$

In our case $\tilde{\sigma}$ is the identity and $\int_\sigma \mathcal{K}\omega = 1$. Then, since $d[\mathcal{K}\omega \restriction \text{comp}(\ell)] = 0$ and $\int_\sigma \mathcal{K}\omega \neq 0$, $\text{comp}(\ell)$ is not simply connected; see Exercise 117, p. 198.

Exercise 141, p. 297 (Periods of a surface). Since T^2 is a group, its fundamental group $\pi_1(T^2, \text{class}(0,0))$ is Abelian, and the set of connected components $\pi_0(\text{Loops}(T^2))$ is in a one-to-one correspondence with $\pi_1(T^2, \text{class}(0,0))$; see Exercise 87, p. 123. Precisely, $\pi_1(T^2, \text{class}(0,0)) = \pi_1(T, [0])^2$, where we simply denote $\text{class}(x)$ by $[x]$, for $x \in \mathbf{R}$ and $\text{class}(x) \in T = \mathbf{R}/\mathbf{Z}$, and then $\text{class}(x,y) = ([x],[y])$. Now, $\pi_1(T,[0]) \simeq \mathbf{Z}$, each class is represented by a loop $t \mapsto [nt]$, with $n \in \mathbf{Z}$. Therefore, each class of loop in T^2 is represented by a loop $\ell_{n,m} : t \mapsto ([nt],[mt])$. Now, let $s \mapsto \sigma_s$ be a loop in $\text{Loops}(T^2)$ based in $\ell_{n,m}$, the map $(t,s) \mapsto \sigma_s(t)$, from $\mathbf{R}^2$ to T^2 has a unique lift $t \mapsto (x_s(t), y_s(t))$ in $\mathbf{R}^2$ — that is, $\sigma_s(t) = ([x_s(t)],[y_s(t)])$ — such that $(x_0(0), y_0(0)) = (0,0)$, (art. 8.25). Since $\sigma_0 = \ell_{n,m}$, $x_0(t) = nt + k$ and $y_0(t) = mt + k'$, but $(x_0(0), y_0(0)) = (0,0)$ implies $k = k' = 0$. Then, $x_0(t) = nt$ and $y_0(t) = mt$. Next, since σ_s is a loop, $\sigma_s(1) = \sigma_s(0)$, that is, $x_s(1) = x_s(0) + k$ and $y_s(1) = y_s(0) + k'$, with $k, k' \in \mathbf{Z}$. Computed for $s = 0$, that gives $k = n$ and $k' = m$, thus $x_s(1) = x_s(0) + n$ and $y_s(1) = y_s(0) + m$. The last condition, $\sigma_0 = \sigma_1 = \ell_{n,m}$, gives $x_1(t) = x_0(t) + k$ and $y_1(t) = y_0(t) + k'$, that is, $x_1(t) = nt + k$ and $y_1(t) = mt + k'$, with $k, k' \in \mathbf{Z}$. Summarized, these conditions write

$$\begin{array}{lll}
x_0(t) = nt & x_1(t) = nt + k & x_s(1) - x_s(0) = n, \\
y_0(t) = mt & y_1(t) = mt + k' & y_s(1) - y_s(0) = m.
\end{array}$$

Now, let us define

$$\xi(u)_s(t) = \begin{pmatrix} [u(nt + sk) + (1-u)x_s(t)] \\ [u(mt + sk') + (1-u)y_s(t)] \end{pmatrix}.$$

With the conditions summarized above, we can check that $\xi(u)_s(0) = \xi(u)_s(1)$ and $\xi(u)_0(t) = \xi(u)_1(t) = \ell_{n,m}$. Thus, ξ is a fixed-ends homotopy of loops connecting $\xi(0) = [s \mapsto \sigma_s]$ to $\xi(1) = [s \mapsto \sigma'_s]$, with $\sigma'_s(t) = ([nt + sk],[mt + sk'])$. Next, since $\mathcal{K}\omega \upharpoonright \mathrm{comp}(\ell_{n,m})$ is closed, the integrals of $\mathcal{K}\omega$ on the loops $[s \mapsto \sigma_s]$ and $[s \mapsto \sigma'_s]$ coincide; see Exercise 116, p. 198. Hence, thanks to Exercise 140, p. 297 ($\diamondsuit$), that gives

$$\begin{aligned}
\int_{[s\mapsto\sigma_s]} \mathcal{K}\omega &= \int_0^1 dt \int_0^1 ds \det \left[D\left(\begin{pmatrix} t \\ s \end{pmatrix} \mapsto \begin{pmatrix} nt+sk \\ mt+sk' \end{pmatrix} \right)_{\binom{t}{s}} \right] \\
&= \int_0^1 dt \int_0^1 ds \det \begin{pmatrix} n & k \\ m & k' \end{pmatrix} \\
&= nk' - mk.
\end{aligned}$$

Therefore,

$$\mathrm{Periods}(\mathcal{K}\omega \upharpoonright \mathrm{comp}(\ell_{n,m})) = \{nk' - mk \mid k, k' \in \mathbf{Z}\}.$$

Note that on the component of the constant loop $\ell_{0,0}$, the periods of $\mathcal{K}\omega$ vanish, which means that $\mathcal{K} \upharpoonright \mathrm{comp}(\ell_{0,0})$ is exact. Also note that, for $(n, m) \neq (0, 0)$,

$$\mathrm{Periods}(\mathcal{K}\omega \upharpoonright \mathrm{comp}(\ell_{n,m})) = \gcd(n, m)\,\mathbf{Z}.$$

Remark finally that, since T^2 is a group, the connected components of $\mathrm{Loops}(T^2)$ form a group for the pointwise addition. For all $(n, m) \in \mathbf{Z}^2$, the map $\phi_{n,m} : \ell \mapsto \ell + \ell_{n,m}$, defined on the connected component of the constant loop $\ell_{0,0}$, is a diffeomorphism from $\mathrm{comp}(\ell_{0,0})$ to $\mathrm{comp}(\ell_{n,m})$, mapping $\ell_{0,0}$ to $\ell_{n,m}$. The computation of the periods shows that $\phi_{n,m}$ is not an automorphism of $\mathcal{K}\omega$.

☞ **Exercise 142, p. 307** (Compact supported real functions, I). Let us first remark that the definition of $\omega(P)_r(\delta r, \delta' r)$ makes sense. Since, by definition of the compact diffeology, for any r_0, there exist an open neighborhood $V \subset U$ and a compact K of $\mathbf{R}$ such that $P(r)$ and $P(r_0)$ coincide outside K, the derivatives

$$t \mapsto \frac{\partial}{\partial r}\frac{\partial P(r)(t)}{\partial t}(\delta r) \quad \text{and} \quad t \mapsto \frac{\partial P(r)(t)}{\partial r}(\delta' r)$$

are compact supported real functions, with their supports in K.

1) Note that $\omega(P)_r$ is antisymmetric. Indeed,

$$\begin{aligned}
\omega(P)_r(\delta r, \delta' r) &= \int_{-\infty}^{+\infty} \frac{\partial}{\partial r}\frac{\partial P(r)(t)}{\partial t}(\delta r) \frac{\partial P(r)(t)}{\partial r}(\delta' r)\, dt \\
&= \int_{-\infty}^{+\infty} \frac{\partial}{\partial t}\left(\frac{\partial P(r)(t)}{\partial r}(\delta r)\right) \frac{\partial P(r)(t)}{\partial r}(\delta' r)\, dt \\
&= \int_{-\infty}^{+\infty} \frac{\partial}{\partial t}\left(\frac{\partial P(r)(t)}{\partial r}(\delta r) \frac{\partial P(r)(t)}{\partial r}(\delta' r)\right) dt \\
&\quad - \int_{-\infty}^{+\infty} \frac{\partial P(r)(t)}{\partial r}(\delta r) \frac{\partial}{\partial t}\left(\frac{\partial P(r)(t)}{\partial r}(\delta' r)\right) dt \\
&= 0 - \int_{-\infty}^{+\infty} \frac{\partial}{\partial r}\frac{\partial P(r)(t)}{\partial t}(\delta' r) \frac{\partial P(r)(t)}{\partial r}(\delta r)\, dt \\
&= -\omega(P)_r(\delta' r, \delta r).
\end{aligned}$$

Now, let $F : V \to U$ be a smooth m-parametrization. Let $s \in V$, $\delta s, \delta' s \in \mathbf{R}^m$, let us denote $r = F(s)$, $\delta r = D(F)_s(\delta s)$ and $\delta' r = D(F)_s(\delta' s)$. We have

$$\begin{aligned}
\omega(P \circ F)_s(\delta s, \delta' s) &= \int_{-\infty}^{+\infty} \frac{\partial}{\partial s}\left[\frac{\partial P(F(s))(t)}{\partial t}\right](\delta s) \frac{\partial P(F(s))(t)}{\partial s}(\delta' s)\, dt \\
&= \int_{-\infty}^{+\infty} \frac{\partial}{\partial r}\left[\frac{\partial P(r)(t)}{\partial t}\right]\left(\frac{\partial r}{\partial s}(\delta s)\right) \frac{\partial P(r)(t)}{\partial r}\left(\frac{\partial r}{\partial s}(\delta' s)\right) dt \\
&= \int_{-\infty}^{+\infty} \frac{\partial}{\partial r}\left[\frac{\partial P(r)(t)}{\partial t}\right](\delta r) \frac{\partial P(r)(t)}{\partial r}(\delta' r)\, dt \\
&= F^*(\omega(P))_s(\delta s, \delta' s).
\end{aligned}$$

Thus, ω satisfies the conditions to be a differential 2-form on X.

2) Denoting

$$r = \begin{pmatrix} s \\ s' \end{pmatrix}, \quad \delta r = \begin{pmatrix} 1 \\ 0 \end{pmatrix}, \quad \delta' r = \begin{pmatrix} 0 \\ 1 \end{pmatrix}, \quad \text{and} \quad P(r) = sf + s'g,$$

we have immediately

$$\begin{aligned}
\frac{\partial}{\partial r}\left[\frac{\partial P(r)(t)}{\partial t}\right](\delta r) &= \frac{\partial}{\partial s}\frac{\partial}{\partial t}\Big(sf(t) + s'g(t)\Big) &= \dot{f}(t), \\
\frac{\partial P(r)(t)}{\partial r}(\delta' r) &= \frac{\partial}{\partial s'}\Big(sf(t) + s'g(t)\Big) &= g(t).
\end{aligned}$$

Then,

$$\omega\left(\begin{pmatrix} s \\ s' \end{pmatrix} \mapsto sf + s'g\right)_{\binom{s}{s'}} \begin{pmatrix} 1 \\ 0 \end{pmatrix} \begin{pmatrix} 0 \\ 1 \end{pmatrix} = \int_{-\infty}^{+\infty} \dot{f}(t) g(t)\, dt = \bar{\omega}(f, g).$$

3) Let $u \in X$ and P be a plot of X. Using the notation above, we have

$$\begin{aligned}
T_u^*(\omega)(P)_r(\delta r, \delta' r) &= \omega(T_u \circ P)_r(\delta r, \delta' r) \\
&= \omega(r \mapsto P(r) + u)_r(\delta r, \delta' r),
\end{aligned}$$

and since

$$\frac{\partial}{\partial r}\Big(P(r) + u\Big) = \frac{\partial P(r)}{\partial r},$$

we get $\omega(r \mapsto P(r) + u)_r(\delta r, \delta' r) = \omega(r \mapsto P(r))_r(\delta r, \delta' r)$, that is, $T_u^*(\omega) = \omega$.

4) The space X is contractible. Indeed, the map $s \mapsto sf$, $s \in \mathbf{R}$, is a (smooth) deformation retraction connecting the constant map $f \mapsto 0$ to $\mathbf{1}_X$. Thus, X is null-homotopic. Then, since the holonomy group is a homomorphic image of $\pi_1(X)$ (art. 9.7, item 2), we get $\Gamma = \{0\}$.

5) Since Γ vanishes, the path moment map $\Psi(p)$, for a path p connecting f to g, depends only on the ends f and g, we can chose the path $p : s \mapsto sg + (1-s)f$. Let F be a plot of G, that is, a plot of X. The definition (art. 9.2, (♡)) gives then

$$\begin{aligned}
\Psi(p)(F)_r(\delta r) &= \int_0^1 \omega\left[\begin{pmatrix} s \\ u \end{pmatrix} \mapsto T_{F(u)}(p(s+t))\right]_{\binom{s=0}{u=r}} \begin{pmatrix} 1 \\ 0 \end{pmatrix} \begin{pmatrix} 0 \\ \delta r \end{pmatrix} dt \\
&= \int_0^1 \omega\left[\begin{pmatrix} s \\ u \end{pmatrix} \mapsto p(s+t) + F(u)\right]_{\binom{s=0}{u=r}} \begin{pmatrix} 1 \\ 0 \end{pmatrix} \begin{pmatrix} 0 \\ \delta r \end{pmatrix} dt.
\end{aligned}$$

Let us introduce

$$x = \begin{pmatrix} s \\ r \end{pmatrix}, \quad x_0 = \begin{pmatrix} 0 \\ r \end{pmatrix}, \quad \delta x_0 = \begin{pmatrix} 1 \\ 0 \end{pmatrix}, \quad \text{and} \quad \delta' x_0 = \begin{pmatrix} 0 \\ \delta r \end{pmatrix}.$$

Then, the integrand $\mathcal{I}$ in the above formula rewrites

$$\begin{aligned}
\mathcal{I} &= \omega\left[\begin{pmatrix} s \\ u \end{pmatrix} \mapsto p(s+t) + F(u)\right]_{\left(\begin{smallmatrix} s=0 \\ u=r \end{smallmatrix}\right)} \begin{pmatrix} 1 \\ 0 \end{pmatrix} \begin{pmatrix} 0 \\ \delta r \end{pmatrix} \\
&= \omega\,[x \mapsto p(s+t) + F(u)]_{x_0}\,(\delta x_0, \delta' x_0) \\
&= \int_{-\infty}^{+\infty} \left\{ \frac{\partial}{\partial x}\Big((s+t)\dot{g}(\tau) + (1-s-t)\dot{f}(\tau) + \dot{F}(u)(\tau)\Big)\Big|_{x=x_0} \delta x_0 \right. \\
&\times \left. \frac{\partial}{\partial x}\Big((s+t)g(\tau) + (1-s-t)f(\tau) + F(u)(\tau)\Big)\Big|_{x=x_0} \delta' x_0 \right\} d\tau \\
&= \int_{-\infty}^{+\infty} \left\{ \frac{\partial}{\partial s}\Big((s+t)\dot{g} + (1-s-t)\dot{f} + \dot{F}(r)\Big)\Big|_{s=0} \right. \\
&\times \left. \frac{\partial}{\partial r}\Big(tg(\tau) + (1-t)f(\tau) + F(r)(\tau)\Big)\delta r \right\} d\tau \\
&= \int_{-\infty}^{+\infty} \left\{ \Big(\dot{g}(\tau) - \dot{f}(\tau)\Big) \frac{\partial F(r)(\tau)}{\partial r}\delta r \right\} d\tau.
\end{aligned}$$

We get finally

$$\begin{aligned}
\Psi(p)(F)_r(\delta r) &= \int_0^1 \left[\int_{-\infty}^{+\infty} \left\{ \Big(\dot{g}(\tau) - \dot{f}(\tau)\Big) \frac{\partial F(r)(\tau)}{\partial r}\delta r \right\} d\tau\right] dt \\
&= \int_{-\infty}^{+\infty} \left\{ \Big(\dot{g}(\tau) - \dot{f}(\tau)\Big) \frac{\partial F(r)(\tau)}{\partial r}\delta r \right\} d\tau.
\end{aligned}$$

☞ **Exercise 143, p. 310** (Compact supported real functions, II). Let ψ be the 2-points moment map associated with the action of G.

1) Thanks to Exercise 142, p. 307, it is clear that

$$\psi(f, g) = \mu(g) - \mu(f) \quad \text{with} \quad \mu(f)(F)_r(\delta r) = \int_{-\infty}^{+\infty} \dot{f}(t) \frac{\partial F(r)(t)}{\partial r}\delta r\, dt,$$

where F is a plot of G, that is, a plot of X. We have just to check that $\mu(f)$ is an element of $\mathcal{G}^*$, that is, invariant by G. Let $u \in X$, we have

$$\begin{aligned}
T_u^*(\mu(f))(F)_r(\delta r) &= \mu(T_u \circ F)_r(\delta r) \\
&= \mu(r \mapsto F(r) + u)_r(\delta r) \\
&= \int_{-\infty}^{+\infty} \dot{f}(t) \frac{\partial [F(r)(t) + u(t)]}{\partial r}\delta r\, dt \\
&= \int_{-\infty}^{+\infty} \dot{f}(t) \frac{\partial F(r)(t)}{\partial t}\delta r\, dt \\
&= \mu(f)(F)_r(\delta r).
\end{aligned}$$

Thus, μ is indeed a primitive of ψ.

2) Thanks to (art. 9.10, item 3), modulo a coboundary, a Souriau cocycle is given by $\theta(u) = \psi(f_0, T_u(f_0))$, where f_0 is element of X. We can choose $f_0 = 0$, which gives $\theta(u) = \psi(0, T_u(0)) = \psi(0, u) = \mu(u) - \mu(0) = \mu(u)$. Thus, $\theta = \mu$. Now,

a coboundary is defined by $\Delta c = [u \mapsto \mathrm{Ad}_*(u)(c) - c]$. But G is Abelian, thus $\mathrm{Ad} = \mathbf{1}_G$, hence there is no nontrivial coboundary except 0. Therefore, since $\mu \neq 0$, θ is not trivial.

Exercise 144, p. 320 (Compact supported real functions, III). First of all the moment map μ computed in Exercise 143, p. 310, is injective. Indeed, since μ is linear we have just to solve the equation $\mu(f) = 0$, that is,

$$0 = \int_{-\infty}^{+\infty} \dot{f}(t) \frac{\partial F(r)(t)}{\partial r} \delta r \; dt$$

for all n-plots $F : U \to X$, $n \in \mathbf{N}$, for all $r \in U$ and all $\delta r \in \mathbf{R}^n$. Let us choose $F(r) = rg$, with $g \in X$, $r \in \mathbf{R}$, and $\delta r = 1$. Then,

$$0 = \int_{-\infty}^{+\infty} \dot{f}(t) g(t) \, dt \quad \text{for all} \quad g \in X.$$

Thanks to the fundamental lemma of variational calculus, since f is smooth, $\dot{f} = 0$, that is, $f = \mathrm{cst}$. But f is compact supported, then $f = 0$. Hence, μ is injective. Now, X is obviously the quotient of itself by $u \mapsto T_u(0) = u$. Therefore, (X, ω) is a symplectic homogeneous space.

Exercise 145, p. 327 (The classical moment map). Let Ψ be the paths moment map of G on (M, ω). For all $p \in \mathrm{Paths}(M)$, $\Psi(p)$ is a 1-form on G. Thus, $\Psi(p)$ is characterized by its values on the 1-plots $F : t \mapsto g_t$ (art. 6.37). But since $\Psi(p)$ is a left-invariant 1-form, $\Psi(p)$ is characterized by its values on the 1-plots pointed at $\mathbf{1}_G$, that is, $F(0) = \mathbf{1}_G$. Next, since G is a Lie group, every path F, pointed at $\mathbf{1}_G$, is tangent to a homomorphism, thus we can assume that $F \in \mathrm{Hom}^\infty(\mathbf{R}, G)$, and then $F(t) = \exp(tZ)$, with $Z \in T_{\mathbf{1}_G}(G)$. According to (art. 9.20), since for $t = 0$ and $\delta t = 1$,

$$\delta p(s) = [D(F(0))(p(s))]^{-1} \left. \frac{\partial F(t)(p(s))}{\partial t} \right|_{t=0} (1) = Z_M(p(s)),$$

we get

$$\Psi(p)(F)_0(1) = \int_0^1 \omega_{p(s)}(\dot{p}(s), Z_M(p(s))) \, ds.$$

Now, let us assume that the action of G is Hamiltonian, that is, $\Psi(p)(F)_0(1) = \mu(m')(F)_0(1) - \mu(m)(F)_0(1)$, with $m' = p(1)$ and $m = p(0)$. With our notation, that gives $\Psi(p)(F)_0(1) = \mu_Z(m') - \mu_Z(m)$. Let us apply this computation to the path $p_s(t) = p(st)$. After a change of variable, we get

$$\mu_Z(m_s) = \int_0^s \omega_{p(t)}(\dot{p}(t), Z_M(p(t))) \, dt + \mu_Z(m)$$

with $m_s = p(s)$. The derivative of this identity, with respect to s, for $s = 0$, gives then

$$\left. \frac{\partial \mu_Z(m_s)}{\partial s} \right|_{s=0} = \omega(\delta m, Z_M(m)), \quad \text{with} \quad \delta m = \dot{p}(m).$$

Therefore, with obvious notation, μ_Z is the solution of the differential equation $\omega(Z_M, \cdot) = -d[\mu_Z](\cdot)$, that is, $i_{Z_M}(\omega) = -d\mu_Z$. Finally, decomposing Z in a basis $\{\xi^i\}_{i=1}^N$ of the tangent space $T_{\mathbf{1}_G}(G)$, $Z = \sum_{i=1}^N Z_i \, \xi^i$, gives $\mu_Z(m) = \sum_{i=1}^N Z_i \, \mu_i(m)$, where $\mu_i = \mu_{\xi^i}$. Then, $\bar{\mu}(m) : Z \mapsto \mu_Z(m)$ belongs to the dual

$[T1_G(G)]^*$, which is identified with the space of momenta $\mathcal{G}^*$. We find again, that way, the classical definition of the moment map $\bar{\mu} : M \to [T1_G(G)]^*$, [**Sou70**].

Exercise 146, p. 328 (The cylinder and SL(2, R)). First of all, let X and X′ be two vectors of $\mathbf{R}^2$ and $M \in SL(2, \mathbf{R})$, $\omega(MX, MX') = \det(M)\omega(X, X') = \omega(X, X')$, thus SL(2, R) preserves ω. Let us next check that SL(2, R) is transitive on $\mathbf{R}^2 - \{0\}$. Let $X = (x, y) \neq (0, 0)$. If $x \neq 0$ or $y \neq 0$, then

$$M = \begin{pmatrix} x & \frac{1}{1+y} \\ y & \frac{1+y}{x} \end{pmatrix} \text{ or } \begin{pmatrix} x & \frac{x-1}{y} \\ y & 1 \end{pmatrix} \in SL(2, \mathbf{R}) \quad \text{and} \quad M\begin{pmatrix} 1 \\ 0 \end{pmatrix} = \begin{pmatrix} x \\ y \end{pmatrix}.$$

Hence, $SL(2, \mathbf{R})$ is transitive. Now, since $\mathbf{R}^2$ is simply connected, the action of $SL(2, \mathbf{R})$ is Hamiltonian (art. 9.7), and since this action has a fixed point, it is exact (art. 9.10, Note 2). Consider now the path γ_X, connecting 0 to X, and the 1-plot $F_\sigma : s \mapsto e^{s\sigma}$, with $\sigma \in \mathfrak{sl}(2, \mathbf{R})$. The application of (art. 9.20, (♢)) gives

$$\Psi(\gamma_X)(F_\sigma)_0(1) = \int_0^1 \omega(\dot{\gamma}_X(t), \delta\gamma_X(t))\, dt = \int_0^1 \omega(X, t\sigma X)\, dt = \frac{1}{2}\omega(X, \sigma X),$$

thanks to (art. 9.20, (♡)), and to

$$\delta\gamma_X(t) = \left.\frac{\partial e^{s\sigma}(\gamma_X(t))}{\partial s}\right|_{s=0}(1) = t\sigma X.$$

And we deduce the expression of the moment map μ, the constant is fixed with $\mu(0) = 0$. It is not difficult then to check that the preimages of μ_M are the pairs $\pm X$. Actually, the image of μ_M is a 1-sheet hyperboloid in $\mathbf{R}^3 \simeq \mathfrak{sl}(2, \mathbf{R})$.

Exercise 147, p. 350 (The moment of imprimitivity). First of all let us check the variance of Taut by the action of $\mathcal{C}^\infty(X, \mathbf{R})$. Let f be a smooth real function defined on X, and let us denote by $Q \times P$ a plot of $X \times \Omega^1(X)$. We have $\bar{f}^*(\mathrm{Taut})(P \times Q)_r = \mathrm{Taut}(\bar{f} \circ (Q \times P))_r = (P(r) - df)(Q)_r = P(r)(Q)_r - df(Q)_r = \mathrm{Taut}(Q \times P)_r - df(\mathrm{pr}_1 \circ (Q \times P))_r = \mathrm{Taut}(Q \times P)_r - \mathrm{pr}_1^*(df)(Q \times P)_r$. Thus, $\bar{f}^*(\mathrm{Taut}) = \mathrm{Taut} - \mathrm{pr}_1^*(df)$. Now let us check that this action is compatible with the value relation. Let (x, α) and (x', α') be two elements of $X \times \Omega^1(X)$ such that $\mathrm{value}(\alpha)(x) = \mathrm{value}(\alpha')(x')$, that is, $x = x'$ and, for every plot Q of X centered at x, $\alpha(Q)_0 = \alpha'(Q)_0$. Then, $(\alpha - df)(Q)_0 = (\alpha' - df)(Q)_0$ and $\mathrm{value}(\alpha - df)(x) = \mathrm{value}(\alpha)(x) - \mathrm{value}(df)(x)$, or $(\alpha - df)(x) = \alpha(x) - df(x)$. Thus, the action of $\mathcal{C}^\infty(X, \mathbf{R})$ projects on T^*X as the action $\bar{f} : (x, a) \mapsto a - df(x)$. Now, since $\bar{f}^*(\mathrm{Taut}) = \mathrm{Taut} - \mathrm{pr}_1^*(df)$, clearly $\bar{f}^*(\mathrm{Liouv}) = \mathrm{Liouv} - \pi^*(df)$. Put differently, $\bar{f}^*(\mathrm{Liouv}) = \mathrm{Liouv} - dF(f)$ with $F \in \mathcal{C}^\infty(\mathcal{C}^\infty(X, \mathbf{R}), \mathcal{C}^\infty(T^*X, \mathbf{R}))$ and $F(f) = \pi^*(f) = f \circ \pi$.

Let us denote by $R(x, a)$ the orbit map $f \mapsto a - df(x)$. Let p be a path in T^*X such that $p(0) = (x_0, a_0)$ and $p(1) = (x_1, a_1)$. By definition, for $\omega = d\mathrm{Liouv}$, $\Psi(p) = \hat{p}^*(\mathcal{K}\omega) = \hat{p}^*(\mathcal{K}d\mathrm{Liouv})$. Now, applying the property of the Chain-Homotopy operator $\mathcal{K} \circ d + d \circ \mathcal{K} = \hat{1}^* - \hat{0}^*$, we get

$$\begin{aligned}
\Psi(p) &= \hat{p}^*(\mathcal{K}d\mathrm{Liouv}) \\
&= \hat{p}^*(\hat{1}^*(\mathrm{Liouv}) - \hat{0}^*(\mathrm{Liouv}) - d\mathcal{K}\mathrm{Liouv}) \\
&= (\hat{1} \circ \hat{p})^*(\mathrm{Liouv}) - (\hat{0} \circ \hat{p})^*(\mathrm{Liouv}) - d[(\mathcal{K}\mathrm{Liouv}) \circ \hat{p}] \\
&= R(x_1, a_1)^*(\mathrm{Liouv}) - R(x_0, a_0)^*(\mathrm{Liouv}) - d[f \mapsto \mathcal{K}\mathrm{Liouv}(\hat{p}(f))].
\end{aligned}$$

Let us consider first the term $[f \mapsto \mathcal{K}\mathrm{Liouv}(\hat{p}(f))]$. Let $p(t) = (x_t, a_t)$, then $\hat{p}(f) = [t \mapsto (x_t, a_t - df(x_t))]$. Thus,

$$\begin{aligned}
\mathcal{K}\mathrm{Liouv}(\hat{p}(f))) &= \int_0^1 a_t[s \mapsto x_s]_{s=t}\, dt - \int_0^1 df[t \mapsto x_t]\, dt \\
&= \int_0^1 a_t[s \mapsto x_s]_{s=t}\, dt - [f(x_1) - f(x_0)].
\end{aligned}$$

Hence,

$$\begin{aligned}
d[f \mapsto \mathcal{K}\mathrm{Liouv}(\hat{p}(f))] &= d\left\{ f \mapsto \int_0^1 a_t[s \mapsto x_s]_{s=t}\, dt - [f(x_1) - f(x_0)] \right\} \\
&= -d[f \mapsto f(x_1) - f(x_0)].
\end{aligned}$$

Let us compute $R(x, a)^*(\mathrm{Liouv})$, for (x, a) in T^*X. Let $P : U \to \mathcal{C}^\infty(X, \mathbf{R})$ be a plot, we have

$$\begin{aligned}
R(x, a)^*(\mathrm{Liouv})(P) &= \mathrm{Liouv}(R(x, a) \circ P) \\
&= \mathrm{Liouv}(r \mapsto \overline{P(r)}(x, a)) \\
&= \mathrm{Liouv}(r \mapsto a + d[P(r)](x)) \\
&= (a + d[P(r)](x))(r \mapsto x). \\
&= 0.
\end{aligned}$$

The last equality happens because the 1-form $a + d[P(r)](x)$ is evaluated on the constant plot $r \mapsto x$, and every form evaluated on a constant plot vanishes. We get finally

$$\Psi(p) = d[f \mapsto f(x_1)] - d[f \mapsto f(x_0)].$$

Now, clearly, $\Psi(p) = \psi(p(0), p(1))$, with $p(0) = (x_0, a_0)$ and $p(1) = (x_1, a_1)$. Then, the action of $\mathcal{C}^\infty(X, \mathbf{R})$ is Hamiltonian, $\Gamma = \{0\}$, and for the 2-points moment map, we have $\psi((x_0, a_0), (x_1, a_1)) = \mu(x_1, a_1) - \mu(x_0, a_0)$, with

$$\mu : (x, a) \mapsto d[f \mapsto f(x)].$$

Noting δ_x the real function $f \mapsto f(x)$, $\mu(x, a) = d\delta_x$. Let us now check the invariance of μ. Note that, for every $h \in \mathcal{C}^\infty(X, \mathbf{R})$, $\delta_x \circ L(h) = [f \mapsto f(x) + h(x)]$. Thus, for all $h \in \mathcal{C}^\infty(X, \mathbf{R})$, $\hat{h}^*(\mu)(x, a) = \hat{h}^*(d\delta_x) = d(\delta_x \circ L(h)) = d[f \mapsto f(x) + h(x)] = d[f \mapsto f(x)] = d\delta_x = \mu(x, a)$. Hence, μ is invariant, and it is a primitive of the 2-points moment map ψ. Therefore, the Souriau class of the action of $\mathcal{C}^\infty(X, \mathbf{R})$ on T^*X vanishes.

Afterword

I was a student of Jean-Marie Souriau, working on my doctoral dissertation, when he introduced « diffeology ». I remember well, we used to gather for a seminar at that time — the beginning of the 1980s — every Tuesday, at the Center for Theoretical Physics, at Marseille's Luminy campus. Jean-Marie was trying to generalize his quantization procedure to a certain kind of coadjoint orbits of infinite dimensional groups of diffeomorphisms. He wanted to regard these groups of diffeomorphisms as Lie groups, like everybody, but he also wanted to avoid topological finessing, feeling that that was not essential for this goal. He invented then a lighter « differentiable » structure on groups of diffeomorphisms. These groups quickly became autonomous objects. I mean, he gave up groups of diffeomorphisms for abstract groups, equipped with an abstract differential structure. He called them « *groupes différentiels* », this was the first name for the future diffeological groups.

Differential spaces **are born**. Listening to Jean-Marie talking about his differential groups, I had the feeling that these structures, the axiomatics of differential groups, could be easily extended to any set, not necessarily groups, and I remember a particularly hot discussion about this question in the Luminy campus cafeteria. It was during a break in our seminar. We were there, the whole group: JMS (as we call him), Jimmy Elhadad, Christian Duval, Paul Donato, Henry-Hugues Fliche, Roland Triay, and myself. Souriau denied the interest of considering anything other than orbits of differential groups (Souriau was really, but really, « group-oriented »), and I decided when I had the time — I was working on the classification of SO(3)-symplectic manifolds which has nothing to do with diffeology — to generalize his axiomatics for any sets. But I never got the opportunity to do it. Sometime later, days or weeks, I don't remember exactly, he outlined the general theory of « *espaces différentiels* » as he called them. I would have liked to do it, anyway... I must say that, at that time, these constructions appeared to us, his students, as a fine construction, but so general that it could not turn out into great results, it could give at most some intellectual satisfaction. We were dubious. I decided to forget differential spaces and stay focused on « real maths », the classification of SO(3)-symplectic manifolds. I went to Moscow, spent a year there, and came back with a complete classification in dimension 4 and some general results in any dimension. This work represented for me a probable doctoral thesis. It was the first global classification theorem in symplectic geometry after the homogeneous case, the famous Kirillov-Kostant-Souriau theorem, which states that any homogeneous symplectic manifold is a covering of some coadjoint orbit. But Jean-Marie didn't pay any attention to my work, looking away from it, as he was completely absorbed by his « differential spaces ». I was really disappointed, I thought that this work deserved to become my doctorate. At the same time, Paul Donato gave a general construction of the universal covering for any quotient of « differential groups », that is, the universal covering of any homogeneous « differential space ». This construction became his doctoral thesis. I decided then to give up, for a moment, symplectic geometry and to get into the world of differential spaces, since it was the only subject about which JMS was able, or willing, to talk at that time.

The coming of the irrational torus. It was the year 1984, we were taking part in a conference about symplectic geometry, in Lyon, when we decided, together with Paul, to test diffeology on the *irrational torus*, the quotient of the 2-torus by an irrational line. This quotient is not a manifold but remains a diffeological space, moreover a diffeological group. We decided to call it T_α, where α is the slope of the line. The interest for this example came, of course, from the Denjoy-Poincaré flow about which we heard so much during this conference. What had diffeology to say about this group, for which topology is completely dry? We used the techniques worked out by Paul and computed its homotopy groups, we found $\mathbf{Z} + \alpha\mathbf{Z} \subset \mathbf{R}$ for the fundamental group and zero for the higher ones. The real line $\mathbf{R}$ itself appeared as the universal covering of T_α. I remember how we were excited by this computation, as we didn't believe really in the capabilities of diffeology for saying anything serious about such « singular » spaces or groups. Don't forget that differential spaces had been introduced for studying infinite dimensional groups and not singular quotients. We continued to explore this group and found that, as diffeological space, T_α is characterized by α, up to a conjugation by $\mathrm{GL}(2, \mathbf{Z})$, and we found that the components of the group of diffeomorphisms of T_α distinguish the cases where α is quadratic or not. It became clear that diffeology was not such a trivial theory and deserved to be more developed. At the same time, Alain Connes introduced the first elements of noncommutative geometry and applied them to the irrational flow on the torus — our favorite example — and his techniques didn't give anything more (in fact less) than the diffeological approach, which we considered more in the spirit of ordinary geometry. We were in a good position to know the application of Connes' theory on irrational flows as he had many fans, in the Center for Theoretical Physics at that time, developing his ideas.

All in all, this example convinced me that diffeology was a good tool, not as weak as it seemed to be. And I decided to continue to explore this path. The result of the computation of the homotopy group of T_α made me think that everything was as if the irrational flow was a true fibration of the 2-torus: the fiber $\mathbf{R}$ being contractible the homotopy of the quotient T_α had to be the same as the total space T^2, and one should avoid Paul's group specific techniques to get it. But, of course, T_α being topologically trivial it could not be an ordinary locally trivial fibration. I decided to investigate this question and, finally, gave a definition of diffeological fiber bundles, which are not locally trivial, but locally trivial along the plots — the smooth parametrizations defining the diffeology. It showed two important things for me: The first one was that the quotient of a diffeological group by any subgroup is a diffeological fibration, and thus $T^2 \to T_\alpha$. The second point was that diffeological fibrations satisfy the exact homotopy sequence. I was done, I understood why the homotopy of T_α, computed with the techniques elaborated by Paul, gave the homotopy of T^2, because of the exact homotopy sequence. I spent one year on this job, and I returned to Jean-Marie with that and some examples. He agreed to listen to me and decided that it could be my dissertation. I defended it in November 1985, and became since then completely involved in the diffeology adventure.

Differential, differentiable, or diffeological spaces? The choice of the wording « differential spaces » or « differential groups » was not very happy, because « differential » is already used in maths and has some kind of usage, especially « differential groups » which are groups with an operation of derivation. This was quoted often to us. I remember Daniel Kastler insisting that JMS change this name. From time to time we tried to find something else, without success. Finally, it was during the defense of Paul's thesis, if memory serves me right, when Van Est suggested the word « *difféologie* » like « *topologie* » as a replacement for « *différentiel* ». We found the word accurate and we decided to use it, and « *espaces différentiels* » became « *espaces difféologiques* ». There was a damper, however, « *différentiel* » as well as « *topologique* » have four spoken syllables when « *difféologique* » has five. Anyway, I used and abused this new denomination, many friends laughed at me, and one of them once told me, Your « *dix fées au logis* » — which means "ten fairies at home" — since then, there is no time when I say diffeology without thinking of these ten fairies waiting at home... Later, Daniel Bennequin pointed out to me that Kuo-Tsai Chen, in his work, *Iterated path integrals* [**Che77**] in the 1970s, defined « differentiable spaces » which looked a lot like « diffeological spaces ». I got to the library, read Chen's paper and drew a rapid (but unfounded) conclusion that our « diffeological spaces » were just equivalent to Chen's « differentiable spaces », with a slight difference in the definition. I was very disappointed, I was working on a subject I thought really new and it appeared to be known and already worked out. I decided to drop « diffeology » for « differentiable » and to give honor to Chen, but my attempt to use Chen's vocabulary was aborted — the word « diffeology » had already moved into practice, having myself helped to popularize it. However, it is good to notice that, although Chen's and Souriau's axiomatics look alike, Souriau's choice is better adapted to the geometrical point of view. Defining plots on open domains, rather than on standard simplices or convex subsets, changes dramatically the scope of the theory.

Last word? I would add some words about the use or misuse of diffeology. Some friends have expressed their skepticism about diffeology, and told me that they are waiting for diffeology to prove something great. Well, I don't know any theory proving anything, but I know mathematicians proving theorems. Let me put it differently: number theory doesn't prove any theorem, mathematicians solve problems raised by number theory. A theory is just a framework to express questions and pose problems, it is a playground. The solutions of these problems depend on the skill of the mathematicians who are interested in them. As a framework for formulating questions in differential geometry, I think diffeology is a very good one, it offers good tools, simple axioms, simple vocabulary, simple but rich objects, it is a stable category, and it opens a wide field of research. Now, I understand my friends, there are so many attempts to extend the usual category of differential geometry, and so many expectations, that it is legitimate to be doubtful. Nevertheless, I think that we now have enough convincing examples, simple or more elaborate, for which diffeology brings concrete and formal results. And this is an encouragement to persist on this path, to develop new diffeological tools, and perhaps to prove some day, some great theorem :).

At the time I began this book, Jean-Marie Souriau was alive and well. He asked me frequently about my progress. He was eager to know if people were buying his theory, and he was happy when I could say sometimes that, yes, some people in Tel Aviv or in Texas mentioned it in some paper or discussed it on some web forum. Now, as I'm finishing this book and writing the last sentences, Jean-Marie is no longer with us. He will not see the book published and complete. It is sad, diffeology *was his last program, in which he had strong expectations regarding geometric quantization. I am not sure if diffeology will fulfill his expectations, but I am sure that it is now a mature theory, and I dedicate this work to his memory. Whether it is the right framework to achieve Souriau's quantization program is still an open question.*

Notation and Vocabulary

Diffeology and diffeological spaces

$\mathbf{R}$	The *real numbers.*
$\mathbf{N}$	The *natural integers.*
$\times$	The product of sets or the product of a number by something.
$\operatorname{def}(f)$	The *set of definition* of the map f.
$\operatorname{val}(f)$	The values of the map f.
$\operatorname{Maps}(X, Y)$	The maps from X to Y.
1_X	The identity map of X.
n-domain	A (nonempty) open set of the vector space $\mathbf{R}^n$.
domain	An arbitrary n-domain, for some n.
$\operatorname{Domains}(\mathbf{R}^n)$	The domains of $\mathbf{R}^n$.
Domains	The domains of the $\mathbf{R}^n$, n running over $\mathbf{N}$.
$\mathfrak{P}(E)$	The *powerset* of the set E, that is, the set of all the subsets of the set E.
$\dim(U)$	The *dimension* of the domain U, that is, if $U \subset R^n$, $\dim(U) = n$.
Parametrization	Any map defined from a domain to some set.
$\operatorname{Param}(X)$	The parametrizations of a set X.
$\operatorname{Param}(U, X)$	The parametrizations of a set X defined on a domain U.
$\mathbf{x}$	A bold lower case letter, a constant map (here with value x).
$D(F)(s)$ or $D(F)_s$ or $\frac{\partial r}{\partial s}$	For any smooth map $F : s \mapsto r$ between real domains, the tangent linear map of F at the point s.
pr_k	From a product, the projection on the k-th factor.
$\mathcal{D}$	Refers to a diffeology.
D1, D2, D3	Refers to the three axioms of diffeology.
$\mathcal{D}(U, X)$	The plots, defined on U, of the space X.
$\mathcal{D}(X, X')$ or $\mathcal{C}^\infty(X, X')$	The smooth maps from X to X'.
$\operatorname{Diff}(X, X')$	The diffeomorphisms from X to X'.
{Set}, {Diffeology} etc.	Names of categories.
$\mathcal{D}_\circ(X)$	The discrete diffeology of X.
$X_\circ$	X equipped with the discrete diffeology.
$\mathcal{D}_\bullet(X)$	The coarse diffeology of X.
$X_\bullet$	X equipped with the coarse diffeology.
$f^*(\mathcal{D}')$	The pullback of the diffeology $\mathcal{D}'$ by f.
$\coprod_{i \in \mathcal{I}} X_i$	The sum (coproduct) of diffeological spaces.

class(x)	The equivalence class of x.
$f_*(\mathcal{D})$	The pushforward of the diffeology $\mathcal{D}$ by f.
$\prod_{i \in \mathcal{I}} X_i$	The product of diffeological spaces.
Paths(X)	The space of paths in X, that is, $\mathcal{C}^\infty(\mathbf{R}, X)$.
Paths$_k$(X)	The space of iterated k-paths in X, that is, $\mathcal{C}^\infty(\mathbf{R}, \mathrm{Paths}_{k-1}(X))$.
δx	Any vector of $\mathbf{R}^n$, when x is a generic point in an n-domain.
$\langle \mathcal{F} \rangle$	The diffeology generated by a family $\mathcal{F}$ of parametrizations.
Nebula($\mathcal{F}$)	The nebula of the family of parametrizations $\mathcal{F}$.
dim(X)	The global dimension of X, for a diffeological space.
Gen(X)	The generating families of the diffeology of X.

Locality and diffeologies

$\mathcal{C}^\infty_{\mathrm{loc}}(X, X')$	The locally smooth maps from X to X'.
$\mathrm{Diff}_{\mathrm{loc}}(X, X')$	The local diffeomorphisms from X to X'.
$=_{\mathrm{loc}}$	Is equal/coincides locally.
germ	The germ of a map.
D-open	Open for the D-topology.
$\dim_x(X)$	The pointwise dimension of X at x.

Diffeological vector spaces

$L^\infty(E, E')$	The smooth linear maps from E to E'.
$\mathcal{H}$, $\mathcal{H}_{\mathbf{R}}$, $\mathcal{H}_{\mathbf{C}}$	The standard Hilbert space, on $\mathbf{R}$, on $\mathbf{C}$.
$X \cdot Y$	The scalar or Hermitian product.

Modeling spaces, manifolds, etc.

$\mathcal{S}_{\mathbf{R}}$, $\mathcal{S}_{\mathbf{C}}$	The infinite Hilbert sphere, in $\mathcal{H}_{\mathbf{R}}$ or $\mathcal{H}_{\mathbf{C}}$.
$\mathcal{P}$, $\mathcal{P}_{\mathbf{C}}$	The infinite projective space.
$\mathbf{H}_n$, $\mathbf{K}_n$	The half n-space or the n-corner.

Homotopy of diffeological spaces

$\hat{0}(\gamma)$, $\hat{1}(\gamma)$, ends(γ)	Starting point, ending point, and ends of a path γ.
Paths(X, A, B)	The paths in X, starting in A and ending in B.
Loops(X)	Loops in X.
Loops(X, x), Loops$_n$(X, x)	Loops and iterated loops in X, based at x.
$\gamma \vee \gamma'$	Concatenation of paths.
rev(γ)	The reverse path of γ.
stPaths, stLoops	Stationary paths, loops.
$\gamma^\star$	Smashed path, making it stationary.
$\pi_0(X)$, $\pi_k(X, x)$	The components of X and k-th-homotopy groups based at x.
$\pi_k(X, A, x)$	Relative homotopy pointed-sets/groups.
$\Pi(X)$	The Poincaré groupoid of X.

Cartan-De Rham calculus

$E * F$ or $L(E, F)$	The space of linear maps from E to F.
E^*	Dual vector space E, that is, $E * \mathbf{R}$.
Vector — Covector	An element of a vector space — An element of its dual.
$\otimes$, $\wedge$	Tensor product, exterior product.
$\mathrm{sgn}(\sigma)$	Signature of the permutation σ.
$L^\infty(E, F)$, E^*_∞	Smooth linear maps, smooth dual.
$M^*(A)$	Pullback of a covariant tensor A by a linear map M.
vol, $\mathrm{vol}_{\mathcal{B}}$	Volume, canonical volume of a basis $\mathcal{B}$.
$\Lambda^k(E)$	The space of all k-linear forms on a vector space E.
$\Omega^k(X)$	The space of differentiable k-forms on a diffeological space X.
$\Lambda^k_x(X)$	The space of the values of the k-forms of X at the point x.
$\alpha(P)$	The value of the differential form α on the plot P.
$f^*(\alpha)$	The pullback of the differential form α by the smooth map f.
$d\alpha$	The *exterior derivative* of the differential form α.
θ	The canonical 1-form on $\mathbf{R}$ or on an irrational torus $\mathbf{R}/\Gamma$.
δP, $\delta\sigma$, δc	The *variation* of a plot P, a cube σ, a chain c.
$\alpha\rfloor\delta\sigma$	The *contraction* of a k-form α by a variation of a cube σ, a differential $(k-1)$-form on $\mathrm{def}(\sigma)$.
i_F	The *contraction* by an arc of diffeomorphisms F.
$\pounds_F$	The Lie derivative by an arc of diffeomorphisms F.
$\mathrm{Cub}_p(X)$, $C_p(X)$	The smooth p-cubes, p-chains, in X.
$Z_\star(X, \cdot)$, $Z^\star(X, \cdot)$	The cycles/cocycles groups with coefficients.
$B_\star(X, \cdot)$, $B^\star(X, \cdot)$	The boundary/coboundary groups with coefficients.
$H_\star(X, \cdot)$, $H^\star(X, \cdot)$	The homology/cohomology groups with coefficients.
$\int_\sigma \alpha$, $\int_c \alpha$	The integral of a differential form on a cube/cubic-chain.
Magma	A set A equipped with an internal operation $\star$.
$\mathcal{K}$	The Chain-Homotopy operator.
$L(\gamma)$, $R(\gamma)$	Pre- or post-concatenation by γ.

Diffeological groups

$\mathrm{Hom}^\infty(G, G')$	The smooth homomorphisms from G to G'.
G^{Ab}	The Abelianized group $G/[G, G]$.
$R(x)$ or $\hat{x}$	The orbit map from G to X.
$L(g)$, $R(g)$	The left/right multiplication by g in G.
$\mathrm{Ad}(g)$	The adjoint action of g on G, $\mathrm{Ad}(g) = L(g) \circ R(g^{-1})$.
Coset	The left or right orbits of a subgroup in a group.
$\mathcal{G}^*$	The *space of momenta* of a diffeological group G.
Ad_*, $\mathrm{Ad}_*^{\Gamma,\theta}$	The coadjoint action of g on $\mathcal{G}^*$, on $\mathcal{G}^*/\Gamma$ with cocycle θ.

Diffeological fiber bundles

src, trg, χ The source, target and characteristic maps on a groupoid.

$f^*(T)$ The pullback by f of the total space T of a projection.

$T \times_G E$ or $T \times_\rho E$ An associate bundle.

Θ A connection on a principal bundle.

HorPaths(Y, Θ) The horizontal paths in Y, for the connection Θ.

T_Γ A *torus* $\mathbf{R}^n/\Gamma$, Γ discrete (diffeologically) and generator of $\mathbf{R}^n$.

$\mathcal{R}(X)$ The diffeological group of $(\mathbf{R}, +)$-principal bundles over X.

T_α The *torus of periods* of a closed form α.

Symplectic diffeology

Ψ, Ψ_ω The *moment of paths*, the index ω is for *universal.*

ψ, ψ_ω The 2-*points moment map.*

μ, μ_ω A 1-*point moment map.*

θ, θ_ω The *lack of equivariance* of a moment map, a group cocycle in $\mathcal{G}^*/\Gamma$.

σ, σ_ω The *Souriau class*, the class of the cocycles θ, θ_ω.

Bibliography

[Arn65] Vladimir I. Arnold, *Small denominators, mapping of the circumference onto itself*, Isvestja A.N. SSSR series math., vol. 25, 1965.

[Arn80] Vladimir I. Arnold, *Chapitres supplémentaires à la théorie des équations différentielles*, MIR, Moscou, 1980.

[BaHo09] John C. Baez and Alexander E. Hoffnung, *Convenient Categories of Smooth Spaces*, 2009. `http://arXiv.org/abs/0807.1704`.

[Ban78] Augustin Banyaga, *Sur la structure du groupe des difféomorphismes qui préservent une forme symplectique*, Comment. Math. Helv., vol. 53, pp. 174–227, 1978.

[BaWe97] Sean Bates and Alan Weinstein, *Lectures on the geometry of quantization*, Berkeley Mathematics Lecture Notes, American Mathematical Society, Providence, RI, 1997.

[BFBF09] Daniel Bennequin, Ronit Fuchs, Alain Berthoz, Tamar Flash, *Movement Timing and Invariance Arise from Several Geometries*, PLoS Comput. Biol., vol. 5, **7**, 2009.

[BeGo72] Marcel Berger and Bernard Gostiaux, *Cours de géométrie*, PUF, Armand Colin, Paris, 1972.

[BFW10] Christian Blohmann, Marco C. B. Fernandes, and Alan Weinstein, *Groupoid Symmetry And Constraints In General Relativity. 1: Kinematics*, 2010. `http://arxiv.org/abs/1003.2857v1`.

[Bom67] Jan Boman, *Differentiability of a function and of its compositions with functions of one variable*, Mathematica Scandinavica, vol. 20, pp. 249–268, 1967.

[Boo69] William M. Boothby, *Transitivity of the automorphisms of certain geometric structures*, Trans. Amer. Math. Soc., vol. 137, pp. 93–100, 1969.

[Bou55] Nicolas Bourbaki, *Espaces vectoriels topologiques*, Eléments de mathématiques, Hermann, Paris, 1955.

[Bou61] Nicolas Bourbaki, *Topologie générale*, Eléments de mathématiques, Hermann, Paris, 1961.

[Bou72] Nicolas Bourbaki, *Théorie des ensembles*, Eléments de mathématiques, Hermann, Paris, 1972.

[Bou82] Nicolas Bourbaki, *Variétés différentielles et analytiques*, Eléments de mathématiques, Diffusion CCLS, Paris, 1982.

[BoCh67] Zenon I. Borevitch and Igor R. Chafarevitch, *Théorie des nombres*, Monographies Internationales de mathématiques Modernes, Gauthier-Villars, Paris, 1967.

[Bot78] Raoul Bott, *On some formulas for the characteristic classes of group actions*, Differential Topology, Foliations and Gelfand-Fuchs Cohomology, Lect. Notes in Math., vol. 652, Springer Verlag, 1978.

[BoGa07] Charles P. Boyer and Krzysztof Galicki, *Sasakian Geometry*, Department of Mathematics and Statistics, University of New Mexico, Albuquerque, N.M. 87131, 2007.

[Bry93] Jean-Luc Brylinski, *Loop spaces, characteristic classes and geometric quantization*, Birkäuser, 1993.

[Car87] Pierre Cartier, *Jacobiennes généralisées, monodromie unipotente et intégrales itérées*, Séminaire Bourbaki – 40ème année, vol. 687, 1987–88.

[Che77] Kuo-Tsai Chen, *Iterated path integrals*, Bull. Amer. Math. Soc., vol. 83, **5**, pp. 831–879, 1977.

[Die70a] Jean Dieudonné, *Eléments d'analyse*, vol. I, Gauthiers-Villars, Paris, 1970.

[Die70c] Jean Dieudonné, *Eléments d'analyse*, vol. III, Gauthiers-Villars, Paris, 1970.

[Doc76] Manfred P. Do Carmo, *Differential geometry of curves and surfaces*, Prentice-Hall Inc. Englewood Cliffs, New Jersey, 1976.

[Dnl99] Simon K. Donaldson, *Moment maps and diffeomorphisms*, Asian Journal of Math., vol. 3, pp. 1–16, 1999.

[Don84] Paul Donato, *Revêtement et groupe fondamental des espaces différentiels homogènes*, Thèse de doctorat d'état, Université de Provence, Marseille, 1984.

[Don88] Paul Donato, *Géométrie des orbites coadjointes des groupes de difféomorphismes*, in Proc. Lect. Notes in Math. vol. 1416 pp. 84–104 1988

[Don94] Paul Donato, $\mathrm{Diff}(\mathrm{S}^1)$ *as coadjoint orbit in the Virasoro space of moments*, Journal of Geometry and Physics, vol. 13, pp. 299–305, 1994.

[DoIg85] Paul Donato and Patrick Iglesias, *Exemple de groupes difféologiques : flots irrationnels sur le tore*, C.R. Acad. Sci. Paris, vol. 301, **4**, 1985.

[DNF82] Boris Doubrovine, Serguei Novikov and Anatoli Fomenko, *Géométrie contemporaine*, MIR, Moscou, 1982.

[EOM02] *Encyclopedia of Mathematics*, Springer Verlag, New York-Heidelberg-Berlin, 2002.

[FuRo81] Dimitry B. Fuchs and Vladimir A. Rohlin, *Premier cours de topologie, Chapitres géométriques*, MIR, Moscou, 1981.

[God40] Kurt Gödel, *The consistency of the axiom of choice and of the generalised continuum-hypothesis with the axioms of set theory*, Princeton University Press, Princeton–New Jersey, 1940.

[God64] Roger Godement, *Topologie algébrique et théorie des faisceaux*, Hermann, Paris, 1964.

[GuPo74] Victor Guillemin and Alan Pollack, *Differential topology*, Prentice Hall, New Jersey, 1974.

[Her79] Michael R. Hermann *Sur la conjugaison différentiable des difféomorphismes du cercle à des rotations*, Publication Mathématiques, I.H.E.S. Bures sur Yvette, 1979.

[HoYo61] John G. Hocking and Gail S. Young, *Topology*, Dover Publication, 1961.

[Hu59] Sze-Tsen Hu, *Homotopy Theory*, Academic Press, New York, 1959.

[Igl85] Patrick Iglesias, *Fibrations difféologiques et homotopie*, Thèse de doctorat d'État, Université de Provence, Marseille, 1985.

[Igl86] Patrick Iglesias, *Difféologie d'espace singulier et petits diviseurs*, C.R. Acad. Sci. Paris, vol. 302, pp. 519–522, 1986.

[Igl87] Patrick Iglesias, *Connexions et difféologie*, Aspects dynamiques et topologiques des groupes infinis de transformation de la mécanique, Hermann, Paris, vol. 25, pp. 61–78, 1987.

[Igl87b] Patrick Iglesias, *Bicomplexe cohomologique des espaces différentiables*, Preprint CPT-87/P.2052, Marseille, France, 1987.

[Igl91] Patrick Iglesias, *Les SO(3)-variétés symplectiques et leur classification en dimension 4*, Bull. Soc. Math. France, vol. 119, **4**, pp. 371–396, 1991.

[Igl95] Patrick Iglesias, *La trilogie du Moment*, Ann. Inst. Fourier, vol. 45, 1995.

[Igl98] Patrick Iglesias, *Les origines du calcul symplectique chez Lagrange*, L'Enseignement Mathématique, vol. 44, pp. 257–277, 1998.

[IgLa90] Patrick Iglesias and Gilles Lachaud, *Espaces différentiables singuliers et corps de nombres algébriques*, Ann. Inst. Fourier, Grenoble, vol. 40, **1**, pp. 723–737, 1990.

[IKZ05] Patrick Iglesias, Yael Karshon and Moshe Zadka, *Orbifolds as diffeologies*, Trans. Am. Math. Soc., vol. 362, **6**, pp. 2811–2831, 2010.

[IZK10] Patrick Iglesias-Zemmour and Yael Karshon, *Smooth Lie groups actions are diffeological subgroups*, Proc. Am. Math. Soc., vol. 140, pp. 731–739, 2012.

[Kak43] Shizuo Kakutani, *Topological properties of the unit sphere of a Hilbert space*, Proc. Imp. Acad. Tokyo, **19**, pp. 269–271, 1943.

[KaMe85] Mikhail I. Kargapolov and Iuri I. Merzliakov, *Elements de la théorie des groupes*, MIR, Moscou, 1985.

[KaZo12] Yael Karshon and Masrour Zoghi, *Orbifold groupoids and their underlying diffeology*, In preparation, 2012.

[Kir76] Alexandre A. Kirillov, *Elements of the theory of representations*, Berlin, Springer Verlag, 1976.

[Kle74] Felix Klein, *Le programme d'Erlangen*, Discours de la méthode, Gauthier-Villars, Paris, 1974.

[KoNo63] Shoshichi Kobayashi and Katsumi Nomizu, *Foundations of differential geometry*, Interscience Publishers, 1963.

[YKS10] Yvette Kosmann-Schwarzbach, *The Noether Theorems*, Sources and Studies in the History of Mathematics and Physical Sciences, Springer Verlag, 2010.

[Kos70] Bertram Kostant, *Orbits and quantization theory*, in Proc. International Congress of Mathematicians, 1970-1971.

[Lee06] John M. Lee, *Introduction to Smooth Manifolds*, Graduate Texts in Mathematics, Springer Verlag, New York, 2006.

[Les03] Joshua Leslie, *On a Diffeological Group Realization of Certain Generalized Symmetrizable Kac-Moody Lie Algebras*, Journal of Lie Theory, vol. 13, **3**, 2003.

[McK87] Kirill Mackenzie, *Lie groupoids and Lie algebroids in differential geometry*, London Mathematical Society Lecture Notes Series, **124**, Cambridge University Press,Cambridge, UK, 1987.

[McL71] Saunders Mac Lane, *Categories for the working mathematician*, Graduate Texts in Mathematics, Springer Verlag, New York-Heidelberg-Berlin, 1971.

[McL75] Saunders Mac Lane, *Homology*, Graduate Texts in Mathematics Springer Verlag, New York-Heidelberg-Berlin, 1975.

[Man70] Youri I. Manine, *Cours de géométrie algébrique*, Moscow State University Ed., 1970.

[MaWe74] Jerrold Marsden and Alan Weinstein, *Reduction of symplectic manifolds with symmetry*, Rep. Math. Phys., vol. 5, pp. 121–130, 1974.

[Mil74] John Milnor, *Caracteristic classes*, Annals of Math. Studies, Princeton University Press, vol. 76, 1974.

[Mos66] Jurgen Moser, *A rapidly convergent iteration method and non-linear differential equations II*, Ann. Scuola Norm. di Pisa, vol. III, **20**, 1966.

[Omo86] Stephen Malvern Omohundro, *Geometric Perturbation Theory in Physics*, World Scientific, 1986.

[Sat56] Ichiro Satake, *On a generalization of the notion of manifold*, Proceedings of the National Academy of Sciences, vol. 42, pp. 359–363, 1956.

[Sat57] Ichiro Satake, *The Gauss-Bonnet theorem for V-manifolds*, Journal of the Mathematical Society of Japan, vol. 9, **4**, pp. 464–492, 1957.

[Sch11] Urs Schreiber, *Differential cohomology in a cohesive ∞-topos*, Habilitation thesis, Universität Hamburg, Germany, 2011.

[Sch75] Gerald Schwarz, *Smooth Functions Invariant under the Action of a Compact Lie Group*, Topology, vol. 14, pp. 63–68, 1975.

[Sik67] Roman Sikorski, *Abstract covariant derivative*, Colloquium Mathematicum, vol. 18, pp. 151–172., 1967.

[Smi66] Wolfgang J. Smith, *The De Rham theorem for generalized spaces*, Tohôku Math. Journ., vol. 18, **2**, 1966.

[Sou64] Jean-Marie Souriau, *Calcul linéaire*, P.U.F. Paris, 1964.

[Sou70] Jean-Marie Souriau, *Structure des systèmes dynamiques*, Dunod, Paris, 1970.

[Sou83] Jean-Marie Souriau, *Géométrie globale du problème à deux corps*, Modern Developments in Analytical Mechanics, Accademia della Scienza di Torino, pp. 369–418, 1983.

[Sou80] Jean-Marie Souriau, *Groupes différentiels*, Lect. Notes in Math., Springer Verlag, New York, vol. 836, pp. 91–128, 1980.

[Sou84] Jean-Marie Souriau, *Groupes différentiels et physique mathématique*, Collection travaux en cours, Hermann, Paris, pp. 73–119, 1984.

[Sta10] Andrew Stacey, *Comparative Smootheology*, Theory and Applications of Categories, vol. 25, **4**, pp. 64–117, 2011.

[Tyc35] Andrei Tychonoff, *Ein Fixpunktsatz*, Math. Ann., vol. 111, pp. 767–776, 1935.

[Wey39] Hermann Weyl, *Classical Groups*, Princeton University Press, New Jersey, 1939–1946.

[Wht78] Georges W. Whitehead, *Elements of Homotopy Theory*, vol. 61, Graduate Texts in Math, Springer Verlag, New York, 1978.

[Whi43] Hassler Whitney, *Differentiable even functions*, Duke Mathematics Journal, vol. 10, pp. 159–160, 1943.

[Zie96] François Ziegler, *Théorie de Mackey symplectique, in Méthode des orbites et représentations quantiques*, Thèse de doctorat d'Université, Université de Provence, Marseille, 1996. `http://arxiv.org/pdf/1011.5056v1.pdf`.